INTRODUCTION TO POLYMERS

Third Edition

INTRODUCTION TO POLYMERS

Third Edition

Robert J. Young and Peter A. Lovell

CRC Press is an imprint of the
Taylor & Francis Group, an **informa** business

CRC Press
Taylor & Francis Group
6000 Broken Sound Parkway NW, Suite 300
Boca Raton, FL 33487-2742

Printed in the United States of America on acid-free paper
Version Date: 20110511

International Standard Book Number: 978-0-8493-3929-5 (Paperback)

Visit the Taylor & Francis Web site at
http://www.taylorandfrancis.com

and the CRC Press Web site at
http://www.crcpress.com

Contents

PART II Characterization of Polymers

PART III Phase Structure and Morphology of Bulk Polymers

Preface to the Third Edition

The second edition of this book was published at the start of a decade during which the discipline of polymer science saw several major developments come to fruition, including methods for effecting 'living' radical polymerization, dendrimer and hyperbranched polymer syntheses, use of metallocene catalysts for the polymerization of olefins, ring-opening methathesis polymerization, methods for the synthesis of inherently conducting polymers and their use in electronic devices, advanced high-performance polymer composites and specialized biomedical applications (both structural and functional). Theoretical understanding of polymers matured further during this time with scaling concepts, reptation theory, mechanisms for toughening polymers, and understanding of phase separation behaviour of multicomponent polymer systems becoming established in mainstream polymer science. Since the beginning of the new millennium, these advances have evolved to the point where they are now routine and used by the many scientists and engineers working on polymers in both academia and industry. Furthermore, during the past 20 years, the significance of developments in polymer science has been recognized by award of several polymer-related Nobel Prizes in both chemistry and physics. Over the same period, the use of polymers has continued to grow and diversify, particularly into medicine, electronics and aerospace applications.

Thus, the content of the second edition had become somewhat dated by midway through the past decade and preparation of a third edition was an imperative. In planning the new edition, we have taken the opportunity not only to include the science which underpins the important new developments and to bring the subject matter up to date, but also to completely reorganize the book into a more visibly coherent form that, we hope, readers will appreciate, whether the book is used as a basis for teaching or for learning the fundamental aspects of contemporary polymer science. The original 5 chapters have been reorganized into 25 self-contained chapters grouped into 4 parts that focus on (1) concepts and synthesis, (2) characterization, (3) structure and morphology, and (4) properties. Whilst retaining the overall balance between these themes, the content of each previous chapter has been reconsidered in detail, revised, reorganized where appropriate and expanded significantly to include the new developments. Anyone who is familiar with the second edition will recognize the previous material, but also should instantly feel the freshness and accessibility of the new structure.

Each of the chapters on polymer synthesis (Part I) has been expanded to include the most important new developments at a depth designed to inform the selection of appropriate method(s) for the synthesis of particular polymer structures and to facilitate the understanding of polymerization mechanisms and kinetics. Some of the more obvious new topics are 'living' radical polymerization, various methods for the synthesis of conducting polymers, strategies for the synthesis of dendrimers and hyperbranched polymers, metallocene polymerization and strategies for block copolymer synthesis. There also are many other less obvious additions, such as modern methods for the evaluation of rate coefficients in radical polymerization, catalytic chain transfer, non-linear radical polymerizations, free-radical ring-opening polymerization, supramolecular polymers and graft copolymer synthesis. Additionally, polymerization mechanisms have been made more explicit by showing electron movements.

The coverage of polymer characterization (Part II) has been expanded to show the complete derivations of the Flory–Huggins theory and the related theory for phase separation behaviour of polymer solutions, and to give a more thorough description of dynamic light scattering. New topics such as diffusion, solution behaviour of polyelectrolytes and field-flow fractionation methods have been added. And there is a major expansion of the coverage of spectroscopic methods, to which is devoted a complete chapter with new topics that include UV-visible spectroscopy, Raman spectroscopy, use of NMR spectroscopy for the determination of detailed structure, including sequence distributions, and methods of mass spectrometry.

The topics under structure and morphology (Part III) have been rearranged so that the amorphous state is covered (more logically) ahead of the crystalline state. Completely new material on macromolecular dynamics and reptation, on liquid crystalline polymers and on thermal analysis, has been included. In addition, a complete new chapter on multicomponent polymer systems has been added to reflect the growth in importance of polymer blends and block copolymers, and the maturity in the understanding of such materials. Many of the diagrams and micrographs have been updated to more clearly highlight features of polymer morphology.

The topics under polymer properties (Part IV) also benefit from the breakdown into separate chapters and contain the most obvious changes compared to the second edition, with two completely new topics dealt with in chapters on composites and on electrical properties. The previous material has been reorganized into five chapters, which now include topics such as effects of chain entanglements, swelling of elastomers, impact behaviour, ductile fracture and a more thorough coverage of the rubber-toughening of brittle plastics.

Although the new edition expands upon the previous material and adds many new topics and concepts, the philosophy of the book is unchanged and is designed for teaching and learning at both undergraduate and postgraduate levels, as well as for scientists in industry and research. As before, it is written, as far as possible, to be self-contained with most equations derived fully and topics cross-referenced between chapters where appropriate. The new structure should assist learning and the teaching of specific subjects. At the end of each chapter, a list of further reading material is provided to assist the reader in expanding their knowledge of the subject, and sets of problems are given to test understanding, particularly of numerical aspects.

The preparation of the third edition has been a major undertaking that has been fitted in amongst many other conflicting commitments, both private and professional, and its completion would not have been possible without the help and support of other people. The authors would like to thank their colleagues, former students and friends from around the world who have provided material for the new edition and given helpful comments on certain aspects of the manuscript. Finally, they again would like to thank their families for the support they have shown through the long period of time that it has taken to prepare this new edition.

Robert J. Young
Peter A. Lovell
School of Materials
The University of Manchester
Manchester, United Kingdom

Part I

Concepts, Nomenclature and Synthesis of Polymers

1 Concepts and Nomenclature

1.1 THE ORIGINS OF POLYMER SCIENCE AND THE POLYMER INDUSTRY

Polymers have existed in natural form since life began, and those such as DNA, RNA, proteins and polysaccharides play crucial roles in plant and animal life. From the earliest times, man has exploited naturally-occurring polymers as materials for providing clothing, decoration, shelter, tools, weapons, writing materials and other requirements. However, the origins of today's polymer industry commonly are accepted as being in the nineteenth century when important discoveries were made concerning the modification of certain natural polymers.

In 1820, Thomas Hancock discovered that when masticated (i.e. subjected repeatedly to high shear forces), natural rubber becomes more fluid making it easier to blend with additives and to mould. Some years later, in 1839, Charles Goodyear found that the elastic properties of natural rubber could be improved, and its tackiness eliminated, by heating with sulphur. Patents for this discovery were issued in 1844 to Goodyear, and slightly earlier to Hancock, who christened the process vulcanization. In 1851, Nelson Goodyear, Charles' brother, patented the vulcanization of natural rubber with large amounts of sulphur to produce a hard material more commonly known as hard rubber, ebonite or vulcanite.

Cellulose nitrate, also called nitrocellulose or gun cotton, first became prominent after Christian Schönbein prepared it in 1846. He was quick to recognize the commercial value of this material as an explosive, and within a year gun cotton was being manufactured. However, more important to the rise of the polymer industry, cellulose nitrate was found to be a hard elastic material which was soluble and could be moulded into different shapes by the application of heat and pressure. Alexander Parkes was the first to take advantage of this combination of properties and in 1862 he exhibited articles made from Parkesine, a form of plasticized cellulose nitrate. In 1870, John and Isaiah Hyatt patented a similar but more easily processed material, named celluloid, which was prepared using camphor as the plasticizer. Unlike Parkesine, celluloid was a great commercial success.

In 1892, Charles Cross, Edward Bevan and Clayton Beadle patented the 'viscose process' for dissolving and then regenerating cellulose. The process was first used to produce viscose rayon textile fibres, and subsequently for the production of cellophane film.

The polymeric materials described so far are semi-synthetic since they are produced from natural polymers. Leo Baekeland's phenol–formaldehyde 'Bakelite' resins have the distinction of being the first fully synthetic polymers to be commercialized, their production beginning in 1910. The first synthetic rubber to be manufactured, known as methyl rubber, was produced from 2,3-dimethylbutadiene in Germany during World War I as a substitute, albeit a poor one, for natural rubber.

Although the polymer industry was now firmly established, its growth was restricted by the considerable lack of understanding of the nature of polymers. For over a century, scientists had been reporting the unusual properties of polymers, and by 1920, the common belief was that they consisted of physically-associated aggregates of small molecules. Few scientists gave credence to the viewpoint so passionately believed by Hermann Staudinger, that polymers were composed of very large molecules containing long sequences of simple chemical units linked together by covalent bonds. Staudinger introduced the word 'macromolecule' to describe polymers, and during the 1920s, vigorously set about proving his hypothesis to be correct. Particularly important were his studies of the synthesis, structure and properties of polyoxymethylene and of polystyrene, the results from which left little doubt as to the validity of the macromolecular viewpoint. Staudinger's hypothesis was further substantiated by the crystallographic studies of natural polymers reported by Herman Mark and Kurt Meyer, and by the classic work of Wallace Carothers on the preparation of polyamides and polyesters. Thus by the early 1930s, most scientists were convinced of the macromolecular structure of polymers. During the following 20 years, work on polymers increased enormously: the first journals devoted solely to their study were published and most of the fundamental principles of *Polymer Science* were established. The theoretical and experimental work of Paul Flory was prominent in this period, and for his long and substantial contribution to Polymer Science, he was awarded the Nobel Prize for Chemistry in 1974. In 1953, Staudinger had received the same accolade in recognition of his pioneering work.

Not surprisingly, as the science of macromolecules emerged, a large number of synthetic polymers went into commercial production for the first time. These include polystyrene, poly(methyl methacrylate), nylon 6.6, polyethylene, poly(vinyl chloride), styrene–butadiene rubber, silicones and polytetrafluoroethylene, as well as many others. From the 1950s onwards, regular advances, too numerous to mention here, have continued to stimulate both scientific and industrial progress, and as the discipline of polymer science progresses into the twenty-first century there is increasing emphasis on the development of more specialized, functional polymers for biomedical, optical and electronic applications.

Whilst polymer science undoubtedly is now a mature subject, its breadth and importance continue to increase and there remain many demanding challenges awaiting scientists who venture into this fascinating multidisciplinary science.

1.2 BASIC DEFINITIONS AND NOMENCLATURE

Several important terms and concepts must be understood in order to discuss fully the synthesis, characterization, structure and properties of polymers. Most of these will be defined and discussed in detail in subsequent chapters. However, some are of such fundamental importance that they must be defined at the outset.

In strict terms, a *polymer* is a *substance* composed of molecules which have long sequences of one or more species of atoms or groups of atoms linked to each other by primary, usually covalent, bonds. The emphasis upon substance in this definition is to highlight that although the words polymer and *macromolecule* are used interchangeably, the latter strictly defines the molecules of which the former is composed.

Macromolecules are formed by linking together *monomer* molecules through chemical reactions, the process by which this is achieved being known as *polymerization*. For example, polymerization of ethylene yields polyethylene, a typical sample of which may contain molecules with 50,000 carbon atoms linked together in a chain. It is this long chain nature which sets polymers apart from other materials and gives rise to their characteristic properties.

1.2.1 Skeletal Structure

The definition of macromolecules presented up to this point implies that they have a *linear* skeletal structure which may be represented by a chain with two ends. Whilst this is true for many macromolecules, there are also many with *non-linear* skeletal structures of the type shown in Figure 1.1.

Cyclic polymers (*ring polymers*) have no chain ends and show properties that are quite different to their linear counterparts. *Branched polymers* have side chains, or *branches*, of significant length which are bonded to the main chain at *branch points* (also known as *junction points*), and are characterized in terms of the number and size of the branches. *Network polymers* have three-dimensional structures in which each chain is connected to all others by a sequence of junction points and other chains. Such polymers are said to be *crosslinked* and are characterized by their *crosslink density*, or *degree of crosslinking*, which is related directly to the number of junction points per unit volume.

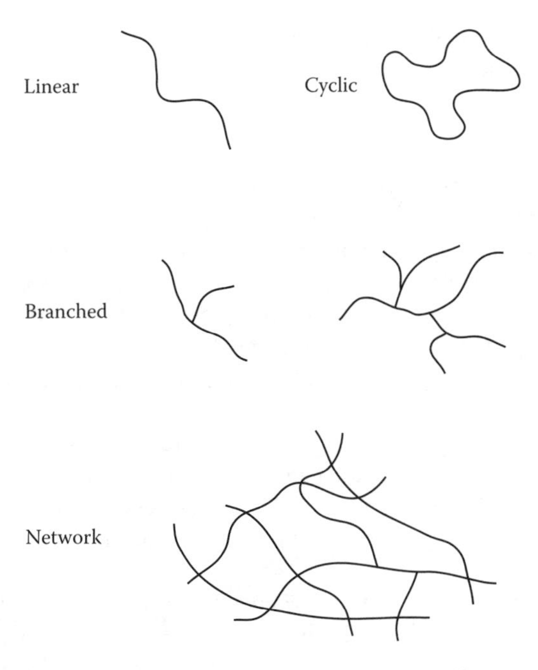

FIGURE 1.1 Skeletal structures representative of linear, cyclic and non-linear polymers.

Branched and network polymers may be formed by polymerization, or can be prepared by linking together (i.e. *crosslinking*) pre-existing chains.

These variations in skeletal structure give rise to major differences in properties. For example, linear polyethylene has a melting point about 20 °C higher than that of branched polyethylene. Unlike linear and branched polymers, network polymers do not melt upon heating and will not dissolve, though they may swell considerably in compatible solvents. The importance of crosslink density has already been described in terms of the vulcanization (i.e. sulphur-crosslinking) of natural rubber. With low crosslink densities (i.e. low levels of sulphur) the product is a flexible elastomer, whereas it is a rigid material when the crosslink density is high.

In addition to these more conventional skeletal structures, there has been growing interest in more elaborate skeletal forms of macromolecules. Of particular interest are *dendrimers*, which are highly branched polymers of well-defined structure and molar mass, and *hyperbranched polymers*, which are similar to dendrimers but have a much less well-defined structure and molar mass. Simple depictions of their skeletal structures are shown in Figure 1.2. Research into these types of polymers intensified during the 1990s and they are now beginning to find applications which take advantage of their unusual properties. For example, because of their high level of branching, they are extremely crowded but as a consequence have voids and channels within the molecule and have a

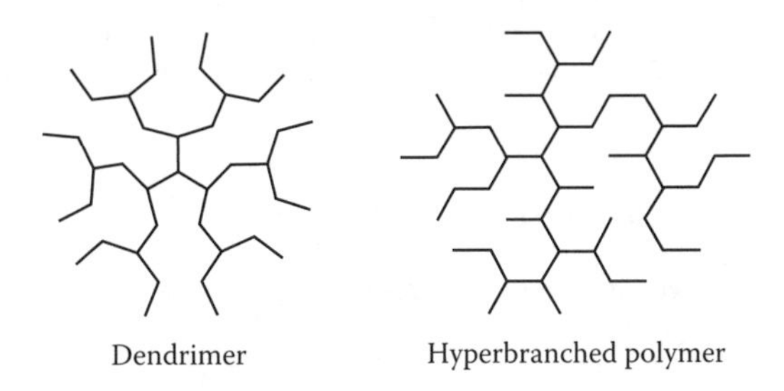

FIGURE 1.2 Skeletal structures representative of dendrimers and hyperbranched polymers.

large number of end groups around their periphery that can be functionalized, leading to therapeutic applications such as in targeted drug delivery.

1.2.2 Homopolymers

The formal definition of a homopolymer is a polymer derived from one species of monomer. However, the word *homopolymer* often is used more broadly to describe polymers whose structure can be represented by multiple repetition of a single type of *repeat unit* which may contain one or more species of *monomer unit*. The latter is sometimes referred to as a *structural unit*.

The chemical structure of a polymer usually is represented by that of the repeat unit enclosed by brackets. Thus the hypothetical homopolymer ~~A–A–A–A–A–A–A–A~~ is represented by $\text{-}\!\!\left[\text{A}\right]\!\!_n\text{-}$ where n is the number of repeat units linked together to form the macromolecule. Table 1.1 shows the chemical structures of some common homopolymers together with the monomers from which they are derived and some comments upon their properties and uses. It should be evident that slight differences in chemical structure can lead to very significant differences in properties. Entries (9) and (10) in Table 1.1 are examples of homopolymers for which the repeat unit contains two monomer (structural) units; for each of the other examples, the repeat unit and the monomer unit are the same.

The naming of polymers, or envisaging of the chemical structure of a polymer from its name, is often an area of difficulty. At least in part this is because most polymers have more than one correct name, the situation being further complicated by the variety of trade names which also are used to describe certain polymers. The approach adopted here is to use names which most clearly and simply indicate the chemical structures of the polymers under discussion.

The names given to the polymers in Table 1.1 exemplify elementary aspects of nomenclature. Thus source-based nomenclature places the prefix 'poly' before the name of the monomer, the monomer's name being contained within parentheses unless it is a simple single word. In structure-based nomenclature, the prefix 'poly' is followed in parentheses by words which describe the chemical structure of the repeat unit. This type of nomenclature is used for polymers (9) and (10) in Table 1.1.

1.2.3 Copolymers

The formal definition of a *copolymer* is a polymer derived from more than one species of monomer. However, in accordance with the use of the word homopolymer, it is common practice to use a structure-based definition. Thus the word copolymer more commonly is used to describe polymers whose molecules contain two or more different types of repeat unit. Hence polymers (9) and (10) in Table 1.1 usually are considered to be homopolymers rather than copolymers.

There are several categories of copolymer, each being characterized by a particular arrangement of the repeat units along the polymer chain. For simplicity, the representation of these categories will be illustrated by copolymers containing only two different types of repeat unit (A and B).

Statistical copolymers are copolymers in which the sequential distribution of the repeat units obeys known statistical laws (e.g. Markovian). *Random copolymers* are a special type of statistical copolymer in which the distribution of repeat units is truly random (some words of caution are necessary here because older textbooks and scientific papers often use the term random copolymer to describe both random and non-random statistical copolymers). A section of a truly random copolymer is represented below

~~B–B–B–A–B–B–A–B–A–A~~

Alternating copolymers have only two different types of repeat units and these are arranged alternately along the polymer chain

~~A–B–A–B–A–B–A–B–A–B~~

TABLE 1.1
Some Common Homopolymers

Monomers	Polymer	Comments
1. Ethylene $CH_2{=}CH_2$	Polyethylene (PE) $-[CH_2-CH_2]_n-$	Moulded objects, tubing, film, electrical insulation, e.g. 'Alkathene', 'Lupolen'.
2. Propylene $CH_2{=}CH(CH_3)$	Polypropylene (PP) $-[CH_2-CH(CH_3)]_n-$	Similar uses to PE; lower density, stiffer, e.g. 'Propathene', 'Novolen'.
3. Tetrafluoroethylene $CF_2{=}CF_2$	Polytetrafluoroethylene (PTFE) $-[CF_2-CF_2]_n-$	Mouldings, film, coatings; high temperature resistance, chemically inert, excellent electrical insulator, very low coefficient of friction; expensive, e.g. 'Teflon', 'Fluon'.
4. Styrene $CH_2{=}CH(C_6H_5)$	Polystyrene (PS) $-[CH_2-CH(C_6H_5)]_n-$	Cheap moulded objects, e.g. 'Styron', 'Hostyren'. Modified with rubbers to improve toughness, e.g. high-impact polystyrene (HIPS) and acrylonitrile–butadiene–styrene copolymer (ABS). Expanded by volatilization of a blended blowing agent (e.g. pentane) to produce polystyrene foam.
5. Methyl methacrylate $CH_2{=}C(CH_3)(C{=}O)OCH_3$	Poly(methyl methacrylate) (PMMA) $-[CH_2-C(CH_3)(C{=}O)OCH_3]_n-$	Transparent sheets and mouldings; used for aeroplane windows; more expensive than PS, e.g. 'Perspex', 'Diakon', 'Lucite', 'Oroglass', 'Plexiglas'.
6. Vinyl chloride $CH_2{=}CH(Cl)$	Poly(vinyl chloride) (PVC) $-[CH_2-CH(Cl)]_n-$	Water pipes and gutters, bottles, gramophone records; plasticized to make PVC leathercloth, raincoats, flexible pipe and hose, toys, sheathing on electrical cables, e.g. 'Darvic', 'Welvic', 'Vinoflex', 'Hostalit'.

(continued)

TABLE 1.1 (continued)
Some Common Homopolymers

Monomers	Polymer	Comments
7. Vinyl acetate $CH_2{=}CH{-}O{-}C({=}O){-}CH_3$	Poly(vinyl acetate) $\left[CH_2{-}CH(O{-}C({=}O){-}CH_3)\right]_n$	Surface coatings, adhesives, chewing gum.
8. Ethylene oxide 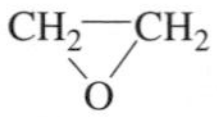 $CH_2{-}CH_2$ (with O bridging, three-membered ring)	Poly(ethylene oxide) (PEO) $\left[CH_2{-}CH_2{-}O\right]_n$	Water-soluble packaging films, textile sizes, thickeners, e.g. 'Carbowax'.
9. Ethylene glycol $HO{-}CH_2{-}CH_2{-}OH$ and terephthalic acid 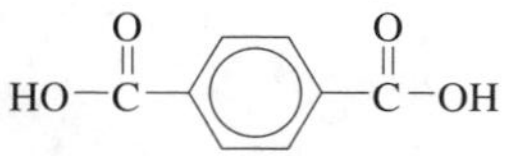 $HO{-}C({=}O){-}C_6H_4{-}C({=}O){-}OH$	Poly(ethylene terephthalate) (PET)[a] 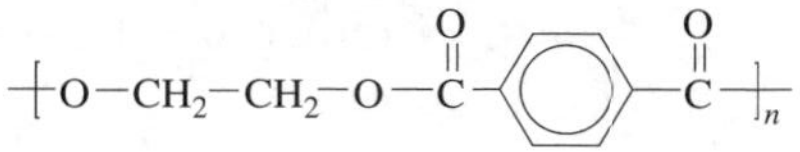 $\left[O{-}CH_2{-}CH_2{-}O{-}C({=}O){-}C_6H_4{-}C({=}O)\right]_n$	Textile fibres, film, bottles, e.g. 'Terylene', 'Dacron', 'Melinex', 'Mylar'.
10. Hexamethylene diamine $H_2N{-}(CH_2)_6{-}NH_2$ and sebacic acid $HO{-}C({=}O){-}(CH_2)_8{-}C({=}O){-}OH$	Poly(hexamethylene sebacamide) (nylon 6.10)[a] $\left[N(H){-}(CH_2)_6{-}N(H){-}C({=}O){-}(CH_2)_8{-}C({=}O)\right]_n$ (6 carbons; 10 carbons)	Mouldings, fibres, e.g. 'Ultramid 6.10'.

[a] The polymer has two monomer units in the repeat unit.

Statistical, random and alternating copolymers generally have properties which are intermediate to those of the corresponding homopolymers. Thus by preparing such copolymers, it is possible to combine the desirable properties of the homopolymers into a single material. This is not normally possible by blending because most homopolymers are immiscible with each other.

Block copolymers are linear copolymers in which the repeat units exist only in long sequences, or *blocks*, of the same type. Two common block copolymer structures are represented below and usually are termed AB di-block and ABA tri-block copolymers

A–A–A–A–A–A–A–A–A–A–B–B–B–B–B–B–B–B–B

A–A–A–A–A–A–A–A–B–B–B–B–B–B–B–B–A–A–A–A–A–A–A–A

Graft copolymers are branched polymers in which the branches have a different chemical structure to that of the main chain. In their simplest form, they consist of a main homopolymer chain with branches of a different homopolymer

```
                B–B–B–B–B–B
                |
                B
                |
A–A–A–A–A–A–A–A–A–A–A–A–A–A–A–A–A–A–A–A
                          |
                          B
                          |
                          B–B–B–B–B–B–B–B
```

In distinct contrast to the types of copolymers described earlier, block and graft copolymers usually show properties characteristic of each of the constituent homopolymers. They also have some unique properties that arise because the chemical linkage(s) between the homopolymer sequences prevent them from acting entirely independently of each other.

The current principles of nomenclature for copolymers are indicated in Table 1.2 where A and B represent source- or structure-based names for these repeat units. Thus, a statistical copolymer of ethylene and propylene is named poly(ethylene-*stat*-propylene), and an ABA tri-block copolymer of styrene (A) and isoprene (B) is named polystyrene-*block*-polyisoprene-*block*-polystyrene. In certain cases, additional square brackets are required. For example, an alternating copolymer of styrene and maleic anhydride is named poly[styrene-*alt*-(maleic anhydride)].

1.2.4 Classification of Polymers

The most common way of classifying polymers is outlined in Figure 1.3 where they are first separated into three groups: *thermoplastics*, *elastomers* and *thermosets*. Thermoplastics are then further separated into those which are crystalline and those which are amorphous (i.e. non-crystalline). This method of classification has an advantage in comparison to others since it is based essentially upon the underlying molecular structure of the polymers.

Thermoplastics, often referred to just as plastics, are linear or branched polymers which become liquid upon the application of heat. They can be moulded (and remoulded) into virtually any shape using processing techniques such as injection moulding and extrusion, and now constitute by far the largest proportion of the polymers in commercial production. Generally, thermoplastics do not crystallize easily upon cooling to the solid state because this

TABLE 1.2
Principles of Nomenclature for Copolymers

Type of Copolymer	Example of Nomenclature
Unspecified	Poly(A-*co*-B)
Statistical	Poly(A-*stat*-B)
Random	Poly(A-*ran*-B)
Alternating	Poly(A-*alt*-B)
Block	PolyA-*block*-polyB
Graft[a]	PolyA-*graft*-polyB

[a] The example is for polyB branches on a polyA main chain.

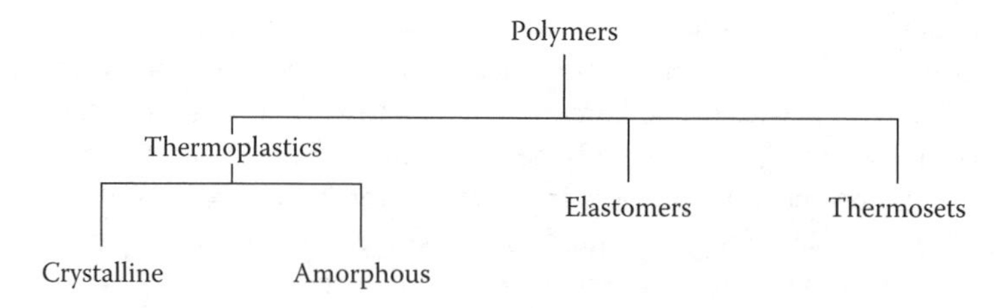

FIGURE 1.3 Classification of polymers.

requires considerable ordering of the highly coiled and entangled macromolecules present in the liquid state. Those which do crystallize invariably do not form perfectly crystalline materials but instead are *semi-crystalline* with both crystalline and amorphous regions. The crystalline phases of such polymers are characterized by their *melting temperature* T_m, above which such polymers can be converted into artefacts by conventional polymer-processing techniques such as extrusion, injection moulding and compression moulding.

Many thermoplastics are, however, completely amorphous and incapable of crystallization, even upon annealing. Amorphous polymers (and amorphous phases of semi-crystalline polymers) are characterized by their *glass transition temperature* T_g, the temperature at which they transform abruptly from the *glassy state* (hard) to the *rubbery state* (soft). This transition corresponds to the onset of chain motion; below T_g the polymer chains are unable to move and are 'frozen' in position. Both T_m and T_g increase with increasing chain stiffness and increasing forces of intermolecular attraction.

It is a common misnomer that completely amorphous polymers 'melt'; they do not (because they have no ordered phases, there is nothing to melt!) and may simply be considered as reducing steadily in viscosity as temperature increases above T_g until the viscosity becomes low enough for so-called melt processing.

Elastomers are crosslinked rubbery polymers (i.e. rubbery networks) that can be stretched easily to high extensions (e.g. 3× to 10× their original dimensions) and which rapidly recover their original dimensions when the applied stress is released. This extremely important and useful property is a reflection of their molecular structure in which the network is of low crosslink density. The rubbery polymer chains become extended upon deformation but are prevented from permanent flow by the crosslinks, and driven by entropy, spring back to their original positions on removal of the stress. The word rubber, often used in place of elastomer, preferably should be used for describing rubbery polymers that are not crosslinked.

Thermosets normally are rigid materials and are network polymers in which chain motion is greatly restricted by a high degree of crosslinking. As for elastomers, they are intractable once formed and degrade rather than become fluid upon the application of heat. Hence, their processing into artefacts is often done using processes, such as compression moulding, that require minimum amounts of flow.

1.3 MOLAR MASS AND DEGREE OF POLYMERIZATION

Many properties of polymers show a strong dependence upon the size of the polymer chains, so it is essential to characterize their dimensions. This normally is done by measuring the *molar mass* M of a polymer which is simply the mass of 1 mol of the polymer and usually is quoted in units of g mol^{-1} or kg mol^{-1}. The term 'molecular weight' is still often used instead of molar mass, but is not preferred because it can be somewhat misleading. It is really a dimensionless quantity, the relative molecular mass, rather than the weight of an individual molecule which is of course a very small quantity (e.g. ~10^{-19} to ~10^{-18} g for most polymers). By multiplying the numerical value of molecular weight by the specific units g mol^{-1} it can be converted into the equivalent value of molar mass. For example, a molecular weight of 100,000 is equivalent to a molar mass of 100,000 g mol^{-1} which in turn is equivalent to a molar mass of 100 kg mol^{-1}.

For network polymers the only meaningful molar mass is that of the polymer chains existing between junction points (i.e. *network chains*), since the molar mass of the network itself essentially is infinite.

The molar mass of a homopolymer is related to the *degree of polymerization x*, which is the number of repeat units in the polymer chain, by the simple relation

$$M = xM_0 \tag{1.1}$$

where M_0 is the molar mass of the repeat unit. Equation 1.1 also can be used for copolymers, but M_0 needs to be replaced by M_0^{cop}, which is the mean repeat unit molar mass of the copolymer taking into account its composition

$$M_0^{\text{cop}} = \sum X_j M_0^j \tag{1.2}$$

where

X_j is the mole fraction, and
M_0^j the molar mass of repeat units of type 'j' in the copolymer.

1.3.1 Molar Mass Distribution

With very few exceptions, polymers consist of macromolecules (or network chains) with a range of molar masses. Since the molar mass changes in intervals of M_0, the distribution of molar mass is discontinuous. However, for most polymers, these intervals are extremely small in comparison to the total range of molar mass and the distribution can be assumed to be continuous, as exemplified in Figure 1.4.

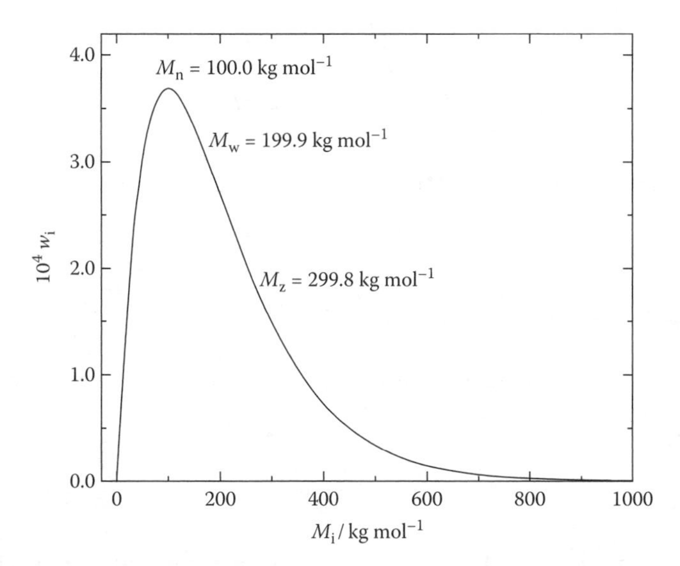

FIGURE 1.4 A typical weight-fraction molar mass distribution curve (for a polymer with the most probable distribution of molar mass and a repeat unit molar mass of 100 g mol^{-1}).

1.3.2 Molar Mass Averages

Whilst a knowledge of the complete molar mass distribution is essential in many uses of polymers, it is convenient to characterize the distribution in terms of molar mass averages. These usually are defined by considering the discontinuous nature of the distribution in which the macromolecules exist in discrete fractions 'i' containing N_i molecules of molar mass M_i.

The *number-average molar mass* $\bar{M}_n$ is defined as the sum of the products of the molar mass of each fraction multiplied by its mole fraction, i.e.

$$\bar{M}_n = \sum X_i M_i \tag{1.3}$$

where X_i is the mole fraction of molecules of molar mass M_i and is given by the ratio of N_i to the total number of molecules. Therefore it follows that

$$\bar{M}_n = \frac{\sum N_i M_i}{\sum N_i} \tag{1.4}$$

showing this average to be the arithmetic mean of the molar mass distribution. It often is more convenient to use weight fractions rather than numbers of molecules. The weight fraction w_i is defined as the mass of molecules of molar mass M_i divided by the total mass of all the molecules present, i.e.

$$w_i = \frac{N_i M_i}{\sum N_i M_i} \tag{1.5}$$

from which it can be deduced that

$$\sum\left(\frac{w_i}{M_i}\right) = \frac{\sum N_i}{\sum N_i M_i} \tag{1.6}$$

Combining Equations 1.4 and 1.6 gives $\bar{M}_n$ in terms of weight fractions

$$\bar{M}_n = \frac{1}{\sum (w_i/M_i)} \tag{1.7}$$

The *weight-average molar mass* $\bar{M}_w$ is defined as the sum of the products of the molar mass of each fraction multiplied by its weight fraction, i.e.

$$\bar{M}_w = \sum w_i M_i \tag{1.8}$$

By combining this equation with Equation 1.5, $\bar{M}_w$ can be expressed in terms of the numbers of molecules

$$\bar{M}_w = \frac{\sum N_i M_i^2}{\sum N_i M_i} \tag{1.9}$$

The ratio $\bar{M}_w/\bar{M}_n$ must by definition be greater than unity for a *polydisperse* polymer and is known as the *polydispersity* or *heterogeneity index* (often referred to as *PDI*). Its value often is used as a measure of the breadth of the molar mass distribution, though it is a poor substitute for knowledge of the complete distribution curve. Typically $\bar{M}_w/\bar{M}_n$ is in the range 1.5–2.0, though there are many polymers which have smaller or very much larger values of polydispersity index. A perfectly *monodisperse* polymer would have $\bar{M}_w/\bar{M}_n = 1.00$.

It should be noted that in 2009 IUPAC (the international organization which defines nomenclature and terminology in pure and applied fields of chemistry) criticized use of the long-established terms 'monodisperse' (because it is a self-contradictory word) and 'polydisperse'/'polydispersity' (because they are tautologous). Paraphrasing their comments, they recommended that a polymer sample composed of a single macromolecular species should be called a *uniform polymer* (instead of monodisperse) and a polymer sample composed of macromolecular species of differing molar masses a *non-uniform polymer* (instead of polydisperse). They further recommended that polydispersity should be replaced by a new term, *dispersity* (given the symbol $Đ$), such that $Đ_M$ is the molar mass dispersity ($= \bar{M}_w/\bar{M}_n$), $Đ_X$ is the degree-of-polymerization dispersity ($= \bar{x}_w/\bar{x}_n$) and for most polymers $Đ = Đ_M = Đ_X$. It is much too early to know whether these new terms will become embedded in the vocabulary of, and future publications from, the world polymer community, but they are used from here onwards in this book because the recommendations are sensible and based on use of good English.

Higher molar mass averages than $\bar{M}_w$ sometimes are quoted. For example, certain methods of molar mass measurement (e.g. sedimentation equilibrium and dynamic light scattering) yield the *z-average molar mass* $\bar{M}_z$, which is defined as follows

$$\bar{M}_z = \frac{\sum N_i M_i^3}{\sum N_i M_i^2} = \frac{\sum w_i M_i^2}{\sum w_i M_i} \tag{1.10}$$

In addition, averages with more complex exponents are obtained from other methods of polymer characterization (e.g. from dilute solution viscometry and sedimentation measurements).

Degree of polymerization averages are of more importance than molar mass averages in the theoretical treatment of polymers and polymerization, as will be highlighted in the subsequent chapters. For homopolymers they may be obtained simply by dividing the corresponding molar mass average by M_0. Thus the *number-average* and *weight-average degrees of polymerization* are given by

$$\bar{x}_n = \frac{\bar{M}_n}{M_0} \tag{1.11}$$

and

$$\bar{x}_w = \frac{\bar{M}_w}{M_0} \tag{1.12}$$

The same equations can be applied to calculate $\bar{x}_n$ and $\bar{x}_w$ for copolymers by replacing M_0 with M_0^{cop}.

PROBLEMS

1.1 A sample of polystyrene is found to have a number-average molar mass of 89,440 g mol^{-1}. Neglecting contributions from end groups, calculate the number-average degree of polymerization of this sample.

Assuming that the sample has a molar mass dispersity of 1.5, calculate its weight-average molar mass.

1.2 Calculate the mean repeat unit molar mass for a sample of poly[ethylene-*stat*-(vinyl acetate)] that comprises 12.9 wt% vinyl acetate repeat units. Given that its number-average molar mass is 39,870 g mol^{-1}, calculate the number-average degree of polymerization of the copolymer.

1.3 Three mixtures were prepared from three very narrow molar mass distribution polystyrene samples with molar masses of 10,000, 30,000 and 100,000 g mol^{-1} as indicated below:

(a) Equal numbers of molecules of each sample
(b) Equal masses of each sample
(c) By mixing in the mass ratio 0.145:0.855 the two samples with molar masses of 10,000 and 100,000 g mol^{-1}

For each of the mixtures, calculate the number-average and weight-average molar masses and comment upon the meaning of the values.

FURTHER READING

General Historical and Introductory Reading

Note that although many of these textbooks use old terminology and are out of print, they do provide very good, simple introductions to polymer science and can be found in libraries.

Campbell, I.M., *Introduction to Synthetic Polymers*, 2nd edn., Oxford University Press, New York, 2000.
Elias, H.-G., *Mega Molecules*, Springer-Verlag, New York, 1987.
Greenwood, C.T. and Banks, W., *Synthetic High Polymers*, Oliver & Boyd, Edinburgh, U.K., 1968.
ICI Plastics Division, *Landmarks of the Plastics Industry*, Kynoch Press, Birmingham, U.K., 1962.
Kaufman, M., *The First Century of Plastics – Celluloid and Its Sequel*, The Plastics Institute, London, U.K., 1963.
Mandelkern, L., *An Introduction to Macromolecules*, 2nd edn., Springer-Verlag, New York, 1983.
Mark, H.F., *Giant Molecules*, Time-Life International, New York, 1970.
Morawetz, H., *Polymers – The Origins and Growth of a Science*, John Wiley & Sons, New York, 1985.
Nicholson, J.W., *The Chemistry of Polymers*, Royal Society of Chemistry, London, U.K., 1991.
Tonelli, A.E. and Srinivasarao, M., *Polymers from the Inside Out: An Introduction to Macromolecules*, John Wiley & Sons, New York, 2001.
Treloar, L.R.G., *Introduction to Polymer Science*, Wykeham Publications, London, 1970.

Macromolecular Nomenclature

Barón, M., Bikales, N., Fox, R., and Work, W., Macromolecular nomenclature and terminology: A brief history of IUPAC activities, *Chemistry International*, **24(6)**, 10 (2002).
Jenkins, A.D. and Loening, K.L., Nomenclature, in *Comprehensive Polymer Science*, Allen, G. and Bevington, J.C. (eds), Chap. 2, Vol. 1 (Booth, C. and Price, C., eds), Pergamon Press, Oxford, U.K., 1989.
Metanomski, W.V., *Compendium of Macromolecular Nomenclature*, Blackwell Science, Oxford, U.K., 1991.

2 Principles of Polymerization

2.1 INTRODUCTION

The most basic requirement for polymerization is that each molecule of monomer must be capable of being linked to two (or more) other molecules of monomer by chemical reaction, i.e. monomers must have a *functionality* of two (or higher). Given this relatively simple requirement, there are a multitude of chemical reactions and associated monomer types that can be used to effect polymerization. Consequently, the number of different synthetic polymers that have been prepared is extremely large and many can be formed by more than one type of polymerization. Hence, the number of individually different polymerization reactions that have been reported is extraordinarily large. To discuss each of these would be an enormous task which fortunately is not necessary since it is possible to categorize most polymerization reactions into a relatively small number of classes of polymerization, each class having distinctive characteristics.

The purpose of this chapter is to set out some of the most important guiding principles of polymerization. These principles are built upon in the following chapters on polymerization, which give details of the chemistry and kinetics of the most important types of polymerization in use today.

2.2 CLASSIFICATION OF POLYMERIZATION REACTIONS

The classification of polymerization reactions used in the formative years of polymer science was due to Carothers and is based upon comparison of the molecular formula of a polymer with that of the monomer(s) from which it was formed. *Condensation polymerizations* are those which yield polymers with repeat units having fewer atoms than are present in the monomers from which they are formed. This usually arises from chemical reactions which involve the elimination of a small molecule (e.g. H_2O, HCl). *Addition polymerizations* are those which yield polymers with repeat units having identical molecular formulae to those of the monomers from which they are formed. Table 1.1 contains examples of each class: the latter two examples are condensation polymerizations involving elimination of H_2O, whereas the others are addition polymerizations.

Carothers' method of classification was found to be unsatisfactory when it was recognized that certain condensation polymerizations have the characteristic features of typical addition polymerizations and that some addition polymerizations have features characteristic of typical condensation polymerizations. A better basis for classification is provided by considering the underlying polymerization mechanisms, of which there are two general types. Polymerizations in which the polymer chains grow step-wise by reactions that can occur between any two molecular species are known as *step-growth polymerizations*. Polymerizations in which a polymer chain grows only by reaction of monomer with a reactive end-group on the growing chain are known as *chain-growth polymerizations*, and usually require an initial reaction between the monomer and an *initiator* to start the growth of the chain.

The modern preference is to simplify these names to *step polymerization* and *chain polymerization*, and this practice will be used here. The essential differences between these classes of polymerization are highlighted in Table 2.1 which illustrates for each mechanism the reactions involved in growth of the polymer chains to a degree of polymerization equal to eight.

In step polymerizations the degree of polymerization increases steadily throughout the reaction, but the monomer is rapidly consumed in its early stages (e.g. when $\bar{x}_n = 10$ only 1% of the monomer

TABLE 2.1
A Schematic Illustration of the Fundamental Differences in Reaction Mechanism between Step Polymerization and Chain Polymerization[a]

Formation of	Step Polymerization	Chain Polymerization
Dimer	o+o → o–o	I+o → I–o
		I–o+o → I–o–o
Trimer	o–o+o → o–o–o	I–o–o+o → I–o–o–o
Tetramer	o–o–o+o → o–o–o–o	I–o–o–o+o → I–o–o–o–o
	o–o+o–o → o–o–o–o	
Pentamer	o–o–o–o+o → o–o–o–o–o	I–o–o–o–o+o → I–o–o–o–o–o
	o–o+o–o–o → o–o–o–o–o	
Hexamer	o–o–o–o–o+o → o–o–o–o–o–o	I–o–o–o–o–o+o → I–o–o–o–o–o–o
	o–o+o–o–o–o → o–o–o–o–o–o	
	o–o–o+o–o–o → o–o–o–o–o–o	
Heptamer	o–o–o–o–o–o+o → o–o–o–o–o–o–o	I–o–o–o–o–o–o+o → I–o–o–o–o–o–o–o
	o–o+o–o–o–o–o → o–o–o–o–o–o–o	
	o–o–o+o–o–o–o → o–o–o–o–o–o–o	
Octomer	o–o–o–o–o–o–o+o → o–o–o–o–o–o–o–o	I–o–o–o–o–o–o–o+o → I–o–o–o–o–o–o–o–o
	o–o+o–o–o–o–o–o → o–o–o–o–o–o–o–o	
	o–o–o+o–o–o–o–o → o–o–o–o–o–o–o–o	
	o–o–o–o+o–o–o–o → o–o–o–o–o–o–o–o	

[a] I = initiator species, o = molecule of monomer and repeat unit, – = chemical link.

remains unreacted). All the polymer chains continue to grow throughout the reaction as the conversion of functional groups into chain links increases. Carothers was quick to recognize that in order to attain even moderately high degrees of polymerization ($\bar{x}_n > 100$), the extent of reaction of functional groups needs to be extremely high (greater than 99.9%), something that he set about achieving by careful design of laboratory apparatus. This key feature of step polymerizations also highlights the importance of using *clean* reactions in which contributions from side reactions are completely absent or negligibly small.

By contrast, in chain polymerizations, high degrees of polymerization are attained at low monomer conversions, the monomer being consumed steadily throughout the reaction. After its growth has been initiated, each polymer chain forms rapidly by successive additions of molecules of monomer to the reactive site at the chain end. In many chain polymerizations, more than 1000 repeat units are added to a single propagating chain in less than a second and the activity of the chain is lost (i.e. its activity dies and it can no longer propagate) after only a fraction of a second or a few seconds of chain growth. As the percentage conversion of monomer into polymer increases, it is simply the number of polymer molecules formed that increases; the degree of polymerization of those polymer molecules already formed does not change.

2.3 MONOMER FUNCTIONALITY AND POLYMER SKELETAL STRUCTURE

The *functionality* of a monomer is best defined as the number of chain links it can give rise to, because it is not necessarily equal to the number of functional groups present in the monomer, i.e. it is not always immediately obvious from the chemical structure of the monomer. The first seven examples in Table 1.1 are of polymers formed by chain polymerizations of olefinic monomers which contain a single C=C double bond; these monomers have a functionality of two because each C=C double bond gives rise to two chain links. The eighth example is of a ring-opening chain polymerization of an epoxide group which also gives rise to two chain links; hence ethylene oxide also has a functionality of two. The remaining two examples in Table 1.1 are step polymerizations of monomers that have functional groups which are mutually reactive towards each other; in these examples, each monomer possesses two functional groups and has a functionality of two because each functional group can give rise to only a single chain link by reaction with the complementary functional group (e.g. in example (9) of Table 1.1, a $-CO_2H$ group reacts with a $-OH$ group to give a single ester chain link with loss of H_2O). Thus all the examples in Table 1.1 are of *linear polymerizations* of monomers with a functionality of two.

If a monomer has a functionality greater than two, then this will lead to the formation of branches and possibly to the formation of a network polymer, depending on the particular polymerization and

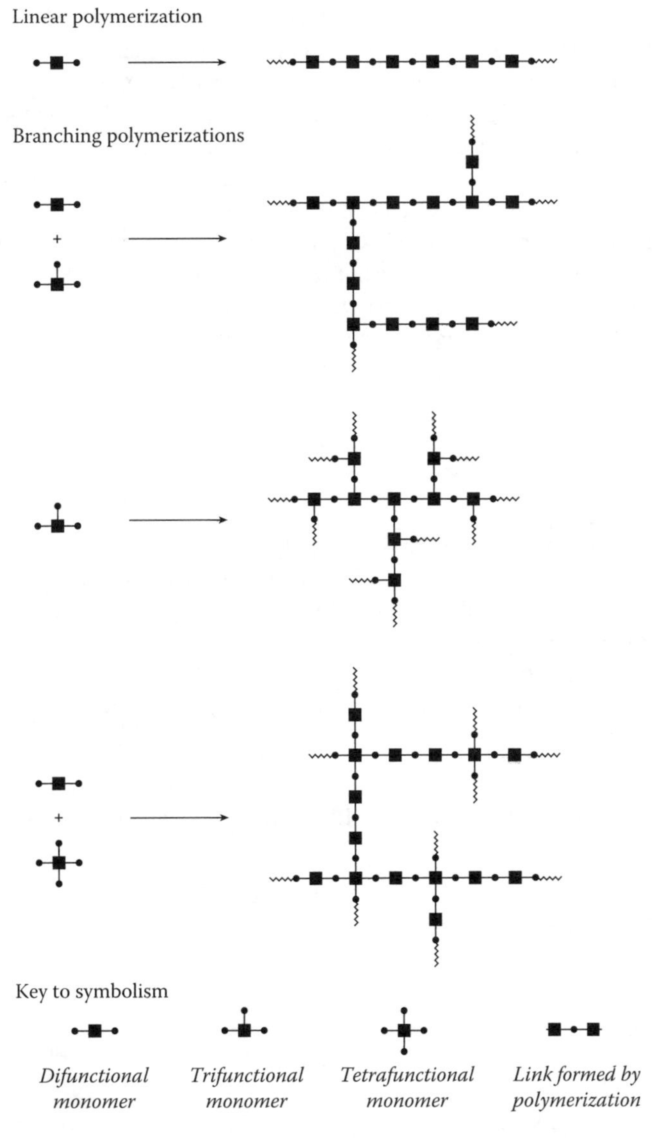

FIGURE 2.1 Schematic representations of how polymer skeletal structure is affected by monomer functionality.

reactant stoichiometry. The effect of monomer functionality is illustrated schematically in Figure 2.1. Although this depiction of the effect of monomer functionality is very much an oversimplification, it nevertheless demonstrates the importance of functionality. The formation of dendrimers and hyperbranched polymers is achieved using monomers with functionalities of three or higher, but in a way that is controlled so that a network polymer cannot form.

2.4 FUNCTIONAL GROUP REACTIVITY AND MOLECULAR SIZE: THE PRINCIPLE OF EQUAL REACTIVITY

Chemical reactions proceed as a consequence of collisions during encounters between mutually-reactive functional groups. At each encounter, the functional groups collide repeatedly until they either diffuse apart or, far more rarely, react. Under normal circumstances, the reactivity of a functional group depends upon its collision frequency and not upon the collision frequency of the molecule to which it is attached. As molecular size increases, the rate of molecular diffusion decreases, leading to larger time intervals between encounters (i.e. to fewer encounters per unit time). This effect is compensated by the greater duration of each encounter giving rise to a larger number of functional group collisions per encounter. Hence the reactivity of a functional group can be expected to be approximately independent of molecular size.

Mathematical analysis of polymerizations is simplified greatly by assuming that the intrinsic reactivity of a functional group is independent of molecular size and unaffected by the reaction of other functional group(s) in the molecule of monomer from which it is derived. This principle of *equal reactivity of functional groups* was proposed by Flory who demonstrated its validity for functional groups in many step polymerizations by examining the kinetics of model reactions. Similarly, analysis of the kinetics of chain polymerizations shows that it is reasonable to assume that the reactivity of the active species at the chain end is independent of the degree of polymerization.

PROBLEMS

2.1 Polymerization of vinyl chloride produces poly(vinyl chloride) of high molar mass at low monomer conversions, whereas polymerization of ethylene glycol with terephthalic acid produces poly(ethylene terephthalate) of high molar mass only at very high conversions of the –OH and $-CO_2H$ groups. (See entries (6) and (9) in Table 1.1 for the chemical structures of the monomers and polymers.)

For each polymerization, explain what the observations indicate in terms of the type of polymerization taking place.

2.2 The following rate coefficient k_{ester} data were obtained from studies of the model esterification reactions shown below. (Data taken from Bhide, B.V. and Sudborough, J.J., *J. Indian Inst. Sci.*, 8A, 89, 1925.)

$$H{-}(CH_2)_x{-}\underset{\underset{O}{\|}}{C}{-}OH + HO{-}CH_2CH_3 \xrightarrow[\text{HCl, 25 °C}]{k_{ester}} H{-}(CH_2)_x{-}\underset{\underset{O}{\|}}{C}{-}O{-}CH_2CH_3 + H_2O$$

x	1	2	3	4	5	8	9	11	13	15	17
k_{ester} / 10^{-4} dm^3 mol^{-1} s^{-1}	22.1	15.3	7.5	7.5	7.4	7.5	7.4	7.6	7.5	7.7	7.7

Discuss briefly the significance of these data for kinetics analysis of polyester formation.

2.3 Describe the effects on the *skeletal* structure of the polymer produced, if some ethylene glycol dimethacrylate were included in a polymerization of methyl methacrylate (see entry (5) of Table 1.1).

$$\begin{array}{c} \quad CH_3 \qquad\qquad\qquad\qquad\qquad CH_3 \\ CH{=}\overset{|}{C}{-}\underset{\|}{C}{-}O{-}(CH_2)_2{-}O{-}\underset{\|}{C}{-}\overset{|}{C}{=}CH_2 \\ \qquad\quad O \qquad\qquad\qquad\qquad O \end{array}$$

Ethylene glycol dimethacrylate

FURTHER READING

Cowie, J.M.G. and Arrighi, V., *Polymers: Chemistry and Physics of Modern Materials*, 3rd edn., CRC Press, Boca Raton, FL, 2007.

Flory, P.J., *Principles of Polymer Chemistry*, Cornell University Press, Ithaca, NY, 1953.

Odian, G., *Principles of Polymerization*, 4th edn., Wiley-Interscience, New York, 2004.

3 Step Polymerization

3.1 INTRODUCTION

Step polymerizations are, in most cases, logical extensions of simple organic chemistry linking reactions to chain formation and were the first types of polymerizations to be understood from a fundamental perspective, the work of Carothers during the 1920s and 1930s being seminal in this respect. In this chapter, the most important types of step polymerization are introduced and exemplified. The mechanisms of the reactions are not shown because they are identical to the well-established mechanisms of the equivalent small molecule reactions which can be found in any university-level textbook on organic chemistry.

Most step polymerizations involve reactions which produce links that contain a heteroatom and so the polymers normally are grouped into generic classes according to the type of links created in the polymerization. Some of the most common classes of polymers named according to the linking group in the chain backbone are shown in Table 3.1.

The majority of step polymerizations are based on quite simple linking reactions and, for linear step polymerizations, the greatest challenge is the need to take the reactions to extremely high conversions in order to produce chains of sufficient length to realize useful properties. This is very demanding and, as a consequence, most polymers produced by linear step polymerization have molar masses in the range 10–100 kg mol^{-1}, but more often 15–30 kg mol^{-1}, which is much lower than for polymers prepared by chain polymerizations (typically 50–10,000 kg mol^{-1}).

Although this chapter is much shorter than the overall coverage of chain polymerization, this is because of the relative simplicity in the chemistry of step polymerizations and the diversity and relative complexity of the different types of chain polymerization (for which the mechanistic chemistry is presented in the following chapters because it is not covered in standard textbooks on organic chemistry). Thus, despite the shorter coverage in terms of the number of pages devoted to step polymerization, it is important to emphasize that step polymerizations are used extensively for commercial production of polymers, including many of those encountered in everyday life, such as polyesters, polyamides, polyurethanes, polycarbonate and epoxy resins, as well as many high-performance polymers (e.g. Kevlar, PEEK).

3.2 LINEAR STEP POLYMERIZATION

Step polymerizations involve successive reactions between pairs of mutually-reactive functional groups which initially are provided by the monomer(s). The number of functional groups present on a molecule of monomer is of crucial importance, as can be appreciated by considering the formation of ester linkages from the condensation reaction of carboxylic acid groups with hydroxyl groups. Acetic acid and ethyl alcohol are *monofunctional* compounds which upon reaction together yield ethyl acetate with the elimination of water

$$CH_3{-}\underset{\underset{O}{\|}}{C}{-}OH + HO{-}CH_2CH_3 \longrightarrow CH_3{-}\underset{\underset{O}{\|}}{C}{-}O{-}CH_2CH_3 + H_2O$$

TABLE 3.1
Some Common Classes of Polymer Named according to the Heteroatom-Containing Linking Group in the Chain Backbone

Class of Polymer	Structure of Linking Group
Polyether	$\sim\!\sim -\text{O}- \sim\!\sim$
Polysulphide	$\sim\!\sim -\text{S}- \sim\!\sim$
Polyester	$\sim\!\sim -\overset{\overset{\displaystyle \text{O}}{\|}}{\text{C}}-\text{O}- \sim\!\sim$
Polycarbonate	$\sim\!\sim -\text{O}-\overset{\overset{\displaystyle \text{O}}{\|}}{\text{C}}-\text{O}- \sim\!\sim$
Polyamide	$\sim\!\sim -\overset{\overset{\displaystyle \text{O}}{\|}}{\text{C}}-\underset{\underset{\displaystyle \text{H}}{\vert}}{\text{N}}- \sim\!\sim$
Polyurethane	$\sim\!\sim -\underset{\underset{\displaystyle \text{H}}{\vert}}{\text{N}}-\overset{\overset{\displaystyle \text{O}}{\|}}{\text{C}}-\text{O}- \sim\!\sim$
Polyurea	$\sim\!\sim -\underset{\underset{\displaystyle \text{H}}{\vert}}{\text{N}}-\overset{\overset{\displaystyle \text{O}}{\|}}{\text{C}}-\underset{\underset{\displaystyle \text{H}}{\vert}}{\text{N}}- \sim\!\sim$

but because ethyl acetate is incapable of further reaction, a polymer chain cannot form. Now consider the reaction between terephthalic acid and ethylene glycol, both of which are *difunctional*

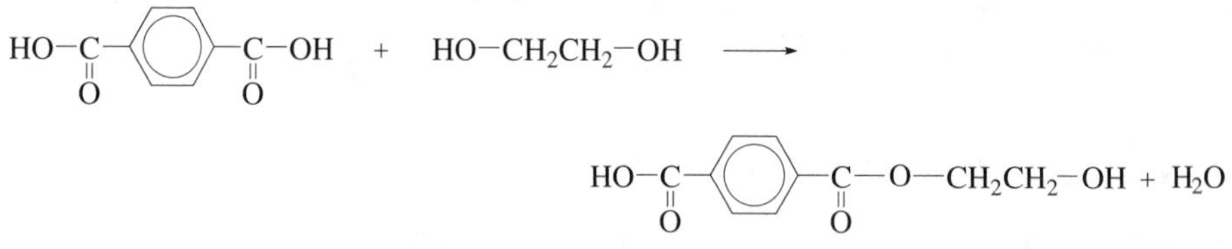

$$\text{HO}-\underset{\underset{\displaystyle \text{O}}{\|}}{\text{C}}-\text{C}_6\text{H}_4-\underset{\underset{\displaystyle \text{O}}{\|}}{\text{C}}-\text{OH} + \text{HO}-\text{CH}_2\text{CH}_2-\text{OH} \longrightarrow$$

$$\text{HO}-\underset{\underset{\displaystyle \text{O}}{\|}}{\text{C}}-\text{C}_6\text{H}_4-\underset{\underset{\displaystyle \text{O}}{\|}}{\text{C}}-\text{O}-\text{CH}_2\text{CH}_2-\text{OH} + \text{H}_2\text{O}$$

The product of their reaction is an ester which possesses one carboxylic acid end-group and one hydroxyl end-group (i.e. it also is difunctional). This *dimer*, therefore, can react with other molecules of terephthalic acid, ethylene glycol or dimer leading to the formation of *difunctional trimers* or *difunctional tetramer*. Growth of linear polymer chains then proceeds via further condensation reactions in the manner indicated for step polymerization in Table 2.1. Hence *linear step polymerizations* involve reactions of *difunctional monomers*. If a trifunctional monomer were included, reaction at each of the three functional groups would lead to the formation of a branched polymer and may ultimately result in the formation of a network. For example, if terephthalic acid were reacted with glycerol, $HOCH_2CH(OH)CH_2OH$, the product would be a non-linear polyester. It follows that polymerizations involving monomers of functionality greater than two will produce non-linear polymers, as was shown schematically in Figure 2.1 and is covered in detail in Section 3.3.

3.2.1 Polycondensation

Step polymerizations that involve reactions in which small molecules are eliminated are termed *polycondensations.*

3.2.1.1 Synthesis of Polyesters, Polyamides and Polyethers by Polycondensation

The formation of linear *polyesters* from reaction of carboxylic acid or acid halide groups with alcohol groups, as described in Section 3.2, is a classic example of polycondensation and may be represented more generally by

$$n\ \mathrm{HO{-}\underset{\|\atop O}{C}{-}R_1{-}\underset{\|\atop O}{C}{-}OH} + n\ \mathrm{HO{-}R_2{-}OH} \longrightarrow \mathrm{HO{+}\underset{\|\atop O}{C}{-}R_1{-}\underset{\|\atop O}{C}{-}O{-}R_2{-}O{+}_n H} + (2n-1)\,\mathrm{H_2O}$$

$$n\ \mathrm{X{-}\underset{\|\atop O}{C}{-}R_1{-}\underset{\|\atop O}{C}{-}X} + n\ \mathrm{HO{-}R_2{-}OH} \longrightarrow \mathrm{X{+}\underset{\|\atop O}{C}{-}R_1{-}\underset{\|\atop O}{C}{-}O{-}R_2{-}O{+}_n H} + (2n-1)\,\mathrm{HX}$$

$\mathrm{X = Cl\ or\ Br}$

where R_1 and R_2 represent any divalent group (usually hydrocarbon). Reaction between carboxylic acid and alcohol groups is slow and the reactions have to be performed at moderate-to-high temperatures (80–300 °C), usually in the presence of an acid catalyst. The reaction between terephthalic acid and ethylene glycol shown in Section 3.2 produces poly(ethylene terephthalate), the most widely used polyester. Carboxylic acid halides are much more reactive and their reaction with alcohol groups to produce polyesters proceeds at room temperature, so they often are used in simple laboratory syntheses of polyesters.

Reactions of the type shown above are referred to as RA_2+RB_2 *step polymerizations* where R is any divalent group and A and B represent the mutually-reactive functional groups.

Polyesters also can be prepared from single monomers which contain both types of functional group, i.e. ω-hydroxy carboxylic acids (but not ω-hydroxy carboxylic acid halide monomers because they cannot be synthesized due to the high reactivity of carboxylic acid halide groups towards hydroxyl groups)

$$n\ \mathrm{HO{-}R{-}\underset{\|\atop O}{C}{-}OH} \longrightarrow \mathrm{H{+}O{-}R{-}\underset{\|\atop O}{C}{+}_n OH} + (n-1)\,\mathrm{H_2O}$$

With each condensation reaction, the polymer chain grows but remains an ω-hydroxy carboxylic acid and so can react further. This is an example of an *ARB step polymerization*. The use of monomers of this type has the advantage that, provided they are pure, an exact stoichiometric equivalence of the two functional groups is guaranteed. As described quantitatively in Section 3.2.3.1, very slight excesses of one monomer in a RA_2+RB_2 polymerization reduce significantly the attainable degree of polymerization because the polymer chains become terminated with functional groups derived from the monomer present in excess (e.g. both end-groups are ultimately of type B if RB_2 is in excess). Since these functional groups are unreactive towards each other, further growth of the chains is not possible.

Polyamides can be prepared by polycondensations analogous to those used to prepare polyesters, the hydroxyl groups simply being replaced by amine groups, e.g.

$$n\ \mathrm{HO{-}\underset{\|\atop O}{C}{-}R_1{-}\underset{\|\atop O}{C}{-}OH} + n\ \mathrm{H_2N{-}R_2{-}NH_2} \longrightarrow \mathrm{HO{+}\underset{\|\atop O}{C}{-}R_1{-}\underset{\|\atop O}{C}{-}\overset{H}{N}{-}R_2{-}\overset{H}{N}{+}_n H} + (2n-1)\,\mathrm{H_2O}$$

$$n\,X-\underset{\underset{O}{\|}}{C}-R_1-\underset{\underset{O}{\|}}{C}-X + n\,H_2N-R_2-NH_2 \longrightarrow X\left[\underset{\underset{O}{\|}}{C}-R_1-\underset{\underset{O}{\|}}{C}-\overset{\overset{H}{|}}{N}-R_2-\overset{\overset{H}{|}}{N}\right]_n H + (2n-1)\,HX$$

$X = Cl$ or Br

$$n\,H_2N-R-\underset{\underset{O}{\|}}{C}-OH \longrightarrow H\left[\overset{\overset{H}{|}}{N}-R-\underset{\underset{O}{\|}}{C}\right]_n OH + (n-1)\,H_2O$$

Aliphatic polyamides are called *nylons*, the most important of which are prepared by RA_2+RB_2 polymerization, e.g. nylon 6.6 ($R_1=(CH_2)_4$; $R_2=(CH_2)_6$) and nylon 6.10 ($R_1=(CH_2)_8$; $R_2=(CH_2)_6$), where the first number in the name is the number of carbon atoms in the diamine monomer and the second number is the number of carbon atoms in the diacid (or diacid halide) monomer. Control of the number of carbon atoms in each structural unit is important because this determines the extent to which the amide links in separate chains can align to form hydrogen bonds (see Section 17.8.3.1).

The formation of *polyethers* by dehydration of diols is one of relatively few examples of RA_2 polycondensation

$$n\,HO-R-OH \longrightarrow HO\left[R-O\right]_n H + (n-1)\,H_2O$$

but the polymerization is not very controlled. A much better method for synthesis of polyethers is the RA_2+RB_2 polycondensation of dihalides with dialkoxides

$$n\,X-R_1-X + n\,\overset{+}{Mt}\,\overset{-}{O}-R_2-\overset{-}{O}\,\overset{+}{Mt} \longrightarrow X-R_1\left[O-R_2-O-R_1\right]_{n-1} O-R_2-\overset{-}{O}\,\overset{+}{Mt} + (2n-1)\,MtX$$

$X = Cl, Br, I$
$Mt = Li, Na, K$

However, most polyethers are prepared by the ring-opening polymerization of epoxides, which is described in detail in Chapter 7.

3.2.1.2 Synthesis of Engineering and High-Performance Polymers by Polycondensation

Polymers with aromatic groups (e.g. 1,4-phenylene units) in the chain backbone tend to have improved mechanical properties and greater resistance to degradation on exposure to heat or radiation, effects which arise from the stiffening effect and high stability of aromatic groups. Such polymers are most easily prepared by step polymerization and find use in engineering and high-performance applications. Some specific examples of important aromatic polymers prepared by polycondensation are shown in Table 3.2.

3.2.1.3 Synthesis of Conducting Polymers by Polycondensation

Some important conducting polymers (Section 25.3.4) can be prepared by polycondensation. Several polycondensation routes to *trans*-poly(1,4-phenylene vinylene) have been developed, some of which are shown below.

(i) $RA_2 + RB_2$ polyaddition via Wittig coupling

$$n\,\overset{-}{Cl}\,\overset{+}{P}Ph_3CH_2-C_6H_4-CH_2\overset{+}{P}Ph_3\,\overset{-}{Cl} + n\,OHC-C_6H_4-CHO \xrightarrow[20\,°C]{\text{LiOEt catalysis}} \left[C_6H_4-CH{=}CH\right]_n$$

TABLE 3.2
Some Engineering and High-Performance Aromatic Polymers Prepared by Polycondensation

Polycondensation	Comments
$n\ HO-C_6H_4-C(CH_3)_2-C_6H_4-OH + n\ Cl-C(=O)-Cl \longrightarrow [-O-C_6H_4-C(CH_3)_2-C_6H_4-O-C(=O)-]_n + (2n-1)\ HCl$ Polycarbonate (PC)	Mouldings and sheet; transparent and tough; used for safety glasses, screens and glazing, e.g. 'Lexan', 'Merlon'
$n\ KO-C_6H_4-S(=O)_2-C_6H_4-Cl \longrightarrow [-O-C_6H_4-S(=O)_2-C_6H_4-]_n + (n-1)\ KCl$ Polyethersulphone (PES)	Mouldings, coatings, membranes, e.g. 'Victrex PES'
$n\ KO-C_6H_4-OK + n\ F-C_6H_4-C(=O)-C_6H_4-F \longrightarrow [-O-C_6H_4-O-C_6H_4-C(=O)-C_6H_4-]_n + (2n-1)\ KF$ Polyetheretherketone (PEEK)	Mouldings, composites, bearings, coatings; very high continuous use temperature (260 °C), e.g. 'Victrex PEEK'
$n\ Cl-C_6H_4-Cl + n\ Na_2S \longrightarrow [-C_6H_4-S-]_n + (2n-1)\ NaCl$ Poly(phenylene sulphide) (PPS)	Mouldings, composites, coatings, e.g. 'Ryton', 'Tedur,' 'Fortron'
$n\ Cl-C(=O)-C_6H_4-C(=O)-Cl + n\ H_2N-C_6H_4-NH_2 \longrightarrow [-C(=O)-C_6H_4-C(=O)-N(H)-C_6H_4-N(H)-]_n + (2n-1)\ HCl$ Poly(*p*-phenylene terephthalamide) (PPTA)	High-modulus fibres, e.g. 'Kevlar', 'Twaron'

(continued)

TABLE 3.2 (continued)
Some Engineering and High-Performance Aromatic Polymers Prepared by Polycondensation

Polycondensation	Comments
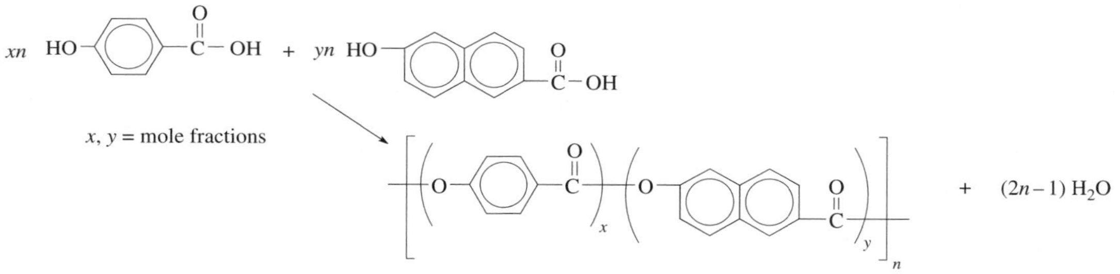 xn HO–C₆H₄–COOH + yn HO–C₁₀H₆–COOH → + $(2n-1)\ H_2O$; x, y = mole fractions. Poly[(1,4-benzoate)-*co*-(2,6-naphthoate)]	Liquid crystalline polyester for mouldings and high-modulus fibres, e.g. 'Vectra', 'Vetran'
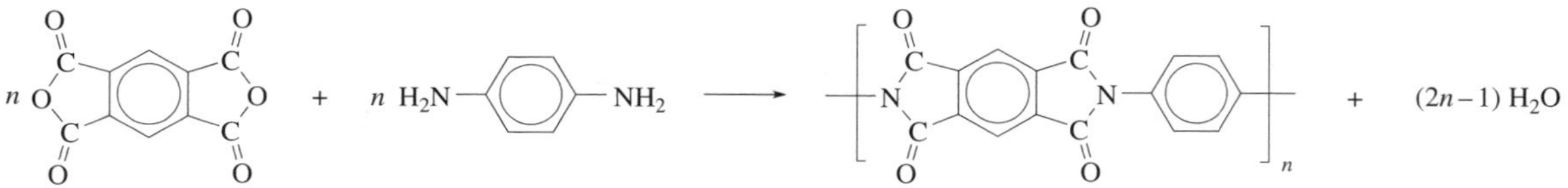 n (pyromellitic dianhydride) + n H_2N–C₆H₄–NH_2 → + $(2n-1)\ H_2O$. A polyimide	Films, coatings, adhesives, laminates, e.g. 'Kapton', 'Vespel'
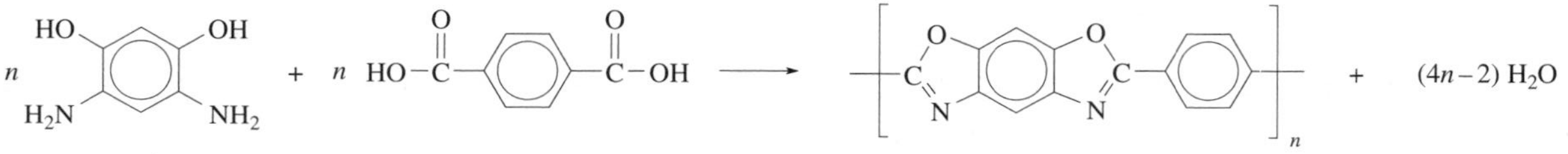 n (HO)₂C₆H₂(NH_2)₂ + n HO–C(=O)–C₆H₄–C(=O)–OH → + $(4n-2)\ H_2O$. Poly(*p*-phenylene benzoxazole) (PBO)	High-modulus fibres, e.g. 'Zylon'

(ii) $RA_2 + RB_2$ polyaddition via Heck coupling

$$n\ \mathrm{Br{-}C_6H_4{-}Br} + n\ \mathrm{CH_2{=}CH_2} \xrightarrow[100\ ^\circ\mathrm{C}]{\mathrm{Pd(OCOMe)_2\ catalysis}} \left[\mathrm{C_6H_4{-}CH{=}CH}\right]_n$$

(iii) RA_2 polyaddition via McMurry coupling

$$n\ \mathrm{OHC{-}C_6H_4{-}CHO} \xrightarrow[85\ ^\circ\mathrm{C}]{\mathrm{TiCl_3/Zn/Cu\ catalysis}} \left[\mathrm{C_6H_4{-}CH{=}CH}\right]_n$$

Regioregular poly(3-alkyl-2,5-thiophene)s can be synthesized by ARB polyaddition of semi-Grignard reagents of 2,5-dibromo-3-alkylthiophenes

$$2n\ \mathrm{BrMg{-}C_4HS(R){-}Br} \xrightarrow[-5-25\ ^\circ\mathrm{C}]{\mathrm{Ni\{Ph_2P(CH_2)_3PPh_2\}Cl_2\ catalysis}} \left[\mathrm{C_4HS(R){-}C_4HS(R)}\right]_n$$

R = alkyl

The mechanisms of the reactions described in this section are quite complex, so are not considered here, but they are described in reviews and in more advanced textbooks on organic chemistry.

3.2.1.4 Synthesis of Polysiloxanes by Polycondensation

Polysiloxanes (also known simply as *siloxanes*) are unusual in that they have a completely inorganic backbone of –Si–O– bonds, which gives them high thermal stability. Synthesis of polysiloxanes by polycondensation is achieved through hydrolysis of highly-reactive dichlorodialkylsilanes

$$n\ \mathrm{Cl{-}Si(R_1)(R_2){-}Cl} + (n+1)\ \mathrm{H_2O} \longrightarrow \mathrm{HO}\left[\mathrm{Si(R_1)(R_2){-}O}\right]_n\mathrm{H} + (2n\ \mathrm{HCl})$$

where R_1 and R_2 can be alkyl (e.g. methyl) or aryl (e.g. phenyl) groups and may be different or the same. The polycondensation is unusual in that after partial hydrolysis of the monomer via

$$\mathrm{Cl{-}Si(R_1)(R_2){-}Cl} + \mathrm{H_2O} \longrightarrow \mathrm{Cl{-}Si(R_1)(R_2){-}OH} + \mathrm{HCl}$$

$$\mathrm{Cl{-}Si(R_1)(R_2){-}OH} + \mathrm{H_2O} \longrightarrow \mathrm{HO{-}Si(R_1)(R_2){-}OH} + \mathrm{HCl}$$

$RA_2 + RB_2$ polymerization

$$Cl-\underset{R_2}{\overset{R_1}{\underset{|}{\overset{|}{Si}}}}-Cl + HO-\underset{R_2}{\overset{R_1}{\underset{|}{\overset{|}{Si}}}}-OH \longrightarrow Cl-\underset{R_2}{\overset{R_1}{\underset{|}{\overset{|}{Si}}}}-O-\underset{R_2}{\overset{R_1}{\underset{|}{\overset{|}{Si}}}}-OH + HCl$$

$RA_2 + ARB$ and $RB_2 + ARB$ polymerization

$$Cl-\underset{R_2}{\overset{R_1}{\underset{|}{\overset{|}{Si}}}}-Cl + HO-\underset{R_2}{\overset{R_1}{\underset{|}{\overset{|}{Si}}}}-Cl \longrightarrow HO-\underset{R_2}{\overset{R_1}{\underset{|}{\overset{|}{Si}}}}-O-\underset{R_2}{\overset{R_1}{\underset{|}{\overset{|}{Si}}}}-Cl + HCl$$

$$HO-\underset{R_2}{\overset{R_1}{\underset{|}{\overset{|}{Si}}}}-OH + Cl-\underset{R_2}{\overset{R_1}{\underset{|}{\overset{|}{Si}}}}-OH \longrightarrow HO-\underset{R_2}{\overset{R_1}{\underset{|}{\overset{|}{Si}}}}-O-\underset{R_2}{\overset{R_1}{\underset{|}{\overset{|}{Si}}}}-OH + HCl$$

and RA_2 polymerization

$$HO-\underset{R_2}{\overset{R_1}{\underset{|}{\overset{|}{Si}}}}-OH + HO-\underset{R_2}{\overset{R_1}{\underset{|}{\overset{|}{Si}}}}-OH \longrightarrow HO-\underset{R_2}{\overset{R_1}{\underset{|}{\overset{|}{Si}}}}-O-\underset{R_2}{\overset{R_1}{\underset{|}{\overset{|}{Si}}}}-OH + H_2O$$

can occur simultaneously. In order to control the degree of polymerization attained upon complete hydrolysis (as indicated in the general equation above), it is usual to include monofunctional chlorosilanes, e.g. chlorotrimethylsilane would lead to a polysiloxane with unreactive trimethylsilyl end groups

$$CH_3-\underset{CH_3}{\overset{CH_3}{\underset{|}{\overset{|}{Si}}}}-O-\left[\underset{R_2}{\overset{R_1}{\underset{|}{\overset{|}{Si}}}}-O\right]_n-\underset{CH_3}{\overset{CH_3}{\underset{|}{\overset{|}{Si}}}}-CH_3$$

The most important monomer is dichlorodimethylsilane ($R_1 = R_2 = CH_3$) which produces poly(dimethyl siloxane) (PDMS)

$$\left[\underset{CH_3}{\overset{CH_3}{\underset{|}{\overset{|}{Si}}}}-O\right]_n$$

PDMS

A massive range of commercial grades of PDMS are available, many of which are copolymers prepared by including low levels of other dichlorosilanes as comonomers, e.g. dichlorodiphenylsilane ($R_1 = R_2 = Ph$), which provides repeat units that further enhance thermal stability, and dichloromethylvinylsilane ($R_1 = CH_3$; $R_2 = CH{=}CH_2$), which provides C=C bonds for use in subsequent crosslinking. Polycondensation is used to prepare low-moderate molar mass grades of PDMS, which are liquids or soft solids. High molar mass grades for use as elastomers tend to be produced by ring-opening polymerization of cyclic siloxanes, as described in Chapter 7.

3.2.2 Polyaddition

Step polymerizations in which the monomers react together without the elimination of other molecules are termed *polyadditions*. In contrast to polycondensation, there are relatively few important polymers prepared by polyaddition, the most important being polyurethanes and epoxy resins.

3.2.2.1 Synthesis of Linear Polyurethanes and Polyureas by Polyaddition

Linear *polyurethanes* are prepared by RA_2+RB_2 polyaddition of diisocyanates with diols

$$n\,O{=}C{=}N{-}R_1{-}N{=}C{=}O \;+\; n\,HO{-}R_2{-}OH \longrightarrow \left[\overset{O}{\overset{\|}{C}}{-}\underset{H}{\underset{|}{N}}{-}R_1{-}\underset{H}{\underset{|}{N}}{-}\overset{O}{\overset{\|}{C}}{-}O{-}R_2{-}O\right]_n$$

which proceeds quite rapidly at room temperature because isocyanate groups are highly reactive; nevertheless, catalysts often are used to increase the rate of polymerization. The most commonly used diisocyanates are

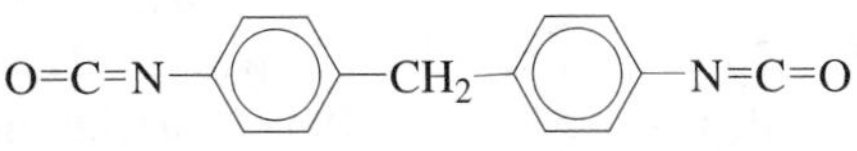

$O{=}C{=}N{-}C_6H_4{-}CH_2{-}C_6H_4{-}N{=}C{=}O$

bis(4-isocyanatophenyl)methane,
methylene diphenylene diisocyanate (MDI)

$O{=}C{=}N{-}(CH_2)_6{-}N{=}C{=}O$

1,6-diisocyanatohexane,
hexamethylene diisocyanate (HDI)

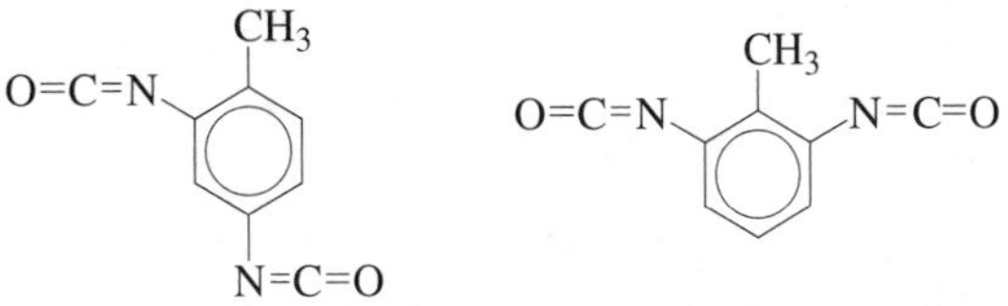

CH_3, $O{=}C{=}N$, $N{=}C{=}O$ (2,4-isomer); CH_3, $O{=}C{=}N$, $N{=}C{=}O$ (2,6-isomer)

diisocyanatotoluene, toluene diisocyanate (TDI)
(mixed 2,4- and 2,6- isomers; 80% 2,4- in commercial TDI)

An extremely wide range of diols can be used, including simple diols and low molar mass α,ω-dihydroxy polyethers, e.g.

$$HO{-}(CH_2)_y{-}OH \qquad HO{-}(CH_2)_x{-}\overset{R}{\overset{|}{C}H}\left[O{-}(CH_2)_x{-}\overset{R}{\overset{|}{C}H}\right]_{n-1}OH$$

e.g. $y = 2,3,4,6$ e.g. $x = 1$, R = H or CH_3; $x = 3$, R = H

Historically, α,ω-dihydroxy polyesters also have been used, but polyether diols are now preferred. The choice of diisocyanate and diol determines the properties of the polyurethane, which can range from being a rigid solid (if low molar mass diols are used) to a rubbery material (e.g. if poly(propylene oxide) diols with molar masses of about 2–8 kg mol^{-1} are used). Many commercial linear polyurethanes are *segmented copolymers* prepared using a mixture of short- and long-chain diols (see Section 9.4.1).

The analogous reaction of diisocyanates with diamines yields *polyureas*

$$n\,O{=}C{=}N{-}R_1{-}N{=}C{=}O \;+\; n\,H_2N{-}R_2{-}NH_2 \longrightarrow \left[\overset{O}{\overset{\|}{C}}{-}\underset{H}{\underset{|}{N}}{-}R_1{-}\underset{H}{\underset{|}{N}}{-}\overset{O}{\overset{\|}{C}}{-}\underset{H}{\underset{|}{N}}{-}R_2{-}\underset{H}{\underset{|}{N}}\right]_n$$

and proceeds at very high rates, such that intimate mixing of the monomers requires specialist high-speed mixing equipment. The diisocyanates used are the same as for preparation of polyurethanes. In addition to simple aliphatic amines and α,ω-diamino polyethers akin to the diols used in polyurethane synthesis, aromatic diamines also are used. By reacting diisocyanates with diols and diamines, poly(urethane-*co*-urea)s are produced.

3.2.2.2 Other Polymers Prepared by Polyaddition

As stated above, besides polyurethanes and polyureas, relatively few polymers are prepared by polyaddition. Two further types of polyaddition will be considered here.

The Diels–Alder reaction of dienes with dienophiles is a 4+2 cycloaddition

R_1 R R_2 R Δ R_1 R R_2 R

where R_1, R_2 and R represent substituent groups, and carbon and hydrogen atoms are omitted for clarity. The reaction proceeds at temperatures of 25–150 °C and works best when R_1 and R_2 are electron donating and R is electron withdrawing (or vice versa), but it is reversible and the retro-reaction occurs at high temperatures. A wide variety of polymers have been prepared by Diels–Alder polyaddition, of which two examples will be given. Bis(2-buta-1,3-dienyl)methylacetal undergoes RA_2+RB_2 polyaddition with 1,4-benzoquinone to give a soluble amorphous Diels–Alder polymer

n O O + n O O 100–150 °C O O O O O O O O $n-1$

A more complex example involves the Diels–Alder reaction of 2-vinylbuta-1,3-diene with 1,4-benzoquinone, which first produces a monomeric adduct that then undergoes ARB polyaddition at higher temperature

O O + 25 °C O O O O 145 °C O O O O O O O O O O O O O O $2n$ O O O O O O O O $n+1$

The polymer is insoluble except at very low degrees of polymerization and the chain is completely rigid with a more crooked contour than indicated here because there are two possible orientations for the addition reaction; the structure shown arises from alternation of the orientation. Polymers of this type are called *ladder polymers*, based on the ladder-like nature of their backbone, and have improved thermal stability because two main-chain bonds must break for chain scission to occur.

Diels–Alder polymerizations largely have remained curiosities, but the RA_2+RB_2 polyaddition of diamines with bismaleimides has been commercialized. A generic example is shown below

n' [bismaleimide: N—Ar—N] + n $H_2N—Ar'—NH_2$

Excess (i.e. $n' > n$)

[maleimide-terminated oligomer: N—Ar—N … NH—Ar'—NH … N—Ar—N … NH—Ar'—NH … N—Ar—N]$_{n-1}$

where Ar and Ar′ are aromatic groups (e.g. Ar = Ar′ = —C$_6$H$_4$—CH_2—C$_6$H$_4$—). As shown above, using an excess of bismaleimide produces oligomers with maleimide end-groups, which are known generically as *bismaleimide (BMI) resins.* This is only one of many methods for synthesis of BMI resins and a very wide range of structurally different BMI resins are available commercially for use as matrixes in composites and as high-performance adhesives. The BMI resins are converted to crosslinked materials *in situ* during processing by a second-stage reaction, e.g. free-radical polymerization of the C=C bonds in the maleimide end-groups or further reaction with polyfunctional amines.

3.2.3 Theoretical Treatment of Linear Step Polymerization

The principle of equal reactivity of functional groups (described in Section 2.4) is fundamental to simplification of the theoretical treatment of step polymerization and was shown to be valid for the most common polymerizations by the experimental studies of Flory. On this basis, step polymerization involves random reactions occurring between any two mutually-reactive molecular species. Intrinsically, each of the possible reactions is equally probable and their relative preponderances depend only upon the relative numbers of each type of molecular species (i.e. monomer, dimer, trimer, etc.). This assumption of equal reactivity is implicit in each of the theoretical treatments of step polymerization that follow.

3.2.3.1 Carothers Theory

Carothers developed a simple method of analysis for predicting the molar mass of polymers prepared by step polymerization. He recognized that the number-average degree of polymerization *with respect to monomer units* is given by the relation

$$\bar{x}_n = \frac{N_0}{N} \tag{3.1}$$

where

N_0 is the number of molecules present initially

N is the number of molecules remaining after a time t of polymerization

Assuming that there are equal numbers of mutually-reactive functional groups, $\bar{x}_n$ can be related to the *extent of reaction p* at time *t* which is given by

$$p = \frac{\text{Number of functional groups that have reacted}}{\text{Number of functional groups present initially}}$$

and is the probability that any functional group present initially has reacted. Since the total number of molecules decreases by one for each pair-wise reaction between functional groups

$$p = \frac{N_0 - N}{N_0} \quad \text{which rearranges to} \quad \frac{N_0}{N} = \frac{1}{1-p} \tag{3.2}$$

Combining Equations 3.1 and 3.2 gives the Carothers equation

$$\bar{x}_n = \frac{1}{1-p} \tag{3.3}$$

This equation is applicable to $RA_2 + RB_2$, ARB and RA_2 polymerizations in which there is an exact stoichiometric balance in the numbers of mutually-reactive functional groups. The equation highlights the need to attain very high extents of reaction of functional groups in order to produce polymers with useful physical properties. Normally, degrees of polymerization of the order of 100 or above are required, hence demanding values of $p \geq 0.99$. This clearly demonstrates the necessity for using monomers of high purity and chemical reactions that are either highly efficient or can be forced towards completion.

The number-average molar mass $\bar{M}_n$ is related to $\bar{x}_n$ by

$$\bar{M}_n = \bar{M}_0 \bar{x}_n$$

where $\bar{M}_0$ is the mean molar mass of a monomer unit and is given by

$$\bar{M}_0 = \frac{\text{Molar mass of the repeat unit}}{\text{Number of monomer units in the repeat unit}}$$

Slight stoichiometric imbalances significantly limit the attainable values of $\bar{x}_n$. Consider a $RA_2 + RB_2$ polymerization in which RB_2 is present in excess. The ratio of the numbers of the two different types of functional group (A and B) present initially is known as the *reactant ratio r*, and for linear step polymerization is always defined so that it is less than or equal to one. Thus for the reaction under consideration

$$r = \frac{N_A}{N_B} \tag{3.4}$$

where N_A and N_B are, respectively, the numbers of A and B functional groups present initially. Since there are two functional groups per molecule

$$N_0 = \frac{N_A + N_B}{2}$$

which upon substitution for N_A from Equation 3.4 gives

$$N_0 = \frac{N_B(1+r)}{2} \tag{3.5}$$

It is common practice to define the extent of reaction p in terms of the functional groups present in minority (i.e. A groups in this case). On this basis

$$\begin{aligned}\text{number of unreacted A groups} &= N_A - pN_A \\ &= rN_B(1-p) \\ \text{number of unreacted B groups} &= N_B - pN_A \\ &= N_B(1-rp)\end{aligned}$$

so that

$$N = \frac{rN_B(1-p) + N_B(1-rp)}{2}$$

i.e.

$$N = \frac{N_B(1+r-2rp)}{2} \tag{3.6}$$

Substitution of Equations 3.5 and 3.6 into Equation 3.1 yields the more *general Carothers equation*

$$\bar{x}_n = \frac{1+r}{1+r-2rp} \tag{3.7}$$

of which Equation 3.3 is the special case for $r=1$. Table 3.3 gives values of $\bar{x}_n$ calculated using Equation 3.7 and reveals the dramatic reduction in $\bar{x}_n$ when r is less than unity. Thus only very slight stoichiometric imbalances can be tolerated if useful polymers are to be formed, the corollary of which is that r must be controlled with great accuracy. It is now absolutely clear that, in order to control r with the necessary precision, the monomers used in linear step polymerizations must be of very high purity and that the linking reactions must be clean. Assuming these criteria are satisfied,

TABLE 3.3
Variation of $\bar{x}_n$ with p and r according to Equation 3.7

	$\bar{x}_n$ at				
r	$p=0.90$	$p=0.95$	$p=0.99$	$p=0.999$	$p=1.000$[a]
1.000	10.0	20.0	100.0	1000.0	∞
0.999	10.0	19.8	95.3	666.8	1999.0
0.990	9.6	18.3	66.8	166.1	199.0
0.950	8.1	13.4	28.3	37.6	39.0
0.900	6.8	10.0	16.1	18.7	19.0

[a] As $p \to 1, \bar{x}_n \to \frac{(1+r)}{(1-r)}$

Equation 3.7 can be used to exert control of molar mass by using slight imbalances in stoichiometry to place an upper limit on $\bar{x}_n$ for reactions taken to very high conversions of functional groups.

Equation 3.7 also is applicable to reactions in which a monofunctional compound is included to control $\bar{x}_n$, e.g. RA_2+RB_2+RB or ARB+RB. All that is required is to re-define the reactant ratio

$$r = \frac{N_A}{N_B + 2N_{RB}}$$

where

N_A and N_B are, respectively, the initial numbers of A and B functional groups from the difunctional monomer(s)

N_{RB} is the number of molecules of RB present initially.

The factor of 2 is required because one RB molecule has the same quantitative effect in limiting $\bar{x}_n$ as one excess RB_2 molecule.

3.2.3.2 Statistical Theory

The theory of Carothers is restricted to prediction of number-average quantities. In contrast, simple statistical analyses based upon the random nature of step polymerization allow prediction of degree of polymerization distributions. Such analyses were first described by Flory.

For simplicity RA_2+RB_2 and ARB polymerizations in which there is exactly equivalent stoichiometry will be considered here. The first stage in the analysis is to calculate the probability $P(x)$ of existence of a molecule consisting of exactly x monomer units at time t when the extent of reaction is p. A molecule containing x monomer units is created by the formation of a sequence of $(x-1)$ linkages. The probability that a particular sequence of linkages has formed is the product of the probabilities of forming the individual linkages. Since p is the probability that a functional group has reacted, the probability of finding a sequence of two linkages is p^2, the probability of finding a sequence of three linkages is p^3 and the probability of finding a sequence of $(x-1)$ linkages is $p^{(x-1)}$. For a molecule to contain exactly x monomer units, the xth (i.e. last) unit must possess a terminal unreacted functional group. The probability that a functional group has not reacted is $(1-p)$ and so

$$P(x) = (1-p)p^{(x-1)} \tag{3.8}$$

Since $P(x)$ is the probability that a molecule chosen at random contains exactly x monomer units, it must also be the *mole fraction* of x-mers. If the total number of molecules present at time t is N, then the total number N_x of x-mers is given by

$$N_x = N(1-p)p^{(x-1)} \tag{3.9}$$

Often N cannot be measured and so is eliminated by substitution of the rearranged form of Equation 3.2, $N=N_0(1-p)$, to give

$$N_x = N_0(1-p)^2 p^{(x-1)} \tag{3.10}$$

which is an expression for the number of molecules of degree of polymerization x in terms of the initial number of molecules N_0 and the extent of reaction p.

The weight fraction w_x of x-mers is given by

$$w_x = \frac{\text{Total mass of molecules with degree of polymerization } x}{\text{Total mass of all the molecules}}$$

Thus, neglecting end groups

$$w_x = \frac{N_x(x\bar{M}_0)}{N_0\bar{M}_0} = \frac{xN_x}{N_0} \tag{3.11}$$

Combining Equations 3.10 and 3.11 gives

$$w_x = x(1-p)^2 p^{(x-1)} \tag{3.12}$$

Equations 3.8 and 3.12 define what is known as the *most probable* (or *Flory* or *Flory–Schulz*) *distribution*, the most important features of which are illustrated by the plots shown in Figure 3.1. Thus the mole fraction $P(x)$ decreases continuously as the number of monomer units in the polymer chain increases, i.e. at all extents of reaction, the mole fraction of monomer is greater than that of any other species. In contrast, the weight fraction distribution shows a maximum at a value of x that is very close to $\bar{x}_n$. As the extent of reaction increases, the maximum moves to higher values of x and the weight fraction of monomer becomes very small.

Knowledge of the distribution functions (i.e. $P(x)$ and w_x) enables molar mass averages to be evaluated. From Equation 1.3 the number-average molar mass may be written as

$$\bar{M}_n = \sum P(x)M_x$$

Recognizing that $M_x = x\bar{M}_0$ and substituting for $P(x)$ using Equation 3.8 gives

$$\bar{M}_n = \sum x\bar{M}_0(1-p)p^{(x-1)}$$

$$\text{i.e. } \bar{M}_n = \bar{M}_0(1-p)\sum xp^{(x-1)}$$

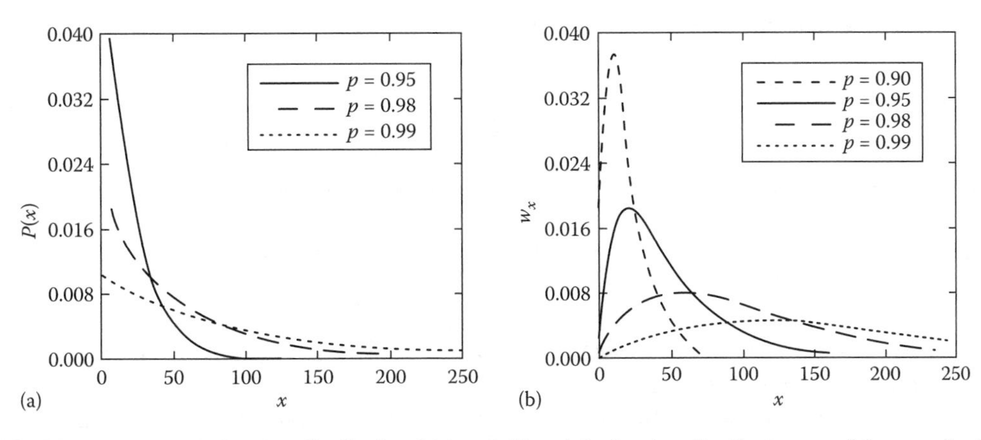

FIGURE 3.1 (a) Mole-fraction distribution $P(x)$ and (b) weight-fraction distribution w_x of degree of polymerisation x with respect to monomer units for various extents of reaction p in a linear step polymerisation with a reactant ratio $r=1$. Note that the values of number-average degree of polymerisation $\bar{x}_n$ with respect to monomer units corresponding to $p=0.90$, 0.95, 0.98 and 0.99 are, respectively: 10, 20, 50 and 100. (Data taken from Flory, *J. Am. Chem. Soc.*, 58, 1877, 1936.)

Using the mathematical relation

$$\sum_{x=1}^{\infty} xp^{(x-1)} = (1-p)^{-2} \quad \text{for } p < 1$$

the equation for $\bar{M}_n$ reduces to

$$\bar{M}_n = \frac{\bar{M}_0}{(1-p)} \tag{3.13}$$

Since $\bar{x}_n = \bar{M}_n/\bar{M}_0 = 1/(1-p)$ this equation is equivalent to the Carothers Equation 3.3, but this time it has been derived from purely statistical considerations.

The application of Equation 1.8 enables the weight-average molar mass to be written as

$$\bar{M}_w = \sum w_x M_x$$

Using Equation 3.12 it follows that

$$\bar{M}_w = \bar{M}_0(1-p)^2 \sum x^2 p^{(x-1)}$$

and another mathematical relation

$$\sum_{x=1}^{\infty} x^2 p^{(x-1)} = (1+p)(1-p)^{-3} \quad \text{for } p < 1$$

leads to

$$\bar{M}_w = \bar{M}_0 \frac{(1+p)}{(1-p)} \tag{3.14}$$

and hence to the weight-average degree of polymerization

$$\bar{x}_w = \frac{(1+p)}{(1-p)} \tag{3.15}$$

Combining Equations 3.13 and 3.14 gives an equation for the molar mass dispersity $\bar{M}_w/\bar{M}_n$

$$\frac{\bar{M}_w}{\bar{M}_n} = 1 + p \tag{3.16}$$

which for most linear polymers prepared by step polymerization is close to 2 (since high values of p are required to form useful polymers).

The mole fraction and weight fraction distributions for step polymerizations in which there is a stoichiometric imbalance are similar to those just derived for the case of exactly equivalent stoichiometry. Thus all linear step polymerizations lead to essentially the same form of molar mass distribution.

Before closing this section, it must again be emphasized that the degrees of polymerization given are with respect to monomer units *and not* repeat units.

3.2.3.3 Kinetics of Step Polymerization

The assumption of equal reactivity of functional groups also greatly simplifies the kinetics of step polymerization since a single rate coefficient applies to each of the step-wise reactions. It is usual to define the overall rate of reaction as the rate of decrease in the concentration of one or other of the functional groups, i.e. in general terms for equimolar stoichiometry

$$\text{Rate of reaction} = -\frac{\mathrm{d}[\mathrm{A}]}{\mathrm{d}t} = -\frac{\mathrm{d}[\mathrm{B}]}{\mathrm{d}t}$$

Most step polymerizations involve bimolecular reactions, which often are catalysed. Thus, neglecting elimination products in polycondensations, the general elementary reaction is

$$\sim\!\!\sim\mathrm{A} + \mathrm{B}\sim\!\!\sim + \text{catalyst} \longrightarrow \sim\!\!\sim\mathrm{AB}\sim\!\!\sim + \text{catalyst}$$

and so the rate of reaction is given by

$$-\frac{\mathrm{d}[\mathrm{A}]}{\mathrm{d}t} = k'[\mathrm{A}][\mathrm{B}][\text{Catalyst}] \tag{3.17}$$

where k' is the rate coefficient for the reaction. Since the concentration of a true catalyst does not change as the reaction proceeds, it is usual to simplify the expression by letting $k = k'[\text{Catalyst}]$ giving

$$-\frac{\mathrm{d}[\mathrm{A}]}{\mathrm{d}t} = k[\mathrm{A}][\mathrm{B}] \tag{3.18}$$

For equimolar stoichiometry $[\mathrm{A}] = [\mathrm{B}] = c$ and Equation 3.18 becomes

$$-\frac{\mathrm{d}c}{\mathrm{d}t} = kc^2$$

This equation may be integrated by letting $c = c_0$ at $t = 0$

$$\int_{c_0}^{c} -\frac{\mathrm{d}c}{c^2} = \int_0^t k\mathrm{d}t$$

and gives

$$\frac{1}{c} - \frac{1}{c_0} = kt$$

which may be rewritten in terms of the extent of reaction by recognizing that $c_0/c = N_0/N$ and applying Equation 3.2

$$\frac{1}{(1-p)} - 1 = c_0kt \tag{3.19}$$

This equation also applies to reactions which proceed in the absence of catalyst, though the rate coefficient is different and obviously does not include a term in catalyst concentration.

Certain step polymerizations are self-catalysed, that is, one of the types of functional group also acts as a catalyst (e.g. carboxylic acid groups in a polyesterification). In the absence of an added catalyst the rate of reaction for such polymerizations is given by

$$-\frac{d[A]}{dt} = k''[A][B][A] \tag{3.20}$$

assuming that the A groups catalyse the reaction. Again letting [A] = [B] = c, Equation 3.20 becomes

$$-\frac{dc}{dt} = k''c^3$$

which upon integration over the same limits as before gives

$$\frac{1}{c^2} - \frac{1}{c_0^2} = 2k''t$$

or in terms of the extent of reaction

$$\frac{1}{(1-p)^2} - 1 = 2c_0^2k''t \tag{3.21}$$

Equations 3.19 and 3.21 have been derived assuming that the reverse reaction (i.e. depolymerisation) is negligible. This is satisfactory for many polyadditions, but for reversible polycondensations, it requires the elimination product to be removed continuously as it is formed. The equations have been verified experimentally using step polymerizations that satisfy this requirement, as is shown by the polyesterification data plotted in Figure 3.2. These results further substantiate the validity of the principle of equal reactivity of functional groups.

3.2.4 Ring Formation

A complication not yet considered is the *intramolecular* reaction of terminal functional groups on the same molecule. This results in the formation of cyclic molecules (i.e. rings), e.g. in the preparation of a polyester

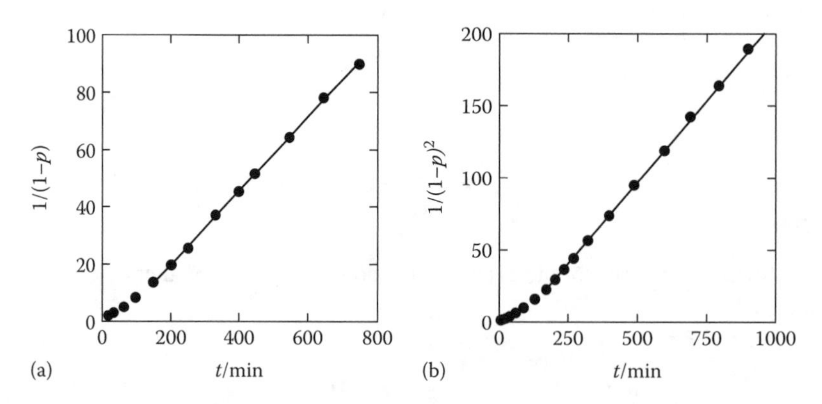

FIGURE 3.2 (a) Plot of $1/(1-p)$ as a function of time for the polymerization of diethylene glycol with adipic acid using p-toluene sulphonic acid as a catalyst at 109 °C and (b) plot of $1/(1-p)^2$ as a function of time for the polymerization of diethylene glycol with adipic acid at 166 °C. (Data taken from Flory, *J. Am. Chem. Soc.*, 61, 3334, 1939.)

$$\text{HO}\sim\sim\sim\overset{\overset{\text{O}}{\|}}{\text{C}}-\text{OH} \xrightarrow{-\text{H}_2\text{O}} \text{cyclic}\ \underset{\underset{\text{O}}{\|}}{\text{C}}-\text{O}$$

The ease of ring formation depends strongly upon the number of atoms linked together in the ring. For example, 5-, 6- and, to a lesser extent, 7-membered rings are stable and often form in preference to linear polymer. For the self-condensation of ω-hydroxy carboxylic acids $HO—(CH_2)_i—CO_2H$ when $i=3$ only the monomeric lactone is produced.

$$\text{HO}-(\text{CH}_2)_3-\overset{\overset{\text{O}}{\|}}{\text{C}}-\text{OH} \longrightarrow \begin{matrix}\text{CH}_2-\text{CH}_2\\ |\quad\quad |\\ \text{CH}_2\quad \text{O}\\ \diagdown\ \diagup\\ \text{C}\\ \|\\ \text{O}\end{matrix} + \text{H}_2\text{O}$$

When $i=4$, some polymer is produced in addition to the corresponding monomeric lactone, and when $i=5$, the product is a mixture of polymer with some of the monomeric lactone.

Normally, 3- and 4-membered rings and 8- to 11-membered rings are unstable due to bond-angle strain and steric repulsions between atoms crowded into the centre of the ring, respectively, and usually are not formed. Whilst 12-membered and larger rings are more stable and can form, their probability of formation decreases as the ring size increases. This is because the probability of the two ends of a single chain meeting decreases as their separation (i.e. the chain length) increases. Thus large rings rarely form.

Ring formation disturbs the form of the molar mass distribution and reduces the ultimate molar mass. However, since linear polymerization is a bimolecular process and ring formation is a unimolecular process, it is possible to greatly promote the former process relative to the latter by using high monomer concentrations. This is why many step polymerizations are performed in bulk (i.e. using only monomer(s) plus catalysts in the absence of a solvent).

3.2.5 Linear Step Polymerization Processes

Although step polymerizations can be carried out in a solvent that dissolves the monomers and the polymer to be produced, finding suitable solvents can be difficult because the polymers often are semi-crystalline and of low solubility. The isolation of the polymer from the solvent also can prove difficult. Hence, most step polymerizations are performed by reacting liquid monomers together in the absence of a solvent.

The preceding sections highlight the many constraints upon the formation of high molar mass polymers by linear step polymerization. Special polymerization systems often have to be developed to overcome these constraints and are exemplified here by systems developed for the preparation of polyesters and polyamides.

Ester interchange (or *transesterification*) reactions commonly are employed in the production of polyesters, the most important example being the preparation of poly(ethylene terephthalate). The direct polyesterification reaction of terephthalic acid with ethylene glycol indicated in Table 1.1 is complicated by the high melting point of terephthalic acid (in fact it sublimes at 300 °C before melting) and its low solubility. Thus poly(ethylene terephthalate) is prepared in a two-stage process. The first stage involves formation of bis(2-hydroxyethyl)terephthalate either by reaction of dimethylterephthalate with an excess of ethylene glycol (i.e. via ester interchange)

$$\text{CH}_3\text{O}-\overset{\overset{\text{O}}{\|}}{\text{C}}-\langle\bigcirc\rangle-\overset{\overset{\text{O}}{\|}}{\text{C}}-\text{OCH}_3 + (2+x)\,\text{HOCH}_2\text{CH}_2\text{OH} \xrightarrow{147\text{–}197\ ^\circ\text{C}}$$

$$\text{HOCH}_2\text{CH}_2\text{O}-\overset{\overset{\text{O}}{\|}}{\text{C}}-\langle\bigcirc\rangle-\overset{\overset{\text{O}}{\|}}{\text{C}}-\text{OCH}_2\text{CH}_2\text{OH} + 2\,\text{CH}_3\text{OH} + x\,\text{HOCH}_2\text{CH}_2\text{OH}$$

or more commonly nowadays by direct esterification of terephthalic acid with an excess of ethylene glycol

$$\mathrm{HO{-}\overset{O}{\overset{\|}{C}}{-}C_6H_4{-}\overset{O}{\overset{\|}{C}}{-}OH} + (2+x)\,\mathrm{HOCH_2CH_2OH} \xrightarrow{227\text{–}257\ ^\circ\mathrm{C}}$$

$$\mathrm{HOCH_2CH_2O{-}\overset{O}{\overset{\|}{C}}{-}C_6H_4{-}\overset{O}{\overset{\|}{C}}{-}OCH_2CH_2OH} + 2\,\mathrm{H_2O} + x\,\mathrm{HOCH_2CH_2OH}$$

The methanol or water produced during these first-stage reactions is removed as it is formed. On completion of the first stage, the reaction temperature is raised to about 277 °C so that the excess ethylene glycol and the ethylene glycol produced by further ester interchange reactions can be removed, and so that the polymer is formed above its melting temperature (265 °C)

$$n\,\mathrm{HOCH_2CH_2O{-}\overset{O}{\overset{\|}{C}}{-}C_6H_4{-}\overset{O}{\overset{\|}{C}}{-}OCH_2CH_2OH} \longrightarrow$$

$$\mathrm{HOCH_2CH_2O}\left[\mathrm{\overset{O}{\overset{\|}{C}}{-}C_6H_4{-}\overset{O}{\overset{\|}{C}}{-}OCH_2CH_2O}\right]_n\mathrm{H} + (n-1)\,\mathrm{HOCH_2CH_2OH}$$

Thus by using ester interchange reactions and removing the ethylene glycol produced, the need for strict stoichiometric control is eliminated.

The preferred method for preparing aliphatic polyamides from diamines and diacids is *melt polymerization* of the corresponding nylon salt. For example, in the preparation of nylon 6.6, hexamethylene diamine and adipic acid are first reacted together at low temperature to form hexamethylene diammonium adipate (nylon 6.6 salt) which then is purified by recrystallisation. The salt is heated gradually up to about 277 °C to effect melt polymerization and maintained at this temperature whilst removing the water produced as steam

$$\underset{\text{Hexamethylene diamine}}{n\,\mathrm{H_2N{-}(CH_2)_6{-}NH_2}} + \underset{\text{Adipic acid}}{n\,\mathrm{HO{-}\overset{O}{\overset{\|}{C}}{-}(CH_2)_4{-}\overset{O}{\overset{\|}{C}}{-}OH}}$$

$$\downarrow$$

$$n\,\mathrm{\overset{+}{H_3N}{-}(CH_2)_6{-}\overset{+}{N}H_3\overset{-}{O}{-}\overset{O}{\overset{\|}{C}}{-}(CH_2)_4{-}\overset{O}{\overset{\|}{C}}{-}\overset{-}{O}}$$

$$\downarrow$$

$$\mathrm{H}\left[\mathrm{NH{-}(CH_2)_6{-}NH{-}\overset{O}{\overset{\|}{C}}{-}(CH_2)_4{-}\overset{O}{\overset{\|}{C}}}\right]_n\mathrm{OH} + (2n-1)\,\mathrm{H_2O}$$

A major advantage of melt polymerization by *salt dehydration* is that the use of a pure salt guarantees exact 1:1 stoichiometry.

A convenient method for preparation of polyesters and polyamides in the laboratory is the reaction of diacid chlorides with diols and diamines, respectively (i.e. Schotten–Baumann reactions).

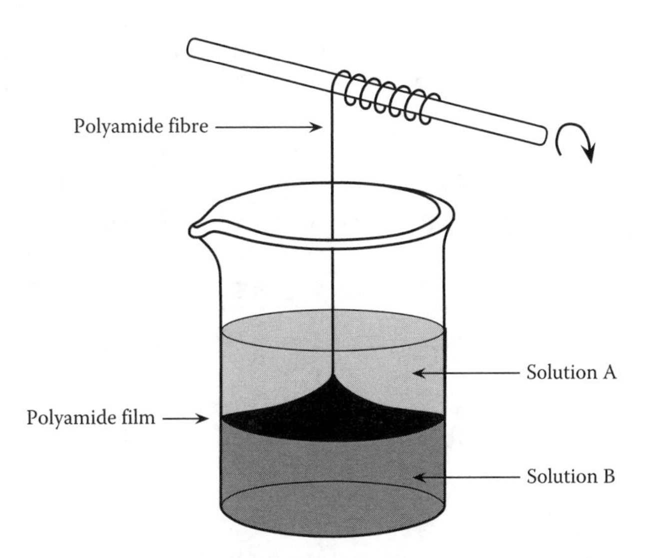

FIGURE 3.3 Schematic illustration of interfacial polymerization to produce a polyamide. A is an aqueous solution of the diamine and B is a solution of the diacid chloride in carbon tetrachloride. The polyamide is produced as a film at the interface between the two solutions and can be drawn off in the form of a fibre. It is usual to include a base (e.g. NaOH) in solution A in order to neutralize the HCl formed by the reaction.

These reactions proceed rapidly at low temperatures and often are performed as *interfacial polymerizations* in which the two reactants are dissolved separately in immiscible solvents which are then brought into contact. The best known example of this is the 'nylon rope trick' where a continuous film of nylon is drawn from the interface as illustrated in Figure 3.3. For example, the preparation of nylon 6.10 would proceed by the following reaction

$$n\,\mathrm{H_2N{-}(CH_2)_6{-}NH_2} \quad + \quad n\,\mathrm{Cl{-}\overset{\displaystyle O}{\overset{\|}{C}}{-}(CH_2)_8{-}\overset{\displaystyle O}{\overset{\|}{C}}{-}Cl}$$

Hexamethylene diamine   Sebacoyl chloride

$$\downarrow$$

$$\mathrm{H}\!\left[\mathrm{NH{-}(CH_2)_6{-}NH{-}\overset{\displaystyle O}{\overset{\|}{C}}{-}(CH_2)_8{-}\overset{\displaystyle O}{\overset{\|}{C}}}\right]_n\!\mathrm{Cl} \quad + \quad (2n-1)\,\mathrm{HCl}$$

The reaction takes place at the organic solvent side of the interface and, because it usually is diffusion-controlled, there is no need for strict control of stoichiometry.

3.3 NON-LINEAR STEP POLYMERIZATION

The inclusion of a monomer with a functionality greater than two has a dramatic effect upon the structure and molar mass of the polymer formed. In the early stages of such reactions, the polymer has a branched structure and, consequently, increases in molar mass much more rapidly with extent of reaction than for a linear step polymerization. As the reaction proceeds, further branching reactions lead ultimately to the formation of complex network structures which have properties that are quite different from those of the corresponding linear polymer. For example, reaction of a dicarboxylic acid $R(COOH)_2$ with a triol $R'(OH)_3$ would lead to structures of the type

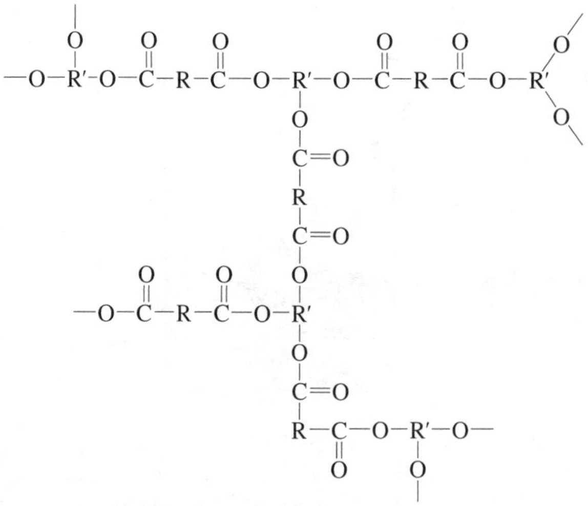

```
      |                                                   /
      O      O      O              O      O            O
      |      ||     ||             ||     ||          /
—O—R'—O—C—R—C—O—R'—O—C—R—C—O—R'
                            |                        \
                            O                         O
                            |                          \
                            C=O
                            |
                            R
                            |
                            C=O
                            |
           O      O         O
           ||     ||        |
     —O—C—R—C—O—R'
                            |
                            O
                            |
                            C=O
                            |
                            R—C—O—R'—O—
                               ||      |
                               O       O
                                       |
```

The point at which the first network molecule is formed is known as the *gel point* because it is manifested by *gelation*, that is, an abrupt change of the reacting mixture from a viscous liquid to a solid gel which shows no tendency to flow. Before dealing with theoretical predictions of the gel point, the chemistry of some important network-forming step polymerizations will be described.

3.3.1 Network Polymers

Network polymers produced by step polymerizations were amongst the first types of synthetic polymers to be commercialized and often are termed *resins*. The polymers are completely intractable and so at the stage when the network chains are generated, the polymerizations must be carried out within a mould to produce the required artefact.

3.3.1.1 Formaldehyde-Based Resins

Formaldehyde-based resins were the first network polymers prepared by step polymerization to be commercialized. They are prepared in two stages. The first involves the formation of a prepolymer of low molar mass which may either be a liquid or a solid. In the second stage, the prepolymer is forced to flow under pressure to fill a heated mould in which further reaction takes place to yield a highly crosslinked, rigid polymer in the shape of the mould. Since formaldehyde is difunctional, in order to form a network polymer the co-reactants must have a functionality f greater than two; those most commonly employed are phenol ($f=3$), urea ($f=4$) and melamine ($f=6$)

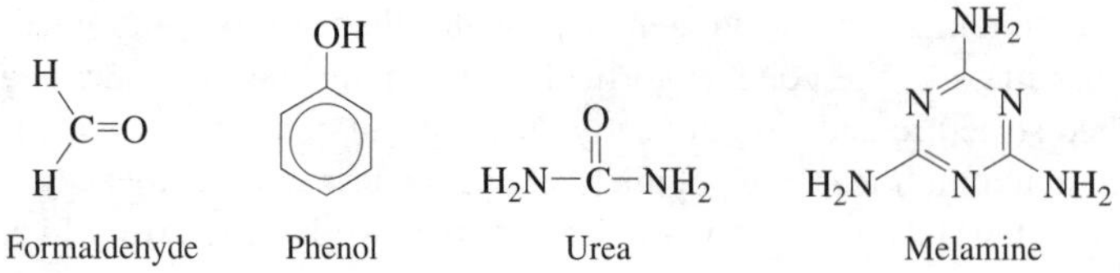

```
                       OH                              NH2
   H                   |                                |
    \               [benzene]        O              N /  \ N
     C=O                             ||              |    ||
    /                           H2N—C—NH2     H2N /  \ N /  \ NH2
   H
```

Formaldehyde Phenol Urea Melamine

The hydroxyl group in phenol activates the benzene ring towards substitution in the 2-, 4- and 6-positions. Upon reaction of phenol with formaldehyde, methylol substituent groups are formed, e.g.

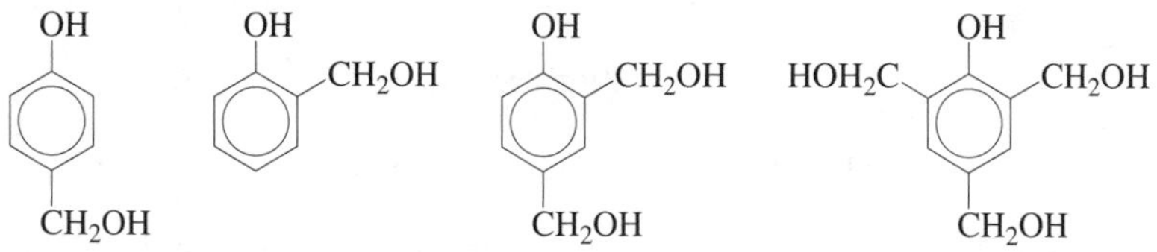

OH, CH_2OH; OH, CH_2OH; OH, CH_2OH, CH_2OH; OH, HOH_2C, CH_2OH, CH_2OH

Further reaction leads principally to the formation of methylene bridges but also to dimethylene ether links

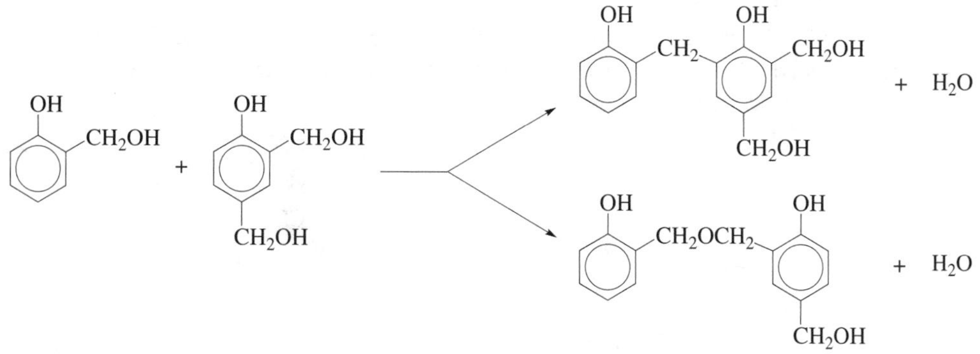

OH, CH_2OH + OH, CH_2OH, CH_2OH → OH, CH_2, OH, CH_2OH, CH_2OH + H_2O ; OH, CH_2OCH_2, OH, CH_2OH + H_2O

There are two types of *phenol–formaldehyde resin*. Those prepared using an excess of formaldehyde with base catalysis are known as *resoles*. The resole prepolymers possess many unreacted methylol groups that upon further heating react to produce the network structure.

Novolaks are prepared using an excess of phenol and acid catalysis, which promotes condensation reactions of the methylol groups. Thus the prepolymers produced contain no methylol groups and are unable to crosslink, e.g.

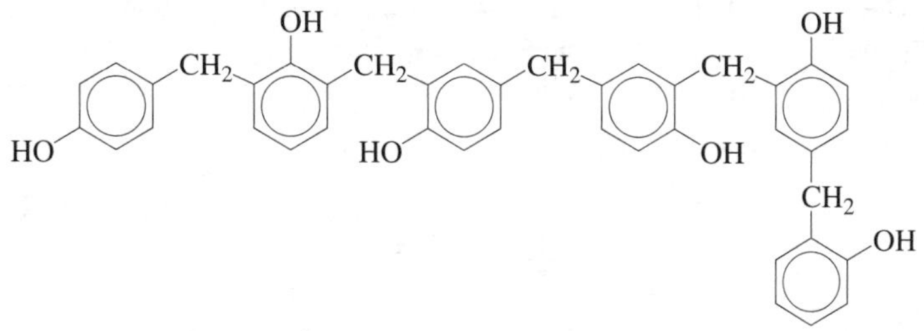

HO, CH_2, OH, CH_2, HO, CH_2, OH, CH_2, OH, CH_2, OH

Normally, they are dried and ground to a powder, then compounded (i.e. mixed) with fillers (e.g. mica, glass fibres, sawdust), colourants and hardeners. Hexamethylenetetramine, with magnesium or calcium oxide as a catalyst, usually is employed as the hardener to activate curing (i.e. crosslinking) when the compound is converted into an artefact by heating in a mould. Most of the crosslinks formed are methylene bridges, though some dimethylene amine ($–CH_2–NH–CH_2–$) links are formed. The fillers are added to reduce the cost and to improve the electrical or mechanical properties of the resin.

The chemistry of *urea–formaldehyde* and *melamine–formaldehyde* resins involves the formation and condensation reactions of *N*-methylol groups, e.g. in general terms

$$-NH_2 + CH_2{=}O \longrightarrow -NH{-}CH_2{-}OH \xrightarrow{HN<} -NH{-}CH_2{-}N< + H_2O$$

$$-NH{-}CH_2{-}OH \xrightarrow{HO-CH_2-NH-} -NH{-}CH_2{-}O{-}CH_2{-}NH- + H_2O$$

The reactions usually are arrested at the prepolymer stage by adjusting the pH of the reacting mixture to slightly alkaline. After blending (e.g. with fillers, pigments, hardeners, etc.) the prepolymers are cured by heating in a mould. The hardeners are compounds (e.g. ammonium sulphamate) which decompose at mould temperatures to give acids that catalyse further condensation reactions.

3.3.1.2 Epoxy Resins

Epoxy resins are low molar mass prepolymers containing epoxide end-groups. The most important are the diglycidyl ether prepolymers prepared by the reaction of excess epichlorohydrin with bisphenol-A in the presence of a base

$$(n+2)\,ClCH_2{-}\underset{\backslash O/}{CH{-}CH_2} + (n+1)\,HO{-}C_6H_4{-}C(CH_3)_2{-}C_6H_4{-}OH \xrightarrow{NaOH}$$

Epichlorohydrin   Bisphenol-A

$$CH_2{-}CH{-}CH_2{-}O{-}\left[C_6H_4{-}C(CH_3)_2{-}C_6H_4{-}O{-}CH_2{-}\underset{OH}{CH}{-}CH_2{-}O\right]_n{-}C_6H_4{-}C(CH_3)_2{-}C_6H_4{-}O{-}CH_2{-}CH{-}CH_2 \quad \text{(terminal epoxide rings)}$$

These prepolymers are either viscous liquids or solids depending upon the value of n. Usually, they are cured by the use of multifunctional amines which undergo a polyaddition reaction with the terminal epoxide groups in the manner indicated below (for a diamine, $f=4$)

$$4\ \sim\!CH{-}CH_2\ (\text{epoxide}) + H_2N{-}R{-}NH_2 \longrightarrow (\sim\!\underset{OH}{CH}{-}CH_2)_2N{-}R{-}N(CH_2{-}\underset{OH}{CH}\!\sim)_2$$

Together, the amine R group and epoxy resin chain length control the mechanical properties of the fully-cured resin. Some examples of polyfunctional amines used to cure epoxy resins are shown below together with their functionalities and common names

$$H_2N-(CH_2)_2-NH-(CH_2)_2-NH_2$$

Diethylenetriamine (DETA), $f = 5$

$$H_2N-(CH_2)_2-NH-(CH_2)_2-NH-(CH_2)_2-NH_2$$

Triethylenetetramine (TETA), $f = 6$

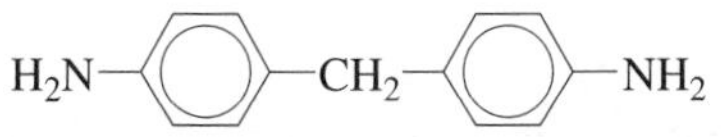

$$H_2N-C_6H_4-CH_2-C_6H_4-NH_2$$

4,4′-Diaminodiphenylmethane (DDM), $f = 4$

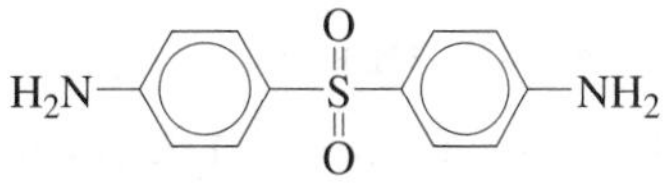

$$H_2N-C_6H_4-SO_2-C_6H_4-NH_2$$

4,4′-Diaminodiphenylsulphone (DDS), $f = 4$

Epoxy resins are characterized by low shrinkage on curing and find use as adhesives, electrical insulators, surface coatings and matrix materials for fibre-reinforced composites.

3.3.1.3 Network Polyurethanes

Polyurethane networks find a wide variety of uses (e.g. elastomers, flexible foams and rigid foams) and usually are prepared by the reaction of diisocyanates with *polyols*, which are branched polyether (or, less commonly, polyester) prepolymers that have hydroxyl end-groups, e.g. reaction of MDI (Section 3.2.2.1) with a poly(propylene oxide) triol

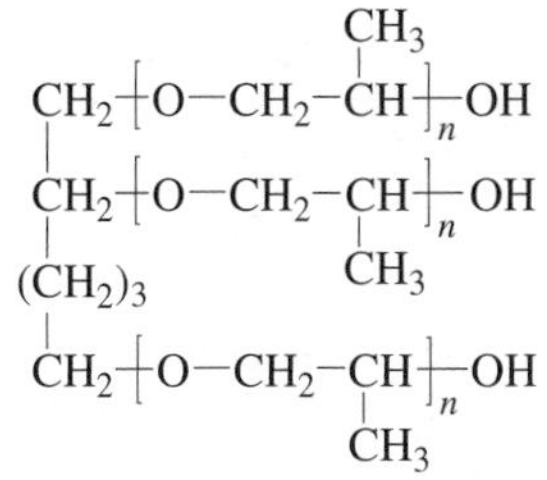

$$\begin{array}{l} CH_2 + O-CH_2-CH(CH_3) +_n OH \\ | \\ CH_2 + O-CH_2-CH(CH_3) +_n OH \\ | \\ (CH_2)_3 \\ | \\ CH_2 + O-CH_2-CH(CH_3) +_n OH \end{array}$$

via the chemistry described in Section 3.2.2.1. Commercial poly(propylene oxide) polylols often have terminal ethylene oxide units (i.e. $-O-CH_2-CH_2-OH$) (see Section 7.3.1) to increase reactivity (i.e. primary $-CH_2OH$ groups are more reactive than the secondary $-CH(CH_3)OH$ groups). The molar mass and functionality of the prepolymer determine the crosslink density and hence the flexibility of the network formed. Typically, polyether polyols have a functionality of 3–6 and those with $\bar{M}_n$ of up to about 1 kg mol^{-1} are used to prepare rigid polyurethanes, whereas flexible polyurethanes are prepared from those with $\bar{M}_n$ of about 2–8 kg mol^{-1}.

Polyurethane foams can be formed by inclusion of a small amount of water, which reacts rapidly with the isocyanate groups to give an unstable carbamic acid that instantly decomposes to produce an amine end-group and carbon dioxide gas, which causes foaming of the polyurethane as it is formed

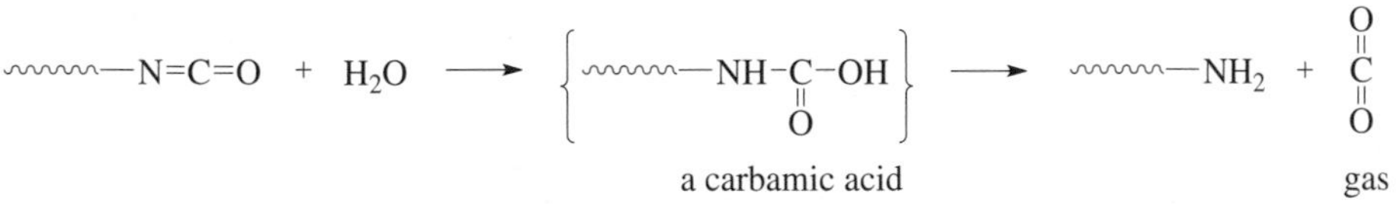

$$\sim\sim-N=C=O + H_2O \longrightarrow \left\{ \sim\sim-NH-\underset{\underset{O}{\|}}{C}-OH \right\} \longrightarrow \sim\sim-NH_2 + O=C=O$$

a carbamic acid — gas

The amine end-group is extremely reactive towards isocyanate groups, so reacts immediately to form a *urea* link

$$\sim\sim\!-NH_2 + O{=}C{=}N\!-\!\sim\sim \longrightarrow \sim\sim\!-\underset{H}{N}-\underset{O}{\overset{\|}{C}}-\underset{H}{N}-\!\sim\sim$$

urea link

Polyurethane foams can also be produced by inclusion of compounds that vaporize due to the heat released by the exothermic reaction of isocyanates groups with hydroxyl groups.

3.3.2 Gelation Theory

Three-dimensionally crosslinked polymers are incapable of macroscopic viscous flow because, at the molecular level, the crosslinks prevent the network chains from flowing past one another. When the first *network molecule* forms in a non-linear polymerization, it encompasses the whole reactant mixture which instantly becomes immobilized; this corresponds to the *gel point*, which clearly is a very important stage in a network-forming polymerization. Much effort has been devoted to prediction of the extent of reaction at the gel point. Here, simple theories for the gel point in non-linear step polymerizations are introduced.

3.3.2.1 Carothers Theory of Gelation

A simple theory for prediction of gel points can be derived using the principles employed in the linear step polymerization theory of Carothers (Section 3.2.3.1). When there is a stoichiometric balance in the numbers of mutually-reactive functional groups, the number-average functionality f_{av} is used and is defined by

$$f_{av} = \frac{\sum N_i f_i}{\sum N_i}$$

where N_i is the initial number of molecules of monomer i which has functionality f_i. Thus, if there are N_0 molecules present initially, the total number of functional groups present is $N_0 f_{av}$. If at time t there are N molecules present, then the number of functional groups that have reacted is $2(N_0 - N)$ since the number of molecules decreases by one for each link produced by the reaction together of *two* functional groups. Therefore the extent of reaction p, which is the probability that a functional group present initially has reacted, is given by

$$p = \frac{2(N_0 - N)}{N_0 f_{av}}$$

Simplifying and substituting for N/N_0 using Equation 3.1 gives

$$p = \frac{2}{f_{av}}\left(1 - \frac{1}{\bar{x}_n}\right) \tag{3.22}$$

which can be rearranged to

$$\bar{x}_n = \frac{2}{2 - p f_{av}} \tag{3.23}$$

This equation reduces to the simple Carothers Equation 3.3 when $f_{av} = 2.0$. Slight increases in f_{av} above 2.0 give rise to substantial increases in the value of $\bar{x}_n$ attained at specific extents of reaction, as is demonstrated in Table 3.4.

TABLE 3.4
Values of $\bar{x}_n$ for Different Values of p and f_{av} Calculated using Equation 3.23

	$\bar{x}_n$ at				
f_{av}	$p=0.50$	$p=0.70$	$p=0.90$	$p=0.95$	$p=0.99$
2.0	2	3.33	10	20	100
2.1	2.10	3.77	18.18	400	Gelled
2.2	2.22	4.35	100	Gelled	Gelled
2.3	2.35	5.13	Gelled	Gelled	Gelled

If it is postulated that gelation occurs when $\bar{x}_n$ goes to infinity, then it follows from Equation 3.22 that the critical extent of reaction p_c for gelation is given by

$$p_c = \frac{2}{f_{av}} \tag{3.24}$$

It is possible to extend the theory to prediction of p_c when there is an imbalance of stoichiometry. However, this simple method of analysis will not be pursued further here because it is rather inelegant and always yields overestimates of p_c. The approach is fundamentally flawed because it is based upon $\bar{x}_n$, an average quantity, tending to infinity. Molecules with degrees of polymerization both larger and smaller than $\bar{x}_n$ are present, and it is the largest molecules that undergo gelation first.

3.3.2.2 Statistical Theory of Gelation

The basic statistical theory of gelation was first derived by Flory who considered the reaction of an f-functional monomer RA_f ($f>2$) with the difunctional monomers RA_2 and RB_2. It is necessary to define a parameter called the *branching coefficient*, α, which is the probability that an f-functional unit is connected via a chain of difunctional units to another f-functional unit. In other words α is the probability that the general sequence of linkages shown below exists

$${}_{(f-1)}\mathrm{AR{-}AB{-}R{-}B}\left[\mathrm{A-R-AB-R-B}\right]_i\mathrm{A-R-A}_{(f-1)}$$

In order to derive an expression for α from statistical considerations it is necessary to introduce another term, γ, which is defined as the initial ratio of A groups from RA_f molecules to the total number of A groups. Using γ, it is possible to calculate the probabilities for the existence of each of the linkages in the general sequence given above. If the extent of reaction of the A groups is p_A and of the B groups is p_B then

$$\text{probability of }{}_{(f-1)}\mathrm{AR{-}\overrightarrow{AB}{-}R{-}B} = p_A$$

$$\text{probability of }\mathrm{B{-}R{-}\overrightarrow{BA}{-}R{-}A} = p_B(1-\gamma)$$

$$\text{probability of }\mathrm{A{-}R{-}\overrightarrow{AB}{-}R{-}B} = p_A$$

$$\text{probability of }\mathrm{B{-}R{-}\overrightarrow{BA}{-}RA_{(f-1)}} = p_B\gamma$$

where the arrows indicate the linkages under consideration and the direction in which the chain is extending. Thus the probability that the general sequence of linkages has formed is

$$p_A[p_B(1-\gamma)p_A]^i p_B\gamma$$

The branching coefficient is the probability that sequences with all values of i have formed and so is given by

$$\alpha = p_A p_B \gamma \sum_{i=0}^{\infty} \left[p_A p_B (1-\gamma) \right]^i$$

Using the mathematical relation

$$\sum_{i=0}^{\infty} x^i = \frac{1}{(1-x)} \quad \text{for } x < 1$$

then

$$\alpha = p_A p_B \gamma [1 - p_A p_B (1-\gamma)]^{-1} \tag{3.25}$$

Network molecules can form when n chains are expected to lead to more than n chains through branching of some of them. The maximum number of chains that can emanate from the end of a single chain, such as that analysed above, is $(f-1)$ and so the probable number of chains emanating from the chain end is $\alpha(f-1)$. Network molecules can form if this probability is not less than one, i.e. $\alpha(f-1) \geq 1$. Thus the *critical branching coefficient* α_c for the onset of gelation is given by

$$\alpha_c = \frac{1}{(f-1)} \tag{3.26}$$

Substitution of this equation into Equation 3.25 followed by rearrangement gives an expression for the product of the critical extents of reaction at gelation

$$(p_A p_B)_c = \frac{1}{1 + \gamma(f-2)} \tag{3.27}$$

If the *reactant ratio* r is defined as the initial ratio of A groups to B groups then $p_B = rp_A$ and

$$(p_A)_c = [r + r\gamma(f-2)]^{-1/2} \tag{3.28}$$

$$(p_B)_c = r^{1/2}[1 + \gamma(f-2)]^{-1/2} \tag{3.29}$$

When RA_2 molecules are absent (i.e. in a $RA_f + RB_2$ polymerization), $\gamma = 1$ and $(p_A p_B)_c = 1/(f-1)$, in which case $(p_A p_B)_c = \alpha_c$.

3.3.2.3 Validity of the Carothers and Statistical Theories of Gelation

The predictions of the Carothers theory and the statistical theory of gelation can now be compared with each other and with experimental gelation data. Figure 3.4 shows the variation of p, $\bar{x}_n$ and viscosity, η, with time for a reaction between diethylene glycol (a diol), succinic acid (a diacid) and propan-1,2,3-tricarboxylic acid (a triacid). Gelation occurred after about 238 min and was manifested by η becoming infinite. The observed extent of reaction at gelation was 0.894 when $\bar{x}_n$ was only about 25 (i.e. well below the value of infinity assumed in the Carothers theory of gelation), but the statistical theory also is inaccurate and predicts $(p_A p_B)_c = 0.844$. In a similar polymerization carried out using $r = 1.000$ and $\gamma = 0.194$, the gel point occurred at $p = 0.939$ which may be compared to

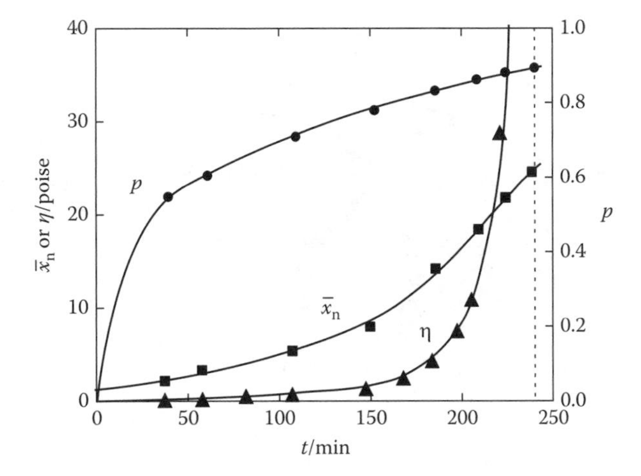

FIGURE 3.4 Variation of p, $\bar{x}_n$ and η with time for a p-toluene sulphonic acid catalysed polymerization at 109 °C of diethylene glycol with succinic acid and propan-1,2,3-tricarboxylic acid in which $r = 1.002$ and $\gamma = 0.404$. The dashed line marks the gel point which occurred at $p = 0.894$. (Data taken from Flory, *J. Am. Chem. Soc.*, 63, 3083, 1941.)

the prediction of $p_c = 0.968$ from the Carothers Equation 3.24 which (as expected) is high, whereas Equation 3.27 gives a value of 0.915, which again is lower than observed. The statistical theory always underestimates p_c because it does not take into account the effect of intramolecular reactions between end groups. These give rise to loops in the polymer structure and the polymerization must proceed to higher extents of reaction in order to overcome this wastage of functional groups. When the effects of intramolecular reaction are eliminated (e.g. extrapolation of p_c values to infinite reactant concentrations; see Section 3.2.4) the statistical theory gives accurate predictions of the gel point. More advanced statistical theories have been developed in which intramolecular reactions and other effects are included; they provide a fundamentally satisfactory basis for accurate prediction of gelation but are beyond the scope of this book.

The theories of gelation presented here can be applied only when it is clear that the assumption of equal reactivity of functional groups is satisfactory. In terms of the polymerizations described earlier, the theories generally are applicable to the formation of polyester and polyurethane networks, but not to the formation of formaldehyde-based resins and epoxy resins. Failure of the principle of equal reactivity for the latter systems results from modification of the reactivity of a particular functional group by reaction of another functional group in the same molecule of monomer.

3.3.3 Dendrimers

Dendrimers are highly branched molecules which (ideally) have a specific, perfectly symmetric skeletal geometry and a unique molar mass. The elegance of their molecular structures is reminiscent of fractals and coral, and is only hinted at by the very simple dendrimer structure shown in Figure 1.2. Synthesis of dendrimers can be traced back to the late 1970s, but it was not until the late 1980s, when organic polymer chemists became interested in their unique properties, that research into dendrimers exploded. Since then, dendrimers have become a major focus for research.

There are many different classes of dendrimer in terms of their design and chemical structure, but all have common features. There is a multifunctional *core* at the centre, connected to successive *generations* of *branching units*, the incomplete spatial packing of which leaves *internal voids and channels* within the dendrimer, and an outer sheath of end groups, termed *surface groups*, that can carry chemical functionality. These features of a dendrimer are depicted in Figure 3.5. The concepts of *core*, *generation*, *branching units* and *surface groups* are best defined through schematic representation of a dendrimer, as shown in Figure 3.6.

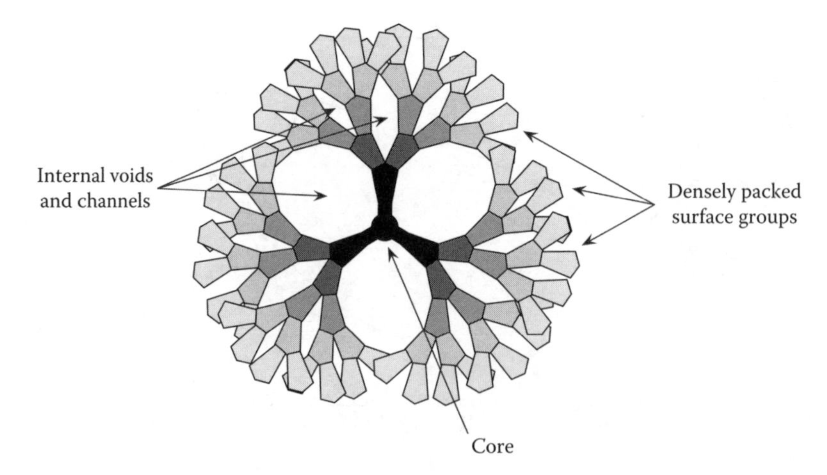

FIGURE 3.5 Dendrimer molecules are characterized by their high concentration of surface groups and the presence of internal voids and channels, as illustrated in this schematic diagram of a fifth-generation dendrimer.

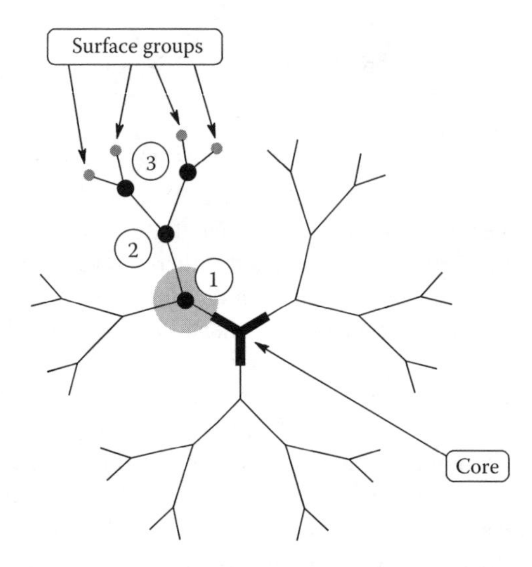

FIGURE 3.6 Schematic representation of a third-generation dendrimer. The *core* is shown in bold lines. The terminology for dendrimers is illustrated for one arm of the dendrimer: the *junction point* in each *branching unit* is highlighted by a solid black dot; the numbers in circles define the *generation numbers*, which increase by one for each successive branching unit; the links associated with the first-generation branching unit are highlighted by the grey circle; and the *surface groups* are indicated by the solid grey dots at the chain ends.

3.3.3 1 Synthesis of Dendrimers

Dendrimers are synthesized one generation at a time in carefully controlled step-wise reactions. Their synthesis, therefore, most closely resembles step polymerization. For example, if a ARB_2 monomer was polymerized onto an RB_3 core molecule whilst preventing ARB_2 self-polymerization and ensuring that each generation of growth was completed before the succeeding generation began to form, dendrimers with structures of the types shown in Figures 3.5 and 3.6 would be formed (i.e. with trifunctional junction points, RB_3 at the core and successive generations of ARB_2 branching units). This would be a *divergent* dendrimer synthesis in which the dendrimer is formed from the core outwards. Before considering how to overcome the challenges in achieving such control, it is important to recognize the spatial constraints on growth of dendrimers in this way.

To create a true dendrimer without imperfections, a single new branching unit must be linked to each surface group of the preceding dendrimer. If the functionality of the core is f_{core} and that of the branching units is f_{br} then $f_{core}(f_{br}-1)^{g-1}$ branching units must be added to produce the gth generation dendrimer from the $(g-1)$-th generation dendrimer (the core may be considered as the *zero*-th generation in this context). Hence, *excluding the core*, the number of branching units N_{br} in a dendrimer of generation g is given by

$$N_{br} = \sum_{x=1}^{g} f_{core}(f_{br}-1)^{x-1} \tag{3.30}$$

Thus N_{br} increases very rapidly with g and the demands on the chemistry required to produce the dendrimer become ever greater with each generation of branching units to be added. This is not the only constraint, however, because spatial restrictions also have a major impact on dendrimer synthesis. Consider the very simple picture of a dendrimer as a spherical entity, then assuming that the lengths of bonds linking each generation are the same (which generally is true), the radius of the dendrimer molecule will increase by the same amount for each generation added and the volume of the dendrimer molecule will increase approximately with g^3, i.e. far less rapidly with g than the increase of N_{br} with g. It is now easy to appreciate that there will come a point at which steric crowding will prevent the formation of further generations. This often is referred to as the *starburst limit*. Calculations suggest this will occur at around the 10th generation for $f_{br}=3$ and the 5th generation when $f_{br}=4$. Crowding also becomes an issue when attempting to draw the chemical structures of dendrimers. Compromises have to be made in order to draw in two dimensions large molecules that fill space in three dimensions; to avoid overlapping of atoms and bonds, correct constraints on bond angles and/or bond lengths must be relaxed and, even then, the drawings are restricted to lower-generation dendrimers to prevent overlaps of bonds and atoms. For this reason, dendrimer structures often are shown schematically with letters or blocks representing parts of the chemical structure.

The simple analysis given above shows clearly that synthesis of dendrimers is highly demanding and becomes rapidly more difficult with each successive generation of branching units to be added. Ultimately, steric crowding restricts further chemical reaction and so it is difficult to synthesize higher-generation dendrimers of perfect structure. Nevertheless, both lower-generation perfect dendrimers and higher-generation imperfect dendrimers are of considerable academic and commercial interest.

Since a large number of different types of linking reaction have been used to produce dendrimers, there are now many different chemical classes of dendrimer, far too numerous to describe here. For details, the reader should consult review articles, some of which are given in Further Reading. The focus here will be to demonstrate the general principles of dendrimer synthesis, but one real example will be considered to highlight the need to use methodical, step-wise chemical reactions.

One of the first, and more important, dendrimers to be synthesized via the *divergent* approach is a poly(amidoamine) (PAMAM), the structure of which is shown in Figure 3.7. The chemistry underlying the synthesis of a PAMAM dendrimer is described here because it illustrates the factors that need to be considered in any dendrimer synthesis. First and foremost, the linking reactions must be highly efficient (to ensure that all surface groups are reacted) and very clean (i.e. free from any significant side reactions). Second, all except the linking functional group in the branching unit (monomer) to be added must be *protected* from reaction so that only one generation of branching units is added. When addition of that generation is complete, the protected

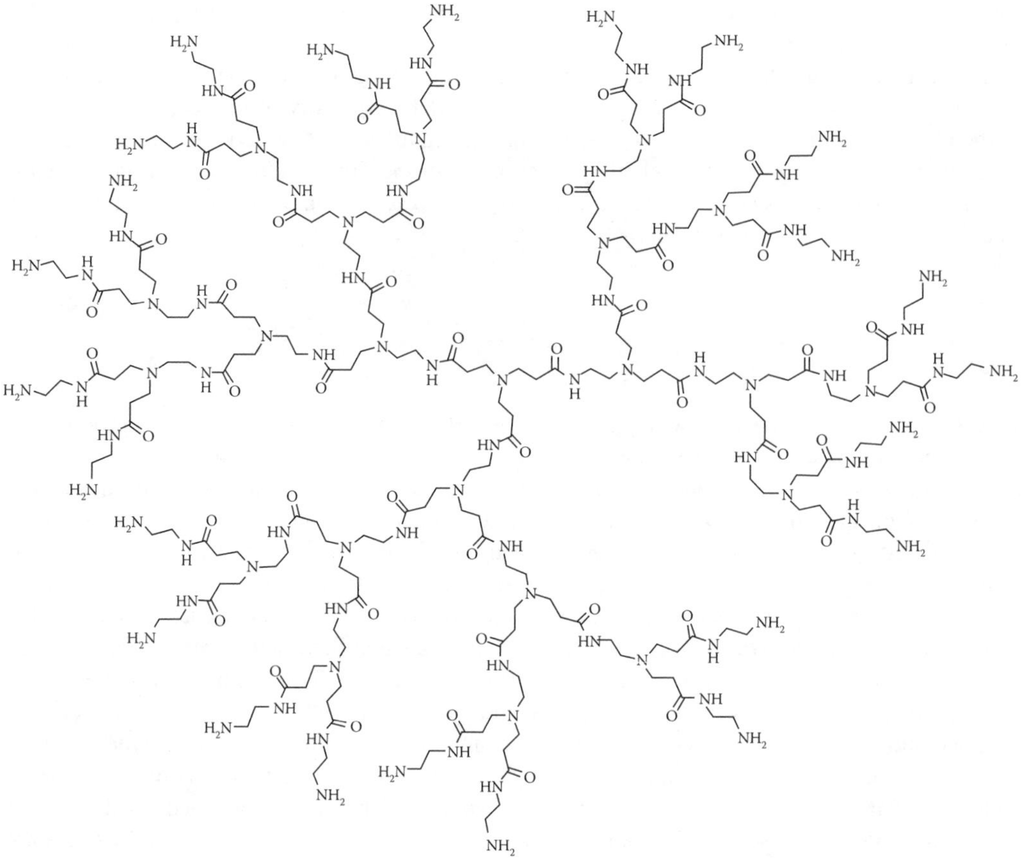

FIGURE 3.7 A third-generation poly(amidoamine) (PAMAM) dendrimer.

functional groups must then be *deprotected*, or *converted*, to a reactive form so that they can be linked to the next generation of branching units. The chemistry used for synthesis of PAMAM dendrimers is

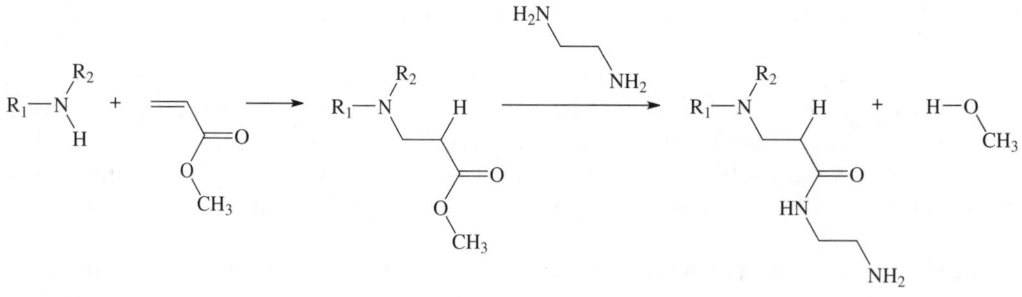

where R_1 and R_2 represent H or a fragment of the dendrimer. The linking reaction involves Michael addition of an amine N–H to the C=C bond of methyl acrylate; the methyl ester group is then subjected to transamidation with 1,2-diaminoethane, displacing methanol from the ester to produce the corresponding amide and leave an amine ($-NH_2$) surface group (i.e. end group). Each of these terminal amine groups then is reacted with two further molecules of methyl acrylate (one per N–H bond) and the sequence of chemical reactions repeated to produce the PAMAM dendrimer. (Note that, due to electron delocalization with the adjacent carbonyl group, the amide N–H is too weakly nucleophilic to add to methyl acrylate; hence, only the amine N–H react.) The starting point (i.e. core) of the PAMAM dendrimer shown in Figure 3.7 is ammonia (NH_3), which adds to three molecules of methyl acrylate (one per N–H) in producing the first-generation dendrimer.

Synthesis of a PAMAM dendrimer provides an example of *divergent* dendrimer synthesis in which a bifunctional branching surface group ($-NH_2$) is created from an inactive monofunctional group ($-CO_2CH_3$) in the activation step prior to linking the next generation of branching units. In other divergent dendrimer syntheses, each branching unit added possesses two or more protected surface groups that are simply deprotected to reveal the reactive groups. As indicated above, many different protection/deprotection chemistries have been used for this purpose, so the reader should consult review articles on dendrimers to gain an appreciation of the wide variety of ways in which this has been achieved.

The divergent approach to dendrimer synthesis was the first to be used and is very effective in producing significant quantities of dendrimer because molar mass increases rapidly with each new generation. However, the divergent approach has the disadvantages that the number of chemical reactions that need to be effected to produce the next-generation dendrimer increases rapidly with generation number and the perfect next-generation dendrimer is difficult to separate from impurities, particularly molecules that do not have a complete generation of new branching units but otherwise are very similar in size and chemical structure to the perfect dendrimer molecules. Thus, higher-generation dendrimers prepared in this way often contain some molecules with imperfections in their structure. Other more efficient and versatile approaches have, therefore, been developed. The *convergent* approach grows the dendrimer from the surface inwards and produces sections of the dendrimer known as *wedges* or *dendrons*. The principles of the divergent and convergent approaches are set out schematically in Figures 3.8 and 3.9. The convergent approach, of course, requires the dendrons to be joined together to produce the dendrimer in a final reaction of their activated *focal point groups* with a multifunctional core molecule, so there is no difference in the total number of chemical reactions required to produce a given dendrimer compared to the divergent approach. The divergent and convergent approaches can also be mixed; for example, the second-generation dendrimer with activated surface groups from the divergent synthesis shown in Figure 3.8 could be reacted with the third-generation dendron with an activated focal point group from the convergent synthesis shown in Figure 3.9 to produce a fifth-generation dendrimer with protected surface groups.

The so-called *double exponential* approach is another important synthetic route to dendrons and is depicted in Figure 3.10. It involves the formation of dendrons by the reaction of the activated surface group form of a branching unit or an already formed dendron with its activated focal point group form. As for the convergent approach, the dendrons produced must be joined together in a final reaction with a multifunctional core molecule to produce the dendrimer. Reduced reactivity of focal point groups due to steric crowding is the principal problem in achieving this coupling and becomes a serious restriction as the starburst limit is approached. Nevertheless, the convergent and double exponential approaches have the major benefit that fewer chemical reactions are required to produce the perfect, next-generation dendron (e.g. only two reactions are involved for each step in the convergent synthesis shown in Figure 3.9) and the dendron produced is easier to separate from impurities because there is a greater difference in molecular size. Even more important, these approaches also facilitate the synthesis of chemically-asymmetric dendrimers by the joining together of dendrons that have different chemical structures and/or different surface groups. Thus the number of possibilities in dendrimer synthesis is vast, thereby providing a rich field for imaginative research.

3.3.3.2 Applications of Dendrimers

Dendrimers have many unusual properties. They are very compact molecules that do not crystallize, so they dissolve easily and their solutions have low viscosity compared to solutions of linear macromolecules of the same molar mass. Thus dendrimers can be used as rheology modifiers. As the

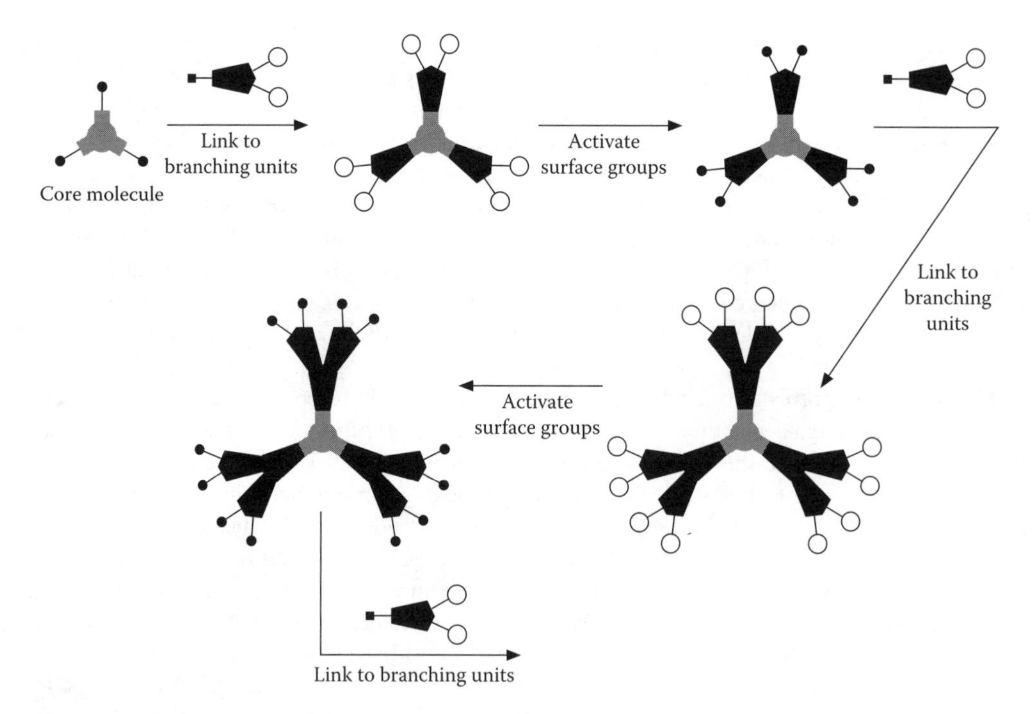

FIGURE 3.8 Schematic representation of the *divergent* strategy for direct synthesis of dendrimers. The elongated solid pentagons represent monomer units. The open circles represent protected surface groups. The small solid squares and circles represent, respectively, the deprotected focal point and surface groups that react with each other to produce the links between generations.

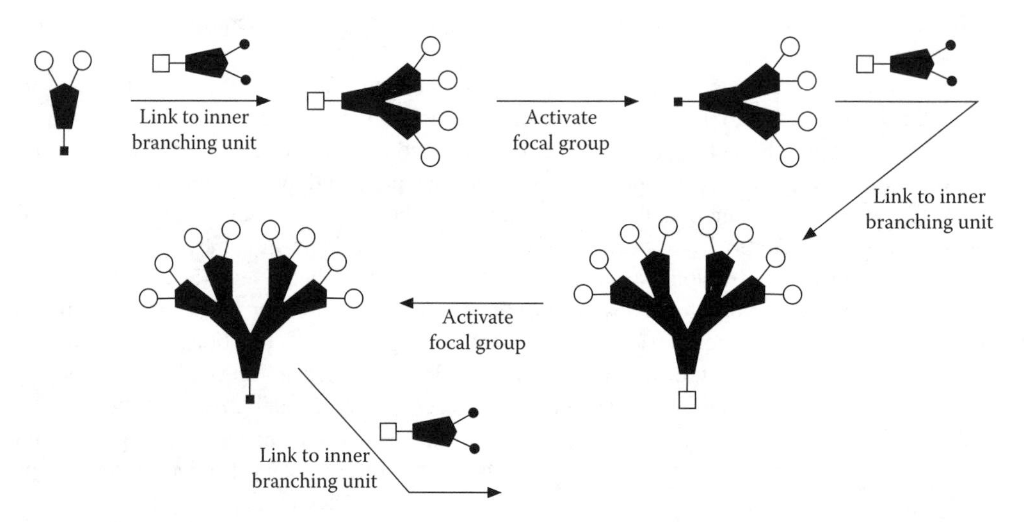

FIGURE 3.9 Schematic representation of the *convergent* strategy for synthesis of *dendrons* that have to be joined together in a final reaction (not shown here) of their activated focal point group with a multifunctional core molecule. The elongated solid pentagons represent monomer units. The open squares and circles represent protected focal point and surface groups, respectively. The small solid squares and circles represent, respectively, the deprotected focal point and surface groups that react with each other to produce the links between generations.

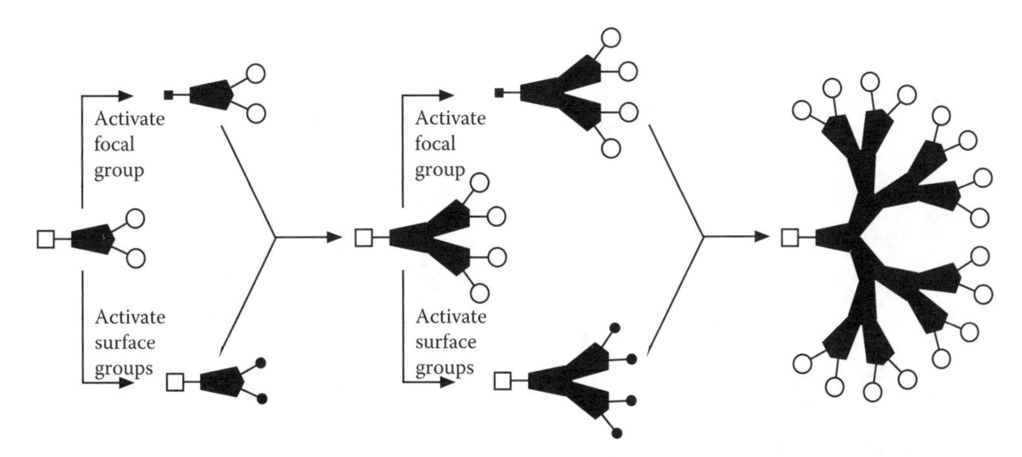

FIGURE 3.10 Schematic representation of the *double exponential* strategy for synthesis of *dendrons* that have to be joined together in a final reaction (not shown here) of their activated focal point group with a multifunctional core molecule. The elongated solid pentagons represent monomer units. The open squares and circles represent protected focal point and surface groups, respectively. The small solid squares and circles represent, respectively, the deprotected focal point and surface groups that react with each other to produce the links between generations.

generation number increases, in order to reduce steric crowding of the surface groups, they become globular in shape and gain the characteristic features indicated in Figure 3.5. The internal voids and channels are dynamic in shape and size, because the links (bonds) in the dendrimer are capable of rotation, but nevertheless define a rather special nanoscale environment that provides opportunities for catalysis, light harvesting, molecular recognition and trapping of guest molecules. The high density of surface functional groups also offers many possibilities for linking other molecules (such as sensors, drugs and genes) to the surface of a dendrimer. There is, therefore, great interest in the use of dendrimers for medical and therapeutic applications. Additionally, there is growing interest in the self-assembly of dendrons to produce highly organized *supramolecular* structures.

3.3.4 Hyperbranched Polymers

The stringent demands for controlled chemistry in the synthesis of dendrimers present a limitation to their exploitation, despite their interesting properties. In many applications, their highly branched, compact skeletal structure with a proliferation of functional end-groups is their most desirable feature and this potentially can be mimicked by *hyperbranched polymers* which also are highly branched but have far from perfect skeletal geometry. Indeed, just like most other types of polymer, hyperbranched polymers are statistical entities, with distributions of molar mass, degree of branching and skeletal structure (i.e. there are many geometric isomers with the same molar mass and degree of branching). The advantage of hyperbranched polymers in comparison to dendrimers is the ease with which they can be prepared, generally via a single polymerization reaction. For example, if a non-linear step polymerization is stopped just before it reaches the gel point, the product will be a hyperbranched polymer. More significantly, ARB_x $(x>2)$ step polymerizations are branching, but incapable of gelation and always produce hyperbranched polymers. Hyperbranched polymers also can be prepared by chain polymerization, but step polymerization methods are more important.

The concept of *branching factor* (or *degree of branching*) is important when describing hyperbranched polymers and is the mole fraction of fully branched and terminal monomer units, so is given by

$$\textit{branching factor} = \frac{N_{\mathrm{fbr}} + N_{\mathrm{term}}}{N_{\mathrm{total}}} \tag{3.31}$$

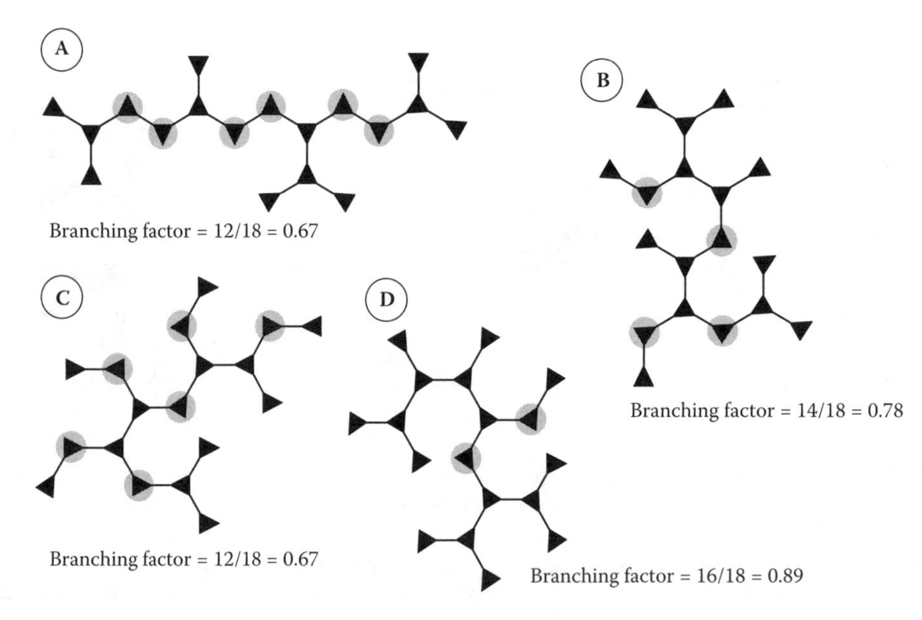

FIGURE 3.11 Schematic diagrams of four branched molecules, each comprising 18 repeat units of a trifunctional monomer which is represented by the solid triangles. The lines represent chemical links between monomer units. To assist calculation of the branching factors for these molecules, the grey circles highlight monomer units that are neither fully branched nor terminal.

where

N_{fbr} is the number of fully branched monomer units
N_{term} is the number of terminal monomer units
N_{total} is the total number of monomer units, including those that are not fully branched

A perfect dendrimer has a branching factor = 1 (with $N_{br} = N_{fbr} + N_{term} = N_{total}$). The branching factor most commonly is measured using nuclear magnetic resonance spectroscopy (see Chapter 15).

Some schematic diagrams of branched polymer structures, each comprising 18 monomer units, are given in Figure 3.11 to illustrate the diversity of skeletal structure and the importance of the branching factor calculation. Although molecules **A** and **C** appear to have quite different extents of branching and molecules **B**, **C** and **D** appear to have similar extents of branching, this is not the case – it is simply the way in which they are drawn that deceives the eye because molecules **A** and **C** have the same value of branching factor and molecule **D** has a much higher branching factor than molecule **C**, with molecule **B** having an intermediate value.

The development of hyperbranched polymers as mimics of dendrimers for functional applications is still in progress in relation to their use, e.g. as hosts for guest molecules and as delivery devices. However, their ease of synthesis and much reduced cost compared to dendrimers has led to more extensive applications in bulk materials (taking advantage of their low viscosities and high miscibility) and also in applications where the high number of functional end-groups can be used to advantage (e.g. their use as high-functionality crosslinking agents in coating applications).

3.3.4.1 Synthesis of Hyperbranched Polymers by Non-Linear Step Polymerization

All non-linear step polymerizations are capable of producing hyperbranched polymers provided that gelation is avoided by limiting conversion or controlling reactant stoichiometry (see Section 3.3.2.2). However, the most important routes are ARB_x step polymerizations where $x > 2$ since they cannot undergo gelation and so can be taken to high extents of reaction. The reason for the absence of gelation is self-evident from the massive stoichiometric imbalance of functional groups inherent in the ARB_x monomer structure. The statistical theory of non-linear step polymerization described in Section 3.3.2.2 can be applied to provide a more rigorous explanation of why these polymerizations do not gel. However, definition of the branching coefficient α needs to be reconsidered, since for an ARB_x system it is simply the probability that an f-functional unit is connected to another f-functional unit, which clearly is given by the extent of reaction of the B groups, p_B. Given that the ratio of B:A functional groups is x, then

$$\alpha = p_B = \frac{p_A}{x}$$

where p_A is the extent of reaction of A groups. Since the monomer functionality $f = x + 1$,

$$\alpha = \frac{p_A}{f-1}$$

Thus the condition for existence of network molecules, i.e. $\alpha(f-1) \geq 1$, cannot be attained because $p_A < 1$ in the absence of intramolecular reaction between A and B groups (which itself will give rise to loops in the hyperbranched polymer structure rather than the intermolecular links necessary for network formation).

Hence, ARB_x step polymerizations are widely used for the preparation of hyperbranched polymers. As for dendrimer synthesis, a wide range of chemical reactions (i.e. mutually-reactive A and B groups) have been used for synthesis of hyperbranched polymers, including all the main classes described for linear step polymerizations in Section 3.2. An example is the self-polycondensation of dihydroxy carboxylic acid monomers, such as 2,2-di(hydroxymethyl)propionic acid

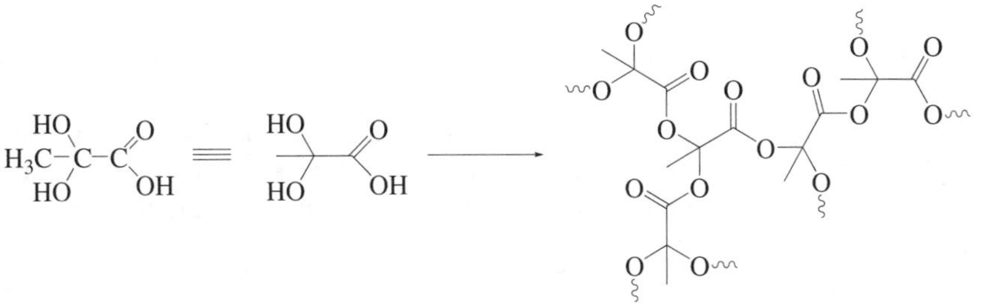

The branching factor for hyperbranched polymers increases with x for high conversion ARB_x step polymerizations and typically is about 0.5 for $x = 2$ and around 0.7 when $x = 4$. Strategies for increasing the branching factor have been developed and typically involve copolymerization of ARB_x branching monomers with RB_y core monomers where $y \geq 3$, which also has the effect of reducing molar mass dispersity. Nevertheless, it remains difficult to produce hyperbranched polymers with branching factors much greater than 0.8 and molar mass dispersities below 2 by non-linear step polymerization.

PROBLEMS

3.1 For each of the following polymerizations, write down a balanced chemical equation showing the structure of the polymer formed, including the repeat unit and the end groups.

(a) $n\ H_2N\!-\!(CH_2)_6\!-\!NH_2 + n\ ClOC\!-\!(CH_2)_8\!-\!COCl \longrightarrow$

(b) 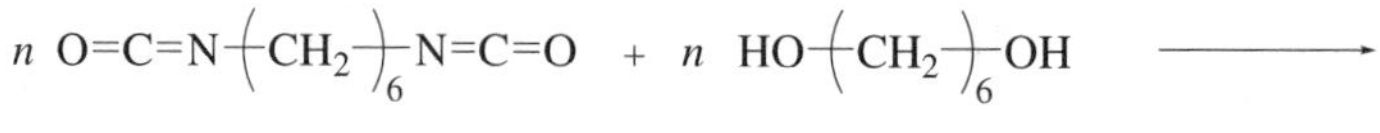 $n\ O{=}C{=}N{+}CH_2{+}_6N{=}C{=}O \ + \ n\ HO{+}CH_2{+}_6OH \longrightarrow$

(c) n 2,5-diamino-1,4-benzenedithiol (HS, SH, H_2N, NH_2 on the benzene ring) $+ \ n\ HO{-}\overset{O}{\overset{\|}{C}}{-}C_6H_4{-}\overset{O}{\overset{\|}{C}}{-}OH \longrightarrow$

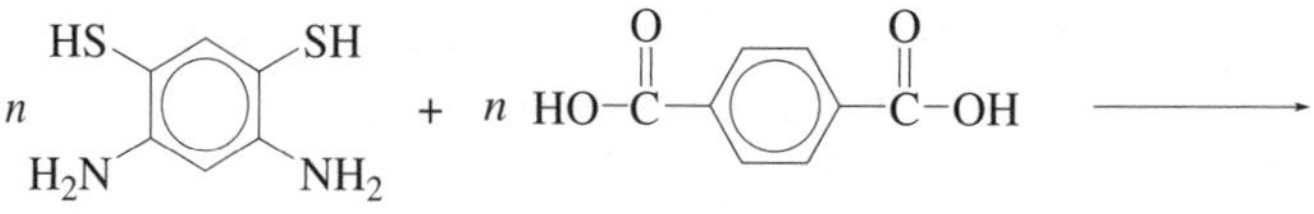

3.2 The repeat unit structures of three polymers, labelled (I)–(III), are shown below. For each polymer, write down the chemical structures of the monomers from which the polymer could be prepared.

(I) 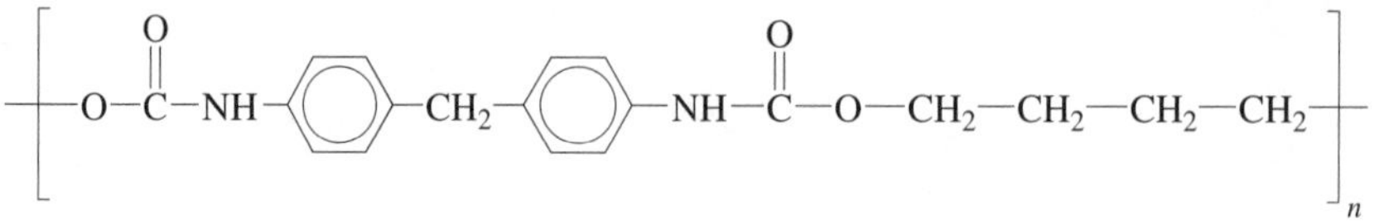 $\left[O{-}\overset{O}{\overset{\|}{C}}{-}NH{-}C_6H_4{-}CH_2{-}C_6H_4{-}NH{-}\overset{O}{\overset{\|}{C}}{-}O{-}CH_2{-}CH_2{-}CH_2{-}CH_2 \right]_n$

(II) 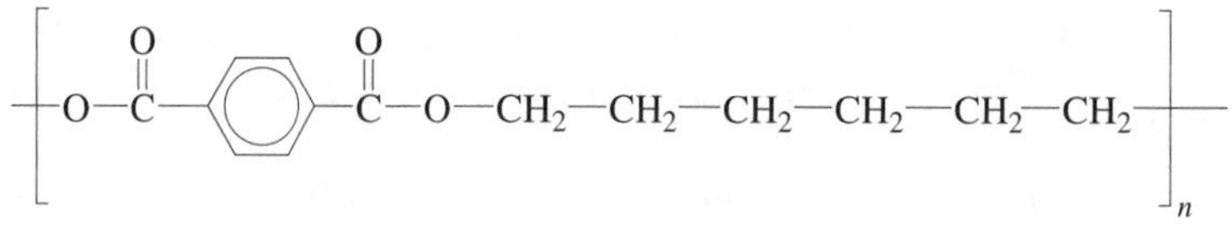 $\left[O{-}\overset{O}{\overset{\|}{C}}{-}C_6H_4{-}\overset{O}{\overset{\|}{C}}{-}O{-}CH_2{-}CH_2{-}CH_2{-}CH_2{-}CH_2{-}CH_2 \right]_n$

(III) 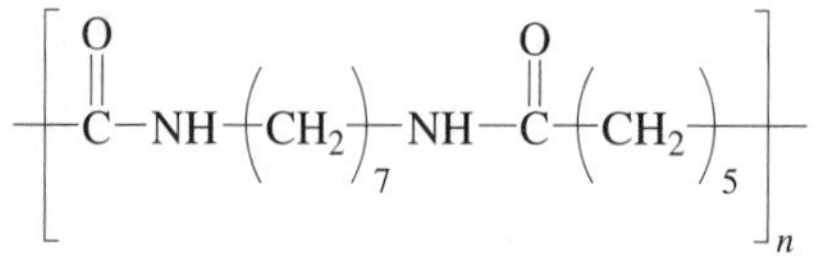 $\left[\overset{O}{\overset{\|}{C}}{-}NH{+}CH_2{+}_7NH{-}\overset{O}{\overset{\|}{C}}{+}CH_2{+}_5 \right]_n$

3.3 (a) Write down a balanced chemical equation for polymerization of a ω-hydroxy carboxylic acid ($HORCO_2H$).

(b) The following data were obtained during the condensation of 12-hydroxystearic acid at 160.5 °C in the molten state by sampling the reaction mixture at various times. The [COOH] was determined for each sample by titrating with ethanolic sodium hydroxide.

t / h	[COOH] / mol dm^{-3}
0	3.10
0.5	1.30
1.0	0.83
1.5	0.61
2.0	0.48
2.5	0.40
3.0	0.34

Determine the rate coefficient for the reaction under these conditions and the order of the reaction. Also suggest whether or not a catalyst was used.

(c) What would be the extent of the reaction after (i) 1 h and (ii) 5 h?

3.4 Neglecting the contribution of end groups to polymer molar mass, calculate the percentage conversion of functional groups required to obtain a polyester with a number-average molar mass of 24,000 g mol^{-1} from the monomer $HO(CH_2)_{14}COOH$.

3.5 A polyamide was prepared by bulk polymerization of hexamethylene diamine (9.22 g) with adipic acid (11.68 g) at 280 °C. Analysis of the whole reaction product showed that it contained 2.6×10^{-3} mol of carboxylic acid groups. Evaluate the number-average molar mass $\bar{M}_n$ of the polyamide, and also estimate its weight-average molar mass $\bar{M}_w$ by assuming that it has the most probable distribution of molar mass.

3.6 One kilogram of a polyester with a number-average molar mass $\bar{M}_n = 10{,}000$ g mol^{-1} is mixed with one kilogram of another polyester with a $\bar{M}_n = 30{,}000$ g mol^{-1}. The mixture then is heated to a temperature at which ester interchange occurs and the heating continued until full equilibration is achieved. Assuming that the two original polyester samples and the new polyester, produced by the ester interchange reaction, have the most probable distribution of molar mass, calculate $\bar{M}_n$ and weight-average molar mass $\bar{M}_w$ for the mixture before and after the ester interchange reaction.

3.7 Consider 1:1 stoichiometric polymerization of 1,4-diaminobutane with sebacoyl chloride.

(a) Write down a balanced chemical equation for the polymerization.

(b) Name the polymer formed by this polymerization.

(c) Using the Carothers equation, calculate the extent of reaction required to produce this polymer with a number-average molar mass $\bar{M}_n = 25.0$ kg mol^{-1}. You should neglect the effects of end groups on molar mass.

(d) Explain briefly the consequences of including 1,1,4,4-tetraminobutane in the polymerization with an exact stoichiometric ratio of 2:1:4 of 1,4-diaminobutane to 1,1,4,4-tetraminobutane to sebacoyl chloride.

3.8 A polycondensation reaction takes place between 1.2 moles of a dicarboxylic acid, 0.4 moles of glycerol (a triol) and 0.6 moles of ethylene glycol (a diol).

(a) Calculate the critical extents of reaction for gelation using (i) the statistical theory of Flory and (ii) the Carothers theory.

(b) Comment on the observation that the measured value of the critical extent of reaction is 0.866.

3.9 Show that the number-average functionality f_{av} for a non-linear $RA_2 + RA_f + RB_2$ step polymerization performed with an overall 1:1 stoichiometry of A and B groups is given by

$$f_{av} = \frac{4}{2 - \{(\gamma/f)(f-2)\}}$$

where $f > 2$ and γ is the initial ratio of A groups from RA_f molecules to the total number of A groups.

3.10 Calculate the branching factor for both of the following hyperbranched oligomeric chains prepared from a trifunctional monomer which is represented by the solid triangles. The lines represent chemical links between monomer units.

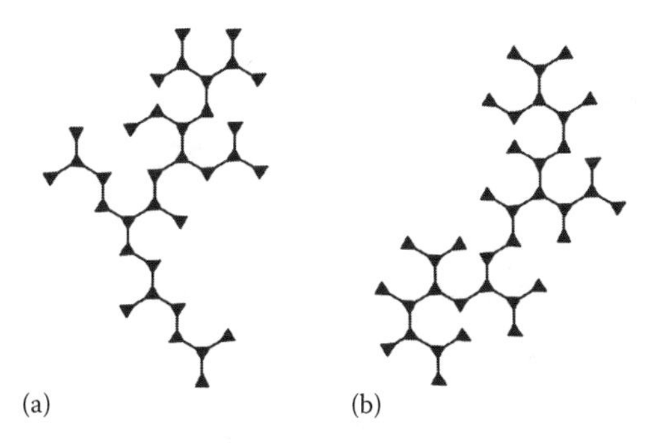

FURTHER READING

General Reading

Allen, G. and Bevington, J.C. (eds), *Comprehensive Polymer Science*, Vol. 5 (Eastmond, G.C., Ledwith, A., Russo, S., and Sigwalt, P., eds), Pergamon Press, Oxford, 1989.

Flory, P.J., *Principles of Polymer Chemistry*, Cornell University Press, Ithaca, NY, 1953.

Lenz, R.W., *Organic Chemistry of Synthetic High Polymers*, Interscience, New York, 1967.

Mark, H.F., Bikales, N.M., Overberger, C.G., and Menges, G. (eds), *Encyclopedia of Polymer Science and Engineering*, Wiley-Interscience, New York, 1985–1989.

Odian, G., *Principles of Polymerization*, 4th edn., Wiley-Interscience, New York, 2004.

Smith, M.B. and March, J., *March's Advanced Organic Chemistry: Reactions, Mechanisms and Structure*, John Wiley & Sons, Hoboken, NJ, 2007.

Stevens, M.P., *Polymer Chemistry: An Introduction*, 3rd edn., Oxford University Press, New York, 1998.

Dendrimers

Aulenta, F., Hayes, W., and Rannard, S., Dendrimers: A new class of nanoscopic containers and delivery devices, *European Polymer Journal*, **39**, 1741 (2003).

Boas, U. and Heegaard, P.M.H., Dendrimers in drug research, *Chemical Society Reviews*, **33**, 43 (2004).

Grayson, S.M. and Fréchet, J.M.J., Convergent dendrons and dendrimers: From synthesis to applications, *Chemical Reviews*, **101**, 3819 (2001).

Matthews, O.A., Shipway, A.N., and Stoddart, J.F., Dendrimers – Branching out from curiosities into new technologies, *Progress in Polymer Science*, **23**, 1 (1998).

Newkome, G.R., Moorefield, C.N., and Vögtle, F., *Dendrimers and Dendrons – Concepts, Syntheses, Applications*, Wiley-VCH, Weinheim, Germany, 2001.

Tomalia, D.A. and Fréchet. J.M. (eds), Dendrimers and dendritic polymers, *Progress in Polymer Science*, **30**, 217 (2005).

Vögtle, F., Richardt, G., and Werner, N., *Dendrimer Chemistry*, Wiley-VCH, Weinheim, Germany (2009).

Hyperbranched Polymers

Gao, C. and Yan, D., Hyperbranched polymers: From synthesis to applications, *Progress in Polymer Science*, **29**, 183 (2004).

Jikei, M. and Kakimoto, M-A., Hyperbranched polymers: A promising new class of materials, *Progress in Polymer Science*, **26**, 1233 (2001).

Yates, C.R. and Hayes, W., Synthesis and applications of hyperbranched polymers, *European Polymer Journal*, **40**, 1257 (2004).

4 Radical Polymerization

4.1 INTRODUCTION TO RADICAL POLYMERIZATION

Free radicals are independently existing species that possess an unpaired electron and normally are highly reactive with short lifetimes. A carbon-based free radical typically has sp^2 hybridization and has a general structure that can be represented in several ways, such as

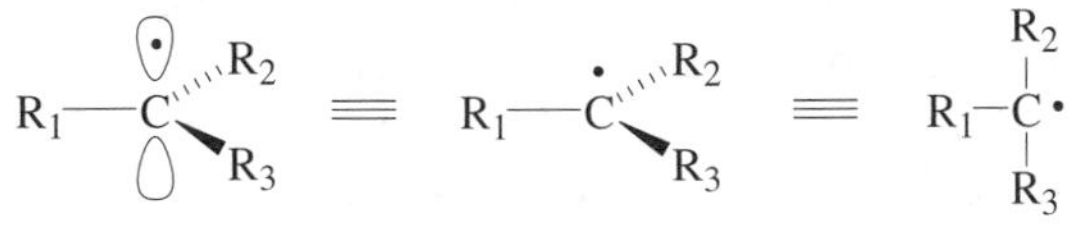

$$R_1-\dot{C}(R_2)(R_3) \equiv R_1-\overset{\bullet}{C}(R_2)(R_3) \equiv R_1-\overset{R_2}{\underset{R_3}{C}}\bullet$$

(i) **(ii)** **(iii)**

where R_1, R_2 and R_3 represent atoms or molecular fragments attached to the carbon atom bearing the unpaired electron. Structure **(i)** is a spatial representation in which the thin lines are in the plane of the paper, the dashed line is at 90° to and protruding behind the plane of the paper and the solid wedge is at 90° to and protruding forward from the plane of the paper; the single residual p-orbital is represented by the two lobes above and below the C atom and contains the unpaired electron which is represented by a 'dot' and circulates around both lobes of the p-orbital (but in structure **(i)** is shown at an instant in time when it is present in the upper lobe). Structure **(ii)** is a simplified version of **(i)** in which the lobes of the p-orbital are omitted, and structure **(iii)** is an even simpler non-spatially accurate representation in which correct bond angles are neglected. The latter two representations of free radicals are most common.

Radical polymerizations have been performed since the early days of polymer science and remain the most widely used and versatile types of polymerization for unsaturated monomers (containing C=C bonds) because almost all are susceptible to this form of chain polymerization (see Table 5.1). Each polymer molecule grows by sequential additions of molecules of unsaturated monomer to a terminal free-radical reactive site known as an *active centre*. The growing chain radical attacks the π-bond of a molecule of monomer (considered below as CH_2=CHX for simplicity) causing it to break homolytically.

$$\sim\!\sim CH_2-\underset{X}{\overset{\bullet}{C}H} + CH_2=\underset{X}{CH} \longrightarrow \sim\!\sim CH_2-\underset{X}{CH}-CH_2-\underset{X}{\overset{\bullet}{C}H}$$

The single-headed arrows depict movement of single electrons and are shown to highlight the processes that occur during the addition of monomer. Thus, one electron from the π-bond joins with the unpaired electron from the terminal carbon atom of the chain radical, creating a new bond to the methylene (CH_2) carbon atom of the C=C bond; the remaining π-bond electron moves to the other C=C carbon atom, which becomes the new active radical site. Hence, upon every addition of monomer, the active centre is transferred to the newly-created chain end.

Use of the term *free-radical polymerization* to describe all polymerizations in which the propagating species is a free radical has become less appropriate since the 1990s due to the development of *reversible-deactivation (living/controlled)* radical polymerizations (see Section 4.5) in which the active radical sites are not free for most of the time, but instead are reversibly deactivated and only become active in adding monomer for very short bursts of time. Hence, the generic term has been abbreviated to *radical polymerizations* in recognition of the very important distinction between the long-established, conventional *free*-radical polymerizations and the much newer *reversible-deactivation* radical polymerizations.

4.2 THE CHEMISTRY OF CONVENTIONAL FREE-RADICAL POLYMERIZATION

Free-radical polymerization is the most widely practised method of radical polymerization, and is used almost exclusively for the preparation of polymers from monomers of the general structure $CH_2{=}CR_1R_2$ (see entries (1) and (3)–(7) of Table 1.1). In common with other types of chain polymerization, the reaction can be divided into three distinct basic stages: *initiation*, *propagation* and *termination*. A further process, known as *chain transfer*, can occur in all chain polymerizations and often makes a very significant contribution. In the following four sections, the general chemistry associated with each of these four processes is described in the context of free-radical polymerizations by considering for simplicity the polymerization of a general vinyl monomer, $CH_2{=}CHX$ (where X is the substituent group). Nevertheless, the chemistry is equally applicable to 1,1-disubstituted monomers ($CH_2{=}CXY$, where Y is the second substituent group) and to 1,2-disubstituted monomers (CHX=CHY), but the latter are of low reactivity due to steric hindrance in propagation and do not easily homopolymerize (see Section 9.3.5).

4.2.1 Initiation

Initiation involves creation of the free-radical active centre and usually takes place in two steps. The first is the formation of free radicals from an *initiator* and the second is the addition of one of these free radicals to a molecule of monomer.

There are two principal ways in which free radicals can be formed: (i) *homolytic scission* (i.e. *homolysis*) of a single bond in which the two bonding electrons go one each onto the two atoms associated with the original bond (thereby always producing two free radical species), and (ii) transfer of a single electron to or from an ion or molecule (e.g. in redox reactions) which often are termed *single-electron transfer* processes, some of which produce only a single free-radical species.

Homolysis can be effected by the application of heat (Δ) and there are many compounds that contain weak bonds which undergo *thermolysis* at useful rates above about 50 °C. Compounds containing peroxide (–O–O–) or azo (–N=N–) linkages are particularly useful as *initiators* and undergo homolysis as follows

$$R_1{-}O{-}O{-}R_2 \xrightarrow{\Delta} R_1{-}O^{\bullet} + {}^{\bullet}O{-}R_2$$

$$R{-}N{=}N{-}R \xrightarrow{\Delta} 2R^{\bullet} + N{\equiv}N$$

where R_1 and R_2 can be aromatic, alkyl or H (often with $R_1 = R_2$) and R is aromatic or alkyl. The single-headed arrows again depict movement of single electrons and show how the molecules decompose. Some specific *thermal initiators* are shown below.

$$C_6H_5{-}\overset{O}{\overset{\|}{C}}{-}O{-}O{-}\overset{O}{\overset{\|}{C}}{-}C_6H_5 \xrightarrow{\Delta} 2\,C_6H_5{-}\overset{O}{\overset{\|}{C}}{-}O^{\bullet}$$

Benzoyl peroxide Benzoyloxy radicals

$$(CH_3)_3C{-}O{-}OH \xrightarrow{\Delta} (CH_3)_3C{-}O^{\bullet} + {}^{\bullet}OH$$

t-Butyl hydroperoxide → *t*-Butoxy radical + Hydroxyl radical

$$(CH_3)_2C(CN){-}N{=}N{-}C(CN)(CH_3)_2 \xrightarrow{\Delta} 2\,(CH_3)_2C^{\bullet}(CN) + N{\equiv}N$$

Azobisisobutyronitrile or 2,2'-Azobis(2-cyanopropane) → 2-Cyanopropyl radicals + Nitrogen

Such homolysis reactions are widely used to initiate free-radical polymerizations in the convenient temperature range of 60–90 °C. Many of the primary radicals produced undergo further breakdown before reaction with monomer, for example, β-scissions such as

$$C_6H_5{-}C(=O){-}O^{\bullet} \longrightarrow C_6H_5^{\bullet} + O{=}C{=}O$$

Phenyl radical + Carbon dioxide

$$(CH_3)_2C(CH_3){-}O^{\bullet} \longrightarrow {}^{\bullet}CH_3 + (CH_3)_2C{=}O$$

Methyl radical + Acetone

Homolysis can also be brought about by the action of radiation (usually ultraviolet) rather than heat, i.e. by *photolysis*. For example, the dissociation of azobisisobutyronitrile (via the same chemistry as that shown above) can be effected using ultraviolet radiation, though this is of low photochemical efficiency. Hence, it is more common to use *photochemical initiators* that decompose efficiently when exposed to ultraviolet radiation, such as benzophenone and benzoin derivatives (e.g. benzoin alkyl ethers)

$$C_6H_5{-}C(=O){-}C_6H_5 \xrightarrow{h\nu} C_6H_5{-}C^{\bullet}(=O) + {}^{\bullet}C_6H_5$$

Benzophenone → Benzoyl radical + Phenyl radical

$$C_6H_5{-}C(=O){-}CH(OR){-}C_6H_5 \xrightarrow{h\nu} C_6H_5{-}C^{\bullet}(=O) + {}^{\bullet}CH(OR){-}C_6H_5$$

Benzoin alkyl ether (*R* = *alkyl*) → Benzoyl radical + Alkoxybenzyl radical

An advantage of photolysis is that the formation of free radicals begins at the instant of exposure and ceases as soon as the light source is removed.

Redox reactions often are used when it is necessary to perform free-radical polymerizations at low temperatures (typically below 50 °C). Two *redox initiation* systems will be considered to exemplify these types of reaction. In the first example, shown below, a Fe^{2+} ion donates an electron to the OH oxygen atom of cumyl hydroperoxide causing the adjacent O–O peroxide bond to break homolytically; the OH oxygen atom thereby receives a second electron and becomes negatively charged (giving the hydroxyl ion) and the remaining oxygen atom from the peroxide bond receives the other peroxide bonding electron (giving the cumyloxy radical).

$$C_6H_5-C(CH_3)_2-O-OH + Fe^{2+} \longrightarrow C_6H_5-C(CH_3)_2-O^{\bullet} + \overline{O}H + Fe^{3+}$$

Cumyl hydroperoxide | Ferrous ion | Cumyloxy radical | Hydroxyl ion | Ferric ion

Similar chemistry occurs with all peroxides and hydroperoxides (e.g. R–O–O–R will produce an R–O• radical plus the alkoxide ion, RO^-, and Fe^{3+}) and can be activated using many other transition metals (e.g. the reaction of R–O–O–R with Co^{2+} yields an R–O• radical plus RO^- and Co^{3+} and is widely employed to initiate room temperature curing of the resin matrices in production of glass-fibre reinforced composites; see Section 4.6.2.1). The next example is a completely inorganic redox initiation system that is useful for polymerizations carried out in water

$$^{-}O-SO_2-O-O-SO_2-O^{-} + HO-SO-O^{-} \longrightarrow {}^{-}O-SO_2-O^{-} + {}^{\bullet}O-SO_2-O^{-} + HO-SO-O^{\bullet}$$

Persulphate ion | Bisulphite ion | Sulphate ion | Sulphate radical-anion | Bisulphite radical

In this case, one electron from the lone pair of electrons (negative charge) on the bisulphite ion is donated to one of the peroxide oxygen atoms, causing the peroxide bond to break homolytically, each of the peroxide oxygen atoms receiving one of the two bonding electrons in the process (hence, one finishes with two electrons, to yield the sulphate ion, and the other with an unpaired electron, to yield the sulphate radical-anion); both the sulphate radical-anion and the bisulphite radical are capable of initiating free-radical polymerization.

An *active centre* is created in the second step of initiation, when a free radical $R^{\bullet}$ generated from the initiation system attacks the π-bond of a molecule of monomer via the same chemistry of monomer addition shown in Section 4.1. Two modes of addition are possible

$$R^{\bullet} + CH_2{=}CHX \xrightarrow{(I)} R-CH_2-\dot{C}HX$$

$$R^{\bullet} + CH_2{=}CHX \xrightarrow{(II)} R-CHX-\dot{C}H_2$$

Mode **(I)** predominates because attack at the methylene (CH_2) carbon is less sterically hindered and yields a product radical that is more stable because of the effects of the attached substituent group X (which provides steric stabilization of the radical and often also contributes mesomeric stabilization).

Not all of the radicals formed from the initiator are destined to react with monomer. Some are lost in side reactions such as those for benzoyl peroxide shown below.

$$2\,C_6H_5{-}\overset{O}{\overset{\|}{C}}{-}O^{\bullet} \longrightarrow C_6H_5{-}\overset{O}{\overset{\|}{C}}{-}O{-}C_6H_5 + O{=}C{=}O$$

$$2\,C_6H_5^{\bullet} \longrightarrow C_6H_5{-}C_6H_5$$

$$C_6H_5{-}\overset{O}{\overset{\|}{C}}{-}O{-}O{-}\overset{O}{\overset{\|}{C}}{-}C_6H_5 + C_6H_5^{\bullet} \longrightarrow C_6H_5{-}\overset{O}{\overset{\|}{C}}{-}O{-}C_6H_5 + C_6H_5{-}\overset{O}{\overset{\|}{C}}{-}O^{\bullet}$$

The latter type of reaction is known as *induced decomposition* and makes a significant contribution to wastage of peroxide initiators.

4.2.2 Propagation

Propagation involves growth of a polymer chain by rapid sequential addition of monomer to the active centre, as described in Section 4.1. The time required for each monomer addition typically is of the order of a millisecond and so several thousand additions can take place within a few seconds. As with the second step of initiation, there are two possible modes of propagation, which in the first propagation step following initiation would be

$$R{-}CH_2{-}\underset{X}{\overset{\bullet}{C}H} + CH_2{=}\underset{X}{CH} \xrightarrow{(\mathrm{I})} R{-}CH_2{-}\underset{X}{CH}{-}CH_2{-}\underset{X}{\overset{\bullet}{C}H} \quad \text{Head-to-tail addition}$$

$$R{-}CH_2{-}\underset{X}{\overset{\bullet}{C}H} + CH_2{=}\underset{X}{CH} \xrightarrow{(\mathrm{II})} R{-}CH_2{-}\underset{X}{CH}{-}\underset{X}{CH}{-}\overset{\bullet}{C}H_2 \quad \text{Head-to-head addition}$$

Mode **(I)** again dominates for the same reasons as described in the previous section and the polymer chains formed consist predominantly of head-to-tail sequences of repeat units

$$\sim CH_2{-}\underset{X}{CH}{-}CH_2{-}\underset{X}{CH}{-}CH_2{-}\underset{X}{CH}{-}CH_2{-}\underset{X}{CH}\sim$$

If a head-to-head addition occurs, it will be followed immediately by a tail-to-tail addition to generate the more stable active centre which will then continue to propagate principally via head-to-tail addition. The extent of head-to-head and tail-to-tail additions is immeasurably small for most monomers; mode **(II)** only contributes significantly for the few monomers for which X is small (e.g. similar in size to a H atom) and provides no mesomeric stabilization (e.g. in free-radical polymerization of $CH_2{=}CHF$). Thus, for the purposes of a more general description of radical polymerizations, it is entirely reasonable to neglect mode **(II)** and to assume that propagation proceeds exclusively by mode **(I)** head-to-tail addition and can be represented in its most general form by

$$R\left[CH_2{-}\underset{X}{CH}\right]_{x-1}CH_2{-}\underset{X}{\overset{\bullet}{C}H} + CH_2{=}\underset{X}{CH} \longrightarrow R\left[CH_2{-}\underset{X}{CH}\right]_{x-1}CH_2{-}\underset{X}{CH}{-}CH_2{-}\underset{X}{\overset{\bullet}{C}H}$$

where x is the degree of polymerization prior to the monomer addition.

4.2.3 Termination

In the *termination* stage, the active centre is destroyed irreversibly and propagation ceases. The two most common mechanisms of termination in radical polymerizations involve bimolecular reaction of propagating chains. *Combination* involves the *coupling* together of two growing chains to form a single polymer molecule

$$\text{R}\!\left[\text{CH}_2\!-\!\underset{\text{X}}{\text{CH}}\right]_{x-1}\!\text{CH}_2\!-\!\underset{\text{X}}{\dot{\text{C}}\text{H}} \;+\; \underset{\text{X}}{\dot{\text{C}}\text{H}}\!-\!\text{CH}_2\!\left[\underset{\text{X}}{\text{CH}}\!-\!\text{CH}_2\right]_{y-1}\!\text{R}$$

$$\downarrow$$

$$\text{R}\!\left[\text{CH}_2\!-\!\underset{\text{X}}{\text{CH}}\right]_{x-1}\!\text{CH}_2\!-\!\underset{\text{X}}{\text{CH}}\!-\!\underset{\text{X}}{\text{CH}}\!-\!\text{CH}_2\!\left[\underset{\text{X}}{\text{CH}}\!-\!\text{CH}_2\right]_{y-1}\!\text{R}$$

where x and y are the degrees of polymerization of the respective chain radicals prior to the combination reaction. Note that a *single* dead polymer molecule (with degree of polymerization $= x + y$) is produced with an initiator fragment (the R group) at both chain ends and that the radical coupling reaction also gives rise to a 'head-to-head' linkage. The other common process by which propagating chain radicals terminate is *disproportionation*, which involves abstraction of a hydrogen atom from the penultimate C atom of one growing chain radical by another, the remaining electron from the C–H bond joining with the unpaired electron on the terminal C atom of that chain to create a terminal π-bond

$$\text{R}\!\left[\text{CH}_2\!-\!\underset{\text{X}}{\text{CH}}\right]_{x-1}\!\text{CH}_2\!-\!\underset{\text{X}}{\dot{\text{C}}\text{H}} \;+\; \underset{\text{X}}{\dot{\text{C}}\text{H}}\!-\!\overset{\text{H}}{\text{CH}}\!\left[\underset{\text{X}}{\text{CH}}\!-\!\text{CH}_2\right]_{y-1}\!\text{R}$$

$$\downarrow$$

$$\text{R}\!\left[\text{CH}_2\!-\!\underset{\text{X}}{\text{CH}}\right]_{x-1}\!\text{CH}_2\!-\!\underset{\text{X}}{\text{CH}_2} \;+\; \underset{\text{X}}{\text{CH}}\!=\!\text{CH}\!\left[\underset{\text{X}}{\text{CH}}\!-\!\text{CH}_2\right]_{y-1}\!\text{R}$$

Thus, *two* dead polymer molecules are formed (with degrees of polymerization x and y, respectively), one with a saturated end group and the other with an unsaturated end group, and both with an initiator fragment at the other chain end.

In general, both types of termination reaction take place but to different extents depending upon the monomer and the polymerization conditions. For example, it is found that polystyrene chain radicals terminate principally by combination whereas poly(methyl methacrylate) chain radicals terminate predominantly by disproportionation, especially at temperatures above 60 °C. These observations are more general in that combination tends to dominate termination in polymerizations of vinyl monomers ($CH_2{=}CHX$), whereas disproportionation dominates in polymerizations of α-methylvinyl monomers ($CH_2{=}C(CH_3)X$). This is because the α-CH_3 group provides an additional three C–H bonds from which a H atom can be abstracted

$$\sim\!\text{CH}_2\!-\!\underset{\text{X}}{\overset{\text{CH}_3}{\text{C}}}\!\cdot \;+\; \cdot\underset{\text{X}}{\overset{\overset{\text{H}}{\text{CH}_2}}{\text{C}}}\!-\!\text{CH}_2\!\sim \;\longrightarrow\; \sim\!\text{CH}_2\!-\!\underset{\text{X}}{\overset{\text{CH}_3}{\text{CH}}} \;+\; \underset{\text{X}}{\overset{\text{CH}_2}{\overset{\|}{\text{C}}}}\!-\!\text{CH}_2\!\sim$$

4.2.4 Chain Transfer

Chain transfer reactions occur in most chain polymerizations and are reactions in which the active centre is transferred from the active chain end to another species in the polymerization system. In their most generic form, chain transfer reactions in radical polymerizations can be represented by

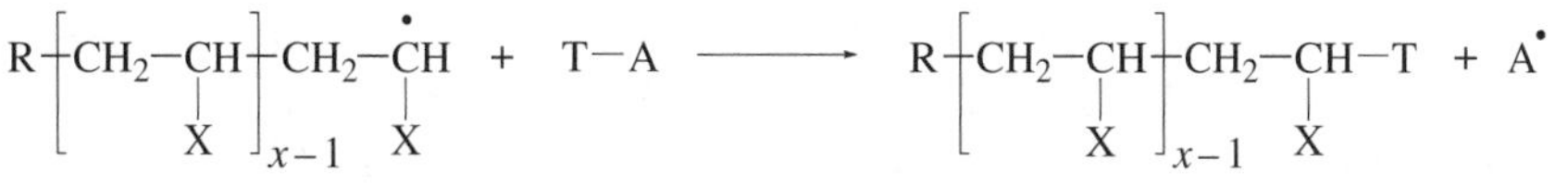

$$\mathrm{R}\!\left[\mathrm{CH_2{-}\underset{X}{\underset{|}{CH}}}\right]_{x-1}\!\mathrm{CH_2{-}\underset{X}{\underset{|}{\dot{C}H}}} + \mathrm{T{-}A} \longrightarrow \mathrm{R}\!\left[\mathrm{CH_2{-}\underset{X}{\underset{|}{CH}}}\right]_{x-1}\!\mathrm{CH_2{-}\underset{X}{\underset{|}{CH}}{-}T} + \mathrm{A^{\bullet}}$$

where T and A are fragments linked by a single bond in a hypothetical molecule TA. The chain radical abstracts T (often a hydrogen or halogen atom) from TA causing homolytic scission of the T–A bond to yield a dead polymer molecule and the radical $A^{\bullet}$, which if sufficiently reactive may then react with a molecule of monomer to initiate the growth of a new chain.

4.2.4.1 Chain Transfer with Small Molecules

Chain transfer can occur with molecules of *initiator*, *monomer*, *solvent* and deliberately added *transfer agents*. Some specific examples are given in Figure 4.1, where for simplicity only the terminal unit of the chain radical is shown.

Clearly, each of these chain transfer reactions leads to death of a propagating chain. The first chain transfer reaction is another example of induced decomposition of a peroxide (described at the end of Section 4.2.1). Chain transfer to monomer is exemplified by the commercially important monomer, vinyl acetate; the product radical reinitiates polymerization but also possesses a terminal C=C double bond which itself can undergo polymerization, leading to formation of *long-chain branches* in the resulting polymer. In the case of toluene (a good solvent for many monomers and polymers), the chain radical abstracts a benzylic H atom (rather than a H atom from the benzene ring) because this yields the benzyl radical which is stabilized mesomerically by delocalization of the unpaired electron around the benzene ring (which cannot occur for the aryl radicals produced by H-abstraction from the benzene ring). The final example is of a molecule, CBr_4, which is so active in chain transfer that it can only be tolerated in very small quantities, otherwise the polymer chains

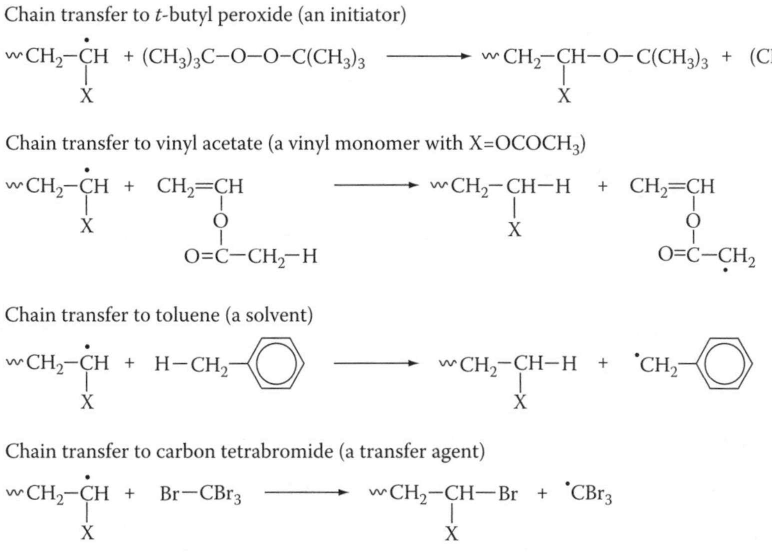

Chain transfer to *t*-butyl peroxide (an initiator)

$$\sim\mathrm{CH_2{-}\underset{X}{\underset{|}{\dot{C}H}}} + \mathrm{(CH_3)_3C{-}O{-}O{-}C(CH_3)_3} \longrightarrow \sim\mathrm{CH_2{-}\underset{X}{\underset{|}{CH}}{-}O{-}C(CH_3)_3} + \mathrm{(CH_3)_3C{-}O^{\bullet}}$$

Chain transfer to vinyl acetate (a vinyl monomer with $X{=}OCOCH_3$)

$$\sim\mathrm{CH_2{-}\underset{X}{\underset{|}{\dot{C}H}}} + \mathrm{CH_2{=}CH{-}O{-}C({=}O){-}CH_2{-}H} \longrightarrow \sim\mathrm{CH_2{-}\underset{X}{\underset{|}{CH}}{-}H} + \mathrm{CH_2{=}CH{-}O{-}C({=}O){-}\dot{C}H_2}$$

Chain transfer to toluene (a solvent)

$$\sim\mathrm{CH_2{-}\underset{X}{\underset{|}{\dot{C}H}}} + \mathrm{H{-}CH_2{-}C_6H_5} \longrightarrow \sim\mathrm{CH_2{-}\underset{X}{\underset{|}{CH}}{-}H} + \mathrm{{}^{\bullet}CH_2{-}C_6H_5}$$

Chain transfer to carbon tetrabromide (a transfer agent)

$$\sim\mathrm{CH_2{-}\underset{X}{\underset{|}{\dot{C}H}}} + \mathrm{Br{-}CBr_3} \longrightarrow \sim\mathrm{CH_2{-}\underset{X}{\underset{|}{CH}}{-}Br} + \mathrm{{}^{\bullet}CBr_3}$$

FIGURE 4.1 Some specific examples of chain transfer to initiator, monomer, solvent and transfer agent.

formed would be too short; such compounds are called *transfer agents* (or *modifiers* or *regulators*) because they can be used effectively in very small quantities to control degree of polymerization.

The consequences of chain transfer to small molecules present in a polymerization system will be described more completely in Section 4.3.5.

4.2.4.2 Chain Transfer to Polymer

Chain transfer to polymer occurs in polymerization of some monomers and leads to the formation of branched polymer molecules, so it can have a major effect on the skeletal structure and properties of the polymers produced. *Intramolecular* chain transfer to polymer (or *backbiting*) reactions give rise to *short-chain branches*, whereas *long-chain branches* result from *intermolecular* chain transfer to polymer. The polymerization of ethylene at high temperature and high pressure (used to manufacture low-density polyethylene, LDPE) is a classic example of the importance of chain transfer to polymer; the chemistry is shown in Figure 4.2. Note that the so-called short-chain branches resulting from backbiting are in fact *n*-butyl, ethyl and 2-ethylhexyl side groups. The short- and long-chain branches in LDPE greatly restrict the ability of the polymer to crystallize and compared to linear polyethylene (see Sections 6.4.7 and 6.5.6), LDPE has a much lower melting point (105–115 °C cf. 135–140 °C) and degree of crystallinity (45%–50% cf. about 90%). The long-chain branches also have a major effect on the rheology of LDPE in the molten state.

For other monomers, chain transfer to polymer depends on the nature of the monomer structure and may proceed via abstraction of an atom (usually H) from either the backbone or a substituent group. For example, in radical polymerizations of acrylic monomers ($CH_2{=}CHCO_2R$ where $R{=}H$ or alkyl), chain transfer to polymer proceeds via H-abstraction from the tertiary backbone C–H bonds. This occurs because the tertiary radical produced is significantly more stable than the 'normal' secondary propagating radical (i.e. the additional group attached to the tertiary carbon atom provides significant steric stabilization of the radical). The intermolecular reaction is illustrated for methyl acrylate in Figure 4.3. The H-abstraction can also occur intramolecularly via six-membered cyclic transition states (in backbiting reactions analogous to those shown in Figure 4.2). In fact, as for ethylene, there is evidence that the intramolecular reaction in polymerization of acrylic monomers is far more frequent than the intermolecular reaction.

In radical polymerization of vinyl acetate, chain transfer to polymer proceeds via H-abstraction from the side group, as shown in Figure 4.4. The reason for this difference compared to methyl acrylate (which is an isomer of vinyl acetate) is the position of the carbonyl group, which for vinyl acetate is attached to the methyl group; thus the product radical from H-abstraction at the methyl group is mesomerically stabilized by electron delocalization with the carbonyl π-bond (this stabilization is not possible for the product from H-abstraction at a backbone tertiary C–H bond). Again chain transfer to polymer can also proceed intramolecularly, though the chemistry will be slightly different from that shown for ethylene because the H-abstraction is from the side group.

Polymers from disubstituted monomers of structure $CH_2{=}CR_1R_2$ have no tertiary backbone C–H bonds, and so chain transfer to polymer proceeds only via the side group and then only if it contains labile H atoms. For some monomers, therefore, chain transfer to polymer is negligible (e.g. in polymerization of methacrylic monomers which have $R_1{=}CH_3$ and $R_2{=}CO_2R$ where $R{=}H$, alkyl or aryl).

Chain transfer to polymer in polymerization of 1,3-dienes is a special case that is considered in Section 4.6.2.2.

4.3 KINETICS OF CONVENTIONAL FREE-RADICAL POLYMERIZATION

For any polymerization, two quantities are of paramount importance: the rate at which monomer is polymerized and the degree of polymerization of the polymer produced. The kinetics of free-radical polymerization can be analysed to obtain equations that predict these quantities by considering the general chemistry indicated in Section 4.2.

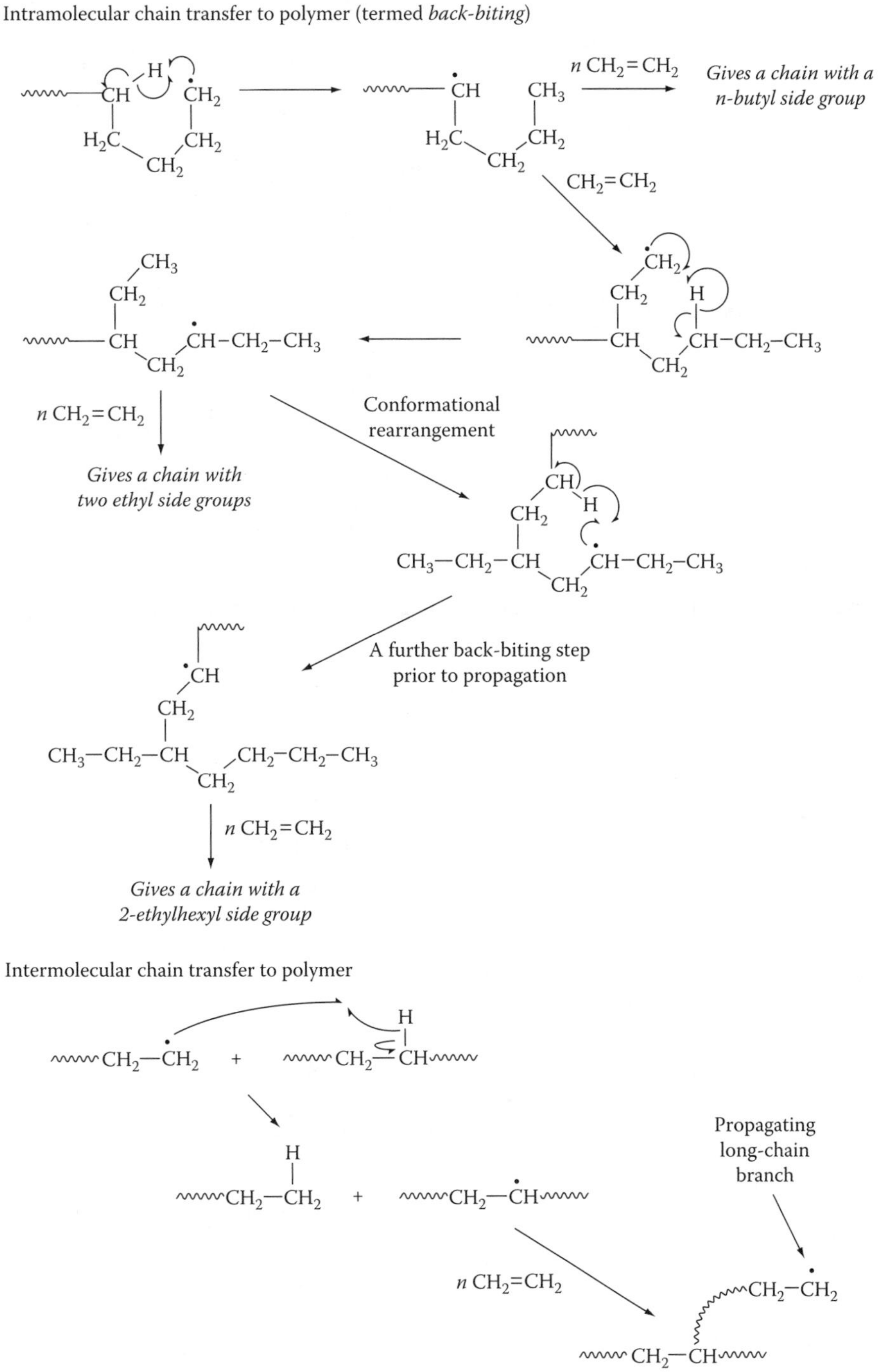

Intramolecular chain transfer to polymer (termed *back-biting*)

$n\,CH_2{=}CH_2$

Gives a chain with a n-butyl side group

$CH_2{=}CH_2$

$n\,CH_2{=}CH_2$

Gives a chain with two ethyl side groups

Conformational rearrangement

A further back-biting step prior to propagation

$n\,CH_2{=}CH_2$

Gives a chain with a 2-ethylhexyl side group

Intermolecular chain transfer to polymer

Propagating long-chain branch

$n\,CH_2{=}CH_2$

FIGURE 4.2 The chemistry of intramolecular and intermolecular chain transfer to polymer in free-radical polymerization of ethylene.

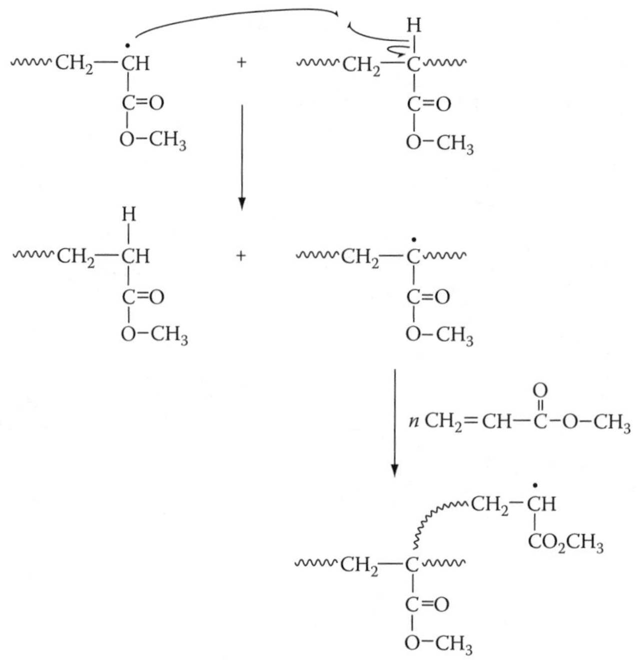

FIGURE 4.3 The chemistry of intermolecular chain transfer to polymer in free-radical polymerization of methyl acrylate.

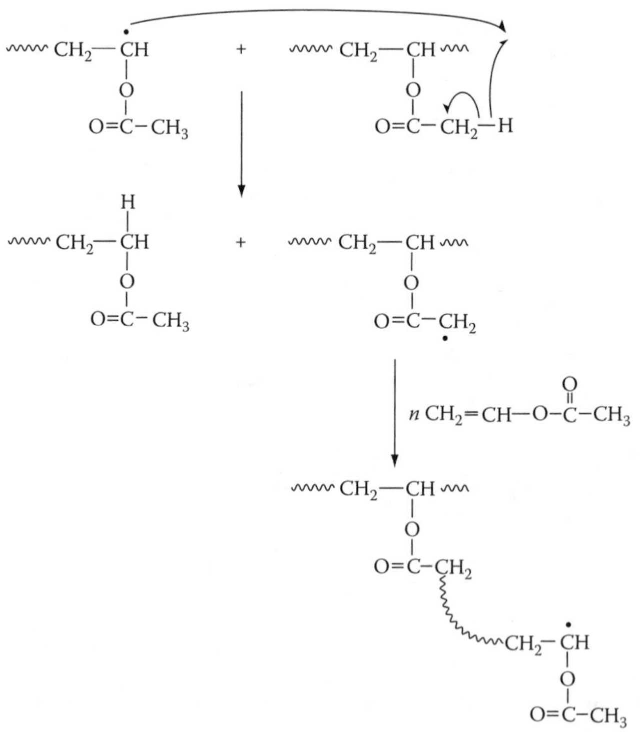

FIGURE 4.4 The chemistry of intermolecular chain transfer to polymer in free-radical polymerization of vinyl acetate.

4.3.1 Rate of Polymerization

The *rate of polymerization* R_p is the rate at which monomer (M) is consumed in polymerization

$$R_p = -\frac{d[M]}{dt} \tag{4.1}$$

In order to obtain a kinetics expression for R_p in terms of experimentally accessible quantities, the kinetics of each stage in free-radical polymerization needs to be considered individually.

Initiation may be represented by the two steps:

$$I \xrightarrow{\text{Slow}} n\,R^{\bullet}$$

$$R^{\bullet} + M \xrightarrow{\text{Fast}} RM_1^{\bullet}$$

where

n is the number of free radicals $R^{\bullet}$ formed upon breakdown of one molecule of the initiator I (usually $n=1$ or 2)

M represents either a molecule of monomer or a monomer unit in a polymer chain

Since the formation of $R^{\bullet}$ from the initiator proceeds much more slowly than the reaction of $R^{\bullet}$ with monomer, the first step is rate-determining and controls the rate R_i of formation of active centres, which then is defined as

$$R_i = \frac{d[R^{\bullet}]}{dt} \tag{4.2}$$

If it is assumed that the rate coefficient k_p for propagation is independent of the length of the growing chain, then *propagation* can be represented by a single general reaction

$$M_i^{\bullet} + M \longrightarrow M_{i+1}^{\bullet}$$

in which the initiator fragment R has been omitted for simplicity. Given that long-chain polymer molecules are formed, the amount of monomer consumed in the initiation stage is negligible compared to that consumed by propagating chains and so the rate of consumption of monomer is given by

$$-\frac{d[M]}{dt} = k_p[M_1^{\bullet}][M] + k_p[M_2^{\bullet}][M] + \ldots + k_p[M_i^{\bullet}][M] + \ldots$$

or

$$-\frac{d[M]}{dt} = k_p[M]([M_1^{\bullet}] + [M_2^{\bullet}] + \ldots + [M_i^{\bullet}] + \ldots)$$

If $[M^{\bullet}]$ is the total concentration of all radical species (i.e. $[M^{\bullet}] = \sum_{i=1}^{\infty}[M_i^{\bullet}]$), then

$$-\frac{d[M]}{dt} = k_p[M][M^{\bullet}] \tag{4.3}$$

Termination can be represented by

$$\mathrm{M}_i^\bullet + \mathrm{M}_j^\bullet \begin{cases} \xrightarrow{k_{tc}} \mathrm{P}_{i+j} \\ \xrightarrow{k_{td}} \mathrm{P}_i + \mathrm{P}_j \end{cases}$$

where k_{tc} and k_{td} are the rate coefficients for combination and disproportionation, respectively. Thus, the overall rate at which radicals are consumed is given by

$$-\frac{d[M^\bullet]}{dt} = 2k_{tc}[M^\bullet][M^\bullet] + 2k_{td}[M^\bullet][M^\bullet]$$

The factors of 2 and use of the total concentration $[M^\bullet]$ of radical species arise because *two* growing chains of *any length* are consumed by each termination reaction. The equation can be simplified to

$$-\frac{d[M^\bullet]}{dt} = 2k_t[M^\bullet]^2 \tag{4.4}$$

where k_t is the overall rate coefficient for termination, given by

$$k_t = k_{tc} + k_{td}$$

At the start of the polymerization, the rate of formation of radicals greatly exceeds the rate at which they are lost by termination. However, $[M^\bullet]$ increases rapidly and so the rate of loss of radicals by termination increases. A value of $[M^\bullet]$ is soon attained at which the latter rate exactly equals the rate of radical formation. The net rate of change in $[M^\bullet]$ is then zero and the reaction is said to be under *steady-state conditions*. In practice, most free-radical polymerizations operate under steady-state conditions for all but the first few seconds. If this were not so and $[M^\bullet]$ continuously increased, then the reaction would go out of control and could lead to an explosion. Assuming that chain transfer reactions have no effect on $[M^\bullet]$ (i.e. that the product radicals from any chain transfer reactions immediately reinitiate a new chain), steady-state conditions are defined by

$$\frac{d[R^\bullet]}{dt} = -\frac{d[M^\bullet]}{dt}$$

and from Equations 4.2 and 4.4

$$R_i = 2k_t[M^\bullet]^2$$

so that the steady-state total concentration of all radical species is given by

$$[M^\bullet] = \left(\frac{R_i}{2k_t}\right)^{1/2} \tag{4.5}$$

This equation can be substituted into Equation 4.3 and combined with Equation 4.1 to give a general expression for the *steady-state rate of polymerization*

$$R_p = -\frac{d[M]}{dt} = k_p[M]\left(\frac{R_i}{2k_t}\right)^{1/2}$$

i.e.

$$R_p = k_p \left(\frac{R_i}{2k_t} \right)^{1/2} [M] \tag{4.6}$$

It must be emphasized that Equation 4.6, and any other equation derived from it, applies only under steady-state conditions.

The initiation stage can now be considered in more detail so that equations for R_i can be derived for substitution into Equation 4.6. The most common method of initiation is thermolysis for which

$$I \xrightarrow{k_d} 2\,R^{\bullet}$$

and the rate of formation of active centres is given by

$$R_i = 2fk_d[I] \tag{4.7}$$

where

k_d is the rate coefficient for initiator dissociation

f, the *initiator efficiency*, is the fraction of primary free radicals $R^{\bullet}$ that successfully initiate polymerization.

The factor of 2 enters because two radicals are formed from one molecule of initiator. Normally, the initiator efficiency is in the range 0.3–0.8 as a consequence of loss of radicals due to side reactions (see Sections 4.2.1 and 4.2.4.1). Substitution of Equation 4.7 into Equation 4.6 gives for *initiation by thermolysis*

$$R_p = k_p \left(\frac{fk_d}{k_t} \right)^{1/2} [M][I]^{1/2} \tag{4.8}$$

When initiation is brought about by *photolysis* of an initiator (i.e. $I \xrightarrow{h\nu} 2R^{\bullet}$)

$$R_i = 2\phi \varepsilon I_o[I] \tag{4.9}$$

and so

$$R_p = k_p \left(\frac{\phi \varepsilon I_o}{k_t} \right)^{1/2} [M][I]^{1/2} \tag{4.10}$$

where

I_o is the intensity of the incident light

ε is the molar absorptivity of the initiator

ϕ is the quantum yield (i.e. the photochemical equivalent of initiator efficiency).

For initiation using the *redox couple* of cumyl hydroperoxide (CumOOH) with a ferrous ion (as shown in Section 4.2.1)

$$R_i = fk_r[\text{CumOOH}][Fe^{2+}] \tag{4.11}$$

where k_r is the rate coefficient for the redox reaction. Note that the front factor in the equation is unity because only one radical is produced. Substituting Equation 4.11 into Equation 4.6 gives

$$R_p = k_p\left(\frac{fk_r}{2k_t}\right)^{1/2}[M][CumOOH]^{1/2}[Fe^{2+}]^{1/2} \tag{4.12}$$

Using the same approach, specific equations for R_p can be derived for any initiation system.

4.3.2 Number-Average Degree of Polymerization

Kinetics analysis can provide only the *number-average degree of polymerization* $\bar{x}_n$ of the polymer produced, which is given by

$$\bar{x}_n = \frac{\text{Moles of monomer consumed in unit time}}{\text{Moles of polymer formed in unit time}} \tag{4.13}$$

On the basis of the simple kinetics scheme analysed in the previous section and *neglecting the effects of chain transfer* (which will be considered in Section 4.3.5),

$$(\bar{x}_n)_0 = \frac{k_p[M][M^\bullet]}{k_{tc}[M^\bullet]^2 + 2k_{td}[M^\bullet]^2}$$

where the subscript zero in $(\bar{x}_n)_0$ indicates that it is the value of $\bar{x}_n$ in the absence of chain transfer. The denominator of the equation takes into account the fact that combination produces one polymer molecule whereas disproportionation produces two. Under steady-state conditions, $[M^\bullet]$ can be substituted using Equation 4.5 to give

$$(\bar{x}_n)_0 = \frac{k_p[M]}{(1+q)k_t^{1/2}(R_i/2)^{1/2}} \tag{4.14}$$

where $q = k_{td}/k_t$ and is the fraction of termination reactions that proceed by disproportionation $(0 \le q \le 1)$. When termination occurs only by combination, $q = 0$, whereas when it occurs only by disproportionation, $q = 1$.

Substitution of Equation 4.7 into Equation 4.14 gives for *initiation by thermolysis*

$$(\bar{x}_n)_0 = \frac{k_p[M]}{(1+q)(fk_d k_t)^{1/2}[I]^{1/2}} \tag{4.15}$$

Similarly, from Equation 4.9, when initiation is by *photolysis*

$$(\bar{x}_n)_0 = \frac{k_p[M]}{(1+q)(\phi\varepsilon I_o k_t)^{1/2}[I]^{1/2}} \tag{4.16}$$

and, from Equation 4.11, when using the *cumyl hydroperoxide/ferrous ion redox couple* for initiation

$$(\bar{x}_n)_0 = \frac{2^{1/2}k_p[M]}{(1+q)(fk_r k_t)^{1/2}[CumOOH]^{1/2}[Fe^{2+}]^{1/2}} \tag{4.17}$$

Historically, the *kinetic chain length* ν has featured in considerations of $\bar{x}_n$ and is the average number of repeat units added to a single chain radical prior to termination. Hence, under steady-state conditions, ν is given by the ratio of the rates of propagation and termination

$$\nu = \frac{k_p[M^\bullet][M]}{2k_t[M^\bullet]^2} = \frac{k_p[M]}{2k_t^{1/2}(R_i/2)^{1/2}}$$

which, from comparison with Equation 4.14, shows that

$$(\bar{x}_n)_0 = \left(\frac{2}{1+q}\right)\nu$$

Thus, $(\bar{x}_n)_0 = \nu$ when termination is exclusively by disproportionation, and $(\bar{x}_n)_0 = 2\nu$ when termination is exclusively by combination.

4.3.3 Features of the Steady-State Equations for R_p and $(\bar{x}_n)_0$

The general prediction from the simple kinetics analysis is that

$$R_p \propto [M][I]^{1/2} \quad \text{and} \quad (\bar{x}_n)_0 \propto [M][I]^{-1/2}$$

Thus, by increasing [M], it is possible to increase both R_p and $(\bar{x}_n)_0$, though there is an upper limit imposed by the bulk concentration of the monomer, $[M]_{bulk}$ (i.e. with no dilution). Increasing [I] gives rise to an increase in R_p but a reduction in $(\bar{x}_n)_0$ (or *vice versa*). Thus, in many important polymerization processes that do not use a solvent (such as bulk and suspension polymerization, see Sections 4.4.1 and 4.4.3), there is a compromise between R_p and $(\bar{x}_n)_0$, which usually is controlled by the need to target a specific value of $(\bar{x}_n)_0$.

The expressions derived for R_p and $(\bar{x}_n)_0$ are said to be *instantaneous equations* because they apply at a *specific instant in time* when [M] and [I] have specific values. As a particular reaction proceeds, both [M] and [I] decrease at individual rates which depend upon the initiator, monomer, temperature and targeted R_p and $(\bar{x}_n)_0$. Clearly, R_p should reduce continuously as conversion increases (ultimately becoming zero), but because $(\bar{x}_n)_0$ depends on the ratio $[M]/[I]^{1/2}$ it can increase, decrease or remain reasonably constant depending on the particular polymerization and conditions. In order to predict the change in R_p and $(\bar{x}_n)_0$ with conversion, the kinetics equations must be integrated, for which purpose it often is assumed that [I] is constant so that the integrations are simplified.

The validity of the expressions for R_p and $(\bar{x}_n)_0$, i.e. of the simple kinetics scheme, will be examined in the following two sections. Experimentally, this is most easily done by measuring the initial values of R_p and $(\bar{x}_n)_0$, and correlating them with the initial concentrations $[M]_0$ and $[I]_0$. There is ample experimental evidence to support the relationship $(R_p)_{initial} \propto [M]_0[I]_0^{1/2}$ for most free-radical polymerizations, but the predictions of $\{(x_n)_0\}_{initial}$ from Equation 4.14 are generally high because the equation does not include the significant effects of chain transfer (which are described quantitatively in Section 4.3.5).

4.3.4 Diffusion Constraints on Rates of Propagation and Termination

Free radicals are highly reactive species that react at very high rates. Thus if one or more of the reacting species has a low rate of diffusion, this can be become the rate-determining process and the reaction is then said to be *diffusion controlled*. In free-radical polymerizations, as a chain radical propagates, it grows in size rapidly and its rate of *translational diffusion* (also termed *centre-of-mass diffusion*) becomes very low. Hence, it is important to understand diffusion constraints on the rates of propagation and termination in radical polymerizations.

Propagation involves reaction of a chain radical with a highly mobile, small molecule of monomer. Although translational diffusion of the chain radical is slow, under most conditions the rate of diffusion of monomer to the active chain end greatly exceeds the chemical rate of propagation, so propagation is not normally diffusion controlled (for exceptions, see the end of this section).

Termination, however, always is diffusion controlled because it is an extremely fast chemical reaction involving two highly reactive chain radicals that have low rates of diffusion. There are three regimes of diffusion that are of importance.

1. *Translational diffusion*: This is movement of the centre of mass of the chain through space. In the absence of entanglements between chains (i.e. when the concentration of already-formed polymer is low, e.g. at low conversions), diffusion corresponds to simple motion of the chain through space. At higher concentrations of already-formed polymer, chains entangle and translational diffusion is very slow and proceeds via *reptation* (see Section 16.4.2). In the timescale of the chemical termination reaction, most chains are effectively frozen in position; only those chains which are short enough to be free from entanglements are able to undergo translational diffusion reasonably freely. Under such conditions, encounters between chain radicals are dominated by those between short-chain (mobile) radicals and long-chain (immobile) radicals, so-called *short–long termination* (the distinction between *short* and *long* being in terms of whether the chain is below or above the critical chain length for entanglement).
2. *Segmental diffusion*: Even when two chain radicals encounter each other, the chemical termination reaction can only take place if the two active chain ends come into intimate contact. In most encounters, this is not the situation initially and segmental diffusion (via bond rotations and local reconfiguration of the chains) is required before the active chain ends come into close enough proximity to react. Due to their high reactivity, there is then a very high probability that the two radicals will undergo the chemical termination reaction.
3. *Reaction diffusion*: At high concentrations of already-formed polymer under conditions where even short-chain radicals are immobilized in the timescale of the chemical termination reaction, the active chain ends can only move by reaction with monomer (i.e. with each addition of monomer, the active end moves the distance associated with the two main-chain bonds just added) and the rate of termination becomes independent of chain length. This is known as *reaction diffusion*.

Therefore, the rate coefficient for termination is controlled by whichever of these diffusion processes is slowest. Under most conditions of polymerization, the rates of segmental and translational diffusion are lowest, so they are the most important processes. Both processes become slower as the concentration of already-formed polymer increases (i.e. as the viscosity of the reaction medium increases), but the effect is much greater on translational diffusion. At low conversions, termination is controlled by segmental diffusion, but as conversion increases (i.e. as the polymer concentration and viscosity increase), there is a change to control by translational diffusion when it is the rate of diffusion of short-chain radicals that becomes the controlling factor in determining the rate of termination. The absolute viscosity of the reaction medium is the major factor in controlling the rate of diffusion of short-chain radicals and increases as the total polymer concentration and the degree of polymerization of the polymer increase. For a given percentage conversion of monomer to polymer, both the polymer concentration and (from Equation 4.14) its degree of polymerization increase with $[M]_0$.

When $[M]_0$ is high, the change from segmental to translational diffusion control of termination by slowly moving short-chain radicals can lead to a large reduction in k_t (by several orders of magnitude), which results in a large increase in R_p and $(\bar{x}_n)_0$ because both depend on $k_t^{-1/2}$ (see Equations 4.6 and 4.14). This is manifested in the phenomenon known as *autoacceleration* (also known as the *Trommsdorff–Norrish effect* or the *gel effect*), which is demonstrated in Figure 4.5 by some

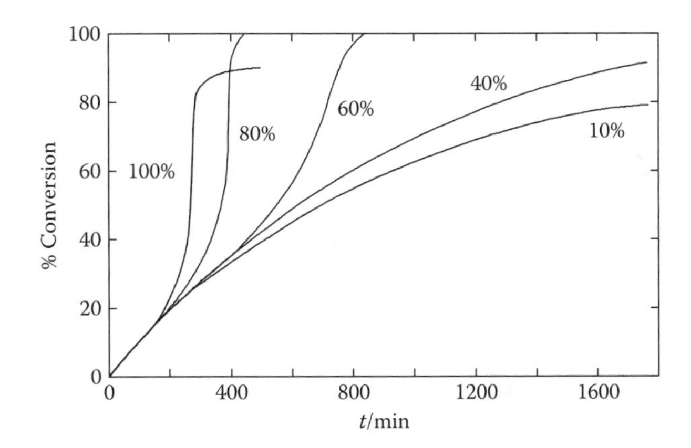

FIGURE 4.5 Effect of initial monomer concentration ($[MMA]_0$ given as a percentage) upon the conversion-time curves for polymerization of MMA at 50 °C using benzoyl peroxide as initiator and benzene as solvent. Note that when $[MMA]_0$ is high, there is a sharp increase in R_p as the conversion of monomer increases and that this was reported to correspond with an increase in the molar mass of the polymer formed. (Data taken from Schultz, G.V. and Harborth, G., *Makromol. Chem.*, 1, 106, 1947.)

experimental conversion-time curves for polymerization of methyl methacrylate (MMA) at different $[MMA]_0$. By plotting conversion against time, the $[MMA]_t$ data are normalized to $[MMA]_0$ (with R_p = slope of the conversion-time curve × $[MMA]_0$) and so all the curves should be the same if the rate coefficients are constant. As would be predicted from the preceding description of diffusion effects, autoacceleration is observed to begin at lower conversions and to be more pronounced as $[MMA]_0$ increases. Because free-radical polymerizations are exothermic, energy is evolved at an increasing rate after autoacceleration begins and if the dissipation of energy is poor there may well be an explosion. Autoacceleration can occur in any free-radical polymerization but is far more prominent for some monomers (e.g. acrylates and methacrylates) than for others (e.g. styrene and vinyl acetate) depending on the value of $k_p/k_t^{1/2}$ at low conversion and on the value of $(\bar{x}_n)_0$ and the extent to which it is reduced by chain transfer reactions. The simplest ways of avoiding autoacceleration are to stop the reaction before translational diffusion of short-chain radicals becomes difficult (i.e. before the onset of autoacceleration) or to use dilute solutions of monomer (which reduces $\bar{x}_n$ and the polymer concentration at a given conversion).

The reaction mixture will become extremely viscous and can even vitrify (i.e. become glass-like) at high conversions in high $[M]_0$ polymerizations of monomers that produce polymers whose (non-plasticized) glass transition temperatures are above the reaction temperature. This is because the unreacted monomer is not present in sufficient quantity to plasticize the polymer enough to reduce its glass transition temperature to a value well below the reaction temperature. If the polymerization mixture vitrifies, short-chain radicals become immobilized and even small molecules, such as monomer, are slow to diffuse. Both propagation and termination then become controlled by reaction diffusion. Such effects are evident in the conversion-time data shown in Figure 4.5. Poly(methyl methacrylate) has a glass transition temperature of 105 °C, whereas the reaction temperature is only 50 °C. For the bulk polymerization, the effect is so severe that the polymerization effectively stops at about 90% conversion (and could only be completed by raising the reaction temperature to well above 105 °C to facilitate diffusion of monomer all the way to 100% conversion).

4.3.5 Effects of Chain Transfer

The values of $\bar{x}_n$ found experimentally often are very much lower than $(\bar{x}_n)_0$ calculated from Equation 4.14 and its analogues, indicating that chain transfer reactions make a significant contribution in terminating the growth of chain radicals. As stated in Section 4.2.4, chain transfer with small molecules may be represented generically by

$$M_i^{\bullet} + T\!-\!A \longrightarrow M_i\!-\!T + A^{\bullet}$$

$$A^{\bullet} + M \longrightarrow AM_1^{\bullet}$$

where T and A are fragments of the molecule undergoing chain transfer and the product radical $A^{\bullet}$ is capable of reinitiating polymerization. If reinitiation is rapid, the rate of polymerization is not affected, but since growth of chain radicals is terminated prematurely, $\bar{x}_n$ is reduced.

All molecular species present in a free-radical polymerization are potential sources of chain transfer. Thus, when applying Equation 4.13, a series of terms are required in the denominator in order to take account of the dead molecules of polymer formed by chain transfer to *monomer*, *initiator* and *solvent*, as well as those formed by combination and disproportionation. Hence

$$\bar{x}_n = \frac{k_p[M][M^{\bullet}]}{k_{tc}[M^{\bullet}]^2 + 2k_{td}[M^{\bullet}]^2 + k_{trM}[M^{\bullet}][M] + k_{trI}[M^{\bullet}][I] + k_{trS}[M^{\bullet}][S]}$$

where k_{trM}, k_{trI} and k_{trS} are respectively the rate coefficients for chain transfer to monomer, initiator and solvent (S). Under steady-state conditions, $[M^{\bullet}]$ is given by Equation 4.5 and the above equation can be reduced to

$$\frac{1}{\bar{x}_n} = \frac{(1+q)k_t^{1/2}(R_i/2)^{1/2}}{k_p[M]} + \frac{k_{trM}}{k_p} + \frac{k_{trI}[I]}{k_p[M]} + \frac{k_{trS}[S]}{k_p[M]}$$

which is known as the *Mayo* or *Mayo–Walling Equation* and normally is written in the form

$$\frac{1}{\bar{x}_n} = \frac{1}{(\bar{x}_n)_0} + C_M + C_I\frac{[I]}{[M]} + C_S\frac{[S]}{[M]} \tag{4.18}$$

where

$(\bar{x}_n)_0$ is given by Equation 4.14 and its analogues

the *transfer constants* C_M, C_I and C_S are the ratios of k_{tr}/k_p for each type of chain transfer reaction.

Provided that all the concentrations, rate coefficients and transfer constants are known accurately, Equation 4.18 gives accurate predictions of $\bar{x}_n$.

Although chain transfer to polymer has a major effect on the skeletal structure of the polymer chains formed, it has no effect on $\bar{x}_n$ because the number of polymer molecules before and after the chain transfer reaction are the same (i.e. one dead polymer chain and one propagating chain). Hence, no term for chain transfer to polymer is needed in Equation 4.18.

4.3.5.1 Determination of Transfer Constants

Equation 4.18 can be applied to low conversion experimental data for $\bar{x}_n$ to determine transfer constants. Advantage can be taken of the fact that azo initiators do not undergo induced decomposition or other chain transfer reactions and have $C_I \approx 0$. Thus, by using an azo initiator in the absence of a solvent (i.e. bulk polymerization), only the $(\bar{x}_n)_0$ and C_M terms contribute to $\bar{x}_n$ and so C_M can be determined from an appropriate extrapolation of low conversion $\bar{x}_n$ data obtained at different $[I]_0$ (e.g. a plot of $(1/\bar{x}_n)_{initial}$ against $[I]_0^{1/2}/[M]_{bulk}$ has intercept C_M at $[I]_0^{1/2}/[M]_{bulk} = 0$). Once C_M is known, C_I can be determined for other initiators (e.g. peroxides) from further bulk polymerizations. Finally, C_S can be obtained by inclusion of the compound under investigation, either as a solvent or in small quantity as an additive if the compound is highly susceptible to chain transfer. By using an azo initiator and varying $[S]_0$ (and hence also $[M]_0$), a bivariate extrapolation is required with

TABLE 4.1
Typical Magnitudes of Transfer Constants Observed in Free-Radical Polymerization of Styrene at 60 °C

Compound	Bond Cleaved (T–A)	Transfer Constant, C_{TA} $(=k_{trTA}/k_p)$
Styrene	H—C(Ph)=CH_2	7×10^{-5}
Benzoyl peroxide	PhCOO—OOCPh	5×10^{-2}
Benzene	H—Ph	2×10^{-6}
Toluene	H—CH_2Ph	12×10^{-6}
Chloroform	H—CCl_3	5×10^{-5}
Carbon tetrachloride	Cl—CCl_3	1×10^{-2}
Carbon tetrabromide	Br—CBr_3	2
Dodecyl mercaptan	H—$SC_{12}H_{25}$	15

$[I]_0^{1/2}/[M]_0$ and $[S]_0/[M]_0$ as the two variables, from which it is possible to extract both C_M and C_S from the low conversion $\bar{x}_n$ data.

Table 4.1 lists for various compounds typical values of transfer constants for polymerization of styrene at 60 °C. The transfer constant for styrene shows that one chain transfer to monomer event occurs for approximately every 10^4 propagation events. Chain transfer to the initiator benzoyl peroxide is significant and is an example of induced decomposition (see Sections 4.2.1 and 4.2.4.1). In contrast, chain transfer to azobisisobutyronitrile is negligible, which is why it is a good initiator for studies of reaction kinetics. The wide variations in transfer constant for the other compounds listed highlight the need for careful choice of the solvent to be used in a free-radical polymerization. For example, carbon tetrachloride would be a poor choice as solvent except for preparation of low molar mass polymers. Compounds with high transfer constants (e.g. carbon tetrabromide and dodecyl mercaptan) cannot be used as solvents because they would prevent propagation; instead, they are employed at low concentrations to control (reduce) $\bar{x}_n$ and are known as *chain transfer agents* or *regulators* or *modifiers*. In general, transfer constants increase as the strength of the bond cleaved decreases and as the stability of the product radical increases.

4.3.6 Catalytic Chain Transfer

Catalytic chain transfer (CCT) is a special type of chain transfer which can be used in polymerization of methacrylates and was first reported in 1975 for free-radical polymerization of MMA using cobalt (II) porphyrin complexes that promote transfer of a hydrogen atom from the propagating chain end to monomer. The mechanism is believed to involve formation of a cobalt (III) hydride complex, which then transfers a hydrogen atom to a molecule of monomer, thus initiating a new chain, as shown in Figure 4.6.

The Co(II) acts as a true catalyst, being oxidized to Co(III) and then regenerated in each cycle. Transfer of the hydrogen atom is so efficient (with transfer constant values typically in the range 100 to 30,000) that even with very low levels of Co(II) (e.g. 10^{-3}–10^{-4} mol dm^{-3}), CCT is the dominant mechanism by which propagating chains terminate and low degrees of polymerization are achieved. Hence, all the polymer chains produced have a C=C bond at the chain end and a very high proportion have the structure

$$\mathrm{H}\!\left[\mathrm{CH_2}-\underset{\underset{\displaystyle \mathrm{OR}}{|}}{\underset{\mathrm{C=O}}{|}}\!\!\overset{\mathrm{CH_3}}{\overset{|}{\mathrm{C}}}\right]_x\!\!-\mathrm{CH_2}-\underset{\underset{\displaystyle \mathrm{OR}}{|}}{\underset{\mathrm{C=O}}{|}}\!\!\overset{\mathrm{CH_2}}{\overset{\|}{\mathrm{C}}}$$

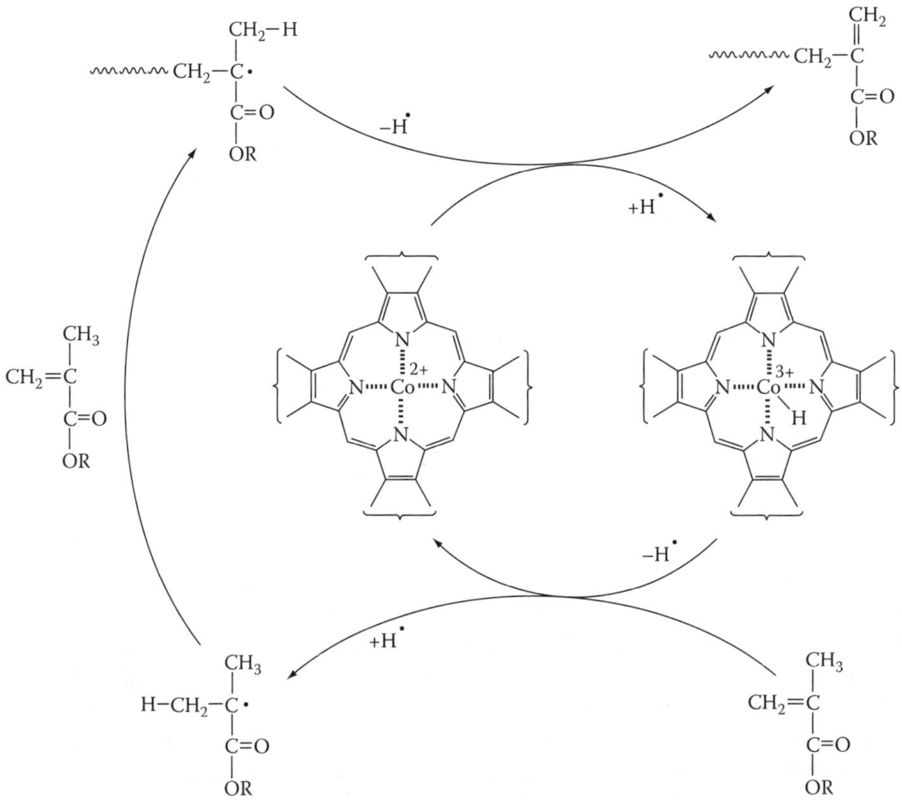

FIGURE 4.6 The mechanism of catalytic chain transfer with cobalt (II) porphyrins in free-radical polymerization of a methacrylate (R = H or alkyl). The Co(II) atom donates an electron to form a Co(III)–H bond through homolytic cleavage of a C–H bond in the terminal α-CH_3 group, the residual electron from the cleaved C–H bond and the chain's terminal unpaired electron then joining together to form a C=C bond at the end of the terminated chain. The Co(III)–H bond then breaks homolytically to regenerate Co(II) and yield $H^\bullet$, which initiates a new chain.

though some will have end groups resulting from primary radicals produced by the initiator. Under conditions where CCT is totally dominant and most chains are initiated by H-atom transfer from Co(III)–H, it has to be recognized that the propagating chain must possess two repeat units in order for CCT to occur (monomer is regenerated if CCT occurs when $x = 1$) and so the number-average degree of polymerization is given by

$$(\bar{x}_n)_{CCT} = 2 + \frac{k_p[M][M^\bullet]}{k_{CCT}[M^\bullet][Co(II)]}$$

where

k_{CCT} is the rate coefficient for CCT

[Co(II)] is the concentration of the cobalt (II) complex.

Thus

$$(\bar{x}_n)_{CCT} = 2 + \frac{[M]}{C_{CCT}[Co(II)]}$$

where C_{CCT} (= k_{CCT}/k_p) is the transfer constant for CCT.

The most important feature of CCT is that it produces low molar mass polymer chains with terminal C=C bonds which can be used as *macromonomers* for homopolymerization or copolymerization (see Section 9.5.2). Thus, CCT is used widely for preparing oligomethacrylate macromonomers, for which purpose cobaloximes nowadays are the preferred CCT agents, for example

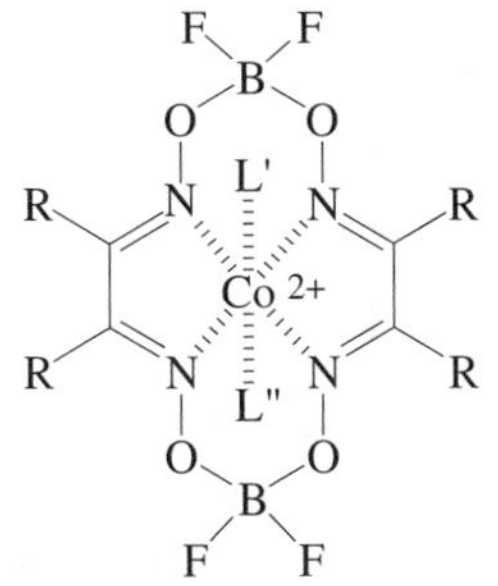

where R is a hydrocarbon group (e.g. CH_3 or Ph) and L′ and L″ are axial ligands that are oriented perpendicular to and in front/behind the plane of the complex (e.g. L′ and L″ = H_2O, CH_3OH, $N(CH_2CH_3)_3$, PPh_3).

CCT also can be used with other monomers that have α-methyl groups from which a hydrogen atom can be abstracted in the same way as for methacrylates (e.g. α-methylstyrene and methacrylonitrile), but is more complex for monomers without α-methyl groups because H-abstraction must then take place from the CH_2 and consumption of the cobalt catalyst through formation of stable Co(III)–C bonds by reaction of Co(II) with the propagating radical is common.

4.3.7 Inhibition and Retardation

Certain substances react with the free-radical active centres to produce species (either radical or non-radical) which are incapable of reinitiating polymerization. If the reaction is highly efficient, polymerization is prevented and the substance is said to be an *inhibitor*. When the reaction is less efficient or yields species that slowly reinitiate polymerization, the rate of polymerization is reduced and the substance is known as a *retarder*. The effect of adding an inhibitor or a retarder to a free-radical polymerization is shown schematically in Figure 4.7.

Nitrobenzene is a retarder for polymerization of styrene and acts via chain transfer reactions; the product radicals are of relatively low reactivity and only slowly add to molecules of styrene.

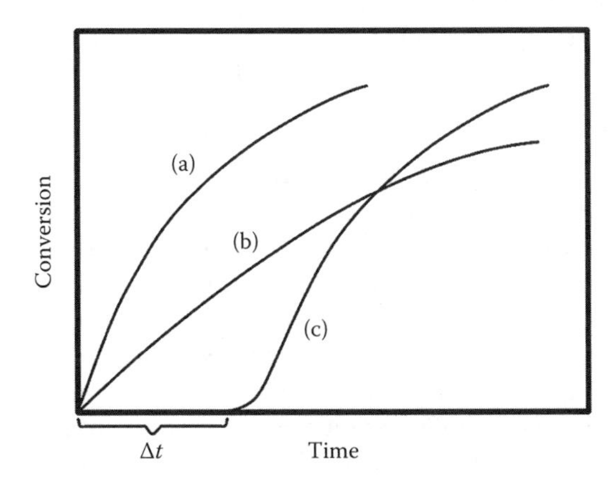

FIGURE 4.7 Schematic examples of inhibition and retardation: (a) normal polymerization; (b) polymerization in the presence of a retarder; and (c) polymerization in the presence of an inhibitor (where Δt is the induction period).

In this case, both the rate and degree of polymerization are reduced. Phenols and quinones inhibit the polymerization of most monomers by scavenging every active centre formed. An *induction period* is observed whilst the molecules of inhibitor are consumed (and hence deactivated); after this period, polymerization proceeds in the normal way.

Oxygen can act as a retarder or an inhibitor through the following chemistry

$$\sim CH_2-\dot{C}H(X) + O{=}O \longrightarrow \sim CH_2-CH(X)-O-\dot{O}$$

Peroxy radical

The product peroxy radical may reinitiate polymerization slowly, or not at all, and so oxygen can act as a retarder or an inhibitor. If reinitiation does occur, this gives rise to chains with weak peroxide (–O–O–) bonds in the backbone which are sites for bond scission leading to premature degradation of the polymer during use (e.g. during processing at high temperatures). Therefore, oxygen must be excluded from all radical polymerizations in order to obtain reproducible results and to eliminate the possibility of introducing weak backbone links. Hence, it is usual to perform radical polymerizations under an inert atmosphere, usually nitrogen.

Inhibitors are added to monomers at low levels to prevent premature polymerization during transportation and storage. Ideally, therefore, monomers should be purified prior to use. On an industrial scale, this is not practical and so monomer is supplied with an inhibitor that is less active in the absence of oxygen (e.g. a phenolic inhibitor), and the level of initiator is increased to overcome any effects of the inhibitor.

Autoinhibition (also termed degradative chain transfer to monomer) is observed in the polymerization of monomers that have easily-abstracted atoms, e.g. for allylic monomers

$$R^{\bullet} + CH_2{=}CH-CH_2X \longrightarrow R-H + CH_2{=}CH-\dot{C}HX$$

Allylic monomer — Allylic radical

where X is usually a functional group (e.g. allyl chloride, X=Cl; allyl alcohol, X=OH; and allyl acetate, $X=OCOCH_3$). The driving force for H-abstraction from allylic monomers is the strong mesomeric stabilization of the allylic product radical, which therefore only slowly reinitiates polymerization. Hence, polymerizations of allylic monomers proceed at very low rates and produce polymers with $\bar{x}_n < 20$ (due to the high value of C_M).

4.3.8 Molar Mass Distribution

The statistical analysis of linear step polymerization (Section 3.2.3.2) can be extended to the prediction of molar mass distributions for polymers prepared by free-radical polymerization. However, the situation is more complex because the growth of an active centre can be terminated by various different reactions and because the molar mass of the polymer formed changes with time due to the different rates of decrease in [M] and [I]. In order to simplify the analysis, [M] and [I] will be taken as constants, i.e. the equations derived will be applicable only at *low conversions*.

It is necessary to define a new parameter, β, as the probability that an active centre will propagate rather than terminate. Thus, β is given by the ratio of the rate of propagation to the sum of the rates of all reactions that an active centre can undergo (i.e. propagation, combination, disproportionation and chain transfer). It is analogous to the parameter p (the extent of reaction) in step polymerization. However, for free-radical polymerization, N is the total number of polymer molecules formed and N_0 is the total number of monomer molecules polymerized (cf. Section 3.2.3.1).

The simplest case to consider is when the growth of an active centre is terminated by *disproportionation* and/or *chain transfer* (though not to polymer). In each case, the length of the dead polymer chain is the same as that of the chain radical immediately prior to its termination. A chain of i repeat units is formed after initiation (i.e. from $RM_1^{\bullet}$) by a sequence of $(i-1)$ propagation reactions. The probability that one of these reactions has taken place is β and the probability that $(i-1)$ successive monomer additions have occurred is the product of the individual probabilities, i.e. $\beta^{(i-1)}$. The probability that the next reaction is termination rather than propagation is $(1-\beta)$ and so the probability $P(i)$ that a chain with degree of polymerization i is formed is given by

$$P(i) = (1-\beta)\beta^{(i-1)} \tag{4.19}$$

Furthermore, since one termination event has taken place for each polymer molecule formed

$$\beta = \frac{N_0 - N}{N_0} \text{ which rearranges to } N = N_0(1-\beta) \tag{4.20}$$

It can be seen that Equations 4.19 and 4.20 are exactly analogous to the linear step polymerization Equations 3.8 and 3.2, respectively (i replaces x and β replaces p). Thus, under these conditions of termination, polymers prepared by free-radical polymerization also have the most probable distribution of molar mass (Section 3.2.3.2) and

$$\bar{M}_w/\bar{M}_n = 1+\beta$$

For the formation of long molecules, $\beta \rightarrow 1$ and $\bar{M}_w/\bar{M}_n \rightarrow 2$ as for linear polymers prepared by step polymerization.

The analysis of termination by *combination* is slightly more complex since two growing chains are terminated by their coupling together. A dead polymer molecule of degree of polymerization i is formed when a growing chain of length j terminates by combination with a growing chain of length $(i-j)$ where j can be any integer from 1 to $(i-1)$. Thus, in this case, the probability $P(i)$ is given by

$$P(i) = \sum_{j=1}^{i-1} P(j)P(i-j)$$

where from Equation 4.19

$$P(j) = (1-\beta)\beta^{(j-1)}$$

$$P(i-j) = (1-\beta)\beta^{(i-j-1)}$$

Substituting these expressions into the equation for $P(i)$ gives

$$P(i) = \sum_{j=1}^{i-1} (1-\beta)^2\beta^{(i-2)}$$

which easily simplifies to the following relation defining $P(i)$ for termination by combination

$$P(i) = (i-1)(1-\beta)^2\beta^{(i-2)} \tag{4.21}$$

Equation 4.21 describes the mole fraction distribution for the polymer formed and may be used to derive the number-average molar mass from

$$\bar{M}_n = \sum P(i)M_i$$

Since $M_i = iM_0$, where M_0 is the molar mass of the repeat unit, then applying Equation 4.21 gives

$$\bar{M}_n = M_0(1-\beta)^2 \sum i(i-1)\beta^{(i-2)}$$

Using the mathematical relation

$$\sum_{i=1}^{\infty} i(i-1)\beta^{(i-2)} = 2(1-\beta)^{-3} \quad \text{for } \beta < 1$$

the equation for $\bar{M}_n$ reduces to

$$\bar{M}_n = \frac{2M_0}{(1-\beta)} \tag{4.22}$$

In order to evaluate the weight-average molar mass $\bar{M}_w$, it is necessary to obtain an expression for the weight fraction w_i of i-mers. Equation 3.11 is applicable but requires N_i, the total number of i-mers. This is given by

$$N_i = NP(i) \tag{4.23}$$

Since two propagating chains are terminated by each combination event

$$\beta = \frac{N_0 - 2N}{N_0} \text{ which rearranges to } N = \frac{N_0(1-\beta)}{2} \tag{4.24}$$

and so from Equations 3.11, 4.21, 4.23 and 4.24

$$w_i = \left(\frac{1}{2}\right) i(i-1)(1-\beta)^3 \beta^{(i-2)}$$

and from

$$\bar{M}_w = \sum w_i M_i \quad \text{with } M_i = iM_0$$

is obtained

$$\bar{M}_w = \left(\frac{1}{2}\right) M_0(1-\beta)^3 \sum i^2(i-1)\beta^{(i-2)}$$

Another mathematical relation

$$\sum_{i=1}^{\infty} i^2(i-1)\beta^{(i-2)} = 2(2+\beta)(1-\beta)^{-4} \quad \text{for } \beta < 1$$

leads to

$$\bar{M}_w = \frac{M_0(2+\beta)}{(1-\beta)} \tag{4.25}$$

Thus, the molar mass dispersity for termination by combination is given by

$$\bar{M}_w/\bar{M}_n = (2+\beta)/2 \tag{4.26}$$

and for the formation of long molecules, $\beta \rightarrow 1$ and so $\bar{M}_w/\bar{M}_n \rightarrow 1.5$. Therefore, the molar mass distribution of the polymer formed is narrower when chain growth is terminated by combination than when it occurs by disproportionation and/or chain transfer to molecules other than polymer molecules.

As stated in Section 4.3.5, chain transfer to polymer has no effect upon $\bar{M}_n$ because there is no change in the number of polymer molecules formed. Furthermore, the molar mass distribution is not affected when the reaction is intramolecular because it simply changes the skeletal structure of that molecule of polymer (giving short-chain branches). However, if chain transfer to polymer proceeds intermolecularly, then some chain radicals are terminated prematurely (i.e. are shorter than they would otherwise have been) and some dead chains become active again and grow further by propagation of a long-chain branch (i.e. they become longer than they would otherwise have been). Hence, although intermolecular chain transfer to polymer has no effect on $\bar{M}_n$, it does lead to broadening of the molar mass distribution and to an increase in the higher molar mass averages (e.g. $\bar{M}_w$ and $\bar{M}_z$).

4.3.9 Determination of Individual Rate Coefficients

In order to use the equations for rate and degree of polymerization predictively, it is necessary to know the values of the individual rate coefficients for the particular polymerization of interest. For initiation by thermolysis f, k_d, k_p and k_t are required to predict R_p. Additionally, values of transfer constants C_{tr} and the termination parameter q are required for prediction of $\bar{x}_n$. Strategies for determination of transfer constants were described in Section 4.3.5.1. The parameter q is rather more difficult to determine, but its value can be calculated from $(\bar{x}_n)_0$ with knowledge of f, k_d, k_p and k_t or (more accurately) from spectroscopic analysis of polymer end groups provided that $\bar{x}_n$ is sufficiently low for the end-group concentrations to be measured reliably (see Chapter 15).

Although C_{tr} data are sufficient for prediction of $\bar{x}_n$ using Equation 4.18, values for k_{tr} are important for more thorough kinetics analyses (e.g. of the chain-length dependence of k_t) and can be calculated from C_{tr} data once k_p is known.

In the following three sections, principles for the measurement of f, k_d, k_p and k_t are described. Most methods make use of bulk polymerizations in order to avoid the complicating effects of chain transfer to solvent. The determination of f and k_d is relatively straightforward, but accurate determination of k_p and k_t is more problematic.

4.3.9.1 Determination of f and k_d

Measurement of the rate of loss of initiator (usually by spectroscopic methods) is sufficient to obtain k_d

$$-\frac{d[I]}{dt} = k_d[I] \tag{4.27}$$

Although it is not possible to determine f directly, the quantity fk_d can be determined and together with the value of k_d this enables f to be calculated. The simplest method for determining fk_d is to measure the rate of consumption of a known quantity of a highly-efficient inhibitor that reacts stoichiometrically with chain radicals; the stable diphenylpicrylhydrazyl radical (DPPH) often is used for this purpose since it reacts rapidly with other radicals by 1:1 coupling and changes from an intense purple colour to colourless upon reaction, so its consumption can be followed by UV-visible spectroscopy (Section 15.3).

$$-\frac{d[\text{DPPH}]}{dt} = 2fk_d[\text{I}]$$

Alternatively fk_d can be determined by quantifying initiator-derived end groups in the polymer produced, although this is difficult because the end groups are at low concentration; spectroscopic methods can be used and the initiator can be labelled (e.g. enriched in ^{13}C or ^{14}C) to make it easier to analyse. In favourable cases, measurement of k_d and fk_d can be achieved simultaneously by performing a reaction inside a spectrometer provided that the initiator and the end groups derived from it give independently distinguishable signals.

Some typical values of k_d for common initiators are given in Table 4.2 together with the corresponding half lives, which are useful to consider when selecting an initiator. The value of f varies quite widely, but is typically in the range 0.3–0.8 and tends to reduce as [M] reduces.

4.3.9.2 Early Approaches to Determining k_p and k_t

In the formative years of polymer science, steady-state and non-steady-state methods were combined in order to determine k_p and k_t. Measurements of the initial steady-state rate of polymerization (e.g. by dilatometry) at known $[M]_0$ and $[I]_0$ enables the ratio $k_p/k_t^{1/2}$ to be evaluated from

TABLE 4.2
Some Typical Magnitudes of the Rate Coefficient k_d for Initiator Dissociation and the Corresponding Initiator Half-Lives

Initiator	Temperature / °C	k_d / s^{-1}	Half-Life / h[a]
2,2′-Azobis(2-cyanopropane)	50	2×10^{-6}	96.3
	60	1×10^{-5}	19.3
1,1′-Azobis(1-cyanocyclohexane)	80	7×10^{-6}	27.5
	95	5×10^{-5}	3.85
Benzoyl peroxide	60	2×10^{-6}	96.3
	80	2×10^{-5}	9.63
tert-Butyl peroxide	80	8×10^{-8}	2407
	100	9×10^{-7}	214
Potassium persulphate[b]	60	3×10^{-6}	64.2
	80	7×10^{-5}	2.75

[a] The half-life is the time for [I] to reduce to $[I]_0/2$, which is given by $\ln(2)/k_d$ from integration of Equation 4.27.

[b] k_d depends on pH.

Equation 4.8 provided that fk_d has been evaluated, as described above. (Although $k_p/k_t^{1/2}$ could also be determined from measurements of $(\bar{x}_n)_0$ using Equations 4.15 and 4.18, the values obtained are unreliable because the calculation requires several parameters to be known with accuracy, including the termination parameter q.) In order to separate k_p and k_t a second relationship between them is needed and was provided by the average lifetime τ of an active centre, i.e. the average time elapsing between creation of an active centre and its termination

$$\tau = \frac{\text{concentration of active centres}}{\text{rate of loss of active centres}} = \frac{[\text{M}^\bullet]}{2k_t[\text{M}^\bullet]^2} = \frac{1}{2k_t[\text{M}^\bullet]}$$

Since $R_p = k_p[\text{M}][\text{M}^\bullet]$, the equation may be written in the form

$$\tau = \frac{k_p[\text{M}]}{2k_t R_p}$$

Thus, simultaneous measurements of τ and R_p at known [M] enable the ratio k_p/k_t to be evaluated. These measurements are made under non-steady-state conditions using photoinitiation in a technique known as the *rotating sector method* which is elegant but experimentally challenging. Nevertheless, knowledge of $k_p/k_t^{1/2}$ and k_p/k_t allows the values of the individual rate coefficients to be calculated.

Although this approach has been used extensively, it is fraught with difficulties and requires great care. The advent of more robust methods has led to the realization that many of the values of k_p and k_t obtained using the approach just described are unreliable. Hence, it is very important to establish the method of determination when taking k_p and k_t values from the literature.

4.3.9.3 Modern Approaches to Determining k_p and k_t

Nowadays, k_p and k_t are determined more directly, often in separate experiments. *Pulsed-laser polymerization* (*PLP*) has emerged as the most reliable method for determination of k_p. PLP involves the use of photochemical initiators activated by exposure to very short, well-defined pulses of intense UV laser radiation (usually the third harmonic 355 nm wavelength from a Nd-YAG laser). The pulses are of consistent intensity, precise frequency ν (typically 0.1–10 Hz), and of a duration (typically 5–10 ns) that is very much shorter than the time between pulses. Each laser pulse generates a burst of new primary radicals. The primary radicals from the first pulse can be considered to give simultaneous, instant initiation of chain radicals (because the duration of the pulse is very short compared to the time between pulses). In the time between the first and second pulses, these chain radicals propagate. Many of these chain radicals undergo essentially instantaneous termination with the next crop of (small, highly mobile) radicals generated by the second pulse. This sequence of events is repeated with every pair of successive laser pulses. Thus, each chain radical may be considered to propagate for a period defined by the time interval between pulses which is precisely $1/\nu$. On this basis, the polymer should have a degree of polymerization given by

$$x = k_p[\text{M}](1/\nu) \tag{4.28}$$

since $k_p[\text{M}]$ is the rate at which a chain radical adds monomer. Since [M] and ν are known, k_p can be easily determined from Equation 4.28 provided that x can be measured accurately. This is not so simple because chains that are initiated by one pulse and then terminated at the next pulse comprise only a proportion of the total polymer formed. Some chain radicals may terminate or undergo chain transfer between pulses and some may survive several pulses before terminating. Theoretical treatments have concluded that the low molar mass point of inflection ($M_{\text{inflection}}$) in the molar mass distribution curve (usually obtained by gel permeation chromatography, see Section 14.3) corresponds to chains that are created and terminated by successive pulses (i.e. that have degree of polymerization $x = M_{\text{inflection}}/M_0$). On this basis, reliable k_p data have been determined for a range of important monomers at several different temperatures, thus enabling Arrhenius parameters (Section 4.3.10) to be determined and, therefore, prediction of k_p for any temperature.

Determination of k_t is rather more difficult, not least because its value is sensitive to (i.e. varies with) the concentration and molar mass distribution of already-formed polymer (see Section 4.3.4). Most methods use the approach of creating a population of chain radicals by photochemical initiation, then turning off the supply of radicals (by stopping irradiation) and monitoring the decay of [M•] with time either directly (by electron spin resonance spectroscopy) or indirectly (e.g. by measuring R_p and using Equation 4.3 with knowledge of [M] and k_p). The data are analysed to obtain k_t using the integrated form of Equation 4.4. An enhanced form of PLP, known as *time-resolved PLP*, has been developed in which the consumption of monomer between laser pulses is measured spectroscopically; this gives access to measurement of both k_p and k_t provided that the quantum yield for photoinitiation is known. Although each of these approaches is capable of yielding accurate values of k_t, the values obtained by different research groups are not necessarily consistent with each other for the reasons indicated above. Thus, great care is required when selecting a value of k_t for use in predicting R_p and $\bar{x}_n$ or when testing more advanced kinetics theories.

Values of Arrhenius parameters and k_p determined for some important monomers using PLP are given in Table 4.3. Values of k_t are not given here because they vary with conversion (see Section 4.3.4) and so no unique value can be assigned; however, the values of k_t at low conversions typically fall in the range 10^5–10^8 dm^3 mol^{-1} s^{-1}.

4.3.10 Effects of Temperature

The temperature dependences of the rate of polymerization and the degree of polymerization arise from the dependences upon temperature, T, of the individual rate coefficients. The latter are described in terms of the appropriate Arrhenius equations

$$k_d = A_d \exp(-E_d/\mathbf{R}T)$$

$$k_p = A_p \exp(-E_p/\mathbf{R}T)$$

$$k_t = A_t \exp(-E_t/\mathbf{R}T)$$

$$k_{tr} = A_{tr} \exp(-E_{tr}/\mathbf{R}T)$$

TABLE 4.3
Some Values of the Arrhenius Collision Factor A_p and Activation Energy E_p for Propagation Determined by PLP for Free-Radical Polymerization, and the Corresponding Rate Coefficients for Propagation k_p at 60 °C

Monomer	A_p / dm^3 mol^{-1} s^{-1}	E_p / kJ mol^{-1}	k_p (60 °C) / dm^3 mol^{-1} s^{-1}
Styrene[a]	4.27×10^7	32.51	341
Methyl methacrylate[b]	2.67×10^6	22.36	833
Ethyl methacrylate[b]	4.06×10^6	23.38	877
n-Butyl methacrylate[b]	3.78×10^6	22.88	978
n-Dodecyl methacrylate[b]	2.50×10^6	21.02	1,266
n-Butyl acrylate[c]	2.21×10^7	17.9	34,507
Vinyl acetate[d]	2.7×10^8	27.84	11,737

[a] Buback, M. et al., *Macromol. Chem. Phys.*, 196, 3267, 1995.
[b] Beuermann, S. et al., *Macromol. Chem. Phys.*, 201, 1355, 2000.
[c] Asua, J.M. et al., *Macromol. Chem. Phys.*, 205, 2151, 2004.
[d] Hutchinson, R.A. et al., *Macromolecules*, 27, 4530, 1994.

where

A_d, A_p, A_t and A_{tr} are the (nominally temperature-independent) *collision factors* and E_d, E_p, E_t and E_{tr} are the *activation energies* for the individual reactions

R is the gas constant.

TABLE 4.4
Typical Ranges for the Activation Energies of the Individual Processes in Free-Radical Polymerization

Process	Activation Energy / kJ mol^{-1}
Initiator dissociation	$110 < E_d < 160$
Propagation	$15 < E_p < 40$
Termination	$2 < E_t < 20$
Chain transfer	$30 < E_{tr} < 80$

Hence, it is the activation energies that determine the sensitivity of the value of a rate coefficient to changes in temperature. Typical values of activation energy for the above processes are given in Table 4.4.

From Equations 4.8 and 4.15 for initiation by thermolysis, $R_p \propto k_p k_d^{1/2} k_t^{1/2}$ and $(\bar{x}_n)_0 \propto k_p / k_d^{1/2} k_t^{1/2}$. Applying the Arrhenius equations, taking logarithms and differentiating with respect to temperature gives

$$\frac{d\{\ln R_p\}}{dT} = \frac{(2E_p + E_d) - E_t}{2\mathbf{R}T^2} \tag{4.29}$$

and

$$\frac{d\{\ln(\bar{x}_n)_0\}}{dT} = \frac{2E_p - (E_d + E_t)}{2\mathbf{R}T^2} \tag{4.30}$$

Since $E_d >> E_p > E_t$ then d{ln R_p}/dT is positive and so the rate of polymerization increases with temperature, but d{ln($\bar{x}_n$)$_0$}/dT is negative and the degree of polymerization decreases. This is because increasing the temperature increases [M•], i.e. more chain radicals exist at a given time, which increases the rate of polymerization (∝[M•]) but gives rise to an even bigger increase in the rate of termination (∝[M•]^2) thereby reducing $(\bar{x}_n)_0$.

The rate coefficient dependences of the kinetics Equations 4.10 and 4.16 for polymerization initiated by photolysis differ from those for thermolysis only in that they have no term in k_d, and so their analysis yields equations that are simply Equations 4.29 and 4.30 with $E_d=0$. Thus, both d{ln R_p}/dT and d{ln($\bar{x}_n$)$_0$}/dT are positive, showing that for photolysis both R_p and $(\bar{x}_n)_0$ increase with temperature (because the propagation rate coefficient increases more rapidly with temperature than does the termination rate coefficient).

The variation of $\bar{x}_n$ with temperature also depends upon the temperature dependence of the transfer constants C_{tr} in Equation 4.18. In general terms, $C_{tr}=k_{tr}/k_p$ and a similar analysis to that for R_p and $(\bar{x}_n)_0$ leads to

$$\frac{d\{\ln C_{tr}\}}{dT} = \frac{E_{tr} - E_p}{\mathbf{R}T^2} \tag{4.31}$$

Normally, $E_{tr} > E_p$ and so transfer constants increase with temperature. Hence, chain transfer becomes more significant as temperature increases and contributes increasingly to the reduction in $\bar{x}_n$.

4.3.10.1 Ceiling Temperature

The predicted continuous increase in R_p with temperature is based upon an assumption that the propagation reaction is irreversible. This assumption is not correct and the reverse reaction, *depropagation*, also can occur, i.e. the propagation step is more accurately represented by

$$M_i^{\bullet} + M \rightleftharpoons M_{i+1}^{\bullet}$$

Under equilibrium conditions

$$\Delta G_p^o = -\mathbf{R}T \ln\{K_p\} \tag{4.32}$$

where ΔG_p^o is the standard Gibbs energy for propagation and K_p is the equilibrium constant which is given by

$$K_p = \frac{[\mathrm{M}^\bullet]_{eq}}{[\mathrm{M}^\bullet]_{eq}[\mathrm{M}]_{eq}} = [\mathrm{M}]_{eq}^{-1} \tag{4.33}$$

where $[\mathrm{M}]_{eq}$ is the equilibrium monomer concentration. Substitution of $\Delta G_p^o = \Delta H_p^o - T\Delta S_p^o$ and Equation 4.33 into Equation 4.32 leads to

$$\ln\{[\mathrm{M}]_{eq}\} = \frac{\Delta H_p^o}{\mathbf{R}T} - \frac{\Delta S_p^o}{\mathbf{R}} \tag{4.34}$$

where ΔH_p^o and ΔS_p^o are, respectively, the standard enthalpy and entropy changes for propagation. Since propagation involves formation of a σ-bond from a less stable π-bond, ΔH_p^o is negative (typically –50 to –100 kJ mol^{-1}), i.e. *polymerization is exothermic*. However, because propagation increases the degree of order in the system, ΔS_p^o also is negative (typically –100 to –120 J K^{-1} mol^{-1}) and opposes propagation. Thus, depropagation is promoted by an increase in temperature and from Equation 4.34 $[\mathrm{M}]_{eq}$ increases with increasing temperature. At the *ceiling temperature* T_c, $[\mathrm{M}]_{eq}$ becomes equal to the concentration of monomer in the bulk monomer, $[\mathrm{M}]_{bulk}$, and propagation does not occur. Hence, a monomer will not polymerize above its T_c. Rearrangement of Equation 4.34 gives

$$T_c = \frac{\Delta H_p^o}{\Delta S_p^o + \mathbf{R}\ln\{[\mathrm{M}]_{bulk}\}} \tag{4.35}$$

Some values of T_c are given in Table 4.5. For most monomers, T_c is well above the temperatures (0–90 °C) normally employed for free-radical polymerization and therefore is not restrictive. However, certain monomers have low values of T_c and can only be polymerized at low temperatures. This generally is the case for 1,1-disubstituted monomers with bulky substituents and results from the reduction in magnitude of ΔH_p^o caused by steric interactions between the substituent groups in adjacent repeat units of the polymer chain (e.g. ΔH_p^o for styrene is approximately –70 kJ mol^{-1} but for α-methylstyrene is about –35 kJ mol^{-1}).

TABLE 4.5
Ceiling Temperatures for Different Monomers (CH_2=CXY)

Monomer	X	Y	Ceiling Temperature, T_c / °C
Styrene	H	Ph	310
α-Methyl styrene	CH_3	Ph	61
Methyl methacrylate	CH_3	$COOCH_3$	220

4.4 FREE-RADICAL POLYMERIZATION PROCESSES

The four most commonly used methods for performing free-radical polymerization are bulk, solution, suspension and emulsion polymerization. The kinetics of the first three of these processes can be described using the theory developed so far and so these processes will be considered first. Emulsion polymerization, however, has quite different kinetics which will be considered in Section 4.4.4.

Although the processes are described below specifically in the context of free-radical polymerization, the general features, advantages and disadvantages also apply to other types of polymerization carried out using these processes.

4.4.1 Bulk Polymerization

Bulk polymerization is the simplest process and involves only the monomer and a monomer-soluble initiator. Thus, [M] has its maximum value, which is defined by the density of the bulk monomer at the polymerization temperature (converted from g cm^{-3} to units of mol dm^{-3} through knowledge of M_0, the molar mass of the monomer). The kinetics theory for free-radical polymerization (Section 4.3) informs us that

$$R_p \propto [M][I]^{1/2} \quad \text{and} \quad (\bar{x}_n)_0 \propto [M][I]^{-1/2}$$

Hence, the high concentration of monomer gives rise to high rates of polymerization and high degrees of polymerization. However, the viscosity of the reaction medium increases rapidly with conversion (i.e. as polymer forms there is less monomer, i.e. less 'solvent'), making it difficult to remove the heat evolved upon polymerization, because of the inefficient stirring, and leading to autoacceleration. Bulk polymerization can, therefore, be difficult to reproduce. Note also that in order to increase the molar mass of the polymer produced, [I] and/or polymerization temperature have to be reduced, which in either case will result in a reduced rate of polymerization. Thus, there has to be compromise between polymer molar mass and speed of production.

The problems of viscosity build-up and autoacceleration normally are avoided by restricting the reaction to low conversions, though on an industrial scale the process economics necessitate recovery and recycling of unreacted monomer. A different complication arises when the polymer is insoluble in its monomer (e.g. acrylonitrile, vinyl chloride) since the polymer precipitates as it forms and the usual kinetics do not apply.

The principal advantage of bulk polymerization is that it minimizes contamination by impurities and produces high molar mass polymer of high purity. For example, it is used to prepare transparent sheets of poly(methyl methacrylate) in a two-stage process. The monomer is first partially polymerized to yield a viscous solution which then is poured into a sheet mould where polymerization is completed at high temperature. This method reduces the problems of heat transfer and shrinkage. The high ratio of surface area to volume of the sheet moulds provides for efficient heat transfer and control of autoacceleration, and the initial partial monomer conversion reduces problems associated with contraction in volume upon polymerization within the sheet mould.

4.4.2 Solution Polymerization

Many of the difficulties associated with bulk polymerization can be overcome if the monomer is polymerized in solution. The solvent must be selected so that it dissolves not only the initiator and monomer, but also the polymer that is to be produced. The solvent needs to be included at a level which reduces the viscosity of the reaction medium so that it can be stirred efficiently across the full conversion range, thus facilitating good heat transfer and eliminating (or at least enabling control of) autoacceleration. However, the inclusion of solvent is not without compromise and can lead to other complications.

The reduced [M] gives rise to decreases in the rate and degree of polymerization. Furthermore, if the solvent is not chosen with care, chain transfer to solvent may be appreciable and can result in a major reduction in the degree of polymerization (see Section 4.3.5 and Table 4.1, from which it is obvious that chloroform, carbon tetrabromide and dodecyl mercaptan most definitely are not be suitable as solvents). Finally, isolation of the polymer requires either

1. evaporation of the solvent, or
2. precipitation of the polymer by adding the solution to a sufficient (typically at least 5×) excess of a non-solvent (such that the final mixture of solvent/non-solvent overall is a non-solvent for the polymer, but still a solvent for any unreacted monomer and initiator), and then collection of the polymer by filtration or centrifugation followed by drying.

Clearly, neither process is very efficient and so tends to be restricted to laboratory work. Thus, commercial use of solution polymerization tends to be restricted to the preparation of polymers for applications that require the polymer to be used in solution (e.g. solvent-borne paints and adhesives; solution spinning of polyacrylonitrile fibres).

4.4.3 Suspension Polymerization

A better way of avoiding the problems of heat transfer associated with bulk polymerization on an industrial scale is to use *suspension polymerization*. This is essentially a bulk polymerization in which the reaction mixture is suspended as tiny droplets in an inert medium. The initiator, monomer and polymer must be insoluble in the suspension medium, which usually is water and so suspension polymerization tends to be used for monomers and polymers that have very low (or no) solubility in water.

A solution of initiator in monomer is prepared and then added to the preheated aqueous suspension medium. Polymerization temperatures are limited by the boiling point of water and so are usually in the range 70–90 °C. Droplets of the organic phase are formed and maintained in suspension by

1. vigorous agitation throughout the reaction, which typically is provided through use of baffled reactors with turbine stirrers, and
2. dispersion stabilizers dissolved in the aqueous phase; these typically are present at levels of 3–5 wt% to monomer and usually are low molar mass water-soluble polymers, such as poly(vinyl alcohol) or hydroxyalkylcelluloses, used in combination with surfactants (surfactants will be discussed in more detail when considering emulsion polymerization in Section 4.4.4).

The low viscosity and high thermal conductivity of the aqueous continuous phase and the high surface area of the dispersed droplets provide for good heat transfer. Each droplet acts as a tiny bulk polymerization reactor for which the normal kinetics apply. At high conversions, autoacceleration can occur but is much better controlled than in bulk polymerization due to the greatly improved heat dissipation.

The droplets are converted directly to polymer which, therefore, is produced in the form of beads (typically 0.05–2 mm diameter) which are isolated easily by filtration or centrifugation provided that they are rigid and not tacky. Thus, the polymerization must be taken to complete conversion (to prevent plasticization and reduction of the polymer glass transition temperature, T_g, by unreacted monomer) and normally is not used to prepare polymers that have T_gs below about 60 °C.

Suspension polymerization is used widely on an industrial scale (e.g. for polymerization of styrene, MMA and vinyl chloride), though care has to be taken to remove the dispersion stabilizers by thorough washing of the beads before drying. Nevertheless, the polymers produced by suspension polymerization usually are contaminated by low levels of the dispersion stabilizers.

4.4.4 Emulsion Polymerization

Another heterogeneous polymerization process of great industrial importance is *emulsion polymerization*. The reaction components differ from those used in suspension polymerization only in that the initiator must not be soluble in monomer but soluble only in the aqueous dispersion medium. Nevertheless, this difference has far-reaching consequences for the mechanism and kinetics of polymerization, and also for the form of the reaction product, which is a colloidally-stable dispersion of particulate polymer in water known as a *latex*. The polymer particles generally have diameters in the range 0.1–1 μm, i.e. about three orders of magnitude smaller than for suspension polymerization. In view of its commercial importance and the differences in mechanism and kinetics compared to the other processes for free-radical polymerization, emulsion polymerization will be considered in some detail.

Anionic surfactants most commonly are used as dispersion stabilizers, typically at levels of 1–5 wt% to monomer. They consist of molecules with hydrophobic hydrocarbon chains at one end of which is a hydrophilic anionic head group and its associated counter-ion (e.g. sodium lauryl sulphate, $CH_3(CH_2)_{11}SO_4^-Na^+$). Due to their hydrophobic tails, they have a low molecular solubility in water and above a certain characteristic concentration, the *critical micelle concentration* (CMC), the surfactant molecules associate into spherical aggregates known as *micelles* which contain of the order of 100 molecules and typically are about 5 nm in diameter. The surfactant molecules in the micelles have their hydrophilic head groups in contact with the water molecules and their hydrocarbon chains pointing inwards to form a hydrophobic core which has the ability to absorb considerable quantities of water-insoluble substances. For example, such surfactants are used in detergents, soaps and washing powders and liquids; in these applications, the surfactant micelles absorb greases and oils into their cores.

When a water-insoluble monomer is added to an aqueous solution containing a surfactant well above its CMC and the mixture is subjected to reasonably vigorous agitation, three phases are established: (i) the aqueous phase in which small quantities of surfactant and monomer are molecularly dissolved; (ii) large (about 1–10 μm diameter) droplets of monomer maintained in suspension by adsorbed surfactant molecules and agitation; and (iii) small (about 5–10 nm diameter) monomer-swollen micelles which are far greater in number (10^{18}–10^{21} dm^{-3}) than the monomer droplets (10^{12}–10^{15} dm^{-3}) but contain a relatively small amount of the total monomer. This is the situation at the beginning of an emulsion polymerization, but in addition, the aqueous phase also contains the initiator which usually is either a persulphate or a redox system (see Section 4.2.1). This situation is depicted in Figure 4.8.

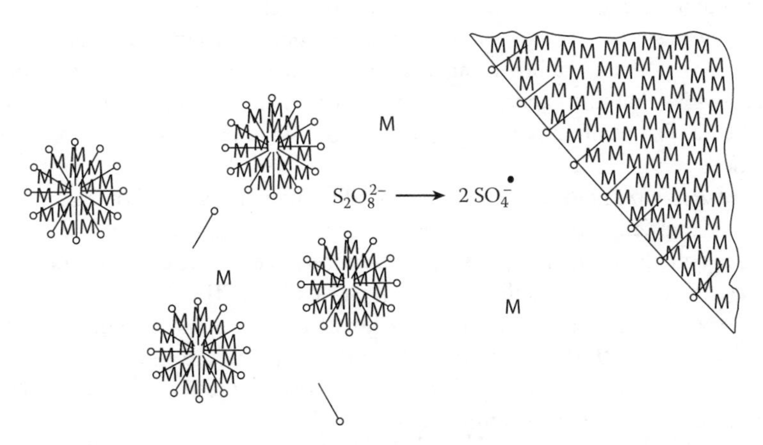

FIGURE 4.8 Schematic diagram showing the three phases that exist at the start of an emulsion polymerization with a surfactant (drumsticks) well above its CMC: (i) the aqueous phase contains small amounts of dissolved monomer (M), surfactant and initiator (persulphate is shown); (ii) monomer-swollen surfactant micelles; and (iii) large droplets of monomer (shown in part only) stabilized by adsorbed surfactant and agitation.

4.4.4.1 Particle Nucleation (Interval I)

Particle nucleation is the first stage of an emulsion polymerization and is known as *Interval 1*. There are several mechanisms by which particles can be formed. Whilst they can all contribute to particle nucleation, usually one mechanism predominates depending on the monomer used and the precise conditions of polymerization. In all cases, the primary free radicals formed from the initiator react with molecules of monomer dissolved in the aqueous phase to produce oligomeric radical species that continue to propagate in the aqueous phase. These oligomeric radicals have several possible fates. They may terminate in the aqueous phase to produce species that are surfactant-like (because the oligomeric chain is hydrophobic and the initiator end-group is hydrophilic), or they may continue to propagate until they reach a critical degree of polymerization z at which they become surface active, or can grow still further until they reach another critical degree of polymerization j at which they become insoluble in the aqueous phase and precipitate from it.

Micellar nucleation (first proposed by Harkins) occurs when the oligomeric radicals (z-mers or higher) adsorb and diffuse into monomer-swollen micelles, thereby initiating polymerization. Entry of oligomeric radicals into monomer droplets is very unlikely because the smaller micelles have a much higher total surface area and also are more efficient at capturing radicals. Each monomer-swollen micelle that captures a radical becomes a particle nucleus within which propagation is supported by absorption of monomer from the aqueous phase, there being concurrent diffusion of monomer from droplets into the aqueous phase to maintain equilibrium. These monomer-swollen polymer-particle nuclei soon exceed the size of the original micelles; the micellar surfactant becomes adsorbed surfactant and is supplemented by additional surfactant in order to maintain the colloidal stability of the particles (which is due to electrostatic and/or steric repulsion between adsorbed surfactant layers). This additional surfactant derives principally from disruption of monomer-swollen micelles that have not undergone initiation, the monomer released being redistributed throughout the system. As the reaction progresses, there comes a point at which micelles are consumed completely, thus signifying the end of particle nucleation. Micellar nucleation can only occur when the surfactant concentration is above CMC and is more likely to predominate for hydrophobic monomers with low solubilities in water (e.g. styrene and n-butyl acrylate).

For monomers with higher solubilities in water (e.g. MMA, methyl acrylate and vinyl acetate), *homogeneous nucleation* is important and can be dominant. In this mechanism, an oligomeric radical continues to propagate in solution in the aqueous phase (i.e. homogeneously) until it becomes a j-mer, at which point the chain radical collapses and becomes a *primary particle* whilst retaining the radical site at its chain end. The other end group derived from the initiator (e.g. SO_4^- from persulphate) provides some colloidal stability but not sufficient; hence the primary particles adsorb surfactant to achieve colloidal stability and absorb monomer so that the chain radical can continue to propagate. Provided that this happens, each primary particle becomes a latex particle. Nucleation ceases when the number of already-formed particles is high enough to ensure that they capture all oligomeric radicals.

Coagulative nucleation is most easily understood as an extension of homogenous nucleation in which the primary particles coagulate with each other until the aggregate particle attains a size at which it becomes colloidally stable and can absorb appreciable quantities of monomer. These aggregate particles are the latex particles that grow in Intervals II and III.

Each of these particle nucleation mechanisms can contribute in most emulsion polymerizations, as summarized in Figure 4.9. However, when the surfactant concentration is below the CMC, only the homogeneous and coagulative mechanisms are possible. If the surfactant concentration is very low, coagulative nucleation will be dominant.

4.4.4.2 Particle Growth (Intervals II and III)

Provided that the particles remain colloidally stable, the number N_p of particles per unit volume of latex (typically 10^{16}–10^{18} particles dm^{-3}) remains constant after the end of Interval I. Polymerization

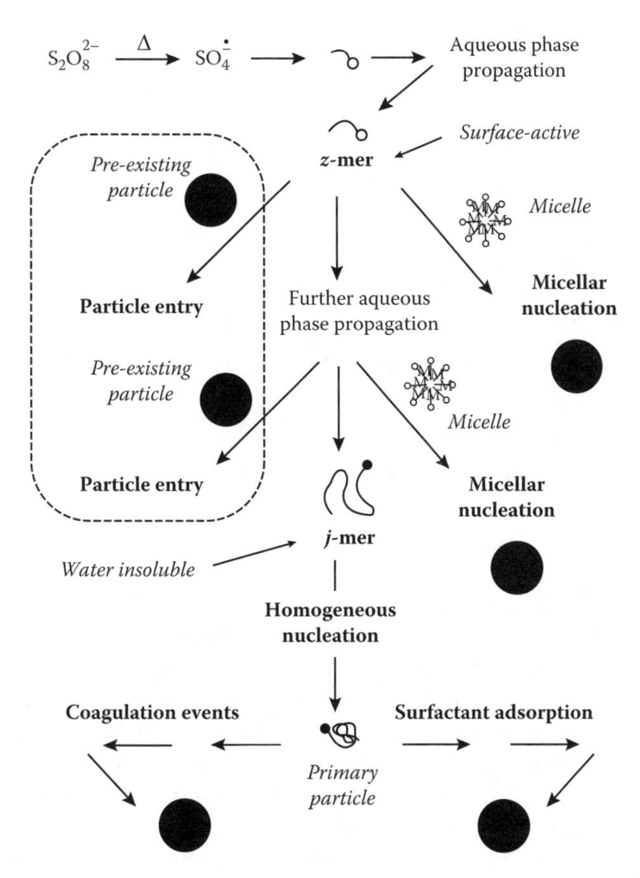

FIGURE 4.9 Schematic diagram of processes that occur during emulsion polymerization and how they lead to particle nucleation, which ceases when the radical entry events become dominant (e.g. due to the very high number of pre-existing particles and/or when all surfactant is adsorbed to pre-existing particles).

within these particles continues, supported by diffusion of monomer through the aqueous phase from the monomer droplets, as described above. Thus, the monomer droplets have only one function, which is to serve as reservoirs of monomer. The rate of monomer diffusion exceeds the rate of polymerization so that the concentration $[M]_p$ of monomer within a particle remains constant. Since N_p is constant, the rate of polymerization also is constant; this period of the polymerization is known as *Interval II*. Eventually, the monomer droplets are exhausted, marking the end of the period of constant rate. Thereafter, in *Interval III*, $[M]_p$ and the rate of polymerization decreases continuously as the remaining monomer present in the particles is polymerized. The three intervals of emulsion polymerization are identified in the schematic conversion-time curve shown in Figure 4.10.

4.4.4.3 Simple Kinetics of Emulsion Polymerization

If $\bar{n}$ is the average number of radicals per latex particle, then the corresponding number of moles of radicals is $\bar{n}/\mathbf{N}_A$, where $\mathbf{N}_A$ is the Avogadro constant. Thus, the rate of polymerization in an 'average' particle is $k_p[M]_p(\bar{n}/\mathbf{N}_A)$ and so the *rate of polymerization per unit volume of latex* is given by

$$R_p = k_p[M]_p(\bar{n}/\mathbf{N}_A)N_p \tag{4.36}$$

where

R_p and k_p have their usual meaning
$[M]_p$ and N_p are as defined in Section 4.4.4.2.

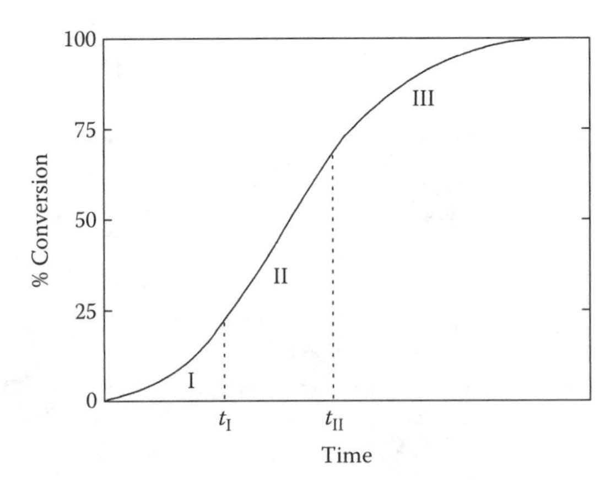

FIGURE 4.10 Schematic diagram showing the variation of conversion with time for a classic emulsion polymerization showing Intervals I, II and III, where t_I and t_{II} are the reaction times at the completion of Intervals I and II, respectively.

The value of $\bar{n}$ can vary widely dependent upon the reaction formulation and the conditions used. Furthermore, during Interval I, N_p is increasing, and during Interval III, $[M]_p$ is decreasing. Although Equation 4.36 is applicable to all stages in an emulsion polymerization, full theoretical treatment is complex because of the need to derive equations for N_p and $\bar{n}$ in terms of experimentally-accessible quantities. This is non-trivial because both N_p and $\bar{n}$ are determined by balances between several competing processes and the reaction mechanism usually is much more complex than indicated in this basic introduction to emulsion polymerization. For example, chain transfer agents often are used to reduce $\bar{x}_n$, small radical species (e.g. product radicals from chain transfer events) can desorb from the particles, and small amounts of water-soluble comonomers are often included.

Only the simplest situation will be considered here (known as *Smith–Ewart Case 2* conditions), for which it is assumed that radical desorption from particles does not occur and that the particles are so small that two radical species can exist independently within a particle only for very short periods of time before reacting together. Under these conditions, a radical cannot escape the particle once captured and termination can be considered to occur immediately upon entry of a second radical species into a particle that already contains one propagating chain radical. The particle then remains dormant until entry of another radical initiates the propagation of a new chain. Therefore, on average, each particle contains one propagating chain radical for half the time of its existence and none for the remaining half. Under these conditions, $\bar{n}$= 1/2 and

$$R_p = \frac{k_p[M]_p N_p}{2\mathbf{N}_A} \tag{4.37}$$

Furthermore, if the molar rate of formation of radical species from the initiator is ρ_i, then the average time interval between successive entries of radicals into a particle is $(N_p/\rho_i \mathbf{N}_A)$. Since each propagating chain radical adds molecules of monomer at a rate $k_p[M]_p$, the number-average degree of polymerization of the polymer formed is given by

$$\bar{x}_n = k_p[M]_p(N_p/\rho_i \mathbf{N}_A) \tag{4.38}$$

provided that chain transfer is negligible and that the propagating chain radical is of much higher degree of polymerization than the oligomeric chain radical that entered and brought about termination. Equations 4.37 and 4.38 are known as the *Smith–Ewart 'Case 2' Equations* and are best applied to Interval II when $[M]_p$ and N_p are constant. They show that both R_p and $\bar{x}_n$ can be increased by increasing N_p (e.g. by using a higher concentration of surfactant), i.e. high molar mass polymer can be formed at high rates of polymerization without compromise, unlike in bulk and suspension polymerization processes (see Section 4.3.3). This is because simultaneously propagating chain radicals are segregated into separate particles and cannot react with each other. Hence the isolation of individual propagating chains into separate particles, termed *compartmentalization*, is the key reason for the unique kinetics of emulsion polymerization.

4.4.4.4 Benefits and Applications of Emulsion Polymerization

Further important features of emulsion polymerization are the excellent heat transfer, the relatively low viscosity of the product latexes at high polymer concentrations, and the ability to control particle morphology (e.g. formation of core-shell particle structures by successive additions of different monomers). Additionally, due to the increasingly stringent environmental legislation on emission of volatile organic compounds (VOCs), water-borne polymers (i.e. latexes) prepared by emulsion polymerization are growing in importance as environmentally friendly alternatives to solvent-borne polymers, especially for coating applications.

Polymers prepared by emulsion polymerization are used either directly in latex form (e.g. emulsion paints, water-borne adhesives, paper coatings, binders for non-woven fabrics, foamed carpet-backings) or after isolation by coagulation or spray drying of the latex (e.g. synthetic rubbers and thermoplastics). In this respect, contamination by inorganic salts and dispersion stabilizers is often the most significant problem.

4.4.4.5 Miniemulsion Polymerization

Miniemulsion polymerization was developed from emulsion polymerization with the objective of controlling N_p by starting the polymerization with a *miniemulsion* that comprises monomer droplets which are sufficiently small (typically 50–300 nm diameter) and large enough in number (10^{16}–10^{18} dm^{-3}) to capture efficiently all radicals. The locus for particle nucleation is then the miniemulsion monomer droplets, each of which (ideally) becomes a polymer particle, and so N_p is defined by the number of miniemulsion droplets N_d present at the start of the polymerization. Once polymer has been formed within a miniemulsion droplet, it becomes a monomer-swollen particle and polymerization continues in much the same way as during Interval III of an emulsion polymerization.

In order to produce miniemulsions, a two-component stabilizer system is required: a surfactant plus a costabilizer compound that is soluble in the monomer, but highly insoluble in water (e.g. hexadecane). A macroemulsion of costabilizer/monomer mixture in water is prepared and then subjected to high shear (e.g. using an ultrasonicator or mechanical homogenizer) to produce the miniemulsion droplets which are colloidally stabilized by adsorbed surfactant molecules (the surfactant being used at levels typical for emulsion polymerization). Compared to the macroemulsion that exists at the outset of an emulsion polymerization, the much smaller droplets of a miniemulsion have much greater surface area and most of the surfactant is consumed in stabilizing the miniemulsion droplets so that there are relatively few (or ideally, no) micelles present. The role of the costabilizer is to create an osmotic pressure that opposes diffusion of monomer out of the droplets, thus preventing the miniemulsion from transforming back into a macroemulsion. The costabilizer is typically used at a level of 1–4 wt% to monomer and can be any compound that is monomer soluble, highly water-insoluble and of relatively low molar mass. Thus low molar mass preformed polymer can be used as the costabilizer and is particularly effective in achieving high droplet nucleation efficiencies ($N_p/N_d \approx 1$).

Although developed in order to gain more direct control of N_p, miniemulsion polymerization has other advantages that arise because there is no need for monomer transport through the aqueous phase and because the droplets can be considered as microreactors. Thus, in addition to being used simply as an alternative to emulsion polymerization for preparation of latexes, it is finding a range of unique applications. For example, it is possible to produce latexes from monomers that are completely insoluble in water and to produce hybrid particles by including water-insoluble additives (e.g. other polymers/resins or dispersions of particulate materials such as pigments) in the monomer prior to its emulsification so that the additives become incorporated into the final particles. Additionally, it is possible to use the general principle to produce latexes of other types of polymers by preparing a miniemulsion from the monomers and then carrying out *in situ* polymerization within the droplets (e.g. epoxy resin latexes have been produced in this way from step polymerization of miniemulsions of epoxy resin monomers).

4.4.4.6 Microgels

Microgels are crosslinked submicron (or multi-micron) size particles prepared by emulsion copolymerization of mono- and multi-functional monomers that produce network polymers, which are insoluble in water under the conditions of polymerization, but which under other conditions are miscible with water. Most commonly, this involves carrying out the emulsion polymerization at temperatures well above the lower critical solution temperature (LCST; see Section 14.2.1) for the network polymer being formed. On cooling of the product latex to below the LCST, the microgels become miscible with, and so absorb, water from the continuous phase of the latex, thereby becoming highly swollen; as for macroscopic networks, microgels swell but do not dissolve due to the crosslinks. Microgel latexes are finding a wide variety of uses, but are particularly relevant to biomedical applications where advantage can be taken of their swelling–deswelling behaviour (e.g. in diagnostics and controlled release). The most important microgel material is crosslinked poly(*N*-isopropylacrylamide) for which the LCST (~32 °C) is close to body temperature.

4.4.5 Strategies for Performing Polymerization Processes

For each of the different types of polymerization process, the procedure by which the polymerization is carried out can have a profound effect upon the kinetics of polymerization and the properties of the resulting polymer. Indeed, polymers with quite different performance characteristics can be produced from the same reaction formulation by appropriate control of the type of process and conditions used. As for many other chemical processes, there are three generic strategies for carrying out a polymerization process.

1. *Batch*: All reactants are added completely to the reaction vessel at the start of the polymerization.
2. *Semi-continuous batch (or semi-batch)*: Only part of the total reaction formulation is introduced at the beginning of the reaction, the remainder being added according to a predetermined schedule during the course of the polymerization.
3. *Continuous*: Reactants are added continuously to the reactor from which product is removed continuously such that there is a balance between the input and output streams.

Batch processes are of limited versatility for manufacture of polymers by free-radical polymerization and mainly find use in academic studies and simple evaluations of reaction formulations. By comparison, semi-batch processes are very versatile and are widely used, both industrially and in academic laboratories. Continuous processes tend to be used when very large volumes of a specific polymer needs to be manufactured, but (unlike semi-batch reactors) are not easily switched from production of one polymer to another.

Description of the effects of process strategy upon polymerization kinetics and polymer properties are beyond the scope of this book, and the reader is recommended to consult textbooks and reviews on polymerization process engineering for more detail.

4.5 REVERSIBLE-DEACTIVATION ('LIVING') RADICAL POLYMERIZATIONS

Chain polymerizations that have no mechanism through which growth of a propagating chain can be terminated (e.g. by termination or chain transfer reactions) are termed *living polymerizations* because the active sites are not lost and will continue to propagate as long as monomer is present (e.g. they will resume propagation if more monomer is added after the initial quantity has been consumed). If all chains can be initiated simultaneously, living polymerization has the major advantage that it will produce polymers with narrow molar mass distributions ($M_w/M_n < 1.1$, see Section 5.3.2.4). Furthermore, living polymerization also gives access to control of copolymer architecture (e.g. block copolymers can be synthesized by polymerizing two different monomers in sequence, see Section 9.4.2). Thus, living polymerization is important in the synthesis of well-defined homopolymers and copolymers.

Living anionic polymerizations were the first living polymerizations to be developed, but are restricted to relatively few types of olefinic monomers (see Section 5.3.2). In contrast, a very wide range of olefinic monomers can be used in radical polymerizations, which is partly why free-radical polymerization is so versatile and commercially important. However, synthesis of well-defined homopolymers and copolymers by free-radical polymerization is constrained by the statistics described in Section 4.3.8 and by chain transfer and termination reactions. Thus, achieving living radical polymerization is highly desirable, but this requires a means of stopping chain radicals from being killed. This problem exercised the minds of polymer chemists for many decades, but during the late 1980s and through the 1990s several strategies were developed for massively reducing termination, though not entirely eliminating it. Their important feature is that they are sufficiently living to facilitate synthesis of well-defined homopolymers and copolymers from the wide range of monomers that are susceptible to radical polymerization. These new types of radical polymerization initially were termed *living radical polymerizations* but because termination and chain transfer reactions are still possible, they began to be referred to as *controlled radical polymerizations.* Both terms are in wide use, but neither captures correctly how these polymerizations operate. In 2009, IUPAC (the international organization which defines nomenclature and terminology in pure and applied fields of chemistry) recommended use of the generic term *reversible-deactivation radical polymerization,* a good name because it captures the essence of all types of (what, up to then, were called) living/controlled radical polymerizations.

The basic principle underlying all so-called living/controlled radical polymerizations is to suppress termination to the extent that it becomes insignificant by reversibly trapping and temporarily deactivating the chain radicals. Although the activation–deactivation cycle is rapid, the chain radicals that are free can still propagate (but also can undergo all other possible reactions, though termination has a much-reduced probability). From this simple picture, it is easy to see why reversible-deactivation radical polymerization is a far better name for these types of radical polymerization.

The various types of reversible-deactivation radical polymerization employ one or other of two general strategies which are illustrated in Figure 4.11 and will be described here in generic terms. The first is shown in Figure 4.11(a) and involves rapid, reversible end-capping of the chain radical by a chain-capping species, where the equilibrium lies far to the left. In this way, $[M^\bullet]$ is reduced massively, causing an enormous reduction in the rate of termination (because this depends upon $[M^\bullet]^2$). Since Equation 4.3 still is applicable and $[M^\bullet]$ is very low, a feature of these types of reversible-deactivation radical polymerizations is that they normally proceed at much lower rates than conventional free-radical polymerizations. A further interesting feature of reversible-deactivation radical polymerizations that proceed via Strategy 1 of Figure 4.11 is that, in principle, they should not suffer from autoacceleration (see Section 4.3.4) because both the rate of polymerization and the number-average degree of polymerization are independent of k_t. The second strategy is shown in Figure 4.11(b) and makes use of highly-efficient chain transfer reactions in which a free chain radical displaces a trapped chain radical from an end-capped species and in the process becomes end-capped. In this case, it is the very high efficiency of the exchange process and the much higher number of trapping agent molecules present compared to the total number of primary radicals produced from the initiator that makes termination negligible. Hence, $[M^\bullet]$ can have values similar to those for conventional free-radical polymerizations. In this case, the normal rate of polymerization equation applies, so rate autoacceleration can occur but usually is mitigated because the polymerizations often are designed to produce polymers with lower molar masses.

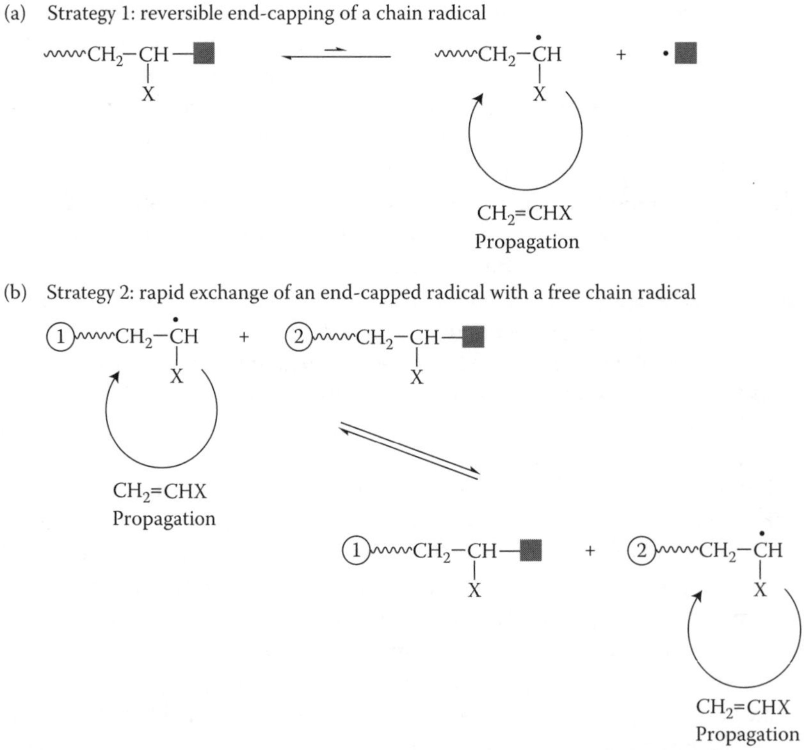

(a) Strategy 1: reversible end-capping of a chain radical

$\sim\sim CH_2-CH(X)-■ \rightleftharpoons \sim\sim CH_2-\dot{C}H(X) + \cdot■$

$CH_2{=}CHX$
Propagation

(b) Strategy 2: rapid exchange of an end-capped radical with a free chain radical

$(1)\sim\sim CH_2-\dot{C}H(X) + (2)\sim\sim CH_2-CH(X)-■ \rightleftharpoons (1)\sim\sim CH_2-CH(X)-■ + (2)\sim\sim CH_2-\dot{C}H(X)$

$CH_2{=}CHX$
Propagation

$CH_2{=}CHX$
Propagation

FIGURE 4.11 The basic principles of the two strategies for reversible-deactivation radical polymerization in which the probability of bimolecular termination between chain radicals is reduced massively due to the highly efficient and reversible radical capture mechanisms, thereby giving the characteristics of living polymerization. In each case, the solid square represents a chain-capping species. In Strategy 2 the two chains are labelled to make it clear that radical exchange takes place.

Both types of reversible-deactivation radical polymerization are *quasi living* because, although at any instant in time most of the chain radicals are in the end-capped dormant state, when released they can undergo propagation, chain transfer and termination as in a conventional free-radical polymerization, though as indicated above, termination has a much lower probability. Whilst reversible end-capping enables the chain radicals to be protected from bimolecular termination (i.e. kept 'alive'), in order to produce polymers with narrow molar mass distributions, it is essential that the rates at which chain radicals are released and recaptured by the trapping agent are both high. Thus, the most important criterion for living-like control in reversible-deactivation radical polymerizations is rapid, reversible exchange between the active and dormant states. Each chain then grows with approximately equal probability in very short bursts of activity and the Poisson distribution of molar mass can be achieved, as in a true living polymerization (see Section 5.3.2.4). Under such conditions, when a chain radical is released from its end-capped state, it typically is free for about 10–100 μs (the *transient radical lifetime*, t_{rad}) before it becomes trapped again. Recognizing that k_p values typically fall in the range from about 400 to 40,000 $dm^3\ mol^{-1}\ s^{-1}$ (see Table 4.3) and that [M] typically is in the range 4–9 mol dm^{-3}, applying Equation 4.28 (with $1/\nu$ replaced by t_{rad}) shows that this timescale corresponds, on average, to the addition of around 0.02–35 repeat units per active period (the value of 0.02 indicating that, on average, 50 activation–deactivation cycles are required before a single repeat unit is added). The radical lifetimes from chain initiation to termination in conventional free-radical polymerizations (see Section 4.3.9.2) also vary considerably, but typically are in the range of around 0.1–10 s, i.e. they are several orders of magnitude greater than t_{rad}.

Because termination is suppressed almost completely in reversible-deactivation radical polymerizations and exchange between the active and dormant states is rapid, they show the characteristic features of true living polymerizations, namely: (i) degree of polymerization increases linearly with conversion and the molar mass distribution is narrow (see Sections 5.3.2.3 and 5.3.2.4) and (ii) chains can be grown first with one monomer and then another to produce block copolymers (see Section 9.4.2). When these features are considered together with the wide range of functional groups that are tolerant of radicals, it is not surprising that there has been an explosion of academic and industrial interest in synthesis of well-defined polymers from a wide range of monomers using methods of reversible-deactivation radical polymerization. This remains a rapidly developing area of polymer chemistry.

In the following sections, the three most important types of reversible-deactivation radical polymerization are described to provide sufficient background for a basic understanding. However, the reader will need to consult the literature, in particular recent reviews and conference proceedings (beyond those given here in Further Reading) for a deeper insight and in order to keep up-to-date.

4.5.1 Nitroxide-Mediated Radical Polymerization

Although the origins of the principles underlying many of the important reversible-deactivation radical polymerizations can be traced to earlier research, it was the development of nitroxide-mediated polymerization (NMP), as a consequence of work by Solomon, Rizzardo and Cacioli at the Commonwealth Scientific and Industrial Research Organisation, Australia (CSIRO) during the mid-1980s, which triggered the rapid evolution of reversible-deactivation radical polymerization during the 1990s.

Nitroxides ($R_1R_2N–O^•$) are *persistent radicals* which do not self-react, but will couple rapidly with carbon-centred radicals and have long been used as radical traps in electron spin resonance spectroscopy (see Section 15.2.1). What the researchers at CSIRO discovered was that by using nitroxides as traps in radical polymerizations they could produce low molar mass polymers in a controlled way. Georges at Xerox extended their work and showed that it is possible to trap propagating radicals reversibly and achieve living-like conditions in radical polymerizations of styrene using 2,2,6,6-tetramethylpiperidinyl-*N*-oxyl (TEMPO) as the nitroxide at 130 °C. During this early period of development, NMP more commonly was called *stable-free-radical-mediated polymerization* (*SFRP*), a more generic term that is used less often nowadays.

The principles of NMP are shown in Figure 4.12 and involve reversible trapping and release of the propagating radical by the nitroxide. In the trapped form, the end-group structure is that of an *alkoxyamine* in which the C–O bond is weak and dissociates to regenerate the chain radical and nitroxide. The chain grows in very short bursts of propagation when it is released from the alkoxyamine. Once dissociated from the alkoxyamine end group, the free nitroxide (being small) can diffuse rapidly and

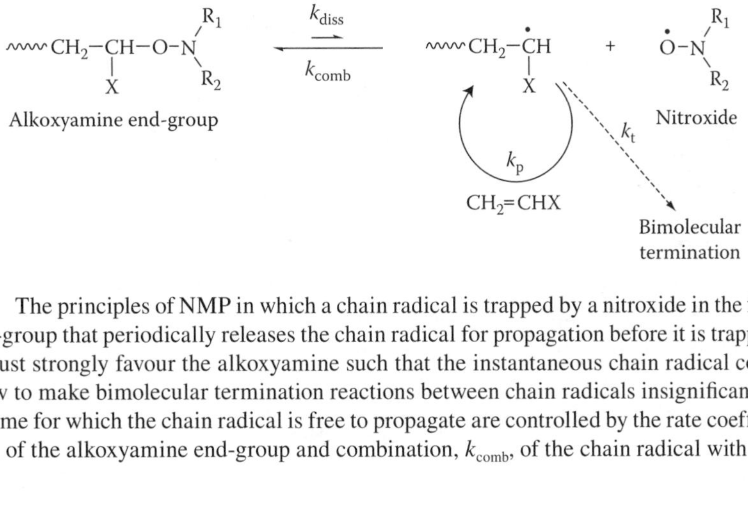

$$\sim\!\sim CH_2{-}CH(X){-}O{-}NR_1R_2 \underset{k_{comb}}{\overset{k_{diss}}{\rightleftharpoons}} \sim\!\sim CH_2{-}\dot{C}H(X) + {}^{\bullet}O{-}NR_1R_2$$

Alkoxyamine end-group; k_p, $CH_2{=}CHX$; k_t, Bimolecular termination; Nitroxide

FIGURE 4.12 The principles of NMP in which a chain radical is trapped by a nitroxide in the form of an alkoxyamine end-group that periodically releases the chain radical for propagation before it is trapped again. The equilibrium must strongly favour the alkoxyamine such that the instantaneous chain radical concentration is sufficiently low to make bimolecular termination reactions between chain radicals insignificant. The equilibrium and the time for which the chain radical is free to propagate are controlled by the rate coefficients for dissociation, k_{diss}, of the alkoxyamine end-group and combination, k_{comb}, of the chain radical with the nitroxide.

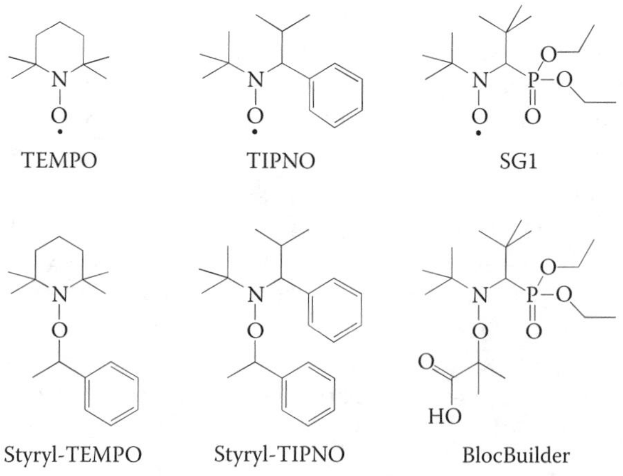

TEMPO TIPNO SG1

Styryl-TEMPO Styryl-TIPNO BlocBuilder

FIGURE 4.13 Chemical structures of some nitroxides and alkoxyamines that are used widely in NMP (for clarity, the carbon and hydrogen atoms are not shown, just the carbon skeleton). The full names for TIPNO and SG1 are, respectively, 2,2,5-trimethyl-4-phenyl-3-azahexane-3-nitroxide and *N*-(2-methylpropyl)-*N*-(1,1-diethylphosphono-2,2-dimethylpropyl)-*N*-oxyl. BlocBuilder is a trade name for what is effectively the alkoxyamine of methacrylic acid and SG1.

so there is a dynamic equilibrium in which a chain radical is likely to be recaptured by a different nitroxide molecule than the one from which it was released when the alkoxyamine end-group dissociated. Because [M•] is very low, bimolecular termination events are massively reduced and rates of polymerization are low. Thus NMP operates according to Strategy 1 of Figure 4.11 and requires control of the dissociation-combination equilibrium such that it lies far to the left and exchange is rapid. For each monomer-nitroxide pairing these conditions are realized over a fairly narrow temperature range.

Although TEMPO is used widely, it is most effective for polymerization of styrene at 125–130 °C and not so appropriate for other types of monomer. From the late 1990s onwards, a number of acyclic nitroxides were developed which are capable of achieving living-like conditions in radical polymerizations of other monomers. The chemical structures of TEMPO and two acyclic nitroxides used widely in NMP are shown in Figure 4.13. An important feature in the structure of TIPNO and SG1 is the hydrogen atom attached to one of the carbon atoms bonded to the nitrogen atom (on its right-hand side in Figure 4.13), because this changes the reactivity and makes NMP possible with styrene and other monomers (principally acrylates, acrylamides and dienes) at lower temperatures (typically 110–120 °C).

In the earlier work on NMP, the nitroxide was included in an otherwise conventional free-radical polymerization formulation in which an initiator was used to generate the chain radicals. An alternative, much better, procedure is to prepare an alkoxyamine ($R_1R_2NO–R_3$) from the nitroxide and (usually) a monomer, and to use the alkoxyamine as the initiator. Figure 4.13 also gives the chemical structures of some important monomeric alkoxyamine initiators of the nitroxides shown. The use of a monomeric alkoxyamine for initiation is much more precise in terms of the control of radical concentration and polymer molar mass. It also makes it easy to use precise excesses of the free nitroxide to exert further control of the dissociation-combination equilibrium, so nowadays use of alkoxyamines as initiators is the standard way of conducting NMP.

4.5.1.1 Kinetics of Nitroxide-Mediated Radical Polymerization

Control of the equilibrium shown in Figure 4.12 is crucial to success in achieving living-like conditions in NMP. This often is achieved by adding some free nitroxide (e.g. 5–20 mol% to the alkoxyamine). The equilibrium constant K_{NMP} is given by

$$K_{NMP} = \frac{[M^{\bullet}][R_1R_2NO^{\bullet}]}{[P-ONR_1R_2]} \tag{4.39}$$

where

$[M^•]$ and $[R_1R_2NO^•]$ are, respectively, the total concentration of free chain radicals (all chain lengths) and the concentration of free nitroxide (which will be greater than $[M^•]$ if excess free nitroxide has been added)

$[P–ONR_1R_2]$ is the total concentration of polymer alkoxyamines of all chain lengths.

Assuming use of an alkoxyamine initiator ($R_1R_2NO–R_3$) and that $[R_1R_2NO^•]$ and $[P–ONR_1R_2]$ do not change during the polymerization, then these concentrations can be equated to the values at the start of polymerization, $[R_1R_2NO^•]_0$ and $[P–ONR_1R_2]_0 \approx [R_3–ONR_1R_2]_0$, in which case

$$[M^•] = \frac{K_{NMP}[R_3 - ONR_1R_2]_0}{[R_1R_2NO^•]_0} \tag{4.40}$$

This may now be substituted into Equation 4.3 to give

$$R_p = -\frac{d[M]}{dt} = K_{NMP}k_p[M]\left(\frac{[R_3 - ONR_1R_2]_0}{[R_1R_2NO^•]_0}\right) \tag{4.41}$$

Thus, R_p should be first order in [M]. Note also that R_p is independent of k_t and so NMP should not suffer from rate autoacceleration. Equation 4.41 can be integrated as follows

$$\int_{[M]_0}^{[M]_t} \frac{d[M]}{[M]} = -K_{NMP}k_p\left(\frac{[R_3 - ONR_1R_2]_0}{[R_1R_2NO^•]_0}\right)\int_0^t dt$$

which is solved easily to give

$$\ln\left(\frac{[M]_0}{[M]_t}\right) = K_{NMP}k_p\left(\frac{[R_3 - ONR_1R_2]_0}{[R_1R_2NO^•]_0}\right)t \tag{4.42}$$

Thus, a plot of $\ln([M]_0/[M]_t)$ versus t should be linear and have a slope from which K_{NMP} can be determined if k_p, $[R_1R_2NO^•]_0$ and $[R_3–ONR_1R_2]_0$ are known. In practice, such plots are linear initially, but tend to show positive curvature as conversion increases. This is due to the *persistent-radical effect* in which $[M^•]$ reduces with conversion due to contributions from bimolecular chain radical termination events, which in turn causes $[R_1R_2NO^•]$ to increase because nitroxides are persistent radicals that cannot be lost by self-combination. Fischer analysed this effect using more advanced methods which are not appropriate to introduce here and obtained the following equation for NMP performed using an alkoxyamine initiator *in the absence of excess added nitroxide*

$$[M^•] = \left(\frac{K_{NMP}[R_3 - ONR_1R_2]_0}{3k_t}\right)^{1/3} t^{-1/3} \tag{4.43}$$

in which k_t appears as a consequence of including bimolecular termination into the kinetics analysis. Substitution of Equation 4.43 into Equation 4.3 gives an equation for R_p which takes into account the persistent-radical effect

$$R_p = -\frac{d[M]}{dt} = k_p[M]\left(\frac{K_{NMP}[R_3 - ONR_1R_2]_0}{3k_t}\right)^{1/3} t^{-1/3} \tag{4.44}$$

Thus, the persistent-radical effect leads to a dependence of R_p upon $k_t^{-1/3}$ and so an autoacceleration effect can occur, but is mitigated by the weaker dependency on k_t (c.f. $R_p \propto k_t^{-1/2}$ for conventional free-radical polymerization) and the dependence on $t^{-1/3}$. Rearranging Equation 4.44 and applying integration limits

$$\int_{[M]_0}^{[M]_t} \frac{d[M]}{[M]} = -k_p \left(\frac{K_{NMP}[R_3 - ONR_1R_2]_0}{3k_t} \right)^{1/3} \int_0^t t^{-1/3} dt$$

which upon integration gives

$$\ln\left(\frac{[M]_0}{[M]_t}\right) = \frac{3k_p}{2}\left(\frac{K_{NMP}[R_3 - ONR_1R_2]_0}{3k_t}\right)^{1/3} t^{2/3} \quad (4.45)$$

Thus, when the persistent-radical effect is operating in the absence of excess added nitroxide, a plot of $\ln([M]_0/[M]_t)$ versus $t^{2/3}$ is predicted to be linear, but K_{NMP} can be determined from the slope only if k_t is known. Such plots usually give better straight lines than plots of Equation 4.42.

The number-average degree of polymerization of the polymer produced is determined simply by the initial molar ratio of the monomer to alkoxyamine and the fractional monomer conversion c

$$\bar{x}_n = \frac{c[M]_0}{[R_3 - ONR_1R_2]_0} \quad (4.46)$$

because each alkoxyamine molecule generates a single chain, i.e. $[P] = [R_3\text{–}ONR_1R_2]_0$. Thus, $\bar{x}_n$ should increase linearly with c, which is a characteristic feature of living polymerizations (see Section 5.3.2.3).

4.5.1.2 Side Reactions in Nitroxide-Mediated Radical Polymerization

One issue in NMP is disproportionation of the chain radical and nitroxide, which leads to loss of the chain radical and conversion of the nitroxide to a hydroxylamine

$$\sim\!\!\sim CH(H)\text{—}\dot{C}H(X) + \dot{O}\text{–}NR_1R_2 \longrightarrow \sim\!\!\sim CH{=}CH(X) + H\text{–}O\text{–}NR_1R_2$$

Clearly this results in loss of the living-like characteristics for that particular chain radical. For styrene and acrylate monomers, this reaction occurs but is not a serious problem in terms of producing polymers of well-defined molar mass and narrow molar mass distribution. However, disproportionation is a serious problem in NMP of methacrylate monomers and involves abstraction of a hydrogen atom from the α-methyl group

$$\sim\!\!\sim CH_2\text{–}\dot{C}(CH_2H)(CO_2R) + \dot{O}\text{–}NR_1R_2 \longrightarrow \sim\!\!\sim CH_2\text{–}C({=}CH_2)(CO_2R) + H\text{–}O\text{–}NR_1R_2$$

Thus, NMP of methacrylates is uncontrolled.

A side reaction unique to styrene is generation of radicals through self-reaction involving three molecules of styrene. This is believed to proceed via an initial Diels–Alder reaction between two molecules of styrene

followed by reaction of the conjugated triene product with another molecule of styrene

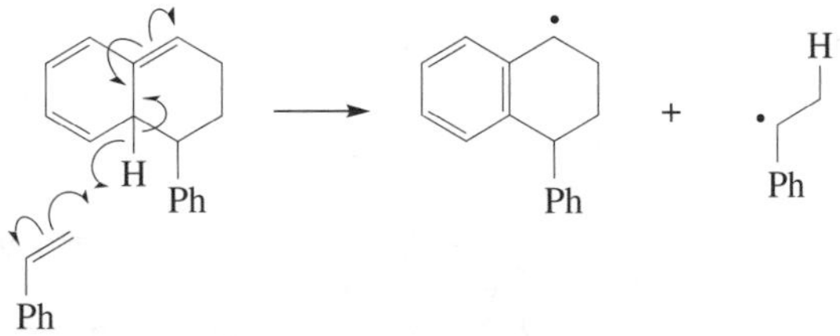

thus generating two radicals, both of which can initiate polymerization. This self-initiation process proceeds at significant rates at higher temperatures (e.g. styrene polymerizes at appreciable rates on heating at 80 °C in the absence of an initiator) and so makes a major contribution in NMP of styrene because this is carried out at 125–130 °C. This is both a problem and a benefit because although, in principle, it creates an imbalance between the chain radical and nitroxide concentrations, in practice it serves to compensate for loss of chain radicals due to bimolecular termination and disproportionation reactions. It does, however, control the rate of polymerization and makes it difficult to predict exactly the progress and control of NMP of styrene.

4.5.2 Atom-Transfer Radical Polymerization

Research into control of cationic polymerizations using transition metals (see Section 5.2.2) led to the discovery, first reported independently by Sawamoto and Matyjaszewski in 1995, of transition-metal-catalysed radical polymerizations that have the characteristics of living polymerization. This type of polymerization is now more commonly termed *atom-transfer radical polymerization (ATRP)*. The principles of such polymerizations are shown in Figure 4.14.

The transition metal (Mt) is in the form of a halide compound that is complexed by ligands and behaves as a true catalyst, but more usually is called an *activator*. Organic halides RX (usually bromides or chlorides) are used as initiators. Initiation proceeds via a single-electron transfer from the metal (Mt^{z+}) to the halogen atom in the R–X bond, causing homolysis of the bond to yield $R^{\bullet}$ and an increase in the oxidation state of the metal atom (to $Mt^{(z+1)+}$) which captures the released halogen atom. Sooner rather than later, propagation of the chain radical is intercepted by the reverse process in which the oxidized transition metal complex donates a halogen atom back to the propagating radical resulting in deactivation of the chain radical through formation of a new C–X bond at the chain end and regeneration of Mt^{z+}. Thus, ATRP employs Strategy 1 of Figure 4.11 and so has features similar to NMP: (i) the chain grows through a series of activation-propagation-deactivation cycles; (ii) the equilibrium must be controlled so that it lies far to the left and exchange is rapid, this being achieved for each monomer by careful choice of the initiator, activator complex and temperature; and (iii) most chains exist in the dormant state, so $[M^{\bullet}]$ is very low, thereby massively reducing the probability of bimolecular termination events, but also reducing the rate of polymerization.

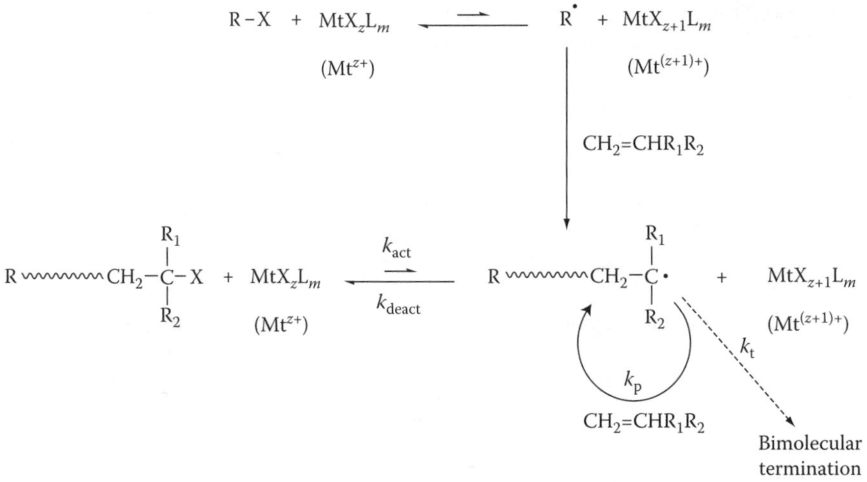

$$\mathrm{R{-}X} + \mathrm{MtX}_z\mathrm{L}_m \rightleftharpoons \mathrm{R}^{\bullet} + \mathrm{MtX}_{z+1}\mathrm{L}_m$$

(Mt^{z+}) $(\mathrm{Mt}^{(z+1)+})$

$\mathrm{CH_2{=}CHR_1R_2}$

$$\mathrm{R}\sim\sim\mathrm{CH_2{-}C(R_1)(R_2){-}X} + \mathrm{MtX}_z\mathrm{L}_m \underset{k_{\mathrm{deact}}}{\overset{k_{\mathrm{act}}}{\rightleftharpoons}} \mathrm{R}\sim\sim\mathrm{CH_2{-}C^{\bullet}(R_1)(R_2)} + \mathrm{MtX}_{z+1}\mathrm{L}_m$$

(Mt^{z+}) $(\mathrm{Mt}^{(z+1)+})$

k_p $\mathrm{CH_2{=}CHR_1R_2}$

k_t Bimolecular termination

FIGURE 4.14 The principles of transition-metal-catalysed radical polymerization, more commonly referred to as ATRP, in which the chain radical is released and captured through single-electron transfer processes involving a transition metal and a C–X bond. RX is an organic halide (the initiator), Mt is a transition metal with initial oxidation state $+z$, X is a halogen, L is a ligand, m molecules of which are coordinated to the transition metal, and R_1 and R_2 are the substituent groups in the monomer. The transition metal acts as a true catalyst in changing its oxidation state reversibly, as indicated in the parentheses under the transition metal complexes. The equilibrium must strongly favour the C–X bond such that the instantaneous chain radical concentration is sufficiently low to make bimolecular termination reactions between chain radicals insignificant. The equilibrium and the time for which the chain radical is free to propagate are controlled by the rate coefficients for activation, k_{act}, and deactivation, k_{deact}, of the chain radical.

4.5.2.1 Kinetics of ATRP

As for NMP, the equilibrium shown in Figure 4.14 must be controlled such that the chain radical concentration is low and exchange is rapid. The equilibrium constant K_{ATRP} is given by

$$K_{\mathrm{ATRP}} = \frac{[\mathrm{M}^{\bullet}][\mathrm{Mt}^{(z+1)+}\mathrm{L}_m]}{[\mathrm{RX}][\mathrm{Mt}^{z+}\mathrm{L}_m]} \tag{4.47}$$

where

[RX] is the concentration of initiator

$[\mathrm{Mt}^{z+}\mathrm{L}_m]$ and $[\mathrm{Mt}^{(z+1)+}\mathrm{L}_m]$ are the concentrations of the complexed transition metal ion in its initial and higher oxidation states, respectively.

Rearranging this equation for the total chain radical concentration

$$[\mathrm{M}^{\bullet}] = K_{\mathrm{ATRP}}[\mathrm{RX}]\left\{\frac{[\mathrm{Mt}^{z+}\mathrm{L}_m]}{[\mathrm{Mt}^{(z+1)+}\mathrm{L}_m]}\right\} \tag{4.48}$$

and then substitution into Equation 4.3 gives the following equation for the rate of polymerization

$$R_{\mathrm{p}} = -\frac{\mathrm{d}[\mathrm{M}]}{\mathrm{d}t} = K_{\mathrm{ATRP}}k_{\mathrm{p}}[\mathrm{M}][\mathrm{RX}]\left\{\frac{[\mathrm{Mt}^{z+}\mathrm{L}_m]}{[\mathrm{Mt}^{(z+1)+}\mathrm{L}_m]}\right\} \tag{4.49}$$

which shows that R_{p} is first order in both [M] and [RX], and that it is independent of k_{t} (thus eliminating autoacceleration, as for NMP in the absence of the persistent-radical effect). Comparison of

Equation 4.49 with the equivalent NMP Equation 4.41 highlights a distinguishing feature of ATRP, namely, that the rate depends not only on the concentrations of initiator and persistent radical (which is $Mt^{(z+1)+}$ in ATRP), but also on the activator concentration. For ATRP, the rate is controlled by the concentration ratio $[Mt^{z+}L_m]/[Mt^{(z+1)+}L_m]$ and so is little affected by changes in the absolute value of $[Mt^{z+}L_m]$. Thus, unlike NMP (which requires $[R_1R_2NO^\bullet]$ to be similar to $[M^\bullet]$), ATRP is truly catalytic in its requirements for Mt^{z+}. However, very low values of $[Mt^{z+}L_m]$ are not useful because the exchange rate would then be much too low; hence, $[Mt^{z+}L_m]$ still needs to be quite high to produce polymers with narrow molar mass distributions. Since each molecule of initiator (RX) produces one polymer chain, the number-average degree of polymerization is given by

$$\bar{x}_n = \frac{c[M]_0}{[RX]_0} \tag{4.50}$$

where

$[M]_0$ and $[RX]_0$ are, respectively, the initial monomer and initiator concentrations
c is the fractional monomer conversion.

Hence, ATRP has the characteristics of living polymerization, with $\bar{x}_n$ being controlled by the ratio $[M]_0/[RX]_0$ and increasing linearly with c.

The persistent-radical effect also is evident in ATRP with $[Mt^{(z+1)+}L_m]$ increasing (and $[Mt^{z+}L_m]$ reducing by an equal amount) as chain radicals are lost through occasional bimolecular termination events. The effect on the rate of polymerization is more complex than for NMP because the rate depends on the ratio $[Mt^{z+}L_m]/[Mt^{(z+1)+}L_m]$ for ATRP. If the amount of Mt^{z+} is lower than the amount of chains that undergo bimolecular termination, eventually the transition metal will exist entirely as $Mt^{(z+1)+}$ and the reaction will stop; this is another reason why catalytic amounts of Mt^{z+} cannot be used.

4.5.2.2 Initiators, Transition Metals and Ligands for ATRP

The choice of initiator, transition metal and coordinating ligand is crucial for achieving living-like conditions in ATRP. For each ATRP system (initiator, activator complex, monomer and reaction medium), there is a relatively narrow optimum temperature range, but this can vary widely from one system to another (e.g. controlled ATRP has been carried out successfully at 20 °C and also at 130 °C for different systems).

The amount of initiator (RX) controls the number of polymer chains produced and so it must be capable of fast, quantitative initiation of chains in order to provide the desired control of molar mass and molar mass distribution. Thus, the R–X bond should be of slightly greater activity towards atom transfer than the resulting dormant polymer chain-end C–X bond. For this reason, the most common initiators are bromides and chlorides that have some structural similarity to the dormant end-unit of the polymer chains to be produced. Some examples are shown in Figure 4.15 and are capable of initiating controlled ATRP for several monomers.

(i) $CH_3-CH(C_6H_5)-X$ X = Br or Cl
(ii) $CH_3-O-C(=O)-CH(CH_3)-X$ X = Br or Cl
(iii) $CH_3CH_2-O-C(=O)-C(CH_3)_2-Br$

FIGURE 4.15 Some commonly used ATRP initiators (R–X): (i) 1-halo-1-phenylethanes (suitable for ATRP of styrene); (ii) methyl 2-halopropionates (suitable for ATRP of acrylates and styrene); and (iii) ethyl 2-bromoisobutyrate (suitable for ATRP of methacrylates and acrylates).

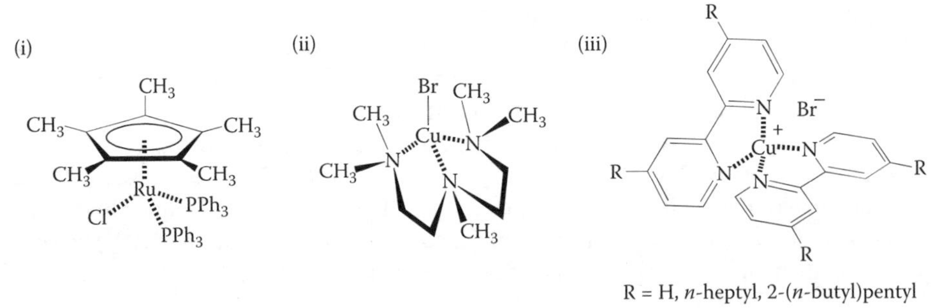

(i) (ii) (iii)

R = H, *n*-heptyl, 2-(*n*-butyl)pentyl

FIGURE 4.16 Some commonly-used ATRP transition-metal-complex activators (MtX_zL_m): (i) the triphenylphosphine complex of ruthenium (II) pentamethylcyclopentadienyl chloride, which is neutral; (ii) the *N,N,N′,N″,N″*-pentamethyldiethylenetriamine complex of copper (I) bromide, which is neutral; and (iii) the 4,4′-disubstitutedbipyridyl complex of copper (I) bromide, which is ionic (if R = H the complex is insoluble; if long alkyl R groups are used, the complex is soluble in common monomers and solvents).

The first reported transition-metal-catalysed radical polymerizations involved use of ruthenium (Ru^{2+}) and copper (Cu^{+}) halide complexes as activators. Several other transition metals also have been used successfully (e.g. iron, nickel, palladium and rhodium), but use of copper-based activators has become dominant. Though not essential, it is normal to use the transition metal halide corresponding to the initiator (e.g. CuBr for RBr and CuCl for RCl initiators). The purpose of the ligands is to coordinate to Mt^{z+}, activating it towards loss of an electron and atom transfer. Nitrogen-based ligands are used extensively for ATRP with Cu^{+} as activator, whereas more complicated activator complexes are needed for other transition metals (e.g. phosphines are often used for ruthenium-based activators). Some ATRP activator complexes are shown in Figure 4.16.

ATRP is much more versatile than NMP and controlled ATRP has been achieved for a wide range of monomers (e.g. styrenics, acrylates, methacrylates, acrylonitrile, acrylamides and vinyl pyridines). The less reactive monomers (such as ethylene, vinyl acetate and vinyl chloride) have proven more difficult and require much more reactive ATRP activators, which brings complications in terms of the low stability of the activator towards oxidation. Acidic protons cannot be tolerated, which means that monomers such as acrylic acid and methacrylic acid can be polymerized only in their ionic form (i.e. as the carboxylate salt); an alternative approach is to use acrylates/methacrylates with ester side groups (e.g. trimethylsiloxyl esters, $-OSi(CH_3)_3$) that can be converted easily to carboxylic acid groups after the polymer has been produced.

4.5.2.3 Alternative Strategies for Initiation of ATRP

Reverse ATRP was introduced to overcome problems associated with the low oxidative stability of the more highly reactive (easily oxidized) activator complexes used in ATRP. The oxidized form of the activator is used instead, together with a conventional free-radical initiator (e.g. azobisisobutyronitrile) so that the ATRP initiator and activator are generated *in situ*, for example

$$\tfrac{1}{2}\,R_1{-}N{=}N{-}R_1 + MtX_{z+1}L_m \xrightarrow{\Delta} R_1^{\bullet} + MtX_{z+1}L_m \rightleftharpoons R_1{-}X + MtX_zL_m$$

ATRP mechanism as shown in Figure 4.14

For control, the concentrations of free-radical initiator and $Mt^{(z+1)+}L_m$ must be comparable. *Simultaneous reverse and normal initiation (SR&NI)* ATRP involves use of a conventional free-radical initiator and a normal ATRP initiator (RX) in the presence of the oxidized form of the activator, $Mt^{(z+1)+}$. The radicals produced from the free-radical initiator are captured by the same process as in reverse ATRP, but also by chain transfer to RX. A further development of this approach is *initiators for continuous activator regeneration (ICAR)* in which a conventional free-radical initiator is used in an otherwise normal ATRP to produce free radicals slowly throughout the polymerization in order to continuously regenerate Mt^{z+} from $Mt^{(z+1)+}$ and counteract the increase in $[Mt^{(z+1)+}L_m]$ due to the persistent-radical effect, which consequently allows lower low levels of transition metal to be used. Radicals generated by thermal self-initiation (see Section 4.5.1.2) can serve a similar purpose in ATRP of styrene at temperatures above 80 °C.

A better approach is to use *activators generated by electron transfer (AGET)* for initiation in which reducing agents (e.g. ascorbic acid) that cannot initiate new chains are used to generate the activator Mt^{z+} from $Mt^{(z+1)+}$ *in situ* in the presence of the normal ATRP initiator (RX). A related approach is to use *activators regenerated by electron transfer (ARGET)* in which the reducing agent is present in large excess. Under such conditions, the activator is continuously regenerated during the polymerization and there always is enough reducing agent to prevent the increase in $[Mt^{(z+1)+}L_m]$ due to the persistent-radical effect, again enabling low levels of transition metal to be used.

4.5.3 Reversible-Addition-Fragmentation Chain-Transfer Radical Polymerization

The researchers at CSIRO in Australia were responsible for another major development in reversible-deactivation radical polymerization with their discovery of reversible-addition-fragmentation chain-transfer radical (RAFT) polymerization, first revealed in 1998 by Le, Moad, Rizzardo and Thang. RAFT polymerization evolved from earlier work at CSIRO on control of radical polymerizations using addition-fragmentation chain-transfer agents. Together, RAFT and ATRP are now the most extensively used types of reversible-deactivation radical polymerization.

RAFT polymerization is, in essence, a normal free-radical polymerization to which is added a highly active dithioester transfer agent (the *RAFT agent*) that fragments during the chain-transfer process to release a new radical and generate a new dithioester species via the addition-fragmentation mechanism shown below

$$R^{\bullet} + S{=}C(Z){-}S{-}A \rightleftharpoons R{-}S{-}\dot{C}(Z){-}S{-}A \rightleftharpoons R{-}S{-}C(Z){=}S + A^{\bullet}$$

where the electron movements are shown for the forward reaction sequence, though it should be noted that the addition-fragmentation process is reversible. In order for this chain-transfer reaction to be facile, the reactivities of the two radicals ($R^{\bullet}$ and $A^{\bullet}$) must be similar and the group Z must activate the exchange process. If $R^{\bullet}$ is a chain radical, it becomes trapped by addition to the RAFT agent and that chain remains dormant until it is released in a further addition-fragmentation chain-transfer process when another chain radical adds to the dithioester end-group. By using a RAFT agent that is very highly active in chain transfer at a level that greatly exceeds the total number of primary radicals generated from the initiator during the course of the whole polymerization, addition-fragmentation chain-transfer to dithioester groups becomes the dominant process by which propagating chains are both generated and captured, thereby making bimolecular termination an event of low probability. Since capture of a chain radical by a dithioester group is reversible and fast, at any instant in time, most chains are in the dormant dithioester form. The overall reaction scheme for RAFT polymerization is shown and described in Figure 4.17. Initially, addition-fragmentation chain-transfer to the RAFT agent dominates, but as the RAFT agent is consumed and the concentration of polymeric RAFT species grows, the latter begin to contribute and eventually become the only dithioester species when the RAFT agent is consumed completely.

4.5.3.1 Kinetics of RAFT Polymerization

If the product radical $A^{\bullet}$ is of similar reactivity to the chain radicals, inclusion of the RAFT agent should have little effect upon the steady-state radical concentration, and the rate of polymerization should be given by Equation 4.6, i.e. it should be similar to that for the equivalent conventional

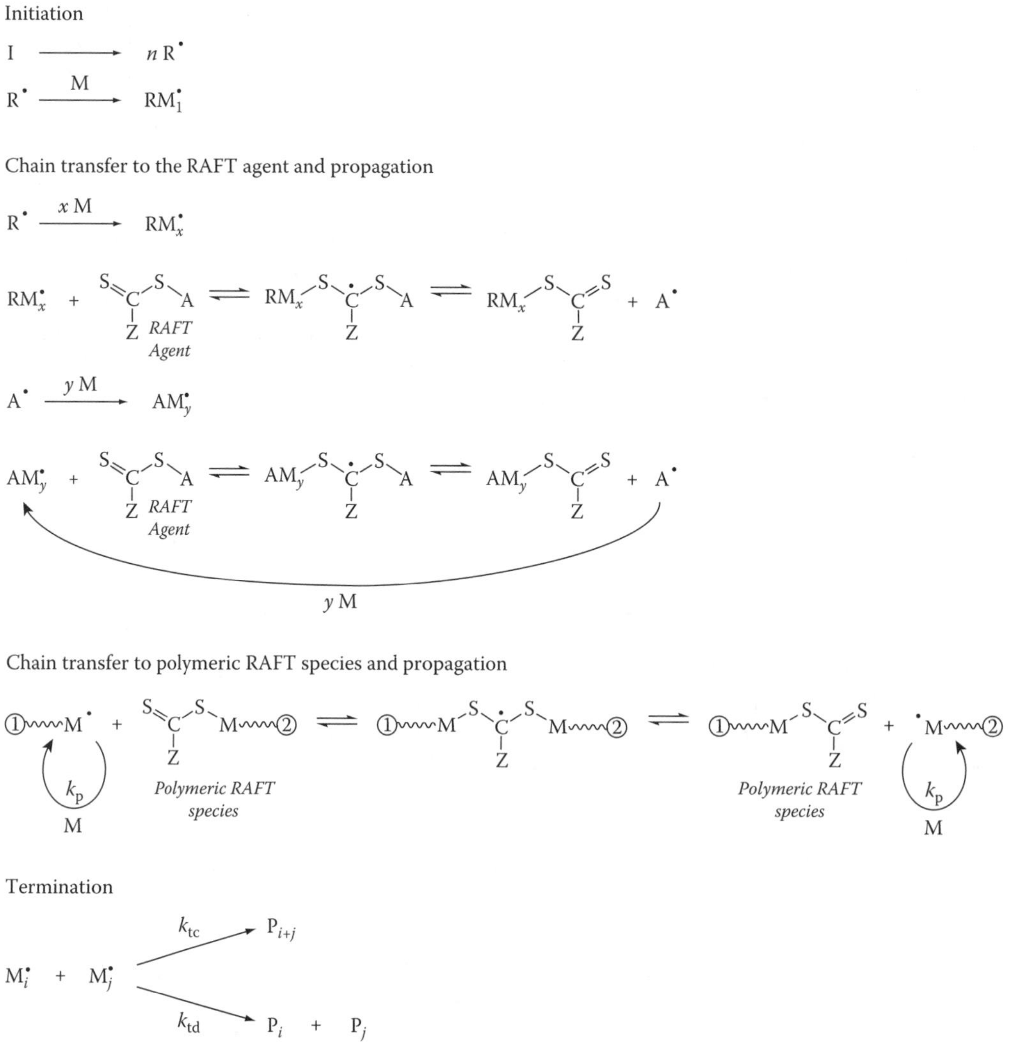

Initiation

$$I \longrightarrow n\,R^{\bullet}$$

$$R^{\bullet} \xrightarrow{M} RM_1^{\bullet}$$

Chain transfer to the RAFT agent and propagation

$$R^{\bullet} \xrightarrow{x\,M} RM_x^{\bullet}$$

$$RM_x^{\bullet} + S{=}C(Z){-}S{-}A \rightleftharpoons RM_x{-}S{-}\dot{C}(Z){-}S{-}A \rightleftharpoons RM_x{-}S{-}C(Z){=}S + A^{\bullet}$$

RAFT Agent

$$A^{\bullet} \xrightarrow{y\,M} AM_y^{\bullet}$$

$$AM_y^{\bullet} + S{=}C(Z){-}S{-}A \rightleftharpoons AM_y{-}S{-}\dot{C}(Z){-}S{-}A \rightleftharpoons AM_y{-}S{-}C(Z){=}S + A^{\bullet}$$

RAFT Agent

($A^{\bullet} \xrightarrow{y\,M} AM_y^{\bullet}$)

Chain transfer to polymeric RAFT species and propagation

$$①{\sim}M^{\bullet} + S{=}C(Z){-}S{-}M{\sim}② \rightleftharpoons ①{\sim}M{-}S{-}\dot{C}(Z){-}S{-}M{\sim}② \rightleftharpoons ①{\sim}M{-}S{-}C(Z){=}S + {}^{\bullet}M{\sim}②$$

Polymeric RAFT species (k_p, M) *Polymeric RAFT species* (k_p, M)

Termination

$$M_i^{\bullet} + M_j^{\bullet} \xrightarrow{k_{tc}} P_{i+j}$$

$$M_i^{\bullet} + M_j^{\bullet} \xrightarrow{k_{td}} P_i + P_j$$

FIGURE 4.17 The chemistry of reversible-addition-fragmentation chain-transfer (RAFT) polymerization in which a dithioester RAFT agent is used to provide control and living-like characteristics to what would otherwise be a conventional free-radical polymerization. Chain radicals are formed via homolysis of a normal free-radical initiator, but are captured with high efficiency through addition to the RAFT agent, followed by fragmentation of the adduct to release the radical A• which itself initiates a new chain. This is the *pre-equilibrium*. When the original RAFT agent has been consumed completely, chain radicals continue to be captured rapidly, but now by the polymeric RAFT end-group species which results in release of another chain radical (as is made clear by the labels on the chain ends). This is the *main equilibrium*. During their short life between release and capture, chain radicals are free to undergo all possible reactions, with propagation being far the most likely, though bimolecular termination still can occur, as can other chain transfer reactions (e.g. to monomer, solvent and polymer), though these are not shown in the scheme.

free-radical polymerization in the absence of the RAFT agent. Thus, RAFT polymerizations proceed at much higher rates than NMP and ATRP (and all other reversible-deactivation radical polymerizations that work via Strategy 1 of Figure 4.11) where living-like conditions are achieved principally by suppressing the radical concentration. This is a major advantage for RAFT polymerization. In practice, however, there often is an induction period in RAFT polymerizations, which quite commonly is followed by a lower rate of polymerization than for the same polymerization in the absence of the RAFT agent. The reasons for the induction period and the subsequent retardation of polymerization are not yet understood, but there is strong evidence that the relatively high stability of the intermediate adduct radical is a key factor. Due to its high stability, the rate of fragmentation may be reduced significantly and there is likely to be a greater probability of the adduct radical undergoing bimolecular radical coupling with a propagating chain radical (leading to termination), both of which would result in a reduction of the concentration of propagating chain radicals and a consequent reduction in the rate of polymerization.

Provided that the RAFT agent is highly efficient so that bimolecular termination makes an insignificant contribution, virtually all the polymer chains produced will have one dithioester end group derived from the original RAFT agent. Under such conditions, the number-average degree of polymerization of the polymer produced is given by

$$\bar{x}_n = \frac{c[M]_0}{[RAFT]_0} \tag{4.51}$$

where

$[M]_0$ and $[RAFT]_0$ are, respectively, the initial monomer and RAFT agent concentrations
c is the fractional monomer conversion

Thus, RAFT also shows the characteristics of a living polymerization with $\bar{x}_n$ controlled by the ratio $[M]_0/[RAFT]_0$ and increasing linearly with c.

Unlike NMP and ATRP, for RAFT polymerization, it is relatively simple to predict the fraction of dead chains produced. This is because the concentration of living chains is equal to $[RAFT]_0$ (assuming the RAFT agent is highly efficient) and, since the dead chains result from bimolecular termination of chain radicals the original source of which are primary radicals produced from the initiator, the concentration of dead chains is given by

$$[\text{dead chains}]_t = 2f\{[I]_0 - [I]_t\} \times \left(\frac{1+q}{2}\right) = (1+q)f\{[I]_0 - [I]_t\} \tag{4.52}$$

in which q is the termination parameter (see Section 4.3.2), f is the initiator efficiency, and $[I]_0$ and $[I]_t$ are the initiator concentrations at the start and after time t, respectively. In Equation 4.52, $2f\{[I]_0-[I]_t\}$ is the cumulative total concentration of chain radicals generated from initiator primary radicals and the factor $(1+q)/2$ is the average number of dead chains produced per chain radical, taking into account the contributions from bimolecular termination by combination and disproportionation. Integration of Equation 4.27 gives $[I]_t=[I]_0 \exp(-k_d t)$ and so the fraction of dead chains is given by

$$\text{fraction of dead chains} = \frac{(1+q)f[I]_0\{1-\exp(-k_d t)\}}{[RAFT]_0 + (1+q)f[I]_0\{1-\exp(-k_d t)\}} \tag{4.53}$$

Given that most polymerizations are carried out over a period of 2–10 h, for which $\{1-\exp(-k_d t)\}$ is typically <0.1, then the criterion for achieving living-like conditions in RAFT polymerization is $[RAFT]_0 \gg [I]_0$. For example, if $[RAFT]_0/[I]_0=10$, $\{1-\exp(-k_d t)\}=0.06$, $q=1$ and $f=0.7$, the percentage of dead chains is predicted to be just 0.8%. Since $[RAFT]_0$ needs to be reduced as the target value of $\bar{x}_n$ increases, $[I]_0$ also needs to be reduced to maintain livingness, the consequence being a reduced rate of polymerization.

4.5.3.2 RAFT Agents

The choice of RAFT agent (i.e. the Z and A groups) is important in achieving living-like conditions. So far, only dithioesters have been considered, but several other classes of RAFT agent can provide living-like radical polymerization: trithiocarbonates (Z=SR), dithiocarbamates ($Z=NR_2$) and

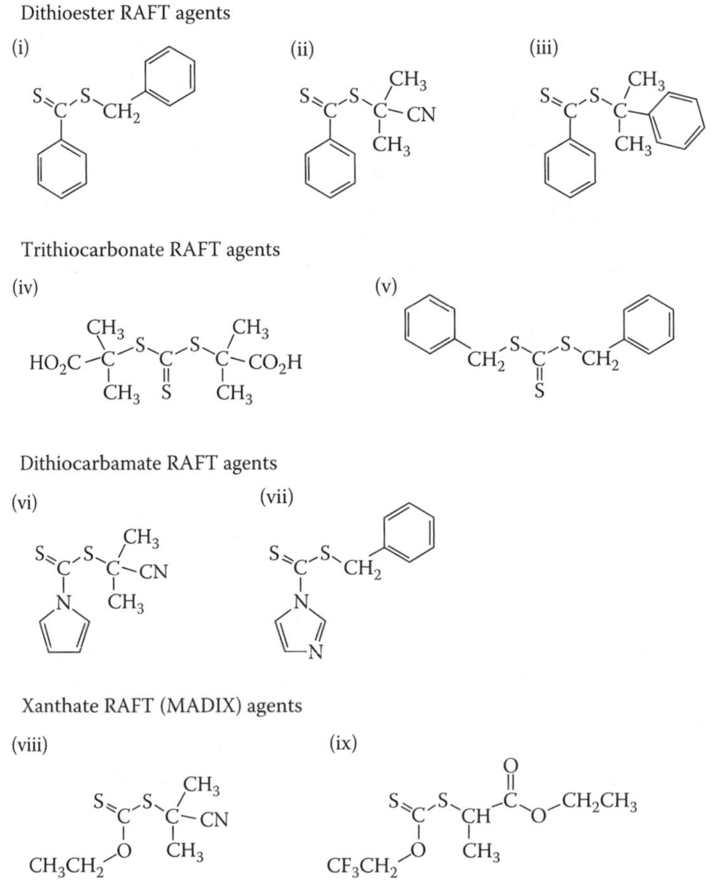

Dithioester RAFT agents

(i) (ii) (iii)

Trithiocarbonate RAFT agents

(iv) (v)

Dithiocarbamate RAFT agents

(vi) (vii)

Xanthate RAFT (MADIX) agents

(viii) (ix)

FIGURE 4.18 The chemical structures of some commonly used RAFT agents.

xanthates (Z=OR). The chemical structures of some RAFT agents from each of the main classes of RAFT agent are shown in Figure 4.18.

Dithiobenzoate RAFT agents (Z=Ph), such as (i)–(iii) in Figure 4.18, are the most widely used dithioesters because the phenyl group stabilizes the intermediate adduct radical by delocalization of the unpaired electron around the benzene ring, thereby activating strongly the addition step. For the symmetric trithiocarbonates (iv) and (v) of Figure 4.18, Z=S–A where A=$C(CH_3)_2CO_2H$ for (iv) and A=CH_2Ph for (v). The main RAFT equilibrium is then

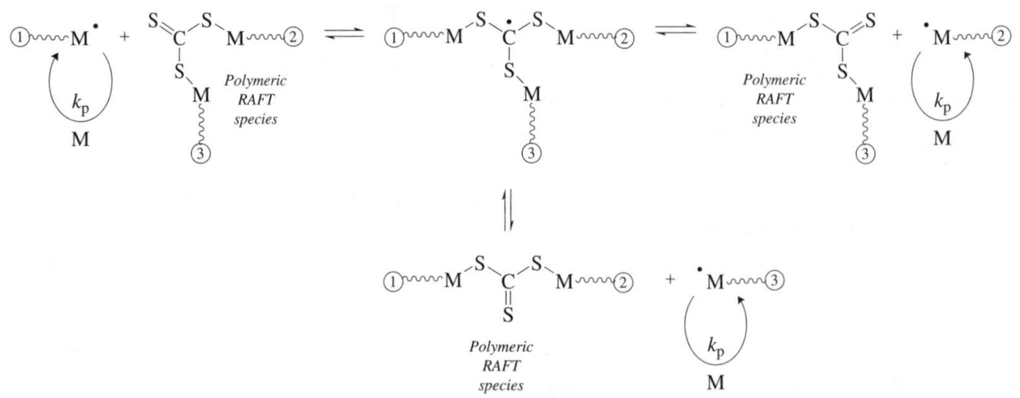

Polymeric RAFT species

k_p

M

in which the chains are labelled to make it clear that fragmentation of the intermediate adduct radical can proceed reversibly in any one of three directions: the reverse direction to release the chain radical that just added, or through release of either of the chain radicals that were trapped in the polymeric RAFT species prior to addition of the chain radical.

Dithiocarbamate RAFT agents, such as (vi) and (vii) in Figure 4.18 (where Z is N-pyrrole and N-imidazole, respectively), are similar in behaviour to dithioesters and react via the mechanism shown in Figure 4.17. Use of xanthates for effecting living-like radical polymerization was discovered by Zard in collaboration with researchers at Rhodia, almost simultaneously with the development of RAFT polymerization; they named these polymerizations *macromolecular design via the interchange of xanthates* (*MADIX*). The mechanism for MADIX polymerization is the same as shown for RAFT polymerization in Figure 4.17 with Z=OR, and so MADIX polymerization is simply a specific form of RAFT polymerization.

The chain-transfer activity of the different classes of RAFT agent fall in the order: dithiobenzoates > trithiocarbonates ≈ dithioalkanoates (Z=alkyl) ≈ N-pyrrole/N-imidazole dithiocarbamates > xanthates > dithiodialkylcarbamates ($Z=NR_2$ where R=alkyl). This order can change depending on the radical leaving group ($A^\bullet$) for which the ease of loss follows the order of radical stability and for the radical leaving groups in the RAFT agents shown in Figure 4.18 is

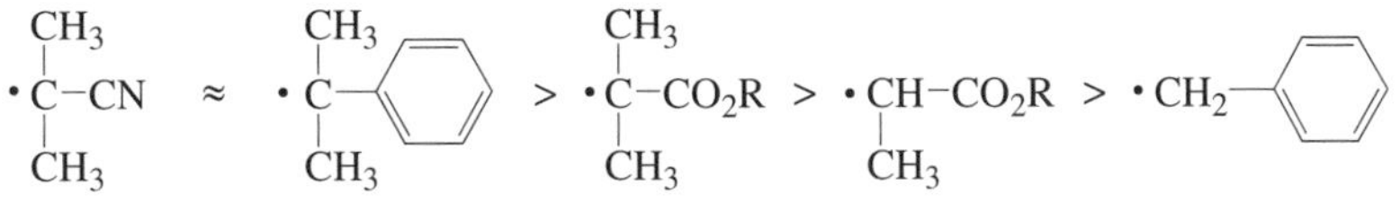

$$\cdot C(CH_3)_2-CN \approx \cdot C(CH_3)_2-C_6H_5 > \cdot C(CH_3)_2-CO_2R > \cdot CH(CH_3)-CO_2R > \cdot CH_2-C_6H_5$$

where R=H, alkyl, aryl.

Because RAFT polymerization is effectively a slight perturbation of normal free-radical polymerization, it can be used with a very wide range of monomers. The more reactive RAFT agents (e.g. dithiobenzoates and trithiocarbonates) are suitable for effecting living-like radical polymerization of more reactive monomers such as styrene, methacrylates, acrylates, acrylamides and vinyl pyridines. Controlled RAFT polymerization of monomers of low reactivity (which consequently give polymeric radicals of very high reactivity, vinyl acetate being an important example) requires less active RAFT agents and so xanthates and dithiodialkylcarbamates are the most suitable RAFT agents for such monomers.

4.6 NON-LINEAR RADICAL POLYMERIZATIONS

The methods of radical polymerization have been considered so far only in terms of homopolymerization. Radical copolymerization is extremely important and considered in detail in Chapter 9 together with all other forms of copolymerization. Nevertheless, it is relevant to consider here radical copolymerizations that involve comonomers which have more than one C=C bond. Such copolymerizations lead to the formation of non-linear polymers and are used in the production of a wide range of materials from crosslinked particles for gel permeation chromatography (see Section 14.3) to crosslinked matrices for fibre-reinforced composites (see Section 24.2). These two examples are representative of the two principal ways of producing crosslinked polymers by radical copolymerization which are considered in more detail in the following sections.

4.6.1 Non-Linear Radical Polymerizations Involving Crosslinking Monomers

As illustrated schematically in Figure 4.19, copolymerization of a mono-olefin with a *crosslinking monomer*, i.e. a monomer that has more than one C=C bond, leads to the formation of branched and network polymers. The junction points in the resulting non-linear polymer are provided by the linking groups between the C=C bonds in the crosslinking monomer. In most cases, crosslinked

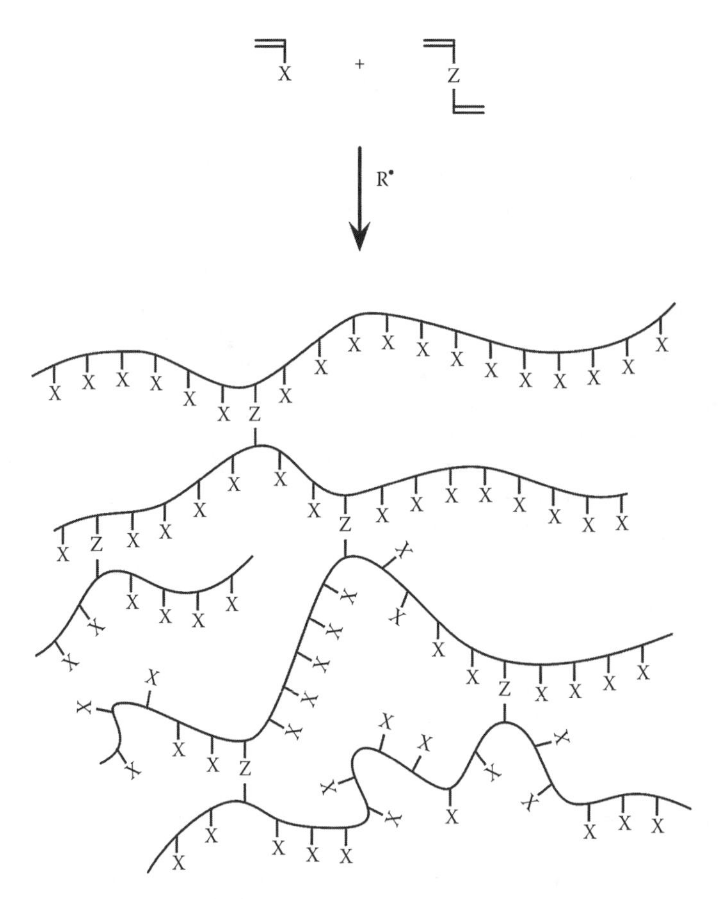

FIGURE 4.19 Formation of non-linear polymers by radical copolymerization is illustrated here for a simple case in which a mono-olefin is copolymerized with a di-olefin. X is the substituent group in the mono-olefin monomer, Z is the linking group in the di-olefin monomer and R• represents primary radicals from a free-radical initiator.

polymers are required and can be formed by using only a low level of the crosslinking monomer. The degree of crosslinking is controlled by the mole fraction of crosslinking monomer, the differences in reactivity of the various types of C=C bond in the monomers employed and the overall conversion. Because crosslinked polymers are intractable, they need to be produced directly in the form required for their application and so it is usual to carry out crosslinking polymerizations in the absence of a solvent (e.g. by bulk, suspension or emulsion polymerization). In some cases, highly branched (soluble) polymers are required (e.g. viscosity modifiers) and can be produced using crosslinking monomers if a chain transfer agent is included at high enough level to prevent formation of a network.

The chemical structures of some common crosslinking monomers are shown in Figure 4.20. Divinylbenzene is used in the synthesis of crosslinked polystyrene beads; such beads are used directly as gels in gel permeation chromatography, but for many other applications (e.g. ion exchange resins, solid-phase peptide synthesis supports) they are subjected to chemical reactions that introduce functional groups to the phenyl side groups in the polystyrene chains. Despite its extensive use as a crosslinking monomer, commercial divinylbenzene is far from ideal. It is synthesized by (usually incomplete) dehydrogenation of mixed diethylbenzene isomers and consequently is a mixture of divinylbenzene isomers and, as supplied, usually contains about 20% of ethylstyrene isomers that possess only a single C=C bond. Furthermore, because the C=C bonds are connected directly to

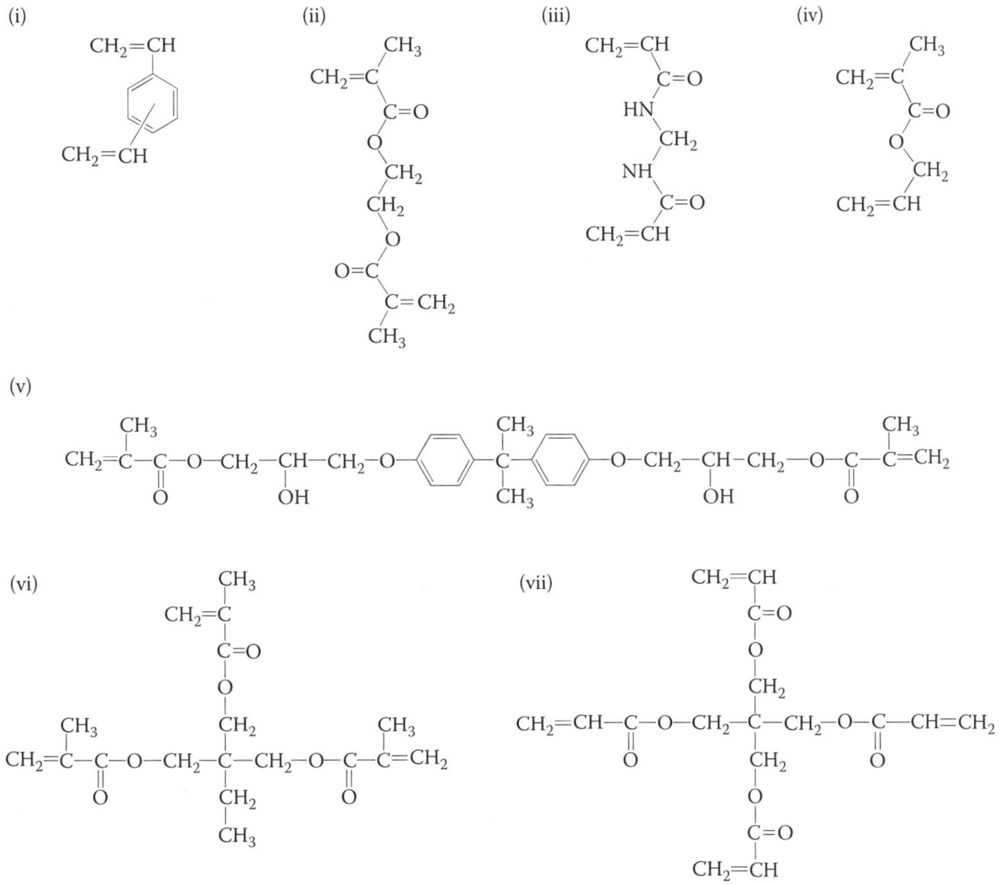

FIGURE 4.20 Common crosslinking monomers used in the synthesis of network polymers by radical polymerization: (i) divinylbenzene (usually supplied as a mixture comprising the 1,3-isomer (~53%) and the 1,4 isomer (~26%) with the remainder being the corresponding ethylstyrene isomers); (ii) ethylene glycol dimethacrylate; (iii) methylene bisacrylamide; (iv) allyl methacrylate; (v) bisphenol-A glycidyl methacrylate; (vi) trimethylolpropane trimethacrylate; and (vii) pentaerythritol tetraacrylate.

the benzene ring, the reactivity of the second C=C bond changes significantly after reaction of the first C=C bond. The two principal divinylbenzene isomers also have different reactivities. Hence, the crosslinking produced is non-uniform. Ethylene glycol dimethacrylate is much better in this respect, because the two C=C bonds are sufficiently far apart not to affect each other's reactivity, and finds extensive use as a crosslinking monomer in acrylate and methacrylate crosslinking polymerizations; it also is used as the crosslinking agent in synthesis of *hydrogels*, which are hydrophilic, crosslinked water-swellable network polymers (e.g. crosslinked poly(2-hydroxyethyl methacrylate) used in soft contact lenses). Bisphenol-A glycidyl methacrylate is more commonly known as BIS-GMA and is a major component in formulations for acrylic dental fillings. Methylene bisacrylamide is used extensively to produce crosslinked polymers from acrylamide and related monomers, e.g. as the crosslinking monomer in production of polyacrylamide electrophoresis gels and in preparation of hydrogels and microgels (see Section 4.4.4.6). Crosslinking monomers of higher functionality (such as (vi) and (vii) in Figure 4.20) are used when higher degrees of crosslinking are required. Allyl methacrylate ((iv) in Figure 4.20) is an example of a *graftlinking monomer*, the characteristic feature of which is the presence of C=C bonds with very different reactivities; in allyl methacrylate, the methacrylate C=C bond is far more reactive than the allyl C=C bond, which has very low reactivity. Graftlinking monomers are inefficient crosslinkers because relatively few of the low reactivity C=C bonds polymerize; however, the reactive C=C bond copolymerizes effectively to leave many unreacted pendent allyl C=C bonds on the polymer chains, some of which do react with newly added monomer. This is used to provide grafting at interfaces between different polymers, e.g. to provide strong interfaces between phases in core-shell particles produced by emulsion polymerization (see Section 4.4.4.4).

4.6.2 Non-Linear Radical Polymerizations Involving Unsaturated Polymers

Polymerization of a monomer in the presence of a polymer which has C=C bonds along, pendent to or terminal to, its backbone leads to branching and crosslinking. There are two situations where this is important: (i) where crosslinking is a desired outcome; (ii) where branching and crosslinking needs to be restricted.

4.6.2.1 Crosslinking of Unsaturated Resins

Crosslinked polymers can be formed by copolymerization of a monomer with a prepolymer that possesses several C=C bonds. This is used in the production of glass-fibre reinforced composites to produce crosslinked matrix polymers *in situ*, e.g. for construction of the hulls for small- to medium-size boats and yachts. The basic principles of crosslinking of unsaturated prepolymers are illustrated schematically in Figure 4.21. The unsaturated prepolymer is supplied as a concentrated solution in the monomer, which acts as a solvent to reduce the viscosity so that the resin can impregnate and wet the glass-fibre reinforcement (see Section 24.2) before free-radical copolymerization of the prepolymer C=C bonds with the monomer generates the crosslinking chains. As the copolymerization progresses, the viscosity increases because the solvent (i.e. monomer) is being converted into polymer of increasingly high molar mass through branch and, ultimately, network formation. Rigid materials usually are required, so monomers that produce homopolymers with glass transition temperatures well above room temperature are used and the reactions are taken to complete monomer conversion. In some cases, a mixture of monomers is employed, including some crosslinking monomers (Figure 4.20). The degree of crosslinking in the final material is controlled by the concentration of C=C bonds in the unsaturated prepolymer, the concentration of prepolymer, the concentration of monomer and, if included, the concentration of crosslinking monomers.

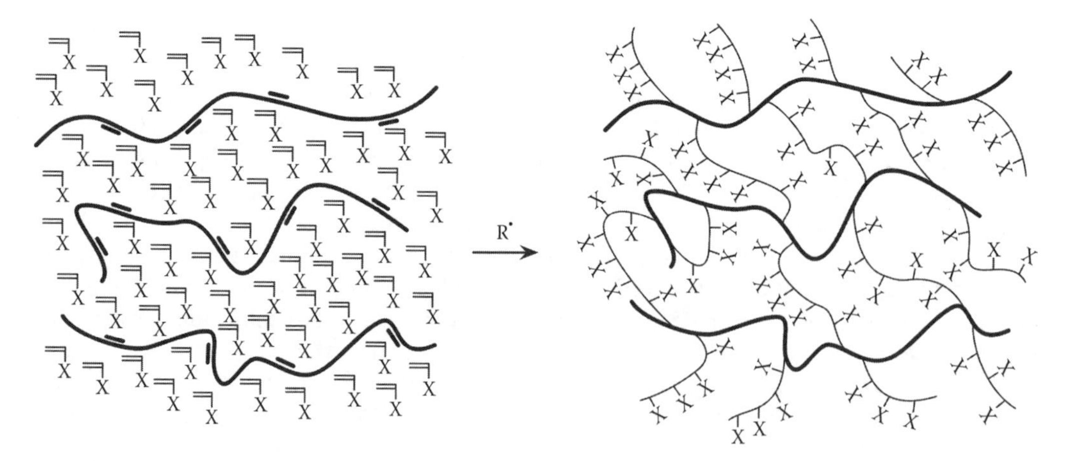

FIGURE 4.21 Schematic representation of free-radical crosslinking copolymerization of an unsaturated prepolymer (bold lines) dissolved in a monomer (=–X). Copolymerization of the monomer with the C=C bonds in the unsaturated prepolymer gives rise to crosslinking.

The most common materials of this type are *polyester resins*, which comprise low molar mass unsaturated polyesters dissolved, typically, in styrene. The unsaturated polyesters are prepared by polycondensation (see Section 3.2.1.1) of saturated diols with a mixture of saturated and unsaturated dicarboxylic acids, the most common of which are shown in Figure 4.22. The structural units in the polyester produced from the unsaturated dicarboxylic acids provide the sites for crosslinking. For many applications, crosslinking needs to proceed at ambient temperature, in which case redox initiation systems are employed (e.g. Co^{2+} with peroxides; see Section 4.2.1) with the redox initiator components being dissolved into the polyester resin just prior to use.

4.6.2.2 Branching and Crosslinking during Polymerization of 1,3-Dienes

Homopolymerization (and copolymerization) of 1,3-diene monomers (CH_2=CR—CH=CH_2) produces unsaturated polymers with one C=C bond per diene repeat unit. Thus, part way through such a polymerization, there exists a mixture of unreacted monomer and highly unsaturated polymer that is very similar to the situation illustrated in Figure 4.21. As conversion of the monomer increases, therefore, there is an increasing probability that the C=C bonds in the polymer already produced will copolymerize with the unreacted monomer to give branches and, eventually, crosslinked (intractable) polymer. The C=C bonds in *cis*-1,4 and *trans*-1,4 repeat units form the majority of the repeat units in the polymer (see Section 6.3) and are of lower reactivity in copolymerization because of steric hindrance from the two bulky chain substituents, one on each carbon atom in the C=C bond. The 1,2 (and 3,4) repeat units are present at much lower levels, but their pendent C=C bonds are more reactive because they have substituents only on one carbon atom in the C=C bond and only one of those substituents is a bulky chain. Thus, despite their low levels, the pendent C=C bonds in 1,2 (and 3,4) repeat units can make a significant contribution to branching and crosslinking. In addition to copolymerization, chain transfer to polymer makes a significant contribution due to the ease of H-abstraction from any of the four labile allylic C–H bonds present in each repeat unit, for example

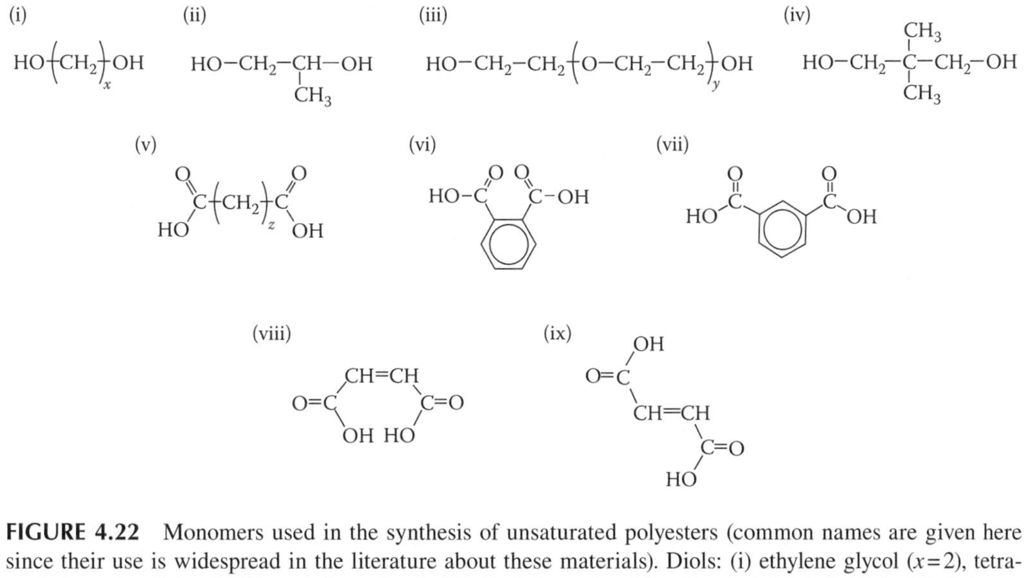

(i) HO$\left(CH_2\right)_x$OH

(ii) HO–CH_2–CH(CH_3)–OH

(iii) HO–CH_2–$CH_2$$\left(O–CH_2–CH_2\right)_y$OH

(iv) HO–CH_2–C(CH_3)$_2$–CH_2–OH

(v) HO(O=)C$\left(CH_2\right)_z$C(=O)OH

(vi) HO–C(=O)–C_6H_4–C(=O)–OH (ortho)

(vii) HO–C(=O)–C_6H_4–C(=O)–OH (meta)

(viii) O=C(OH)–CH=CH–C(OH)=O (cis)

(ix) HO(O=)C–CH=CH–C(=O)OH (trans)

FIGURE 4.22 Monomers used in the synthesis of unsaturated polyesters (common names are given here since their use is widespread in the literature about these materials). Diols: (i) ethylene glycol (x=2), tetramethylene glycol (x=4); (ii) propylene glycol; (iii) diethylene glycol (y=1), triethylene glycol (y=2); (iv) neopentyl glycol. Saturated aliphatic and aromatic dicarboxylic acids: (v) succinic acid (z=2), adipic acid (z=4), sebacic acid (z=8); (vi) phthalic acid (or its anhydride); (vii) isophthalic acid. Unsaturated dicarboxylic acids: (viii) maleic acid (or its anhydride); (ix) fumaric acid.

$$\sim\!\!\sim\!\!\sim\!-\mathrm{CH_2-C(R)=CH-CH_2}-\!\sim\!\!\sim\!\!\sim \longrightarrow \sim\!\!\sim\!\!\sim\!-\dot{\mathrm{C}}\mathrm{H-C(R)=CH-CH_2}-\!\sim\!\!\sim\!\!\sim \;+\; \sim\!\!\sim\!\!\sim\!-\mathrm{H}$$

$CH_2{=}CR{-}CH{=}CH_2$

The branching and crosslinking arising from copolymerization and chain transfer to polymer needs to be controlled, which is achieved in two ways. Inclusion of a highly active transfer agent (usually a mercaptan) limits the extent to which chains propagate through more than one polymer C=C bond because chain transfer to the transfer agent reduces the probability of a propagating chain growing long enough to react with more than a few polymer C=C bonds. The other important control is simply to restrict the monomer conversion to values low enough (typically around 30%) to limit formation of crosslinked polymer.

PROBLEMS

4.1 Calculate the half-life of benzoyl peroxide in benzene at 66 °C given that the rate coefficient k_d for dissociation of this initiator under these conditions is 8×10^{-6} s^{-1}. What will be the change in initiator concentration after 1 h at 66 °C? Comment on the values obtained.

4.2 In a free-radical polymerization that proceeds in the absence of chain transfer, what would be the effect of (a) increasing the monomer concentration [M] four times at constant initiator concentration [I] and (b) increasing [I] four times at constant [M] upon

(i) the total radical concentration at steady state,
(ii) the rate of polymerization, and
(iii) the number-average degree of polymerization.

4.3 Write down a reaction scheme for polymerization of styrene initiated by thermolysis of azobisisobutyronitrile, including both combination and disproportionation as possible modes of termination.

A sample of polystyrene, prepared by bulk polymerization at 60 °C using radioactive (^{14}C) azobisisobutyronitrile (^{14}C-AIBN) as initiator, was found to have a number-average molar mass $\bar{M}_n$ of 1000 kg mol^{-1} and a radioactivity of 6×10^3 counts s^{-1} g^{-1} (measured using a liquid-scintillation counter). Given that the ^{14}C-AIBN has a radioactivity of 6×10^9 counts s^{-1} mol^{-1}, determine the mode of termination that operated during preparation of the polystyrene sample.

4.4 (a) Calculate the concentration of benzoyl peroxide required to prepare polystyrene with a number-average molar mass of 750 kg mol^{-1} by bulk polymerization at 60 °C. Assume: (i) the initiator is 60% efficient; (ii) the rate coefficients for initiator dissociation, propagation and termination are, respectively, 2×10^{-6} s^{-1}, 341 dm^3 mol^{-1} s^{-1} and 4×10^7 dm^3 mol^{-1} s^{-1}; (iii) the density of styrene at the polymerization temperature is 0.88 g cm^{-3}; and (iv) that termination by disproportionation, and chain transfer to initiator and monomer, are negligible.

(b) Calculate initial values of (i) the radical concentration and (ii) the rate of bulk polymerization expected for bulk polymerization of styrene at the concentration of benzoyl peroxide determined in (a).

4.5 Methyl methacrylate was polymerized at a mass concentration of 200 g dm^{-3} in toluene using azobisisobutyronitrile as initiator at a mass concentration of 1.64×10^{-2} g dm^{-3} and a reaction temperature of 60 °C. Calculate the initial rate of polymerization and the molar mass of the poly(methyl methacrylate) formed in the initial stages of the reaction given that the relevant rate coefficients at 60 °C are:

Initiator dissociation, $k_d = 8.5 \times 10^{-6}$ s^{-1}
Propagation, $k_p = 833$ dm^3 mol^{-1} s^{-1}
Termination, $k_t = 9.3 \times 10^6$ dm^3 mol^{-1} s^{-1}
Transfer to monomer, $k_{trM} = 3.93 \times 10^{-3}$ dm^3 mol^{-1} s^{-1}
Transfer to solvent, $k_{trS} = 7.34 \times 10^{-3}$ dm^3 mol^{-1} s^{-1}

Assume that the initiator efficiency $f = 0.7$, that termination by combination is negligible, and that the density of the initial solution of MMA in toluene is 860 g dm^{-3}.

4.6 Assuming that in free-radical polymerization the rate coefficients for initiator dissociation, propagation and termination can be replaced by the appropriate Arrhenius expressions

$$k_d = A_d \exp(-E_d/\mathbf{R}T)$$

$$k_p = A_p \exp(-E_p/\mathbf{R}T)$$

$$k_t = A_t \exp(-E_t/\mathbf{R}T)$$

calculate the changes in the rate of polymerization and the degree of polymerization caused by increasing the temperature of polymerization of styrene in benzene, initiated by azobisisobutyronitrile, from 60 °C to 70 °C given that:

$$E_p = 34 \text{ kJ mol}^{-1}$$

$$E_t = 10 \text{ kJ mol}^{-1}$$

$$E_d = 126 \text{ kJ mol}^{-1}$$

Assume that the concentration of monomer and initiator and the values of f and q remain unchanged when the temperature is increased.

4.7 During interval II of an emulsion polymerization of styrene at 60 °C, there are 2.7×10^{16} particles dm^{-3} of latex, each latex particle containing styrene monomer at a volume fraction of 0.53. Given that, at reaction temperature, the density and rate coefficient for propagation of styrene are 0.88 g cm^{-3} and 341 dm^3 mol^{-1} s^{-1}, respectively, calculate the rate of polymerization in interval II using the Smith–Ewart Case 2 equation.

4.8 A bulk nitroxide-mediated polymerization of styrene is to be carried out at 125 °C using styryl-TEMPO alkoxyamine as the mediator. Calculate: (a) the concentration styryl-TEMPO alkoxyamine required to produce polystyrene with a number-average degree of polymerization of 100 at 60% conversion; and (b) the time required for the polymerization to attain this conversion, taking into account the persistent-radical effect. The values at 125 °C for the parameters required in the kinetics equations are:

Rate coefficient for propagation, $k_p = 2311$ dm^3 mol^{-1} s^{-1}
Rate coefficient for termination, $k_t = 1.6 \times 10^8$ dm^3 mol^{-1} s^{-1}

Equilibrium constant, $K_{NMP} = 2.1 \times 10^{-11}$ mol dm^{-3}
Density of styrene = 0.84 g cm^{-3}

4.9 State, giving your reasoning, the method of radical polymerization you would use to prepare each of the following polymers.
(a) Poly(vinyl acetate) for use in formulating water-borne adhesives.
(b) Poly(methacrylic acid) with a number-average molar mass of 10 kg mol^{-1} and low molar mass dispersity.
(c) Polyacrylonitrile for use in production of acrylic fibres.
(d) Poly(methyl methacrylate) for use in optical applications.
(e) Low molar mass poly(*t*-butyl methacrylate) with terminal C=C bonds.
(f) Polystyrene for use in solvent-borne adhesives.
(g) Poly(methyl methacrylate) beads for use in formulating bone cements.

FURTHER READING

General Reading

Allen, G. and Bevington, J.C. (eds), *Comprehensive Polymer Science*, Vol. 3 (Eastmond, G.C., Ledwith, A., Russo, S. and Sigwalt, P., eds), Pergamon Press, Oxford, U.K., 1989.
Flory, P.J., *Principles of Polymer Chemistry*, Cornell University Press, Ithaca, NY, 1953.
Gridnev, A.A. and Ittel, S.D., Catalytic chain transfer in free-radical polymerizations, *Chemical Reviews*, **101**, 3611 (2001).
Lenz, R.W., *Organic Chemistry of Synthetic High Polymers*, Interscience, New York, 1967.
Mark, H.F., Bikales, N.M., Overberger, C.G. and Menges, G. (eds), *Encyclopedia of Polymer Science and Engineering*, Wiley-Interscience, New York, 1985–89
Matyjaszewski, K. and Davis, T.P. (ed), *Handbook of Radical Polymerization*, John Wiley & Sons, Hoboken, NJ, 2002.
Moad, G. and Soloman, D.H., *The Chemistry of Radical Polymerization*, 2nd edn., Elsevier, Oxford, 2006.
Odian, G., *Principles of Polymerization*, 4th edn., Wiley-Interscience, New York, 2004.
Stevens, M.P., *Polymer Chemistry: An Introduction*, 3rd edn., Oxford University Press, New York, 1998.

Emulsion Polymerization

Fitch, R.M., *Polymer Colloids: A Comprehensive Introduction*, Academic Press, New York, 1997.
Gilbert, R.G., *Emulsion Polymerization: A Mechanistic Approach*, Academic Press, London, 1995.
Lovell, P.A. and El-Aasser, M.S. (eds), *Emulsion Polymerization and Emulsion Polymers*, John Wiley and Sons, Chichester, U.K., 1997.
van Herk, A.M. (ed), *Chemistry and Technology of Emulsion Polymerisation*, Blackwell, Oxford, U.K., 2005.

Reversible-Deactivation Radical Polymerization

Barner-Kowollik, C., Buback, M., Charleux, B., Coote, M.L., Drache, M., Fukuda, T., Goto, A. et al., Mechanism and kinetics of dithiobenzoate-mediated RAFT polymerization. I. The current situation, *Journal of Polymer Science, Part A: Polymer Chemistry*, **44**, 5809 (2006).
Braunecker, W.A. and Matyjaszewski, K., Controlled/living radical polymerization: features, developments, and perspectives, *Progress in Polymer Science*, **32**, 93 (2007).
Fischer, H., The persistent radical effect in controlled radical polymerizations, *Journal of Polymer Science, Part A: Polymer Chemistry*, **37**, 1885 (1999).
Fischer, H., The persistent radical effect: A principle for selective radical reactions and living radical polymerizations, *Chemical Reviews*, **101**, 3581 (2001).
Goto, A. and Fukuda, T., Kinetics of living radical polymerization, *Progress in Polymer Science*, **29**, 329 (2004).
Hawker, C.J., Bosman, A.W., and Harth, E., New polymer synthesis by nitroxide mediated living radical polymerizations, *Chemical Reviews*, **101**, 3661 (2001).

Kamigaito, M., Ando, T., and Sawamoto, M., Metal-catalyzed living radical polymerization, *Chemical Reviews*, **101**, 3689 (2001).

Matyjaszewski, K. and Xia, J., Atom transfer radical polymerization, *Chemical Reviews*, **101**, 2921 (2001).

Moad, G., Rizzardo, E., and Thang, S.H., Living radical polymerization by the RAFT process, *Australian Journal of Chemistry*, **58**, 379 (2005).

Perrier, S. and Takolpuckdee, P., Macromolecular design via reversible addition: fragmentation chain transfer (RAFT)/xanthates (MADIX) polymerization, *Journal of Polymer Science, Part A: Polymer Chemistry*, **43**, 5347 (2005).

5 Ionic Polymerization

5.1 INTRODUCTION TO IONIC POLYMERIZATION

Chain polymerization of olefinic monomers can also be effected via active centres which possess an ionic charge. There are two types of *ionic polymerization*: those in which the active centre is positively charged are termed *cationic polymerizations* and those in which it is negatively charged are termed *anionic polymerizations*. Because the active centre has an ionic charge, these polymerizations are more monomer-specific than radical polymerization and will proceed only with monomers that have substituent groups which can stabilize the active centre (e.g. by inductive and/or mesomeric effects). For cationic active centres of the type $\sim\sim CH_2-\overset{+}{C}HX$, polymerization will proceed if the substituent group X is able to donate electrons and/or delocalize the positive charge. However, for polymerization via anionic active centres (e.g. $\sim\sim CH_2-\overset{-}{C}HX$), the substituent group must be able to withdraw electrons and/or delocalize the negative charge. Thus, although most monomers are susceptible to radical polymerization, cationic and anionic polymerization tend to be mutually exclusive for a given monomer (see Table 5.1). Only when the substituent group has a weak inductive effect and is capable of delocalizing both positive and negative charges will a monomer undergo both cationic and anionic polymerization (e.g. styrene and 1,3-dienes).

There are further important distinctions between free-radical and ionic polymerizations. For example, many ionic polymerizations proceed at very much higher rates than is usual for free-radical polymerization, largely because the concentration of actively propagating chains is very much higher (typically by a factor of 10^4–10^6). A further difference is that a propagating ionic active centre is accompanied by a *counter-ion* of opposite charge. Both the rate of polymerization and the stereochemistry of propagation are influenced by the counter-ion and its degree of association with the active centre. Thus, the polarity of a polymerization solvent and its ability to solvate the counter-ion can have very significant effects upon ionic polymerizations. Furthermore, termination cannot occur by reaction between two ionic active centres because they are of similar charge and, hence, repel each other.

5.2 CATIONIC POLYMERIZATION

Cationic polymerizations have been investigated since the early days of polymer science, but compared to radical polymerizations, they have found relatively few applications. This is, in part, due to the restricted range of monomers that can be used, but also because in conventional cationic polymerizations the very highly reactive carbocation active sites show a strong tendency to undergo chain transfer and rearrangement reactions that limit the degree of polymerization and make it more difficult to control skeletal structure. These issues are discussed in the following section before considering more recent developments, which have led to better control through use of initiation systems that give more stable active centres.

5.2.1 Conventional Cationic Polymerizations

Conventional cationic polymerizations are chain reactions that can be considered in terms of initiation, propagation, chain transfer and termination processes.

TABLE 5.1
Susceptibility of Different Classes of Olefinic Monomer to Radical, Cationic and Anionic Polymerization[a]

Monomer	Structure[b]	Radical	Cationic	Anionic
Ethylene	$CH_2{=}CH_2$	✓	✓	×
1-Alkyl olefins[c] (e.g. propylene, $R_1=CH_3$)	$CH_2{=}C(H)(R_1)$	(×)	(×)	×
1,1-Dialkyl olefins (e.g. isobutylene, $R_1=R_2=CH_3$)	$CH_2{=}C(R_1)(R_2)$	(×)	✓	×
1,3-Dienes (e.g. butadiene, R = H; isoprene, $R=CH_3$)	$CH_2{=}CH{-}CR{=}CH_2$	✓	✓	✓
Styrene (R = H), α-methyl styrene ($R=CH_3$)	$CH_2{=}C(R)(Ph)$	✓	✓	✓
Vinyl halides (e.g. vinyl chloride, X = Cl)	$CH_2{=}C(H)(X)$	✓	×	(×)
Vinyl ethers (e.g. ethyl vinyl ether, $R_1=CH_2CH_3$)	$CH_2{=}CH{-}O{-}R_1$	(×)	✓	×
Vinyl esters (e.g. vinyl acetate, $R_1=CH_3$)	$CH_2{=}CH{-}O{-}C({=}O){-}R_1$	✓	×	(×)
Acrylates (R = H), methacrylates ($R=CH_3$) (e.g. *n*-butyl acrylate, R = H, $R_1=CH_2CH_2CH_2CH_3$; methyl methacrylate, $R=R_1=CH_3$)	$CH_2{=}CR{-}C({=}O){-}O{-}R_1$	✓	×	✓
Acrylamide (R = H), methacrylamide ($R=CH_3$)	$CH_2{=}CR{-}C({=}O){-}NH_2$	✓	×	(×)
Acrylonitrile (R = H), Methacrylonitrile ($R=CH_3$)	$CH_2{=}CR{-}C{\equiv}N$	✓	×	✓

[a] ✓ means susceptible, × means not susceptible, and (×) indicates that whilst the monomer might be expected to polymerize (e.g. on the basis of inductive and/or mesomeric effects) it does not do so successfully because of side reactions (e.g. chain transfer to monomer).
[b] R_1 and R_2 = alkyl, R = H or CH_3, Ph = phenyl, and X = halogen.
[c] Also called α-olefins.

5.2.1.1 Initiation of Cationic Polymerization

Cationic active centres are created by reaction of monomer with electrophiles (e.g. R^+). Protonic acids such as sulphuric acid (H_2SO_4) and perchloric acid ($HClO_4$) can be used as initiators and act by addition of a proton (H^+) to monomer. However, hydrogen halide acids (e.g. HCl or HBr) are not suitable as initiators because the halide counter-ion rapidly combines with the carbocationic active centre to form a stable covalent bond. Lewis acids such as boron trifluoride (BF_3), aluminium chloride ($AlCl_3$) and tin tetrachloride ($SnCl_4$) are the most important 'initiators' but must be used in conjunction with a so-called co-catalyst which can be water, but more commonly is an organic halide

$$BF_3 + H_2O \rightleftharpoons H^+(BF_3OH)^-$$

$$AlCl_3 + RCl \rightleftharpoons R^+(AlCl_4)^-$$

For the BF_3/H_2O system, the second step of initiation for a monomer of general structure $CH_2{=}CR_1R_2$ involves both electrons from the π-bond joining with the H^+ ion to create a new H–CH_2 bond, thereby leaving an empty orbital (i.e. a positive charge) on the carbon bearing the substituents with which the counter-ion then becomes associated. This is shown below

$$H^+(BF_3OH)^- + CH_2{=}C(R_1)(R_2) \longrightarrow H{-}CH_2{-}C^+(R_1)(R_2)\ (BF_3OH)^-$$

More generally, the second step of initiation can be represented by

$$\overset{+}{R}\overset{-}{A} + CH_2{=}C(R_1)(R_2) \longrightarrow R{-}CH_2{-}C^+(R_1)(R_2)\ A^-$$

where R^+ is the electrophile and A^- is the counter-ion.

The term 'co-catalyst' (as used above) is common in the literature on cationic polymerization, but nevertheless is poor. It is clear from the chemistry of initiation that the 'co-catalyst' is in fact the initiator and that the Lewis acid behaves as an activator which also provides a more stable counter-ion.

5.2.1.2 Propagation in Cationic Polymerization

For reasons of carbocation stability (cf. radical polymerization, Sections 4.2.1 and 4.2.2), propagation proceeds predominantly via successive head-to-tail additions of monomer to the active centre.

$$\sim\!CH_2{-}C^+(R_1)(R_2)\ A^- + CH_2{=}C(R_1)(R_2) \longrightarrow \sim\!CH_2{-}C(R_1)(R_2){-}CH_2{-}C^+(R_1)(R_2)\ A^-$$

5.2.1.3 Termination and Chain Transfer in Cationic Polymerization

Unlike in radical polymerization, for which the active centres bear no charge, termination by reaction together of two propagating chains is not possible for ionic polymerizations due to charge repulsion. Instead, growth of individual chains terminates most commonly by either unimolecular rearrangement of the ion pair in which a penultimate C–H bond breaks to release H^+ and generate a terminal C=C bond in the dead molecule of polymer, e.g.

$$\sim\!CH(H){-}C^+(R_1)(R_2)\ A^- \longrightarrow \sim\!CH{=}C(R_1)(R_2) + \overset{+}{H}\overset{-}{A}$$

or chain transfer. Chain transfer to monomer often makes a significant contribution and involves abstraction of a H-atom from the penultimate C–H bond by monomer, resulting in formation of a monomeric carbocation

$$\sim\!CH(H){-}C^+(R_1)(R_2)\ A^- + CH_2{=}C(R_1)(R_2) \longrightarrow \sim\!CH{=}C(R_1)(R_2) + H{-}CH_2{-}C^+(R_1)(R_2)\ A^-$$

In either case, the resulting electrophile becomes associated with the counter-ion A^- and starts the growth of a new polymer chain with little effect on the concentration of actively propagating chains. The terminal C=C bond in the dead polymer chain is of low reactivity due to steric hindrance to reaction arising from 1,2-disubstitution (and the presence of three substituents if both R_1 and $R_2 \neq H$) and so does not easily participate in the polymerization (i.e. it tends to remain unreacted).

Chain transfer to solvent and reactive impurities (e.g. H_2O) can further limit the degree of polymerization, so the choice of solvent and the purity of the reactants is of great importance. Chain transfer to polymer also is significant and results in the formation of branched polymers via chemistry analogous to that described for radical polymerization (see Section 4.2.4.2).

5.2.1.4 Kinetics of Conventional Cationic Polymerization

The exact mechanism of cationic polymerization often depends upon the type of initiator, the structure of the monomer and the nature of the solvent. Often the reaction is heterogeneous because the initiator is only partially soluble in the reaction medium. These features make the formulation of a general kinetics scheme somewhat difficult. Nevertheless, a kinetics scheme based upon the chemistry given in the previous three sections will be analysed, though of course the equations obtained must be applied with discretion.

The following general kinetics scheme will be assumed:

Initiation: $$\overset{+}{R}\overset{-}{A} + M \xrightarrow{k_i} RM_1^{+}\overset{-}{A}$$

Propagation: $$RM_n^{+}\overset{-}{A} + M \xrightarrow{k_p} RM_{n+1}^{+}\overset{-}{A}$$

Ion-pair rearrangement: $$RM_n^{+}\overset{-}{A} \xrightarrow{k_t} RM_n + \overset{+}{H}\overset{-}{A}$$

Chain transfer to monomer: $$RM_n^{+}\overset{-}{A} + M \xrightarrow{k_{trM}} RM_n + HM_1^{+}\overset{-}{A}$$

Assuming that the amount of monomer consumed in initiation is negligible (i.e. long chains are formed), the rate of polymerization is given by

$$R_p = -\frac{d[M]}{dt} = k_p[M][M^+] \tag{5.1}$$

where $[M^+] = \sum_{n=1}^{\infty}[RM_n^+]$. Applying the steady-state condition $d[M^+]/dt = 0$ gives

$$\frac{d[M^+]}{dt} = k_i[R^+A^-][M] - k_t[M^+] = 0$$

which can be rearranged to

$$[M^+] = \left(\frac{k_i}{k_t}\right)[R^+A^-][M]$$

Substitution of this relation into Equation 5.1 gives an expression for the steady-state rate of polymerization

$$R_p = \left(\frac{k_i k_p}{k_t}\right)[R^+A^-][M]^2 \tag{5.2}$$

which has a second-order dependence upon [M] resulting directly from the usually satisfactory assumption that the second step of initiation is rate-determining (cf. free-radical polymerization, Section 4.3.1), i.e. that R^+A^- is present in its fully active form at the start of the polymerization. (Note, however, that under conditions where the formation of R^+A^- is rate-determining for initiation, application of the steady-state condition will lead to a first-order dependence of R_p upon [M].)

If the ionic product of termination by ion-pair rearrangement or chain transfer to monomer is capable of rapidly initiating polymerization, then $[M^+]=[R^+A^-]$ and $R_p=k_p[M][R^+A^-]$.

For each of the above kinetics possibilities, the equilibrium constant for formation of R^+A^- must be taken into account when evaluating $[R^+A^-]$.

Analysis for the number-average degree of polymerization $\bar{x}_n$ is somewhat simpler. Equation 4.13 of Section 4.3.2 again is valid and leads to

$$\bar{x}_n = \frac{k_p[M][M^+]}{k_t[M^+]+k_{trM}[M^+][M]}$$

for the above kinetics scheme where consumption of monomer by processes other than propagation is assumed to be negligible. Notice that the $[M^+]$ term divides out, i.e. $\bar{x}_n$ is independent of the concentration of propagating chains. Inversion and simplification of the equation gives

$$\frac{1}{\bar{x}_n} = \frac{k_t}{k_p[M]} + \frac{k_{trM}}{k_p} \tag{5.3}$$

This is the cationic polymerization equivalent of Equation 4.18. Additional terms can be added to the right-hand side to take account of other chain transfer reactions, e.g. for chain transfer to solvent the additional term is $k_{trS}[S]/k_p[M]$. In each case, $\bar{x}_n$ is independent of initiator concentration and in the absence of chain transfer is given by $(\bar{x}_n)_0=(k_p/k_t)[M]$.

5.2.1.5 Effect of Temperature

Following the method of analysis used in Section 4.3.10, appropriate Arrhenius expressions can be substituted into Equation 5.2. Taking logarithms and then differentiating with respect to temperature gives

$$\frac{d\ln(R_p)}{dT} = \frac{E_i+E_p-E_t}{\mathbf{R}T^2} \tag{5.4}$$

The substituent group effects which give rise to stabilization of the cationic active centre also cause polarization of the C=C bond in the monomer, activating it towards attack by electrophiles. Thus, the activation energies E_i and E_p are relatively small and individually must be less than E_t (and also E_{trM}) if polymer is formed. In view of this, the overall activation energy $E_i+E_p-E_t$ can be positive or negative depending upon the particular system. When the overall activation energy is negative, R_p is observed to increase as the temperature is reduced (e.g. polymerization of styrene in 1,2-dichloroethane using $TiCl_4/H_2O$ as the initiator).

A similar analysis for $(\bar{x}_n)_0$ gives

$$\frac{d\ln(\bar{x}_n)_0}{dT} = \frac{E_p-E_t}{\mathbf{R}T^2} \tag{5.5}$$

Since $E_p < E_t$, then $\mathrm{d}\ln(\bar{x}_n)_0/\mathrm{d}T$ must be negative and so $(\bar{x}_n)_0$ decreases as temperature increases. Analysis of the transfer constants ($C_{tr} = k_{tr}/k_p$) gives the same equation as for free-radical polymerization

$$\frac{\mathrm{d}\ln(C_{tr})}{\mathrm{d}T} = \frac{E_{tr} - E_p}{\mathbf{R}T^2}$$

and since $E_{tr} > E_p$ the transfer constants increase with increasing temperature and so the effect of chain transfer in reducing $\bar{x}_n$ becomes more significant as temperature increases. Thus, $\bar{x}_n$ always decreases if the reaction temperature is increased.

5.2.1.6 Solvent and Counter-Ion Effects

The various states of association between an active centre and its counter-ion can be represented by the following equilibria

$$\underset{\textit{Covalent bond}}{RM_n{-}A} \rightleftharpoons \underset{\textit{Contact ion-pair}}{RM_n^{+}A^{-}} \rightleftharpoons \underset{\textit{Solvent-separated ion-pair}}{RM_n^{+}\,||\,A^{-}} \rightleftharpoons \underset{\textit{Free ions}}{RM_n^{+} + A^{-}}$$

Free carbocationic active centres propagate faster than the contact ion pairs (typically by an order of magnitude) and so as the equilibria shift to the right the overall rate coefficient k_p for propagation increases. Thus, polar solvents which favour ion-pair separation (e.g. dichloromethane) and larger counter-ions (e.g. $SbCl_6^-$), which associate less strongly with cations, give rise to higher values of k_p. Furthermore, an increase in the degree of dissociation of the ions can result in a reduction in the overall rate coefficient k_t for ion-pair rearrangement.

An interesting feature of certain so-called *pseudocationic polymerizations* is that they involve propagation of polarized covalently bonded species, such as in the polymerization of styrene initiated by perchloric acid where propagation is thought to proceed via a rearrangement reaction involving a cyclic transition state.

$$\sim\!CH_2{-}\overset{\delta+}{C}HPh{-}O{-}ClO_2{-}\overset{\delta-}{O} \;\cdots\; \overset{\delta-}{C}H_2{=}\overset{\delta+}{C}HPh \longrightarrow \sim\!CH_2{-}CHPh{-}CH_2{-}CHPh{-}O{-}ClO_3$$

The kinetics scheme analysed in Section 5.2.1.4 naturally leads to formation of polymers with a most probable distribution of molar mass (cf. Section 4.3.8). However, polymers formed by cationic polymerization sometimes have complex (e.g. bimodal) molar mass distributions, which is thought to be due to simultaneous growth of different types of active species that interchange slowly compared to the timescale for propagation and have very different but characteristic values of the rate coefficients for propagation and ion-pair rearrangement.

5.2.1.7 Practical Considerations

The propensity for reactions which terminate the growth of propagating chains greatly restricts the usefulness of conventional cationic polymerization. Usually, it is necessary to perform the reactions below 0 °C in order to produce polymers with suitably high molar masses. If water is employed as an initiator, then it should only be used in stoichiometric quantities with the Lewis acid activator, otherwise chain transfer to water will result in a substantial decrease in molar mass. Thus, all reactants and solvents should be rigorously dried and purified to remove impurities which could take part in chain transfer reactions. Many cationic polymerizations are extremely rapid because the initiator is present in its fully active state from the outset (i.e. [M^+] is high and far greater than [$M^\bullet$] in radical polymerizations); complete conversion of the monomer can be achieved in a matter of seconds. This creates additional problems with respect to the operation of non-steady-state conditions and the control of heat transfer.

The most important monomers are isobutylene, styrene and alkyl vinyl ethers. The latter are highly reactive in cationic polymerization due to mesomeric stabilization of the carbocation by delocalization of the positive charge onto the oxygen atom and despite the electron-withdrawing (-I inductive) effect of the OR side group.

The only cationic polymerization of major technological importance is the synthesis of butyl rubber by copolymerization of isobutylene with small proportions of isoprene at low temperature (+10 °C to −100 °C) using an alkyl halide initiator (e.g. CH_3Cl) with a Lewis acid activator (typically $AlCl_3$) and chlorinated solvents (e.g. CH_3Cl can also be used as the solvent). Isoprene is included because it provides repeat units that enable the (essentially polyisobutylene) rubber to be cross-linked using conventional sulphur vulcanization (Section 21.1.2). Molar mass is controlled mainly by using the polymerization temperature to moderate the extent of chain transfer to monomer, with the highest molar mass polymer being produced at the lowest temperature.

5.2.2 Reversible-Deactivation (Living) Cationic Polymerizations

As discussed in the previous sections, the prevalence of chain transfer and rearrangement reactions in conventional cationic polymerizations makes control of the reactions more difficult and limits the degree of polymerization that can be achieved. Chain transfer to monomer is particularly problematic due to a combination of the high acidity of the penultimate CH_2 hydrogen atoms, arising from the formal positive charge on the adjacent terminal carbocation, and the nucleophilicty of the monomers that are susceptible to cationic polymerization. The discovery of pseudocationic polymerizations (Section 5.2.1.6) indicated that it may be possible to suppress this effect by greatly reducing the charge associated with the terminal carbon atom whilst still allowing propagation to proceed. This requires the active site and the counter-ion to associate closely in bonds that have covalent character, but which are weak enough to dissociate briefly to allow short bursts of propagation. Approaches to achieving this balance were developed during the 1980s, and some are outlined in the following section. Even though chain transfer and rearrangement reactions are not completely eliminated, these cationic polymerizations show characteristics of *living polymerization* (Section 5.3.2) in that the chains grow further if more monomer is added. Significantly, they provided an inspiration for the development of approaches to minimizing contributions from termination in radical polymerizations (Section 4.5) and although they often are referred to as *living cationic polymerizations*, here they have been termed *reversible-deactivation cationic polymerizations* due to the close parallels with reversible-deactivation radical polymerization and the fact that they are not truly living.

5.2.2.1 Initiation Systems for Reversible-Deactivation Cationic Polymerization

The approaches to achieving reversible-deactivation cationic polymerization may be divided into two broad categories, though specific initiation systems need to be designed for each monomer in order to achieve the necessary control. The first approach involves activation of terminal C–X covalent bonds by addition of an electrophilic species, where X most often is a halogen or oxygen atom. This was first demonstrated experimentally using hydrogen iodide as initiator and iodine as the activator in polymerization of alkyl vinyl ethers.

$$CH_2{=}CH(OR) \xrightarrow{HI} H{-}CH_2{-}CH(OR){-}I \xrightarrow{I_2} H{-}CH_2{-}\overset{\delta+}{C}H(OR)\cdots\overset{\delta-}{I}\cdots I_2 \xrightarrow{n\ CH_2=CHOR} H{+}CH_2{-}CH(OR){+}_n CH_2{-}\overset{\delta+}{C}H(OR)\cdots\overset{\delta-}{I}\cdots I_2$$

It was shown subsequently that better control is achieved using ZnI_2 as the activator. Living-like polymerization of isobutylene can be achieved in a similar way using initiating species produced by reaction of tertiary esters, ethers, or alcohols with BCl_3, e.g.

$$Ph{-}C(CH_3)_2{-}O{-}C(=O)CH_3 \xrightarrow{BCl_3} Ph{-}\overset{\delta+}{C}(CH_3)_2\cdots \overset{\delta-}{O}{-}C(CH_3){=}O\cdots BCl_3 \xrightarrow{n\ CH_2=C(CH_3)_2} Ph{-}C(CH_3)_2{+}CH_2{-}C(CH_3)_2{+}_{n-1}CH_2{-}\overset{\delta+}{C}(CH_3)_2\cdots\overset{\delta-}{O}{-}C(CH_3){=}O\cdots BCl_3$$

The other generic approach to achieving living-like conditions makes use of conventional Lewis acid initiation systems (Section 5.2.1.1) but with addition of a Lewis base which modifies the nature of the interaction between the carbocation active centre and the counter-ion. Two examples will be given to illustrate this approach. Polymerization of isobutylene using a tertiary chloride initiator with BCl_3 as the Lewis acid activator and dimethyl sulphoxide as the Lewis base moderator is thought to involve the following stable intermediates

$$^tBu{-}C(CH_3)_2{-}Cl \xrightarrow[(CH_3)_2S=O]{BCl_3} {}^tBu{-}\overset{\delta+}{C}(CH_3)_2\cdots\overset{\delta-}{O}(=S(CH_3)_2)\cdots Cl{-}BCl_3 \xrightarrow{n\ CH_2=C(CH_3)_2} {}^tBu{-}C(CH_3)_2{+}CH_2{-}C(CH_3)_2{+}_{n-1}CH_2{-}\overset{\delta+}{C}(CH_3)_2\cdots\overset{\delta-}{O}(=S(CH_3)_2)\cdots Cl{-}BCl_3$$

Alkyl vinyl ethers can be polymerized using an ester as the initiator, $AlEtCl_2$ as the Lewis acid activator and 1,4-dioxane as the Lewis base moderator, for which the stable intermediates are believed to be

$$CH_3{-}CH(O^iBu){-}O{-}C(=O)CH_3 \xrightarrow[\text{1,4-dioxane}]{AlEtCl_2} CH_3{-}\overset{\delta+}{C}H(O^iBu)\cdots O(\text{dioxane})\cdots\overset{\delta-}{O}{-}C(CH_3){=}O\cdots AlEtCl_2$$

$$\xrightarrow{n\ CH_2=CHOR} CH_3{-}CH(O^iBu){+}CH_2{-}CH(OR){+}_{n-1}CH_2{-}\overset{\delta+}{C}H(OR)\cdots O(\text{dioxane})\cdots\overset{\delta-}{O}{-}C(CH_3){=}O\cdots AlEtCl_2$$

In both examples, electron donation from the lone pairs of electrons on the oxygen atom of the Lewis base generates a positively-charged oxonium ion which acts to stabilize (i.e. reduce the charge at) the terminal carbon atom of the chain. In the absence of the Lewis base, cationic polymerization proceeds rapidly in an uncontrolled way.

Propagation is believed to be dominated by addition of monomer to carbocation active centres (most likely in contact or solvent-separated ion pairs) that are in equilibrium with the stable (dormant) species shown in the reaction schemes above. Hence, the dormant species can be considered to be on the leftmost ('covalent') side of the equilibrium shown in Section 5.2.1.6 and to be the most abundant species present. Nevertheless, exchange between the different states of the active centre is rapid and when terminal carbocation species are released, albeit for very short periods of time, they propagate and can undergo the chain transfer and rearrangement reactions described in Section 5.2.1.3. However, averaged over time, the chain ends exist mostly in the dormant state, which protects the active centre and gives the polymerizations a living-like nature in which the degree of polymerization increases approximately linearly with monomer conversion, allowing polymers of narrow molar mass distribution to be prepared if the initiation step proceeds at a rate similar to or greater than the rate of propagation (Section 5.3.2.4).

5.2.2.2 Kinetics of Reversible-Deactivation Cationic Polymerizations

The kinetics of reversible-deactivation cationic polymerizations often do not conform with the kinetics expected for truly living polymerizations (Section 5.3.2.3), though in favourable cases the number-average degree of polymerization is determined simply by c[M]/[I], where c is the fractional monomer conversion and [I] is the initiator concentration. It is not unusual for the rate of polymerization to be independent of monomer concentration, particularly when the rate is controlled by the rate of equilibration between the dormant and active species.

5.2.2.3 Practical Considerations

The factors discussed in Sections 5.2.1.6 and 5.2.1.7 also are important when planning the conditions to be used for reversible-deactivation cationic polymerization. Solvents need to be chosen carefully, not just in terms of their effect on the equilibrium between active and dormant species, but also with consideration of their ability to interfere with the dormant species. For example, chlorinated solvents (e.g. CH_3Cl and CH_2Cl_2) can react with dormant active centres generated from oxygen-based initiators to give chloride-based counter-ions, which may cause uncontrolled polymerization depending on the monomer and reaction conditions

$$\text{Ph}-\underset{CH_3}{\overset{CH_3}{C}}\!\left[CH_2-\underset{CH_3}{\overset{CH_3}{C}}\right]_{n-1}\!CH_2-\underset{CH_3}{\overset{CH_3}{\overset{\delta+}{C}}}\cdots\overset{\delta-}{O}-\underset{CH_3}{C}(=O\cdots BCl_3) \xrightarrow{CH_2Cl_2} \text{Ph}-\underset{CH_3}{\overset{CH_3}{C}}\!\left[CH_2-\underset{CH_3}{\overset{CH_3}{C}}\right]_{n-1}\!CH_2-\underset{CH_3}{\overset{CH_3}{C^{+}}}\,[BCl_4]^{-}$$

As for conventional cationic polymerizations, it is normal to perform the reactions at low temperatures, typically –30 °C to –80 °C.

5.3 ANIONIC POLYMERIZATION

Anionic polymerizations were studied in the early days of polymer science, but at that time strong bases were used as initiators with solvents that readily undergo chain transfer, greatly limiting the degree of polymerization that could be achieved. The opportunities provided by the absence of inherent termination processes in anionic polymerization were not realized until initiation systems and conditions for effecting living anionic polymerization were established in the mid-1950s. Such polymerizations have since become very important for reasons that will be discussed here in some detail. However, before examining living anionic polymerization, polymerization of styrene in liquid ammonia, initiated by potassium amide will be considered briefly because, although it is of minor importance nowadays, it was one of the first anionic polymerizations to be studied in detail and it provides a further example of kinetics analysis.

5.3.1 Polymerization of Styrene in Liquid NH_3 Initiated by KNH_2

Initiation involves dissociation of potassium amide followed by addition of the amide ion to styrene. Termination occurs by proton abstraction from ammonia (i.e. chain transfer to solvent) and so living polystyrene chains are not formed. Hence, the kinetics scheme to be analysed is

Initiation:
$$KNH_2 \rightleftharpoons K^+ + \bar{N}H_2$$
$$\bar{N}H_2 + CH_2{=}CHPh \xrightarrow{k_i} H_2N{-}CH_2{-}\bar{C}HPh$$

Propagation:
$$H_2N{-}\!\left[CH_2{-}CHPh\right]_{n-1}\!{-}CH_2{-}\bar{C}HPh + CH_2{=}CHPh \xrightarrow{k_p} H_2N{-}\!\left[CH_2{-}CHPh\right]_{n}\!{-}CH_2{-}\bar{C}HPh$$

Chain transfer to solvent (NH_3):
$$H_2N{-}\!\left[CH_2{-}CHPh\right]_{n}\!{-}CH_2{-}\bar{C}HPh + H{-}NH_2 \xrightarrow{k_{trS}} H_2N{-}\!\left[CH_2{-}CHPh\right]_{n}\!{-}CH_2{-}CHPh{-}H + \bar{N}H_2$$

The second step of initiation is rate-determining and so the amide ion produced upon chain transfer to ammonia can initiate polymerization, but at a rate controlled by the rate coefficient k_i for initiation. Therefore, it is normal to consider this chain transfer reaction as a true kinetic-chain termination step so that application of the steady-state condition $d[\sim\!\sim CH_2{-}\bar{C}HPh]/dt=0$ gives

$$k_i[\bar{N}H_2][CH_2{=}CHPh] - k_{trS}[\sim\!\sim CH_2{-}\bar{C}HPh][NH_3]=0$$

which upon rearrangement yields

$$[\sim\!\sim CH_2{-}\bar{C}HPh]=\left(\frac{k_i}{k_{trS}}\right)\frac{[\bar{N}H_2][CH_2{=}CHPh]}{[NH_3]}$$

Thus, assuming that the initiation step consumes a negligible amount of monomer, the steady-state rate of polymerization

$$R_p=-\frac{d[CH_2{=}CHPh]}{dt}=k_p[\sim\!\sim CH_2{-}\bar{C}HPh][CH_2{=}CHPh]$$

is given by

$$R_p=\left(\frac{k_ik_p}{k_{trS}}\right)\frac{[\bar{N}H_2][CH_2{=}CHPh]^2}{[NH_3]} \tag{5.6}$$

Analysis for the number-average degree of polymerization using Equation 4.13 leads to

$$\bar{x}_n=\frac{k_p[CH_2{=}CHPh]}{k_{trS}[NH_3]} \tag{5.7}$$

Examination of Equations 5.6 and 5.7 for the effects of temperature yields equations which are identical in form to Equations 5.4 and 5.5 of Section 5.2.1.5, but in which E_t is replaced by the activation energy E_{trS} for chain transfer to ammonia. Reducing the reaction temperature increases $\bar{x}_n$ (because $E_p - E_{trS} \approx -17$ kJ mol^{-1}) but decreases R_p (because $E_i + E_p - E_{trS} \approx +38$ kJ mol^{-1}). Due to the small magnitude of $E_p - E_{trS}$ and the high concentration of ammonia, chain transfer to ammonia is highly competitive with propagation and only low molar mass polystyrene is formed, even at low temperatures.

5.3.2 Polymerization without Termination—Living Anionic Polymerization

An important feature of anionic polymerization is the absence of inherent termination processes. As for cationic polymerization, bimolecular termination between two propagating chains is not possible due to charge repulsion, but unlike cationic polymerization, termination by ion-pair rearrangement does not occur because it requires the highly unfavourable elimination of a hydride ion. Furthermore, the alkali metal (or alkaline earth metal) counter-ions used have no tendency to combine with the carbanionic active centres to form unreactive covalent bonds. Thus, in the absence of chain transfer reactions, the propagating polymer chains retain their active carbanionic end groups and are truly *living*, unlike reversible-deactivation radical and cationic polymerizations (often referred to as 'living') in which termination is still possible (see Sections 4.5 and 5.2.2). Interest in anionic polymerization grew enormously following the work of Szwarc in the mid-1950s. He demonstrated that, under carefully controlled conditions, carbanionic living polymers could be formed using electron transfer initiation. If more monomer is added after complete conversion of the initial quantity, the chains grow further by polymerization of the additional monomer and will again remain active. Such polymer molecules, which permanently retain their active centres (of whatever type) in chain polymerization and continue to grow so long as monomer is available, are termed *living polymers*.

5.3.2.1 Organometallic Initiators for Living Anionic Polymerization

Organolithium compounds (e.g. butyllithium) are the most widely used organometallic initiators. They are soluble in non-polar hydrocarbons but tend to aggregate in such media, i.e.

$$\{\text{RLi}\}_n \rightleftharpoons n\,\text{RLi}$$

where $n=6$ for ^{n}BuLi and $n=4$ for ^{s}BuLi and ^{t}BuLi. The reactivity of the aggregated species is very much lower than that of the free species and it is the latter which principally are responsible for initiation, e.g.

$$^s\text{Bu}^{-}\ \text{Li}^{+} + \text{CH}_2{=}\text{CR}_1\text{R}_2 \longrightarrow {}^s\text{Bu{-}CH}_2{-}\overset{-}{\text{C}}\text{R}_1\text{R}_2\,\text{Li}^{+}$$

Disaggregation can be brought about by the addition of small quantities of polar solvents which are able to solvate the lithium ions, e.g. tetrahydrofuran (THF). If the reaction is performed in polar solvents such as THF, then initiation is rapid because the organolithium compounds exist as free species with more strongly developed ionic character than in non-polar media.

Organometallic compounds of other alkali metals (e.g. benzylsodium and cumylpotassium) are insoluble in non-polar hydrocarbons and so are used only for reactions performed in polar media.

Propagation proceeds in the usual way with predominantly head-to-tail additions of monomer due to steric and electronic effects on the stability of the carbanionic active centre

$$\sim\text{CH}_2{-}\overset{-}{\text{C}}\text{R}_1\text{R}_2\text{Mt}^{+} + \text{CH}_2{=}\text{CR}_1\text{R}_2 \longrightarrow \sim\text{CH}_2{-}\text{CR}_1\text{R}_2{-}\text{CH}_2{-}\overset{-}{\text{C}}\text{R}_1\text{R}_2\text{Mt}^{+}$$

where Mt^+ is the metallic counter-ion.

5.3.2.2 Electron Transfer Initiation for Living Anionic Polymerization

Electron transfer initiation involves donation of single electrons to molecules of monomer to form monomeric radical-anion species which then couple together to give dicarbanion species that initiate polymerization of the remaining monomer. Whilst alkali metals can be used as electron donors, they are insoluble in most organic solvents so the reaction is heterogeneous and initiation is slow and difficult to reproduce. Homogeneous initiation by electron transfer can be achieved in ether solvents, such as THF, using soluble electron-transfer complexes formed by reaction of alkali metals with aromatic compounds (e.g. naphthalene, biphenyl). Sodium naphthalide (also known as naphthalenide) was one of the first to be used and is formed by the addition of sodium to an ethereal solution of naphthalene. An atom of sodium donates an electron to naphthalene producing the naphthalide radical-anion, which is green in colour and stabilized by resonance, though only one canonical form is shown below

Na + [naphthalene] ⇌ $\overset{+}{Na}$ [naphthalide radical-anion]

Naphthalide radical-anion

The position of the equilibrium depends upon the solvent; in THF it lies completely to the right. The complex initiates polymerization by donation of an electron to monomer

$\overset{+}{Na}$ [naphthalide] + $CH_2{=}CR_1R_2$ ⇌ $\dot{C}H_2{-}\overset{..}{\bar{C}}R_1R_2\,\overset{+}{Na}$ + [naphthalene]

and this equilibrium is shifted to the right by rapid coupling of the monomeric radical anions

$$2\ \dot{C}H_2{-}\overset{..}{\bar{C}}R_1R_2\,\overset{+}{Na} \longrightarrow \overset{+}{Na}\,R_2R_1\overset{..}{\bar{C}}{-}CH_2{-}CH_2{-}\overset{..}{\bar{C}}R_1R_2\,\overset{+}{Na}$$

The green colour rapidly disappears and is replaced by the colour of the carbanionic active centres (e.g. for polymerization of styrene, $R_1{=}H$ and $R_2{=}Ph$, the colour is red). Propagation then proceeds by sequential addition of monomer to both ends of the dicarbanionic species.

5.3.2.3 Kinetics of Living Anionic Polymerization

Regardless of whether the initiator is an organometallic compound or an electron-transfer complex, each initiator species generates a single carbanionic active centre. It is usual to assume that the initiator reacts completely before any of the active centres begin to propagate, i.e. that all of the initiator species exist in active form free in solution and that the rate coefficient k_i for initiation is very much greater than the rate coefficient k_p for propagation. Under these conditions, the total concentration of propagating carbanionic active centres is equal to the concentration $[I]_0$ of initiator used. Thus, in the absence of termination reactions

$$R_p = -\frac{d[M]}{dt} = k_p[I]_0[M] \tag{5.8}$$

This shows the reaction to be pseudo-first-order since for each particular reaction $k_p[I]_0$ is a constant. Equation 5.8 usually is satisfactory for homogeneous polymerizations performed in polar solvents. However, for reactions carried out in non-polar solvents, the effects of slow initiation and of aggregation of both the initiator species and the carbanionic active centres must be taken into account when evaluating the concentration of propagating carbanionic active centres.

The number-average degree of polymerization is given by a slight modification of Equation 4.13 in recognition of the living nature of the polymerization

$$\bar{x}_n = \frac{\text{moles of monomer consumed}}{\text{moles of polymer chains produced}} \tag{5.9}$$

Assuming that no active sites are lost, that is, true living conditions operate, application of Equation 5.9 when the fractional conversion of monomer is c gives

$$\bar{x}_n = \frac{cK[\mathrm{M}]_0}{[\mathrm{I}]_0} \tag{5.10}$$

where

$K = 1$ for initiation by organometallic compounds

$K = 2$ for electron transfer initiation

$[\mathrm{M}]_0$ and $[\mathrm{I}]_0$ are the initial concentrations of monomer and initiator, respectively.

Hence, for a given initiator, $\bar{x}_n$ is controlled simply by the ratio $[\mathrm{M}]_0/[\mathrm{I}]_0$ and increases linearly with monomer conversion, attaining the value $K[\mathrm{M}]_0/[\mathrm{I}]_0$ at complete conversion.

For monomers with low ceiling temperatures (e.g. α-methylstyrene; see Section 4.3.10.1) the propagation–depropagation equilibrium must be considered when applying Equation 5.9 since an appreciable monomer concentration may exist at equilibrium (i.e. $c = 1$ may not be achieved).

5.3.2.4 Molar Mass Distributions of Polymers Produced by Living Polymerization

Analysis of chain polymerization without termination for the molar mass distribution of the polymer formed requires a different approach to that used in Sections 3.2.3.2 and 4.3.8 for the analyses of step polymerization and of chain polymerization with termination, respectively. It is necessary to define a rate parameter Φ as the average rate at which a molecule of monomer is added to a single active centre that can be either an initiator species or a propagating chain. The precise form of Φ is immaterial, though it can be expected to contain a rate coefficient and a term in monomer concentration. By using Φ for both the initiation and propagation steps, it is implicitly assumed that $k_i = k_p$. However, provided that long polymer chains are formed, there is no tangible effect upon the predictions of the analysis if $k_i > k_p$.

At the beginning of the polymerization, there are N initiator species, each of which is equally active. After a time t, N_0 of these initiator species have added no molecules of monomer, N_1 have added one molecule of monomer, N_2 have added two molecules of monomer and, in general, N_x have added x molecules of monomer. Since N_0 decreases by one for each reaction between an initiator species and monomer

$$-\frac{\mathrm{d}N_0}{\mathrm{d}t} = \Phi N_0$$

Rearranging this equation and applying integration limits,

$$\int_{N}^{N_0} \frac{\mathrm{d}N_0}{N_0} = -\int_0^t \Phi \mathrm{d}t$$

which by letting $\nu = \int_0^t \Phi \mathrm{d}t$ may be solved to give

$$N_0 = Ne^{-\nu} \tag{5.11}$$

The rate of change in the numbers of all other species (i.e. N_x for $x = 1, 2, 3, \ldots, \infty$) is given by the following general equation

$$\frac{\mathrm{d}N_x}{\mathrm{d}t} = \Phi N_{x-1} - \Phi N_x \tag{5.12}$$

since an x-mer is formed by addition of monomer to an $(x-1)$-mer and is lost by addition of monomer. Recognizing that $\mathrm{d}\nu = \Phi \mathrm{d}t$, then Equation 5.12 can be rearranged to give the following first-order linear differential equation

$$\frac{\mathrm{d}N_x}{\mathrm{d}\nu} + N_x = N_{x-1}$$

the solution of which gives

$$N_x = e^{-\nu} \int_0^{\nu} e^{\nu} N_{x-1} \mathrm{d}\nu \tag{5.13}$$

Substitution of Equation 5.11 into Equation 5.13 yields $N_1 = N\nu e^{-\nu}$, which can then be substituted back into Equation 5.13 to give $N_2 = N(\nu^2/2)e^{-\nu}$, which also can be substituted into Equation 5.13 to give $N_3 = N(\nu^3/6)e^{-\nu}$. This process can be continued for all values of x, but it soon becomes evident that the solutions have the general form

$$N_x = \frac{N\nu^x e^{-\nu}}{x!} \tag{5.14}$$

or

$$P(x) = \frac{\nu^x e^{-\nu}}{x!} \tag{5.15}$$

where $P(x)$ is the mole fraction of x-mers. Equation 5.15 is identical in form to the frequency function of the Poisson distribution and so *the polymer formed has a Poisson distribution of molar mass.* Neglecting initiator fragments, the number-average molar mass is given by

$$\bar{M}_\mathrm{n} = \sum_{x=1}^{\infty} P(x) M_x$$

where

the molar mass of an x-mer $M_x = xM_0$

M_0 is the molar mass of the monomer.

Thus

$$\bar{M}_n = M_0 e^{-\nu} \frac{\sum_{x=1}^{\infty} x\nu^x}{x!}$$

i.e.

$$\bar{M}_n = M_0 \nu e^{-\nu} \frac{\sum_{x=1}^{\infty} \nu^{x-1}}{(x-1)!}$$

Using the standard mathematical relation $\sum_{x=1}^{\infty} \nu^r / r! = e^{\nu}$, the equation for $\bar{M}_n$ reduces to

$$\bar{M}_n = M_0 \nu \tag{5.16}$$

and so ν is the number-average degree of polymerization.

Since N living polymer molecules are formed, the total mass of polymer is $NM_0\nu$. Thus, the weight-fraction w_x of x-mers is given by

$$w_x = \frac{N_x M_x}{NM_0 \nu}$$

neglecting initiator fragments. Substituting $M_x = xM_0$ and Equation 5.14 gives

$$w_x = \frac{e^{-\nu} \nu^{x-1}}{(x-1)!} \tag{5.17}$$

The weight-average molar mass is given by

$$\bar{M}_w = \sum_{x=1}^{\infty} w_x M_x$$

which upon substitution of Equation 5.17 and $M_x = xM_0$ leads to

$$\bar{M}_w = M_0 e^{-\nu} \frac{\sum_{x=1}^{\infty} x\nu^{x-1}}{(x-1)!}$$

Using the mathematical relation $\sum_{r=1}^{\infty} r\nu^{r-1}/(r-1)! = (\nu+1)e^{\nu}$, the equation for $\bar{M}_w$ becomes

$$\bar{M}_w = M_0(\nu + 1) \tag{5.18}$$

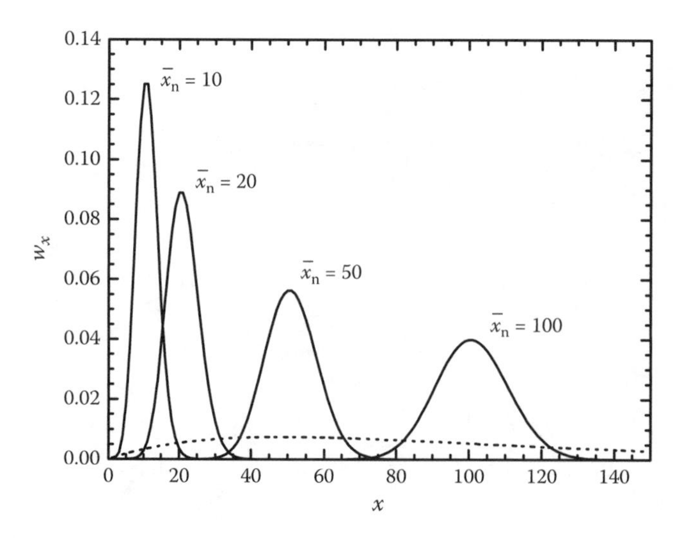

FIGURE 5.1 Weight-fraction Poisson distribution w_x of chain lengths for various values of $\bar{x}_n$ in a polymerization without termination. For comparison, the broken line shows the weight-fraction most probable distribution of chain lengths when $\bar{x}_n = 50$ (see Sections 3.1.3.2 and 4.3.8). Thus, polymerization without termination produces far narrower distributions of chain length than either step polymerization or conventional free-radical polymerization.

Combining Equations 5.16 and 5.18 gives $\bar{M}_w - \bar{M}_n = M_0$ and so the molar mass distribution is extremely narrow, as is revealed by the expression for the molar mass dispersity

$$\frac{\bar{M}_w}{\bar{M}_n} = 1 + \frac{1}{\nu} \tag{5.19}$$

from which it can be seen that $\bar{M}_w/\bar{M}_n \to 1$ as ν $(= \bar{x}_n) \to \infty$. This result arises from the difference (M_0) between $\bar{M}_w$ and $\bar{M}_n$ becoming negligible as molar mass increases and is not due to a further narrowing of the molar mass distribution. In fact, the distribution becomes broader as $\bar{x}_n$ increases, as is shown by the plots of Equation 5.17 given in Figure 5.1, thus highlighting the deficiencies in using $\bar{M}_w/\bar{M}_n$ as an absolute measure of the breadth of a molar mass distribution. Nevertheless, provided that $k_i \geq k_p$, polymerization without termination (i.e. truly living polymerization) always yields polymers with narrow molar mass distributions, and anionic polymerization is widely used for the preparation of such polymer standards.

5.3.2.5 Deactivation of Carbanionic Living Polymers

Most commonly, the active end-groups of carbanionic living polymers are deactivated by reaction with proton donors (usually alcohols) added at the end of the polymerization to give saturated end-groups in the final polymer

$$\sim\!\sim CH_2-\overset{-}{C}R_1R_2\,\overset{+}{Mt} \;+\; R-O-H \longrightarrow \;\sim\!\sim CH_2-CR_1R_2-H \;+\; R-\overset{-}{O}\;\overset{+}{Mt}$$

However, it is also possible to introduce terminal functional groups by making use of reactions that are well established in the field of organic chemistry. For example, carboxylic acid and hydroxyl end-groups can be formed by reactions analogous to those of Grignard reagents

$$\sim\!\sim CH_2-\overset{-}{C}R_1R_2\,Mt^+ + O=C=O \longrightarrow \sim\!\sim CH_2-CR_1R_2-C(=O)-O^-\,Mt^+$$

Dilute acid (H^+)

$$\sim\!\sim CH_2-CR_1R_2-C(=O)OH$$

Carboxylic acid end-group

$$\sim\!\sim CH_2-\overset{-}{C}R_1R_2\,Mt^+ + CH_2-CHR\ (\text{epoxide, }O) \longrightarrow \sim\!\sim CH_2-CR_1R_2-CH_2-CHR-O^-\,Mt^+$$

Dilute acid (H^+)

$$\sim\!\sim CH_2-CR_1R_2-CH_2-CHR-OH$$

Alcohol end-group

R = H gives $\sim\!\sim CH_2-OH$

R = CH_3 gives $\sim\!\sim CH(CH_3)-OH$

$$\sim\!\sim CH_2-\overset{-}{C}R_1R_2\,Mt^+ + CHR=O \longrightarrow \sim\!\sim CH_2-CR_1R_2-CHR-O^-\,Mt^+$$

Dilute acid (H^+)

$$\sim\!\sim CH_2-CR_1R_2-CHR-OH$$

Alcohol end-group

R = H gives $\sim\!\sim CH_2-OH$

R = CH_3 gives $\sim\!\sim CH(CH_3)-OH$

R = Ph gives $\sim\!\sim CH(Ph)-OH$

Thus, polymer molecules which possess either one or two functionalized end-groups can be prepared by controlled deactivation of living polymers formed using either organometallic initiators or electron transfer initiation, respectively. Furthermore, these molecules can be prepared with narrow molar mass distributions and predefined molar masses (by application of Equation 5.10).

5.3.2.6 Solvent and Counter-Ion Effects in Living Anionic Polymerizations

A carbanionic active centre and its counter-ion exist in a series of states of association similar to those described for cationic polymerization in Section 5.2.1.6. However, in anionic polymerization, the ion pairs are much tighter because the counter-ions are considerably smaller than those employed in cationic polymerization. Consequently, the ion pairs propagate at rates which typically are two to four orders of magnitude lower than those for propagation of free carbanionic active centres. The reactivity of ion pairs is greatly enhanced by the use of a more polar solvent which increases the ion-pair separation. Ion pairs can be disrupted by using polar solvents that are able to solvate the counter-ion, e.g. coordination of Li^+ ions by the lone pairs of electrons on the oxygen atoms of 1,2-dimethoxyethane

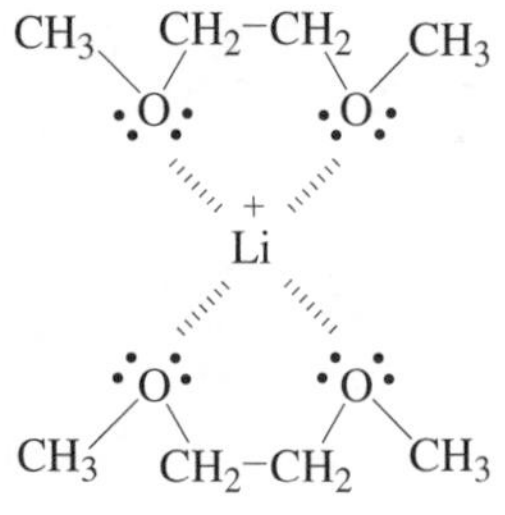

```
CH3   CH2–CH2   CH3
   \  /      \  /
   •O•        •O•
      ''', + ,'''
          Li
      ,''' '''',
   •O•        •O•
   /  \      /  \
CH3   CH2–CH2   CH3
```

thereby releasing the free polymeric carbanion active centre. When solvation is absent or weak (e.g. benzene or 1,4-dioxan), the rate coefficient for ion-pair propagation increases as the size of the counter-ion increases (i.e. $K^+ > Na^+ > Li^+$) due to the consequent increase in the separation of the ions. However, in polar solvating solvents (e.g. tetrahydrofuran or 1,2-dimethoxyethane), the opposite trend often is observed because the smaller counter-ions are more strongly solvated.

Generally, in non-polar solvents, free ions and solvent-separated ion pairs are not present and the contact ion pairs are aggregated. Furthermore, initiation can be slow relative to propagation, usually due to stronger aggregation of the initiator species than the propagating ion pairs, and leads to the formation of polymers with molar mass distributions which are broader than the Poisson distribution. Since the activation energy for propagation is positive, the rates of polymerization increase as the reaction temperature is increased.

In polar solvating solvents, the formation of solvent-separated ion pairs from contact ion-pairs is exothermic whereas the formation of free ions from solvent-separated ion-pairs is essentially athermal. Therefore, the effect of reducing the reaction temperature is to increase the concentration of solvent-separated ion-pairs relative to the concentration of the less reactive contact ion-pairs and the rate of polymerization increases. In contrast to the ion-pairs, the reactivity of free carbanionic active centres is not greatly enhanced by an increase in the strength of solvation of the counter-ions.

The existence of several kinds of propagating species (e.g. free ions and ion pairs) with vastly different rates of propagation could lead to the formation of polymers with broad or complex molar mass distributions. However, the rates of interconversion between the different species normally are greater than their rates of propagation and so there is no significant effect upon the molar mass distribution. Thus polymers prepared by anionic polymerization using fast initiation have narrow molar mass distributions typically with $\bar{M}_w/\bar{M}_n < 1.05$.

The solvent and counter-ion also can influence the stereochemistry of propagation and these effects are considered in Sections 6.2 and 6.3.

5.3.2.7 Practical Considerations for Living Anionic Polymerization

The high reactivity and low concentration of carbanionic active centres makes anionic polymerization very susceptible to inhibition by trace quantities of reactive impurities (e.g. H_2O, CO_2, O_2). Thus all reactants and solvents must be rigorously purified and the reactions must be carried out under inert conditions in scrupulously clean sealed apparatus. High vacuum techniques are often used for this purpose. Monomers with side groups that have acidic hydrogens must be used with those groups in a protected form and the groups then de-protected after the polymer has been produced. For example, monomers with pendent hydroxyl groups can be used as trimethylsilyl ethers and those with carboxylic acid groups can be used as trimethylsilyl or *tert*-butyl esters

$$n\ CH_2{=}C(CH_3)C(=O)O(CH_2)_2OSi(CH_3)_3 \xrightarrow[\text{(ii) } CH_3OH]{\text{(i) RLi}} R{-}[CH_2{-}C(CH_3)(C(=O)O(CH_2)_2OSi(CH_3)_3)]_n{-}H \xrightarrow[\text{aq. HCl in THF}]{\Delta} R{-}[CH_2{-}C(CH_3)(C(=O)O(CH_2)_2OH)]_n{-}H + n\ HO{-}Si(CH_3)_3$$

$$n\ CH_2{=}C(CH_3)C(=O)OSi(CH_3)_3 \xrightarrow[\text{(ii) } CH_3OH]{\text{(i) RLi}} R{-}[CH_2{-}C(CH_3)(C(=O)OSi(CH_3)_3)]_n{-}H \xrightarrow[\text{aq. HCl in THF}]{\Delta} R{-}[CH_2{-}C(CH_3)(C(=O)OH)]_n{-}H + n\ HO{-}Si(CH_3)_3$$

$$n\ CH_2{=}C(CH_3)C(=O)OC(CH_3)_3 \xrightarrow[\text{(ii) } CH_3OH]{\text{(i) RLi}} R{-}[CH_2{-}C(CH_3)(C(=O)OC(CH_3)_3)]_n{-}H \xrightarrow{\Delta} R{-}[CH_2{-}C(CH_3)(C(=O)OH)]_n{-}H + n\ CH_2{=}C(CH_3)_2$$

Anionic polymerization of polar monomers is complicated by side reactions that involve the polar groups. For example, in the polymerization of acrylates and methacrylates, the initiator species and propagating active centres can react with the C=O groups in the monomer (or polymer)

$$A^- + CH_2{=}C(R){-}C(=O){-}O{-}R_1 \rightleftharpoons CH_2{=}C(R){-}C(A)(O^-){-}O{-}R_1 \rightleftharpoons CH_2{=}C(R){-}C(=O){-}A + R_1{-}O^-$$

The side reactions lead to loss of initiator, termination of chain growth and formation of polymers with broad molar mass distributions. For methyl methacrylate ($R{=}R_1{=}CH_3$), the side reactions essentially can be eliminated by using polar solvents, low temperatures, bulky initiators (for which reaction with the C=O group is sterically hindered) and large counter-ions, e.g. by polymerization in tetrahydrofuran at –75 °C using cumylcaesium as initiator.

The ability to produce polymers of well-defined structure using anionic polymerization is of great importance and despite the above difficulties it is widely used for this purpose. Thus, polymers with narrow molar mass distributions, terminally-functionalized polymers and, perhaps most important of all, well-defined block copolymers can be prepared using living anionic polymerization (see Section 9.4.2.1).

5.4 GROUP-TRANSFER POLYMERIZATION

Group-transfer polymerization (*GTP*) was first reported in 1983 by Webster and his coworkers at DuPont, and is most suitable for polymerization of methacrylates. Propagation involves reaction of a terminal silyl ketene acetal with monomer by Michael addition during which the silyl group transfers to the added monomer thus creating a new terminal silyl ketene acetal group.

Polymerization is initiated by monomeric silyl ketene acetals and normally is catalysed by anions (e.g. F^-, HF_2^-, $MeCO_2^-$, $PhCO_2^-$) but also can be catalysed by Lewis acids (e.g. $ZnBr_2$, Al^iBu_2Cl), though their action is less well understood and thought to involve activation of the monomer rather than the silyl ketene acetal. The following is an example of group-transfer polymerization of a methacrylate monomer initiated by dimethylketene methyl trimethylsilyl acetal and catalysed by tris(dimethylamino)sulphonium bifluoride

$$Me_2C{=}C(OMe)(OSiMe_3) + n\,CH_2{=}C(Me)CO_2R \xrightarrow{(Me_2N)_3\overset{+}{S}\,HF_2^-} Me{-}C(Me)(CO_2Me){-}[CH_2{-}C(Me)(CO_2R)]_{n-1}{-}CH_2{-}C(Me){=}C(OR)(OSiMe_3)$$

5.4.1 Mechanism of GTP

Originally, propagation in GTP was believed to proceed via the following *associative* mechanism in which the anion (represented here as a general nucleophile, Nu^-) activates transfer of the trimethylsilyl group by association with the silicon atom

$$\sim CH_2{-}C(Me){=}C(OMe){-}O{-}SiMe_3 \underset{}{\overset{Nu^-}{\rightleftharpoons}} \sim CH_2{-}C(Me){=}C(OMe){-}O{-}SiMe_3\cdots Nu$$

$$\xrightarrow{CH_2{=}C(Me){-}C({=}O){-}OR} \sim CH_2{-}C(Me)(\overset{+}{C}(OMe){-}O{-}SiMe_3\cdots Nu){-}CH_2{-}\overset{-}{C}(Me){-}C({=}O){-}OR$$

$$\longrightarrow \sim CH_2{-}C(Me)(C(OMe){=}O){-}CH_2{-}C(Me){=}C(OR){-}O{-}SiMe_3\cdots Nu$$

$$\overset{-Nu^-}{\rightleftharpoons} \sim CH_2{-}C(Me)(C(OMe){=}O){-}CH_2{-}C(Me){=}C(OR){-}O{-}SiMe_3$$

$$\equiv \sim CH_2{-}C(Me)(C({=}O)OMe){-}CH_2{-}C(Me){=}C(OMe){-}O{-}SiMe_3$$

However, as studies of GTP grew, it became clear that this associative mechanism was inconsistent with many of the observations (e.g. ester enolates, C=C(OR)–O$^-$, act as both initiators and catalysts for GTP, and increasing the anion concentration does not increase the rate of polymerization, but instead eventually poisons the reaction). There is now a very strong body of evidence for a *dissociative* mechanism in which the free enolate ion is responsible for propagation

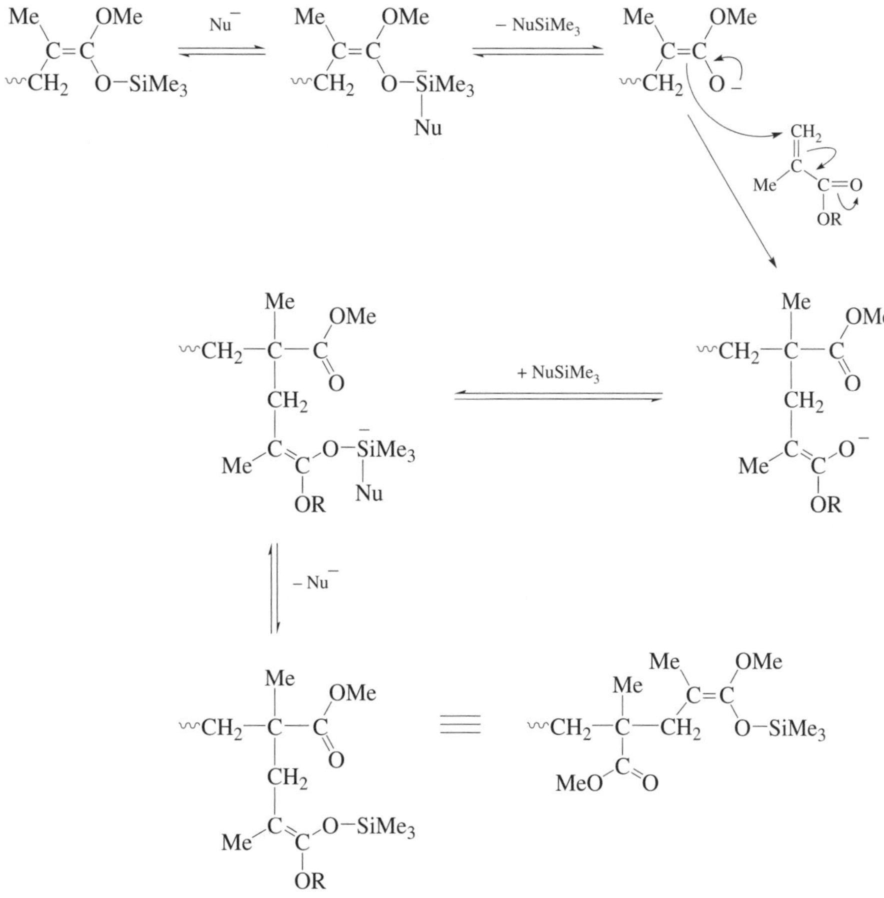

Although only a single propagation reaction is shown here, several monomer additions may occur before $NuSiMe_3$ adds back to the enolate. The operation of this dissociative mechanism shows that GTP is effectively a special form of anionic polymerization.

5.4.2 Practical Considerations and Uses of GTP

GTP is terminated rapidly by compounds containing an active hydrogen and so must be performed under dry conditions using reactants and solvents which have been rigorously dried and purified. If these precautions are taken, then living polymers can be formed and in favourable cases the Poisson distribution of molar mass is obtained (Section 5.3.2.4). Monomers with active hydrogens (e.g. methacrylic acid, 2-hydroxyethyl methacrylate) can be protected using trimethylsilyl derivatives (i.e. $-CO_2SiMe_3$, $-OSiMe_3$), which can be displaced easily by hydrolysis after the polymer is formed (see Section 5.3.2.7).

GTP usually is performed at 50–80 °C, i.e. at much higher temperatures than for living anionic polymerization. This is possible because the side reaction shown in Section 5.3.2.7, which kills anionic polymerization at such temperatures, is the first step of a chain transfer process in GTP of methacrylates. Backbiting also occurs via a similar process to produce a terminal cyclohexanone ring structure

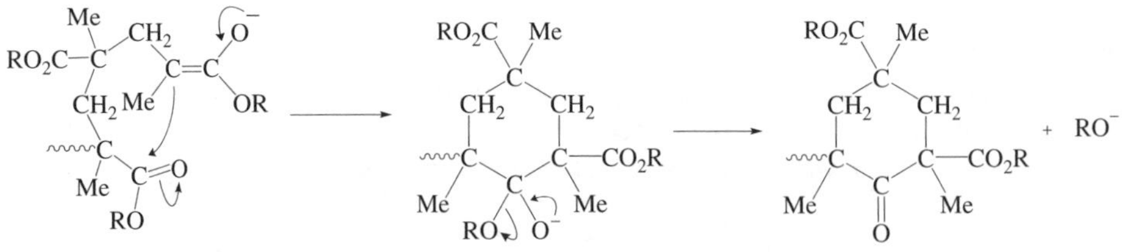

In both cases, the alkoxide ion released can react with silyl ketene acetal end-groups to give the alkyltrimethylsilylether ($ROSiMe_3$) and release the enolate ion for further propagation. However, the backbiting reaction gives rise to loss of active centres in GTP of acrylates because the acidic tertiary hydrogen atom in the cyclohexanone ring is abstracted by alkoxide to give an alcohol (ROH) and a stable enolate which does not propagate

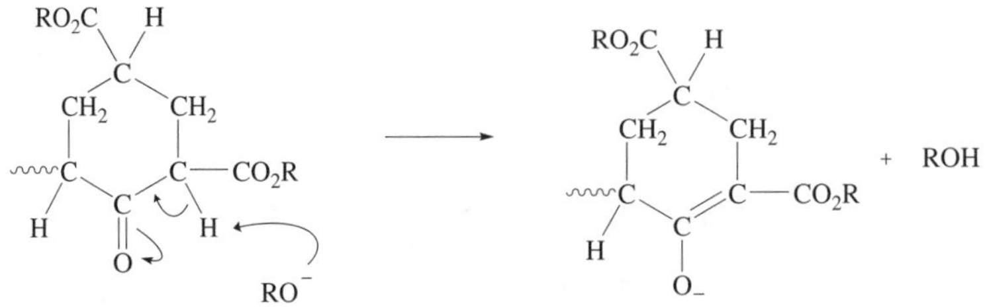

Acrylonitrile and methacrylonitrile polymerize so rapidly by GTP that the reactions are difficult to control. Thus, although polymers have been prepared from acrylates, acrylonitriles and other similar monomers, GTP is most suited to the preparation of low molar mass (<50 kg mol^{-1}) methacrylate polymers, in particular functionalized homopolymers and copolymers. This will now be illustrated.

GTP can be terminated by addition of a proton source (e.g. water or dilute acid) or by coupling of two active species (e.g. with a dihalide). The essence of the chemistry of these reactions is shown below.

Protonation:

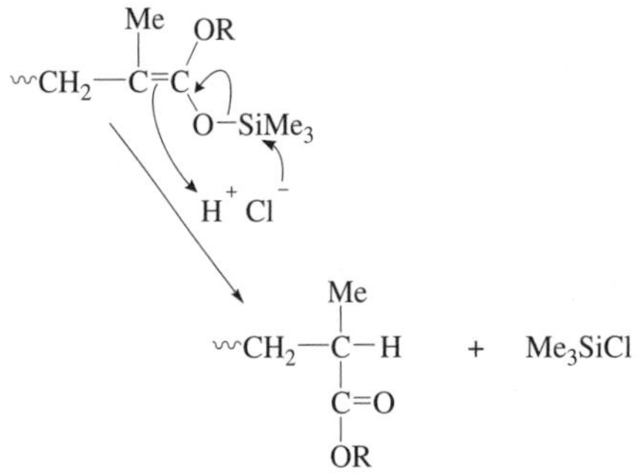

Coupling:

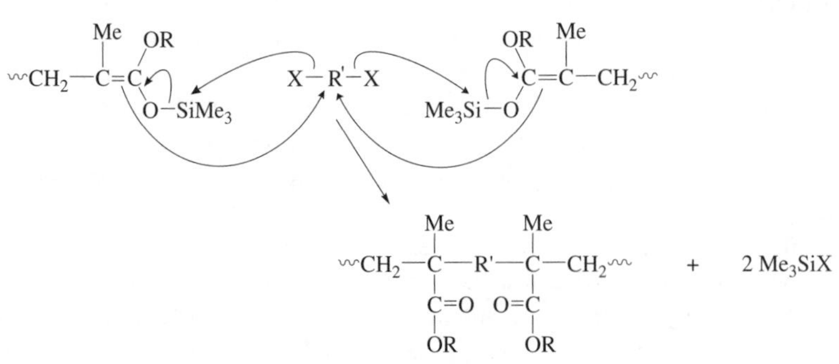

where the π-electrons in the C=C bond move to form the new bond (to H or R′), causing the O–Si bond to break with the bonding electrons moving to create the π-bond in the resulting C=O group. Together with the use of initiators containing protected functional groups, these termination reactions facilitate the preparation of terminally-functional polymers, e.g. poly(methyl methacrylate) with terminal carboxylic acid groups may be prepared as follows

$$\mathrm{Me_2C{=}C(OSiMe_3)_2} + n\,\mathrm{CH_2{=}C(Me)CO_2R} \xrightarrow{(\mathrm{Me_2N})_3\overset{+}{\mathrm{S}}\,\mathrm{HF_2^-}} \mathrm{Me_3SiO{-}C({=}O){-}C(Me)_2{+}CH_2{-}C(Me)(CO_2Me){+}_{n-1}CH_2{-}C(Me){=}C(OMe)(OSiMe_3)}$$

$$\xrightarrow[(-\mathrm{Me_3SiCl})]{\text{aq. HCl}} \mathrm{HO{-}C({=}O){-}C(Me)_2{+}CH_2{-}C(Me)(CO_2Me){+}_{n-1}CH_2{-}C(Me)(CO_2Me){-}H}$$

$$\xrightarrow[\text{(ii) aq. HCl }(-\mathrm{Me_3SiCl})]{\text{(i) } \mathrm{BrCH_2{-}C_6H_4{-}CH_2Br}\ (-\mathrm{Me_3SiBr})} \mathrm{HO{-}C({=}O){-}C(Me)_2{+}CH_2{-}C(Me)(CO_2Me){+}_{n-1}CH_2{-}C(Me)(CO_2Me){-}CH_2{-}C_6H_4{-}CH_2{-}C(Me)(CO_2Me){-}CH_2{+}C(Me)(CO_2Me){-}CH_2{+}_{n-1}C(Me)_2{-}C({=}O){-}OH}$$

5.4.3 Aldol GTP

In an extension of the principles of GTP, DuPont developed *aldol group-transfer polymerization* in which a silyl vinyl ether is polymerized using an aldehyde as the initiator and a Lewis acid as catalyst to give a living silylated poly(vinyl alcohol), e.g.

$$\mathrm{Ph{-}C(H){=}O} + n\,\mathrm{CH_2{=}C(H){-}O{-}SiMe_2{}^tBu} \xrightarrow{\mathrm{ZnBr_2}} \mathrm{Ph{-}C(H)(OSiMe_2{}^tBu){+}CH_2{-}C(H)(OSiMe_2{}^tBu){+}_{n-1}CH_2{-}C(H){=}O}$$

Thus, the active species is an aldehyde group, and propagation is believed to proceed via a mechanism of the type

$$\sim\!\!CH_2\!-\!CH\!=\!O \xrightarrow{\text{Cat}} \sim\!\!CH_2\!-\!CH\!=\!O\cdots\text{Cat} \;+\; CH_2\!=\!CH\!-\!O\!-\!SiMe_2{}^tBu$$

$$\longrightarrow \sim\!\!CH_2\!-\!CH(O^-\cdots\text{Cat})\!-\!CH_2\!-\!C^+H\!-\!O\!-\!SiMe_2{}^tBu \xrightarrow{-\text{Cat}} \sim\!\!CH_2\!-\!CH(O\!-\!SiMe_2{}^tBu)\!-\!CH_2\!-\!CH\!=\!O$$

in which the catalyst (Cat) activates the aldehyde for reaction and the silyl group is transferred from monomer to the previously terminal aldehyde oxygen atom thus generating a new terminal aldehyde group. Hence, the mechanism of aldol GTP has similarities to reversible-deactivation cationic polymerization (Section 5.2.2). At the end of the polymerization, the silyl ether groups can be converted to hydroxyl groups by treatment with an acid to produce poly(vinyl alcohol).

PROBLEMS

5.1 Styrene was polymerized at a mass concentration of 208 g dm^{-3} in ethylene dichloride using sulphuric acid as the initiator at 25 °C. Given that the rate coefficients for propagation, ion-pair rearrangement and transfer to monomer are 7.6 dm^3 mol^{-1} s^{-1}, 4.9×10^{-2} s^{-1} and 0.12 dm^3 mol^{-1} s^{-1}, respectively, calculate the molar mass of the polystyrene formed early in the reaction.

5.2 For polymerization of styrene in tetrahydrofuran at 25 °C using sodium naphthalide as initiator, the rate coefficient for propagation is 550 dm^3 mol^{-1} s^{-1}. If the initial mass concentration of styrene is 156 g dm^{-3} and that of sodium naphthalide is 3.02×10^{-2} g dm^{-3}, calculate the initial rate of polymerization and, for complete conversion of the styrene, the number-average molar mass of the polystyrene formed. Comment upon the expected value of the molar mass dispersity ($\bar{M}_w/\bar{M}_n$) and the stereoregularity of the polystyrene produced.

5.3 Describe, showing the chemistry of the reactions, how you would prepare linear poly(n-butyl methacrylate) with carboxylic acid groups at both chain ends by (a) anionic polymerization and (b) group-transfer polymerization.

FURTHER READING

Allen, G. and Bevington, J.C. (eds), *Comprehensive Polymer Science*, Vol. 3 (Eastmond, G.C., Ledwith, A., Russo, S., and Sigwalt, P., eds), Pergamon Press, Oxford, U.K., 1989.

Lenz, R.W., *Organic Chemistry of Synthetic High Polymers*, Interscience, New York, 1967

Mark, H.F., Bikales, N.M., Overberger, C.G., and Menges, G. (eds), *Encyclopedia of Polymer Science and Engineering*, Wiley-Interscience, New York, 1985–1989.

Matyjaszewski, K. and Sigwalt, P., Unified approach to living and non-living cationic polymerization of alkenes, *Polymer International*, **35**, 1 (1994).

Odian, G., *Principles of Polymerization*, 4th edn., Wiley-Interscience, New York, 2004.

Sawamoto, M., Modern cationic vinyl polymerization, *Progress in Polymer Science*, **16**, 111 (1991).

Stevens, M.P., *Polymer Chemistry: An Introduction*, 3rd edn., Oxford University Press, New York, 1998.

Webster, O.W., Group transfer polymerization: mechanism and comparison with other methods for controlled polymerization of acrylic monomers, *Advances in Polymer Science*, **167**, 1 (2004).

6 Stereochemistry and Coordination Polymerization

6.1 INTRODUCTION TO STEREOCHEMISTRY OF POLYMERIZATION

In addition to the effects of skeletal structure and of the chemical composition of the repeat units, the properties of a polymer are strongly influenced by its molecular microstructure. Variations in the geometric and configurational arrangements of the atoms in the repeat unit, and the distribution of these different spatial arrangements for the repeat units along the chain, are of particular importance.

Different molecular microstructures arise from there being several possible stereochemical modes of propagation. The possibility of head-to-tail and head-to-head placements of the repeat units has been considered already, with the observation that for both steric and energetics reasons the placement is *regioselective*, giving almost exclusively head-to-tail placements for most polymers. Therefore, only head-to-tail placements will be considered in this chapter. The factors which influence the stereochemistry of propagation for mono-olefins and 1,3-dienes will be described in generic terms before considering the extent to which they can be controlled in radical and ionic polymerizations. More specialized methods of polymerization which involve strong coordination of monomer during propagation will then be introduced since they provide much greater constraints on the stereochemistry and are capable of producing highly stereoregular polymers.

6.2 TACTICITY OF POLYMERS

Chiral is the term used to describe objects which are non-superimposable on their mirror image (e.g. human hands and feet). Molecular chirality is of great importance and is used, for example, by nature to control biochemistry. The simplest chiral molecules have an sp^3-hybridized carbon atom to which four different groups are attached, which makes the carbon atom *asymmetric*. For polymers prepared from monomers of the general structure $CH_2{=}CXY$, where X and Y are two different substituent groups, there are two distinct configurational arrangements of the repeat unit

and

where ▼ and ⁝ indicate bonds which are extending above and below the plane of the paper, respectively. These two *stereoisomers* of the repeat unit cannot be interchanged by bond rotation and exist because the substituted sp^3-hybridized carbon atom is asymmetric. Unlike simple organic compounds with asymmetric carbon atoms, the stereoisomers indicated above show no significant optical activity because the two polymer chain residues attached to the asymmetric carbon atom are almost identical. Nevertheless, the existence of two isomeric forms of the repeat unit, and in particular their distribution along the polymer chain, are of great significance. In *isotactic* polymers, all the repeat units have the same configuration, whereas in *syndiotactic* polymers the configuration alternates from one repeat unit to the next. *Atactic* polymers have an irregular or random placement of the two configurations. These three stereochemical forms are shown for short segments of polymer chains in Figure 6.1.

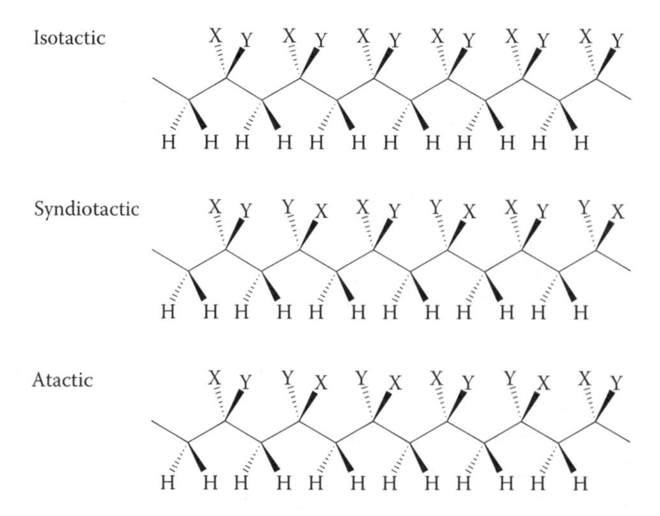

FIGURE 6.1 Different stereochemical forms of polymers derived from monomers of the type CH_2=CXY.

Polypropylene (X = H, Y = CH_3) provides a good example of the importance of tacticity. The commercial material is essentially isotactic and due to its regular structure is crystalline (~65%). It is the crystalline regions that give rise to the good mechanical properties of commercial polypropylene. In contrast, atactic polypropylene is unable to crystallize because of its irregular structure and is a soft, wax-like amorphous material which has no useful mechanical properties.

The tacticity of a polymer is controlled by the stereochemistry of propagation, some elementary aspects of which are illustrated in Figure 6.2. The terminal active centres of propagating chains in free-radical, cationic and anionic polymerizations can be considered to be sp^2 hybridized, the remaining p-orbital containing one, none and two electrons, respectively. This form of hybridization is normal for free radicals and carbocations, but for carbanionic active centres is a consequence of resonance with the substituent group(s) (a requirement for which is coplanarity and a change from the normal sp^3 to sp^2 hybridization). Thus, in each case, there is a planar arrangement of the groups about the terminal active carbon atom and so its configuration in the resulting polymer molecule is determined by the way in which monomer adds to it in a propagation step. As indicated in Figure 6.2, the orientation of the substituent groups on the terminal active carbon atom relative to the orientation of those on the asymmetric carbon atom of the penultimate repeat unit, and the face of the planar active centre to which the molecule of monomer adds, are of great importance. Usually, steric and/or electronic repulsion between similar substituent groups results in a slight preference for syndiotactic rather than isotactic placements. This preference is accentuated by reducing the reaction temperature, and highly syndiotactic polymers can be formed by ionic polymerization in a polar solvent at low temperature (e.g. anionic polymerization of methyl methacrylate initiated by 9-fluorenyllithium at −78 °C in tetrahydrofuran). In contrast, the relatively high temperatures normally employed for radical polymerizations result in the formation of essentially atactic polymers which, as a consequence of their irregular microstructure, do not crystallize.

Highly isotactic polymers can be prepared by ionic polymerization if there is strong coordination of the counter-ion with the terminal units in the polymer chain and with the incoming molecule of monomer. However, this is difficult to achieve with non-polar monomers and usually requires the monomer to have polar substituent groups, which can act as sites for strong coordination (e.g. cationic polymerization: vinyl ethers; anionic polymerization: methacrylate esters). In order to prepare highly isotactic polymers from such polar monomers, the reaction must be carried out at low temperature in a non-polar solvent using an initiator which yields a small counter-ion so that ion-pair association is promoted (e.g. solvent: toluene at −78 °C; cationic initiator: boron

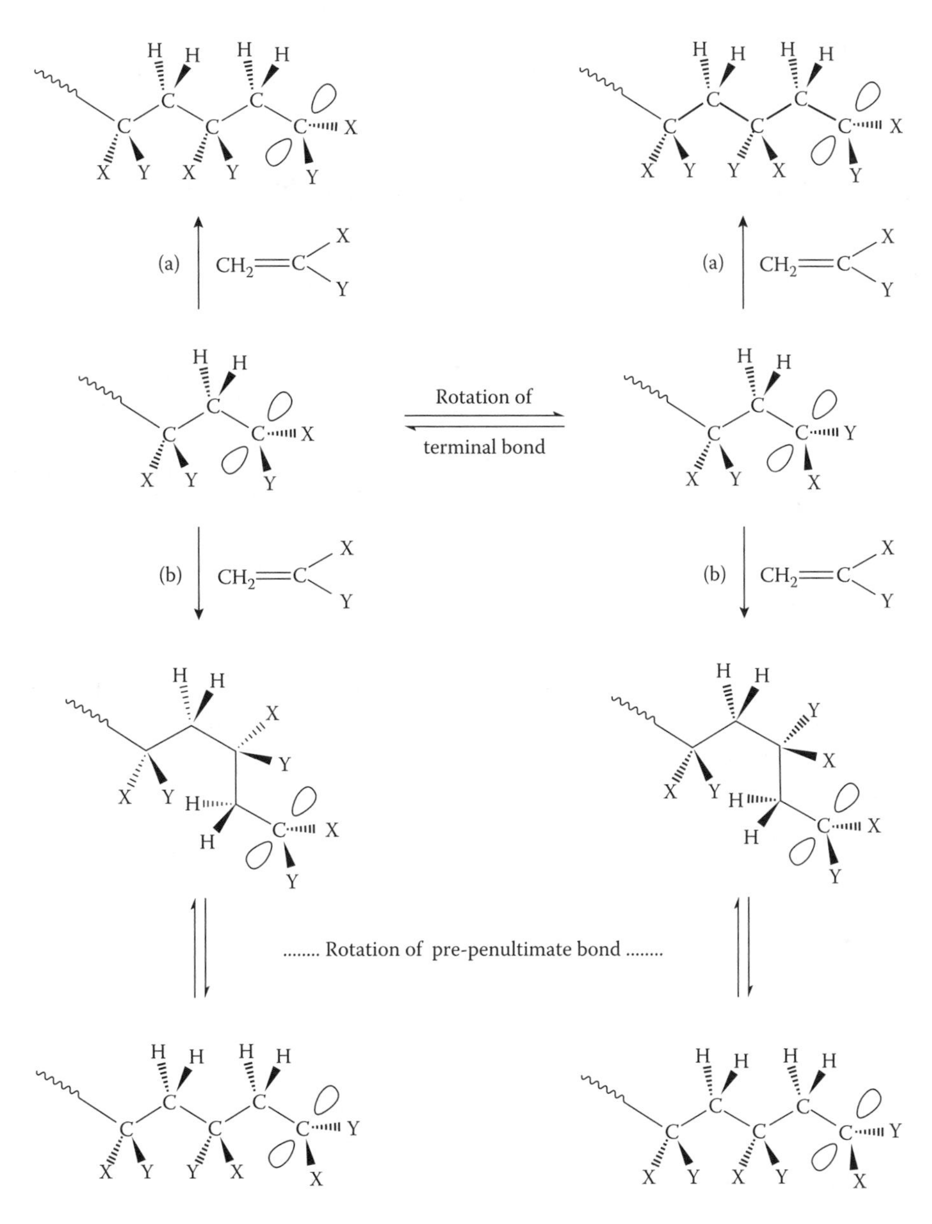

FIGURE 6.2 Elementary features of the stereochemistry of propagation showing the consequences of rotation about the pre-penultimate C—C bond in the active chain and of addition of monomer (a) from above and (b) from below the face of the planar sp^2-hybridized active carbon atom for which the p-orbital contains either a single electron (radical polymerization), no electrons (cationic polymerization), or two electrons (anionic polymerization).

trifluoride etherate; anionic initiator: 1,1-diphenylhexyllithium). The coordination is easily disrupted (e.g. by addition of a small quantity of a polar solvent) resulting in loss of the stereochemical control and formation of predominantly syndiotactic polymer. Isotactic polymers, however, can be prepared from non-polar monomers by polymerizations involving coordination to transition metals (see Sections 6.4 and 6.5).

Polymers with more complex tacticities are formed from monomers of the general structure XCH=CHY since each backbone carbon atom is asymmetric. However, since these monomers do not readily form homopolymers, they will not be considered here.

The complications of tacticity are absent in polymers prepared from monomers of the type $CH_2{=}CX_2$ because they contain no asymmetric backbone carbon atoms and therefore must be stereoregular.

6.3 GEOMETRIC ISOMERISM IN POLYMERS PREPARED FROM CONJUGATED DIENES

The most important conjugated dienes are the following 1,3-dienes

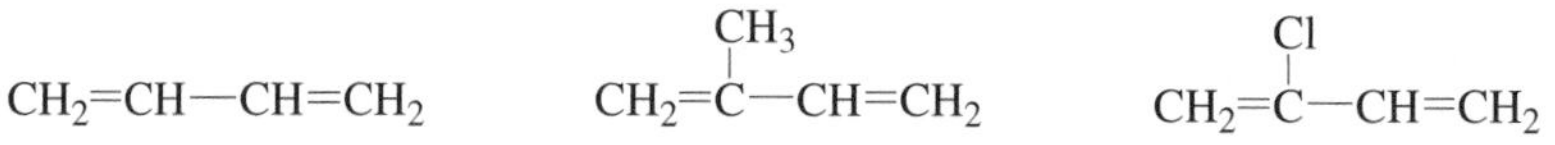

$CH_2{=}CH{-}CH{=}CH_2$	$CH_2{=}C(CH_3){-}CH{=}CH_2$	$CH_2{=}C(Cl){-}CH{=}CH_2$
Butadiene	Isoprene	Chloroprene

which have the general structure $CH_2{=}CR{-}CH{=}CH_2$. There are four basic modes for addition of such 1,3-dienes to a growing polymer chain and these are shown in Table 6.1 (for butadiene there are only three modes because the 1,2- and 3,4-additions are identical since R = H).

The importance of repeat unit isomerism in poly(1,3-dienes) is very clearly demonstrated by the naturally occurring polyisoprenes. *Gutta percha* and *balata* are predominantly *trans*-1,4-polyisoprene, and due to their regular structure are able to crystallize, which causes them to be hard, rigid materials. However, *natural rubber* is *cis*-1,4-polyisoprene, which has a less symmetrical structure that does not allow easy crystallization under normal conditions and so is an amorphous rubbery material. The difference in regularity between these structures is shown below schematically for chain segments containing four head-to-tail repeat units.

cis-1,4-polyisoprene

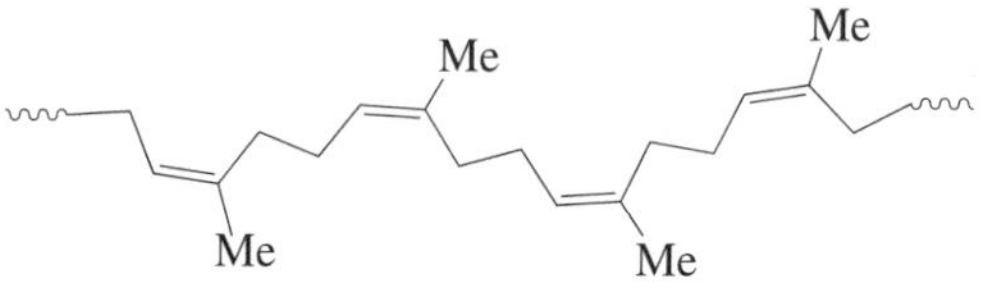

Me Me Me Me

trans-1,4-polyisoprene

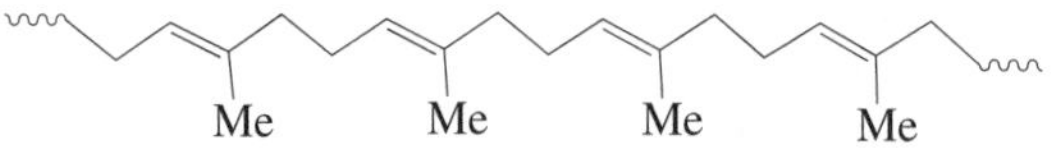

Me Me Me Me

Table 6.2 shows the proportions of the different repeat units in homopolymers of butadiene and isoprene prepared using various polymerization conditions. The factors which are of importance in determining these proportions are as follows.

1. The conformation of the 1,3-diene molecule when it adds to the growing chain, since this at least initially is retained in the new active unit formed by its addition. In the absence of specific effects, the molecules exist mainly in the transoid conformation R, which is more stable than the cisoid R and leads to a preponderance of initially *trans*-active units.
2. The relative stabilities of the various structures for the active unit.
3. For 1,4-addition, the rate of isomerization between the *cis*- and *trans*- forms of an active unit relative to their individual rates of propagation. Transformation of one form into the other results from the combined effects of resonance and bond rotation, e.g.

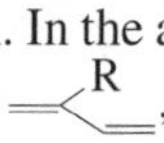

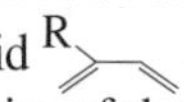

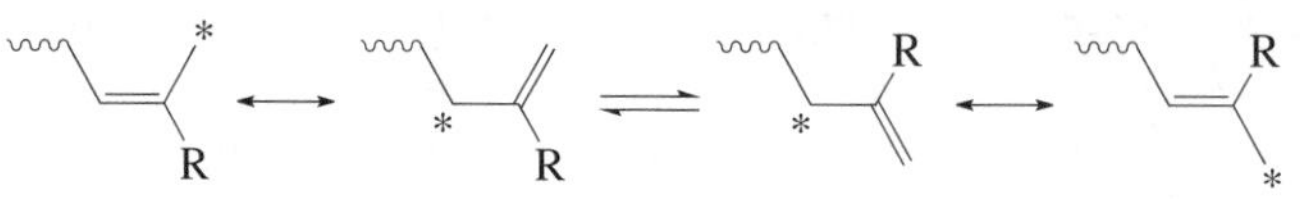

$$* \; R \longleftrightarrow * \; R \rightleftharpoons * \; R \longleftrightarrow R \; *$$

where * represents a single electron or a positive or negative charge.

TABLE 6.1
Basic Modes for Addition of 1,3-Dienes ($CH_2{=}CR{-}CH{=}CH_2$) to a Growing Polymer Chain (M_n^*)[a]

Mode of Addition	Product of Addition	Repeat Unit Structure
1,2-addition	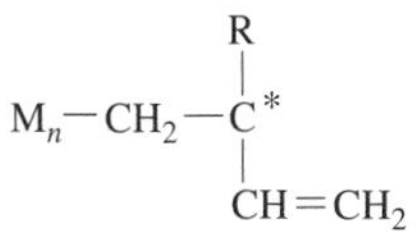 $M_n{-}CH_2{-}C^*(R)(CH{=}CH_2)$	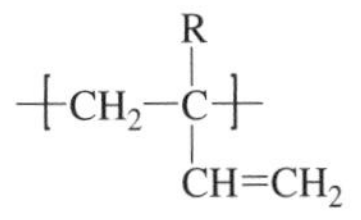 $\text{+}CH_2{-}C(R)(CH{=}CH_2)\text{+}$
3,4-addition	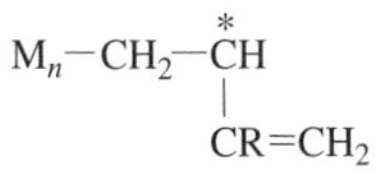 $M_n{-}CH_2{-}C^*H(CR{=}CH_2)$	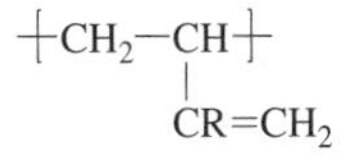 $\text{+}CH_2{-}CH(CR{=}CH_2)\text{+}$
cis-1,4-addition	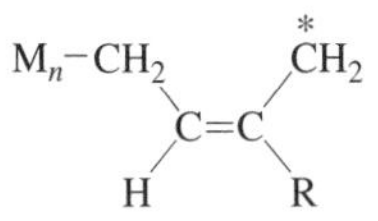 $M_n{-}CH_2(H)C{=}C(R)C^*H_2$ (cis)	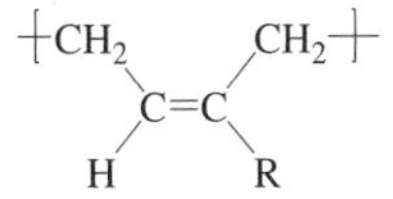 $\text{+}CH_2(H)C{=}C(R)CH_2\text{+}$ (cis)
trans-1,4-addition	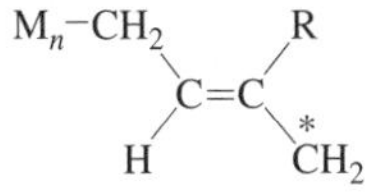 $M_n{-}CH_2(H)C{=}C(R)C^*H_2$ (trans)	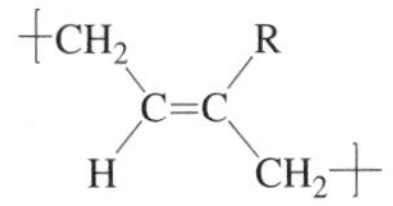 $\text{+}CH_2(H)C{=}C(R)CH_2\text{+}$ (trans)

[a] For each mode, there is the possibility of head-to-head or head-to-tail placement, and for 1,2- and 3,4-addition, the additional complication of isotactic or syndiotactic placement.

TABLE 6.2
Molecular Microstructures of Butadiene and Isoprene Homopolymers Prepared Using Various Polymerization Conditions

		Microstructure (Mole Fractions)			
Monomer	**Polymerization Conditions**	***cis*-1,4**	***trans*-1,4**	**1,2**	**3,4**
Butadiene	Free radical at −20 °C	0.06	0.77	0.17	—
Butadiene	Free radical at 100 °C	0.28	0.51	0.21	—
Butadiene	Anionic in hexane with Li^+ counter-ion at 20 °C	0.68	0.28	0.04	—
Butadiene	Anionic in diethyl ether with Li^+ counter-ion at 0 °C	0.08	0.17	0.75	—
Isoprene	Free radical at −20 °C	0.01	0.90	0.05	0.04
Isoprene	Free radical at 100 °C	0.23	0.66	0.05	0.06
Isoprene	Anionic in cyclohexane with Li^+ counter-ion at 30 °C	0.94	0.01	0.00	0.05
Isoprene	Anionic in diethyl ether with Li^+ counter-ion at 20 °C	← 0.35 →		0.13	0.52

In free-radical polymerization, there are no special effects and the polymers obtained have a high proportion of *trans*-1,4 repeat units which increase in number at the expense of *cis*-1,4 repeat units as the reaction temperature is reduced. The preference for *trans*-1,4 addition is more pronounced for isoprene due to the presence of the methyl substituent group.

Anionic polymerization in a non-polar solvent using Li^+ as the counter-ion leads to formation of polymers with high proportions of *cis*-1,4 repeat units. Under these conditions, as the monomer adds to the growing chain, it is held in the cisoid conformation by strong coordination to the small Li^+ counter-ion and so the active units initially are in the *cis*- form. Also, the electron density in the active unit is greatest at the terminal carbon atom, thus favouring 1,4 propagation. As long as the monomer concentration is sufficiently high, the rate of propagation of the *cis*-form of the active unit exceeds its rate of isomerization and *cis*-1,4 propagation predominates. The rate of *cis*- to *trans*-isomerization of the active unit in the polymerization of isoprene is much lower than for butadiene and gives rise to the very high *cis*-1,4 content of polyisoprene prepared in this way.

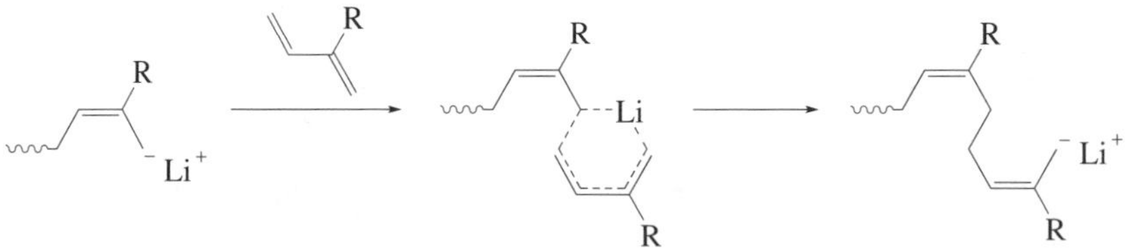

If anionic polymerization is performed in non-polar solvents using counter-ions other than Li^+ or in polar solvents (regardless of the counter-ion), stereochemical control is lost and the proportion of 1,4-repeat units is reduced considerably. High proportions of 1,2-addition and 3,4-addition occur, respectively, for butadiene and isoprene, partly because coordination effects are much weaker but also because in polar solvents the electron density in the active unit is greatest at the carbon atom in the γ-position relative to the terminal carbon atom.

Cationic polymerization is of little use for the preparation of homopolymers from conjugated dienes because side reactions lead to cyclic structures in the polymer chain and loss of a significant proportion of the expected residual unsaturation.

6.4 ZIEGLER–NATTA COORDINATION POLYMERIZATION

The use of ionic polymerizations for preparation of highly stereoregular polymers is restricted to specific monomers, in particular polar monomers. Generally, this method is not appropriate for non-polar monomers because they require stronger coordination than can be achieved with the counter-ions used in ionic polymerizations.

Chromium trioxide based catalysts supported on silica were reported by Phillips Petroleum in 1952 and are able to polymerize ethylene via a coordination mechanism that produces high-density, linear polyethylene (HDPE), which has a higher density because it attains much higher degrees of crystallinity than the highly branched (low density) polyethylene (LDPE) produced by free-radical polymerization (see Section 4.2.4.2). Supported catalysts of very high activity for the polymerization of ethylene have since been prepared from chromates and also from chromacene. However, these Phillips-type catalysts do not give stereochemical control in polymerization of α-olefins and so are used principally for production of HDPE.

In 1953, Ziegler reported the preparation of linear polyethylene by polymerization of ethylene using catalysts prepared from aluminium alkyl compounds and transition metal halides. Natta quickly recognized, and pursued, the potential of this new type of polymerization for the preparation of stereoregular polymers. By slightly modifying the catalysts used in Ziegler's work, he was able to prepare highly isotactic linear crystalline polymers from non-polar α-olefins (e.g. propylene). The enormous academic and industrial importance of these discoveries was recognized in 1963 by the joint award to Ziegler and Natta of the Nobel Prize for Chemistry.

6.4.1 Ziegler–Natta Catalysts

Usually, *Ziegler–Natta catalysts* are broadly defined in terms of their preparation, which involves reacting compounds (commonly halides) of groups IV–VIII transition metals (e.g. Ti, V, Cr, Zr) with organometallic compounds (e.g. alkyls, aryls or hydrides) of groups I–III metals (e.g. Al, Mg, Li). This definition is in fact too broad since not all such reactions yield catalysts suitable for preparing stereoregular polymers. Nevertheless, for each monomer, there is a wide range of catalysts that are suitable.

The catalysts which are useful for preparation of isotactic polymers are *heterogeneous*, i.e. they are insoluble in the solvent, or diluent, in which they are prepared. Their activity and stereoregulating ability are greatly affected by the components, and method, used for their preparation. For example, the α-form of $TiCl_3$ can be used to prepare catalysts suitable for the synthesis of isotactic polypropylene, whereas the β-form yields catalysts which give no stereochemical control. If α-$TiCl_3$ is reacted with $AlEt_2Cl$, it gives a catalyst of lower activity but much higher stereospecificity than that obtained from its reaction with $AlEt_3$. The inclusion of electron donors such as Lewis bases (e.g. ethers, ketones and esters) during preparation of the catalyst also can improve stereospecificity, often but not always with a loss of activity. Ball milling of the catalyst usually improves its activity, not only by increasing the surface area available but also by inducing crystal–crystal transformations.

In the search for higher efficiency, *supported Ziegler–Natta catalysts* have been developed in which the transition metal is either bonded to or occupies lattice sites in the support material. Magnesium compounds are widely used as supports (e.g. $Mg(OH)_2$, $Mg(OEt)_2$, $MgCl_2$). For example, catalysts with both high activity and high stereospecificity can be obtained from $TiCl_4$ supported on $MgCl_2$, which has been ball-milled in the presence of aromatic esters (e.g. ethyl benzoate).

Ziegler–Natta catalysts that are soluble in the solvent in which they are prepared (i.e. *homogeneous*) are of limited use because in general they do not provide stereochemical control. Nevertheless, there are some notable exceptions. For example, syndiotactic polypropylene can be prepared at low temperatures (e.g. −78 °C) using soluble catalysts based upon vanadium compounds (e.g. $VCl_4 + AlEt_3$). In addition, homogeneous catalysts prepared from benzyl derivatives of Ti and Zr have yielded isotactic polypropylene but are of low activity.

The factors which control catalyst activity and stereospecificity will not be considered here since they are complex and not completely understood. Furthermore, although there is strong evidence that propagation occurs by monomer insertion at a metal–carbon bond, there still is no single definitive mechanism for propagation in Ziegler–Natta polymerizations. In the following sections, two mechanisms which are representative of those that have been postulated will be described for heterogeneous catalysts prepared by reaction of α-$TiCl_3$ with trialkylaluminium (AlR_3).

6.4.2 Propagation: Monomer Insertion at Group I–III Metal–Carbon Bonds

A number of mechanisms have been proposed for propagation by insertion of monomer at groups I–III metal–carbon bonds after initial polarization of the monomer by coordination to the transition metal. Since both metals are involved, these often are termed *bimetallic mechanisms*. An example is the mechanism proposed by Natta in which the active site is an electron-deficient bridge complex formed by reaction between a surface Ti atom and AlR_3. Propagation may be represented by

$$\mathrm{Ti}\overset{R}{\underset{Cl}{\lozenge}}\mathrm{Al} \xrightarrow{n+1\ CH_2{=}CHX} \mathrm{Ti}\overset{CH_2-CHX{+}CH_2-CHX{+}_n R}{\underset{Cl}{\lozenge}}\mathrm{Al}$$

where peripheral ligands (i.e. Cl for the Ti atom and R for the Al atom) are omitted for clarity. The proposed mechanism is shown below and involves initial coordination of monomer to the Ti atom. This is followed by cleavage of the Ti–C bridging bond and polarization of the monomer in a six-membered cyclic transition state. The molecule of monomer then inserts into the Al–C bond and the bridge reforms.

$$P = \left(CH_2{-}CHX \right)_n R$$

6.4.3 Propagation: Monomer Insertion at Transition Metal–Carbon Bonds

A *monometallic mechanism* proposed by Cossee and Arlman is the most widely accepted mechanism in which propagation occurs by insertion of monomer at transition metal–carbon bonds. They recognized that for electrical neutrality in α-$TiCl_3$ crystals the octahedrally coordinated surface Ti atoms must have Cl vacancies (i.e. empty d-orbitals) and proposed that the active sites are surface Ti atoms which have been alkylated by reaction with AlR_3. The overall propagation reaction is represented by

$$R{-}Ti{-}\square \xrightarrow{n+1\ CH_2{=}CHX} CHX\left(CH_2{-}CHX\right)_n R{-}CH_2{-}Ti{-}\square$$

where the peripheral Cl ligands have been omitted for clarity and —□ indicates an empty d-orbital. Details of the mechanism are presented below. After initial coordination of the monomer at the vacant d-orbital, it is inserted into the Ti–C bond via a cyclic transition state and migration of the chain, since this requires the least movement of atoms. The polymer chain then migrates back to its original position thus preserving the stereochemical constraints associated with the specific nature of the original active site at the catalyst surface.

$$P = \left(CH_2{-}CHX \right)_n R$$

Although the final back migration of the chain is generally considered to be essential for formation of isotactic polymer, it has been the subject of much discussion. The most widely accepted view is that the original position of the chain is far less crowded and that relief of steric crowding provides the driving force for back migration. However, it also is possible that the two positions for the chain give identical constraints to monomer coordination and insertion, such that the back migration is not necessary, i.e. isotactic sequences would be produced regardless of whether or not the back migration occurs. This is difficult to establish for Ziegler–Natta catalysts because they are heterogeneous, but it certainly is the situation for metallocene catalysts (see Section 6.5.2).

6.4.4 Propagation: Mechanistic Overview

Whilst the precise mechanism of propagation is not known, there are certain mechanistic features which now are widely accepted on the basis of experimental evidence.

1. Monomer initially is coordinated at vacant d-orbitals of transition metal atoms at the catalyst surface.
2. The orientation of a coordinated molecule of monomer is determined by its steric and electronic interactions with the ligands around the transition metal atom. One particular orientation is of lowest energy.
3. The propagation step is completed by insertion of the coordinated molecule of monomer into a metal–carbon bond.
4. The orientation of the molecule of monomer as it inserts into a metal–carbon bond determines the configuration of the asymmetric carbon atom in the newly-formed terminal repeat unit.
5. Isotactic polymer is formed when the preferred orientation for coordination of monomer is of much lower energy than other possible orientations; each successive molecule of monomer then adopts the same preferred orientation as it undergoes coordination and then insertion.
6. The mechanism of monomer insertion always leads to the formation of linear polymer chains, irrespective of the detailed stereochemistry.

The weight of experimental evidence is in favour of monometallic mechanisms involving successive insertions of monomer into a transition metal–carbon bond. In addition, it is found that the methylene carbon atom from the monomer always is bonded to the transition metal atom (i.e. Cat$-CH_2-CHX\{CH_2-CHX\}_n$R and not Cat$-CHX-CH_2\{CHX-CH_2\}_n$R where Cat represents the catalyst surface).

6.4.5 Termination of Chain Growth

Although there is no inherent termination reaction, several types of transfer reaction are possible and these terminate the growth of a propagating chain. Some of the more common ones are summarized below.

(i) Internal hydride transfer

$$\text{Cat}-CH_2-\underset{X}{CH}\{CH_2-\underset{X}{CH}\}_n R \longrightarrow \text{Cat}-H + CH_2{=}\underset{X}{C}\{CH_2-\underset{X}{CH}\}_n R$$

(ii) Chain transfer to monomer

$$\text{Cat}-CH_2-\underset{X}{CH}\{CH_2-\underset{X}{CH}\}_n R + CH_2{=}\underset{X}{CH} \longrightarrow \text{Cat}-CH_2-\underset{X}{CH_2} + CH_2{=}\underset{X}{C}\{CH_2-\underset{X}{CH}\}_n R$$

(iii) Chain transfer to the organometallic compound (MtR_m)

$$\text{Cat}-CH_2-\underset{X}{CH}\{CH_2-\underset{X}{CH}\}_n R + MtR_m \longrightarrow \text{Cat}-R + R_{m-1}Mt\{CH_2-\underset{X}{CH}\}_{n+1} R$$

(iv) Chain transfer to compounds (H–T) with active hydrogen(s)

$$\text{Cat}-CH_2-\underset{X}{CH}\{CH_2-\underset{X}{CH}\}_n R + H-T \longrightarrow \text{Cat}-T + CH_3-\underset{X}{CH}\{CH_2-\underset{X}{CH}\}_n R$$

where Cat and R have their usual meaning, Mt is a group I–III metal of oxidation number m (e.g. if Mt = Al, $m = +3$) and T is a molecular fragment bonded to an active hydrogen atom.

Under normal conditions of polymerization, internal hydride transfer is negligible and termination of propagating chains is dominated by chain transfer processes. Polymer molar mass often is controlled by using hydrogen as a chain transfer agent (i.e. via process (iv) with T = H).

6.4.6 Kinetics

The kinetics of Ziegler–Natta polymerization are complicated by the heterogeneous nature of the reaction and so will only be considered in simple outline here. The rate of polymerization is given by

$$R_p = k_p C_p^* \theta_M \tag{6.1}$$

where

k_p is the rate coefficient for propagation

C_p^* is the concentration of active catalyst sites

θ_M is the fraction of these sites at which monomer is adsorbed.

Note that for heterogeneous kinetics, k_p has the dimensions of s^{-1}. Usually, θ_M is expressed in terms of a standard adsorption isotherm (e.g. Langmuir) and is assumed to have an equilibrium value which depends upon competition between the monomer, the organometallic compound and other species (e.g. hydrogen) for adsorption at the active catalyst sites. Catalyst reactivity often is specified in terms of its *activity*, which normally is reported as the mass of polymer produced per unit mass of transition metal per unit time. Commercial supported Ziegler–Natta catalysts typically have activities in the range 1500–3000 kg of polymer per gram of Ti per hour. When comparing activities for catalysts based on different transition metals, the values need to be adjusted to account for the differences in their atomic weights so that they are per mole of transition metal.

A general equation for the number-average degree of polymerization $\bar{x}_n$ can be obtained by application of Equation 4.13

$$\bar{x}_n = \frac{k_p C_p^* \theta_M}{k_{ht} C_p^* + k_{trM} C_p^* \theta_M + k_{trA} C_p^* \theta_A + k_{trH_2} C_p^* \theta_{H_2}}$$

where

k_{ht}, k_{trM}, k_{trA} and k_{trH_2} are the rate coefficients for internal hydride transfer and for chain transfer to monomer, organometallic compound and hydrogen, respectively (all the rate coefficients having dimensions of s^{-1})

θ_A and θ_{H_2} are the respective fractions of the active catalyst sites at which the organometallic compound and hydrogen are adsorbed.

This equation can be inverted and simplified to yield a general Mayo–Walling equation for Ziegler–Natta polymerization

$$\frac{1}{\bar{x}_n} = \frac{k_{ht}}{k_p \theta_M} + \frac{k_{trM}}{k_p} + \frac{k_{trA} \theta_A}{k_p \theta_M} + \frac{k_{trH_2} \theta_{H_2}}{k_p \theta_M} \tag{6.2}$$

At high concentrations of monomer (i.e. high θ_M) in the absence of hydrogen, Equation 6.2 takes the limiting form $\bar{x}_n = k_p/k_{trM}$.

Due to the heterogeneous nature of Ziegler–Natta catalysts, there are differences in activity between individual active sites on the same catalyst surface and so they are said to be *multi-site catalysts*. Furthermore, termination processes can alter the nature of a given active site and thereby modify its activity (see Section 6.4.5). Clearly, polymer chains will grow more rapidly at sites of high activity than at sites of low activity and so the rate coefficients in Equations 6.1 and 6.2 must be regarded as average quantities. A consequence of the differences in activity between individual active sites is that the polymer formed has a broad distribution of molar mass (typically, $5 < \bar{M}_w/\bar{M}_n < 30$).

6.4.7 Practical Considerations

In general, organometallic compounds are highly reactive and many ignite spontaneously upon exposure to the atmosphere. For this reason, Ziegler–Natta catalysts are prepared and used under inert, dry conditions, typically employing hydrocarbons (e.g. cyclohexane, heptane) as solvents and diluents. Normally, polymerization is carried out at temperatures in the range 50–150 °C with the general observation that rates of polymerization increase but stereospecificity decreases as temperature increases. Most catalysts have some active sites which do not yield stereoregular polymer. Thus, when preparing crystalline isotactic poly(α-olefins), it often is necessary to remove amorphous atactic polymer from the product by solvent extraction.

The three general types of processes used for polymerization of ethylene and α-olefins employ supported heterogeneous catalysts and are the solution, slurry and gas-phase processes. *Solution processes* operate at high temperatures (>130 °C) so that as the polymer forms it dissolves in the hydrocarbon solvent used. At the lower temperatures (50–100 °C) used in *slurry processes,* the polymer is insoluble in the hydrocarbon diluent and precipitates as it forms to give a dispersion (or slurry) of polymer in the diluent. Advances in catalyst technology led to a major increase in the use of *gas-phase processes* which have the distinct advantage of not requiring a solvent or diluent. These processes involve dispersion of the particulate catalyst in gaseous monomer and operate at low temperatures and pressures. Each of these processes are used for commercial production of high-density (linear) polyethylene (HDPE), isotactic polypropylene and copolymers of ethylene with low to medium levels of α-olefins, which are referred to as linear low-density polyethylenes (LLDPE) to distinguish them from the low-density (highly branched) polyethylene (LDPE) homopolymers produced by free-radical polymerization (see Section 4.2.4.2).

Ziegler–Natta catalysts also can be used for preparation of stereoregular polymers from 1,3-dienes. For example, polyisoprene with 96–97% *cis*-1,4 content (i.e. synthetic 'natural rubber') can be prepared using catalysts obtained from $TiCl_4 + Al^iBu_3$.

Attempts have been made to prepare stereoregular polymers from polar monomers (e.g. vinyl chloride, methyl methacrylate) using modified Ziegler–Natta catalysts, but without success. When polymerization does occur, it yields non-stereospecific polymer and is thought to proceed by free-radical mechanisms.

Shirakawa reported in 1974 the fortuitous discovery that films of high molar mass polyacetylene fibrils could be prepared by Ziegler–Natta polymerization of gaseous acetylene at the surface of a highly concentrated $Ti(O^nBu)_4$–$AlEt_3$ catalyst in toluene. This marked a turning point in realizing the conductivity of polyacetylene in a useful form and led ultimately to award of the 2000 Nobel Prize for Chemistry to Shirakawa, Heeger and MacDiarmid for their discovery and development of conducting polymers (see Section 25.3.4).

6.5 METALLOCENE COORDINATION POLYMERIZATION

Metallocenes are organometallic sandwich compounds of a transition metal (Mt) with two cyclopentadienyl (Cp) anions. The simplest metallocenes have the general structure Cp_2Mt

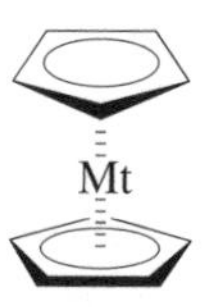

in which the transition metal has an oxidation state of +2 (e.g. ferrocene, where Mt = Fe and titanocene, where Mt = Ti) and ııııııı represents a coordinate bond. Metallocene derivatives are common and have additional groups attached to the metal in a higher oxidation state and/or the cyclopentadienyl ring (e.g. zirconocene dichloride Cp_2ZrCl_2).

Metallocene derivatives were first investigated as catalysts for olefin polymerization during the period when Ziegler–Natta polymerization was being developed, in part because they are soluble (i.e. homogeneous catalysts) and easier to study, but this early work was not very successful. However, the work of Kaminsky during the 1970s and 1980s led to discovery that the reaction of methylaluminoxane with titanocenes and zirconocenes produces high-activity catalysts for homogenous coordination polymerization of ethylene and α-olefins. This discovery triggered rapid development of metallocene catalysts for commercial production of polyolefins during the 1990s.

6.5.1 Metallocene Catalysts

Titanocene and zirconocene derivatives have received most attention as catalysts for effecting polymerization of olefins. By far the most important metallocene catalysts are based upon zirconocene dichloride derivatives which are activated by reaction with methylaluminoxane (MAO) and so such catalyst systems will be the focus here. MAO is formed by controlled hydrolysis of trimethylaluminium and, although it is known to be oligomeric, its precise structure has proven difficult to elucidate. MAO is represented most commonly as a linear oligomer

$$CH_3-\left[\overset{CH_3}{\overset{|}{Al}}-O\right]_x-\overset{CH_3}{\overset{|}{Al}}-CH_3$$

where $x \approx 3$–20, but cyclic oligomer, oligomeric 2D-ladder and oligomeric 3D-cluster structures also have been proposed, the latter being thought to be the most important. The effects of MAO are complex because it not only methylates the zirconocene to produce the polymerization catalyst site, but also undergoes several other reactions which include activation, deactivation and reactivation of the catalyst site. These reactions will be considered in the following section.

6.5.2 Mechanism of Polymerization with Zirconocene:MAO Catalysts

The catalyst sites are formed from zirconocene dichlorides by methylation on reaction with MAO (which can displace either of the two Cl atoms, though only displacement of the lower one is shown here)

$$(C_p^*)_2ZrCl_2 + MAO \rightleftharpoons (C_p^*)_2Zr(Cl)CH_3 + (H_3C)(Cl)Al-O-$$

where C_p^* represents any cyclopentadienyl anion, including those with substituents on the ring. As for Ziegler–Natta polymerization, the mechanism of propagation with zirconocene:MAO catalysts has been the subject of debate, but there is now consensus that the MAO reacts further to activate the catalyst site by formation of an ion pair that (similar to ionic polymerization) can exist as independent ions, a solvent-separated ion pair, a contact ion pair or with strong coordination (and covalent character), the latter two states being depicted below

$$(C_p^*)_2Zr(Cl)CH_3 + MAO \longrightarrow (C_p^*)_2Zr^{\delta+}(CH_3)\cdots{}^{\delta-}[MAO-Cl] \rightleftharpoons (C_p^*)_2Zr^{+}(CH_3)\ [MAO-Cl]^-$$

A large excess of MAO is needed (e.g. molar ratios of Al:Zr above 3000), which has been explained partly in terms of there being only a few Al atoms in MAO that have the attached groups required for activation of the zirconocene, but also because of the other reactions of MAO mentioned earlier. MAO is known to deactivate catalyst sites through a hydrogen-transfer process that eliminates methane

$$(C_p^*)_2Zr(Y)CH_2-H + MAO \rightleftharpoons (C_p^*)_2Zr(Y)CH_2-Al(CH_3)-O- + CH_4$$

where Y represents the [MAO–Cl] counter-ion species and the state of ionization of the Zr–Y bond is neglected. The $-Zr-CH_2-Al-$ species are formed rapidly, but are inactive as polymerization catalysts. Hence, MAO is thought to undergo a further reaction which regenerates the active site

$$\mathrm{Cp^*_2Zr(Y)(-CH_2-Al(CH_3)-O-)} + \mathrm{MAO} \rightleftharpoons \mathrm{Cp^*_2Zr(Y)(CH_3)} + \left\{-O-Al(CH_3)-CH_2-Al(CH_3)-O-\right\}$$

Besides these reactions, MAO also is thought to scavenge impurities.

Propagation proceeds by coordination of monomer to the activated Zr atom, followed by insertion of monomer into the Zr–C bond. Propagation is regioselective, with monomer molecules almost always being inserted such that the methylene carbon atom from the monomer is bonded to the Zr atom. However, the precise mechanism of coordination and propagation is not established with certainty, in part because the intermediate states are difficult to detect, but also because it depends upon the monomer and the nature of the ligands around the Zr atom. Since most metallocene coordination polymerizations are carried out in hydrophobic media, ion-pair separation will be small and any significant separation of the ions is likely to be transient. Nevertheless, there is consensus that the active site is cationic at the point of monomer insertion and that the mechanism involves an initial displacement of the counter-ion by coordinated monomer, which then is followed by monomer insertion through migration of the existing chain because this minimizes the atomic movements necessary for insertion (as in the monometallic mechanism of Ziegler–Natta polymerization; see Section 6.4.3).

$$\mathrm{Cp^*_2Zr(Y)(CH_2-CHXP)} \xrightleftharpoons{CH_2=CHX} \mathrm{Cp^*_2Zr^+(CH_2=CHX)(CH_2-CHXP)}\ Y^- \rightleftharpoons [\text{4-centre transition state}]\ Y^- \rightleftharpoons \mathrm{Cp^*_2Zr^+(\square)(CH_2-CHX-CH_2-CHXP)}\ Y^- \rightleftharpoons \mathrm{Cp^*_2Zr(Y)(CH_2-CHX-CH_2-CHXP)}$$

$Y = [MAO-Cl]$

$P = \left(CH_2-CHX\right)_n R$

The various postulated mechanistic pathways differ mainly in terms of the direction of approach of monomer to the active site and the degree of dissociation of the counter-ion, i.e. the first stage in the mechanism. The mechanism is shown above for the most likely situation in which there is a brief, incomplete dissociation of the counter-ion to an outer-sphere location that is just far enough away from the Zr atom to facilitate monomer coordination and insertion. The ion separation may be sufficiently long for a sequence of several monomer insertions to take place before the ions re-associate. Note that, unlike the monometallic mechanism of Ziegler–Natta polymerization, the metallocene propagation mechanism does not involve a final back migration of the chain to its original position. Thus, irrespective of the detailed mechanism, propagation can be represented by

$$\mathrm{Cp^*_2Zr(Y)(R)} \xrightarrow{n+1\ \mathrm{CH_2{=}CHX}} \mathrm{Cp^*_2Zr(Y)}\!\left[\mathrm{CH_2{-}CHX}\!\left(\mathrm{CH_2{-}CHX}\right)_n\mathrm{R}\right]$$

where $n = 0$ or an even integer

$$\mathrm{Cp^*_2Zr(Y)}\!\left[\mathrm{CH_2{-}CHX}\!\left(\mathrm{CH_2{-}CHX}\right)_n\mathrm{R}\right]$$

where n = an odd integer

where $R = CH_3$ for zirconocene:MAO catalysts. The mechanisms of propagation for other metallocene catalysts are thought to be similar to that for zirconocenes.

6.5.3 Control of Propagation Stereochemistry with Zirconocenes

In order to achieve control of stereochemistry, the active site must be selective in controlling the orientation of monomer as it coordinates to the cationic Zr atom. The zirconocenes considered so far are not able to do this because the substituted cyclopentadienyl anions rotate to give different conformers which present different stereochemical constraints around the Zr atom and direct different modes of monomer insertion; the rotation happens on a timescale that is short compared to propagation and so atactic polymer is produced. Bulky substituents on the cyclopentadienyl anions can drastically reduce the rate of rotation and have been used to produce *oscillating zirconocenes* that adopt distinct stereochemical conformations for periods of time that are long enough for several successive monomer insertions. For example, catalysts produced from bis(2-phenylindene)ziconocene dichloride are in slow equilibrium between two *anti*-conformers, which are selective and produce isotactic placements, and one *syn*-conformer, which is not selective

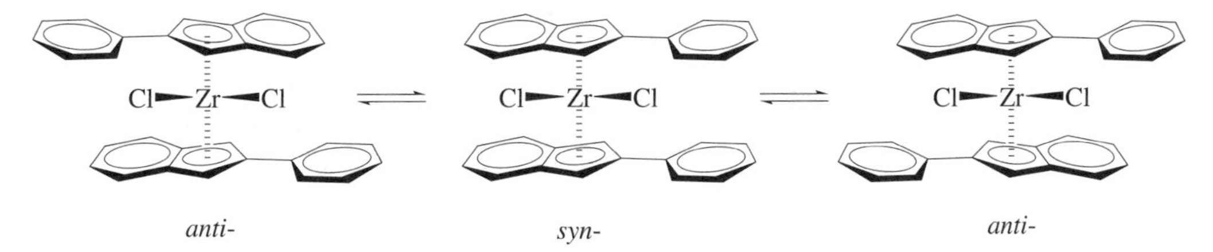

and so produce polymer chains with blocks of isotactic and atactic sequences (so-called *stereoblock polymers*), the length of the blocks increasing as the polymerization temperature reduces (i.e. as the rate of rotation diminishes).

Full stereochemical control most commonly is achieved using *ansa*-zirconocenes in which two substituted cyclopentadienyl anions are linked together by a bridge that prevents rotation of the rings. Examples are catalysts derived from the *ansa*-bisindenylzirconocene dichloride with a $-CH_2CH_2-$ bridge, which has the following three forms

R- *S*- *meso*-

The chiral catalysts produced from the *R*- and *S*-forms are distinct compounds and selective; each is able to produce isotactic polymer, as is a racemic mixture of these forms (because each form acts independently in the mixture). In contrast, catalysts produced from the *meso*-form are non-selective and produce atactic polymer. In catalysts derived from the *R*- and *S*-forms, the two active positions are identical (corresponding to the Cl atom positions of the *ansa*-bisindenylzirconocene dichloride) and so give rise to the same constraints on monomer coordination and insertion (the two positions are said to be *homotopic*). Hence, when the chain moves from one position to the other with each successive monomer insertion, the constraints remain the same and so isotactic sequences are formed. Through careful design of the *ansa*-zirconocene it is possible to produce catalysts in which the two active positions are *enantiotopic* and provide exactly opposite constraints on monomer coordination and insertion, such that syndiotactic polymer is formed as a consequence of the change in position of the chain with each successive monomer insertion. Catalyst activity can be increased by using a shorter bridge (e.g. $Si(CH_3)_2$), but the resulting more open catalyst site usually has lower stereospecificity which then needs to be regained by use of more bulky ligands. Extensive research into *ansa*-zirconocene catalysts of this kind led to their commercialization during the 1990s for production of highly stereospecific polymers from α-olefins.

As for Ziegler–Natta coordination polymerization, there is no inherent termination reaction, but internal hydride transfer and chain transfer to monomer reactions, identical to those shown for Ziegler–Natta polymerization in Section 6.4.5 (i) and (ii), lead to termination of the growth of individual chains. Both reactions produce polymer chains with terminal C=C bonds that (unlike in Ziegler–Natta polymerizations) can participate in propagation with metallocene catalysts of higher activity and/or at higher temperatures and lead to a (usually small) proportion of long-chain branches. For α-olefins, internal hydride transfer is promoted following inverted monomer insertion (i.e. in which the substituted carbon atom from the monomer becomes attached to the Zr atom), which although infrequent, occurs to a greater extent in metallocene polymerizations than in Ziegler–Natta polymerizations.

6.5.4 Kinetics of Metallocene Polymerization

Since the active sites are created at the outset of polymerization and metallocene catalysts are homogeneous, the kinetics of metallocene polymerization should be more easily predicted than for heterogeneous Ziegler–Natta polymerization. Simple analysis gives the following equation for the rate of polymerization

$$R_p = k_p C_p^*[M] \tag{6.3}$$

where

k_p is the rate coefficient for propagation (with its usual dimensions of $dm^3\ mol^{-1}\ s^{-1}$ because the polymerization is homogeneous, cf. Section 6.4.6)

C_p^* is the concentration of active catalyst sites

[M] is the monomer concentration.

In practice, the order with respect to [M] often falls between one and two. The reasons for this remain the subject of debate. One rationale is that the monomer insertion step is triggered by intimate approach of a second molecule of monomer, a hypothesis that can be interpreted in different ways. For example, if it is considered that the active site is created by monomer coordination and insertion of that monomer molecule occurs only when triggered by an incoming second molecule of monomer, then $C_p^* = C_p[M]$ where C_p is the concentration of metallocene:activator catalyst sites (e.g. Cp_2ZrCH_3:MAO–Cl sites), which leads to $R_p \propto [M]^2$. Experimentally observed fractional orders between one and two have been rationalized on the basis that in the absence of a second molecule of monomer, propagation still proceeds (with $R_p \propto [M]$), but at a lower rate than when a second molecule of monomer is present. In this case, simple analysis gives

$$R_p = k_p^{slow} C_p[M] + k_p^{fast} C_p[M]^2 \tag{6.4}$$

where k_p^{slow} and k_p^{fast} are the propagation rate coefficients for propagation in the presence and absence of a second molecule of monomer, respectively, so that the balance between the two modes of propagation will determine the apparent overall order of reaction with respect to [M]. The actual situation would be more complex than implied by Equation 6.4 because each propagation step changes the nature of the active site (e.g. monomer may or may not still be coordinated). Another interpretation is that the active site is in equilibrium between active and inactive states. The reader will need to consult reviews, papers and conference proceedings on metallocene polymerization (beyond those given here in Further Reading) to keep up-to-date with developments in understanding.

Prediction of number-average degree of polymerization may also be treated simply and, by analogy with derivation of Equation 6.2 for Ziegler–Natta polymerization, the Mayo–Walling equation for metallocene polymerization is obtained by replacing adsorbed fractions with the corresponding concentrations

$$\frac{1}{\bar{x}_n} = \frac{k_{ht}}{k_p[M]} + \frac{k_{trM}}{k_p} + \frac{k_{trH_2}[H_2]}{k_p[M]} \tag{6.5}$$

where k_{ht}, k_{trM} and k_{trH_2} are the rate coefficients for internal hydride transfer and for chain transfer to monomer and hydrogen (with the rate coefficients now having their usual dimensions for homogeneous reactions, i.e. the latter two now have dimensions of $dm^3\ mol^{-1}\ s^{-1}$; cf. Section 6.4.6). The term in chain transfer to hydrogen is included because, as for Ziegler–Natta polymerization, hydrogen sometimes is used as a transfer agent to control $\bar{x}_n$ in metallocene polymerizations. Prediction of $\bar{x}_n$ becomes significantly more complex when the effects of an incoming second molecule of monomer are significant.

For most metallocenes, the catalyst dissolves in the reaction medium and each molecule of catalyst is the same. Hence, even though the reactivity of each site may undergo transient changes, for homogeneous metallocene catalysts the individual catalyst sites may be considered equivalent; for this reason, metallocene catalysts are referred to as *single-site catalysts*. This results in the formation of polymers with much narrower molar mass distributions ($2 < \bar{M}_w/\bar{M}_n < 5$) than is possible using heterogeneous (multi–site) Ziegler–Natta catalysts. Furthermore, because each catalyst molecule provides a catalyst site, metallocene catalysts are of higher activity than heterogeneous Ziegler–Natta catalysts in which only a proportion of the total number of transition metal atoms are present in reactive sites. For example, Kaminsky's original work on a simple homogeneous Cp_2ZrCl_2:MAO catalyst revealed a remarkably high activity of ~40 tonnes per gram of Zr per hour for polymerization of ethylene. However, zirconocene catalyst activity is reduced significantly by steric hindrance from the substituents needed to achieve stereochemical control and activities in the range 100–3000 kg of polymer per gram of Zr per hour are more typical.

6.5.5 Other Metallocene and Metallocene-Related Catalysts

In addition to the zirconocene:MAO systems, other catalyst systems have been developed. Important alternatives to MAO include pentafluorophenyl borane derivatives, which are chemically robust and have greater resistance to hydrolysis. For example, highly active zirconocene catalysts are produced by reaction of dimethylzirconocenes with trityl tetra(pentafluorophenyl)borane because this gives a more weakly held anion than from MAO

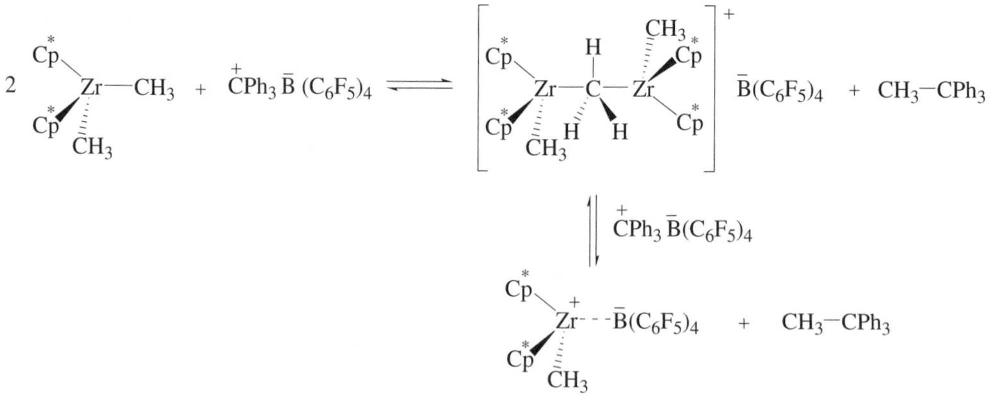

The discovery of zirconocene:MAO catalysts also triggered rapid development of related catalysts, several of which have been commercialized. Dow Chemical developed *ansa*-monocyclopentadienyl amido group IV transition metal catalysts, which have been termed *constrained geometry catalysts*, such as those derived from

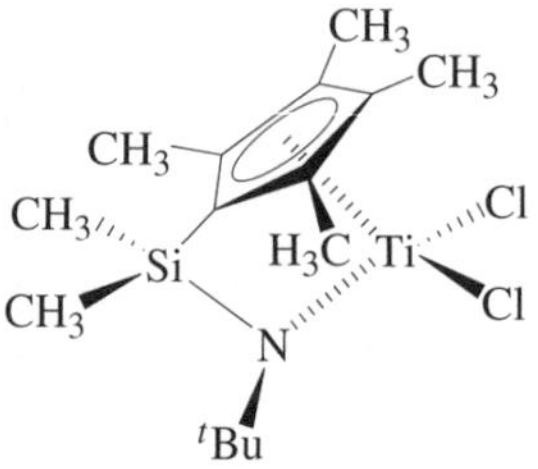

through activation with MAO or organoboranes. Catalysts of this type have a more open active site and so have high activity but do not give good control of stereochemistry in polymerizations of α-olefins. Nevertheless, they are very effective for polymerization of ethylene and are able to copolymerize ethylene with the polyethylene chains that have terminal C=C bonds which are formed by occasional internal hydride transfer reactions during polymerization, thereby producing 'linear' polyethylene with controlled levels of long-chain branching, the level increasing as polymerization temperature increases.

Other catalysts have been developed specifically for polymerization of ethylene, such as MAO-activated late transition metal complexes with bidentate diimine ligands and MAO-activated iron complexes with tridentate 2,6-diiminopryridyl ligands, examples of which are

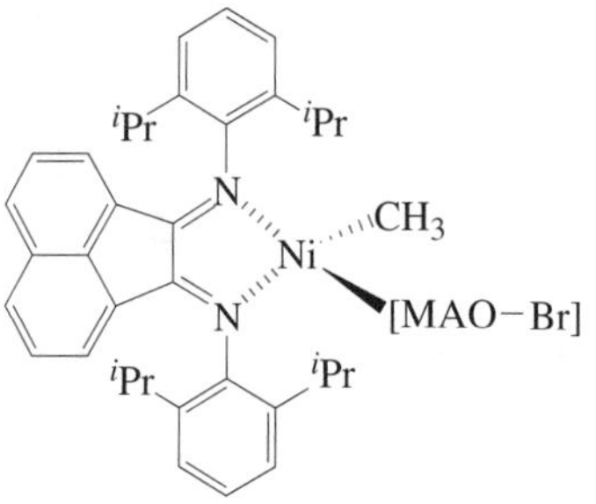

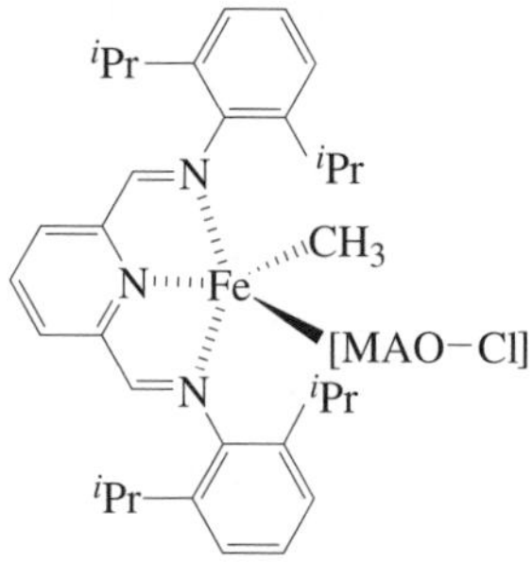

Diimine complexes of Ni and Pd produce polyethylene with a high proportion of methyl side groups, which result from a relatively high propensity for internal hydride transfer with retention of

coordination to the terminal C=C bond generated, followed by simultaneous insertion and reverse hydride transfer to the terminal CH_2

where P represents a polymer chain. The extent of this reaction reduces on increasing the ethylene concentration (because the probability of propagation increases), but increases with polymerization temperature and is far greater with Pd than Ni. Thus, a wide range of different grades of polyethylene with different levels of methyl side groups can be produced. Another feature of the 2,6-diiminopryridyl complexes of iron is that they are tolerant of polar groups and so can be used to copolymerize ethylene with methacrylates and acrylonitrile.

6.5.6 Practical Considerations

Metallocene coordination polymerizations normally are performed at temperatures in the range 10–100 °C in gas phase or in liquid hydrocarbon media. Although homogeneous solution polymerizations often are used in the laboratory, commercial exploitation of metallocene polymerization demanded use of existing reactors and processes designed for heterogeneous supported Ziegler–Natta catalysts. Supported metallocene catalysts were developed for this purpose, mostly using particulate silica and alumina as the support. Several methods have been explored for preparation of supported metallocene catalysts, but by the most effective and important involve initial attachment of MAO to silica support particles followed by addition of the metallocene to form the catalyst sites on the particle surfaces. In most cases, the heterogeneous metallocene catalyst sites obtained via this method show considerable uniformity and still behave as single-site catalysts, producing polymers with similar or greater control of microstructure and similar molar mass distribution as the equivalent homogeneous metallocene catalyst. However, their activity often is reduced significantly as a consequence of the more restricted access to the catalyst site, though this also has a benefit in that the rates of deactivation and termination reactions (e.g. internal hydride transfer and chain transfer to monomer) also are reduced, thereby giving access to production of higher molar mass polymer than is possible with the equivalent homogeneous catalyst. One further benefit of using supports is that the excess of MAO required is greatly reduced; for zirconocenes, typically a molar ratio of Al:Zr in the range 100–500 is sufficient, whereas in homogeneous zirconocene polymerizations ratios in the range 3,000–10,000 are necessary. This also is significant commercially because MAO is relatively expensive.

More active catalysts are used for polymerization of ethylene since the steric constraints which are necessary for control of stereochemistry are not needed. Grades of polyethylene produced using metallocene catalysts at higher temperatures (e.g. 150 °C) have some long-chain branches which result from copolymerization of ethylene with terminally-unsaturated polymer chains displaced from catalyst sites earlier in the reaction by internal hydride transfer and chain transfer to monomer reactions. This occurs because, unlike Ziegler–Natta catalysts, metallocene catalyst sites are more open and accessible and can accommodate larger substituent groups on the C=C bond. Thus, metallocene catalysts also are more effective in producing copolymers of ethylene with α-olefins; for example, higher proportions of comonomer can be included with more uniform sequence distributions than is possible in production of LLDPE by Ziegler–Natta coordination polymerization (Section 6.4.7).

Although Ziegler–Natta catalysts are more easily able to produce very highly isotactic polymers from α-olefins, metallocene catalysts provide much greater flexibility in the control of configurational sequence distribution. By careful design of the metallocene catalyst, it is possible to produce highly isotactic polymer or highly syndiotactic polymer, as well as more complex sequences such as the stereoblock polymers described in Section 6.5.3. Metallocene catalysts also are more capable of controlling the stereochemistry of polymerization of cyclic olefins.

PROBLEMS

6.1 For each of the following monomers, write down each of the possible stereochemical forms of the repeat unit that can be formed during chain polymerization.

(a) $CH_2{=}C(H)({}^{n}Bu)$

(b) $CH_2{=}C(Et)(Et)$

(c) $CH_2{=}CCl{-}CH{=}CH_2$

6.2 A Ziegler–Natta coordination polymerization of propylene was carried out at 50 °C using a $MgCl_2$-supported $TiCl_4$:$AlEt_3$ catalyst activated with ethyl benzoate, *p*-cresol and methyl *p*-toluate. The initial rate of polymerization was 4.55×10^{-4} mol dm^{-3} s^{-1} and the isotactic polypropylene produced early in the polymerization had a number-average molar mass of 320.9 kg mol^{-1}. Using the rate coefficient data given below and assuming that chain transfer to $AlEt_3$ is negligible, calculate the fraction of active sites at which propylene was adsorbed and the concentration of active catalyst sites for this polymerization. If some hydrogen had been included in the polymerization, what would have been the effect on the molar mass of the polypropylene produced?

Rate coefficient for propagation, $k_p = 133$ s^{-1}
Rate coefficient for internal hydride transfer, $k_{ht} = 9.7 \times 10^{-3}$ s^{-1}
Rate coefficient for chain transfer to monomer, $k_{trM} = 7.2 \times 10^{-3}$ s^{-1}

6.3 Outline the methods and conditions of *homopolymerization* you would use to prepare the following polymers, giving reasons for your choices.

(a) Isotactic poly(but-1-ene)
(b) Isotactic poly(methyl methacrylate)
(c) Polyethylene with occasional methyl side groups

FURTHER READING

Allen, G. and Bevington, J.C. (eds), *Comprehensive Polymer Science*, Vol. 4 (Eastmond, G.C., Ledwith, A., Russo, S., and Sigwalt, P., eds), Pergamon Press, Oxford, 1989.

Baugh, L.S and Canich, J.M.M. (eds), *Stereoselective Polymerization with Single-Site Catalysts*, CRC Press, Boca Raton, FL, 2008.

Bochmann, M., Kinetic and mechanistic aspects of metallocene polymerisation catalysts, *Journal of Organometallic Chemistry*, **689**, 3982 (2004).

Boor, J., *Ziegler–Natta Catalysts and Polymerizations*, Academic Press, New York, 1979.

Coates, G.W., Precise control of polyolefin stereochemistry using single-site metal catalysts, *Chemical Reviews*, **100**, 1223 (2000).

Chum, P.S. and Swogger, K.W., Olefin polymer technologies – history and recent progress at the Dow Chemical Company, *Progress in Polymer Science*, **33**, 797 (2008).

Gibson, V.C. and Spitzmesser, S.K., Advances in non-metallocene olefin polymerization catalysis, *Chemical Reviews*, **103**, 283 (2003).

Kaminsky, W., Olefin polymerization catalyzed by metallocenes, *Advances in Catalysis*, **46**, 89 (2001).

Kaminsky, W., The discovery of metallocene catalysts and their present state of the art, *Journal of Polymer Science, Part A: Polymer Chemistry*, **42**, 3911 (2004).

Kaminsky, W. and Sinn, H.-W. (eds), *Transition Metals and Organometallics as Catalysts for Olefin Polymerization*, Springer, Berlin, Germany, 1988.
Lenz, R.W., *Organic Chemistry of Synthetic High Polymers*, Interscience, New York, 1967.
Mark, H.F., Bikales, N.M., Overberger, C.G., and Menges, G. (eds), *Encyclopedia of Polymer Science and Engineering*, Wiley-Interscience, New York, 1985–1989.
Odian, G., *Principles of Polymerization*, 4th edn., Wiley-Interscience, New York, 2004.
Scheirs, J. and Kaminsky, W., *Metallocene-Based Polyolefins – Preparation, Properties and Technology*, Vol. 1, John Wiley & Sons, Chichester, England, 1999.
Scheirs, J. and Kaminsky, W., *Metallocene-Based Polyolefins – Preparation, Properties and Technology*, Vol. 2, John Wiley & Sons, Chichester, England, 2000.
Stevens, M.P., *Polymer Chemistry: An Introduction*, 3rd edn., Oxford University Press, New York, 1998.

7 Ring-Opening Polymerization

7.1 INTRODUCTION TO RING-OPENING POLYMERIZATION

Polymers with the general structure

$$-\!\!\left[\text{R}-\text{Z}\right]\!\!-_n$$

where $-\text{Z}-$ is a linking group (e.g., ether, $-\text{O}-$; ester, $-\text{O}\overset{\text{O}}{\overset{\|}{\text{C}}}-$; amide, $-\text{NH}\overset{\text{O}}{\overset{\|}{\text{C}}}-$)

can be prepared either by step polymerization (Sections 3.2.1 and 3.2.2) or by *ring-opening polymerization* of the corresponding cyclic monomer, i.e.

$$n\ \text{cyclo}(\text{R}-\text{Z}) \longrightarrow -\!\!\left[\text{R}-\text{Z}\right]\!\!-_n$$

Additionally, ring-opening polymerization can be used to prepare polymers, which cannot easily be prepared by other methods, e.g. poly(phosphazene)s. Some important ring-opening polymerizations are listed in Table 7.1. Note that the elongated, curled bonds shown above and in Table 7.1 are for drawing purposes only and allow for more compact representation of the molecules; the actual bonds are normal except for the effects of ring strain, e.g.

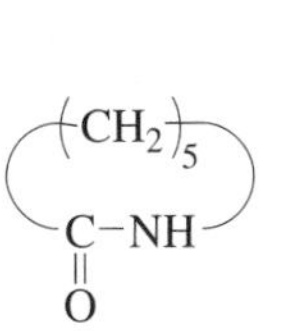

$$\text{cyclo}\left[-(\text{CH}_2)_5-\text{C}(=\text{O})-\text{NH}-\right] \quad \text{represents} \quad \text{seven-membered ring: } \text{CH}_2\text{–CH}_2\text{–CH}_2\text{–CH}_2\text{–CH}_2\text{–N(H)–C(=O)–}$$

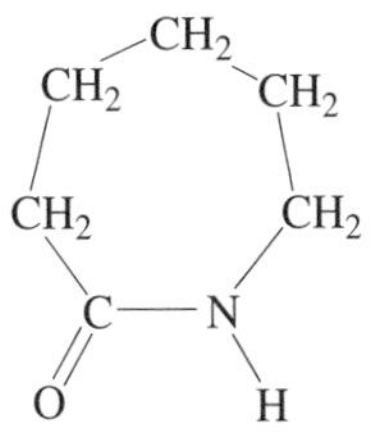

The driving force for ring-opening of cyclic monomers is the relief of bond-angle strain and/or steric repulsions between atoms crowded into the centre of the ring (cf. Section 3.2.4). Therefore, as for other types of polymerization, the enthalpy change for ring-opening is negative. Relief of bond-angle strain is most important for 3- and 4-membered rings, whereas for 8- to 11-membered rings it is the relief of steric crowding that matters. These enthalpic effects are much smaller for five-, six- and seven-membered rings (especially six-membered) and such monomers are more difficult to polymerize. Ring-opening polymerization requires an initiator and in most cases proceeds by chain polymerization mechanisms which most commonly involve sequential additions of monomer to cationic or anionic active centres. However, there are more specialized ring-opening polymerizations which operate by non-ionic mechanisms, e.g. via radical or coordination polymerization. In addition, some ring-opening polymerizations have complex mechanisms in which the ring opens to give a monomer that undergoes polycondensation (e.g. commercial production of nylon 6 (Table 7.1) is effected by hydrolysis of caprolactam to produce $H_2N(CH_2)_5CO_2H$, which then undergoes ARB polycondensation in parallel with ring-opening polymerization of the lactam). Hence the precise mechanism of ring-opening polymerization depends greatly upon the initiator, monomer and polymerization conditions. For this reason, it is not possible to treat generally

TABLE 7.1
Some Important Ring-Opening Polymerizations

Monomer	Polymer
$n\ CH_2{-}CH_2$ (ring via O) Ethylene oxide	$\text{-}[CH_2{-}CH_2{-}O]_n\text{-}$ Poly(ethylene oxide)
$n\ CH_2{-}CH(CH_3)$ (ring via O) Propylene oxide	$\text{-}[CH_2{-}CH(CH_3){-}O]_n\text{-}$ Poly(propylene oxide)
n ring of $(CH_2)_5$ and $C(=O){-}O$ Caprolactone	$\text{-}[(CH_2)_5{-}O{-}C(=O)]_n\text{-}$ Polycaprolactone
n ring of $(CH_2)_5$ and $C(=O){-}NH$ Caprolactam	$\text{-}[(CH_2)_5{-}NH{-}C(=O)]_n\text{-}$ Polycaprolactam (nylon 6)
n/x ring of $(Si(CH_3)_2{-}O)_x$ Normally *either* hexamethylcyclotrisiloxane, D_3 ($x=3$) *or* octamethylcyclotetrasiloxane, D_4 ($x=4$)	$\text{-}[Si(CH_3)_2{-}O]_n\text{-}$ Poly(dimethylsiloxane)
$n/3$ ring of $(PCl_2{=}N)_3$ Hexachlorocyclotriphosphazene	$\text{-}[PCl_2{=}N]_n\text{-}$ Poly(dichlorophosphazene)

the ring-opening polymerization of all cyclic monomers. In the subsequent sections, some specific ring-opening polymerizations are considered in order to illustrate the more important applications and the different types of mechanism that can operate. Most steps in the mechanisms are reversible, but the reactions often are run under conditions of *kinetic control* whereby the forward reactions are forcibly favoured and so only forward reaction arrows are shown here.

7.2 CATIONIC RING-OPENING POLYMERIZATION

The initiators used in cationic ring-opening polymerizations are of the same type as those used for cationic polymerization of ethylenic monomers (Section 5.2.1.1), e.g. strong protonic acids (e.g. H_2SO_4, CF_3SO_3H, CF_3CO_2H), and Lewis acids used in conjunction with 'co-catalysts' (e.g. $Ph_3\overset{+}{C}\,[PF_6]^-$, $CH_3\overset{+}{C}O\,[SbF_6]^-$). For simplicity, the initiator will be generally represented as R^+A^- except where a particular type of initiator is used almost exclusively.

7.2.1 Cationic Ring-Opening Polymerization of Epoxides

Epoxides, also known as *oxiranes*, are three-membered ring cyclic ethers and are very susceptible to ring-opening polymerization due to the high level of ring strain (see the first two examples in Table 7.1). In the *cationic ring-opening polymerization of ethylene oxide*, initiation takes place by addition of R^+ to the epoxide oxygen atom to yield a cyclic oxonium ion (I), which is in equilibrium with the corresponding open-chain carbocation (II)

$$\overset{+}{R}\overset{-}{A} + \underset{\ddot{O}}{CH_2{-}CH_2} \longrightarrow R{-}\overset{+}{\ddot{O}}\langle CH_2{-}CH_2 \rangle \; \overset{-}{A} \rightleftharpoons R{-}\ddot{O}{-}CH_2{-}\overset{+}{C}H_2\overset{-}{A}$$

(I) (II)

In this reaction, the two dots represent lone pairs of electrons, in this case those on oxygen atoms, and the curly arrows indicate electron movements with the double-headed arrows indicating movement of two electrons. Thus, in the addition step, one lone pair from the epoxide oxygen atom provides both electrons for linking to R^+, leaving the oxygen atom with only one lone pair, and thus a positive charge, in the resulting oxonium ion (I). Breaking of the C–O bond in (I), with both electrons moving to become a second lone pair on the oxygen atom, leads to breaking of the oxonium ion ring and leaves the carbon atom with an empty orbital, i.e. a positive charge, in (II); the reverse direction of the equilibrium proceeds via the exact opposite process in which an oxygen lone pair provides both electrons for bonding to the positively charged terminal carbon atom. Both (I) and (II) can propagate: (I) via ring-opening of the cyclic oxonium ion upon nucleophilic attack at a ring carbon atom by the epoxide oxygen atom in another molecule of monomer, and (II) via addition to monomer in a reaction similar to the initiation step. These reactions are depicted below

$$R{+}O{-}CH_2{-}CH_2{+}_n\overset{+}{\ddot{O}}\langle CH_2{-}CH_2\rangle\;\overset{-}{A} \xrightarrow{CH_2{-}CH_2\,(\ddot{O})} R{+}O{-}CH_2{-}CH_2{+}_{n+1}\overset{+}{\ddot{O}}\langle CH_2{-}CH_2\rangle\;\overset{-}{A}\quad (\mathrm{I'})$$

$$R{+}O{-}CH_2{-}CH_2{+}_n\ddot{O}{-}CH_2{-}\overset{+}{C}H_2\overset{-}{A} \xrightarrow{CH_2{-}CH_2\,(\ddot{O})} (\mathrm{I'}) \rightleftharpoons R{+}O{-}CH_2{-}CH_2{+}_{n+1}\ddot{O}{-}CH_2{-}\overset{+}{C}H_2\overset{-}{A}\quad (\mathrm{II'})$$

where only lone pairs on oxygen atoms in the terminal units are shown. In each case, the initial product of propagation has a terminal cyclic oxonium ion (I′) formed from the newly added molecule of monomer; as for the initiation stage, this will be in equilibrium with the open-chain carbocation (II′).

Termination can occur via a rearrangement reaction of the terminal unit involving elimination of H^+ to give an unsaturated C=C end group

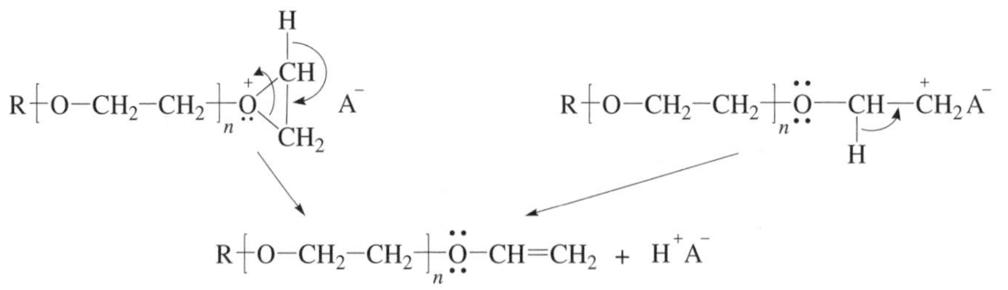

Also, when less stable counter-ions are employed, termination can occur via ion-pair rearrangement, e.g. with $[AlCl_4]^-$ as the counter-ion.

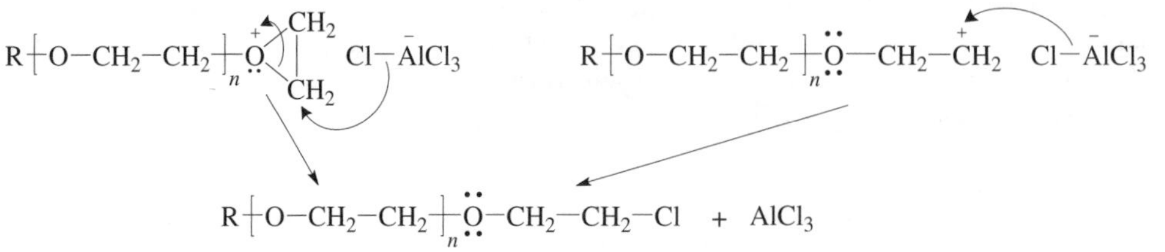

Intramolecular and intermolecular chain transfer to polymer are significant. The former are backbiting processes which lead to the formation of rings, whereas the latter can be considered as interchange reactions and can take place with both linear chain and ring molecules (see Figure 7.1). These reactions give rise to *ring-chain equilibria* which are a characteristic feature of many ring-opening polymerizations (i.e. linear-chain molecules are in equilibrium with ring molecules).

This polymerization is further complicated by other modes of propagation (e.g. via a rearranged form of II′, i.e. $\sim O-\overset{+}{C}H-CH_3$ or via terminal –OH groups if an alcohol is present or if H^+ is the initiating species). Thus, even for ethylene oxide, cationic ring-opening polymerization is quite complex. For epoxide monomers with a substituent group (e.g. propylene oxide), the complexity is greatly increased because it is possible to have head-to-tail or head-to-head placements of repeat units each of which possesses an asymmetric carbon atom that has two possible configurations. The head-to-tail stereoregular forms of such polyethers may be compared to those for ethylenic polymers (Section 6.2). When each asymmetric carbon atom has the same configuration, i.e. in the *isotactic* form, the substituent groups (R) alternate from one side to the other of the planar fully extended backbone.

Isotactic

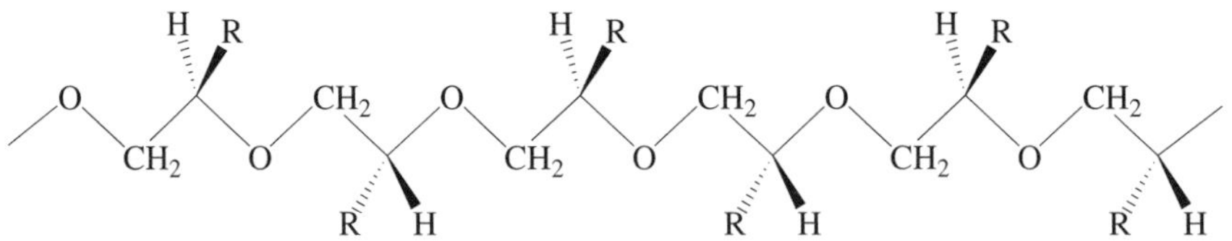

whereas in the *syndiotactic* form the configurations alternate and all the substituent groups are on the same side

Syndiotactic

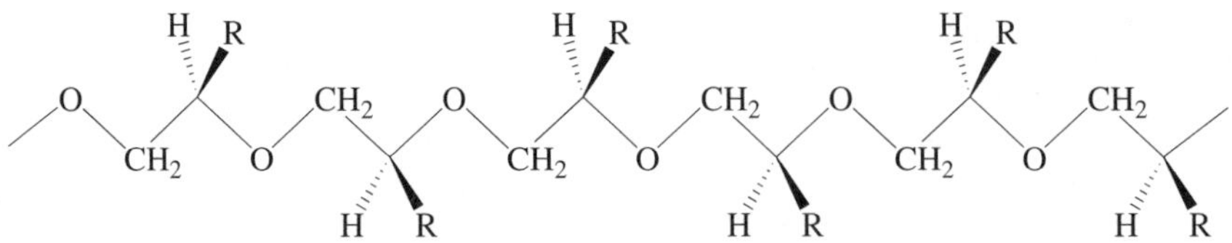

Intramolecular chain transfer to polymer

CH_2 A^- $\overset{+}{O}$ CH_2

~CH_2—CH_2—Ö—CH_2—CH_2~

CH_2—Ö~ A^- $\overset{+}{CH_2}$

~CH_2—CH_2—Ö—CH_2—CH_2~

CH_2—Ö~ CH_2

~CH_2—CH_2—$\overset{+}{O}$—CH_2—CH_2~ A^-

CH_2–CH_2 O

~CH_2—CH_2—$\overset{+}{O}$(CH_2)(CH_2) A^- + CH_2—O~ CH_2 :Ö—CH_2—CH_2~

CH_2—Ö~ CH_2 :Ö—CH_2—CH_2~

+ ~CH_2—$\overset{+}{CH_2}$ A^-

Intermolecular chain transfer to polymer

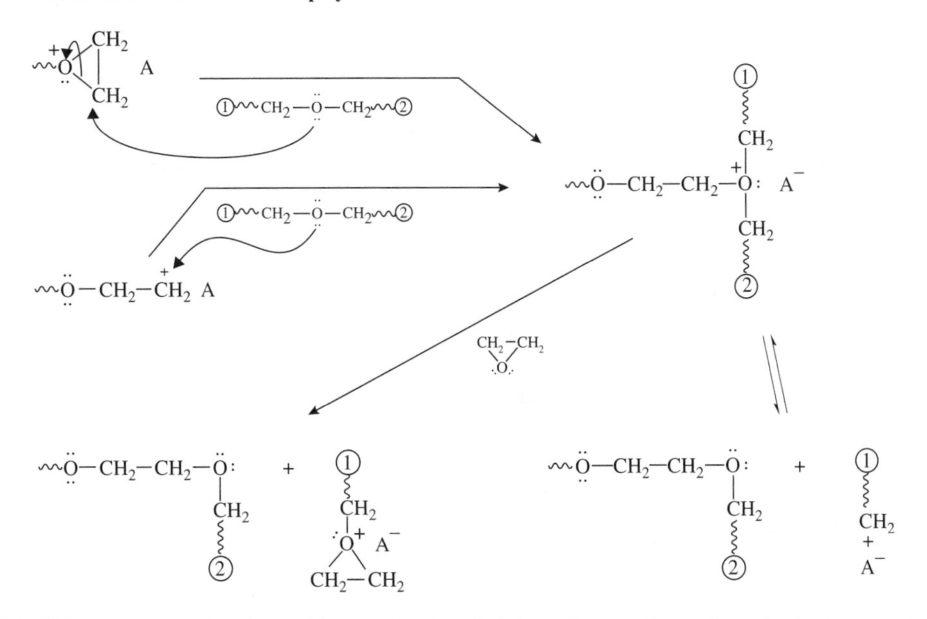

FIGURE 7.1 Intramolecular and intermolecular chain transfer to polymer in cationic ring-opening polymerization of ethylene oxide. The electron movements in the second steps for intermolecular chain transfer to polymer are equivalent to those shown for the second steps in the intramolecular reaction. The circled numbers, ① and ②, are for reference purposes (in a cyclic chain ① and ② would be joined together).

Thus, the positions of the substituent groups relative to the plane of the backbone are opposite to those in isotactic and syndiotactic ethylenic polymers (cf. Figure 6.1). Furthermore, the polyethers are optically active because each repeat unit has a truly asymmetric carbon atom. However, cationic ring-opening polymerization of an optically pure monomer (e.g. H, R, O) does not result in the formation of isotactic polymer since species of type II′ have planar carbocation active centres (sp^3-hybridized, with an empty residual p-orbital) that usually are stabilized by the substituent group and make a more significant contribution to propagation than for ethylene oxide

Rotation of terminal bond

Thus, the specific configuration of the asymmetric carbon atom in the monomer is lost when the planar carbocationic active centre is formed.

7.2.2 Cationic Ring-Opening Polymerization of Lactones

Lactones are cyclic esters with the general structure

and undergo cationic ring-opening polymerization via two principal mechanisms. The dominant mechanism for unsubstituted lactones (i.e. R = H) involves addition of R^+ to the carbonyl oxygen atom to form an oxonium ion, followed by scission of the O–CHR bond upon nucleophilic attack by another molecule of monomer

in which the counter-ion has been omitted from the scheme for simplicity. This mechanism of propagation may be represented by

For substituted lactones (e.g. R = alkyl), this mechanism results in an inversion of the configuration of the asymmetric carbon atom, –CHR– (i.e. the configuration in the polyester is opposite to that in the monomer). However, for such substituted lactones, a second mechanism of propagation is more probable due to the inductive and steric effects of the substituent. This mechanism involves initial addition of R^+ to the ring oxygen atom, followed by scission of the CO—O bond upon nucleophilic attack by another molecule of monomer

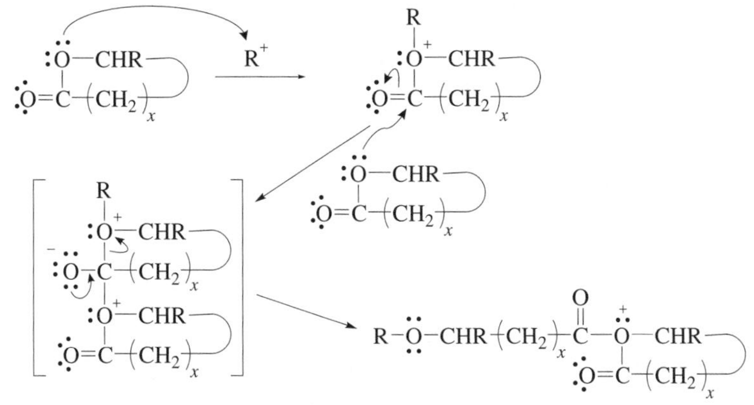

in which the counter-ion again has been omitted and the large square brackets indicate an intermediate species in the reaction. Propagation via this mechanism may be represented by

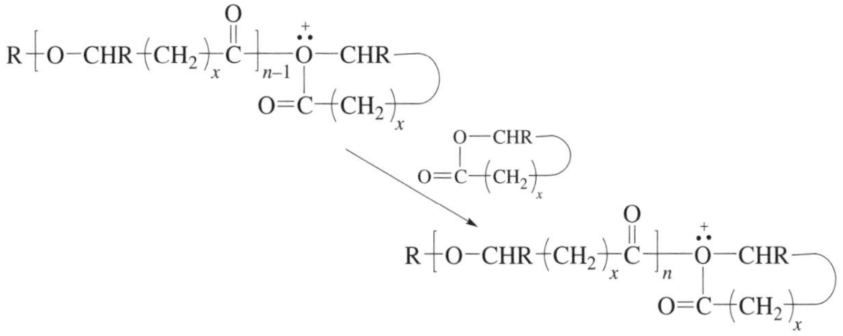

and results in retention of the configuration of the asymmetric carbon atom.

Interchange and backbiting chain transfer to polymer reactions take place in the polymerization of lactones in much the same way as for epoxides. Thus ring-chain equilibria are established and are especially important at high conversions of monomer.

7.2.3 Cationic Ring-Opening Polymerization of Lactams

Cationic ring-opening polymerization of *lactams* (i.e. cyclic amides) usually is initiated with acids which can protonate either the oxygen or nitrogen atom of the lactam. Initiation is believed to involve protonation of the lactam carbonyl oxygen (because it is more basic), which promotes attack of the carbonyl carbon by the lone pair of electrons on the amide nitrogen of another molecule of lactam to give an adduct that rearranges to produce a protonated acyl lactam which then transfers H^+ to another molecule of lactam

$$:\ddot{O}{=}C{-}(CH_2)_x{-}NH{-}CHR \ (\text{ring}) \xrightarrow{H^+} H{-}\overset{+}{\ddot{O}}{=}C{-}(CH_2)_x{-}NH{-}CHR \ (\text{ring})$$

(attack by the ring N–H of $(CH_2)_x{-}C{=}\ddot{O}$: / CHR–$\ddot{N}$H)

$$\left[\ (CH_2)_x{-}C{=}\ddot{O}:\ /\ CHR{-}\overset{+}{N}H{-}\ H{-}\ddot{O}{-}C{-}(CH_2)_x{-}NH{-}CHR\ \right] \rightleftharpoons \left[\ (CH_2)_x{-}C{=}\ddot{O}:\ /\ CHR{-}N:\ H{-}\ddot{O}{-}C{-}(CH_2)_x{-}\overset{+}{N}H_2{-}CHR\ \right]$$

$$\longrightarrow (CH_2)_x{-}C{=}\ddot{O}:\ /\ CHR{-}\ddot{N}{-}C(=\overset{+}{\ddot{O}}{-}H){-}(CH_2)_x{-}CHR{-}\ddot{N}H_2$$

$$+\ :\ddot{O}{=}C{-}(CH_2)_x{-}\ddot{N}H{-}CHR \longrightarrow (CH_2)_x{-}C{=}\ddot{O}:\ /\ CHR{-}\ddot{N}{-}C(=O){-}(CH_2)_x{-}CHR{-}\ddot{N}H_2 + H{-}\overset{+}{\ddot{O}}{=}C{-}(CH_2)_x{-}\ddot{N}H{-}CHR$$

The terminal –NH_2 group in the acyl lactam acts as the nucleophile for attack on another molecule of protonated lactam leading to chain growth via the same chemistry as in the initiation step, with retention of the configuration of the asymmetric carbon atom –CHR– in substituted lactams and releasing H^+ each time to protonate (i.e. activate) another molecule of lactam for addition.

$$(CH_2)_x{-}C{=}\ddot{O}:\ /\ CHR{-}\ddot{N}{-}C(=O){-}(CH_2)_x{-}CHR{-}\ddot{N}H_2 + H{-}\overset{+}{\ddot{O}}{=}C{-}(CH_2)_x{-}\ddot{N}H{-}CHR$$

$$\longrightarrow \left[\ (CH_2)_x{-}C{=}\ddot{O}:\ /\ CHR{-}\ddot{N}{-}C(=O){-}(CH_2)_x{-}CHR{-}\overset{+}{N}H_2{-}\ H{-}\ddot{O}{-}C{-}(CH_2)_x{-}\ddot{N}H{-}CHR\ \right]$$

$$\rightleftharpoons \left[\ (CH_2)_x{-}C{=}\ddot{O}:\ /\ CHR{-}\ddot{N}{-}C(=O){-}(CH_2)_x{-}CHR{-}\ddot{N}H{-}\ H{-}\ddot{O}{-}C{-}(CH_2)_x{-}\overset{+}{N}H_2{-}CHR\ \right]$$

$$\longrightarrow (CH_2)_x{-}C{=}\ddot{O}:\ /\ CHR{-}\ddot{N}{-}C(=O){-}(CH_2)_x{-}CHR{-}NH{-}C(=\overset{+}{\ddot{O}}{-}H){-}(CH_2)_x{-}CHR{-}NH_2$$

$$+\ :\ddot{O}{=}C{-}(CH_2)_x{-}\ddot{N}H{-}CHR \longrightarrow (CH_2)_x{-}C{=}\ddot{O}:\ /\ CHR{-}\ddot{N}{-}C(=O){-}(CH_2)_x{-}CHR{-}NH{-}C(=\ddot{O}:){-}(CH_2)_x{-}CHR{-}NH_2 + H{-}\overset{+}{\ddot{O}}{=}C{-}(CH_2)_x{-}\ddot{N}H{-}CHR \longrightarrow$$

7.2.4 Cationic Ring-Opening Polymerization of Cyclic Siloxanes

High molar mass polysiloxanes are prepared by cationic and anionic ring-opening polymerization of cyclic siloxanes comprising either three or four repeat units (e.g. D_3 or D_4 in Table 7.1).

The chemistry of cationic ring-opening polymerization is shown in Figure 7.2 and is initiated with strong protonic acids which ring-open the cyclic siloxane. The subsequent polymerization is complex

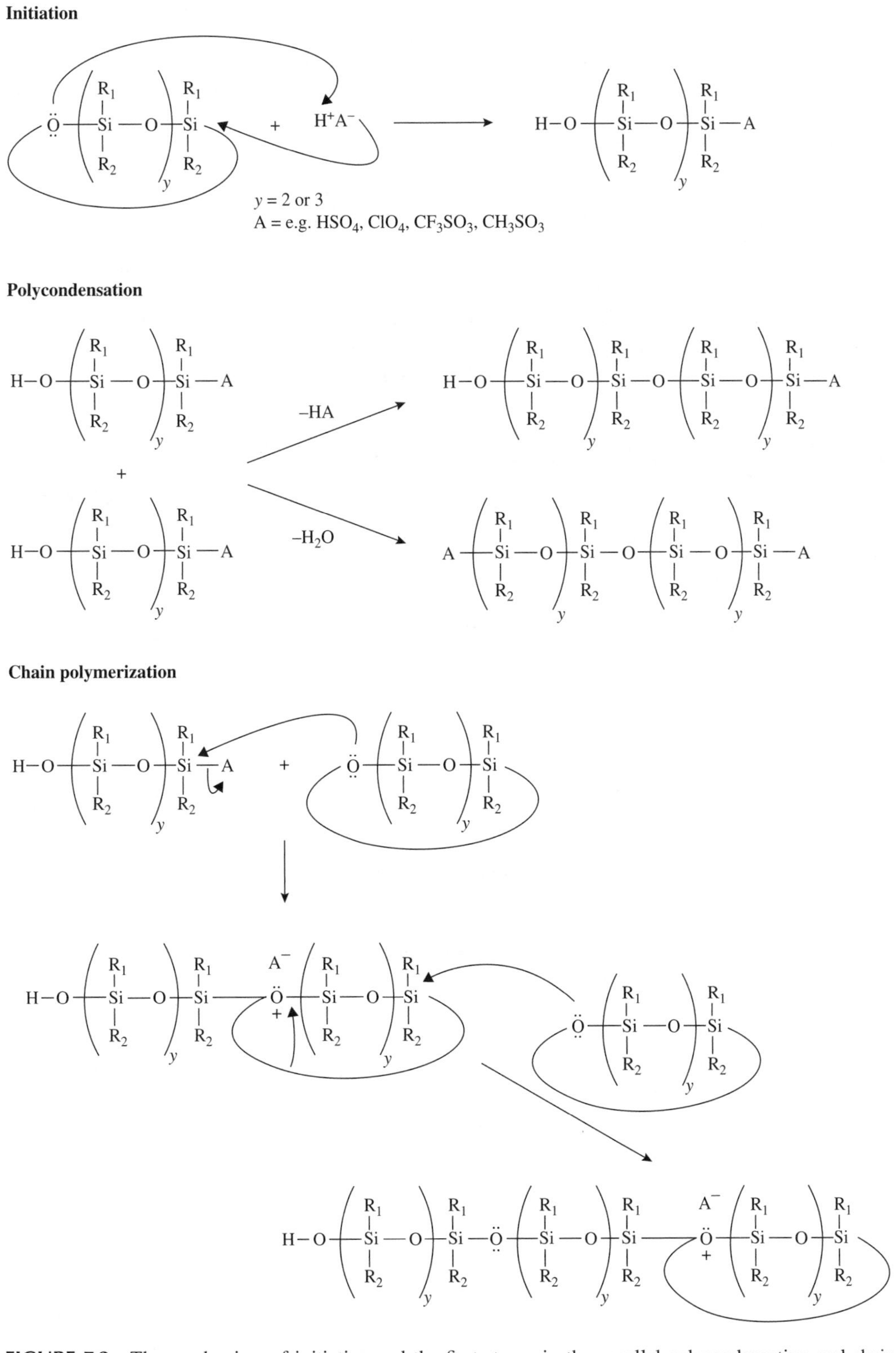

Initiation

$$\text{cyclo-}\left[\ddot{O}{-}\left(Si(R_1)(R_2){-}O\right)_y{-}Si(R_1)(R_2)\right] + H^+A^- \longrightarrow H{-}O{-}\left(Si(R_1)(R_2){-}O\right)_y{-}Si(R_1)(R_2){-}A$$

$y = 2$ or 3

A = e.g. HSO_4, ClO_4, CF_3SO_3, CH_3SO_3

Polycondensation

$$H{-}O{-}\left(Si(R_1)(R_2){-}O\right)_y{-}Si(R_1)(R_2){-}A \;+\; H{-}O{-}\left(Si(R_1)(R_2){-}O\right)_y{-}Si(R_1)(R_2){-}A$$

$$\xrightarrow{-HA} H{-}O{-}\left(Si(R_1)(R_2){-}O\right)_y{-}Si(R_1)(R_2){-}O{-}\left(Si(R_1)(R_2){-}O\right)_y{-}Si(R_1)(R_2){-}A$$

$$\xrightarrow{-H_2O} A{-}\left(Si(R_1)(R_2){-}O\right)_y{-}Si(R_1)(R_2){-}O{-}\left(Si(R_1)(R_2){-}O\right)_y{-}Si(R_1)(R_2){-}A$$

Chain polymerization

$$H{-}O{-}\left(Si(R_1)(R_2){-}O\right)_y{-}Si(R_1)(R_2){-}A + \text{cyclo-}\left[\ddot{O}{-}\left(Si(R_1)(R_2){-}O\right)_y{-}Si(R_1)(R_2)\right]$$

$$\downarrow$$

$$H{-}O{-}\left(Si(R_1)(R_2){-}O\right)_y{-}Si(R_1)(R_2){-}\overset{+}{\ddot{O}}\,A^-\text{-cyclo-}\left[\left(Si(R_1)(R_2){-}O\right)_y{-}Si(R_1)(R_2)\right] + \text{cyclo-}\left[\ddot{O}{-}\left(Si(R_1)(R_2){-}O\right)_y{-}Si(R_1)(R_2)\right]$$

$$\longrightarrow H{-}O{-}\left(Si(R_1)(R_2){-}O\right)_y{-}Si(R_1)(R_2){-}\ddot{O}{-}\left(Si(R_1)(R_2){-}O\right)_y{-}Si(R_1)(R_2){-}\overset{+}{\ddot{O}}\,A^-\text{-cyclo-}\left[\left(Si(R_1)(R_2){-}O\right)_y{-}Si(R_1)(R_2)\right]$$

FIGURE 7.2 The mechanism of initiation and the first stages in the parallel polycondensation and chain polymerization mechanisms in cationic ring-opening polymerization of cyclic siloxanes for which R_1 and R_2 can be alkyl or aryl.

and proceeds via parallel polycondensation and chain polymerization mechanisms. Interchange and backbiting chain transfer reactions also occur and so ring-chain equilibria are established. Despite its complexity, synthesis of high molar mass polysiloxanes by cationic ring-opening polymerization is commercially important and has the advantage that it can be used to polymerize siloxanes with substituents that have functional groups which are reactive towards the strong bases needed to effect anionic ring-opening polymerization (Section 7.3.4).

7.3 ANIONIC RING-OPENING POLYMERIZATION

A very wide range of initiators have been used to effect anionic ring-opening polymerization and these include alkali metals (e.g. Na, K), inorganic bases (e.g. NaOH, KOH, $NaNH_2$), metal alkoxides (e.g. $LiOCH_3$, NaOEt), metal alkyls and hydrides (e.g. Li^nBu, NaH) and electron transfer complexes (e.g. sodium naphthalide). Only use of the first three types of initiator will be considered here. As for cationic ring-opening polymerization, a few important examples will be used to illustrate the chemistry.

7.3.1 ANIONIC RING-OPENING POLYMERIZATION OF EPOXIDES

Polymerization of epoxides by inorganic bases or metal alkoxides (both represented by Mt^+B^-) proceeds as follows

Initiation:

$$\overset{+}{Mt}\,\overset{-}{B} + CH_2\text{—}CHR\ (\text{epoxide, ring O}) \longrightarrow B\text{—}CH_2\text{—}CHR\text{—}\ddot{\underset{\cdot\cdot}{O}}{:}^-\ Mt^+$$

B = e.g. OH, OCH_3, OEt

Propagation:

$$B\text{+}CH_2\text{—}CHR\text{—}O\text{+}_{n-1}CH_2\text{—}CHR\text{—}\ddot{\underset{\cdot\cdot}{O}}{:}^-\ Mt^+ + CH_2\text{—}CHR\ (\text{epoxide})$$

$$\longrightarrow B\text{+}CH_2\text{—}CHR\text{—}O\text{+}_{n}CH_2\text{—}CHR\text{—}\ddot{\underset{\cdot\cdot}{O}}{:}^-\ Mt^+$$

For substituted epoxides (i.e. $R \neq H$), nucleophilic attack of monomer by the active anion takes place at the least sterically hindered CH_2 carbon atom to give the head-to-tail structures shown above. Also, the configuration of the asymmetric carbon atom (—CHR—) in the monomer is retained in the polymer.

There is no inherent termination reaction and so the polymeric anions remain active, i.e. they are living. By using HO^- for initiation and deliberately terminating the polymerization by adding a proton donor (e.g. a dilute acid) when monomer conversion is complete, it is possible to produce linear polyethers with hydroxyl groups at both ends (the metal counter-ion is omitted from the scheme for simplicity)

$$H\bar{O} + n\ CH_2\text{—}CHR\ (\text{epoxide}) \longrightarrow HO\text{+}CH_2\text{—}CHR\text{—}O\text{+}_{n-1}CH_2\text{—}CHR\text{—}\bar{O}$$

$$\xrightarrow{\text{dil. } H_2SO_4} HO\text{+}CH_2\text{—}CHR\text{—}O\text{+}_{n-1}CH_2\text{—}CHR\text{—}OH$$

Extending this principle, if multifunctional alkoxides are used for initiation, then it is possible to produce multiarmed polyethers in which each arm has a hydroxyl end group, e.g. (again omitting the counter-ion)

$$\begin{array}{l} CH_2-O^- \\ CH-O^- \\ (CH_2)_3 \\ CH_2-O^- \end{array} + 3n\,CH_2\!-\!CHR\,(\text{epoxide}) \longrightarrow \begin{array}{l} CH_2-O\!\left[CH_2-CHR-O\right]_{n-1}\!CH_2-CHR-O^- \\ CH-O\!\left[CH_2-CHR-O\right]_{n-1}\!CH_2-CHR-O^- \\ (CH_2)_3 \\ CH_2-O\!\left[CH_2-CHR-O\right]_{n-1}\!CH_2-CHR-O^- \end{array}$$

$$\downarrow \text{dil. } H_2SO_4$$

$$\begin{array}{l} CH_2-O\!\left[CH_2-CHR-O\right]_{n-1}\!CH_2-CHR-OH \\ CH-O\!\left[CH_2-CHR-O\right]_{n-1}\!CH_2-CHR-OH \\ (CH_2)_3 \\ CH_2-O\!\left[CH_2-CHR-O\right]_{n-1}\!CH_2-CHR-OH \end{array}$$

Such protocols are the basis for synthesis of the poly(propylene oxide) *polyols* used in preparation of polyurethanes (see Sections 3.2.2.1 and 3.3.1.3). Before terminating the polymerization, it is usual to add a stoichiometric amount of ethylene oxide after polymerization of the propylene oxide is complete in order to add ethylene oxide units which provide more reactive primary hydroxyl terminal groups.

Chain transfer to monomer can be a significant side reaction if the substituent group possesses hydrogen atoms on the α-carbon atom, e.g. for propylene oxide ($R=CH_3$)

$$\sim CH_2-CH(CH_3)-\ddot{O}:^-\ Mt^+ \ +\ H-CH_2-CH\!-\!CH_2\,(\text{epoxide}) \longrightarrow \sim CH_2-CH(CH_3)-\ddot{O}-H \ +\ CH_2=CH-CH_2-\ddot{O}:^-\ Mt^+$$

which produces allylic alkoxide ions that can initiate polymerization to produce chains with C=C end groups. Thus, most poly(propylene oxide) polyols are contaminated by a small proportion of chains with terminal unsaturation resulting from the chain transfer to monomer and re-initiation reactions. Additionally, intermolecular (interchange) and intramolecular (backbiting) chain transfer to polymer reactions can occur but usually are less significant than in cationic ring-opening polymerization.

7.3.2 Anionic Ring-Opening Polymerization of Lactones

Lactones polymerize via nucleophilic attack at the ring carbonyl group followed by scission of the CO–O bond, the mechanism of which is shown below, specifically for initiation

$$\overset{+}{Mt}\overset{-}{B} + \begin{array}{c} :\ddot{O}-CHR \\ \ddot{O}=C-(CH_2)_x \end{array} \longrightarrow \left[\begin{array}{c} :\ddot{O}-CHR \\ ^{-}:\ddot{O}-C-(CH_2)_x \\ B \end{array} \right] \longrightarrow B-\overset{O}{\overset{\|}{C}}-(CH_2)_x-CHR-\ddot{O}:^{-}\ \overset{+}{Mt}$$

where the large square brackets indicate an intermediate species in the reaction. Propagation proceeds via the same mechanism and can be represented more generally by

$$B\left[\overset{O}{\overset{\|}{C}}-(CH_2)_x-CHR-O\right]_{n-1}\overset{O}{\overset{\|}{C}}-(CH_2)_x-CHR-\ddot{O}:^{-}\ \overset{+}{Mt} + \begin{array}{c} :\ddot{O}-CHR \\ \ddot{O}=C-(CH_2)_x \end{array}$$

$$\longrightarrow B\left[\overset{O}{\overset{\|}{C}}-(CH_2)_x-CHR-O\right]_{n}\overset{O}{\overset{\|}{C}}-(CH_2)_x-CHR-\ddot{O}:^{-}\ \overset{+}{Mt}$$

The configuration of the asymmetric carbon atom —CHR— in a substituted monomer is retained in the polymer. An exception to this general mechanism is the polymerization of β-propiolactone ($x=1$, R=H) by weak bases (e.g. $CH_3COO^- Na^+$) in which nucleophilic attack takes place at the O–CH_2 carbon atom causing scission of this bond and formation of carboxylate ion active species (i.e. $\sim CHR-(CH_2)_x-\overset{O}{\overset{\|}{C}}-O^-$).

7.3.3 Anionic Ring-Opening Polymerization of Lactams

Polymerization of lactams usually is initiated with alkali metals or strong bases and proceeds via a mechanism which initially involves formation of a lactamate ion

$$\begin{array}{c} H\ddot{N}-CHR \\ :\ddot{O}=C-(CH_2)_x \end{array} \xrightarrow{Mt} \begin{array}{c} \overset{+}{Mt}\ ^{-}:\ddot{N}-CHR \\ :\ddot{O}=C-(CH_2)_x \end{array} + \tfrac{1}{2}H_2$$

$$\begin{array}{c} H\ddot{N}-CHR \\ :\ddot{O}=C-(CH_2)_x \end{array} \xrightarrow{\overset{+}{Mt}\overset{-}{B}} \begin{array}{c} \overset{+}{Mt}\ ^{-}:\ddot{N}-CHR \\ :\ddot{O}=C-(CH_2)_x \end{array} + BH$$

The lactamate ion is stabilized by resonance

$$-\overset{\ddot{O}}{\overset{\|}{C}}-\overset{-}{\ddot{N}}- \longleftrightarrow -\overset{:\ddot{O}:^{-}}{\overset{|}{C}}=\ddot{N}-$$

and relatively slowly attacks the carbonyl carbon atom in a molecule of monomer causing scission of the CO—NH bond to produce a highly reactive terminal $-\bar{N}H$ ion which rapidly abstracts H from another molecule of monomer

$$\left[\text{monomer lactamate} + \text{lactam}\right] \rightarrow \left[\text{tetrahedral intermediate}\right] \rightarrow \text{(CH}_2)_x\text{C(=O)–N(CHR)–C(=O)–(CH}_2)_x\text{–CHR–}\bar{\text{N}}\text{H Mt}^+ + \text{HN–CHR–(CH}_2)_x\text{–C=O} \rightarrow$$

$$\text{(CH}_2)_x\text{–C(=O)–N(CHR)–C(=O)–(CH}_2)_x\text{–CHR–NH–H} + \text{Mt}^+\ \bar{\text{N}}\text{–CHR–(CH}_2)_x\text{–C=O}$$

(III)

The ring carbonyl carbon atom in the product (III) is much more strongly activated towards nucleophilic attack than that in the monomer because of the second carbonyl group bonded to the ring nitrogen atom (i.e. (III) is an *N*-acyllactam). Propagation proceeds via similar chemistry

$$\text{H}{\left[\text{NH–CHR–(CH}_2)_x\text{–C(=O)}\right]}_{n-1}\text{N(CHR–(CH}_2)_x\text{–C=O)} + \text{Mt}^+\ \bar{\text{N}}\text{–CHR–(CH}_2)_x\text{–C=O} \rightarrow [\text{intermediate}] \rightarrow$$

$$\text{H}{\left[\text{NH–CHR–(CH}_2)_x\text{–C(=O)}\right]}_{n-1}\bar{\text{N}}(\text{Mt}^+)\text{–CHR–(CH}_2)_x\text{–C(=O)–N(CHR–(CH}_2)_x\text{–C=O)} + \text{HN–CHR–(CH}_2)_x\text{–C=O} \rightarrow$$

$$\text{H}{\left[\text{NH–CHR–(CH}_2)_x\text{–C(=O)}\right]}_{n}\text{N(CHR–(CH}_2)_x\text{–C=O)} + \text{Mt}^+\ \bar{\text{N}}\text{–CHR–(CH}_2)_x\text{–C=O}$$

in which the monomeric lactamate ion attacks the ring carbonyl carbon atom of the terminal *N*-acyllactam causing scission of the CO—N bond to produce a $-\bar{\text{N}}\text{H}$ ion which abstracts H^+ from a molecule of monomer to form another monomeric lactamate ion. It also is possible for the $-\bar{N}H$ ion to attack a *N*-acyllactam end group on another propagating chain, which leads to the formation of branches. For each propagation step, the configuration of the asymmetric carbon atom on a substituted monomer is retained in the polymer.

Induction periods often are observed in the polymerization of lactams whilst the concentration of (III) increases. These can be eliminated by inclusion of acylating agents (e.g. acid halides and anhydrides) that react rapidly with the lactam to form monomeric *N*-acyllactams (IV) which then propagate

where X = Cl or $O-\overset{O}{\overset{\|}{C}}-R'$

(IV)

7.3.4 Anionic Ring-Opening Polymerization of Cyclic Siloxanes

Cyclic siloxanes undergo living anionic ring-opening polymerization via a mechanism similar to that for epoxides. Although metal alkyls can be used as initiators, silanolates more commonly are used and can be formed *in situ* by reaction of the cyclic siloxane with an alkali metal hydroxide.

$y = 2$ or 3

R_1, R_2 = alkyl, aryl

Mt = Li, Na, K, NH_4, PH_4

Propagation then proceeds via similar chemistry

Although chain growth is shown here from just one chain end, propagation will occur from both silanolate end groups. Exchange and backbiting reactions are possible, depending on reaction conditions, but narrow molar mass distribution polysiloxanes can be prepared by making propagation the dominant process, e.g. by using the cyclic trimer ($y=2$), because it is more reactive, and by stopping the polymerization before the monomer concentration becomes low enough for the other processes to become significant.

7.4 FREE-RADICAL RING-OPENING POLYMERIZATION

The mechanisms of the two main types of free-radical ring-opening polymerization are shown generically in Figure 7.3. For efficient ring-opening, the cyclic monomers should have a terminal C=C bond that acts as the site for initial attack, sufficient ring strain to promote ring-opening, and the ability to form thermodynamically stable C=C or C=X double bonds, where X is a heteroatom. Ring-opening competes with 'normal' free-radical polymerization of the C=C bond and so high temperatures (e.g. 80–150 °C) typically are used to promote the ring-opening mechanism. Peroxide initiators tend to be used since they have more appropriate half-lives than azo initiators at these temperatures (see Table 4.2). Ring-opening also is promoted by Z groups which stabilize the propagating radical (e.g. Ph, CN, CO_2R).

Ring-opening polymerization of *cyclic ketene acetals* (type I monomers) is most important and produces polyesters, the extent of ring-opening versus 'normal' C=C bond polymerization depending upon monomer structure and polymerization temperature, as is exemplified by the data in Table 7.2. Thus, free-radical ring-opening polymerization is a convenient method for synthesis of aliphatic polyesters and is particularly effective for monomers with $x=4$ or 5. However, copolymers are formed when both modes of propagation occur, e.g. polymerization of the monomer with $x=2$ at lower temperatures produces

$$n\,CH_2{=}C\langle\begin{matrix}O-CH_2\\ |\\ O-CH_2\end{matrix}\rangle \longrightarrow \left[\left(CH_2-\overset{O}{\overset{\|}{C}}-O-(CH_2)_2\right)_p\left(CH_2-\underset{\underset{CH_2-CH_2}{O\quad O}}{C}\right)_q\right]_n$$

where p and q are the mole fractions of the two types of repeat unit (from Table 7.2, at 60 °C $p \approx q \approx 0.5$).

Polyamides also can be produced in this way; e.g. the following polymerization

$$n\,CH_2{=}C\langle\begin{matrix}O-CH_2\\ |\\ NH-CH_2\end{matrix}\rangle \longrightarrow \left[CH_2-\overset{O}{\overset{\|}{C}}-NH-(CH_2)_2\right]_n$$

gives 100% ring-opening product at 80 °C, i.e. ring-opening is more favourable than for cyclic ketene acetals. However, polythioesters are not produced cleanly, e.g.

$$n\,CH_2{=}C\langle\begin{matrix}O-CH_2\\ |\\ S-CH_2\end{matrix}\rangle \longrightarrow \left[CH_2-\overset{O}{\overset{\|}{C}}-S-(CH_2)_2\right]_n$$

gives just 45% ring-opening product at 160 °C and only 15% at 120 °C. The differences have been explained on the basis of the order of thermodynamic stability of the link produced (amide > ester > thioester).

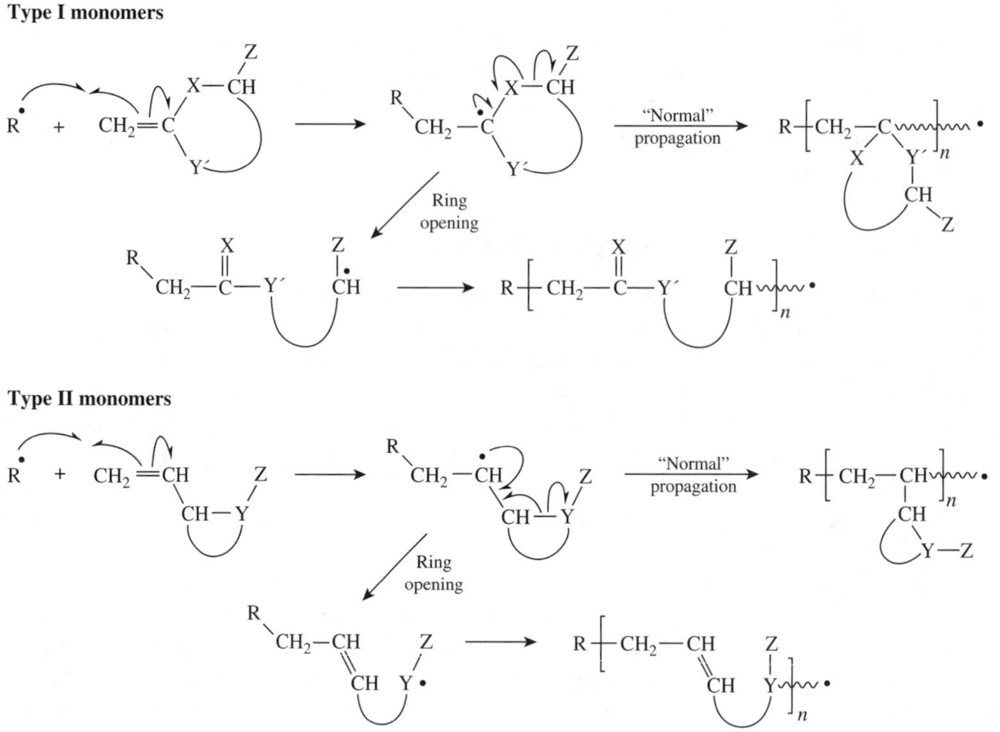

Type I monomers

Type II monomers

FIGURE 7.3 The mechanisms of free-radical polymerization of type I and type II cyclic monomers, where $R^{\bullet}$ is a primary radical from the initiator, X is a heteroatom (usually O), Y′ is a heteroatom (e.g. O or NH), Y is either a carbon atom or a heteroatom (e.g. Y–Z = SO_2), and Z is a substituent atom or group (often simply H). Ring-opening polymerization competes with 'normal' free-radical polymerization, but is promoted by increasing the temperature of polymerization and by Z groups that can confer stability to radicals.

TABLE 7.2
Effect of Ring Size and Temperature on the Extent of Ring-Opening Polymerization of Cyclic Ketene Acetals

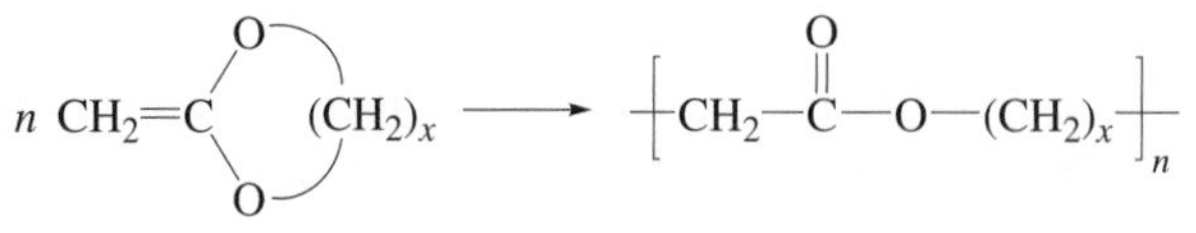

$$n\ CH_2{=}C\langle O_2(CH_2)_x\rangle \longrightarrow \left[CH_2{-}\overset{O}{\overset{\|}{C}}{-}O{-}(CH_2)_x \right]_n$$

x	Temperature / °C	Ring-Opening Polymer / %
2	160	100
2	125	83
2	60	50
3	130	85
4	≥25	100
5	≥25	100

Ring-opening polymerization of type II monomers is less important. Two examples are given below to illustrate the kinds of polymers that can be produced.

$$n\ \underset{\text{(cyclopropane ring: CH–C(X)}_2\text{–CH}_2)}{CH_2{=}CH{-}CH} \xrightarrow[80\,°C]{PhCO_2{-}O_2CPh} \left[CH_2{-}CH{=}CH{-}CH_2{-}\underset{X}{\overset{X}{C}} \right]_n$$

where X = Cl, CN

$$n\ CH_2{=}CH{-}\underset{(CH_2)_x}{CH}{-}S(=O)_2 \xrightarrow[65\text{–}75\,°C]{PhCO_2{-}O_2CPh} \left[CH_2{-}CH{=}CH{-}(CH_2)_x{-}\underset{O}{\overset{O}{S}} \right]_n$$

x = 2, 3, 4

In both polymerizations, 100% ring-opening product is produced, principally due to substituent-group stabilization of the propagating radicals, which are, respectively,

$$\sim\sim CH_2{-}CH{=}CH{-}CH_2{-}\underset{X}{\overset{X}{C}}\bullet \quad \text{and} \quad \sim\sim CH_2{-}CH{=}CH{-}(CH_2)_x{-}\underset{O}{\overset{O}{S}}\bullet$$

7.5 RING-OPENING METATHESIS POLYMERIZATION

Olefin (or *alkene*) *metathesis* is a disproportionation reaction that is catalysed by transition metal complexes and may be represented generally by

$$2\ R_1{-}CH{=}CH{-}R_2 \rightleftharpoons R_1{-}CH{=}CH{-}R_1 + R_2{-}CH{=}CH{-}R_2$$

Although olefin metathesis has been known since the mid-1950s, it was not until the 1970s that it began to be studied more extensively following the discovery by Chauvin that the mechanism involved a metal-carbon double bond. Shrock evaluated many different metals for effecting olefin metathesis, culminating in the development of highly active molybdenum-based catalysts, but it was Grubbs who developed more stable ruthenium-based catalysts of high activity that have become standard for practical applications of olefin metathesis, including polymerization. Chauvin, Shrock and Grubbs were jointly awarded the 2005 Nobel Prize for Chemistry for their contributions to the development of olefin metathesis.

Although there was recognition, as early as the 1960s, that the reaction could be applied to the preparation of polymers by *ring-opening metathesis polymerization* (*ROMP*) of cycloalkenes and bicycloalkenes, it was not until the advent of Grubbs catalysts that ROMP became a more widely used method for polymer synthesis. A simple example is the ROMP of cyclopentene

$$n\ \text{(cyclopentene)} \longrightarrow \left[CH_2{-}CH_2{-}CH_2{-}CH{=}CH \right]_n$$

7.5.1 Chemistry of ROMP

ROMP is similar to Ziegler–Natta polymerization (Section 6.4) in that it involves coordination of monomer to transition metal complexes that can be formed by reaction of transition metal compounds with metal alkyls or Lewis acids. However, in contrast to Ziegler–Natta systems, ROMP involves metal carbenes and the reactions normally are performed using soluble (i.e. homogeneous) catalysts. Catalysts based on titanium, tungsten, molybdenum, rhenium and ruthenium have been used to effect ROMP, but nowadays the most important catalysts are those based on ruthenium, as developed in the work of Grubbs, two examples of which are shown below

Cy = cyclohexyl

Mes = mesityl

These ruthenium-based catalysts are stable in air, tolerant of polar functional groups and effect ROMP at moderate temperatures (typically 25–50 °C).

The active species in propagation is a transition metal carbene with a vacant d-orbital created through temporary loss of a phosphine ligand. For polymerization of a cycloalkene by ruthenium-based catalysts, the overall propagation reaction can be represented by

$$(R_3P)(L)Cl_2Ru{=}CHR' \xrightarrow{n\ \text{cyclo-}C_4H_6Z} (R_3P)(L)Cl_2Ru{=}CH{-}[CH_2{-}Z{-}CH_2{-}CH{=}CH]_n{-}R'$$

where PR_3 is a phosphine ligand (e.g. PCy_3 or PPh_3), L represents a general ligand, R′ is the metallocarbene substituent (e.g. Ph), and Z is the linking group in the cyclic monomer (e.g. $Z{=}CH_2$ in cyclopentene). A general mechanism for propagation in ROMP is shown below and involves coordination of the cycloalkene C=C bond to the transition metal through replacement of a ligand, which is believed to occur via a dissociative process in which departure of the ligand (PR_3 for the ruthenium catalysts shown above) occurs first and is followed by coordination of the cycloalkene at the vacated d-orbital (—□). An unstable metallocyclobutane intermediate then forms via a 2 + 2 cycloaddition and breaks down to give a new transition metal carbene and a new C=C bond which decoordinates upon migration of the carbene to the original site of the metallocarbene, the ligand then being recaptured.

$$L_xMt{=}CH{-}Pol \underset{+PR_3}{\overset{-PR_3}{\rightleftharpoons}} L_{x-1}Mt(\square){=}CH{-}Pol \rightleftharpoons \ldots \rightleftharpoons L_xMt{=}CH{-}CH_2{-}Z{-}CH_2{-}CH{=}CH{-}Pol$$

Mt = Mo, Re, Ru, Ti, W

$$\text{Pol} = {-}[CH_2{-}Z{-}CH_2{-}CH{=}CH]_n{-}R'$$

Thus, the ring-opening reaction proceeds by cleavage of the C=C bond in the monomer.

Intermolecular chain transfer to polymer can occur, leading to broadening of the molar mass distribution. For example

$$L_xMt{=}CH{+}CH_2{-}Z{-}CH_2{-}CH{=}CH{+}_m CH_2{-}Z{-}CH_2{-}CH{=}CH{+}CH_2{-}Z{-}CH_2{-}CH{=}CH{+}_p R'$$

$$R'{+}CH{=}CH{-}CH_2{-}Z{-}CH_2{+}_q CH{=}MtL_x$$

$$\swarrow$$

$$L_xMt{=}CH{+}CH_2{-}Z{-}CH_2{-}CH{=}CH{+}_m CH_2{-}Z{-}CH_2{-}CH{=}CH{+}CH_2{-}Z{-}CH_2{-}CH{=}CH{+}_q R'$$

$$+$$

$$R'{+}CH{=}CH{-}CH_2{-}Z{-}CH_2{+}_p CH{=}MtL_x$$

which also could proceed in the opposite sense to produce one chain with two active metallocarbene end groups and another with none

$$L_xMt{=}CH{+}CH_2{-}Z{-}CH_2{-}CH{=}CH{+}_m CH_2{-}Z{-}CH_2{-}CH{=}CH{+}CH_2{-}Z{-}CH_2{-}CH{=}CH{+}_p R'$$

$$L_xMt{=}CH{+}CH_2{-}Z{-}CH_2{-}CH{=}CH{+}_q R'$$

$$\swarrow$$

$$L_xMt{=}CH{+}CH_2{-}Z{-}CH_2{-}CH{=}CH{+}_m CH_2{-}Z{-}CH_2{-}CH{=}MtL_x$$

$$+$$

$$R'{+}CH{=}CH{-}CH_2{-}Z{-}CH_2{+}_p CH{=}CH{+}CH_2{-}Z{-}CH_2{-}CH{=}CH{+}_q R'$$

Chain transfer to polymer also can proceed intramolecularly (i.e. via backbiting) to give one linear chain of reduced length and one cyclic chain, e.g.

$$R'{+}CH{=}CH{-}CH_2{-}Z{-}CH_2{+}_m CH{=}CH{-}CH_2{-}Z{-}CH_2{\sim\sim}$$

$$L_xMt{=}CH{-}CH_2{-}Z{-}CH_2{-}CH{=}CH{\sim\sim}$$

$$\downarrow$$

$$R'{+}CH{=}CH{-}CH_2{-}Z{-}CH_2{+}_m CH{=}MtL_x \quad + \quad \begin{matrix} CH{-}CH_2{-}Z{-}CH_2{\sim\sim} \\ \| \\ CH{-}CH_2{-}Z{-}CH_2{-}CH{=}CH{\sim\sim} \end{matrix}$$

As in all chain polymerizations, the contributions from chain transfer to polymer can be reduced through use of lower polymerization temperatures.

The metallocarbene end groups in ROMP remain active (i.e. they are living) and will continue to propagate if more monomer is added. Hence, it is normal to remove them on completion of polymerization by addition of a chain-terminating species. For example, for the ruthenium catalysts shown above, it is usual to add ethyl vinyl ether which displaces the chain and gives an inactive ruthenium complex

$$(R_3P)(L)Cl_2Ru{=}C(H){+}CH_2{-}Z{-}CH_2{-}CH{=}CH{+}_n R' \xrightarrow{CH_2{=}C(H)OEt} (R_3P)(L)Cl_2Ru{=}C(H)OEt \; + \; CH_2{=}CH{+}CH_2{-}Z{-}CH_2{-}CH{=}CH{+}_n R'$$

7.5.2 Applications of ROMP

ROMP is used commercially to prepare polymers from norbornene

n → —CH=CH— $_n$

and from dicyclopentadiene

n → —CH=CH— $_n$

The latter polymerization results in the formation of a crosslinked polymer due to participation of the pendant C=C bonds in the polymerization. ROMP also has been used in a convenient route for the synthesis of inherently-conductive polyacetylene (see Sections 8.4 and 25.3.5).

It should be borne in mind that the polymer structures shown above are simplistic since there often are many stereochemical forms of the repeat units. In certain cases, highly stereoregular polymers can be formed, e.g. ROMP of racemic 1-methylnorbornene by rhenium pentachloride gives a polymer with a head-to-tail *cis*-syndiotactic microstructure

n —Me → Me Me $_n$

Use of catalysts that give rapid initiation (such as the catalyst with the *N*-heterocyclic carbene ligand shown in Section 7.5.1) and lower temperatures (to suppress chain transfer reactions) facilitates ROMP with the characteristics of living polymerization (Section 5.3.2). Under such conditions, ROMP can be used to prepare polymers with narrow molar mass distributions and for synthesis of block copolymers (Section 9.4.2).

PROBLEMS

7.1 Write down the structure of the cyclic monomer, and outline the type of ring-opening polymerization you would use to prepare each of the following polymers.

(a) $\left[-\mathrm{Si}(\mathrm{CH_3})(\mathrm{Ph})-\mathrm{O}- \right]_n$

(b) $\left[-\mathrm{CF_2}-\mathrm{CF}(\mathrm{CF_3})-\mathrm{O}- \right]_n$

(c) $\left[-\mathrm{CH{=}CH}-(\mathrm{CH_2})_3- \right]_n$

(d) $\left[-(\mathrm{CH_2})_3-\mathrm{O}-\mathrm{C}(=\mathrm{O})- \right]_n$

(e) $\left[-\mathrm{CH{=}CH}-\mathrm{CH_2}-\mathrm{CCl_2}-\mathrm{CH_2}- \right]_n$

7.2 Discuss the reasons why the structure of polyethers prepared by ring-opening polymerization of cyclic ethers is better controlled when using anionic initiation rather than cationic initiation. Hence, by analogy with ring-opening polymerization of epoxides, give the structure of the cyclic monomer and the conditions you would use to prepare the polymer shown below.

$$HO\left[CH_2-CH_2-CH_2-CH_2-O\right]_n H$$

7.3 Give two ring-opening polymerization methods that use different cyclic monomers for synthesis of the following polymer. In each case, write down the structure of the cyclic monomer and outline the type of ring-opening polymerization used.

$$\left[\left(CH_2\right)_3 NH-\overset{O}{\overset{\|}{C}}\right]_n$$

FURTHER READING

General Reading

Allen, G. and Bevington, J.C. (eds), *Comprehensive Polymer Science*, Vols. 3 and 4 (Eastmond, G.C., Ledwith, A., Russo, S., and Sigwalt, P., eds), Pergamon Press, Oxford, 1989.

Brunelle, D.J. (ed), *Ring-Opening Polymerization: Mechanisms, Catalysis, Structure, Utility*, Hanser, New York, 1993.

Dubois, P., Coulembier, O., and Raquez, J.-M. (eds), *Handbook of Ring-Opening Polymerization*, Wiley-VCH, Weinheim, Germany, 2009.

Frisch, K.C. and Reegen, S.L. (eds), *Ring-Opening Polymerization Kinetics and Mechanisms of Polymerization*, Vol. 2, Marcel Dekker, New York, 1969.

Ivin, K.J. and Saegusa, T. (eds), *Ring-Opening Polymerization*, Vol. 1–3, Elsevier, London, 1984.

Kubisa, P. and Penczek, S., Cationic activated monomer polymerization of heterocyclic monomers, *Progress in Polymer Science*, **24**, 1409 (1999).

Lenz, R.W., *Organic Chemistry of Synthetic High Polymers*, Interscience, New York, 1967.

Mark, H.F., Bikales, N.M., Overberger, C.G., and Menges, G. (eds), *Encyclopedia of Polymer Science and Engineering*, Wiley-Interscience, New York, 1985–1989.

Odian, G., *Principles of Polymerization*, 4th edn., Wiley-Interscience, New York, 2004.

Penczek, S., Cypryk, M., Duda, A., Kubisa, P., and Slomkowski, S., Living ring-opening polymerizations of heterocyclic monomers, *Progress in Polymer Science*, **32**, 247 (2007).

Stevens, M.P., *Polymer Chemistry: An Introduction*, 3rd edn., Oxford University Press, New York, 1998.

Free-Radical Ring-Opening Polymerization

Bailey, W.J., Chou, J.L., Feng, P.-Z., Issari, B., Kuruganti, V., and Zhou, L.-L., Recent advances in free-radical ring-opening polymerization, *Journal of Macromolecular Science, Part A: Pure and Applied Chemistry*, **25**, 781 (1988).

Ring-Opening Metathesis Polymerization

Bielawski, C.W. and Grubbs, R.H., Living ring-opening metathesis polymerization, *Progress in Polymer Science*, **32**, 1 (2007).

Trnka, T.M. and Grubbs, R.H., The development of $L_2X_2Ru{=}CHR$ olefin metathesis catalysts: an organometallic success story, *Accounts of Chemical Research*, **34**, 18 (2001).

8 Specialized Methods of Polymer Synthesis

8.1 INTRODUCTION

There are many specialized methods of polymer synthesis that do not fit easily into the mainstream types of polymerization described in Chapters 3 through 7. The purpose of this chapter is to provide a simple introduction to some that currently are of considerable interest, each type being described briefly with some examples of its use. The treatment is far from exhaustive, so reviews and original papers will need to be consulted in order to gain a deeper understanding and knowledge. There also are many other types of specialized polymerization which are not covered here, such as template polymerization, polymerization in clathrates, cyclopolymerization and plasma polymerization, but for which information can be found in reviews and advanced textbooks.

8.2 SOLID-STATE TOPOCHEMICAL POLYMERIZATION

There are many monomers which are capable of undergoing polymerization in the solid (most commonly crystalline) state, though in general the mechanisms of such polymerizations are not clearly understood. In many cases, the lattice structure of the crystalline monomer is considerably disrupted upon polymerization and, therefore, is not maintained in the polymer produced. Nevertheless, in favourable cases, it is possible to prepare highly-oriented polymers by solid-state polymerization.

Certain polyamides and polyesters can be prepared by solid-state polycondensation step polymerization, usually by heating the appropriate monomer or monomer salt. For example, nylon 11 can be prepared by heating crystals of 11-aminoundecanoic acid (melting point = 188 °C) at 160 °C under vacuum. Polyaddition step polymerization of conjugated dialkene monomers can be induced in the solid state by exposure of the monomer to ultraviolet radiation. For example, irradiation of crystalline 2,5-distyrylpyrazine yields quantitatively a highly-crystalline linear cyclobutane polymer

$$n\ \mathrm{Ph{-}CH{=}CH{-}C_4H_2N_2{-}CH{=}CH{-}Ph} \xrightarrow{h\nu} \left[\mathrm{C_4H_2N_2{-}CH(Ph){-}CH{-}CH{-}CH(Ph)}\right]_n$$

A number of ethylenic monomers and cyclic monomers undergo chain polymerization in the solid state when exposed to high-energy radiation (e.g. γ-radiation). Examples include acrylamide, acrylic acid, acrylonitrile, vinyl acetate, styrene, 1,3,5-trioxane, propriolactone and hexamethylcyclotrisiloxane. These polymerizations may involve free-radical and/or ionic species, but in most cases the precise mechanisms are not known.

The period of greatest research activity in solid-state polymerization was the 1960s. More recently, there has been considerable interest in the *solid-state polymerization of diacetylene* (i.e. *diyne*) *single crystals* which yield macroscopic polymer single crystals virtually free of defects. These polymerizations can be induced thermally or by irradiation (ultraviolet or high-energy radiation) and are said to be *topochemical* since the direction of chain growth is defined by the geometry

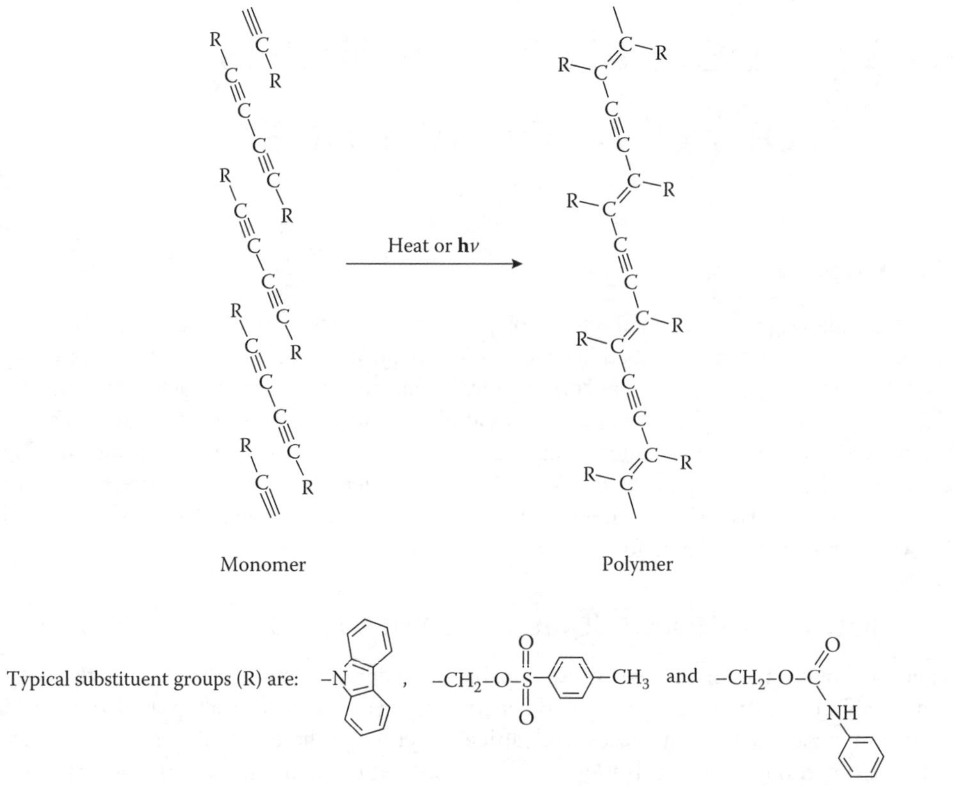

FIGURE 8.1 Schematic representation of the solid-state topochemical polymerization of a symmetrical disubstituted diacetylene monomer.

and symmetry of the monomer crystal lattice structure. The polymerization of a symmetrical disubstituted diacetylene monomer is illustrated schematically in Figure 8.1. The reaction is believed to proceed via carbene species, though free-radical species also may be involved. The result is a direct transition from monomer molecules to polymer without major movements of the atoms involved and so the three-dimensional order of the monomer crystal lattice is, to a large extent, maintained. The polymers produced are completely crystalline, and the polymer molecules are linear, stereoregular, of very high degree of polymerization and lie within the polymer crystals in a chain-extended conformation. Photographs of some polydiacetylene single crystals are shown in Figure 17.11.

Polymers which contain diacetylene units in the chain have been prepared by step polymerization with diacetylene diols, e.g. segmented polyurethanes (Section 9.4.1)

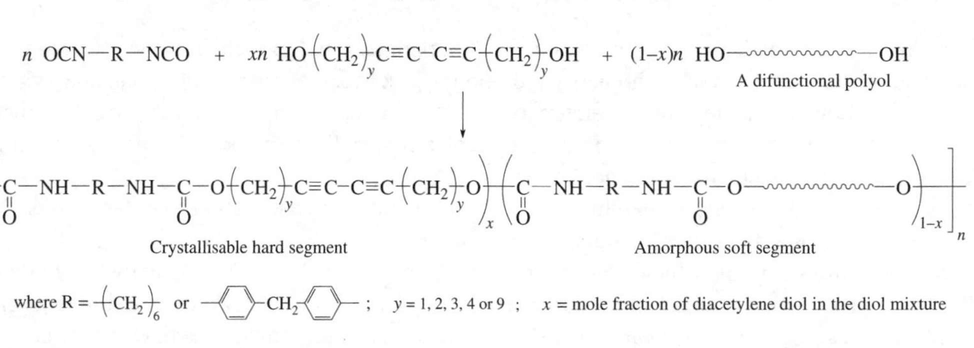

$$n\ \mathrm{OCN{-}R{-}NCO} + xn\ \mathrm{HO{+}CH_2{+}_y C{\equiv}C{-}C{\equiv}C{+}CH_2{+}_y OH} + (1-x)n\ \mathrm{HO{\sim\sim\sim}OH}$$

A difunctional polyol

$$\downarrow$$

$$\left[\left(\underset{\substack{\|\\ O}}{C}\mathrm{-NH-R-NH-}\underset{\substack{\|\\ O}}{C}\mathrm{-O{+}CH_2{+}_y C{\equiv}C{-}C{\equiv}C{+}CH_2{+}_y O}\right)_x\left(\underset{\substack{\|\\ O}}{C}\mathrm{-NH-R-NH-}\underset{\substack{\|\\ O}}{C}\mathrm{-O{\sim\sim\sim}O}\right)_{1-x}\right]_n$$

Crystallisable hard segment — Amorphous soft segment

where R = $+CH_2+_6$ or $-C_6H_4-CH_2-C_6H_4-$; y = 1, 2, 3, 4 or 9 ; x = mole fraction of diacetylene diol in the diol mixture

and aromatic polyesters (Section 3.2.1.1)

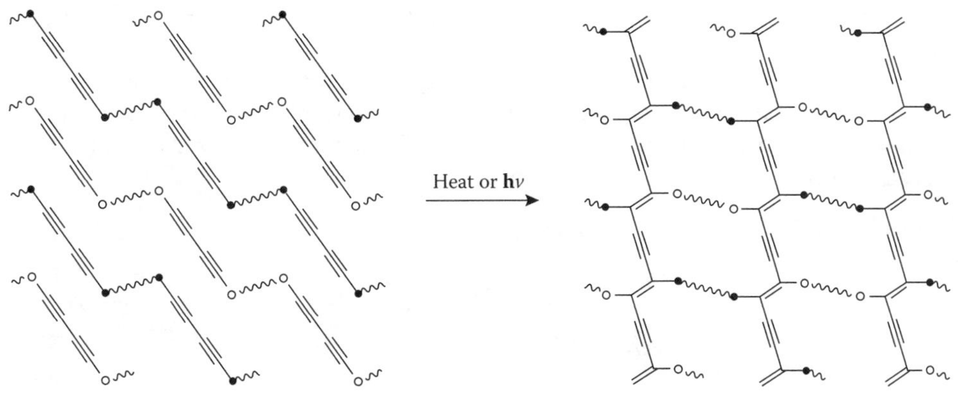

Heat or hν

FIGURE 8.2 Schematic representation of the solid-state topochemical cross-polymerization in crystalline phases of a conjugated diacetylene-containing polymer. The solid and open dots are included to help highlight the position of the original chains in the crosslinked polymer.

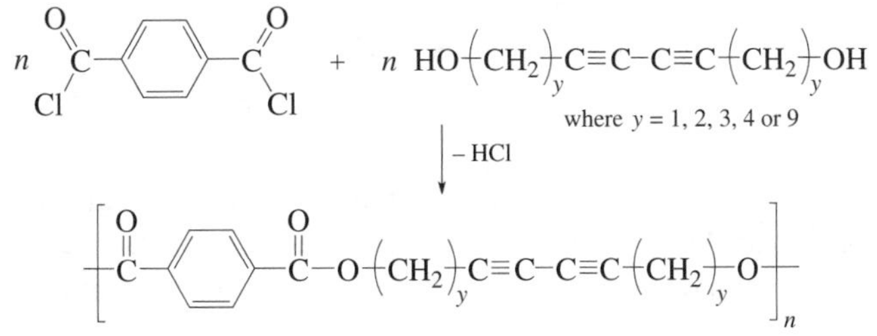

$$n\ \text{ClOC}-C_6H_4-\text{COCl} + n\ \text{HO}\!\left(CH_2\right)_y\!C{\equiv}C{-}C{\equiv}C\!\left(CH_2\right)_y\!\text{OH} \quad \text{where } y = 1, 2, 3, 4 \text{ or } 9$$

$$\xrightarrow{-\text{HCl}} \left[\!-\overset{O}{\overset{\|}{C}}-C_6H_4-\overset{O}{\overset{\|}{C}}-O\!\left(CH_2\right)_y\!C{\equiv}C{-}C{\equiv}C\!\left(CH_2\right)_y\!O-\right]_n$$

Provided that the diacetylene units in the crystalline phases of the polymer have the necessary alignment, solid-state topochemical polymerization can occur on heating or on exposure to ultraviolet light or high-energy radiation and gives rise to *cross-polymerization* of the diacetylene-containing repeat units in the crystalline phases. Solid-state topochemical cross-polymerization of a diacetylene-containing polymer is shown schematically in Figure 8.2.

8.3 POLYMERIZATION BY OXIDATIVE COUPLING

Several important polymers are prepared by oxidative coupling reactions that involve radical and/or radical-ion intermediates. These reactions are effected either chemically or electrochemically, the latter method being particularly useful for synthesis of films of conducting polymers. Some of the more important applications of polymerization by oxidative coupling will be presented here. In many cases, the precise mechanism of polymerization is uncertain, so mechanisms will be given only where there is a reasonable consensus.

8.3.1 Polymerization of Phenols by Oxidative Coupling

Poly(arylene oxide)s are produced from the corresponding phenol by chemically-induced oxidative coupling polymerization in the presence of oxygen and a Cu(I) catalyst

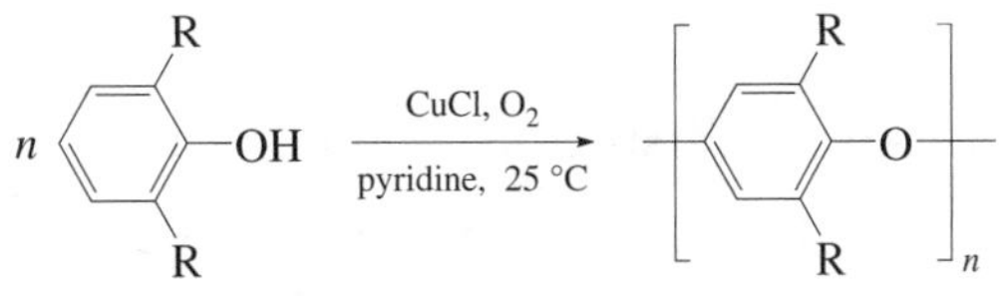

$$n\ \text{(2,6-R}_2C_6H_3)\text{-OH} \xrightarrow[\text{pyridine, 25 °C}]{\text{CuCl, }O_2} \left[\!-\text{(2,6-R}_2C_6H_2)-O-\right]_n$$

where R = H for poly(1,4-phenylene oxide) and R = CH_3 for poly(2,6-dimethyl-1,4-phenylene oxide). The latter polymer is produced commercially and, rather confusingly, often is referred to simply as poly(phenylene oxide) with the acronym PPO, even when supplied as a blend with polystyrene.

The mechanism of polymerization involves abstraction of hydrogen atoms from phenolic OH (end-)groups by oxygen, catalysed by a Cu(I) complex (or possibly a Cu(II) complex formed in situ), followed by radical substitution of a hydrogen atom at the 4-position in a phenoxy (end-)group, which again involves oxygen abstraction of a hydrogen atom

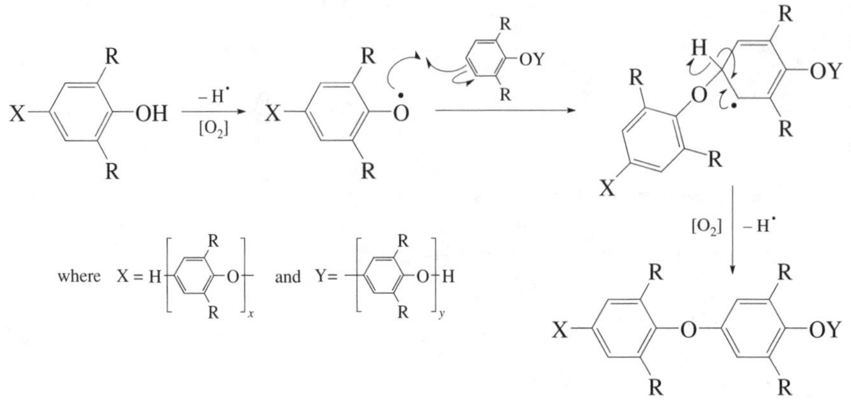

At the beginning of the reaction $x=y=0$ and so the initial reactions are between molecules of the phenol monomer, but as the chain length increases, the reaction can be between any species, i.e. the reaction has the characteristics of step polymerization. Interchange displacement reactions also are possible

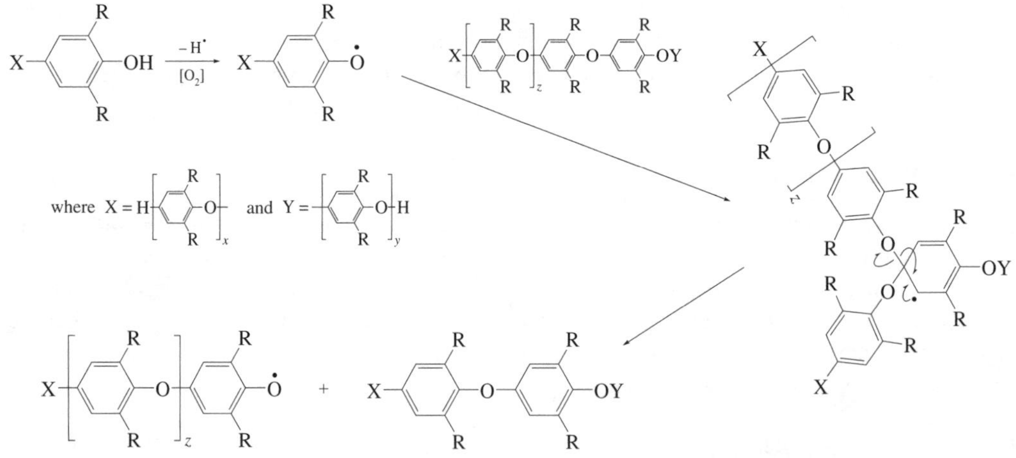

When $R=CH_3$, branching can occur via H-abstraction from the benzylic C–H bonds of the CH_3 groups.

8.3.2 Polymerization of Aniline, Pyrrole and Thiophene by Oxidative Coupling

Several conducting polymers can be prepared by chemical or electrochemical oxidative coupling reactions, the most important being polyaniline, polypyrrole and polythiophene (see Section 25.3.4).

Chemical oxidation of aniline was first reported in the nineteenth century, but its significance was not recognized fully until the inherent conductivity of polyaniline began to be exploited. Many different methods and oxidising agents have been used to effect the polymerization, but most commonly acidic aqueous solutions of aniline are oxidized using aqueous ammonium persulphate. Polyaniline is insoluble in water and so it precipitates as it forms to give either a powder or, if the polymerization is carried out in the presence of a colloid stabilizer, a colloidal dispersion of polyaniline particles. Polymerization in the presence of preformed polyaniline particles can be used to produce fibrous polyaniline. Electrochemical oxidation of aniline can yield polyaniline as a powder, a film, or as fibres, depending on the procedure used, and has the advantage of not requiring the use of an oxidizing agent or colloid stabilizer (which contaminate the polyaniline produced by chemical oxidation methods). The mechanism of oxidative coupling varies according to the method used, but there is reasonable agreement that it involves coupling of aminyl radical-cation intermediates formed by loss of an electron from the lone pair of electrons on the nitrogen atom of an amino group. Any two species can react together, i.e. it is a step polymerization. An outline of the mechanism is shown below up to the formation of trimer

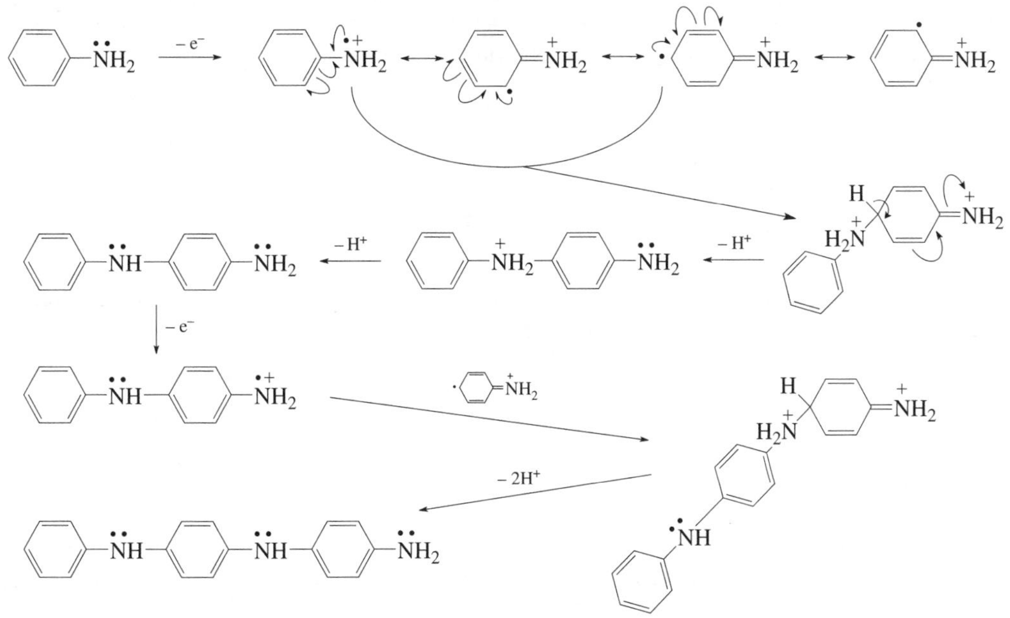

The mechanism is much more complex than indicated here because oxidation of the secondary amine linking groups to radical-cations, and ultimately to imine links, also occurs. Hence, there are many different forms of polyaniline, the structures of which can be represented quite generally in terms of combinations of the fully reduced (amine form) and fully oxidized (imine form)

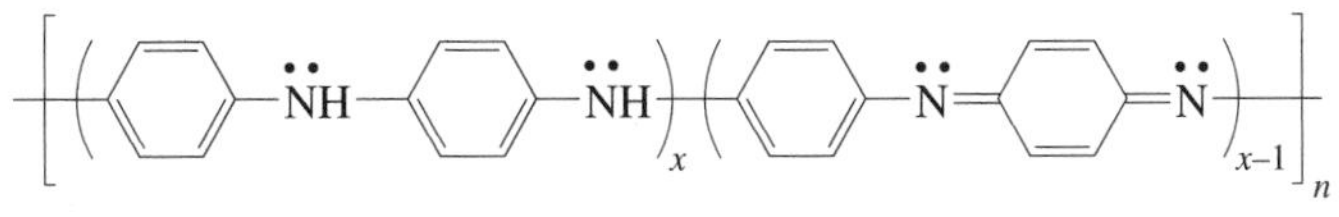

where x is the mole fraction of the amine form. The fully reduced amine form of polyaniline ($x=1$) is known as *leucoemeraldine*, the fully oxidized imine form ($x=0$) as *pernigraline* and the half-oxidized form ($x=0.5$) as *emeraldine*.

Polypyrrole and polythiophene have structural similarity and can be prepared by oxidative coupling of pyrrole and thiophene, respectively,

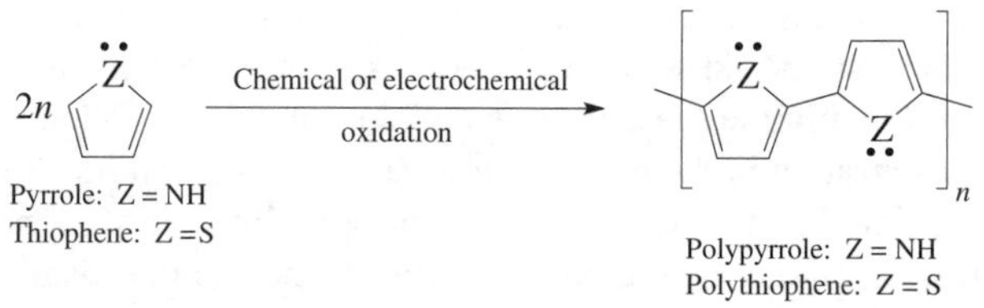

Chemical or electrochemical oxidation

Pyrrole: Z = NH
Thiophene: Z = S

Polypyrrole: Z = NH
Polythiophene: Z = S

in reactions that are akin to those used for synthesis of polyaniline. Aqueous solutions of pyrrole or thiophene most commonly are polymerized at room temperature using ferric chloride as the oxidising agent, the polypyrrole or polythiophene precipitating as it forms to give a powder or, if carried out in the presence of colloid stabilizer, a colloidal dispersion of particles in much the same way as in the chemical oxidative coupling synthesis of polyaniline. Electrochemical oxidation of pyrrole or thiophene solutions can give particles or films of polymer, again in the same way as for polyaniline synthesis. The mechanism of these reactions remains uncertain, but there is reasonable evidence for the involvement of radical-cation intermediates formed as follows

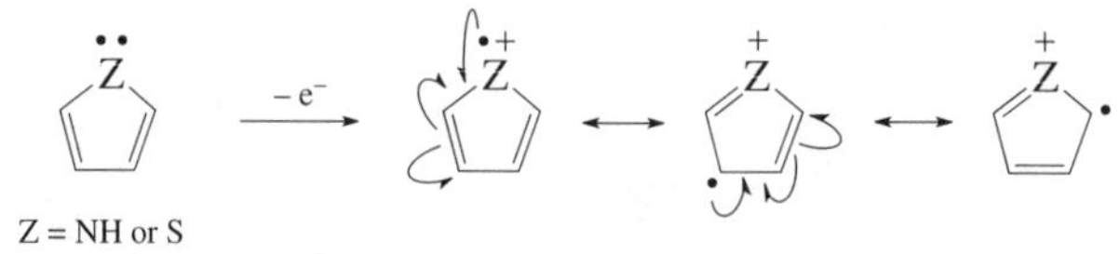

Z = NH or S

The last resonance hybrid is the most stable and undergoes radical coupling followed by loss of H^+

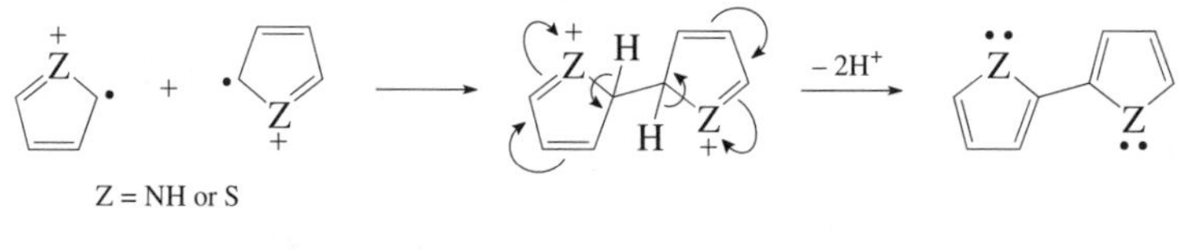

Z = NH or S

Chain growth can occur between any species and so has the characteristics of a step polymerization, which may be represented more generally by

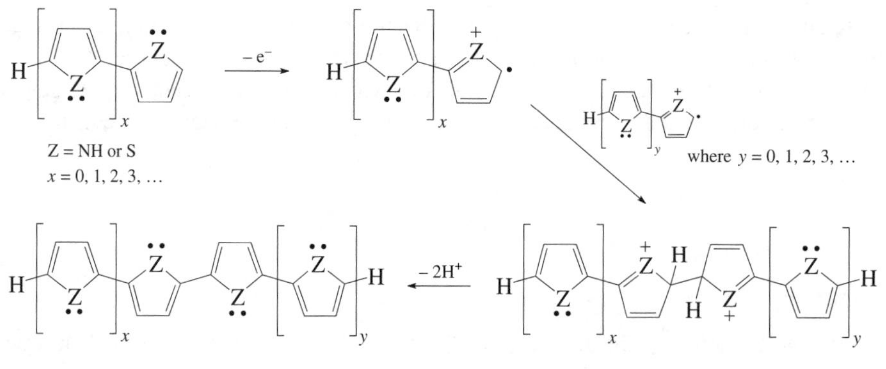

Z = NH or S
$x = 0, 1, 2, 3, \ldots$

where $y = 0, 1, 2, 3, \ldots$

In addition to this radical-cation mechanism, mechanisms involving free-radical and cationic intermediates have been proposed.

An important conducting polythiophene derivative prepared by oxidative coupling is poly(3,4-ethylenedioxythiophene) (PEDOT)

$2n$ — $FeCl_3$ or $(NH_4)_2S_2O_8$, oxidation → PEDOT

8.4 PRECURSOR ROUTES TO INTRACTABLE POLYMERS

Some polymers are completely insoluble and incapable of softening or melting (they are said to be *intractable*), most often due to the polymer having a very high degree of crystallinity and a melting point that is above the temperature at which the polymer degrades. Such polymers cannot be transformed into films, fibres, or solid shapes by conventional melt or solution processing methods. To overcome this problem, these polymers can be prepared via a *precursor route* in which the intractable polymer is produced *in situ* from a precursor polymer that has been processed into the desired final form for the intractable polymer. Conversion of the precursor polymer into the intractable polymer invariably proceeds thermally via elimination of a small, volatile molecule which must escape the precursor polymer without disrupting its physical form. The principal issue with precursor routes is ensuring complete removal of these small molecules from the target intractable polymer. Hence, precursor routes are most suited to producing the intractable polymer in the form of films or fibres (because they have very high surface-to-volume ratios which facilitate easier escape of the small molecule). They are particularly useful for preparation of conducting polymer films because many conducting polymers (Section 25.3.4) are intractable. Precursor routes to poly(1,4-phenylene), poly(1,4-phenylene vinylene) and polyacetylene, all of which are conducting polymers, will be considered to illustrate the principles.

poly(1,4-phenylene) poly(1,4-phenylene vinylene) polyacetylene

One precursor route to poly(1,4-phenylene) is to prepare and then dehydrogenate poly(cyclohexa-1,3-diene), as shown below

n — Ziegler–Natta or free-radical polymerization → reduction →

but the stereochemical control in the precursor polymer synthesis is not sufficiently good to produce poly(1,4-phenylene) cleanly. A slightly different and much more effective approach that produces high-quality poly(1,4-phenylene) is to prepare poly[*cis*-5,6-bis(trimethylsiloxy)cyclo-1,3-diene] by coordination polymerization using bis(allyl)trifluoroacetatonickel(II) ($(ANiTFA)_2$) as catalyst, followed by its conversion to poly[*cis*-5,6-bis(acetoxy)cyclo-1,3-diene], which is soluble and so can be purified and cast into films before the final thermal elimination of acetic acid

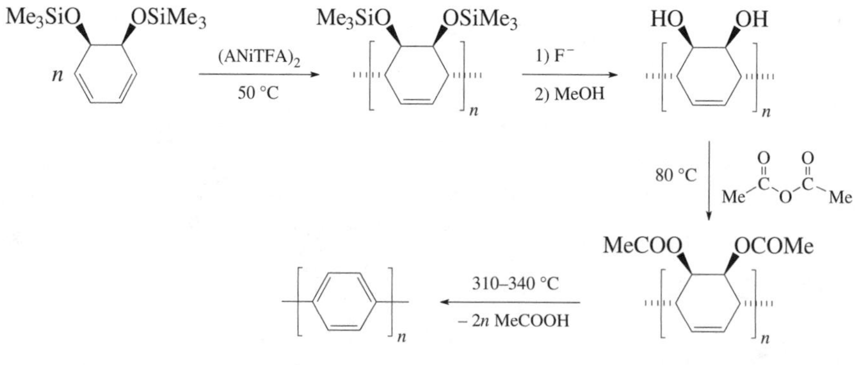

The elimination in the final step proceeds via a concerted intramolecular rearrangement

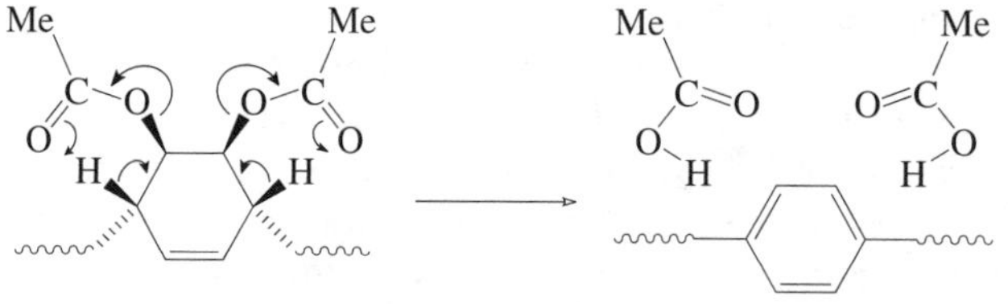

and can be catalysed using Lewis acids or protonic acids. A similar precursor route has been developed for synthesis of *trans*-poly(1,4-phenylene vinylene), by ring-opening metathesis polymerization (ROMP; Section 7.5) of the bis(methoxycarbonyl) ester of bicyclo[2.2.2]octa-5,8-diene-*cis*-2,3-diol to give a soluble precursor polymer that can be purified before casting into a film and heating to eliminate methyl carbonic acid via a mechanism identical to that shown above

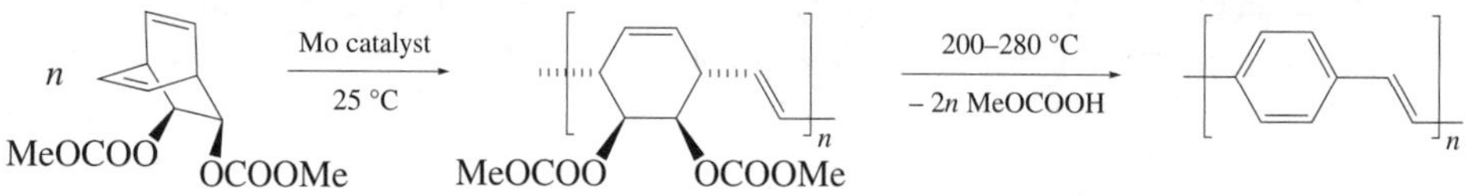

There are many other precursor routes to *trans*-poly(1,4-phenylene vinylene). For example, one that has gained importance involves base-induced polymerization of a 1,4-xylylene intermediate formed *in situ* from α,α'-bis(dialkylsulphonium chloride)-1,4-xylenes (via an uncertain chain polymerization mechanism) to produce a water-soluble sulphonium polyelectrolyte precursor, which after casting into a film, can be converted to *trans*-poly(1,4-phenylene vinylene) by heating

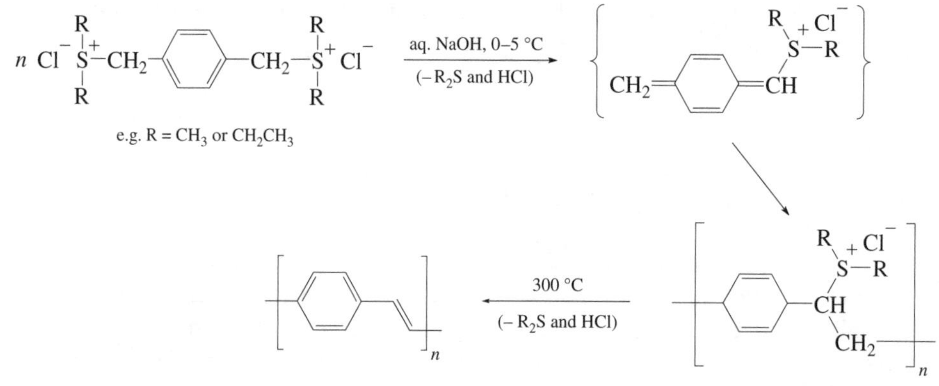

Mild conditions are used in the first-stage polymerization in order to prevent partial conversion to 1,4-phenylene vinylene repeat units because they reduce greatly the solubility of the precursor polymer.

An important precursor route to polyacetylene employs ROMP of substituted cyclobutenes to give a soluble precursor polymer that undergoes a retro Diels–Alder reaction on heating to generate *cis*-polyacetylene, which upon further heating isomerizes to the *trans*-form required for use in conducting polymer applications. This is known as the *Durham route* because it was first developed by Feast and coworkers at Durham University. The original Durham route is shown below

F_3C CF_3 — WCl_6+SnMe_4, 20 °C → F_3C CF_3

50–120 °C

trans-polyacetylene ← 210 °C — *cis*-polyacetylene + F_3C CF_3

ROMP proceeds exclusively at the cyclobutene C=C bond because the ring is highly strained, and the cyclohexene C=C bonds in the monomer are of much lower reactivity in ROMP. Compared to direct syntheses of polyacetylene by coordination or metathesis polymerization of acetylene (which leads to contamination by catalyst residues), this route to polyacetylene has the advantage of enabling the precursor polymer to be purified by dissolution and re-precipitation before it is formed into a film and converted to polyacetylene.

8.5 SUPRAMOLECULAR POLYMERIZATION (POLYASSOCIATION)

Supramolecular chemistry is the chemistry of entities and materials that are created by association of at least two, but more often large numbers of, molecules through intermolecular, non-covalent interactions, such as donor-acceptor interactions, metal ion coordination, or electrostatic interactions. Lehn was a central figure in the establishment of supramolecular chemistry and received the Nobel Prize for Chemistry in 1987 in recognition of his work. Supramolecular assembly is a massive subject and so only a brief introduction will be given here, specifically in relation to preparation of *supramolecular polymers* through assembly of monomeric species into chains, driven by intermolecular interactions, which most commonly take the form of multiple hydrogen bonds.

The formation of a linear supramolecular polymer has the characteristics of linear step polymerization (Sections 2.2 and 3.2) and can be represented schematically as follows

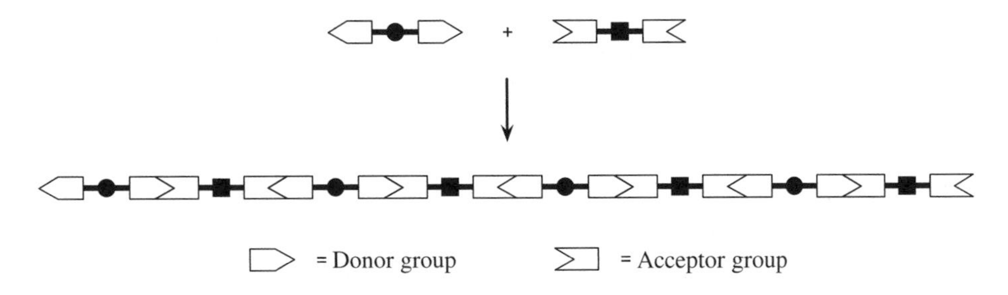

The process by which polymer chains are formed through intermolecular associations has been termed *supramolecular polymerization* or *polyassociation*. Examples of supramolecular polymerizations that involve the formation of three, four and six hydrogen bonds per link are shown in Table 8.1. The supramolecular polymers produced often are designed to self-organize into larger entities, such as liquid crystals (Section 17.5); e.g. the supramolecular polymer shown in example (i) of Table 8.1

TABLE 8.1
Examples of Supramolecular Polymer Synthesis via (i) Triple, (ii) Quadruple and (iii) Sextuple Hydrogen Bond Formation

(i) Supramolecular polymerization through formation of three hydrogen bonds per chain link

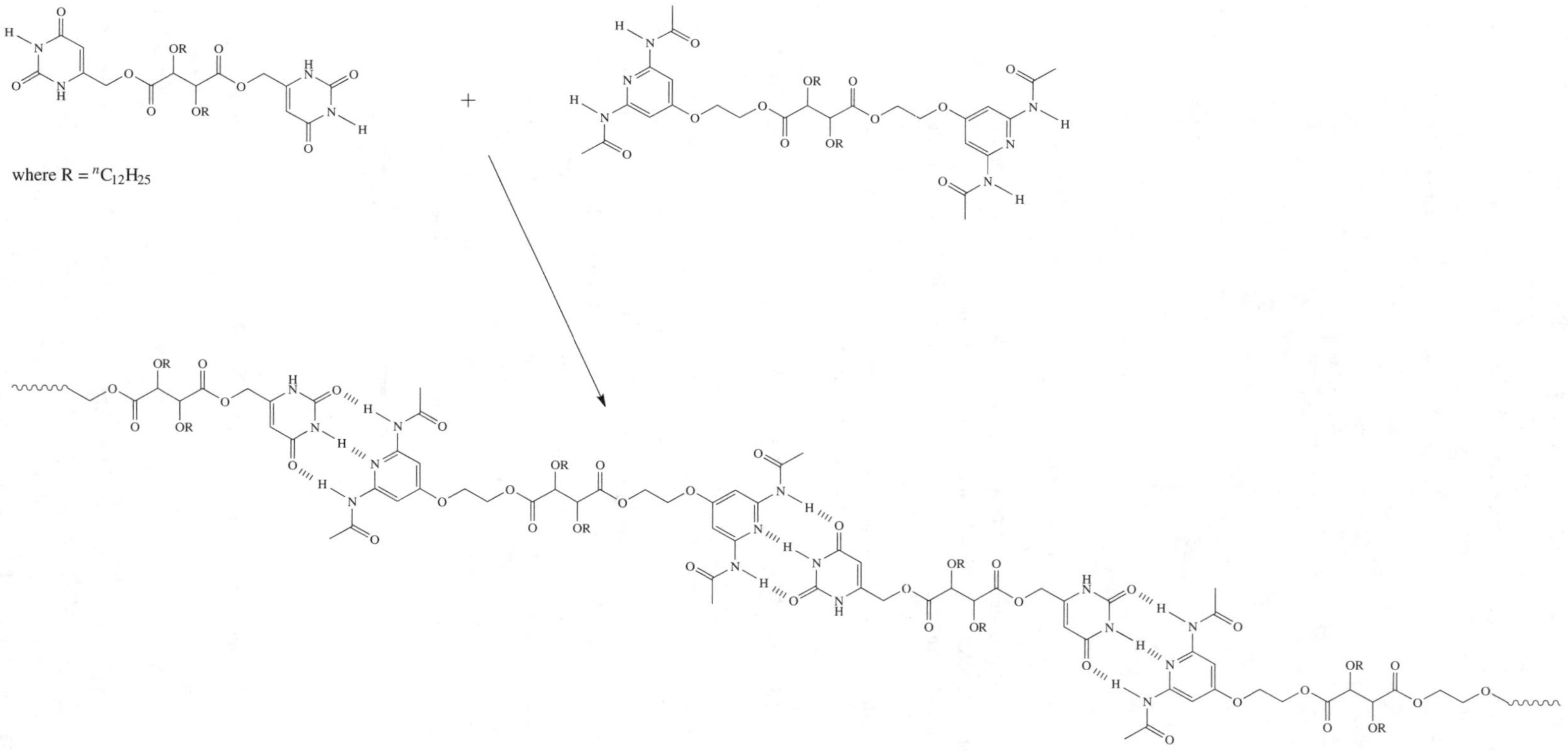

+

where R = $^{n}C_{12}H_{25}$

(continued)

TABLE 8.1 (continued)
Examples of Supramolecular Polymer Synthesis via (i) Triple, (ii) Quadruple and (iii) Sextuple Hydrogen Bond Formation

(ii) Supramolecular polymerization through formation of four hydrogen bonds per chain link

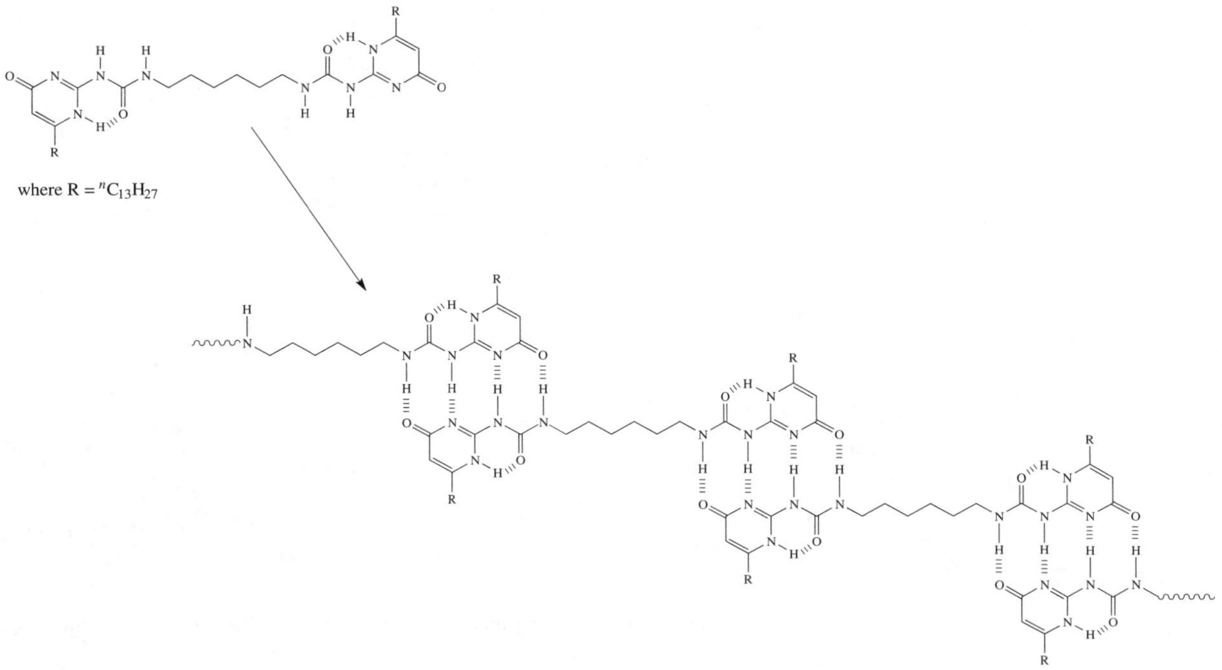

where R = $^{n}C_{13}H_{27}$

(continued)

TABLE 8.1 (continued)
Examples of Supramolecular Polymer Synthesis via (i) Triple, (ii) Quadruple and (iii) Sextuple Hydrogen Bond Formation

(iii) Supramolecular polymerization through formation of six hydrogen bonds per chain link

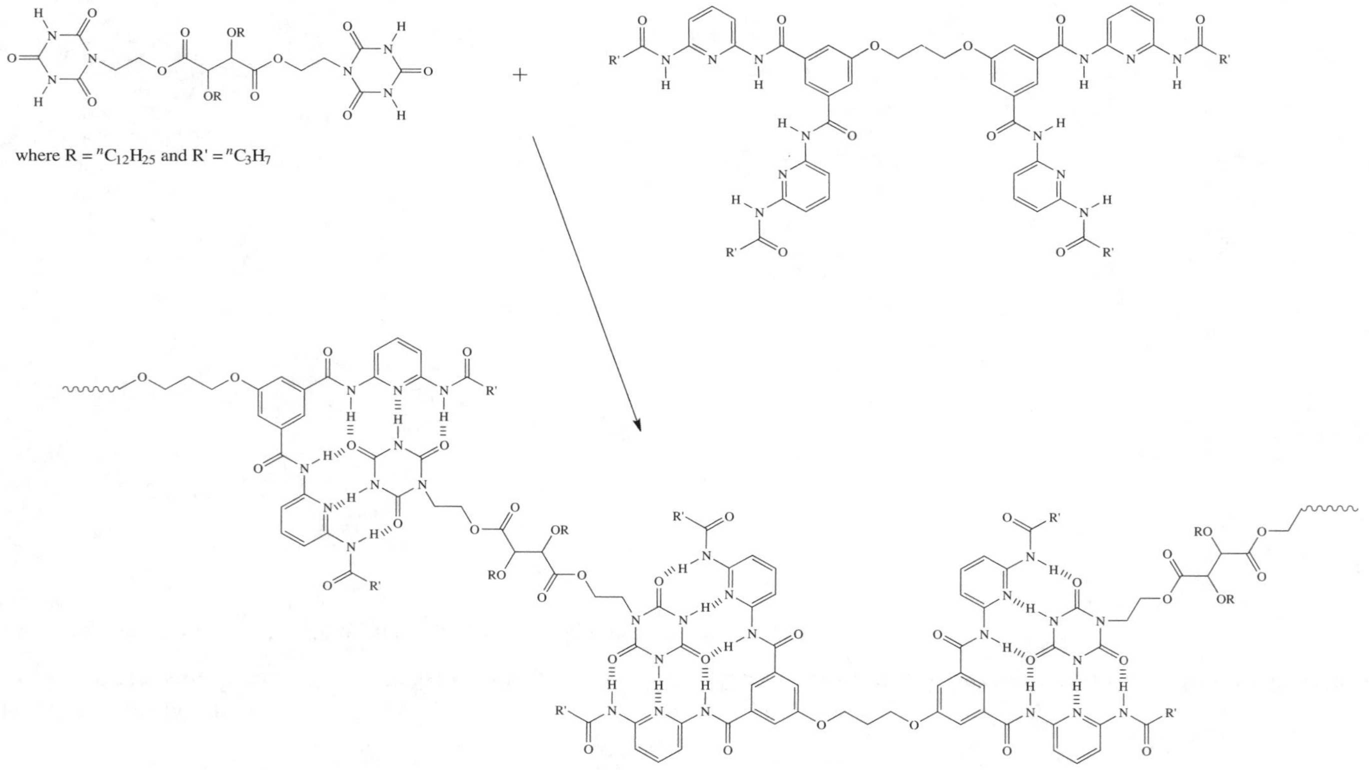

where R = $^{n}C_{12}H_{25}$ and R' = $^{n}C_{3}H_{7}$

displays liquid crystallinity over a very wide range of temperature when the chirality of the tartaric acid residue in each monomer is *meso*.

Multiple hydrogen bonds of the type shown in Table 8.1 greatly increase the stability of the links because all the hydrogen bonds in a link have to dissociate simultaneously for the link to break. Hence, supramolecular polymers often show behaviour that is characteristic of long-chain molecules (e.g. giving solutions of high viscosity (Sections 10.4.2 and 13.2) and being able to be processed into solid artefacts with useful mechanical properties), just as for covalently linked polymers. Nevertheless, because the chains are held together by reversible interactions, supramolecular polymers have an additional dynamic contribution to their properties which arises from their ability to revert to the original monomeric molecules when their environment (temperature and/or solvent) favours full dissociation of the links. Supramolecular polymers also can reorganize their link sequence, molar mass and molar mass distribution under particular conditions due to the dynamic nature of their chain links. This dynamic behaviour is unique to supramolecular polymers and creates opportunities to design polymers, for example, that have reduced 'melt' viscosity during processing, or that can be used as self-healing materials, or which can respond to their environment.

PROBLEMS

8.1 Discuss, in general terms, the criteria for effective solid-state topochemical polymerization.

8.2 An oxidative coupling polymerization of 2,6-dimethylphenol (0.1 mol) is allowed to proceed to almost complete conversion of the monomer, after which phenol (0.1 mol) is added and the polymerization allowed to proceed to complete conversion of both monomers. Discuss the structure of the polymer produced from this process.

8.3 Show how the following molecules (I and II, where $R = {}^{n}C_{12}H_{25}$) would interact with each other. Using simple diagrams to represent the molecules, sketch the form of the supramolecular assembly produced from their interactions.

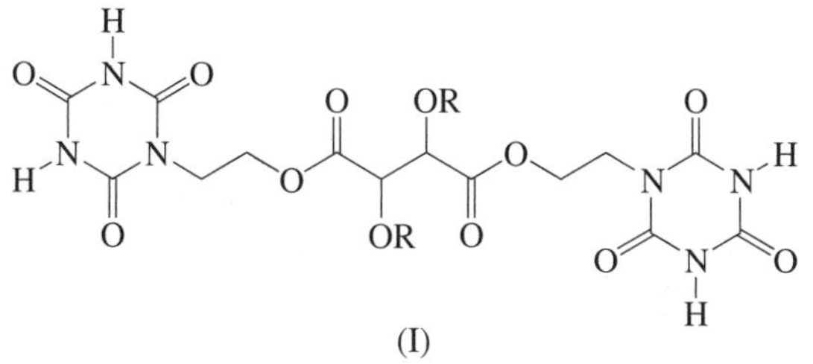

H O N O O OR O H N N O O N N H H O OR O O N O H

(I)

NH_2 N H_2N R N NH_2

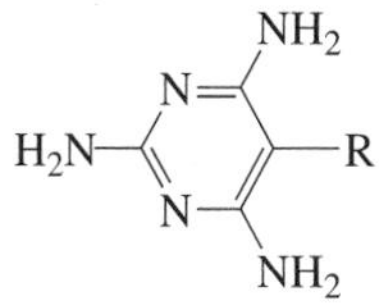

(II)

FURTHER READING

General Reading

Allen, G. and Bevington, J.C. (eds), *Comprehensive Polymer Science*, Vols. 4 and 5 (Eastmond, G.C., Ledwith, A., Russo, S., and Sigwalt, P., eds), Pergamon Press, Oxford, 1989.

Lenz, R.W., *Organic Chemistry of Synthetic High Polymers*, Interscience, New York, 1967.

Odian, G., *Principles of Polymerization*, 4th edn., Wiley-Interscience, New York, 2004.

Stevens, M.P., *Polymer Chemistry: An Introduction*, 3rd edn., Oxford University Press, New York, 1998.

Mark, H.F., Bikales, N.M., Overberger, C.G., and Menges, G. (eds), *Encyclopedia of Polymer Science and Engineering*, Wiley-Interscience, New York, 1985–1989.

Synthesis of Conducting Polymers

Chan, H.S.O. and Ng, S.C., Synthesis, characterization and applications of thiophene-based functional polymers, *Progress in Polymer Science*, **23**, 1167 (1998).

Syed, A.A. and Dinesan, M.K., Polyaniline – a novel polymeric material, *Talanta*, **38**, 815 (1991).

Vernitskaya, T.V. and Efimov, O.N., Polypyrrole: a conducting polymer; its synthesis, properties and applications, *Russian Chemical Reviews*, **66**, 443 (1997).

Supramolecular Polymerization

Bouteiller, L., Assembly via hydrogen bonds of low molar mass compounds into supramolecular polymers, *Advances in Polymer Science*, **207**, 79 (2007).

Brunsveld, L., Folmer, B.J.B., Meijer, E.W., and Sijbesma, R.P., Supramolecular polymers, *Chemical Reviews*, **101**, 4071 (2001).

Lehn, J.-M., Supramolecular polymer chemistry – scope and perspectives, *Polymer International*, **51**, 825 (2002).

Lehn, J.-M., Dynamers: dynamic molecular and supramolecular polymers, *Progress in Polymer Science*, **30**, 814 (2005).

9 Copolymerization

9.1 INTRODUCTION

Copolymers comprise more than one type of repeat unit and their properties are controlled by the nature and proportions of the repeat units together with their sequence distribution along the chain, as outlined in Section 1.2.3 and discussed in Chapters 16 through 18. Thus, it is important to understand how to control copolymer composition and repeat unit sequence distribution.

Simultaneous polymerization of a mixture of monomers is the simplest form of copolymerization and is used to prepare statistical copolymers whose properties are intermediate to those of the corresponding homopolymers. A very high proportion of commercial polymers are copolymers produced in this way and even many commercial 'homopolymers' are in fact copolymers with small proportions of other types of repeat unit included to modify their properties. The control of the statistical step and chain copolymerizations used to prepare such copolymers is considered in Sections 9.2 and 9.3.

In block and graft copolymers, homopolymer chains are connected together by chemical bonds. The homopolymer blocks usually are immiscible, a consequence of which is that block and graft copolymers show properties characteristic of the homopolymer constituents as well as unique properties resulting from chemical linking of the homopolymer blocks (see Chapter 18). Synthesis of block and graft copolymers requires special methods, the most important strategies for which are considered in Sections 9.4 and 9.5 with illustrative examples that demonstrate the most important principles.

9.2 STEP COPOLYMERIZATION

Some simple aspects of statistical step copolymerization will be described here through consideration of linear copolymerizations. The principles introduced are equally applicable to non-linear step copolymerization which, therefore, will not be considered separately.

The simplest linear step copolymerizations are of the general type ARB + AR′B or $RA_2 + R'B_2 + R''B_2$. For example, reaction of hexamethylene diamine with a mixture of adipic and sebacic acids yields a copolyamide containing both nylon 6.6 and nylon 6.10 repeat units.

Most step copolymerizations are taken to high extents of reaction in order to produce copolymers with suitably high molar masses (Sections 3.2.3.1 and 3.2.3.2). A consequence of this is that the overall compositions of the copolymers obtained correspond to those of the comonomer mixtures used to prepare them. However, it must be borne in mind that the sequence distribution of the different repeat units along the copolymer chains is an important factor controlling the properties of a copolymer and that the distribution is affected by differences in monomer reactivity.

When the mutually reactive functional groups (i.e. A and B) have reactivities that are the same for each monomer, the probability of reaction of a particular monomer depends only upon the mole fraction of functional groups that it provides. Most commonly, this situation obtains when the different monomers containing the same functional groups are of similar structure, as for adipic acid and sebacic acid in the example given above. Under these conditions, *random copolymers* which have the most probable distribution of molar mass are formed and the reaction kinetics described in Section 3.2.3.3 apply.

Step copolymerizations which involve two (or more) monomers containing the same type of functional group, but of different reactivity, are more complex and do not yield random copolymers. The monomer containing the more reactive functional groups reacts preferentially and so is incorporated into the oligomeric chains formed early in the reaction in higher proportion than its initial mole fraction in the comonomer mixture. The less reactive monomer participates in the copolymerization to an increasing extent as the concentration of the more reactive monomer is depleted by preferential reaction. Thus the copolymer molecules formed contain significant sequences of the same repeat units, i.e. they have a 'blocky' structure, and the length of these homopolymer sequences increases as the difference in monomer reactivity increases. Such blocky copolymer structures are often formed when performing step copolymerizations of the type $RA_2 + R'B_2 + R''C_2$ where the functional groups of type B and C are unreactive towards each other and only react with the functional groups of type A. For example, reaction of a dicarboxylic acid with a mixture of a diol and a diamine yields a blocky poly(ester-*co*-amide) because the amine groups are more reactive than the hydroxyl groups. However, more subtle differences in reactivity also lead to blockiness in the sequence distribution, e.g. primary hydroxyl groups ($-CH_2OH$) are more reactive than secondary ($-CHROH$), which is why poly(propylene oxide) polyols used in polyurethane formation normally are 'tipped' with ethylene oxide units to make them of equal reactivity to the primary hydroxyl groups of chain extenders (see Sections 3.2.2.1, 3.3.1.3 and 9.4.1).

9.3 CHAIN COPOLYMERIZATION

As for step copolymerization, differences in monomer reactivity in statistical chain copolymerization affect the sequence distribution of the different repeat units in the copolymer molecules formed. The most reactive monomer again is incorporated preferentially into the copolymer chains but, because high molar mass copolymer molecules are initiated, propagate and terminate in short timescales, even at low overall conversions of the comonomers, the high molar mass copolymer molecules formed can have compositions which differ significantly from the composition of the initial comonomer mixture. Also in contrast to step copolymerization, theoretical prediction of the relative rates at which the different monomers add to a growing chain is more firmly established. In the next section, a general theoretical treatment of chain copolymerization of two monomers (so-called binary copolymerization) is presented and introduces an approach that can be applied to derive equations for more complex chain copolymerizations involving three or more monomers.

9.3.1 Copolymer Composition Equation

In order to predict the composition of the copolymer formed at a particular instant in time during a binary chain copolymerization, it is necessary to construct a kinetics model of the reaction. The simplest model will be analysed here and is the *terminal model* which assumes that the reactivity of an active centre depends only upon the terminal monomer unit on which it is located. It is further assumed that the amount of monomer consumed in reactions other than propagation is negligible and that copolymer molecules of high molar mass are formed. Thus, for the copolymerization of monomer A with monomer B, only two types of active centre need be considered

$$\sim A^* \text{ and } \sim B^*$$

where the asterisks represent the active centres. The theory to be developed is applicable to any type of chain polymerization for which the above assumptions are satisfactory (e.g. the asterisk represents an unpaired electron for free-radical copolymerizations, a positive charge for cationic polymerizations, a negative charge for anionic polymerizations or the active site in coordination and ring-opening chain polymerizations).

The ~A* and ~B* active centres propagate by addition of either an A or B monomer molecule and so there are four possible propagation reactions, each with its own rate coefficient

$$\sim\text{A}^* + \text{A} \xrightarrow{k_{\text{AA}}} \sim\text{AA}^*$$

$$\sim\text{A}^* + \text{B} \xrightarrow{k_{\text{AB}}} \sim\text{AB}^*$$

$$\sim\text{B}^* + \text{B} \xrightarrow{k_{\text{BB}}} \sim\text{BB}^*$$

$$\sim\text{B}^* + \text{A} \xrightarrow{k_{\text{BA}}} \sim\text{BA}^*$$

The reactions with rate coefficients k_{AA} and k_{BB} are known as *homopropagation* reactions and those with rate coefficients k_{AB} and k_{BA} are called *cross-propagation* reactions. Hence the rate of consumption of monomer A is given by

$$-\frac{\text{d[A]}}{\text{d}t} = k_{\text{AA}}[\text{A}^*][\text{A}] + k_{\text{BA}}[\text{B}^*][\text{A}] \tag{9.1}$$

where [A*] and [B*] are the total concentrations of propagating chains with terminal A-type and B-type active centres, respectively. Similarly, the rate of consumption of monomer B is given by

$$-\frac{\text{d[B]}}{\text{d}t} = k_{\text{BB}}[\text{B}^*][\text{B}] + k_{\text{AB}}[\text{A}^*][\text{B}] \tag{9.2}$$

At any instant in time during the reaction, the ratio of the amount of monomer A to monomer B being incorporated into the copolymer chains is obtained by dividing Equation 9.2 into Equation 9.1

$$\frac{\text{d[A]}}{\text{d[B]}} = \frac{[\text{A}]}{[\text{B}]}\left\{\frac{k_{\text{AA}}[\text{A}^*]/[\text{B}^*] + k_{\text{BA}}}{k_{\text{BB}} + k_{\text{AB}}[\text{A}^*]/[\text{B}^*]}\right\} \tag{9.3}$$

An expression for the ratio [A*]/[B*] is obtained by applying steady-state conditions to [A*] and [B*], i.e.

$$\frac{\text{d[A}^*]}{\text{d}t} = 0 \quad \text{and} \quad \frac{\text{d[B}^*]}{\text{d}t} = 0$$

In terms of the creation and loss of active centres of a particular type, the contribution of initiation and termination reactions is negligible compared to that of the cross-propagation reactions. Thus

$$\frac{\text{d[A}^*]}{\text{d}t} = k_{\text{BA}}[\text{B}^*][\text{A}] - k_{\text{AB}}[\text{A}^*][\text{B}]$$

and

$$\frac{\text{d[B}^*]}{\text{d}t} = k_{\text{AB}}[\text{A}^*][\text{B}] - k_{\text{BA}}[\text{B}^*][\text{A}]$$

Application of the steady-state condition to either of these equations leads to

$$\frac{[\mathrm{A}^*]}{[\mathrm{B}^*]} = \frac{k_{\mathrm{BA}}[\mathrm{A}]}{k_{\mathrm{AB}}[\mathrm{B}]} \tag{9.4}$$

Substituting Equation 9.4 into Equation 9.3 and simplifying yields one form of the *copolymer composition equation*

$$\frac{\mathrm{d}[\mathrm{A}]}{\mathrm{d}[\mathrm{B}]} = \frac{[\mathrm{A}]}{[\mathrm{B}]}\left(\frac{r_{\mathrm{A}}[\mathrm{A}]+[\mathrm{B}]}{[\mathrm{A}]+r_{\mathrm{B}}[\mathrm{B}]}\right) \tag{9.5}$$

where r_A and r_B are the respective *monomer reactivity ratios* defined by

$$r_{\mathrm{A}} = \frac{k_{\mathrm{AA}}}{k_{\mathrm{AB}}} \quad \text{and} \quad r_{\mathrm{B}} = \frac{k_{\mathrm{BB}}}{k_{\mathrm{BA}}} \tag{9.6}$$

Equation 9.5 gives the *molar ratio* of A-type to B-type repeat units in the copolymer formed at any instant (i.e. a very small interval of time) during the copolymerization when the monomer concentrations are [A] and [B]. Often it is more convenient to express compositions as mole fractions. The *mole fraction* f_A of monomer A in the comonomer mixture is [A]/([A]+[B]) and that of monomer B is $f_B = 1 - f_A$. The mole fraction F_A of A-type repeat units in the copolymer formed at a particular instant in time is d[A]/(d[A]+d[B]) and that of B-type repeat units is $F_B = 1 - F_A$. Addition of unity to both sides of Equation 9.5 allows it to be rearranged in terms of F_A (or F_B), f_A and f_B. The following *copolymer composition equations* are obtained

$$F_{\mathrm{A}} = \frac{r_{\mathrm{A}}f_{\mathrm{A}}^2 + f_{\mathrm{A}}f_{\mathrm{B}}}{r_{\mathrm{A}}f_{\mathrm{A}}^2 + 2f_{\mathrm{A}}f_{\mathrm{B}} + r_{\mathrm{B}}f_{\mathrm{B}}^2} \tag{9.7}$$

and

$$F_{\mathrm{B}} = \frac{r_{\mathrm{B}}f_{\mathrm{B}}^2 + f_{\mathrm{A}}f_{\mathrm{B}}}{r_{\mathrm{A}}f_{\mathrm{A}}^2 + 2f_{\mathrm{A}}f_{\mathrm{B}} + r_{\mathrm{B}}f_{\mathrm{B}}^2} \tag{9.8}$$

These equations apply to all types of chain copolymerization and enable prediction of copolymer composition from the comonomer composition and the monomer reactivity ratios.

9.3.2 Monomer Reactivity Ratios and Copolymer Composition/Structure

Monomer reactivity ratios are important quantities since for a given instantaneous comonomer composition, they control the overall composition of the copolymer formed at that instant and also the sequence distribution of the different repeat units in the copolymer. From Equation 9.6, they are the ratios of the homopropagation to the cross-propagation rate coefficients for the active centres derived from each respective monomer. Thus if $r_A > 1$ then ~A* prefers to add monomer A (i.e. it prefers to homopolymerize), whereas if $r_A < 1$, ~A* prefers to add monomer B and hence copolymerize. Similarly, r_B describes the behaviour of monomer B. Plots of Equation 9.7 for different r_A, r_B pairs are shown in Figure 9.1.

Random copolymers with $F_A = f_A$ (for all values of f_A) are formed when $r_A = r_B = 1$, i.e. when the probability of adding monomer A is equal to the probability of adding monomer B for both types of

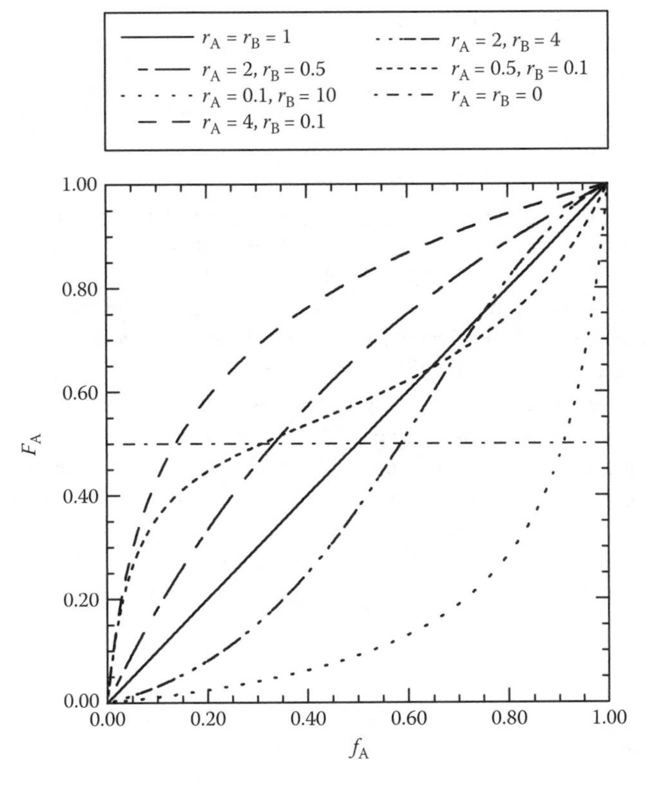

FIGURE 9.1 Plots of Equation 9.7 showing the variation of copolymer composition F_A with comonomer composition f_A for different pairs of r_A, r_B values.

active centre. There are very few copolymerizations which approximate to this condition and those that do involve copolymerization of monomers with very similar structures (e.g. free-radical copolymerization of tetrafluoroethylene with chlorotrifluoroethylene). More commonly, one monomer, assumed here to be monomer A, is more reactive than the other, and both types of active centre prefer to add the more reactive monomer. In terms of reactivity ratios, this gives rise to $r_A > 1$ with $r_B < 1$. On this basis, it is easy to appreciate why there are no simple copolymerizations for which $r_A > 1$ with $r_B > 1$.

Ideal copolymerization is a special case of $r_A > 1$, $r_B < 1$ (or $r_A < 1$, $r_B > 1$) copolymerization for which $r_A r_B = 1$. The name derives from the similarity of the F_A versus f_A curves to boiling point diagrams for ideal liquid–liquid mixtures. Under these conditions

$$r_A = \frac{1}{r_B} \quad \text{i.e.,} \; \frac{k_{AA}}{k_{AB}} = \frac{k_{BA}}{k_{BB}}$$

which means that the *relative rates* at which the two monomers are incorporated into the copolymer chains are the same for both types of active centre (i.e. ~A* and ~B*). Thus, even though one monomer is more reactive and $F_A \neq f_A$, the sequence distribution of the repeat units in the copolymer formed is random. Substitution of $r_B = r_A^{-1}$ into Equation 9.7 gives a simplified *copolymer composition equation for ideal copolymerization*

$$F_A = \frac{r_A f_A}{r_A f_A + f_B} \tag{9.9}$$

Ideal copolymerization occurs when there are no specific effects controlling either one or more of the four propagation reactions, since the relative rates of reaction of the two monomers then depend only upon their absolute relative reactivities. In many copolymerizations $r_A > 1$, $r_B < 1$ (or vice versa) with $r_A r_B \neq 1$ (usually $r_A r_B < 1$). Such copolymerizations give F_A versus f_A curves that are similar to those for ideal copolymerization but skewed towards copolymer compositions richer in the repeat units derived from the more reactive monomer.

The higher the ratio r_A/r_B for $r_A > 1$, $r_B < 1$ copolymerizations, the longer are the continuous sequences of A-type repeat units and the shorter are the continuous sequences of B-type repeat units in the copolymer molecules formed. When $r_A \gg 1$ with $r_B \ll 1$, there is a tendency towards consecutive homopolymerization of the two monomers. The molecules formed early in the reaction have very long sequences of A-type repeat units with the occasional B-type repeat unit (i.e. they are essentially molecules of homopolymer A). Later in the reaction, when monomer A has been consumed virtually completely, the very high concentration of residual monomer B leads to the formation of copolymer molecules that are essentially homopolymer B.

Azeotropic copolymerization occurs when $r_A < 1$ with $r_B < 1$ and when $r_A > 1$ with $r_B > 1$, though the latter of these conditions rarely is observed in practice for the reasons indicated above. The name again comes from analogy with boiling point diagrams of liquid–liquid mixtures, since the F_A versus f_A curves are characterized by their intersection of the $F_A = f_A$ line at one point which corresponds to the *azeotropic composition* $(f_A)_{azeo}$. Substituting $F_A = f_A = (f_A)_{azeo}$ into Equation 9.7 leads to

$$(f_A)_{azeo} = \frac{1 - r_B}{2 - r_A - r_B} \tag{9.10}$$

As the product $r_A r_B$ decreases, there is an increasing tendency towards alternation in the additions of monomer molecules to the propagating chains. The extreme case of azeotropic copolymerization is $r_A = r_B = 0$ and always produces perfectly alternating copolymers, irrespective of the value of f_A (i.e. $F_A = 0.50$ for $0 < f_A < 1$), because the homopropagation reactions do not occur.

9.3.3 Copolymer Composition Drift

For a given pair of comonomers, the value of F_A for the copolymer formed early in the reaction is determined by the initial value of f_A via Equation 9.7. For most copolymerizations, $F_A \neq f_A$ and one monomer is consumed preferentially causing f_A to change as the overall monomer conversion increases. Since Equation 9.7 is applicable to each increment of conversion, the change in f_A gives rise to a variation in F_A with conversion. This is known as *copolymer composition drift* and leads to copolymers which consist of copolymer molecules with significantly different compositions. This broadening of the *distribution of copolymer composition* beyond that arising from the normal statistical variation of copolymer composition about F_A at any specific value of f_A, clearly becomes more significant as the overall monomer conversion increases.

In $r_A > 1$, $r_B < 1$ copolymerizations f_A (and hence F_A) decreases with conversion as monomer A is consumed preferentially. Eventually, monomer A is consumed virtually completely leaving mainly unreacted monomer B (i.e. f_A eventually approaches zero) and so, thereafter, essentially homopolymer of monomer B is formed.

In $r_A < 1$, $r_B < 1$ azeotropic copolymerizations f_A changes with conversion until it becomes equal to either zero or unity and the corresponding homopolymer is formed from then onwards. (For the unrealistic $r_A > 1$, $r_B > 1$ azeotropic condition f_A would change with conversion until it is equal to $(f_A)_{azeo}$ and copolymer with $F_A = (f_A)_{azeo}$ would be formed thereafter.)

For many applications, the tolerance for copolymer composition drift is small and its control is essential. The strategies most commonly used for this purpose are as follows.

1. The overall monomer conversion is limited (usually to ≤5%) in order to reduce the drift in f_A.
2. Additional quantities of the monomer that are consumed preferentially are fed to the reaction vessel at a controlled rate during the copolymerization in order to maintain f_A constant.
3. *Starve-feeding* of the comonomer mixture to the reaction vessel (i.e. feeding at rate below the potential rate of polymerization) in order to achieve very high instantaneous conversions (close to 100%), thereby ensuring that f_A does not vary during the copolymerization. This is widely practiced in preparing structured (e.g. core-shell) latex particles by emulsion polymerization (Section 4.4.4).

9.3.4 Evaluation of Monomer Reactivity Ratios

In order to use the copolymer composition equations predictively, it is necessary to know the values of r_A and r_B. The most simple method for their evaluation involves determination of the compositions (i.e. F_A) of the copolymers formed at low conversions (to minimize composition drift) in a series of copolymerizations performed using known initial comonomer compositions (i.e. f_A). The data can be analysed in several ways. The *Fineman–Ross* method was one of the most widely used and employs a rearranged version of Equation 9.5 with d[A]/d[B] = F_A/F_B (which is applicable at very low conversion)

$$\frac{x}{y}(y-1)=\left(\frac{x^2}{y}\right)r_A - r_B \tag{9.11}$$

where $x = f_A/f_B$ and $y = F_A/F_B$. For each experimental data set, the left-hand side of Equation 9.11 is plotted against x^2/y. The data points are then fitted to a straight line, from which the slope gives r_A and the intercept r_B. This method was shown subsequently to give values of monomer reactivity ratios that are sensitive to the indexing (i.e. which monomer is assigned the label A), thereby highlighting serious flaws in the method. These arise from plotting complex functions of f_A, f_B, F_A and F_B (involving ratios, squares, differences and products that amplify errors in the experimental data), coupled with non-uniform spreading of the data across the axes (which leads to errors arising from linear least squares fitting). The *Kelen–Tudos* method came much later and is a great improvement upon the Fineman–Ross analysis, employing the following expression

$$\frac{(x/y)(y-1)}{\alpha+(x^2/y)}=\left\{\frac{(x^2/y)}{\alpha+(x^2/y)}\right\}\left\{r_A+\frac{r_B}{\alpha}\right\}-\frac{r_B}{\alpha} \tag{9.12}$$

which reduces the amplification of errors and, in particular, includes an arbitrary data spreading factor α which is recommended to have the value given by

$$\alpha=\sqrt{F_A^{\min}F_A^{\max}}$$

where $F_A^{\min}$ and $F_A^{\max}$ are the minimum and maximum values of F_A in the set of experimental data. Equation 9.12 can be reduced to the form

$$\eta=\left\{r_A+\frac{r_B}{\alpha}\right\}\xi-\frac{r_B}{\alpha}$$

in which $\eta=\left[(x/y)(y-1)\right]/\left[\alpha+(x^2/y)\right]$ and $\xi=\left\{(x^2/y)/\left[\alpha+(x^2/y)\right]\right\}$. Thus a plot of η against ξ gives a straight line which can be extrapolated to both $\xi = 0$ and $\xi = 1$, thereby yielding $-r_B/\alpha$ and r_A as the respective η intercepts.

Although the Kelen–Tudos method has yielded much more reliable data, it still is not statistically robust. There now are much more sophisticated and statistically valid methods for evaluation of reactivity ratios, the most favoured being the *error-in-variables* method but this requires knowledge of all experimental errors and so is far more time consuming. Description of these methods is beyond the scope of this book, so the reader should consult up-to-date advanced reviews of copolymerization for details of their basis and use.

9.3.5 Free-Radical Copolymerization

Many commercially important copolymers are prepared by free-radical copolymerization of ethylenic monomers, e.g. styrene–butadiene rubber (SBR), acrylonitrile–butadiene–styrene copolymer (ABS), ethylene–vinyl acetate copolymer (EVA) and acrylonitrile–butadiene copolymer (nitrile rubber).

Most free-radical copolymerizations can be categorized into either one or the other of the following two types: (i) $r_A > 1$, $r_B < 1$ (or $r_A < 1$, $r_B > 1$) copolymerization and (ii) $r_A < 1$, $r_B < 1$ azeotropic copolymerization. This is evident from Table 9.1 which gives the reactivity ratios and their products for some representative free-radical copolymerizations.

The reactivity of a monomer is strongly dependent upon the ability of its substituent group(s) to stabilize the corresponding polymeric radical. This is because the greater is the stability of the polymeric radical, the more readily it is formed by reaction of the monomer. Thus reactive monomers have substituent groups which stabilize the polymeric radical by delocalization of the unpaired electron (i.e. by resonance). Hence a monomer of high reactivity yields a polymeric radical of low reactivity and vice versa. The order in which common substituent groups provide increasing resonance stabilization is

$$-C_6H_5 > -CH{=}CH_2 > -\overset{O}{\overset{\|}{C}}-R \approx -C{\equiv}N \approx -\overset{O}{\overset{\|}{C}}-O-R > -R > -Cl \approx -O-\overset{O}{\overset{\|}{C}}-R \approx -O-R$$

where R is an alkyl group. Vinyl monomers ($CH_2{=}CHX$) tend to be of lower reactivity than the corresponding 1,1-disubstituted monomers ($CH_2{=}CXY$) because the resonance effects of the substituents tend to be additive and because secondary radicals (i.e. $\sim CH_2-\dot{C}HX$) generally are of

TABLE 9.1
Some Typical Values of Reactivity Ratios for Free-Radical Copolymerization at 60 °C

Monomer A	Monomer B	r_A	r_B	$r_A r_B$
Styrene	Butadiene	0.78	1.39	1.08
Styrene	Methyl methacrylate	0.52	0.46	0.24
Styrene	Methyl acrylate	0.75	0.18	0.14
Styrene	Acrylonitrile	0.40	0.04	0.02
Styrene	Maleic anhydride	0.02	0	0
Styrene	Vinyl chloride	17	0.02	0.34
Vinyl acetate	Vinyl chloride	0.23	1.68	0.39
Vinyl acetate	Acrylonitrile	0.06	4.05	0.24
Vinyl acetate	Styrene	0.01	55	0.55
Methyl methacrylate	Methyl acrylate	1.69	0.34	0.57
Methyl methacrylate	*n*-Butyl acrylate	1.8	0.37	0.67
Methyl methacrylate	Acrylonitrile	1.20	0.15	0.18
Methyl methacrylate	Vinyl acetate	20	0.015	0.30
trans-Stilbene	Maleic anhydride	0.03	0.03	0.001

lower stability than tertiary radicals (i.e. $\sim CH_2-\dot{C}XY$) due to steric shielding of the active site being greater for the latter. In contrast, 1,2-disubstituted monomers (CHX=CHY) are of low reactivity due to steric hindrance of the propagation reaction by the 2-substituent in the monomer. Although such monomers homopolymerize only with great difficulty, they can be copolymerized with vinyl and 1,1-disubstituted monomers because of the reduced steric hindrance in the cross-propagation reactions; however, they always have low reactivity ratios.

In $r_A > 1$, $r_B < 1$ copolymerizations, the ratio r_A/r_B increases (or for $r_A < 1$, $r_B > 1$ copolymerization, the ratio r_A/r_B decreases) as the difference between the reactivities of the monomers increases (cf. reactivity ratios for copolymerization of vinyl acetate with vinyl chloride, acrylonitrile and styrene in Table 9.1). Thus, the most simple copolymerizations tend to be between monomers with similar reactivities. However, this general statement requires qualification because of the influence of steric effects (such as those described above) and polar effects. The latter are of great importance and are most evident in azeotropic copolymerizations. In general, as the difference between the electron densities of the C=C bonds (i.e. polarities) of the two monomers increases, the alternating tendency increases (i.e. r_Ar_B decreases; cf. the data in Table 9.1 for copolymerization of styrene with methyl methacrylate, methyl acrylate, acrylonitrile and maleic anhydride). Strong polar effects can negate the effects of steric hindrance, as is demonstrated by the alternating copolymerization of stilbene with maleic anhydride. Since specific steric and polar effects operate in most copolymerizations, there are relatively few copolymerizations which closely approximate to ideal behaviour (i.e. $r_Ar_B = 1$), the majority having $r_Ar_B < 1$.

9.3.5.1 The *Q–e* Scheme

The complexity of copolymerization makes theoretical prediction of reactivity ratios rather difficult. Nevertheless, the semi-empirical *Q–e* scheme often is used for estimating reactivity ratios and also provides an approximate ranking of the reactivities and polarities of monomers. The basis of the scheme is that the rate coefficient k_{pm} for reaction of a polymeric radical (p) with a monomer (m) is given by

$$k_{pm} = P_p Q_m \exp(-e_p e_m) \tag{9.13}$$

where P_p and Q_m are measures of the reactivities of the polymeric radical and the monomer, respectively, and e_p and e_m originally were considered to be measures of the 'electrostatic charges' associated with the polymeric radical and the monomer, respectively. Expressions for k_{AA}, k_{AB}, k_{BB} and k_{BA} can be generated by the application of Equation 9.13 to each propagation reaction. By taking the appropriate ratios of these expressions (see Equation 9.6) and assuming that the charge associated with a polymeric radical is equal to the charge associated with the monomer from which it is derived, the following equations for the reactivity ratios are obtained

$$r_A = \left(\frac{Q_A}{Q_B}\right)\exp\left[-e_A(e_A - e_B)\right] \tag{9.14}$$

$$r_B = \left(\frac{Q_B}{Q_A}\right)\exp\left[-e_B(e_B - e_A)\right] \tag{9.15}$$

where $e_A = e_{pA} = e_{mA}$ and $e_B = e_{pB} = e_{mB}$. Thus the terms in polymeric radical reactivity have divided out and the equations essentially relate reactivity ratios to the reactivities (Q_A and Q_B) and electrostatic charges (e_A and e_B) of the C=C bonds of the two monomers. The prediction of alternating tendency

$$r_Ar_B = \exp[-(e_A - e_B)^2] \tag{9.16}$$

is in accord with experimental observations since $r_A r_B$ decreases as $(e_A - e_B)$ increases. Equation 9.16 also predicts that ideal copolymerization will occur when $e_A = e_B$ since this gives $r_A r_B = 1$.

In order to develop the Q–e scheme, styrene was chosen as the reference monomer and arbitrarily assigned values of Q and e. Initially, the Q and e values of other monomers were evaluated from pairs of experimentally determined r_A, r_B values for copolymerization with styrene. These values were then refined after consideration of experimentally determined r_A, r_B values for copolymerizations not involving styrene and the 'best-fit' Q–e values for each monomer recorded for prediction of pairs of r_A, r_B values using Equations 9.14 and 9.15.

Table 9.2 gives Q–e values for some important monomers and includes those currently accepted for the reference monomer (styrene). The values of Q and e increase with increasing monomer reactivity and increasing electron deficiency of the C=C bond, respectively. Negative values of e indicate that the C=C bond is electron rich.

TABLE 9.2
Values of Q and e for Some Important Monomers

Monomer	Q	e
Isoprene	3.33	−1.22
Butadiene	2.39	−1.05
Styrene	1.00	−0.80
Methyl methacrylate	0.74	0.40
Acrylonitrile	0.60	1.20
Ethyl acrylate	0.52	0.22
Maleic anhydride	0.23	2.25
Vinyl chloride	0.044	0.20
Vinyl acetate	0.026	−0.22

The theoretical basis of the Q–e scheme has been criticized in several respects, the most important being (i) the consideration of polarity effects in terms of electrostatic charges and the assumption of equal charges for a monomer and its corresponding polymeric radical are unrealistic, (ii) alternating tendency results from polarization phenomena in the transition state of propagation and not from interactions of permanent charges (e.g. free-radical copolymerizations normally are not affected by changes in the dielectric constant of the reaction medium), and (iii) steric effects are not directly taken into account, though they are incorporated indirectly into the experimentally evaluated Q–e values. Thus the Q–e scheme cannot be used to give rigorous quantitative predictions. Nevertheless, it does provide a useful guide to free-radical copolymerization behaviour.

9.3.6 Reversible-Deactivation Radical Copolymerization

Simultaneous copolymerization of two or more monomers by reversible-deactivation radical polymerization is quite different to the equivalent free-radical copolymerization due to the living nature of the copolymerization, which determines that all the copolymer molecules grow with the same probability and so should have the same overall composition and sequence distribution of repeat units. Only when the reactivity ratios are reasonably close to unity will statistical copolymers be formed. If the reactivity ratios deviate from unity, such that in a normal free-radical polymerization composition drift would be expected, then that drift manifests itself in the sequence of repeat units because the copolymer chains simply continue to grow and no new chains are formed. The copolymers formed are termed *gradient* or *tapered* since the repeat unit sequence begins with a higher proportion of repeat units from the more reactive monomer (A) and finishes with a higher proportion of repeat units from the less reactive monomer (B), as depicted below where R is an initiator fragment and the solid square represents the end-cap.

R–A–A–A–B–A–A–■

↘

R–A–A–A–B–A–A–B–A–A–B–A–B–■

↘

R–A–A–A–B–A–A–B–A–A–B–A–B–A–B–B–A–B–B–A–■

↘

R–A–A–A–B–A–A–B–A–A–B–A–B–A–B–B–A–B–B–A–B–B–B–A–B–B–B–B–■

Once the chain radical is released, the relative rates of addition of the monomers should simply be controlled by the reactivity ratios through Equation 9.5. Nevertheless, effects of differences in the values of the rate coefficients for activation and deactivation of the active site for the different types of end repeat unit have been observed and can even affect the ability to achieve living-like conditions. Hence, the type of reversible-deactivation radical polymerization used and the specific nature of the end-capping agent are important.

Nitroxide-mediated copolymerization is restricted by the limitations of nitroxide-mediated polymerization (see Section 4.5.1) and tends to be useful only for copolymerizations involving high proportions of styrene.

Atom-transfer radical copolymerization and reversible-addition-fragmentation chain-transfer copolymerization are more versatile (see Sections 4.5.2 and 4.5.3). Using these methods of copolymerization, it is possible to produce statistical, alternating and gradient copolymers from a range of comonomers including styrene, methacrylates, acrylates, acrylonitrile and maleic anhydride. The type of sequence distribution achieved is dependent upon the reactivity ratios, which often are similar to those for normal free-radical copolymerization. Copolymerizations of styrene tend to produce more strongly tapered sequence distributions due to the greater disparities between the reactivity ratios.

9.3.7 Ionic Copolymerization

In comparison to radical copolymerizations, there are relatively few ionic copolymerizations which yield copolymers containing significant proportions of repeat units from each of the monomers involved. This is because the ionic charge associated with the active centre emphasizes substituent group effects and gives rise to much greater differences in monomer reactivities. In most cases, one monomer is much more reactive than the other and so has a strong tendency to homopolymerize, i.e. $r_A \gg 1$, $r_B \ll 1$ or vice versa (see Table 9.3). The formation of copolymers is best accomplished using comonomers which are of very similar structure (e.g. anionic copolymerization of styrene with *para*-methylstyrene).

A further distinguishing feature of ionic copolymerizations is the strong dependence of reactivity ratios upon the nature of the solvent and the counter-ion (and hence the initiator). These effects can be dramatic and can lead to a reversing of the order of relative monomer reactivities (cf. reactivity ratios for anionic copolymerization of styrene with butadiene and isoprene in different solvents given in Table 9.3).

TABLE 9.3
Some Typical Values of Reactivity Ratios for Cationic and Anionic Copolymerization of Styrene (Monomer A)

Type of Copolymerization	Temperature / °C	Initiator	Solvent	Monomer B	r_A	r_B	r_Ar_B
Cationic	−90	$AlCl_3$	Dichloromethane	Isobutylene	0.24	1.79	0.43
	0	BF_3	Nitroethane	Chloroprene	33.0	0.15	4.95
	0	$TiCl_4$	Carbon tetrachloride	*para*-Methoxystyrene	0.05	46	2.30
Anionic	−78	^{s}BuLi	Tetrahydrofuran	Butadiene	11.0	0.04	0.44
	25	^{s}BuLi	Tetrahydrofuran	Butadiene	4.0	0.3	1.20
	25	^{s}BuLi	Benzene	Butadiene	0.04	10.8	0.43
	25	^{n}BuLi	Tetrahydrofuran	Isoprene	9.0	0.10	0.90
	30	^{n}BuLi	Benzene	Isoprene	0.26	10.6	2.76
	20	^{s}BuLi	Benzene	*para*-Methylstyrene	0.74	1.10	0.81

With the exception of block copolymer synthesis (Section 9.4.2.1), butyl rubber is the only copolymer of major industrial importance prepared by ionic copolymerization (see Section 5.2.1.7).

9.3.8 Ziegler–Natta Coordination Copolymerization

For many years, Ziegler–Natta coordination copolymerization of ethylene with propylene and non-conjugated dienes (such as hexa-1,4-diene and 5-ethylidene-2-norbornene) has been used to prepare an important class of rubbers known collectively as EPDM rubbers. Additionally, copolymerization of ethylene with small proportions of higher α-olefins (such as but-1-ene, hex-1-ene and oct-1-ene) is commercially important for production of a range of copolymers known as linear low-density polyethylenes (LLDPE), their lower densities (i.e. lower degrees of crystallinity) and lower melting temperatures resulting from the disruption of polyethylene crystallization by the comonomer repeat units.

Although the generalized copolymer composition Equations 9.7 and 9.8 can be applied to copolymerizations performed using Ziegler–Natta catalysts, there are several features of such reactions which make it less than appropriate. The theory assumes there to be only one type of active centre (e.g. an unpaired electron for free-radical copolymerization), whereas heterogeneous Ziegler–Natta catalysts possess a range of active surface sites with different activities and stereochemical selectivities. Thus the reactivity ratios obtained are average values and are dependent upon the precise methods and conditions employed for the preparation of the catalyst and for copolymerization. Furthermore, attempts to predict copolymer composition drift using the simple theory often are thwarted by time-dependent changes in the catalytic activities of the active surface sites (e.g. those arising from transfer reactions, see Section 6.4.5).

Despite the limited applicability of simple copolymerization theory, it is useful for establishing relative monomer reactivities and the effects of changing catalyst composition. For example, it is found that the relative reactivity of α-olefins decreases as the size of the substituent group increases and that the relative reactivity of ethylene in copolymerization with propylene generally is increased by changing from vanadium- to titanium-based catalysts. Often, very large differences in reactivity ratios are observed for copolymerization of ethylene with higher α-olefins, e.g. r_A(ethylene) > 50 with r_B(but-1-ene) < 0.1.

9.3.9 Metallocene Coordination Copolymerization

One of the many advantages of metallocene polymerization is that the reactivity ratios for copolymerization of ethylene with α-olefins are more favourable, enabling preparation of copolymers with much more uniform sequence distributions and facilitating incorporation of higher levels of α-olefins than is possible with Ziegler–Natta coordination copolymerization. More important, however, is that, because of the more uniform sequence distribution of the α-olefin repeat units, the α-olefins can be used at lower levels than in Ziegler–Natta coordination copolymerizations to achieve the same effects in reducing crystallinity and melting point of the copolymer compared to polyethylene homopolymer, thereby producing LLDPE of enhanced performance; the copolymers also have much narrower molar mass distributions, which has benefits for the polymer rheology. These features of ethylene/α-olefin copolymerizations are a further consequence, and benefit, of the single-site nature of metallocene catalysts (see Section 6.5.4) and enable more realistic analysis of the copolymerizations using the generalized copolymer composition Equations 9.7 and 9.8.

Copolymers of propylene with other α-olefins also can be produced using metallocene catalysts. Other comonomers that have been copolymerized successfully with ethylene and with propylene include cycloolefins, conjugated and non-conjugated dienes, and styrene. The metallocene-related constrained geometry catalysts (see Section 6.5.5) have proven to be very effective in producing ethylene/styrene copolymers with high proportions of well-distributed styrene repeat units.

9.3.10 Other Types of Chain Copolymerization

The other types of chain polymerization described in earlier chapters (notably group-transfer polymerization and the different types of ring-opening polymerization) can be used to prepare statistical copolymers, but such copolymerizations have received relatively little attention. In view of this, they will not be considered here and the reader is referred to more advanced texts.

9.4 BLOCK COPOLYMER SYNTHESIS

There are four principal strategies for synthesis of block copolymers.

1. Linking polymer chains together through mutually-reactive end-groups. If the polymer chains are α,ω-functionalized, such reactions are akin to step polymerizations where the monomers are end-functionalized prepolymers.
2. Polymerization of different monomers in sequence. In the simplest case, this involves using the living polymer formed from one monomer as the initiator for polymerization of a second monomer.
3. Polymerization of one monomer, followed by conversion of the active end-group to a form that initiates a different type of polymerization to which the second monomer added is susceptible.
4. Use of reactive end-groups on a polymer chain to initiate formation of another polymer by chain polymerization.

The best control of block copolymer architecture and block length is achieved using 'living' polymerizations to produce each of the individual blocks. The most important methods for synthesis of block copolymers are described in following sections with examples that illustrate the general principles.

9.4.1 Synthesis of Segmented and Alternating Copolymers by Step Polymerization

By using as comonomers, low molar mass prepolymers with terminal functional groups, step copolymerization can be used to prepare *segmented copolymers* which are akin to alternating block copolymers, but with less well-defined block lengths than normally is associated with block copolymers. Nevertheless, segmented copolymers prepared in this way are of commercial importance. Two examples will be presented to illustrate the concepts.

The ester interchange reaction of dimethylterephthalate, $CH_3OOC-C_6H_4-COOCH_3$, with poly(oxytetramethylene)diol, $H\!-\!\!\left[O\!-\!(CH_2)_4\right]_m\!\!-OH$, and ethylene glycol, $HO-CH_2-CH_2-OH$, yields poly[poly(oxytetramethylene)-*block*-poly(ethylene terephthalate)] a simplified structure of which is shown below

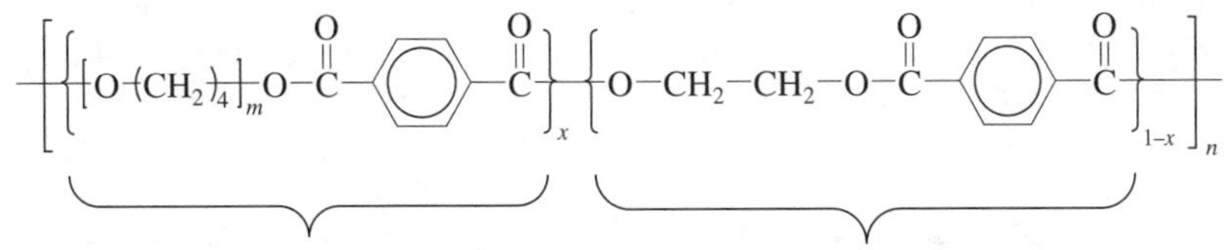

$$\left[\left[\left[O-(CH_2)_4\right]_m-O-\overset{O}{\overset{\|}{C}}-C_6H_4-\overset{O}{\overset{\|}{C}}\right]_x\left[O-CH_2-CH_2-O-\overset{O}{\overset{\|}{C}}-C_6H_4-\overset{O}{\overset{\|}{C}}\right]_{1-x}\right]_n$$

Polyether-based repeat unit (block or segment) — Polyester repeat unit (block or segment)

where x is the mole fraction of the polyether-based repeat units. Since, in this case, the hydroxyl groups are of equal reactivity, the sequence distribution of the two types of repeat unit will be random and the probability of a given sequence of i similar repeat units of type '1' will be given by $[(m_1/(m_1+m_2)]^i$ where m_1 and m_2 are the mole fractions of the corresponding diols in the comonomer mixture. Whilst an isolated polyether-based repeat unit can be considered a 'block' (because it comprises a prepolymer chain), an isolated polyester repeat unit cannot; only sequences of several polyester repeat units can be considered as polyester blocks because (unlike isolated polyester repeat units) these longer sequences can associate and crystallize together in the bulk material. For block copolymers prepared by such step copolymerizations, it is usual to call the blocks *segments* and the copolymers *segmented copolymers*. In the above example, the polyether-based blocks are called *soft segments* because they provide an amorphous rubbery phase, and the polyester blocks are called *hard segments* because they provide a rigid crystalline phase within the bulk material.

Many commercial polyurethanes are segmented copolymers prepared by reaction of a diisocyanate with a prepolymer polyol (hydroxyl group functionality ≥ 2) and a short-chain diol (e.g. butan-1,4-diol), e.g.

$$n\ \mathrm{OCN{-}R{-}NCO} + nx\ \mathrm{HO{\sim\sim\sim}OH} + n(1-x)\ \mathrm{HO{-}R'{-}OH} \longrightarrow \left[\left(\mathrm{O{\sim\sim\sim}O{-}\overset{O}{\overset{\|}{C}}{-}\underset{H}{\underset{|}{N}}{-}R{-}\underset{H}{\underset{|}{N}}{-}\overset{O}{\overset{\|}{C}}}\right)_x\left(\mathrm{O{-}R'{-}O{-}\overset{O}{\overset{\|}{C}}{-}\underset{H}{\underset{|}{N}}{-}R{-}\underset{H}{\underset{|}{N}}{-}\overset{O}{\overset{\|}{C}}}\right)_{1-x}\right]_n$$

Prepolymer-based soft segment   Polyurethane hard segment

where x is the mole fraction of the prepolymer diol in the diol mixture. Most commonly, aliphatic polyester or poly(propylene oxide) polyols are used and often have functionalities > 2, thus giving rise to the formation of segmented copolyurethane networks. As for the polyester example shown above, in the bulk material, the soft segments provide an amorphous rubbery phase, and the polyurethane hard segments a rigid crystalline phase, which has a high melting point due to the high extent of hydrogen bonding between urethane groups. By changing the length of the prepolymer polyol chains and the proportions of the prepolymer polyol and short-chain diol, it is possible to produce highly stiff materials (e.g. using low prepolymer polyol content and/or short prepolymer polyol chains) or soft materials (e.g. using higher contents of longer-chain prepolymer polyols) or anything in between. This is why segmented copolyurethanes are ubiquitous and find use in such a wide variety of applications.

The same principles can be applied to the synthesis of alternating block copolymers by step polymerization of low molar mass α,ω-functionalized di-block copolymers prepared, for example, by living anionic polymerization (Section 9.4.2.1). This strategy facilitates synthesis of perfectly alternating block copolymers in which each block has a pre-determined degree of polymerization and a narrow molar mass distribution. For example,

$$n\ \mathrm{HO{\sim\sim\sim}\!\blacksquare\!\!\blacksquare\!{-}\overset{O}{\overset{\|}{C}}{-}OH} \xrightarrow{-\mathrm{H_2O}} \left[\mathrm{O{\sim\sim\sim}\!\blacksquare\!\!\blacksquare\!{-}\overset{O}{\overset{\|}{C}}}\right]_n$$

where the wavy and solid lines represent homopolymer blocks. The principal limitations are the low concentration of reactive end-groups, which makes the rates of reaction rather low, though this can be improved through use of *click chemistry* (Section 9.4.3.1), and the inability to control the molar mass distribution of the alternating copolymer, which has the form of the most probable distribution because it is governed by the normal statistics of step polymerization (Section 3.2.3.2).

9.4.2 Synthesis of Block Copolymers by Sequential Polymerization

The most convenient way of preparing block copolymers by chain polymerization is through the use of living polymers. This involves the formation of living homopolymer molecules by polymerization of the first monomer (A) under carefully controlled conditions. The terminal active sites of these molecules are then used to initiate polymerization of a different monomer (B), thus extending the molecules by formation of a second homopolymer block to produce an AB di-block copolymer. This general synthetic strategy has been applied to the preparation of block copolymers with well-defined structures using radical, anionic, cationic, group transfer, Ziegler–Natta, metallocene and ring-opening methods of 'living' chain polymerization. In each case, the order in which the monomers are polymerized can be of crucial importance because the living polymer formed from the first monomer must be capable of efficiently initiating polymerization of the second monomer.

9.4.2.1 Synthesis of Block Copolymers by Living Anionic Polymerization

Until the advent of reversible-deactivation radical polymerization, anionic living polymerization provided the most important chain polymerization route to block copolymers. By careful control of the reaction conditions in anionic polymerization, it is possible to produce block copolymers with blocks of pre-defined molar mass, narrow molar mass distribution and controlled stereochemistry. In order to achieve this, polymerization of the first monomer must be complete before addition of the second monomer.

Anionic living polymerization is particularly important for production of block copolymers suitable for use as thermoplastic elastomers (Section 18.3.3), e.g. ABA tri-block copolymers in which homopolymer A is glassy (e.g. polystyrene) and homopolymer B is rubbery (e.g. polyisoprene). Such block copolymers can be prepared by polymerizing styrene using ^{s}BuLi, then adding isoprene to form living di-block copolymer molecules which finally are coupled together by their reaction with a dihalide compound (e.g. dichloromethane or dichlorodimethylsilane). The preparation is performed in a non-polar solvent (e.g. benzene) in order to ensure that styrene is the less reactive of the two monomers and so can be used in the first stage. A schematic representation of the sequence of reactions is shown below

$$^{s}\text{BuLi} \xrightarrow{\text{Styrene}} {}^{s}\text{Bu}\text{▬}^{-}\ \text{Li}^{+} \xrightarrow{\text{Isoprene}} {}^{s}\text{Bu}\text{▬}\sim\sim\sim^{-}\ \text{Li}^{+}$$

$$\xrightarrow{\text{X–R–X}} {}^{s}\text{Bu}\text{▬}\sim\sim\sim\text{R}\sim\sim\sim\text{▬}{}^{s}\text{Bu} + 2\text{LiX}$$

where X = Cl or Br, ▬ represents a polystyrene block and ∼∼∼ represents a polyisoprene block. More complex block copolymer structures can be produced by coupling with polyfunctional halide compounds (e.g. RX_4 would yield a four-armed star-shaped block copolymer). An alternative approach is to produce the ABA tri-block copolymer directly in two sequential growth steps by using a difunctional metal alkyl initiator, such as that formed by reaction of *tert*-butyl lithium with 1,3-diisopropenylbenzene

$$2\ {}^{t}\text{Bu}^{-}\text{Li}^{+} + \text{CH}_2{=}\text{C(CH}_3\text{)–C}_6\text{H}_4\text{–C(CH}_3\text{)}{=}\text{CH}_2 \longrightarrow \text{Li}^{+}\ {}^{-}\text{C(CH}_3\text{)(CH}_2{}^{t}\text{Bu)–C}_6\text{H}_4\text{–C(CH}_3\text{)(CH}_2{}^{t}\text{Bu)}^{-}\ \text{Li}^{+}$$

The degree of polymerization of the central block then needs to be large enough to make the initiator residue in the centre of the block insignificant in terms of its effect on the properties of the block.

Another alternative method for preparation of ABA tri-block copolymers involves the use of an electron transfer initiator (e.g. sodium naphthalide) in a polar solvent, such as tetrahydrofuran. Under these conditions, isoprene is of lower reactivity than styrene, which allows the isoprene to be polymerized first to produce dicarbanionic living polyisoprene molecules which then are used to polymerize styrene before terminating the reaction

$$\text{Isoprene} \xrightarrow[\text{initiation}]{\text{Electron transfer}} \text{Mt}^{+}\ {}^{-}\!\sim\!\sim\!\sim\!\sim\!\sim\!\sim^{-}\ \text{Mt}^{+}$$

$$\xrightarrow{\text{Styrene}} \text{Mt}^{+}\ {}^{-}\!\text{—}\!\sim\!\sim\!\sim\!\sim\!\sim\!\sim\!\text{—}^{-}\ \text{Mt}^{+}$$

$$\xrightarrow{\text{ROH}} \text{H—}\!\sim\!\sim\!\sim\!\sim\!\sim\!\sim\!\text{—H} + 2\,\text{RO}^{-}\,\text{Mt}^{+}$$

where Mt⁺ is a metal ion. This route is less satisfactory for these particular block copolymers because a high proportion of the undesirable 1,2- and 3,4-additions occur upon anionic polymerization of isoprene in a polar solvent (Section 6.3).

Use of difunctional metal alkyl initiators or electron transfer initiation also facilitates synthesis of α,ω-functionalized ABA tri-block copolymers through the use of the chemistry described in Section 5.3.2.5 at the final termination stage. For example, carboxylic acid end-groups can be introduced by terminating chain growth with carbon dioxide

$$\text{Mt}^{+}\ {}^{-}\!\text{—}\!\sim\!\sim\!\sim\!\sim\!\sim\!\sim\!\text{—}^{-}\ \text{Mt}^{+}$$

$$\xrightarrow{CO_2} \text{Mt}^{+}\ {}^{-}\text{O(O=)C—}\!\sim\!\sim\!\sim\!\sim\!\sim\!\sim\!\text{—C(=O)O}^{-}\ \text{Mt}^{+}$$

$$\xrightarrow{\text{dil. } H_2SO_4} \text{HO(O=)C—}\!\sim\!\sim\!\sim\!\sim\!\sim\!\sim\!\text{—C(=O)OH} + 2\,HSO_4^{-}\,\text{Mt}^{+}$$

When preparing block copolymers from styrene and methacrylates by anionic living polymerization, styrene must be polymerized first since it is the less reactive monomer and gives polymeric carbanions that are of lower stability than those from methacrylates. It is common to extend the polystyrene living chains with a single repeat unit of 1,1-diphenylethene before adding the methacrylate monomer because this increases the bulkiness of the terminal carbanion, thereby reducing its attack at the ester group in the methacrylate monomer (Section 5.3.2.7).

In order to form block copolymers, it is essential that the living end-groups are retained throughout the reactions and so they must be protected from even trace quantities of reactive impurities (Section 5.3.2.7). This requires rigorous control of the reaction environment and the use of high-purity monomers and reagents. Blocks can be produced from monomers which have side groups with acidic hydrogens (e.g. hydroxyl or carboxylic acid groups) by protecting those groups and then de-protecting them after the block copolymer has been formed (Section 5.3.2.7).

9.4.2.2 Synthesis of Block Copolymers by Reversible-Deactivation (Living) Cationic Polymerization

The development of reversible-deactivation cationic polymerizations in which the carbocation active centres are captured for most of the time in dormant species (Section 5.2.2) enables the active end-group to be preserved and gives these polymerizations a living character. Their discovery led naturally to the synthesis of block copolymers using the same principles as described for living anionic polymerization. However, the restricted range of monomers (principally isobutylene, styrenes and vinyl ethers) and the need to find initiation systems that give control of polymerization for both monomers reduces the importance of this route to block copolymers.

9.4.2.3 Synthesis of Block Copolymers by Reversible-Deactivation (Living) Radical Polymerization

As discussed in Sections 4.1 and 5.1, most olefinic monomers are capable of radical polymerization because it is tolerant of most types of functional groups and the propagating species do not have formal charges. Hence the development of living-like reversible-deactivation radical polymerizations facilitated synthesis of block copolymers through the application of the principles described in Section 9.4.2.1, but with far less stringent experimental conditions and requirements for monomer purity, and with a greater choice of monomers. All types of reversible-deactivation radical polymerization have been used to prepare block copolymers that have blocks of well-defined molar mass and narrow molar mass distribution, though, unlike living anionic polymerizations, it is not possible to control block stereochemistry. In many cases, the stereochemistry is not important and the ease of carrying out these polymerizations with a wide range of monomers is the dominant driver for their use. Hence, methods of reversible-deactivation radical polymerization have been used extensively for synthesis of a very wide range of block copolymers for use in a large number of application areas, including synthesis of block copolymers *in situ* as starting points for further polymerization, e.g. synthesis of amphiphilic di-block copolymers as colloid stabilizers in emulsion polymerizations (Section 4.4.4). As in reversible-deactivation cationic polymerization, there is a need to find initiator systems that provide control for two different monomers.

Nitroxide-mediated polymerization (NMP; Section 4.5.1) with TEMPO as the nitroxide has been successful in producing polystyrene-polyacrylate block copolymers, but often the polyacrylate blocks have substantial molar mass dispersity because TEMPO does not properly control polymerization of acrylates. The development of acyclic nitroxides, such as TIPNO and SG1 (Section 4.5.1), has greatly improved the ability to prepare well-defined block copolymers by NMP since they are able to control polymerization of styrene and acrylates. For the reasons described in Section 4.5.1.2, NMP of methacrylates is not controlled and so they cannot be used, but some other important monomers have been used successfully, including isoprene and vinyl pyridines. As always, the sequence in which the monomers are polymerized is important and the monomer which gives the most reactive alkoxyamine should be used first to give a macroinitiator from which the next block is grown. For example, in the synthesis of polystyrene-*block*-poly(*n*-butyl acrylate) using TIPNO as the nitroxide, the poly(*n*-butyl acrylate) block should be grown first.

Atom-transfer radical polymerization (ATRP; Section 4.5.2) and reversible-addition-fragmentation chain-transfer (RAFT) polymerization (Section 4.5.3) have been used much more widely for the synthesis of block copolymers. The range of block copolymers prepared are far too numerous to detail here, so a few examples will be described to illustrate the versatility of these methods and some of the strategies for achieving efficient initiation of the second block. In addition to block copolymers prepared by ATRP from the more common monomers (such as styrenes, acrylate and methacrylate esters, acrylonitrile and vinyl pyridines), there are many examples of block copolymers synthesized by ATRP or RAFT from functional monomers, several of which cannot be used in living anionic polymerizations. For example, well-defined block copolymers have been synthesized with blocks prepared from hydroxyalkyl methacrylates, 2-(dialkylamino)ethyl methacrylates, perfluorinated methacrylate esters and salts of 3-sulphopropyl methacrylate and 4-vinylbenzoic acid, such as

$CH_2{=}C(CH_3){-}C({=}O){-}O{-}CH_2{-}CH_2{-}OH$ $\quad$ $CH_2{=}C(CH_3){-}C({=}O){-}O{-}CH_2{-}CH_2{-}N(CH_3)_2$ $\quad$ $CH_2{=}C(CH_3){-}C({=}O){-}O{-}(CF_2)_7{-}CF_3$ $\quad$ $CH_2{=}C(CH_3){-}C({=}O){-}O{-}CH_2{-}CH_2{-}CH_2{-}SO_2{-}O^{-}\ K^{+}$ $\quad$ $CH_2{=}CH{-}C_6H_4{-}C({=}O){-}O^{-}\ Na^{+}$

As described in Section 4.5.2.2, acidic hydrogen atoms interfere in ATRP and so monomers containing acid groups need to be polymerized either in protected form (e.g. as the trimethylsilyl ester) and de-protected after polymerization or in ionized form (as for the acid monomers shown above). RAFT does not suffer from this restriction (monomers with acidic hydrogen atoms can be used directly) and also can be used to polymerize monomers of low reactivity (e.g. vinyl acetate), so is slightly more versatile than ATRP.

In order to produce block copolymers with blocks of well-defined length and narrow molar mass distribution, the rate of initiation of the second block needs to be comparable to or, preferably, higher than the rate of propagation for that block and so the sequence of polymerization again is important. Monomer reactivity in ATRP is in the order acrylonitrile > methacrylates > styrene ≈ acrylates, so blocks should be formed in that order, using the least reactive monomer to form the second block. When monomer reactivities are similar, it is possible to prepare ABA and ABC tri-block copolymers directly by sequential ATRP (e.g. from different methacrylate monomers). An alternative route to ABA tri-block copolymers is the use of difunctional ATRP initiators (i.e. X–R–X where X = Cl or Br). When the target block copolymer architecture requires monomers of different reactivity to be used in order of increasing reactivity, halogen exchange from Br to Cl can be used to facilitate efficient initiation of the second block because the equilibrium constants for dissociation of C–Cl bonds are much lower than for C–Br bonds; an example is shown in Figure 9.2.

In RAFT polymerization, efficient initiation of the second monomer requires that the polymeric radical from the first-formed block is a better leaving group in the addition-fragmentation equilibrium (Section 4.5.3 and Figure 4.17) than the polymeric radical from the second-formed block. For example, for block copolymers formed from styrene and methacrylates or methacrylates and acrylates, the polymethacrylate block should be formed first and then grown by addition of styrene or the acrylate. As for ATRP, ABA and ABC tri-block copolymers can be prepared by sequential RAFT polymerization if the monomers give polymeric radicals of similar leaving group ability or can be produced directly from difunctional RAFT agents provided that the polymerization sequence permits this. Switchable RAFT agents are being developed for synthesis of block copolymers from monomers with large differences in reactivity (e.g. vinyl acetate and methacrylates), for example, *N*-(4-pyridinyl)-*N*-methyldithiocarbamates are of low activity and so are effective RAFT agents for vinyl acetate polymerization, but can be made more reactive by protonation of the pyridinyl nitrogen atom such that they control RAFT polymerization of acrylates, methacrylates and styrene. Hence, poly(methyl methacrylate)-*block*-poly(vinyl acetate) can be prepared by use of the protonated RAFT agent to polymerize methyl methacrylate, followed by deprotonation of the pyridinyl nitrogen atom using a strong base and then RAFT polymerization of vinyl acetate

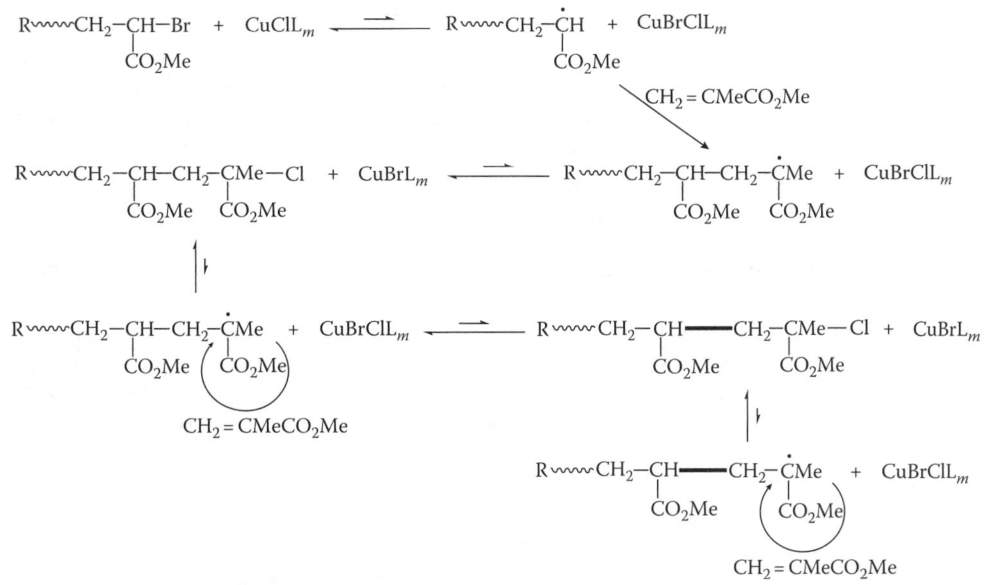

FIGURE 9.2 An example of how halogen exchange can be used to facilitate efficient initiation of a poly(methyl methacrylate) block from a poly(methyl acrylate) ATRP macroinitiator; m is the number of ligands L associated with the Cu, ∿∿∿ represents a poly(methyl acrylate) chain and ▬▬ represents a poly(methyl methacrylate) chain. Poly(methyl acrylate) with terminal C–Br bonds is first prepared using an R–Br initiator with CuBr, then isolated and used as a macroinitiator for ATRP of methyl methacrylate in the presence of CuCl at a molar equivalent to the poly(methyl acrylate) C–Br end groups. The halogen exchange favours the more stable C–Cl bond from which the rate of propagation of methyl methacrylate is much lower than for propagation from the C–Br bond and so initiation of the second block is faster than the subsequent propagation.

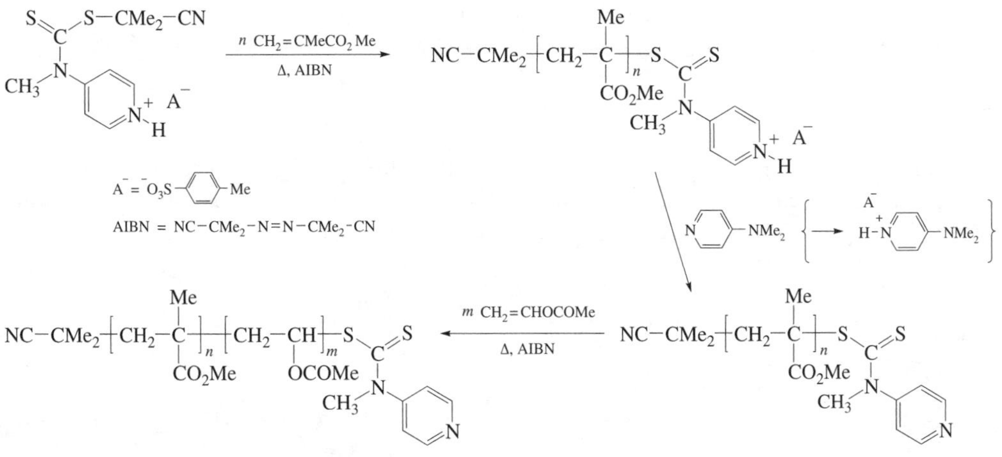

$A^- = {}^-O_3S-C_6H_4-Me$

AIBN = $NC-CMe_2-N=N-CMe_2-CN$

9.4.2.4 Synthesis of Block Copolymers by Other Methods of Living Polymerization

Many ring-opening polymerizations have the characteristics of living polymerization (Chapter 7) and have been used to prepare block copolymers. Some examples of the use of ionic ring-opening polymerizations and ring-opening metathesis polymerization are given in Section 9.4.2.5. Group-transfer polymerization also has been used to prepare block copolymers, though this is most effective for all-methacrylate block copolymers. Coordination polymerizations (Chapter 6) can be used to prepare block copolymers when living-like conditions are achieved, which is more easily accomplished with metallocene catalyst systems.

9.4.2.5 Synthesis of Block Copolymers by Active-Centre Transformation

Methods involving transformation from one type of living polymerization to another were first developed for use in preparing block copolymers when the second monomer to be polymerized is not susceptible to the type of chain polymerization used to form the first block. The use of such active-centre transformations has grown considerably with the development of newer methods of polymerization, such as ring-opening metathesis polymerization and reversible-deactivation radical polymerizations. There are essentially two strategies.

1. Converting end groups of commercially-available homopolymers into functional groups that are then used to initiate polymerization of a monomer to form the next block.
2. Transforming living end-groups of a first-stage polymerization homopolymer into functional groups that then act as initiators of polymerization in a second-stage polymerization to form the next block. This can be done *in situ*, but in most cases is done via isolation of the intermediate end-functional homopolymer followed by its use in a separate second-stage polymerization.

Some specific examples of each approach will be given to illustrate the principles.

Commercially available poly(ethylene oxide) has either one alkyl ether end-group and one hydroxyl end-group or two hydroxyl end-groups depending upon whether it was synthesized using, respectively, alkoxide or hydroxide-initiated ring-opening polymerization of ethylene oxide (see Section 7.3.1). The hydroxyl end-groups have been used widely to create initiating groups at one or both chain ends. For example, an ATRP initiator end-group can be created and used to produce a block copolymer by polymerization of sodium methacrylate

$$CH_3O\!\left[CH_2-CH_2-O\right]_n\!CH_2-CH_2-OH \xrightarrow{Br-\overset{O}{\overset{\|}{C}}-\underset{CH_3}{\underset{|}{C}H}-Br} CH_3O\!\left[CH_2-CH_2-O\right]_n\!CH_2-CH_2-O-\overset{O}{\overset{\|}{C}}-\underset{CH_3}{\underset{|}{C}H}-Br$$

$$\xrightarrow[\text{CuBr/ bipyridyl ligand, } 90\ ^\circ C]{p\ CH_2=\underset{O^-\,Na^+}{\underset{|}{\underset{C=O}{\underset{|}{C}}}}(CH_3)} CH_3O\!\left[CH_2-CH_2-O\right]_n\!CH_2-CH_2-O-\overset{O}{\overset{\|}{C}}-\underset{CH_3}{\underset{|}{C}H}\!\left[CH_2-\overset{CH_3}{\overset{|}{\underset{O^-\,Na^+}{\underset{|}{\underset{C=O}{\underset{|}{C}}}}}}\right]_p\!Br$$

Commercially available poly(dimethyl siloxane) with hydroxy end-groups (e.g. prepared via step polymerization of dichlorodimethylsiloxane, see Section 3.2.1.4) can been converted into a difunctional RAFT agent and then used to produce an ABA tri-block copolymer by polymerization of *N,N*-dimethylacrylamide

$$HO-\underset{CH_3}{\overset{CH_3}{Si}}-O\!\left[\underset{CH_3}{\overset{CH_3}{Si}}-O\right]_n\!\underset{CH_3}{\overset{CH_3}{Si}}-OH \xrightarrow{HO-\overset{O}{\overset{\|}{C}}-CH_2-S-\overset{S}{\overset{\|}{C}}-S-CH_2Ph}$$

$$PhCH_2-S-\overset{S}{\overset{\|}{C}}-S-CH_2-\overset{O}{\overset{\|}{C}}-O-\underset{CH_3}{\overset{CH_3}{Si}}-O\!\left[\underset{CH_3}{\overset{CH_3}{Si}}-O\right]_n\!\underset{CH_3}{\overset{CH_3}{Si}}-O-\overset{O}{\overset{\|}{C}}-CH_2-S-\overset{S}{\overset{\|}{C}}-S-CH_2Ph$$

$$\xrightarrow[(CH_3)_3\underset{CN}{C}-N=N-\underset{CN}{C}(CH_3)_2,\ 60\ ^\circ C]{2p\ CH_2=\underset{N(CH_3)_2}{\underset{|}{\underset{C=O}{\underset{|}{C}H}}}}$$

$$R\!\left[CH_2-\underset{N(CH_3)_2}{\underset{|}{\underset{C=O}{\underset{|}{C}H}}}\right]_p\!S-\overset{S}{\overset{\|}{C}}-S-CH_2-\overset{O}{\overset{\|}{C}}-O-\underset{CH_3}{\overset{CH_3}{Si}}-O\!\left[\underset{CH_3}{\overset{CH_3}{Si}}-O\right]_n\!\underset{CH_3}{\overset{CH_3}{Si}}-O-\overset{O}{\overset{\|}{C}}-CH_2-S-\overset{S}{\overset{\|}{C}}-S\!\left[\underset{N(CH_3)_2}{\underset{|}{\underset{O=C}{\underset{|}{C}H}}}-CH_2\right]_p\!R$$

where $R = C(CH_3)_2CN$ or CH_2Ph.

One principle for transforming living anionic polymerization into NMP has been demonstrated by first polymerizing butadiene using ^{s}Bu, then terminating chain growth by addition of an aldehyde derivative of TEMPO (where the living polybutadiene carbanions add to the aldehyde carbonyl carbon atom). The resulting TEMPO alkoxyamine end-functionalized polybutadiene is isolated and then used as the initiator in a separate second-stage polymerization of styrene

$$^s\text{BuLi} + n\,\text{CH}_2{=}\text{CH}{-}\text{CH}{=}\text{CH}_2 \xrightarrow{-78\,°\text{C}} {}^s\text{Bu}{-}[\text{CH}_2{-}\text{CH}{=}\text{CH}{-}\text{CH}_2]_{n-1}{-}\text{CH}_2{-}\text{CH}{=}\text{CH}{-}\overset{-}{\text{CH}_2}\,\text{Li}^+$$

(i) OHC–C₆H₄–CH(CH₃)–O–N(TEMPO); (ii) CH₃OH

$$^s\text{Bu}{-}[\text{CH}_2{-}\text{CH}{=}\text{CH}{-}\text{CH}_2]_{n-1}{-}\text{CH}_2{-}\text{CH}{=}\text{CH}{-}\text{CH}_2{-}\text{CH(OH)}{-}\text{C}_6\text{H}_4{-}\text{CH(CH}_3){-}\text{O}{-}\text{N}<$$

$$\xrightarrow[120\,°\text{C}]{p\ \text{CH}_2=\text{CH(Ph)}}$$

$$^s\text{Bu}{-}[\text{CH}_2{-}\text{CH}{=}\text{CH}{-}\text{CH}_2]_{n-1}{-}\text{CH}_2{-}\text{CH}{=}\text{CH}{-}\text{CH}_2{-}\text{CH(OH)}{-}\text{C}_6\text{H}_4{-}\text{CH(CH}_3){-}[\text{CH}_2{-}\text{CH(Ph)}]_p{-}\text{O}{-}\text{N}<$$

where carbon and hydrogen atoms have been omitted from the ring structures for reasons of clarity (as also has been done in the following examples).

Transformation of ATRP into cationic ring-opening polymerization has been demonstrated by first polymerizing styrene using a difunctional bromo-initiator, then conversion of the bromo end-groups into perchlorate groups which can be used to effect polymerization of tetrahydrofuran

$$\text{Br}{-}\text{CH(Et)}{-}\text{C(=O)}{-}\text{O}{-}\text{CH}_2{-}\text{CH}_2{-}\text{O}{-}\text{C(=O)}{-}\text{CH(Et)}{-}\text{Br}$$

$$\xrightarrow[110\,°\text{C}]{2n\ \text{CH}_2=\text{CH(Ph)}}$$

$$\text{Br}{-}\text{CH(Ph)}{-}\text{CH}_2{-}[\text{CH(Ph)}{-}\text{CH}_2]_{n-1}{-}\text{CH(Et)}{-}\text{C(=O)}{-}\text{O}{-}\text{CH}_2{-}\text{CH}_2{-}\text{O}{-}\text{C(=O)}{-}\text{CH(Et)}{-}[\text{CH}_2{-}\text{CH(Ph)}]_{n-1}{-}\text{CH}_2{-}\text{CH(Ph)}{-}\text{Br}$$

$$\downarrow 2\ \text{AgClO}_4$$

$$\text{O}_3\text{ClO}{-}\text{CH(Ph)}{-}\text{CH}_2{-}[\text{CH(Ph)}{-}\text{CH}_2]_{n-1}{-}\text{CH(Et)}{-}\text{C(=O)}{-}\text{O}{-}\text{CH}_2{-}\text{CH}_2{-}\text{O}{-}\text{C(=O)}{-}\text{CH(Et)}{-}[\text{CH}_2{-}\text{CH(Ph)}]_{n-1}{-}\text{CH}_2{-}\text{CH(Ph)}{-}\text{OClO}_3 + 2\ \text{AgBr}$$

$$\rightleftharpoons$$

$$\overset{-}{\text{ClO}_4}\,\overset{+}{\text{CH}}\text{(Ph)}{-}\text{CH}_2{-}[\text{CH(Ph)}{-}\text{CH}_2]_{n-1}{-}\text{CH(Et)}{-}\text{C(=O)}{-}\text{O}{-}\text{CH}_2{-}\text{CH}_2{-}\text{O}{-}\text{C(=O)}{-}\text{CH(Et)}{-}[\text{CH}_2{-}\text{CH(Ph)}]_{n-1}{-}\text{CH}_2{-}\overset{+}{\text{CH}}\text{(Ph)}\,\overset{-}{\text{ClO}_4}$$

$$\downarrow -78\,°\text{C},\ 2p\ \text{THF}$$

$$\overset{-}{\text{ClO}_4}\ \text{THF}{-}\overset{+}{\text{O}}{-}[(\text{CH}_2)_4{-}\text{O}]_{p-1}{-}\text{CH(Ph)}{-}\text{CH}_2{-}[\text{CH(Ph)}{-}\text{CH}_2]_{n-1}{-}\text{CH(Et)}{-}\text{C(=O)}{-}\text{O}{-}\text{CH}_2{-}\text{CH}_2{-}\text{O}{-}\text{C(=O)}{-}\text{CH(Et)}{-}[\text{CH}_2{-}\text{CH(Ph)}]_{n-1}{-}\text{CH}_2{-}\text{CH(Ph)}{-}[\text{O}{-}(\text{CH}_2)_4]_{p-1}{-}\overset{+}{\text{O}}\text{-THF}\ \overset{-}{\text{ClO}_4}$$

A more unusual example is the anionic ring-opening polymerization of ethylene oxide, followed by transformation of the end group into an initiator for ring-opening metathesis polymerization of norbornene

$$Ph_2CHLi + n\ \underset{O}{CH_2{-}CH_2} \longrightarrow Ph_2CH{-}\!\left[CH_2{-}CH_2{-}O\right]_{n-1}\!CH_2{-}CH_2{-}O^{-}\ Li^{+}$$

$$\xrightarrow{Br{-}CH_2{-}C_6H_4{-}CH{=}CH_2} Ph_2CH{-}\!\left[CH_2{-}CH_2{-}O\right]_{n-1}\!CH_2{-}CH_2{-}O{-}CH_2{-}C_6H_4{-}CH{=}CH_2$$

$$\xrightarrow{(H)(Et)C{=}Ru(PCy_3)_2Cl_2} Ph_2CH{-}\!\left[CH_2{-}CH_2{-}O\right]_{n-1}\!CH_2{-}CH_2{-}O{-}CH_2{-}C_6H_4{-}CH{=}Ru(PCy_3)_2Cl_2$$

$$\xrightarrow{p\ \text{norbornene}} Ph_2CH{-}\!\left[CH_2{-}CH_2{-}O\right]_{n-1}\!CH_2{-}CH_2{-}O{-}CH_2{-}C_6H_4{-}CH{=}\!\left[CH{-}C_5H_8{-}CH{=}\right]_{p-1}\!CH{-}C_5H_8{-}CH{=}Ru(PCy_3)_2Cl_2$$

A further development in the synthesis of block copolymers, that has the same underlying principle as active-centre transformation, is the use of initiators with dual functionality that can initiate two mechanistically-different types of polymerization. An example of such a dual-functional initiator is shown below

$$Br{-}C(CH_3)_2{-}\overset{O}{\overset{\|}{C}}{-}O{-}CH_2{-}\underset{Ph}{CH}{-}O{-}N(\text{TEMPO})$$

in which the bromo end-group can be used to initiate ATRP (at lower temperature, e.g. 60–90 °C) and the TEMPO end-group can be used to initiate NMP (at higher temperature, e.g. 120–130 °C).

It must be borne in mind that these examples represent only a small fraction of the transformations that have been implemented successfully. The reader needs to consult up-to-date reviews on living polymerizations and block copolymer synthesis to gain a more complete view of what has been achieved using transformation approaches.

9.4.3 Synthesis of Block Copolymers by Coupling of Polymer Chains

Many polymerizations are capable of producing polymers with well-defined end-groups. Some examples are: RA_2+RB_2 step polymerization taken to complete conversion with a slight excess of RB_2 will give polymer with B groups at both chain ends; chain polymerization using initiators that possess a functional group; and living polymerizations terminated using molecules that generate a specific end functionality. Hence, it is possible to produce block copolymers by reacting functional end-groups of one polymer with complementary functional end-groups of another polymer.

Some examples of this principle for block copolymer synthesis have been considered already. At the end of Section 9.4.1, the synthesis of alternating block copolymers by step polymerization of α,ω-functionalized di-block copolymers was described. The first route to ABA tri-block copolymers described in Section 9.4.2.1 provides another example, in which two living AB di-block copolymers were linked together to terminate the polymerization.

Many different linking reactions taken from mainstream organic chemistry have been used for the purpose of coupling end-functionalized homopolymers. The need to achieve high conversions in these linking reactions in order to produce block copolymers free from contamination by unreacted homopolymers is challenging, especially because the concentration of the functional groups is low, and demands the use of highly efficient, clean reactions. Most of the long-established reactions of organic chemistry do not satisfy these criteria and their use normally gives block copolymers contaminated by homopolymers, which in favourable cases can be removed from the block copolymer by selective solvent extraction. However, during the first decade of this century a group of extremely efficient, clean linking reactions have been developed, which are collectively known as *click reactions* because they proceed rapidly and without significant side reactions. There now are several types of click reaction available for linking molecules together efficiently and their use has become widespread. One use of *click chemistry* in polymer science is for linking polymer chains together to produce block and graft copolymers.

9.4.3.1 Synthesis of Block Copolymers by Click-Coupling of Homopolymer Chains

The concept of click chemistry was introduced in 2001 by Kolb, Finn and Sharpless who set out a number of demanding criteria that reactions must satisfy in order to be classified as click reactions. The most important criteria for a click reaction are: broad in scope, rapid, mild conditions, tolerance to other functional groups, no side reactions, high selectivity and very high yielding. The *Huisgen 1,3-dipolar cycloaddition reaction* of an alkyne with an azide was the first type of click chemistry to find extensive use and proceeds rapidly at or just above room temperature in aqueous and organic media to produce 1,2,3-triazole linking groups as follows

$$R_1-C\equiv C-H \quad + \quad \overset{-}{N}=\overset{+}{N}=N-R_2 \;\leftrightarrow\; N\equiv\overset{+}{N}-\overset{-}{N}-R_2 \xrightarrow{Cu(I)} \text{1,4-adduct (principal product)} + \text{1,5-adduct (minor product)}$$

1,4-adduct principal product

1,5-adduct minor product

Although Cu(I) catalysis is used most often, other catalysts for the reaction are Ru, Pd(II), Pt(II) and Ni(II). The 1,4-adduct predominates, but in either case, the substituent groups R_1 and R_2 are linked together. Clearly, use of this click reaction for coupling homopolymer chains together requires one homopolymer chain to have an alkyne end-group and the other an azide end-group

$$\text{——}C\equiv CH \quad + \quad N_3\text{〰〰} \xrightarrow{Cu(I)} \text{——(triazole: C=C(H)–N(〰)–N=N)〰〰}$$

The introduction of these end groups requires the polymers to have functional end-groups which can be reacted further to give an alkyne or azide end-group. Reaction of a functional end-group after polymer synthesis using simple organic chemistry reactions is common, some examples of which are

$$\text{——}\overset{O}{\overset{\|}{C}}-X \xrightarrow[(-HX)]{H-Y-CH_2-C\equiv CH} \text{——}\overset{O}{\overset{\|}{C}}-Y-CH_2-C\equiv CH$$

where X = Cl, Br or OH and Y = O or NH

$$\sim\sim\sim\sim X \xrightarrow[(-NaX)]{NaN_3} \sim\sim\sim\sim N_3$$

where X = Cl, Br or OTs $\left\{ Ts = -\overset{O}{\underset{O}{\overset{\|}{\underset{\|}{S}}}}-C_6H_4-CH_3 \right\}$

$$\sim\sim\sim\sim\overset{O}{\overset{\|}{C}}-X \xrightarrow[(-HX)]{H-Y-(CH_2)_i-N_3} \sim\sim\sim\sim\overset{O}{\overset{\|}{C}}-Y-(CH_2)_i-N_3$$

where X = Cl, Br or OH and Y = O or NH; $i = 2$ or 3

In addition, it is possible to use appropriate initiating or terminating agents in living polymerizations to introduce the alkyne or azide end-group. For example, the following ATRP initiators have been used to produce polymer chains with either alkyne or azide end-groups

$$CH\equiv C-CH_2-O-\overset{O}{\overset{\|}{C}}-\underset{CH_3}{\underset{|}{\overset{CH_3}{\overset{|}{C}}}}-Br \qquad N_3-(CH_2)_3-O-\overset{O}{\overset{\|}{C}}-\underset{CH_3}{\underset{|}{\overset{CH_3}{\overset{|}{C}}}}-Br$$

which is possible because the alkyne and azide groups do not interfere with radical polymerization, though in ATRP care has to be taken to avoid Cu(I)-catalysed oxidative coupling of alkyne groups.

Use of click chemistry in polymer science has revolved mainly around the Huisgen click reaction, but other click reactions are of growing importance, the most significant being the free radical 1:1 addition of a thiol to an alkene, the so-called *thiol-ene reaction*

$$R_1-SH + CH_2=CH-R_2 \longrightarrow R_1-S-CH_2-CH_2-R_2$$

which can be effected either photochemically by irradiation with ultra-violet light or thermally by heating in the presence of a free-radical source (e.g. an azo 'initiator' such as AIBN). Although the Huisgen click reaction continues to be used extensively, use of the thiol-ene click reaction is growing rapidly.

Although applications of click chemistry in block and graft copolymer synthesis are the focus here, it is important to be aware that click chemistry is used much more extensively than this in polymer science, especially in the synthesis of functional polymers where click reactions are used to attach 'active' species to polymer chains (e.g. attachment of receptor or masking groups to polymer chains used in biomedical and therapeutic applications). This is a rapidly growing field of research, so the reader should consult the ongoing literature and up-to-date reviews in order to gain a full appreciation of the opportunities that click chemistry has created and how they are being exploited.

9.4.4 Synthesis of Non-Linear Block Copolymers

So far, all the methods described have focused upon synthesis of linear block copolymers with well-defined block architectures (AB, ABA, etc.). Non-linear block copolymers are produced using the same synthesis methods as for linear block copolymers, either by employing an initiator with a functionality ≥ 3 or by linking together block copolymer chains with linking agents that have a functionality ≥ 3. In this way, multi-armed star-like block copolymers can be produced. For example, a three-armed star AB di-block copolymer can be produced either by sequential living polymerization from a trifunctional initiator or by linking together terminally-functional AB di-block copolymer chains with a trifunctional linking agent

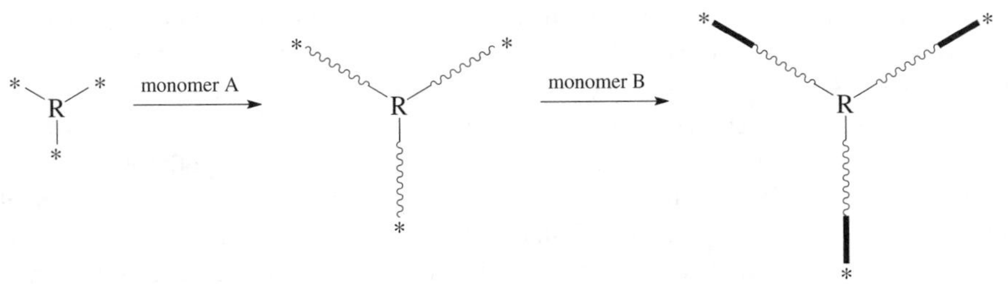

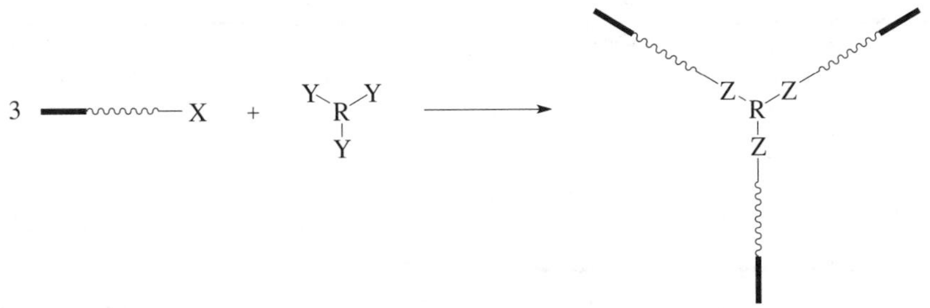

where the X and Y functional groups react to give a Z linking group. Note that the sequence of block formation is generally opposite for these two strategies (e.g. the X end group for the linking reaction is introduced at what was the active end of a living polymer).

9.5 GRAFT COPOLYMER SYNTHESIS

Graft copolymers are the branched equivalents of block copolymers. The methods for their synthesis can be grouped into three broad categories.

1. Activation of groups on a backbone polymer to initiate polymerization of a second monomer, thus forming branches of a different polymer.
2. Copolymerization of the principal backbone-forming monomer(s) with a macromonomer (a prepolymer with terminal polymerizable group; e.g. see Section 4.3.6).
3. Linking end-functionalized chains of one polymer to a backbone polymer that has reactive side groups.

It is worthy of note that the same methods also have been used to graft polymer chains to surfaces, but this aspect of grafting is beyond the scope of this chapter. In the following sections, the principles underlying each of these approaches to graft copolymer synthesis will be illustrated in simple terms. As for the descriptions of block copolymer synthesis, the reader should consult reviews of graft copolymer synthesis to gain a more expansive knowledge and understanding.

9.5.1 Synthesis of Graft Copolymers by Polymerization from a Backbone Polymer

The simplest procedures for polymerization from a backbone polymer involve either its exposure to high-energy radiation (usually ultra-violet light) in the presence of a monomer, or heating a mixture of it with a monomer in the presence of a free-radical initiator. In this way free-radical sites are produced along the backbone polymer chain either by direct interaction of the atoms in the chain with the radiation or via abstraction of atoms (usually hydrogen atoms) from the chain by radical species. The radicals thus created on the backbone polymer initiate polymerization of the added monomer to form the graft chains. Such simple methods have been, and still are, used extensively for grafting because they are simple and versatile, enabling a very wide range of polymers to be used as backbone polymers. Grafting can be enhanced by the presence of groups on the backbone polymer which promote radical abstraction (e.g. see Sections 4.2.4.2 and 4.6.2.2), such groups normally being introduced using a small proportion of an appropriate comonomer in the synthesis of the backbone polymer. Due to the simplicity of this approach, the level of control of graft copolymer architecture is poor (it cannot be used to produce well-defined graft copolymers) and the graft copolymers produced inevitably are contaminated by homopolymer formed via normal free-radical polymerization of the added monomer. The essence of this approach is shown in a very simplified schematic form below with easily abstractable hydrogen atoms present in a side group and directly on the backbone

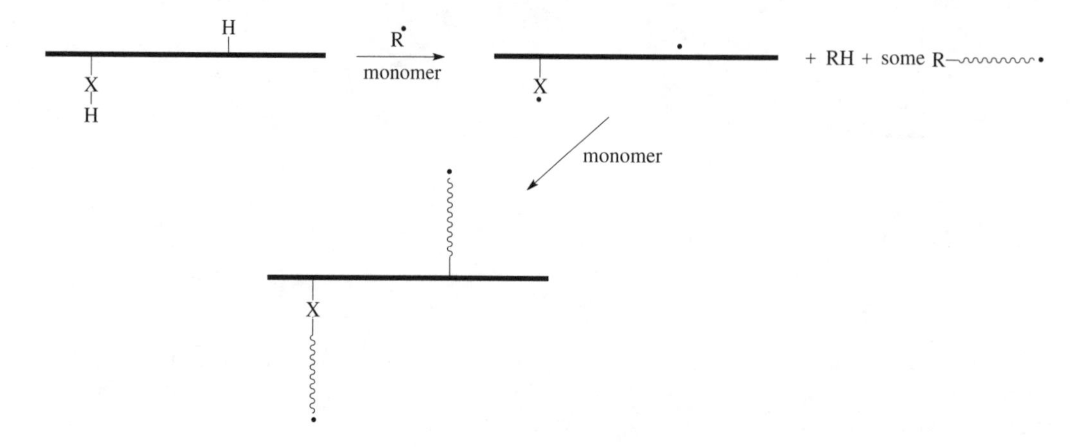

Methods involving the use of backbone polymers which have side groups that can be activated to initiate living polymerization of the added monomer enable much better control of graft copolymer structure and can be designed such that they are not complicated by homopolymer formation. The emergence of reversible-deactivation radical polymerization has greatly added to the ability to produce graft copolymers in this way, facilitating the growth of grafted chains of well-defined molar mass and narrow molar mass distribution. Three monomers that can be used in synthesis of the backbone polymer to introduce side groups which act as initiators of reversible-deactivation radical polymerization are shown below

$CH_2{=}CH{-}(CH_2)_6{-}O{-}CH(Ph){-}O{-}N<$ (2,2,6,6-tetramethylpiperidine ring)

$CH_2{=}CH{-}O{-}C({=}O){-}CH_2{-}Cl$

$CH_2{=}C(CH_3){-}C({=}O){-}O{-}CH_2{-}CH_2{-}O{-}C({=}O){-}CH(CH_3){-}Br$

the first of which provides pendent TEMPO alkoxyamine groups in the copolymer for subsequent initiation of NMP, the other two monomers providing pendent C–Cl and C–Br groups in the copolymer which are suitable for subsequent initiation of ATRP. The essence of this type of approach is shown schematically below for backbone copolymers prepared using these monomers

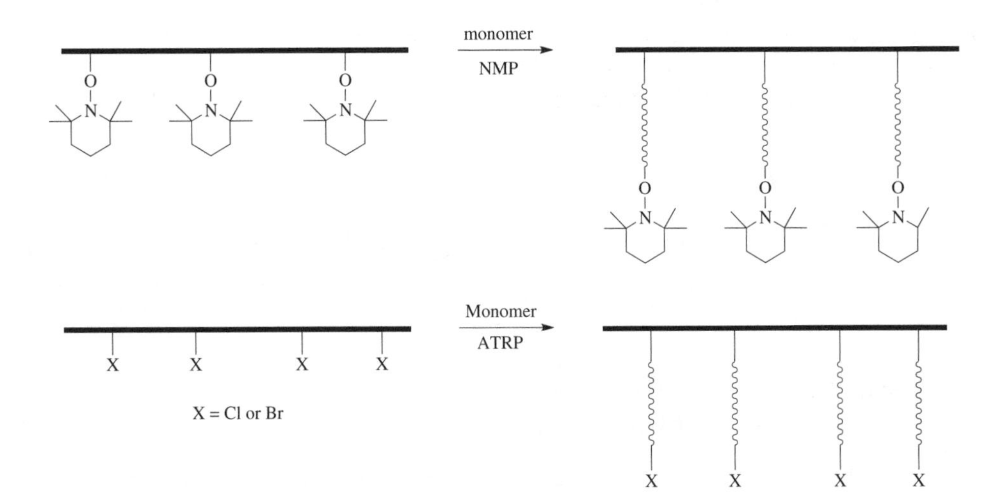

9.5.2 Synthesis of Graft Copolymers by Copolymerization with Macromonomers

An important approach to the preparation of graft copolymers involves the use of *macromonomers*, which are (typically low molar mass) polymers with terminal polymerizable C=C bonds. For example, polymers with a terminal hydroxyl group (e.g. synthesized by ring-opening polymerization of an epoxide or by termination of an anionic living polymerization with formaldehyde) can be reacted with acryloyl or methacryloyl chloride to produce an acrylate or methacrylate macromonomer

$$\sim\sim\sim OH + Cl{-}\overset{O}{\overset{\|}{C}}{-}\overset{R}{\overset{|}{C}}{=}CH_2 \longrightarrow \sim\sim\sim O{-}\overset{O}{\overset{\|}{C}}{-}\overset{R}{\overset{|}{C}}{=}CH_2 + HCl$$

R = H or CH_3 Macromonomer

Graft copolymers are produced by chain copolymerization of the macromonomer with a monomer that provides the main backbone repeat units. This approach enables well-defined graft copolymers to be prepared in terms of the molar mass and molar mass distribution of the branches since these properties are controlled in the synthesis of the macromonomer. The reactivity ratios for macromonomer copolymerization determine the sequence distribution, just as in a normal copolymerization (Sections 9.3.2 and 9.3.3), and usually are similar to those for a conventional monomer with the same kind of C=C bond (e.g. a simple acrylate or methacrylate for the above examples), though steric effects of the macromonomer's polymer chain often reduce the reactivity.

Control of the sequence distribution of grafts is desirable because it can have a major impact upon the properties of the graft copolymer. In reversible-deactivation radical copolymerizations of macromonomers, differences in reactivity ratios lead to gradient sequence distributions in which the grafts (i.e. macromonomer repeat units) are not evenly distributed along the backbone, but instead are increasingly clustered towards one end of the backbone as the disparity in reactivity ratios increases. For example, the schematic gradient copolymer shown in Section 9.3.6 would have the following structure if monomer B is a macromonomer

R−A−A−A−B−A−A−B−A−A−B−A−B−A−B−B−A−B−B−A−B−B−B−A−B−B−B−B−■

Taking this to the extreme, block copolymers which have a homopolymer block from monomer A and a homopolymer block from macromonomer B can be prepared by sequential reversible-deactivation radical polymerization if the macromonomer can be homopolymerized (which usually is the case if the chain in the macromonomer is relatively short)

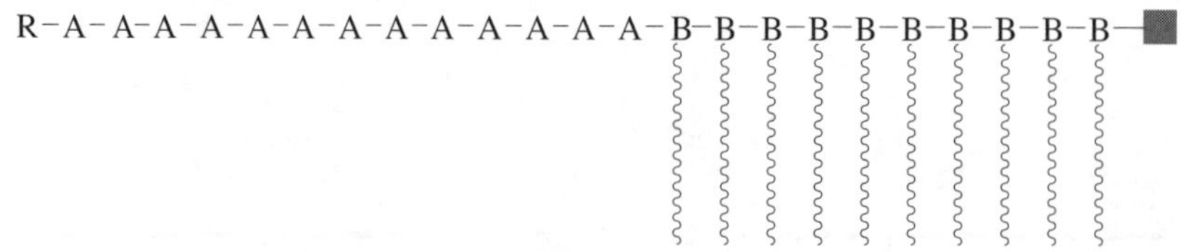

R–A–A–A–A–A–A–A–A–A–A–A–A–A–B–B–B–B–B–B–B–B–B–B–■

9.5.3 Synthesis of Graft Copolymers by Coupling of Polymer Chains to a Backbone Polymer

The practice of linking polymer chains to reactive side-groups on a backbone polymer has not been so extensively used, but has gained considerable impetus with the development of click chemistry. Acrylate and methacrylate monomers with pendent alkyne or azide groups that can be used to prepare copolymers onto which graft chains can be attached by the Huisgen reaction are shown below

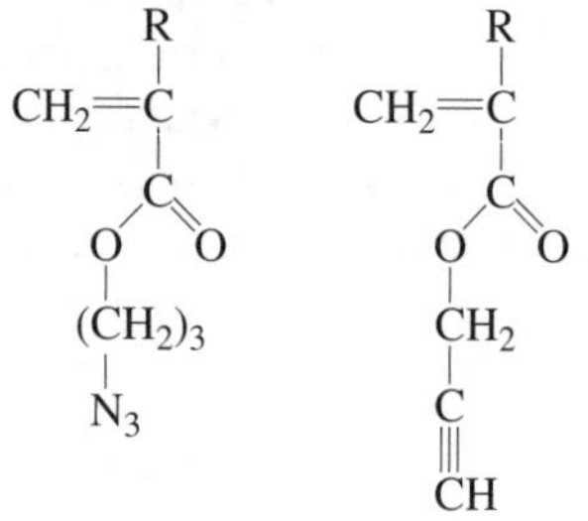

where R = H or CH_3. Synthesis of polymer chains with a terminal alkyne or azide group was described in Section 9.4.3.1 and they provide the necessary complementary functionality for click coupling onto the backbone copolymer. This approach is illustrated schematically below

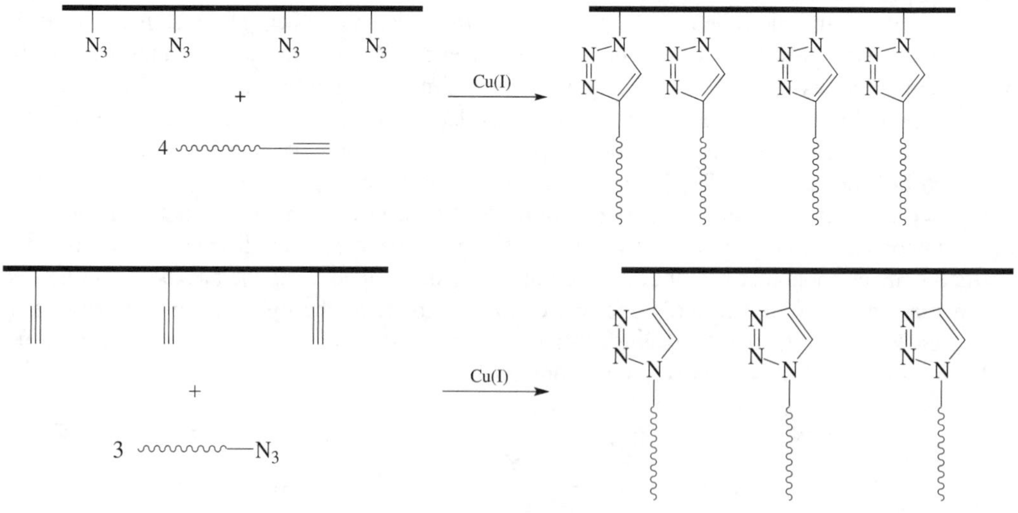

Cu(I)

Cu(I)

This approach also provides an opportunity to attach block copolymer grafts to the backbone polymer.

PROBLEMS

9.1 Table 9.1 gives the reactivity ratios for free-radical copolymerization of styrene with (a) butadiene, (b) methyl methacrylate, (c) methyl acrylate, (d) acrylonitrile, (e) maleic anhydride, (f) vinyl chloride and (g) vinyl acetate. For each of these copolymerizations, use the terminal-model copolymer composition equation to calculate the following.

(i) The composition of the copolymer formed at low conversions of an equimolar mixture of the two monomers.

(ii) The comonomer composition required to form a copolymer consisting of 50 mol% styrene repeat units.

For the azeotropic copolymerizations, calculate the azeotropic composition.

9.2 Using the reactivity ratios given in Table 9.1 and the terminal-model copolymer composition equation to calculate data points, plot the variation of copolymer composition with comonomer composition (as in Figure 9.1) for copolymerization of methyl methacrylate with (a) styrene, (b) methyl acrylate and (c) vinyl acetate. Discuss the curves in relation to control of copolymer composition drift for each of these copolymerizations.

9.3 The initial concentrations of styrene $[S]_0$ and acrylonitrile $[AN]_0$ employed in a series of low conversion free-radical copolymerizations are given in the table below together with the nitrogen contents (%N by weight) of the corresponding poly(styrene-*stat*-acrylonitrile) samples produced.

$[S]_0$ / mol dm^{-3}	3.45	2.60	2.10	1.55
$[AN]_0$ / mol dm^{-3}	1.55	2.40	2.90	3.45
%N in copolymer	5.69	7.12	7.77	8.45

(a) By use of an appropriate plot, evaluate the reactivity ratios for free-radical copolymerization of styrene with acrylonitrile.

(b) Explain why it was necessary to restrict the copolymerizations to low conversions.

9.4 A 0.16 mol dm^{-3} solution (3.50 cm^3) of *s*-butyllithium in toluene was added to a solution of styrene (8.40 g) in toluene (200 cm^3). After complete conversion of the styrene, isoprene (28.00 g) was added. When the isoprene had completely polymerized, the reaction was completed by addition of a 0.10 mol cm^{-3} solution (2.80 cm^3) of dichloromethane in toluene.

(a) Write down the reactions occurring in each stage of this reaction sequence and show the structure of the final polymer.

(b) Evaluate the relevant degrees of polymerization and the corresponding molar masses for the final polymer.

FURTHER READING

Allen, G. and Bevington, J.C. (eds), *Comprehensive Polymer Science*, Vols. 3–5 (Eastmond, G.C., Ledwith, A., Russo, S., and Sigwalt, P., eds), Pergamon Press, Oxford, U.K., 1989.

Davis, K.A. and Matyjaszewski, K. (eds), Statistical, gradient, block and graft copolymers by controlled/living radical polymerizations, *Advances in Polymer Science*, **159**, 1–191 (2002).

Hadjichristidis, N., Pitsikalis, M., and Iatrou, H., Synthesis of block copolymers, *Advances in Polymer Science*, **189**, 1 (2005).

Ito, K. and Kawaguchi, S., Poly(macromonomers): Homo- and copolymerization, *Advances in Polymer Science*, **142**, 1 (1999).

Lenz, R.W., *Organic Chemistry of Synthetic High Polymers*, Interscience, New York, 1967.

Mark, H.F., Bikales, N.M., Overberger, C.G., and Menges, G. (eds), *Encyclopedia of Polymer Science and Engineering*, Wiley-Interscience, New York, 1985–1989.

Moad, G. and Soloman, D.H., *The Chemistry of Radical Polymerization*, 2nd edn., Elsevier, Oxford, 2006.

Odian, G., *Principles of Polymerization*, 4th edn., Wiley-Interscience, New York, 2004.

Pitsikalis, M., Pispas, S., Mays, J.W., and Hadjichristidis, N., Nonlinear block copolymer architectures, *Advances in Polymer Science*, **135**, 1 (1998).

Stevens, M.P., *Polymer Chemistry: An Introduction*, 3rd edn., Oxford University Press, Oxford, U.K., 1998.

Part II

Characterization of Polymers

10 Theoretical Description of Polymers in Solution

10.1 INTRODUCTION

Relationships between the synthesis and molecular properties of polymers (Chapters 3 through 9), and between their molecular and bulk properties (Chapters 16 through 25), provide the foundations of polymer science. In order to establish these relationships, and to test theories, it is essential to accurately and thoroughly characterize the polymers under investigation. Furthermore, use of these relationships to predict and understand the in-use performance of a particular polymer depends upon the availability of good characterization data for that polymer. Thus, polymer characterization is of great importance, both academically and commercially.

Molecular characterization of polymer samples is the focus for Chapters 11 through 15 and involves determination of molar mass averages, molar mass distributions, molecular dimensions, overall compositions, basic chemical structures and detailed molecular microstructures, mostly by analysis of polymers in dilute solution (<20 g dm^{-3}, i.e. <2%w/v). Since the theoretical description of polymers in solution underpins most of these methods, the relevant generic theories will be introduced in this chapter before describing the individual methods in detail in subsequent chapters.

10.2 THERMODYNAMICS OF POLYMER SOLUTIONS

A *solution* can be defined as a homogeneous mixture of two or more substances, i.e. the mixing is on a molecular scale. Under the usual conditions of constant temperature T and pressure P, the thermodynamic requirement for formation of a two-component solution is that the Gibbs free energy G_{12} of the mixture must be less than the sum of the Gibbs free energies G_1 and G_2 of the pure components in isolation. This requirement is defined in terms of the Gibbs free energy of mixing

$$\Delta G_m = G_{12} - (G_1 + G_2)$$

which must be negative (i.e. $\Delta G_m < 0$) for a solution to form. Since Gibbs free energy is related to enthalpy H and entropy S by the standard thermodynamic equation

$$G = H - TS \tag{10.1}$$

a more useful expression for ΔG_m is

$$\Delta G_m = \Delta H_m - T\Delta S_m \tag{10.2}$$

where

ΔH_m is the enthalpy (or heat) of mixing
ΔS_m is the entropy of mixing.

10.2.1 Thermodynamics of Ideal Solutions

Ideal solutions are mixtures of molecules (i) that are identical in size and (ii) for which the energies of like (i.e. 1–1 or 2–2) and unlike (i.e. 1–2) molecular interactions are equal. The latter condition leads to *athermal mixing* (i.e. $\Delta H_m = 0$), which also means that there are no changes in the rotational, vibrational and translational entropies of the components upon mixing. Thus, ΔS_m depends only upon the *combinatorial* (or configurational) entropy change ΔS_m^{comb}, which is positive because the number of distinguishable spatial arrangements of the molecules increases when they are mixed. Hence, ΔG_m is negative and formation of an ideal solution always is favourable. The methods of *statistical mechanics* can be used to derive an equation for ΔS_m^{comb} by assuming that the molecules are placed randomly into cells which are of molecular size and which are arranged in the form of a three-dimensional lattice (represented in two dimensions for cubic cells in Figure 10.1a).

From statistical mechanics, the fundamental relation between the entropy S of an assembly of molecules and the total number Ω of distinguishable degenerate (i.e. of equal energy) arrangements of the molecules is given by the Boltzmann equation

$$S = \mathbf{k} \ln \Omega \tag{10.3}$$

where **k** is the Boltzmann constant. Application of this equation to formation of an ideal solution gives

$$\Delta S_m^{comb} = \mathbf{k}[\ln\Omega_{12} - (\ln\Omega_1 + \ln\Omega_2)] \tag{10.4}$$

where Ω_1, Ω_2 and Ω_{12} are, respectively, the total numbers of distinguishable spatial arrangements of the molecules in the pure solvent, the pure solute and the ideal mixture. Since all the molecules of a pure substance are identical, there is only one distinguishable spatial arrangement of them (i.e. of arranging N identical molecules in a lattice containing N cells). Thus, $\Omega_1 = 1$ and $\Omega_2 = 1$, allowing Equation 10.4 to be reduced to

$$\Delta S_m^{comb} = \mathbf{k} \ln \Omega_{12} \tag{10.5}$$

For ideal mixing of N_1 molecules of solvent with N_2 molecules of solute in a lattice with $(N_1 + N_2)$ cells, the total number of distinguishable spatial arrangements of the molecules is equal to the number of permutations of $(N_1 + N_2)$ objects which fall into two classes containing N_1 identical objects of type 1 and N_2 identical objects of type 2, respectively, i.e.

$$\Omega_{12} = \frac{(N_1 + N_2)!}{N_1!N_2!} \tag{10.6}$$

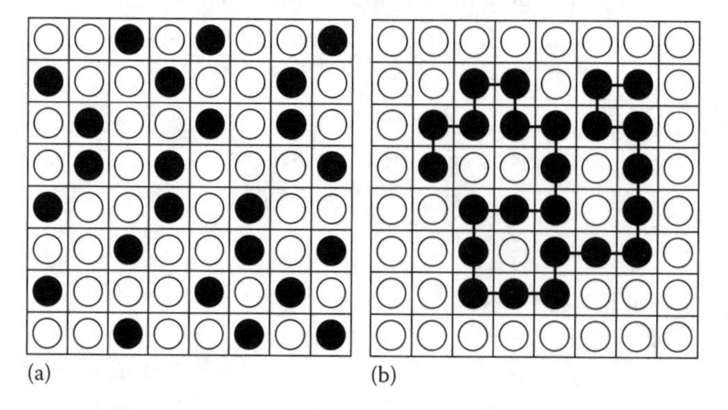

FIGURE 10.1 Schematic representation of a liquid lattice: (a) a mixture of molecules of equal size and (b) a mixture of solvent molecules with a polymer molecule showing the connectivity of polymer segments.

Substitution of Equation 10.6 into Equation 10.5 gives

$$\Delta S_{\mathrm{m}}^{\mathrm{comb}} = \mathbf{k}\ln\left[\frac{(N_1+N_2)!}{N_1!N_2!}\right] \tag{10.7}$$

and introducing Stirling's approximation, $\ln N! = N\ln N - N$ (for large N), leads to

$$\Delta S_{\mathrm{m}}^{\mathrm{comb}} = -\mathbf{k}\left\{N_1\ln\left[\frac{N_1}{(N_1+N_2)}\right]+N_2\ln\left[\frac{N_2}{(N_1+N_2)}\right]\right\} \tag{10.8}$$

It is more usual to write thermodynamic equations in terms of numbers of moles, n, and mole fractions, X, which here are defined by $n_1 = N_1/\mathbf{N_A}$, $n_2 = N_2/\mathbf{N_A}$, $X_1 = n_1/(n_1+n_2)$ and $X_2 = n_2/(n_1+n_2)$, where $\mathbf{N_A}$ is the Avogadro constant. Thus, Equation 10.8 becomes

$$\Delta S_{\mathrm{m}}^{\mathrm{comb}} = -\mathbf{R}[n_1\ln X_1 + n_2\ln X_2] \tag{10.9}$$

since the universal gas constant $\mathbf{R} = \mathbf{k}\mathbf{N}_{\mathrm{A}}$. Hence, for the formation of an ideal solution

$$\Delta G_{\mathrm{m}} = \mathbf{R}T[n_1\ln X_1 + n_2\ln X_2] \tag{10.10}$$

This important equation provides the fundamental basis from which standard thermodynamic relationships for ideal solutions can be derived (e.g. Raoult's law). However, since relatively few solutions behave ideally, the simple lattice theory requires modification to make it more generally applicable. For mixtures of small molecules, non-ideality invariably is due to non-athermal mixing (i.e. $\Delta H_{\mathrm{m}} \neq 0$) and so requires the effects of non-equivalent intermolecular interactions to be taken into account. In contrast, polymer solutions show major deviations from ideal solution behaviour, even when $\Delta H_{\mathrm{m}} = 0$. For example, the vapour pressure of solvent above a polymer solution invariably is very much lower than predicted from Equation 10.10. The failure of the simple lattice theory to give realistic predictions of the thermodynamic properties of polymer solutions arises from the assumption that the solvent and solute molecules are identical in size. Flory and Huggins independently proposed a modified lattice theory which takes account of (i) the large differences in size between solvent and polymer molecules and (ii) intermolecular interactions. This theory is described in an up-to-date form in the following section.

10.2.2 Flory–Huggins Theory

The theory sets out to predict ΔG_{m} for the formation of polymer solutions by considering the polymer molecules to be chains of *segments*, each segment being equal in size to a solvent molecule. The number x of segments in the chain defines the size of a polymer molecule and may be taken as the ratio of the molecular (or molar) volumes of polymer and solvent (hence x is proportional to, but not necessarily equal to, the degree of polymerization). On this basis, it is possible to place solvent molecules and/or polymer molecules in a three-dimensional lattice consisting of identical cells, each of which is the same size as a solvent molecule and has z cells as first neighbours (z being known as the lattice coordination number, the value of which is not important for prediction of ΔG_{m}, but may be considered to be an integer from 6 to 12). Each lattice cell is occupied by either a solvent molecule or a chain segment, and each polymer molecule is placed in the lattice so that its chain segments occupy a continuous sequence of x cells (as indicated in Figure 10.1b).

The first step in the development of the theory is to derive an expression for $\Delta S_{\mathrm{m}}^{\mathrm{comb}}$ when $\Delta H_{\mathrm{m}} = 0$. As for mixing of simple molecules, this involves application of Equation 10.4, but this time with

$\Omega_2 > 1$ because each molecule in a pure amorphous polymer can adopt many different *conformations* (i.e. distinguishable spatial arrangements of the chain of segments). Hence, for polymer solutions

$$\Delta S_{\mathrm{m}}^{\mathrm{comb}} = \mathbf{k} \ln\left(\frac{\Omega_{12}}{\Omega_2}\right) \tag{10.11}$$

The basis of the theory is the formation of a theoretical polymer solution by mixing N_2 polymer molecules, each consisting of x chain segments, with N_1 solvent molecules in a lattice containing $N_1 + xN_2$ (=N) cells which are equal in size to that of a solvent molecule. The N_2 polymer molecules are added one by one to the lattice before adding the solvent molecules. When adding a polymer molecule, the first segment of the chain can be placed in any empty cell. However, for the *connectivity* required to represent a polymer molecule, each successive segment of the remaining $x-1$ segments must be placed in an empty cell adjacent to the previously-placed segment. The individual numbers of possible placements are calculated for each segment of the chain, and by multiplying these numbers together the total number ν of possible conformations of the polymer molecule in the lattice is obtained. This series of calculations is carried out for each of the N_2 polymer molecules as they are added to the lattice. The restrictions imposed by partial occupancy of the lattice are taken into account using a *mean-field approximation* whereby the segments of the previously added polymer molecules are assumed to be distributed uniformly in the lattice (which is necessary to facilitate the mathematical analysis, but is a weakness of the theory that is especially significant for dilute solutions, as will be discussed in Section 10.2.4). Since the N_1 solvent molecules are identical, there is only one distinguishable spatial arrangement of them, namely, one solvent molecule in each of the remaining N_1 empty cells. Thus, the total number of distinguishable spatial arrangements of the mixture of N_2 polymer molecules and N_1 solvent molecules is given by

$$\Omega_{12} = \frac{1}{N_2!}\prod_{\zeta=1}^{N_2} \nu_\zeta \tag{10.12}$$

where Π is the sign for a continuous product (in this case, $\nu_1 \times \nu_2 \times \nu_3 \times \nu_4 \times \ldots \times \nu_{N_2}$). The factor of $1/N_2!$ takes account of the fact that the N_2 polymer molecules are identical and therefore indistinguishable from each other. Consider placement of the ζ-th chain onto the lattice. The previously-placed chains occupy $x(\zeta - 1)$ lattice sites and so the first segment of the ζ-th chain can be placed in any of the remaining $N - x(\zeta - 1)$ lattice sites. The second segment of the ζ-th chain must be placed in an adjacent *vacant* cell. Based on the mean-field approximation and assuming that segments from the ζ-th chain, other than preceding segments in the chain, do not occupy one of the contiguous sites adjacent a segment of the ζ-th chain (which introduces an additional error that is insignificant compared to that from the mean-field approximation), a fraction $[N - x(\zeta - 1)]/N$ of the z cells adjacent to the first segment will be unoccupied and so the number of vacant adjacent cells is $z\{[N - x(\zeta - 1)]/N\}$. Following the same logic, each of the subsequent $x - 2$ segments can be placed in any of $(z-1)$ $\{[N - x(\zeta - 1)]/N\}$ adjacent empty cells (the term $z - 1$ appears because one of the adjacent cells is occupied by the previously-placed segment). Thus

$$\nu_\zeta = \left[N - x(\zeta - 1)\right]\left\{z\frac{\left[N - x(\zeta - 1)\right]}{N}\right\}\left\{(z-1)\frac{\left[N - x(\zeta - 1)\right]}{N}\right\}^{x-2}$$

which simplifies to

$$\nu_\zeta = z(z-1)^{x-2} N^{1-x}\left[N - x(\zeta - 1)\right]^x$$

Substituting this expression into Equation 10.12 and recognising that ζ is the only variable

$$\Omega_{12} = \frac{1}{N_2!}\left\{z(z-1)^{x-2}N^{1-x}\right\}^{N_2}\prod_{\zeta=1}^{N_2}\left[N - x(\zeta-1)\right]^x \tag{10.13}$$

By analogy, the total number of distinguishable spatial arrangements of N_2 chains in the pure amorphous (or liquid) polymer is given by

$$\Omega_2 = \frac{1}{N_2!}\left\{z(z-1)^{x-2}(xN_2)^{1-x}\right\}^{N_2}\prod_{\zeta=1}^{N_2}\left[xN_2 - x(\zeta-1)\right]^x \tag{10.14}$$

because, in this case, $N = xN_2$ (i.e. there are no solvent molecules, so $N_1 = 0$). Although not necessary for derivation of the Flory–Huggins equation, it is instructive to analyse Equations 10.13 and 10.14 fully before considering $\Delta S_{\rm m}^{\rm comb}$. Re-casting Equation 10.14 as the product of two terms is helpful

$$\Omega_2 = \underbrace{\left\{\left[z(z-1)^{x-2}\right]^{N_2}\right\}}_{\omega_{2\rm f}}\underbrace{\left\{\frac{1}{N_2!}(xN_2)^{(1-x)N_2}\prod_{\zeta=1}^{N_2}\left[xN_2 - x(\zeta-1)\right]^x\right\}}_{\omega_{2\rm c}} \tag{10.15}$$

The first term $\omega_{2\rm f}$ is extremely large and simply the total number of conformations of the N_2 chains if they do not encounter each other when being added to the lattice (i.e. each chain has the complete freedom of the lattice, which can therefore be considered to be infinitely large). This highly artificial quantity is massively attenuated by the second term $\omega_{2\rm c}$, which is the fraction of those $\omega_{2\rm f}$ conformations that are allowed when placement of the N_2 chains is confined to a lattice of xN_2 cells (i.e. as in the pure amorphous polymer). Expanding $\omega_{2\rm c}$ gives

$$\omega_{2\rm c} = \frac{1}{N_2!}(xN_2)^{(1-x)N_2}x^{xN_2}(N_2!)^x = N_2^{(1-x)N_2}x^{N_2}(N_2!)^{x-1}$$

By taking the natural logarithm of $\omega_{2\rm c}$ and applying Stirling's approximation (Section 10.2.1), $\omega_{2\rm c}$ is simplified to the form

$$\omega_{2\rm c} = \left[\frac{x}{e^{x-1}}\right]^{N_2} \tag{10.16}$$

Equation 10.13 now can be re-cast in terms of $\omega_{2\rm f}$ and $\omega_{2\rm c}$ by combining it with Equation 10.15

$$\Omega_{12} = \omega_{2\rm f}\omega_{2\rm c}\underbrace{\left\{\left(\frac{N}{xN_2}\right)^{(1-x)N_2}\left(\frac{\prod_{\zeta=1}^{N_2}\left[N - x(\zeta-1)\right]^x}{\prod_{\zeta=1}^{N_2}\left[xN_2 - x(\zeta-1)\right]^x}\right)\right\}}_{\omega_{2\rm d}} \tag{10.17}$$

where ω_{2d} is the factor by which Ω_2 is increased when the chains are given greater spatial freedom by dilution with N_1 solvent molecules. Consider first the continuous product term in the numerator of ω_{2d}

$$\prod_{\zeta=1}^{N_2}[N-x(\zeta-1)]^x = x^{xN_2}\left[\left(\frac{N}{x}\right)\left(\frac{N}{x}-1\right)\left(\frac{N}{x}-2\right)\left(\frac{N}{x}-3\right)\left(\frac{N}{x}-2\right)\ldots\left(\frac{N}{x}-(N_2-1)\right)\right]^x$$

which (since $N/x-N_2=N_1/x$) may be re-written (with little error) as

$$\prod_{\zeta=1}^{N_2}[N-x(\zeta-1)]^x = x^{xN_2}\left[\frac{(N/x)!}{(N_1/x)!}\right]^x$$

Substituting this relation into ω_{2d} from Equation 10.17 and expanding the continuous product term in the denominator (as when evaluating ω_{2c}) gives

$$\omega_{2d} = \left(\frac{N}{xN_2}\right)^{(1-x)N_2}\left[\frac{(N/x)!}{N_2!(N_1/x)!}\right]^x$$

By taking the natural logarithm of ω_{2d} and applying Stirling's approximation, ω_{2c} simplifies to

$$\omega_{2d} = \left(\frac{N}{xN_2}\right)^{N_2}\left(\frac{N}{N_1}\right)^{N_1} \tag{10.18}$$

Thus, in summary, from Equations 10.15 and 10.16

$$\Omega_2 = \underbrace{\left\{\left[z(z-1)^{x-2}\right]^{N_2}\right\}}_{\omega_{2f}}\underbrace{\left\{\left[\frac{x}{e^{x-1}}\right]^{N_2}\right\}}_{\omega_{2c}} \tag{10.19}$$

and from Equations 10.17 through 10.19

$$\Omega_{12} = \Omega_2\underbrace{\left[\left(\frac{N}{xN_2}\right)^{N_2}\left(\frac{N}{N_1}\right)^{N_1}\right]}_{\omega_{2d}} \tag{10.20}$$

Substituting Equations 10.19 and 10.20 into Equation 10.11 gives

$$\Delta S_m^{\text{comb}} = \mathbf{k}\ln\left[\left(\frac{N}{N_1}\right)^{N_1}\left(\frac{N}{xN_2}\right)^{N_2}\right] \tag{10.21}$$

Note that z is absent from this equation (i.e. the precise nature of the lattice is unimportant for evaluation of ΔS_m^{comb}). Equation 10.21 simplifies further to give

$$\Delta S_m^{\text{comb}} = -\mathbf{k}[N_1\ln\phi_1 + N_2\ln\phi_2] \tag{10.22}$$

where ϕ_1 and ϕ_2 are the volume fractions of solvent and polymer, respectively, and are given by $\phi_1 = N_1/(N_1 + xN_2)$ and $\phi_2 = xN_2/(N_1 + xN_2)$. Writing Equation 10.22 in terms of the number of moles, n_1 and n_2, of solvent and polymer in the mixture gives

$$\Delta S_{\mathrm{m}}^{\mathrm{comb}} = -\mathbf{R}[n_1 \ln \phi_1 + n_2 \ln \phi_2] \tag{10.23}$$

which may be compared to the corresponding expression for ideal mixing, i.e. Equation 10.9. Thus the only difference is that for polymer solutions, the mole fractions in Equation 10.9 are replaced by the corresponding volume fractions. Equation 10.23 is in fact a more general expression for athermal mixing and reduces to Equation 10.9 when $x = 1$ (i.e. when both components in the mixture have the same *small* molecular size).

Having derived an expression for $\Delta S_{\mathrm{m}}^{\mathrm{comb}}$, the second step in the development of Flory–Huggins theory is to introduce a term which accounts for the effects of intermolecular interactions. In the original theory, this was considered only in terms of an enthalpy change. However, the term subsequently was modified in recognition that there must be an entropy change associated with the non-randomness induced by interactions. Thus, the effects of interactions will be treated here in terms of a contact Gibbs free energy change, $\Delta G_{\mathrm{m}}^{\mathrm{contact}}$. The calculation is simplified by restriction to first neighbour interactions on the basis that the forces between uncharged molecules are known to decrease rapidly with their distance of separation. Three types of contact need to be considered, each with its own Gibbs free energy of interaction:

Type of Contact	Gibbs Free Energy of Interaction
Solvent–solvent	g_{11}
Segment–segment	g_{22}
Solvent–segment	g_{12}

For every two solvent–segment contacts formed on mixing, one solvent–solvent and one segment–segment contact of the pure components are lost. Thus, the Gibbs free energy change Δg_{12} for the formation of a single solvent–segment contact is given by

$$\Delta g_{12} = g_{12} - \tfrac{1}{2}(g_{11} + g_{22})$$

If p_{12} is the total number of solvent–segment contacts in the solution, then

$$\Delta G_{\mathrm{m}}^{\mathrm{contact}} = p_{12} \Delta g_{12} \tag{10.24}$$

With the exception of segments at chain ends, each chain segment has two connected segmental neighbours. Hence there are $(z-2)x+2$ lattice sites adjacent to each polymer molecule. When x is large, $(z-2)x \gg 2$ and the total number of lattice sites adjacent to all the polymer molecules can be taken as $N_2(z-2)x$. On the basis of the mean-field approximation, a fraction ϕ_1 of these sites will be occupied by solvent molecules. Thus

$$p_{12} = N_2(z-2)x\phi_1$$

from which x can be eliminated to give

$$p_{12} = (z-2)N_1\phi_2 \tag{10.25}$$

because $xN_2\phi_1 = N_1\phi_2$. Combining Equations 10.24 and 10.25 leads to

$$\Delta G_{\mathrm{m}}^{\mathrm{contact}} = (z-2)N_1\phi_2\Delta g_{12} \tag{10.26}$$

The lattice parameter z and the Gibbs free energy change Δg_{12} are not easily accessible, and so are eliminated by introducing a single new parameter, χ, which is commonly known as the *Flory–Huggins polymer–solvent interaction parameter*, and is defined by

$$\chi = \frac{(z-2)\Delta g_{12}}{\mathbf{k}T} \tag{10.27}$$

Thus χ is a temperature-dependent dimensionless quantity which characterizes polymer–solvent interactions and can be expressed more simply in the form

$$\chi = a + \frac{b}{T} \tag{10.28}$$

where a and b are temperature-independent quantities. More generally, χ is given as the sum of enthalpy χ_H and entropy χ_S components

$$\chi = \chi_H + \chi_S \tag{10.29}$$

from which it is simple to show that $\chi_H = b/T$ and $\chi_S = a$, and that

$$\chi_H = -T\left(\frac{\partial\chi}{\partial T}\right) \quad \text{and} \quad \chi_S = \frac{\partial(T\chi)}{\partial T} \tag{10.30}$$

Combining Equations 10.26 and 10.27 together with the relations $n_1 = N_1/\mathbf{N_A}$ and $\mathbf{R} = \mathbf{kN_A}$ leads to

$$\Delta G_{\mathrm{m}}^{\mathrm{contact}} = \mathbf{R}Tn_1\phi_2\chi \tag{10.31}$$

Now all that is required is to combine $\Delta S_{\mathrm{m}}^{\mathrm{comb}}$ (Equation 10.23) and $\Delta G_{\mathrm{m}}^{\mathrm{contact}}$ (Equation 10.31) as follows

$$\Delta G_{\mathrm{m}} = \Delta G_{\mathrm{m}}^{\mathrm{contact}} - T\Delta S_{\mathrm{m}}^{\mathrm{comb}}$$

to give the *Flory–Huggins equation for the Gibbs free energy of mixing*

$$\Delta G_{\mathrm{m}} = \mathbf{R}T[n_1 \ln \phi_1 + n_2 \ln \phi_2 + n_1\phi_2\chi] \tag{10.32}$$

Using the Flory–Huggins equation, it is possible to account for the equilibrium thermodynamic properties of polymer solutions, particularly the large suppression of solvent vapour pressure and elevation of solvent boiling point (Sections 11.3 and 11.4), phase-separation and fractionation behaviour (Section 14.2), melting-point depressions in crystalline polymers (Section 17.8.3), and swelling of networks (Section 21.5.1). However, whilst the theory is able to predict general trends, precise agreement with experimental data is not achieved. The deficiencies of the theory result from the limitations both of the model and of the assumptions employed in its derivation. The use of a single type of lattice for pure solvent, pure polymer and their mixtures is unrealistic and requires there to be no volume change upon mixing. The mathematical procedure used to calculate the total number of possible conformations of a polymer molecule in the lattice does not exclude self-intersections of the chain, which clearly is physically unrealistic. Furthermore, the use of a mean-field approximation to facilitate this calculation for placement of a polymer molecule in a partly-filled lattice is satisfactory only when the volume fraction ϕ_2 of polymer is high. In view of this, Flory–Huggins theory is least satisfactory for dilute polymer solutions because the polymer molecules in such solutions are isolated from each other by regions of pure solvent, i.e. the segments are not distributed uniformly throughout the lattice. Additionally, the interaction parameter χ introduced to account for the effects of contact interactions, is not a simple parameter. Equation 10.29 resulted from a recognition that χ contains both enthalpy and entropy contributions, and χ also has been shown to depend upon ϕ_2. Despite these shortcomings, Flory–Huggins lattice theory was a major step forward towards understanding the thermodynamics of polymer solutions, and is the basis of many other theories which are used widely in practice. Hence, the theory will be developed further in subsequent sections, but the above limitations must be borne in mind when applying the resulting equations. Some of the more refined theories developed since the advent of Flory–Huggins lattice theory will be given brief consideration, but in general are beyond the scope of this book.

10.2.3 Partial Molar Quantities and Chemical Potential

In thermodynamics, it often is necessary to distinguish between *extensive* and *intensive* properties. The value of an extensive property changes with the amount of material in the system, whereas the value of an intensive property is independent of the amount of material. Examples of extensive properties are mass, volume, *total* free energy, enthalpy and entropy, whereas intensive properties include pressure, temperature, density and *molar* free energy, enthalpy and entropy. Generally, systems at equilibrium are defined in terms of their intensive properties.

For multicomponent systems such as polymer solutions, it is important to know how thermodynamic parameters (represented by Z) vary when a small amount of one component (represented by i) is added to the system whilst maintaining the pressure, temperature and amounts of all other components (represented by j) of the system constant. This requires the use of *partial molar quantities* such as that represented by $\bar{Z}_i$ which gives the rate at which Z changes with the number n_i of moles of component i and is defined by the partial differential

$$\bar{Z}_i = \left(\frac{\partial Z}{\partial n_i}\right)_{P,T,nj} \tag{10.33}$$

The partial molar quantities most commonly encountered in the thermodynamics of polymer solutions are partial molar volumes $\bar{V}_i$ and partial molar Gibbs free energies $\bar{G}_i$. The latter quantities are called *chemical potentials* μ_i, which therefore are defined by

$$\mu_i = \left(\frac{\partial G}{\partial n_i}\right)_{P,T,nj} \tag{10.34}$$

In single-component systems, such as pure solvents, the partial molar quantities are identical to the corresponding molar quantities.

For a system at equilibrium, the chemical potential μ_i of component i is related to its activity a_i in the system by the standard thermodynamic relation

$$\mu_i - \mu_i^{\mathrm{o}} = \mathbf{R}T \ln a_i \tag{10.35}$$

where

μ_i^{o} is the chemical potential of component i in its standard state

$\mu_i - \mu_i^{\mathrm{o}}$ is the partial molar Gibbs free energy change, $\overline{\Delta G_i} = (\partial \Delta G/\partial n_i)_{P,T,n_j}$ for formation of the system.

Chemical potential difference equations for the solvent and polymer in polymer solutions can be obtained by partial differentiation of the Flory–Huggins expression for ΔG_m (Equation 10.32). Consider the solvent first

$$\mu_1 - \mu_1^o = \left(\frac{\partial \Delta G_m}{\partial n_1}\right)_{P,T,n_2} = \mathbf{R}T \frac{\partial}{\partial n_1}\{n_1 \ln \phi_1 + n_2 \ln \phi_2 + n_1 \phi_2 \chi\}_{P,T,n_2}$$

Taking into account the fact that ϕ_1 and ϕ_2 are functions of n_1 and n_2, using the standard mathematical procedures for differentiation of products gives

$$\mu_1 - \mu_1^o = \mathbf{R}T \left\{\ln \phi_1 + n_1 \frac{\partial \ln \phi_1}{\partial n_1} + n_2 \frac{\partial \ln \phi_2}{\partial n_1} + \phi_2 \chi + n_1 \chi \frac{\partial \phi_2}{\partial n_1}\right\}_{P,T,n_2} \quad (10.36)$$

From the definitions of $\phi_1 = n_1/(n_1 + xn_2)$ and $\phi_2 = xn_2/(n_1 + xn_2)$, it follows that

$$\phi_2 = 1 - \phi_1 \text{ and so } \left\{\frac{\partial \phi_2}{\partial n_1}\right\}_{P,T,n_2} = -\left\{\frac{\partial \phi_1}{\partial n_1}\right\}_{P,T,n_2} = -\left[\frac{(n_1 + xn_2) - n_1}{(n_1 + xn_2)^2}\right] = -\frac{\phi_2}{(n_1 + xn_2)}$$

$$\left\{\frac{\partial \ln \phi_1}{\partial n_1}\right\}_{P,T,n_2} = \frac{1}{\phi_1}\left\{\frac{\partial \phi_1}{\partial n_1}\right\}_{P,T,n_2} = \frac{\phi_2}{\phi_1 (n_1 + xn_2)}$$

$$\left\{\frac{\partial \ln \phi_2}{\partial n_1}\right\}_{P,T,n_2} = \frac{1}{\phi_2}\left\{\frac{\partial \phi_2}{\partial n_1}\right\}_{P,T,n_2} = -\frac{1}{\phi_2}\left\{\frac{\partial \phi_1}{\partial n_1}\right\}_{P,T,n_2} = -\frac{1}{(n_1 + xn_2)}$$

which upon substitution into Equation 10.36, with simplification, gives for the solvent

$$\mu_1 - \mu_1^o = \mathbf{R}T\left[\ln \phi_1 + \left(1 - \frac{1}{x}\right)\phi_2 + \chi \phi_2^2\right] \quad (10.37)$$

Similarly, for the polymer

$$\mu_2 - \mu_2^o = \left(\frac{\partial \Delta G_m}{\partial n_2}\right)_{P,T,n_1} = \mathbf{R}T \frac{\partial}{\partial n_2}\{n_1 \ln \phi_1 + n_2 \ln \phi_2 + n_1 \phi_2 \chi\}_{P,T,n_1}$$

and so

$$\mu_2 - \mu_2^o = \mathbf{R}T \left\{n_1 \frac{\partial \ln \phi_1}{\partial n_2} + \ln \phi_2 + n_2 \frac{\partial \ln \phi_2}{\partial n_2} + n_1 \chi \frac{\partial \phi_2}{\partial n_2}\right\}_{P,T,n_1} \quad (10.38)$$

Once again, from the definitions of ϕ_1 and ϕ_2

$$\phi_2 = 1 - \phi_1 \text{ and so } \left\{\frac{\partial \phi_2}{\partial n_2}\right\}_{P,T,n_1} = -\left\{\frac{\partial \phi_1}{\partial n_2}\right\}_{P,T,n_1} = -\left[\frac{-xn_1}{(n_1 + xn_2)^2}\right] = \frac{x\phi_1}{(n_1 + xn_2)}$$

$$\left\{\frac{\partial \ln\phi_1}{\partial n_2}\right\}_{P,T,n_1} = \frac{1}{\phi_1}\left\{\frac{\partial \phi_1}{\partial n_2}\right\}_{P,T,n_1} = -\frac{x}{(n_1 + xn_2)}$$

$$\left\{\frac{\partial \ln\phi_2}{\partial n_2}\right\}_{P,T,n_1} = \frac{1}{\phi_2}\left\{\frac{\partial \phi_2}{\partial n_2}\right\}_{P,T,n_1} = -\frac{1}{\phi_2}\left\{\frac{\partial \phi_1}{\partial n_2}\right\}_{P,T,n_1} = \frac{x\phi_1}{\phi_2(n_1 + xn_2)}$$

which upon substitution into Equation 10.38, with simplification, gives for the polymer

$$\mu_2 - \mu_2^{\mathrm{o}} = \mathbf{R}T[\ln\phi_2 - (x-1)\phi_1 + x\chi\phi_1^2] \tag{10.39}$$

and per polymer chain segment, i.e. $(\mu_2 - \mu_2^0)/x$

$$\mu_{\mathrm{s}} - \mu_{\mathrm{s}}^{\mathrm{o}} = \mathbf{R}T\left[\left(\frac{\ln\phi_2}{x}\right) - \left(1 - \frac{1}{x}\right)\phi_1 + \chi\phi_1^2\right] \tag{10.40}$$

Equations 10.37, 10.39 and 10.40 are of fundamental importance. The number of chain segments x appears in these equations entirely as a consequence of its presence in the denominator of the volume fraction definitions, which for polymers with chain-length dispersity are given by

$$\phi_1 = \frac{n_1}{n_1 + \sum_i x_i n_i} = \frac{n_1}{n_1 + \bar{x}_{\mathrm{n}} n_2}$$

$$\phi_2 = \frac{\sum_i x_i n_i}{n_1 + \sum_i x_i n_i} = \frac{\bar{x}_{\mathrm{n}} n_2}{n_1 + \bar{x}_{\mathrm{n}} n_2}$$

because $\bar{x}_{\mathrm{n}} = \sum_i x_i n_i \Big/ \sum_i n_i$ and $\sum_i n_i = n_2$. Thus, in the definitions of ϕ_1 and ϕ_2 for polymers with chain-length dispersity, x is replaced simply by its number-average $\bar{x}_{\mathrm{n}}$ and, consequently, $\bar{x}_{\mathrm{n}}$ also replaces x in each of the derived equations.

Equations 10.37, 10.39 and 10.40 can be combined with Equation 10.35 to obtain expressions for the corresponding activities of the components, which then can be used to calculate theoretical activities for comparison with those determined by experiment. In the following section, the theory will be extended in this way for dilute polymer solutions.

10.2.4 Dilute Polymer Solutions

When analysing the thermodynamic properties of polymer solutions, it is sufficient to consider only one of the components, which for reasons of simplicity normally is the solvent. Thus, application of Equation 10.35 gives

$$\mu_1 - \mu_1^{\mathrm{o}} = \mathbf{R}T\ln a_1 \tag{10.41}$$

which can be separated into *ideal* and *non-ideal* (or *excess*) contributions by recognizing that $a_1 = \gamma_1 X_1$ where γ_1 and X_1 are, respectively, the activity coefficient and mole fraction of the solvent. Hence

$$(\mu_1 - \mu_1^{\mathrm{o}})^{\mathrm{ideal}} = \mathbf{R}T \ln X_1 \tag{10.42}$$

$$(\mu_1 - \mu_1^{\mathrm{o}})^{\mathrm{E}} = \mathbf{R}T \ln \gamma_1 \tag{10.43}$$

where the superscript E indicates an *excess* contribution.

The methods for molar mass characterization use dilute polymer solutions, typically $<20\,\mathrm{g\ dm^{-3}}$, for which $n_1 \gg xn_2$ and $n_1 \gg n_2$, leading to

$$\phi_2 = \frac{xn_2}{(n_1 + xn_2)} \approx \frac{xn_2}{n_1} \tag{10.44}$$

$$X_2 = \frac{n_2}{(n_1 + n_2)} \approx \frac{n_2}{n_1} \tag{10.45}$$

Hence, for dilute polymer solutions both ϕ_2 and X_2 are small and related to each other by $X_2 = \phi_2/x$. Under these conditions, the logarithmic terms in Equations 10.37 and 10.42 can be approximated by series expansion to give

$$\ln \phi_1 = \ln(1 - \phi_2) = -\phi_2 - \frac{\phi_2^2}{2} - \frac{\phi_2^3}{3} - \ldots \tag{10.46}$$

and

$$\ln X_1 = \ln(1 - X_2) = -X_2 - \frac{X_2^2}{2} - \frac{X_2^3}{3} - \ldots$$

which upon substitution of $X_2 = \phi_2/x$ gives

$$\ln X_1 = -\left(\frac{\phi_2}{x}\right) - \frac{(\phi_2/x)^2}{2} - \frac{(\phi_2/x)^3}{3} - \ldots \tag{10.47}$$

Since ϕ_2 is small and x is large, only the first two terms of Equation 10.46 and the first term of Equation 10.47 need to be retained. This leads to

$$\mu_1 - \mu_1^{\mathrm{o}} = \frac{-\mathbf{R}T\phi_2}{x} + \mathbf{R}T\left(\chi - \frac{1}{2}\right)\phi_2^2 \tag{10.48}$$

from Equation 10.37 and

$$(\mu_1 - \mu_1^{\mathrm{o}})^{\mathrm{ideal}} = \frac{-\mathbf{R}T\phi_2}{x} \tag{10.49}$$

from Equation 10.42. Comparison of these equations shows that for dilute polymer solutions, Flory–Huggins theory predicts the excess contribution to be

$$(\mu_1 - \mu_1^{\mathrm{o}})^{\mathrm{E}} = \mathbf{R}T\left(\chi - \frac{1}{2}\right)\phi_2^2 \tag{10.50}$$

in which the quantity $-\mathbf{R}T\phi_2^2/2$ arises from the connectivity of polymer chain segments and the quantity $\mathbf{R}T\chi\phi_2^2$ results from contact interactions. When $\chi = 1/2$, these two effects just exactly compensate each other, $(\mu_1 - \mu_1^{\mathrm{o}})^{\mathrm{E}} = 0$ and the dilute polymer solution behaves ideally, i.e. as if $\Delta H_{\mathrm{m}} = 0$ and the polymer chain segments are not connected. Flory introduced the term *theta conditions* to describe this ideal state of a dilute polymer solution and developed the concept further by defining the excess partial molar enthalpy $\overline{\Delta H_1}^{\mathrm{E}}$ and entropy $\overline{\Delta S_1}^{\mathrm{E}}$ of mixing as follows

$$\overline{\Delta H_1}^{\mathrm{E}} = \mathbf{R}T\kappa\phi_2^2 \tag{10.51}$$

$$\overline{\Delta S_1}^{\mathrm{E}} = \mathbf{R}\psi\phi_2^2 \tag{10.52}$$

where κ and ψ are enthalpy and entropy parameters, respectively. Since

$$(\mu_1 - \mu_1^{\mathrm{o}})^{\mathrm{E}} = \overline{\Delta G_1}^{\mathrm{E}} = \overline{\Delta H_1}^{\mathrm{E}} - T\overline{\Delta S_1}^{\mathrm{E}} \tag{10.53}$$

it is easy to show that

$$\kappa - \psi = \chi - \frac{1}{2} \tag{10.54}$$

and that theta conditions occur at a particular temperature θ, known as the *theta* (or *Flory*) *temperature*, for which $(\mu_1 - \mu_1^{\mathrm{o}})^{\mathrm{E}} = 0$. Thus

$$\overline{\Delta H_1}^{\mathrm{E}} = \theta\overline{\Delta S_1}^{\mathrm{E}} \tag{10.55}$$

This equation suggests that the enthalpic interactions produce a proportionate change in the excess entropy, and reveals that $\overline{\Delta H_1}^{\mathrm{E}}$ and $\overline{\Delta S_1}^{\mathrm{E}}$ (or alternatively κ and ψ) must have the same sign (i.e. either both positive or both negative). Equation 10.55 also provides an alternative definition of the theta temperature, as being the proportionality constant relating $\overline{\Delta H_1}^{\mathrm{E}}$ to $\overline{\Delta S_1}^{\mathrm{E}}$. On this basis, the origin of the variation of θ from one polymer–solvent system to another clearly is evident.

Substituting Equations 10.51 and 10.52 into 10.55 gives

$$\kappa = \left(\frac{\theta}{T}\right)\psi \tag{10.56}$$

which when combined with Equations 10.50 and 10.54 leads to

$$(\mu_1 - \mu_1^{\mathrm{o}})^{\mathrm{E}} = \mathbf{R}T\psi\left[\left(\frac{\theta}{T}\right) - 1\right]\phi_2^2 \tag{10.57}$$

Equations 10.50 and 10.57 reveal that $(\mu_1 - \mu_1^0)^{\mathrm{E}}$ promotes mixing (i.e. is negative) when $\chi < 1/2$ which corresponds to $T > \theta$ for positive ψ and $T < \theta$ for negative ψ. Furthermore, they show that under theta conditions $\chi = 1/2$, $T = \theta$ and the dilute polymer solution behaves as if it were an ideal solution. Some values of θ are given in Table 10.1 for a series of different polymer–solvent systems.

TABLE 10.1
Values of Theta Temperature for Different Polymer–Solvent Systems[a]

Polymer	Solvent	θ / °C
Polyethylene	Biphenyl	125
Polystyrene	Cyclohexane	34
Poly(vinyl acetate)	Methanol	6
Poly(methyl methacrylate)	Butyl acetate	−20
Poly(methyl methacrylate)	Pentyl acetate	41
Poly(vinyl alcohol)	Water	97
Poly(acrylic acid)	1,4-Dioxan	29

[a] The entropy parameter ψ is positive for all of the systems except poly(vinyl alcohol)/water and poly(acrylic acid)/1,4-dioxan for which it is negative.

The theory presented above for dilute polymer solutions is based upon the Flory–Huggins Equation 10.22, which strictly is not valid for such solutions because of the mean-field approximation. Nevertheless, whilst Equations 10.50 and 10.57 do not predict $(\mu_1 - \mu_1^{\text{o}})^{\text{E}}$ accurately, they are of the correct functional form, i.e. the relationships

$$(\mu_1 - \mu_1^{\text{o}})^{\text{E}} \propto \left(\chi - \frac{1}{2}\right)$$

and

$$(\mu_1 - \mu_1^{\text{o}})^{\text{E}} \propto \psi\left[\left(\frac{\theta}{T}\right) - 1\right]$$

also are obtained from the more satisfactory excluded-volume theories (Section 10.3.3). Thus, the Flory–Huggins theory leads to correct conclusions about the general thermodynamic behaviour of dilute polymer solutions, especially with regard to the prediction of ideal behaviour when $\chi = 1/2$ and $T = \theta$. The accuracy of the Flory–Huggins dilute polymer solution theory improves as theta conditions are approached and so the predictions of $(\mu_1 - \mu_1^{\text{o}})^{\text{E}}$ afforded by Equations 10.50 and 10.57 become better as $\chi \rightarrow 1/2$ and $T \rightarrow \theta$.

10.2.5 The Solubility Parameter Approach

In many practical situations, all that is required is a reasonable guide to the compatibility (i.e. miscibility) of specific polymer–solvent systems. Examples include the selection of a solvent for dissolving a particular polymer, the specification of an elastomer which will not swell (i.e. absorb liquid) in applications known to involve contact with certain liquids, and the prediction of possible environmental crazing of a polymer during service (Section 23.4.4). The solubility parameter approach is most useful for this purpose and was first developed by Hildebrand for calculating estimates of the enthalpy of mixing $\Delta H_{\text{m}}^{\text{contact}}$ for mixtures of liquids. The equation employed is

$$\Delta H_{\text{m}}^{\text{contact}} = V_{\text{m}}\phi_1\phi_2(\delta_1 - \delta_2)^2 \tag{10.58}$$

where

V_{m} is the molar volume of the mixture
δ_1 and δ_2 are the solubility parameters of components 1 and 2, respectively.

The *solubility parameter* δ of a liquid is the square root of the energy of vaporization per unit volume and is given by

$$\delta = \left[\frac{(\Delta H_v - \mathbf{R}T)}{V}\right]^{1/2} \tag{10.59}$$

where
ΔH_v is its molar enthalpy of vaporization
V is its molar volume.

The quantity δ^2 is called the *cohesive energy density* (CED) since it characterizes the strength of attraction between the molecules in unit volume. For a volatile liquid, CED, and hence δ, can be determined experimentally by measuring ΔH_v and V.

Equation 10.58 yields only zero or positive values for ΔH_m and predicts that mixing becomes more favourable (i.e. ΔH_m becomes less positive) as the difference between the solubility parameters of the two components decreases, with $\Delta H_m = 0$ when $\delta_1 = \delta_2$. Specific effects, such as hydrogen bonding and charge transfer interactions, can lead to negative ΔH_m but are not taken into account by Equation 10.58 and so a separate qualitative judgement must be made to predict their effect upon miscibility.

Extension of the solubility parameter approach to the prediction of polymer–solvent and polymer–polymer miscibility requires knowledge of δ values for polymers. Because polymers are not volatile, their δ values must be obtained indirectly. Most commonly the δ value is taken as being that of the solvent which gives the maximum degree of swelling for network polymers and the maximum intrinsic viscosity (Section 10.4.2) for soluble polymers. Theoretical estimates of δ can be calculated using *additivity methods*, which are based on the premise that the properties of a substance are determined by a simple summation of contributions from individual parts of its chemical structure

$$\delta = \frac{\rho_p \sum_i f_i F_i}{M_0} \tag{10.60}$$

where
ρ_p is the density of the polymer
f_i is the number of groups of type i present in the repeat unit
F_i is the group molar attraction constant for group i
M_0 is the repeat unit molar mass.

Best-fit values of F_i for small units of chemical structure have been calculated and refined from measured values of δ for a large number of different chemicals and polymers (initially using data from measured ΔH_v and V for small molecules). Tables of F_i values were published first by Small and then refined by Hoy, but the concept of using additivity for prediction of polymer properties has been extended greatly since then and is now documented most thoroughly by van Krevelen (see Further Reading).

Prediction of δ for poly(methyl methacrylate) (PMMA) will be considered here to illustrate the principles. The repeat unit of PMMA is

$$\left[CH_2 - C(CH_3)(C(=O)-O-CH_3) \right]_n$$

Taking into account the numbers of each type of group within the structure and using values of F_i published by Hoy:

Group	f_i	F_i / (J cm^3)$^{1/2}$ mol^{-1}	f_iF_i / (J cm^3)$^{1/2}$ mol^{-1}
$-CH_2-$	1	268	268
$>C<$	1	65	65
$-CH_3$	2	303	606
$-C(=O)-O-$	1	669	669

Thus, the total molar attraction for the PMMA repeat unit is given by

$$\sum_i f_iF_i = 268 + 65 + 606 + 669 = 1608 \text{ (J cm}^3)^{1/2} \text{ mol}^{-1}$$

For PMMA, $\rho_p = 1.18$ g cm^{-3} and $M_0 = 100.12$ g mol^{-1}. Hence, using Equation 10.60, the additivity calculation predicts that $\delta_{PMMA} = (1.18 \text{ g cm}^{-3} \times 1608 \text{ (J cm}^3)^{1/2} \text{ mol}^{-1})/100.12 \text{ g mol}^{-1} = 19.0 \text{ J}^{1/2} \text{ cm}^{-3/2}$.

When carrying out such calculations, it is important to use F_i values from a single source because they are self-consistent. For example, the sets of F_i values published by Small and Hoy are quite different, but usually give similar predictions of δ.

Some values of δ for common solvents and polymers are given in Table 10.2. Note that the value of δ_{PMMA} predicted from additivity using Hoy's F_i values is good. This is not always the case and simple additivity methods are inadequate when there are significant contributions from strong intermolecular interactions (e.g. from hydrogen bonding). For such polymers, the more advanced methods documented by van Krevelen are necessary to obtain reasonable estimates of δ.

Inspection of the data in Table 10.2 shows that solubility parameter differences correctly indicate that, with the exception of methanol and water (for which hydrogen bonding is a major factor), polystyrene can be dissolved in each of the solvents listed. In general, however, the solubility parameter approach must be treated as a guide to miscibility which is most accurate for mixtures of non-polar amorphous polymers with non-polar solvents. The limitations of the theory already have been mentioned with respect to contributions from hydrogen-bonding interactions and other specific effects. Additionally, since the theory is based upon liquid–liquid mixtures, it does not take into account the need to overcome the enthalpy of crystallization when dissolving crystalline polymers. The latter contribution is the reason why crystalline polymers generally are less soluble than amorphous

TABLE 10.2
Solubility Parameters for Common Solvents and Polymers

Solvent	δ / J$^{1/2}$ cm$^{-3/2}$	Polymer	δ / J$^{1/2}$ cm$^{-3/2}$
Acetone	20.3	Polyethylene	16.4
Carbon tetrachloride	17.6	Polystyrene	18.5
Chloroform	19.0	Poly(methyl methacrylate)	19.0
Cyclohexane	16.8	Polypropylene	17.2
Methanol	29.7	Poly(vinyl chloride)	20.0
Toluene	18.2		
Water	47.9		
Xylene	18.0		

polymers and, more specifically, why polyethylene dissolves in toluene and xylene *only* at high temperatures (i.e. above about 80° C).

By combining Equation 10.58 with Equations 10.29 and 10.31 and recognizing that Equation 10.58 provides only the enthalpic contribution to $\Delta G_{\mathrm{m}}^{\mathrm{contact}}$, it is possible to obtain the following expression for χ_H in terms of solubility parameters:

$$\chi_H = \frac{V_1(\delta_1 - \delta_2)^2}{\mathbf{R}T} \tag{10.61}$$

where V_1 is the molar volume of the solvent (because, based on the Flory–Huggins lattice model, $V_1 = V_{\mathrm{m}}/(n_1 + xn_2)$). Together with χ_S, which typically is about 0.2, Equation 10.61 allows estimates of χ to be calculated, though obviously the estimates are subject to the limitations of solubility parameter theory.

10.3 CHAIN DIMENSIONS

The subject of *chain dimensions* is concerned with relating the sizes and shapes of individual polymer molecules to their chemical structure, chain length and molecular environment. In the analysis of polymer solution thermodynamics given in the previous sections, it was not necessary to consider in detail chain dimensions. However, in order to interpret fully the properties of dilute polymer solutions, it is crucial to consider the behaviour of isolated polymer molecules, in particular their chain dimensions. In this and subsequent sections, some of the more elementary aspects of chain dimensions will be introduced.

The shape of a polymer molecule is to a large extent determined by the effects of its chemical structure upon chain stiffness. Since most polymer molecules have relatively flexible backbones, they tend to be highly coiled and can be represented as random coils. However, as the backbone becomes stiffer, the chains begin to adopt a more elongated worm-like shape and ultimately become rod-like. The theories which follow are concerned only with the chain dimensions of linear flexible polymer molecules. More advanced texts should be consulted for treatments of worm-like and rod-like chains.

10.3.1 FREELY-JOINTED CHAINS

The simplest measure of chain dimensions is the length of the chain along its backbone and is known as the *contour length*. For a chain of n backbone bonds each of length l, the contour length is nl. However, for linear flexible chains it is more usual, and more realistic, to consider the dimensions of the molecular coil in terms of the distance separating the chain ends, i.e. the *end-to-end distance* r (Figure 10.2).

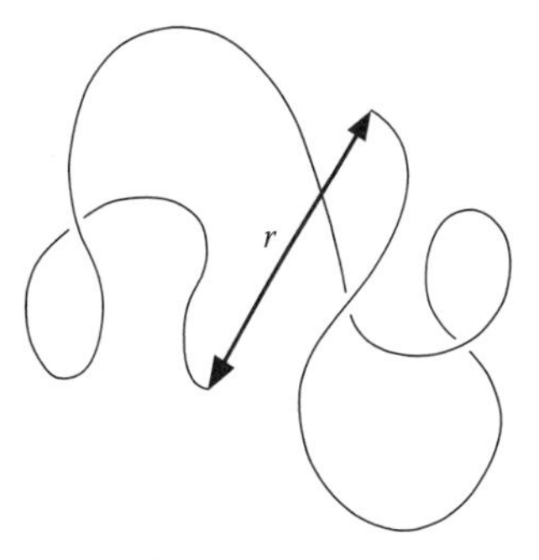

FIGURE 10.2 Schematic representation of a coiled polymer molecule showing the end-to-end distance r.

When considering an isolated polymer molecule, it is not possible to assign a unique value of r because the chain conformation (and hence r) is continuously changing due to rotation of backbone bonds. However, whilst each conformation has a characteristic value of r, certain different conformations give rise to the same value of r. Thus, some values of r are more probable than others and the probability distribution of r can be represented by the *root mean square (RMS) end-to-end distance* $\langle r^2 \rangle^{1/2}$, where $\langle\ \rangle$ indicates that the quantity is averaged over time. The techniques of statistical mechanics can be used to calculate $\langle r^2 \rangle^{1/2}$ but in order to do so a model for the polymer molecule must be assumed. The simplest model is that of a *freely-jointed chain* of n links (i.e. backbone bonds) of length l for which there are no restrictions upon either bond angle or bond rotation. The analysis of this model is a simple extension of the well-known *random-walk* calculation, which was first developed to describe the movement of molecules in an ideal gas. The only difference is that for the freely-jointed chain, each step is of equal length l. A three-dimensional rectangular co-ordinate system is used, as shown in Figure 10.3. One end of the chain is fixed at the origin O and the probability $P(x,y,z)$ of finding the other end within a small volume element $\mathrm{d}x\mathrm{d}y\mathrm{d}z$ at a particular point with co-ordinates (x,y,z) is calculated. The calculation leads to an equation of the form

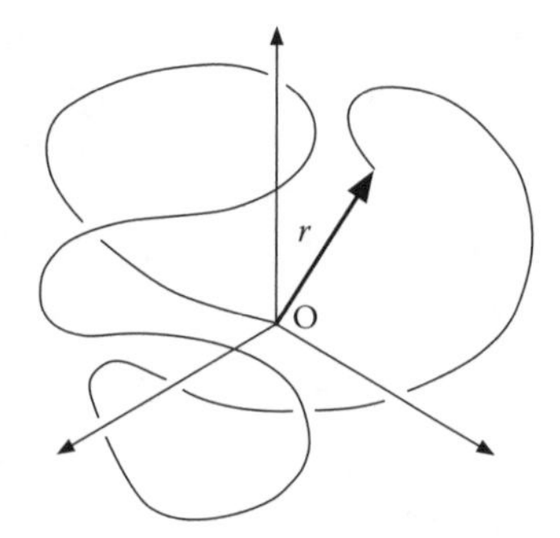

FIGURE 10.3 Diagram showing a coiled polymer molecule of end-to-end distance r in a rectangular co-ordinate system with one chain end fixed at the origin O.

$$P(x,y,z) = W(x,y,z)\mathrm{d}x\mathrm{d}y\mathrm{d}z \tag{10.62}$$

where $W(x,y,z)$ is a *probability density function*, i.e. a probability per unit volume. When $r \ll nl$ and n is large, $W(x,y,z)$ is given by

$$W(x,y,z) = \left(\frac{\beta}{\pi^{1/2}}\right)^3 \exp\left[-\beta^2(x^2+y^2+z^2)\right] \tag{10.63}$$

where $\beta = [3/(2nl^2)]^{1/2}$. Since r^2 is equal to $(x^2+y^2+z^2)$, Equation 10.63 can be simplified to

$$W(x,y,z) = \left(\frac{\beta}{\pi^{1/2}}\right)^3 \exp\left[(-\beta^2 r^2)\right] \tag{10.64}$$

This equation reveals that $W(x,y,z)$ is a *Gaussian* distribution function which has a maximum value at $r=0$, as is shown by the plot of $W(x,y,z)$ given in Figure 10.4a. The co-ordinates (x,y,z) specified by $W(x,y,z)$ define a particular direction from the origin and so correspond only to one of many such co-ordinates, each of which gives rise to an end-to-end distance equal to r but in a different direction. Thus, a more important probability is that of finding one chain end at a distance r in any

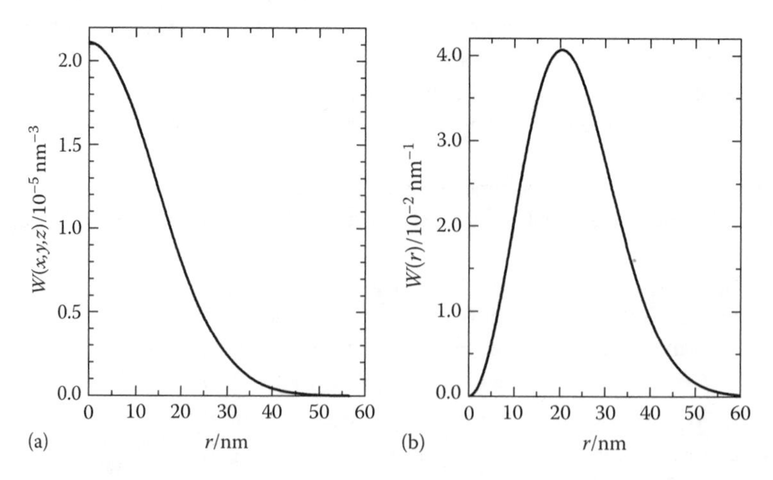

FIGURE 10.4 Plots of (a) the probability density function $W(x,y,z)$ and (b) the radial distribution function $W(r)$ according to Equations 10.64 and 10.66, respectively, for a chain of 10^4 links, each of length 0.25 nm.

direction from the other chain end located at the origin. This is equal to the probability $W(r)\mathrm{d}r$ of finding the chain end in a spherical shell of thickness $\mathrm{d}r$ positioned a radial distance r from the origin. The volume of the shell is $4\pi r^2\mathrm{d}r$ and so

$$W(r)\mathrm{d}r = W(x,y,z)\times 4\pi r^2\mathrm{d}r \tag{10.65}$$

which upon substituting Equation 10.64 for $W(x,y,z)$ leads to the following expression for the *radial distribution function* $W(r)$

$$W(r) = 4\pi\left(\frac{\beta}{\pi^{1/2}}\right)^3 r^2 \exp(-\beta^2 r^2) \tag{10.66}$$

This function is plotted in Figure 10.4b and can be shown by simple differentiation to have a maximum value at $r = 1/\beta$. It also is simple to show that $W(r)$ normalizes to unity, i.e.

$$\int_0^\infty W(r)\mathrm{d}r = 1 \tag{10.67}$$

though this highlights a deficiency of the theory since $W(r)$ is finite when $r > nl$, a condition which physically is not possible (i.e. the end-to-end distance cannot exceed the contour length). However, since $W(r)$ is small for $r > nl$, the errors introduced are negligible.

The mean square end-to-end distance $\langle r^2\rangle$ is the second moment of the radial distribution function and so is defined by the integral

$$\langle r^2\rangle = \int_0^\infty r^2 W(r)\mathrm{d}r \tag{10.68}$$

Combining Equations 10.66 and 10.68 and integrating gives the result

$$\langle r^2\rangle = \frac{3}{2\beta^2} \tag{10.69}$$

Since $\beta^2 = 3/(2nl^2)$, the RMS end-to-end distance is given by

$$\langle r^2\rangle_\mathrm{f}^{1/2} = n^{1/2}l \tag{10.70}$$

where the subscript f indicates that the result is for a freely-jointed chain. This rather simple equation is very important and reveals that $\langle r^2\rangle_\mathrm{f}^{1/2}$ is a factor of $n^{1/2}$ smaller than the contour length. Since n is large, this reveals the highly coiled nature of flexible Gaussian polymer chains. For example, a freely-jointed chain of 10,000 segments, each of length 0.3 nm, will have a fully-extended contour length of 3000 nm whereas its RMS end-to-end distance will be only 30 nm.

In addition to the RMS end-to-end distance, the dimensions of linear chains often are characterized in terms of the RMS distance of a chain segment from the centre of mass of the molecule, which is known simply as the RMS *radius of gyration*, $\langle s^2\rangle^{1/2}$. This quantity has the advantage that it also can be used to characterize the dimensions of branched macromolecules (which have more than two chain ends) and cyclic macromolecules (which have no chain ends). Furthermore, properties of dilute polymer solutions that are dependent upon chain dimensions are controlled by $\langle s^2 \rangle^{1/2}$ rather than $\langle r^2 \rangle^{1/2}$. However, for linear Gaussian chains, $\langle s^2 \rangle^{1/2}$ is related to $\langle r^2 \rangle^{1/2}$ by

$$\langle r^2 \rangle^{1/2} = \frac{\langle s^2 \rangle^{1/2}}{6^{1/2}} \tag{10.71}$$

and so in the theoretical treatment of linear flexible chains, it is usual to consider only $\langle r^2 \rangle^{1/2}$.

10.3.2 Effects of Bond Angle and Short-Range Steric Restrictions

Whilst the freely-jointed chain is a simple model from which to begin prediction of chain dimensions, it is physically unrealistic. The links in real polymer chains are subject to bond angle restrictions and do not rotate freely because of short-range steric interactions. Both of these effects cause the actual $\langle r^2 \rangle^{1/2}$ to be larger than $\langle r^2 \rangle_f^{1/2}$ and are most easily demonstrated by considering the conformations of simple alkanes.

The possible conformations of a molecule of *n*-butane are illustrated schematically in Figure 10.5. Each carbon atom in the molecule is sp^3-hybridized and so is tetrahedral with bond angles of 109.5°. The different conformations result from rotation about the C_2–C_3 bond and have stabilities which depend upon the steric interactions between the methyl groups and hydrogen atoms bonded to C_2 and C_3. The methyl groups are relatively bulky and so the *planar cis* conformation, corresponding to their distance of closest approach, is the least stable. Accordingly, the *planar trans* conformation is the most stable. The potential energy of *n*-butane is given in Figure 10.6 as a function of the angle ϕ through which the C_3–C_4 bond is rotated from the *planar trans* conformation about the plane made by the C_1–C_2–C_3 bonds ($\phi=0$ for *planar trans* and $\phi=180°$ for *planar cis*). The minima correspond to the staggered conformations (*planar trans* and *gauche*$^{\pm}$) in which there is maximum separation of the substituents on C_2 and C_3, i.e. the methyl groups and hydrogen atoms are at 60° to each other. These conformations are shown as Newman projections in Figure 10.7 together with the eclipsed conformations which give rise to the maxima in Figure 10.6. Thus short-range steric interactions are important in determining the relative probabilities of existence of different conformations of a molecule.

The restrictions imposed by short-range steric interactions upon the conformations of a polymer molecule occur at a local level in short sequences of chain segments. The interactions are more complex than in *n*-butane because the conformation about a given chain segment is dependent upon the conformations about the segments to which it is directly connected, i.e. there is an *interdependence* of local chain conformations. This is most easily appreciated by considering the conformations of *n*-pentane in terms of sequences of *planar trans* (*t*) and *gauche*$^{\pm}$ ($g^{\pm}$) conformations. If the

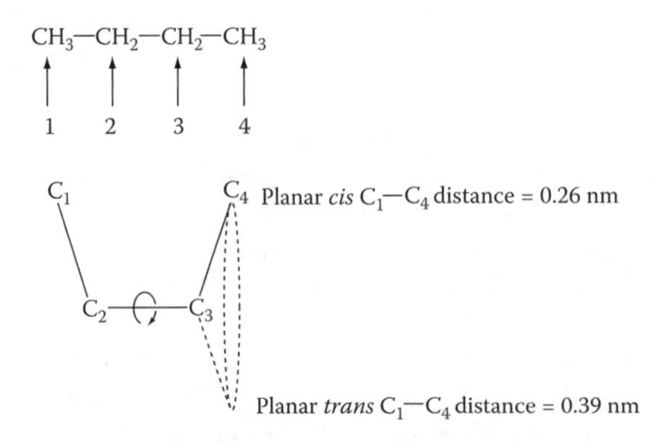

FIGURE 10.5 Effects of bond rotation upon the conformation of *n*-butane.

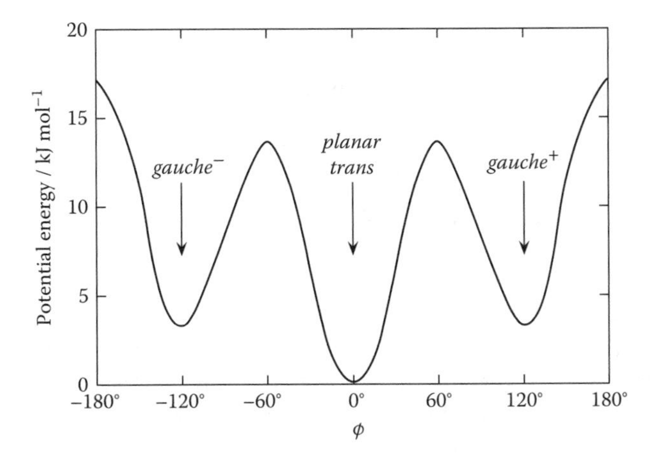

FIGURE 10.6 Potential energy of n-butane as a function of the angle ϕ of bond rotation.

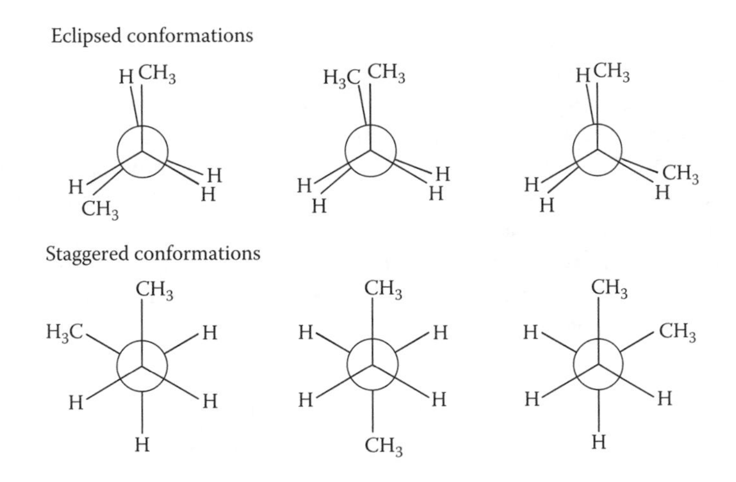

FIGURE 10.7 Newman projections of the eclipsed and staggered conformations of n-butane.

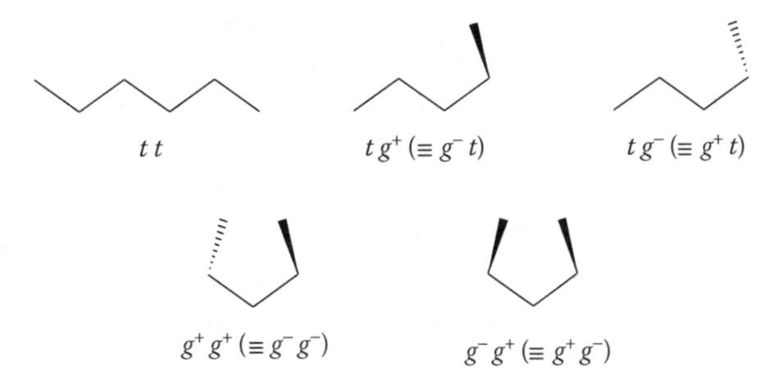

FIGURE 10.8 The five distinct conformations of n-pentane. The solid bonds protrude out from, and the dashed bonds extend behind, the plane of the paper.

three central carbon atoms in n-pentane are fixed in the plane of the paper, then the five distinct conformations of the molecule can be represented in skeletal form, as shown in Figure 10.8. The g^-g^+ and g^+g^- conformations bring the methyl groups into close proximity (with their carbon atoms about 0.25 nm apart) and so are of high energy and low probability in comparison to the other conformations. Such interdependent steric restrictions affect the local chain conformation all along a polymer chain and have a significant effect upon chain dimensions.

The simplest modification to the freely-jointed chain model is the introduction of bond angle restrictions whilst still allowing free rotation about the bonds. This is known as the *valence angle model* and for a polymer chain with backbone bond angles all equal to θ, leads to the following equation for the mean square end-to-end distance

$$\langle r^2 \rangle_{\text{fa}} = nl^2 \left(\frac{1-\cos\theta}{1+\cos\theta} \right) \tag{10.72}$$

where the subscript fa indicates that the result is for chains in which the bonds rotate freely about a fixed bond angle. Since $180° > \theta > 90°$, $\cos\theta$ is negative and $\langle r^2 \rangle_{\text{fa}}^{1/2}$ is greater than $\langle r^2 \rangle_{\text{f}}^{1/2}$. Equation 10.72 is directly applicable to polymers derived from ethylenic monomers because they have C–C backbone bonds with $\theta \approx 109.5°$ for which $\cos\theta \approx -1/3$ and the equation becomes

$$\langle r^2 \rangle_{\text{fa}} = 2nl^2 \tag{10.73}$$

Thus, for such polymers (e.g. polyethylene) bond angle restrictions cause the RMS end-to-end distance to increase by a factor of $2^{1/2}$ (i.e. ~1.4) from that of the freely-jointed chain.

The restrictions upon bond rotation arising from short-range steric interactions are more difficult to quantify theoretically. The usual procedure is to assume that the conformations of each sequence of three backbone bonds are restricted to discrete *rotational isomeric states* corresponding to potential energy minima such as those shown for *n*-butane in Figure 10.6. The effects of interdependent steric interactions of the backbone atoms also can be taken into account in this way (e.g. by excluding g^-g^+ and g^+g^- sequences). For the simplest case of polyethylene, rotational isomeric state theory leads to the following equation

$$\langle r^2 \rangle_0 = nl^2 \left(\frac{1-\cos\theta}{1+\cos\theta} \right) \left(\frac{1+\overline{\cos\phi}}{1-\overline{\cos\phi}} \right) \tag{10.74}$$

where the subscript 0 indicates that the result is for a polymer chain with hindered rotation about a fixed bond angle. The quantity $\overline{\cos\phi}$ is the average value of $\cos\phi$ where ϕ is the angle of bond rotation as defined for *n*-butane (Figures 10.6 and 10.7). When bond rotation is unrestricted, $\overline{\cos\phi}=0$ because all values of ϕ are equally probable causing the positive and negative $\cos\phi$ contributions to cancel each other out and resulting in Equations 10.74 and 10.72 becoming identical. However, due to short-range steric restrictions, values of $|\phi|<90°$ are favoured so that $\overline{\cos\phi}$ is positive and $\langle r^2 \rangle_0^{1/2}$ is greater than $\langle r^2 \rangle_{\text{fa}}^{1/2}$. The presence of side groups on the polymer chain (e.g. phenyl groups for polystyrene) tends to further hinder bond rotation and introduces additional interdependent steric interactions involving the side groups. These effects are very difficult to treat theoretically and so Equation 10.74 usually is written in the more general form

$$\langle r^2 \rangle_0 = \sigma^2 nl^2 \left(\frac{1-\cos\theta}{1+\cos\theta} \right) \tag{10.75}$$

where σ is a *steric parameter* the value of which normally exceeds that of the square root of the $(1+\overline{\cos\phi})/(1-\overline{\cos\phi})$ term in Equation 10.74, and is the factor by which $\langle r^2 \rangle_0^{1/2}$ exceeds $\langle r^2 \rangle_{\text{fa}}^{1/2}$. Since values of σ are rather difficult to calculate, they usually are evaluated from values of $\langle r^2 \rangle_0^{1/2}$ measured experimentally.

TABLE 10.3
Typical Values of σ and C_∞ for Some Common Polymers

Polymer	Temperature / °C	σ	C_∞
Polyethylene	140	1.8	6.8
Isotactic polypropylene	140	1.6	5.2
Poly(vinyl chloride)	25	1.8	6.7
Polystyrene	25	2.3	10.8
Polystyrene	70	2.1	9.2
Poly(methyl methacrylate)	25	2.1	8.6
Poly(methyl methacrylate)	72	1.8	6.6

An idea of the stiffness of a polymer chain can be gained from the ratio $\langle r^2\rangle_0^{1/2}/\langle r^2\rangle_f^{1/2}$, which is the square-root of the *characteristic ratio* C_∞ which is defined as follows

$$C_\infty = \frac{\langle r^2\rangle_0}{nl^2} \tag{10.76}$$

Values of σ and C_∞ for some common polymers are given in Table 10.3. It is found that σ typically has values between 1.5 and 2.5, and that the combination of fixed bond angles and short-range steric interactions causes the RMS end-to-end distances of real polymer chains to be greater than those of freely-jointed chains by factors of 2–3.

It is possible to represent a real polymer chain by an *equivalent freely-jointed chain* of N links each of length b such that the chains have the same contour length (i.e. $nl = Nb$) and the same RMS end-to-end distance (i.e. $\langle r^2\rangle_0^{1/2} = Nb^2$). Thus, in terms of the characteristic ratio

$$N = \frac{n}{C_\infty} \tag{10.77}$$

and

$$b = C_\infty l \tag{10.78}$$

The length b is more commonly known as the *Kuhn segment length* and clearly increases with chain stiffness; it often is used as the basis for defining a chain segment. The *persistence length* is a more a complex quantity which also depends upon chain stiffness; it is defined in words as the average projection of the end-to-end vector on a tangent to the chain contour at a chain end in the limit of infinite chain length, but in more simple terms may be considered as the length from a point on a chain over which correlations in the chain direction with a tangent to that point are lost.

10.3.3 Effects of Long-Range Steric Interactions: Chains with Excluded Volume

The mathematical model for evaluating $\langle r^2\rangle_0^{1/2}$ is that of a series of connected vectors (representing the backbone bonds) which are restricted at a local level to certain allowed orientations relative to each other. The bond vectors are volumeless lines in space and the model places no restrictions upon the relative positions of two bond vectors widely separated in the chain. Thus, the model does not prevent remotely-connected bond vectors occupying the same volume in space, and allows self-intersections of the chain. In a real isolated polymer molecule, each part of the molecule excludes other more remotely-connected parts from its volume. These long-range steric interactions cause the actual RMS end-to-end distance $\langle r^2 \rangle^{1/2}$ to be greater than $\langle r^2 \rangle_0^{1/2}$ and usually are considered in terms of an *excluded volume*. Chain dimensions that correspond to $\langle r^2 \rangle_0^{1/2}$ are unperturbed by the effects of volume exclusion and so are called the *unperturbed dimensions*. The extent to which these dimensions are perturbed in real chains is defined by an *expansion parameter* α_r such that

$$\langle r^2 \rangle^{1/2} = \alpha_r \langle r^2 \rangle_0^{1/2} \tag{10.79}$$

However, expansion of the molecular coil due to volume exclusion is not uniform and is greatest where the segment density is highest. Thus, a different expansion parameter α_s is required to account for expansion from the unperturbed RMS radius of gyration

$$\langle s^2 \rangle^{1/2} = \alpha_s \langle s^2 \rangle_0^{1/2} \tag{10.80}$$

Theoretical calculations give the expansion parameters as power series in the *excluded volume parameter* z which is given by

$$z = \left(\frac{3}{2\pi \langle r^2 \rangle_0} \right)^{3/2} N^2 \beta_e \tag{10.81}$$

where

N is defined by Equation 10.77

β_e is called the *excluded volume integral* and is the volume excluded by one chain segment to another.

For small expansions (i.e. small z), the following relationships are obtained

$$\alpha_r^2 = 1 + \left(\frac{4}{3} \right) z \tag{10.82}$$

$$\alpha_s^2 = 1 + \left(\frac{134}{105} \right) z \tag{10.83}$$

and so

$$\alpha_r > \alpha_s$$

Approximate theories relating expansion parameters α to z give equations of the general form

$$\alpha^5 - \alpha^3 = Kz \tag{10.84}$$

where K is a constant. For large expansions, $\alpha^5 \gg \alpha^3$ and so $\alpha^5 \approx Kz$. Under these conditions, α is dependent upon $n^{1/10}$ because z is proportional to $N^{1/2}$, which in turn is proportional to $n^{1/2}$. Since the unperturbed dimensions are dependent upon $n^{1/2}$, it follows (from Equations 10.79 and 10.80) that the perturbed dimensions of a highly-expanded chain are proportional to $n^{3/5}$.

The expansion parameters for a real chain also embody the effects of interactions of the chain with its molecular environment (e.g. with solvent molecules or other polymer molecules). The simplest case is that of a *pure amorphous polymer.* Each polymer molecule is surrounded by other polymer molecules of the same type. Expansion of a given chain to relieve long-range *intramolecular* steric interactions only serves to create an equal number of *intermolecular* steric interactions with neighbouring chains. These opposing volume exclusion effects exactly counteract each other and so in a pure amorphous polymer, the polymer molecules adopt their unperturbed dimensions (i.e. $\alpha_r = \alpha_s = 1$), as has been proven by neutron scattering experiments (Section 12.4). Since the net energy of interaction is zero, the individual molecular coils interpenetrate each other and become highly entangled.

In *dilute solution*, polymer molecules do not interact with each other and chain expansion depends upon the balance between the intramolecular segment–segment and intermolecular segment–solvent and solvent–solvent interactions. When a polymer is dissolved in a *good solvent*, the isolated polymer molecules expand from their unperturbed dimensions so as to increase the number of segment–solvent contacts (i.e. $\alpha > 1$). In a *poor solvent*, however, the segment–solvent interactions are weak and their Gibbs free energy of interaction may be positive (i.e. unfavourable); the polymer chains contract in order to reduce the number of segment–solvent contacts and so are relatively compact (i.e. values of α are close to unity). Thus, in poor solvents, isolated polymer chains are subjected to two opposing influences: (i) expansion due to unfavourable segment–segment interactions (i.e. volume exclusion), and (ii) contraction due to unfavourable segment–solvent interactions. Under specific conditions in a poor solvent, the energy of segment–solvent interaction is just sufficiently positive to exactly counteract the energy of segment–segment interaction and each polymer molecule assumes its unperturbed dimensions. These conditions correspond to the *theta conditions* introduced in Section 10.2.4 to describe the situation when a dilute polymer solution behaves ideally. Under these conditions, the solvent is said to be a *theta solvent* for the polymer and unperturbed chain dimensions can be measured. If the solvency conditions deteriorate further, the polymer molecules will 'precipitate' to form a separate phase with a very high polymer concentration ($\alpha \approx 1$) rather than undergo extensive contraction ($\alpha < 1$) in the dilute solution (see Section 14.2.2).

The link between chain dimensions and the thermodynamic behaviour of dilute polymer solutions is of great importance and has its origin in the excluded volume integral β_e. This arises because β_e is the volume excluded by one segment to any other segment, which therefore can be part of the same or a different molecule. Theories for β_e lead to equations of the form:

$$\beta_e = C_e\left[\left(\frac{\theta}{T}\right) - 1\right] \tag{10.85}$$

where

C_e is a constant
θ is the theta temperature.

Thus, $\beta_e \rightarrow 0$ as $T \rightarrow \theta$.

Theories of chain expansion use β_e to account for *intramolecular* segment–segment volume exclusion and relate expansion parameters to β_e through the excluded volume parameter z. These theories show that $\alpha_r \rightarrow 1$ and $\alpha_s \rightarrow 1$ as $\beta_e \rightarrow 0$.

In dilute polymer solutions, each polymer molecule excludes all others from its volume. Thus mean-field theories, such as Flory–Huggins theory, are inappropriate and more exact theories of dilute solutions calculate ΔG_m from the volume excluded by one polymer molecule to another, which in turn is calculated using β_e to account for *intermolecular* segment–segment volume exclusion. These theories show that $(\mu_1 - \mu_1^0)^E \rightarrow 0$ as $\beta_e \rightarrow 0$.

It may, therefore, be concluded that when $\beta_e = 0$ polymer molecules assume their unperturbed dimensions and dilute polymer solutions behave ideally.

10.4 FRICTIONAL PROPERTIES OF POLYMER MOLECULES IN DILUTE SOLUTION

Processes such as diffusion, sedimentation and viscous flow involve the motion of individual molecules. When a polymer molecule moves through a dilute solution, it undergoes frictional interactions with solvent molecules. The nature and effects of these frictional interactions depend upon the size and shape of the polymer molecule as modified by its thermodynamic interactions with solvent molecules. Thus, the chain dimensions can be evaluated from measurements of the frictional properties of a polymer molecule.

10.4.1 FRICTIONAL COEFFICIENTS OF POLYMER MOLECULES

Two extremes of the frictional behaviour of polymer molecules can be identified, namely, free-draining and non-draining. A polymer molecule is said to be *free-draining* when solvent molecules are able to flow past each segment of the chain, and *non-draining* when solvent molecules within the coiled polymer chain move with it. These two extremes of behaviour lead to different dependences of the frictional coefficient f_0 of an isolated polymer molecule upon chain length.

Free-draining polymer molecules are considered by dividing them into identical segments each of which has the same frictional coefficient ξ. Since solvent molecules permeate all regions of the polymer coil with equal ease (or difficulty), each segment makes the same contribution to f_0 which therefore is given by

$$f_0 = x\xi \tag{10.86}$$

where x is the number of segments in the chain.

A *non-draining polymer molecule* can be represented by an equivalent impermeable hydrodynamic particle, i.e. one which has the same frictional coefficient as the polymer molecule. Thus, a non-draining random coil can be represented by an equivalent impermeable hydrodynamic sphere of radius R_h. From Stokes' law

$$f_0 = 6\pi\eta_0 R_h \tag{10.87}$$

where η_0 is the viscosity of the pure solvent. By making the reasonable assumption that R_h is proportional to $\langle s^2 \rangle^{1/2}$, then Equation 10.87 can be re-written in the form

$$f_0 = K_0 \alpha_\eta \langle s^2 \rangle^{1/2} \tag{10.88}$$

where

K_0 is a constant for a given system
α_η is the expansion parameter for the hydrodynamic chain dimensions.

Theory shows that for small chain expansions $\alpha_\eta = \alpha_s^{0.81}$, i.e. $\alpha_\eta < a_s$. Since $\langle s^2 \rangle_0^{1/2}$ is proportional to $x^{1/2}$ and for highly expanded coils α_η is approximately proportional to $x^{1/10}$ (from Equation 10.84), Equation 10.88 predicts that

$$f_0 = K_0' x^{a_0} \tag{10.89}$$

where K_0' is another constant and $0.5 < a_0 < 0.6$. This dependence of f_0 upon x can now be contrasted with that for the free-draining polymer molecule, for which f_0 is directly proportional to x.

The frictional behaviour of real polymer molecules comprises both free-draining and non-draining contributions. The free-draining contribution is dominant for very short chains and for highly elongated rod-like molecules, but for flexible (i.e. coiled) chains it decreases rapidly as the chain length increases. Since most polymers consist of long flexible chain molecules, their frictional properties closely approximate to those associated with polymer molecules in the non-draining limit. Therefore, in the following section, this limit will be considered further, specifically in relation to the viscosity of a dilute polymer solution.

10.4.2 Hydrodynamic Volume and Intrinsic Viscosity in the Non-Draining Limit

The viscosity of a suspension of rigid non-interacting spheres is given by the Einstein equation

$$\eta = \eta_0 \left[1 + \left(\frac{5}{2} \right) \phi_2 \right] \tag{10.90}$$

where

η and η_0 are the viscosities of the suspension and the suspension medium, respectively
ϕ_2 is the volume fraction of the spheres.

If the spheres are considered to be impermeable polymer coils of hydrodynamic volume V_h, then

$$\phi_2 = \left(\frac{c}{M} \right) \mathbf{N_A} V_h$$

where

c is the polymer concentration (mass per unit volume)
M is the molar mass of the polymer (mass per mole)
$\mathbf{N_A}$ is the Avogadro constant.

Substituting this equation into Equation 10.90 and simplifying gives

$$\eta_{sp} = \left(\frac{5}{2} \right) \left(\frac{c}{M} \right) \mathbf{N_A} V_h \tag{10.91}$$

where $\eta_{sp} = (\eta - \eta_0)/\eta_0$ and is known as the *specific viscosity*. Non-interaction of the polymer coils requires infinite dilution and this is achieved mathematically by defining a quantity called the *intrinsic viscosity* $[\eta]$ according to the equation

$$[\eta] = \lim_{c \to 0} \left(\frac{\eta_{sp}}{c} \right) \tag{10.92}$$

Thus a more satisfactory form of Equation 10.91 is

$$[\eta] = \left(\frac{5}{2} \right) \frac{\mathbf{N_A} V_h}{M} \tag{10.93}$$

which can be rearranged to give an equation for the hydrodynamic volume of an impermeable polymer molecule

$$V_h = \left(\frac{2}{5}\right)\frac{[\eta]M}{\mathbf{N_A}} \tag{10.94}$$

Thus, the quantity $[\eta]M$ is proportional to V_h and is often, *though incorrectly*, referred to as the hydrodynamic volume.

Assuming that V_h is proportional to $\left(\alpha_\eta \left\langle s^2 \right\rangle_0^{1/2}\right)^3$, then from Equation 10.93

$$[\eta] = \Phi_0^s \alpha_\eta^3 \left(\frac{\left\langle s^2 \right\rangle_0^{3/2}}{M}\right) \tag{10.95}$$

where Φ_0^s is a constant. For non-draining polymer molecules, Φ_0^s is independent of chain structure and chain length, and is dependent only upon the distribution of segments in the molecular coil. The value of Φ_0^s recommended for non-draining Gaussian polymer chains is 3.67×10^{24} mol^{-1} and is based upon theoretical calculations and experimental measurements.

Since $\left\langle s^2 \right\rangle_0$ is directly proportional to the number, x, of chain segments, it is also directly proportional to M. Thus, $\left\langle s^2 \right\rangle_0 / M$ is a constant and Equation 10.95 is more commonly written in the form of the *Flory–Fox Equation*

$$[\eta] = K_\theta \alpha_\eta^3 M^{1/2} \tag{10.96}$$

where

$$K_\theta = \Phi_0^s \left(\frac{\left\langle s^2 \right\rangle_0}{M}\right)^{3/2} \tag{10.97}$$

Furthermore, since α_η is approximately proportional to $x^{1/10}$, and hence to $M^{1/10}$, for highly-expanded flexible chains, the Flory–Fox Equation 10.96 suggests a general relationship of the form

$$[\eta] = KM^a \tag{10.98}$$

where

K and a are constants for a given system, and

a increases in the range $0.5 < a < 0.8$ with the degree of expansion of the molecular coils from their unperturbed dimensions (where $a = 0.5$, i.e. under theta conditions).

This important relationship between $[\eta]$ and M is called the *Mark–Houwink–Sakurada Equation* (or, more commonly, but less correctly, just the Mark–Houwink Equation) and was first proposed on the basis of experimental data.

10.4.3 Diffusion of Polymer Molecules in the Non-Draining Limit

The *translational diffusion coefficient* D of an isolated polymer molecule is related to f_0 by the Einstein equation

$$D = \frac{\mathbf{k}T}{f_0} \tag{10.99}$$

where

k is the Boltzmann constant

T is the temperature.

By combining this equation with Stokes' Equation 10.87, the following relationship is obtained

$$D = \frac{\mathbf{k}T}{6\pi\eta_0 R_\mathrm{h}} \tag{10.100}$$

where R_h is the hydrodynamic radius of the equivalent impermeable sphere. For such non-draining chains, an expression for R_h can be obtained from Equations 10.93 and 10.95

$$R_\mathrm{h} = \left[\left(\frac{3}{10\pi}\right)\left(\frac{\Phi_0^s}{\mathbf{N_A}}\right)\right]^{1/3}\left(\frac{\langle s^2\rangle_0}{M}\right)^{1/2}\alpha_\eta M^{0.5} \tag{10.101}$$

which upon substitution into Equation 10.100 gives

$$D = \left(\frac{\mathbf{k}T}{6\pi\eta_0}\right)\left[\left(\frac{10\pi\mathbf{N_A}}{3\Phi_0^s}\right)\right]^{1/3}\left(\frac{M}{\langle s^2\rangle_0}\right)^{1/2}\alpha_\eta^{-1}M^{-0.5} \tag{10.102}$$

Recognising that the front factor is constant for a given polymer-solvent system and taking into account the dependence of α_η upon M, the equation may be simplified to

$$D = K_D M^{-a_D} \tag{10.103}$$

which is the equivalent of the Mark–Houwink–Sakurada equation for diffusion. The parameters K_D and a_D are constants for a given system and (by analogy with the analysis for $[\eta]$ in the preceding section) the parameter a_D takes values between 0.5 under theta conditions and 0.6 for highly-expanded chains.

10.4.4 Solution Behaviour of Polyelectrolytes

Polyelectrolytes are an important class of polymers in which the chains have a very high proportion of repeat units that bear an ionic charge, a consequence of which is that they are soluble only in water and highly-polar organic solvents. The simplest examples are homopolymers in which an acidic or basic group in the repeat unit has become charged due to reaction, often a simple neutralization. For example, the pendent carboxylic acid groups in poly(acrylic acid) are converted to carboxylate ions on neutralization with a base such as sodium hydroxide to produce an *anionic polyelectrolyte*

$$\left[\mathrm{CH_2{-}CH(COOH)}\right]_n \xrightarrow{n\ \mathrm{NaOH}} \left[\mathrm{CH_2{-}CH(C{=}O)(O^-\ Na^+)}\right]_n$$

A rather different example is the formation of a *cationic polyelectrolyte* when poly(ethylene imine) is reacted with an alkyl halide such as methyl iodide

$$\left[\!-CH_2-CH_2-NH-\right]_n \xrightarrow{n\ CH_3I} \left[\!-CH_2-CH_2-\overset{\overset{I^-}{|}}{\underset{\underset{CH_3}{|}}{\overset{+}{N}H}}-\right]_n$$

Although complete ionization of the chains is indicated here, this is difficult to achieve in practice because as the charge on the polymer chain increases to high levels, it becomes difficult for further charges to be created in close proximity to each other along the chain. Nevertheless, such polyelectrolyte chains are very highly charged. A polyelectrolyte chain will be electrically neutral if all the counter-ions are contained within the coil. However, in solution, although the counter-ions are attracted electrostatically to the chain, there also is an osmotic pressure pulling them away to regions external to the chain where the ions are in low concentration. As a polyelectrolyte solution is diluted, the osmotic pressure grows and an increasing proportion of the counter-ions escape the coil, leaving it with a growing net charge. In these circumstances, the isolated ions on the chain repel each other, causing the chain to expand from its normal Gaussian coil dimensions. This is known as the *polyelectrolyte effect* and is illustrated schematically in Figure 10.9.

The effects of this chain expansion can be dramatic. For example, the normal effect of dilution for solutions of uncharged polymers is to cause the solution viscosity to reduce because there are fewer and fewer macromolecules to restrict the flow of solvent molecules, as is evident from Equation 10.91. In contrast, dilution of an aqueous polyelectrolyte solution leads to massive increases in the solution viscosity because as the solution becomes more dilute the counter-ions diffuse further away from the backbone, thus increasing the net charge on the chain and causing it to expand until ultimately it becomes rod-like at very high dilution (as depicted in Figure 10.9); the enormous increase in chain dimensions more than compensates for the reduction in the number chains in solution and leads to an increase in solution viscosity. This is demonstrated by the upper curve in Figure 10.10.

In the presence of a high concentration of free external ions (from potassium bromide for the lower curve in Figure 10.10), free ions are in high concentration at all concentrations of polyelectrolyte and provide electrostatic shielding of the polyelectrolyte chains, preventing the counter-ions from escaping the chains which, therefore, remain overall neutral. Under these conditions,

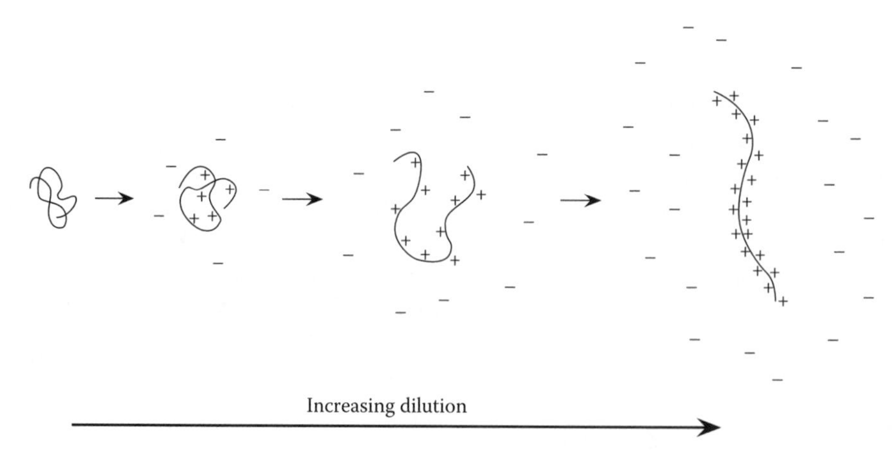

FIGURE 10.9 The *polyelectrolyte effect* in which coil expansion occurs on dilution of a polyelectrolyte chain (cationic in this example) due to diffusion of counter-ions away from the backbone leaving isolated charges on the chain which repel each other.

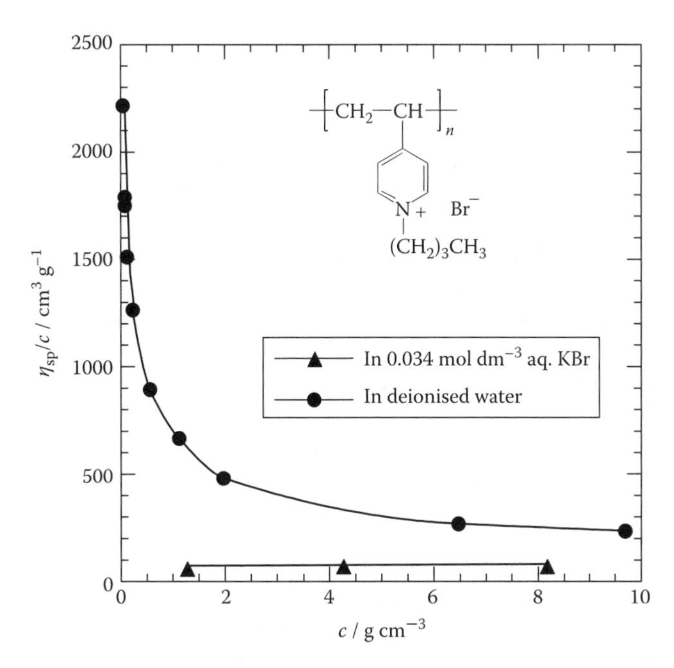

FIGURE 10.10 Variation of reduced viscosity η_{sp}/c with concentration c for aqueous solutions of poly(4-vinyl-*N*-butylpyridinium bromide) at 25 °C. (Data taken from Fuoss, R.M. and Strauss, U.P., *Ann. New York Acad. Sci.*, 51, 836, 1948.)

polyelectrolyte chains behave like uncharged chains, showing a reduction in solution viscosity as concentration decreases. For the lower curve in Figure 10.10, the variation of reduced viscosity with concentration is small, showing that 0.034 mol dm^{-3} aqueous potassium bromide is close to being a theta solvent for poly(4-vinyl-*N*-butylpyridinium bromide) at 25 °C.

Microgels (see Section 4.4.4.6) also can be swollen using the polyelectrolyte effect by, for example, neutralising carboxylic acid groups derived from inclusion of methacrylic (or acrylic) acid in the comonomers used to prepare the microgel. The diameter of a microgel can be increased severalfold by such effects.

PROBLEMS

10.1 From Flory–Huggins theory, the partial molar Gibbs free energy of mixing $\overline{\Delta G_1}$ with respect to the solvent for the formation of a polymer solution is given by

$$\overline{\Delta G_1} = \mu_1 - \mu_1^{\circ} = \mathbf{R}T\left[\ln\phi_1 + \left(1 - \frac{1}{\bar{x}_n}\right)\phi_2 + \chi\phi_2^2\right]$$

where

ϕ_2 is the volume fraction of polymer

$\bar{x}_n$ is the number-average ratio of the molar volume of the polymer to that of the solvent

χ is the Flory–Huggins polymer–solvent interaction parameter.

Use this equation to calculate an estimate of χ for solutions of natural rubber ($\bar{M}_n = 2.5 \times 10^5$ g mol^{-1}) in benzene given that vapour pressure measurements showed that the activity of the solvent in a solution with $\phi_2 = 0.250$ was 0.989. State any assumptions that you make.

10.2 The radial distribution function $W(r)$ of end-to-end distances r for an isolated flexible polymer chain is given by

$$W(r) = 4\pi\left(\frac{\beta}{\pi^{1/2}}\right)^3 r^2 \exp(-\beta^2 r^2)$$

where $\beta = [3/(2nl^2)]^{1/2}$ in which n is the number of links of length l forming the chain.

Show by integration that this distribution function normalizes to unity, and derive equations for (a) the most probable value of r, (b) the root-mean-square value of r and (c) the mean value of r.

Note that for $I(b) = \int_0^\infty x^b \exp(-ax^2)\mathrm{d}x$

$$I(0)=\left(\frac{1}{2}\right)\pi^{1/2}a^{-1/2};\; I(1)=\left(\frac{1}{2}\right)a^{-1};\; I(2)=\left(\frac{1}{4}\right)\pi^{1/2}a^{-3/2};\; I(3)=\left(\frac{1}{2}\right)a^{-2};\; I(4)=\left(\frac{3}{8}\right)\pi^{1/2}a^{-5/2}$$

10.3 For a linear molecule of polyethylene of molar mass 119,980 g mol^{-1}, calculate:

(a) the contour length of the molecule,
(b) the end-to-end distance in the fully-extended molecule, and
(c) the root-mean-square end-to-end distance according to the valence angle model.

In the calculations, end groups can be neglected and it may be assumed that the C–C bonds are of length 0.154 nm and that the valence angles are 109.5°. Comment upon the values obtained. Indicate, giving your reasoning, which of the very large number of possible conformations of the molecule is the most stable.

FURTHER READING

Allen, G. and Bevington, J.C. (eds), *Comprehensive Polymer Science*, Vol. 2 (Booth, C. and Price, C., eds), Pergamon Press, Oxford, U.K., 1989.
Flory, P.J., *Principles of Polymer Chemistry*, Cornell University Press, Ithaca, NY, 1953.
Hiemenz, P.C. and Lodge, T.P., *Polymer Chemistry*, 2nd edn., CRC Press, Boca Raton, FL, 2007.
Mark, H.F., Bikales, N.M., Overberger, C.G., and Menges, G. (eds), *Encyclopedia of Polymer Science and Engineering*, Wiley-Interscience, New York, 1985–1989.
Morawetz, H., *Macromolecules in Solution*, 2nd edn., Wiley-Interscience, New York, 1975.
Munk, P., *Introduction to Macromolecular Science*, Wiley-Interscience, New York, 1989.
Patterson, G., *Physical Chemistry of Macromolecules*, CRC Press, Boca Raton, FL, 2007.
Sun, S.F., *Physical Chemistry of Macromolecules: Basic Principles and Issues*, 2nd edn., Wiley-Interscience, New York, 2004.
Tager, A., *Physical Chemistry of Polymers*, 2nd edn., Mir Publishers, Moscow, Russia, 1978.
Tanford, C., *Physical Chemistry of Macromolecules*, John Wiley & Sons, New York, 1961.

11 Number-Average Molar Mass

11.1 INTRODUCTION TO MEASUREMENTS OF NUMBER-AVERAGE MOLAR MASS

The number-average molar mass $\bar{M}_n$ of a polymer is of great importance and there are several methods for its measurement. The common feature underlying each of these methods is that they count the number of polymer molecules in a given mass of polymer.

Colligative properties are those properties of a solution of a non-volatile solute which depend only upon the number of solute species present in unit volume of the solution and not upon the nature or size of those species. Thus, measurements of the colligative properties of dilute polymer solutions enable $\bar{M}_n$ to be determined for linear and branched homopolymers and copolymers with equal ease. The four important colligative effects are the osmotic pressure of a solution, the lowering of solvent vapour pressure, the elevation of solvent boiling point and the depression of solvent freezing point. These effects and their use in determining $\bar{M}_n$ will be described in Sections 11.2 through 11.4; but before doing so, it is instructive to compare the relative magnitudes of the effects, as indicated by the data in Table 11.1. Inspection of the data shows that osmotic pressure is the only colligative property which can be measured accurately for polymers of high molar mass.

For polymers with well-defined end groups, it also is possible to measure $\bar{M}_n$ by *end-group analysis*, i.e. by counting the number of chain ends. Examples of such measurements are described in Section 11.5.

11.2 MEMBRANE OSMOMETRY

When a solution is separated from the pure solvent by a *semi-permeable membrane*, i.e. a membrane that permits the passage of solvent molecules but not of solute molecules, the solvent molecules always tend to pass through the membrane into the solution. This general phenomenon is known as *osmosis* and the flow of solvent molecules leads to the development of an *osmotic pressure* which at equilibrium just prevents further flow. The *equilibrium osmotic pressure* can be measured using a capillary osmometer, such as that shown schematically in Figure 11.1.

11.2.1 OSMOSIS AND CHEMICAL POTENTIAL

When the pressure P_0 above the solution and solvent compartments is equal, osmosis occurs because the chemical potential μ_1^0 of the pure solvent is higher than the chemical potential μ_1 of the solvent in the solution. Thus, the solvent molecules flow through the membrane in order to reduce their chemical potential. The increase in pressure on the solution side of the membrane caused by this flow has the effect of increasing the chemical potential of the solvent in the solution. At equilibrium, the chemical potential of the solvent in the solution at pressure P is equal to μ_1^0 and so

$$\mu_1^0 = \mu_1 + \int_{P_0}^{P} \left(\frac{\partial \mu_1}{\partial P}\right)_{T,n_1,n_2} \mathrm{d}P \tag{11.1}$$

TABLE 11.1
The Relative Magnitudes of Colligative Effects for Polymers Dissolved in Benzene at a Concentration of 10 g dm^{-3}

$\bar{M}_n$ / g mol^{-1}	Osmotic Pressure[a] / mm Benzene	Lowering of Vapour Pressure / mm Hg	Elevation of Boiling Point / ° C	Depression of Freezing Point / ° C
10,000	288	799×10^{-5}	272×10^{-5}	577×10^{-5}
100,000	29	80×10^{-5}	27×10^{-5}	58×10^{-5}
500,000	6	16×10^{-5}	5×10^{-5}	12×10^{-5}
1,000,000	3	8×10^{-5}	3×10^{-5}	6×10^{-5}

Note: The values are given in practical units and have been calculated assuming ideal solution behaviour.
[a] At 25 °C.

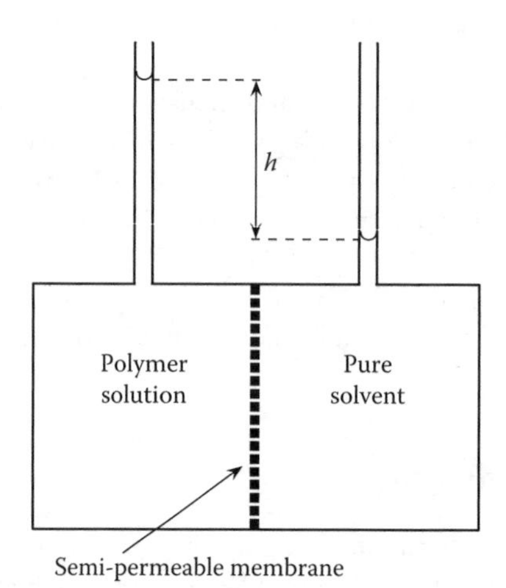

FIGURE 11.1 Schematic representation of a membrane osmometer in which a polymer solution and pure solvent, separated by a semi-permeable membrane, are in equilibrium. The osmotic pressure π is equal to $h\rho_0\mathbf{g}$ where ρ_0 is the solvent density and $\mathbf{g}$ is the acceleration due to gravity.

From Equation 10.34, it follows that

$$\left(\frac{\partial\mu_1}{\partial P}\right)_{T,n_1,n_2} = \frac{\partial}{\partial n_1}\left(\frac{\partial G}{\partial P}\right)_{T,n_1,n_2}$$

Since the solution volume $V = (\partial G/\partial P)_{T,n_1,n_2}$

$$\left(\frac{\partial\mu_1}{\partial P}\right)_{T,n_1,n_2} = \left(\frac{\partial V}{\partial n_1}\right)_{T,P,n_2} = \bar{V}_1 \tag{11.2}$$

where $\bar{V}_1$ is by definition (see Equation 10.33) the partial molar volume of the solvent. Substituting Equation 11.2 into 11.1 and solving the integral gives

$$\mu_1^{\text{o}} = \mu_1 + (P - P_0)\bar{V}_1 \tag{11.3}$$

The pressure difference $(P-P_0)$ is the equilibrium osmotic pressure $\boldsymbol{\pi}$ and so Equation 11.3 can be rewritten in the standard form

$$\mu_1 - \mu_1^{\mathrm{o}} = -\boldsymbol{\pi}\bar{V}_1 \tag{11.4}$$

which is analogous to Equation 10.41. Thus, Equation 11.4 is applicable to any solute/solvent system, regardless of whether or not the system is ideal.

An equation applicable to dilute polymer solutions can be obtained by substituting the Flory–Huggins expression for $\mu_1 - \mu_1^0$ (i.e. substituting Equation 10.48) into Equation 11.4

$$\boldsymbol{\pi} = \frac{\mathbf{R}T\phi_2}{x\bar{V}_1} + \frac{\mathbf{R}T(1/2-\chi)\phi_2^2}{\bar{V}_1} \tag{11.5}$$

For a dilute polymer solution, $\bar{V}_1$ is approximately equal to the molar volume V_1 of the solvent, $\phi_2 = xn_2/(n_1 + xn_2) \approx xn_2/n_1$ and the total solution volume $V \approx n_1V_1$ which when combined gives

$$\frac{\phi_2}{x\bar{V}_1} = \frac{n_2}{V} \tag{11.6}$$

and on substitution into Equation 11.5 leads to

$$\boldsymbol{\pi} = \mathbf{R}T\left(\frac{n_2}{V}\right) + \mathbf{R}T\left(\frac{1}{2}-\chi\right)x^2V_1\left(\frac{n_2}{V}\right)^2 \tag{11.7}$$

This equation shows clearly that $\boldsymbol{\pi}$ depends upon the number of moles of polymer molecules per unit volume of solution (n_2/V).

Assuming that the polymer has molar mass dispersity, $n_2 = \Sigma n_i$ and the total mass of polymer in the solution is given by $m = \Sigma n_iM_i$ where n_i is the total number of moles of polymer molecules of molar mass M_i present in the solution. Thus, from Equation 1.4 it follows that

$$\bar{M}_n = \frac{\sum n_iM_i}{\sum n_i} = \frac{m}{n_2} \tag{11.8}$$

and hence that

$$\frac{n_2}{V} = \left(\frac{m}{V}\right)\left(\frac{n_2}{m}\right) = \frac{c}{\bar{M}_n} \tag{11.9}$$

where c is the polymer concentration (mass per unit volume). Substitution of Equation 11.9 into 11.7 and rearrangement of the resulting equation leads to the following expression for the *reduced osmotic pressure* $\boldsymbol{\pi}/c$

$$\frac{\boldsymbol{\pi}}{c} = \frac{\mathbf{R}T}{\bar{M}_n} + \left(\frac{\mathbf{R}T}{V_1}\right)\left(\frac{1}{2}-\chi\right)\left(\frac{xV_1}{\bar{M}_n}\right)^2 c \tag{11.10}$$

By definition (Section 10.2.2), the number of chain segments is given by

$$x = \frac{V_2}{V_1} \tag{11.11}$$

where V_2 is the molar volume of the polymer. Thus

$$\frac{xV_1}{\bar{M}_n} = \frac{V_2}{\bar{M}_n} = \frac{1}{\rho_2} \tag{11.12}$$

where ρ_2 is the density of the polymer. Substitution of Equation 11.12 into 11.10 gives

$$\frac{\pi}{c} = \frac{\mathbf{R}T}{\bar{M}_n} + \left(\frac{\mathbf{R}T}{V_1\rho_2^2}\right)\left(\frac{1}{2} - \chi\right)c \tag{11.13}$$

on the basis of the Flory–Huggins dilute solution theory. Equation 11.13 shows that for an ideal solution, i.e. under theta conditions $(\chi = 1/2)$

$$\left(\frac{\pi}{c}\right)_\theta = \frac{\mathbf{R}T}{\bar{M}_n} \tag{11.14}$$

and the measurement of π enables an *absolute value* of $\bar{M}_n$ to be calculated (i.e. no simplifying model or instrument calibration is required). For most measurements on dilute polymer solutions, non-ideality is eliminated by extrapolating π/c data to $c = 0$ since

$$\left(\frac{\pi}{c}\right)_{c\to 0} = \frac{\mathbf{R}T}{\bar{M}_n} \tag{11.15}$$

A typical plot of experimental data is shown in Figure 11.2. Thus, $\bar{M}_n$ can be evaluated from the intercept, and from the slope it is possible to calculate an estimate of χ using Equation 11.13. However, since the Flory–Huggins theory strictly is not valid for dilute polymer solutions, Equation 11.13 usually is written in the form of a *virial equation* such as

$$\frac{\pi}{c} = \mathbf{R}T\left[\frac{1}{\bar{M}_n} + A_2c + A_3c^2 + \ldots\right] \tag{11.16}$$

or alternatively

$$\frac{\pi}{c} = \left(\frac{\pi}{c}\right)_{c\to 0}[1 + \Gamma_2 c + \Gamma_3 c^2 + \ldots] \tag{11.17}$$

where A_2 and Γ_2 and A_3 and Γ_3 are known as the second and third virial coefficients, respectively, and quite generally $\Gamma_i = A_i\bar{M}_n$. Virial equations often are used to describe the thermodynamic behaviour of real solutions and real gases, and have the property that at infinite dilution $(c \to 0)$ the behaviour of the solutions or gases becomes ideal. The Flory–Huggins theory in the form of Equation 11.13 gives

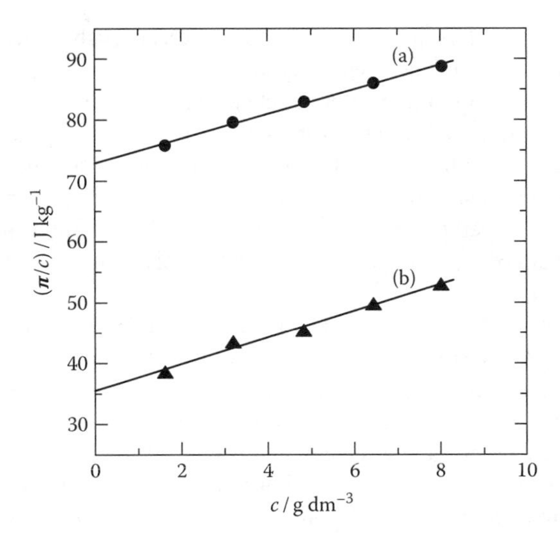

FIGURE 11.2 Plots of π/c against c obtained from measurements at 25 °C on solutions in toluene of samples of polystyrene with $\bar{M}_n$ values of (a) 36.0 and (b) 71.0 kg mol^{-1}. (Data taken from Kamide, K. et al., *Br. Polym. J.*, 15, 91, 1983.)

$$A_2 = \frac{(1/2 - \chi)}{V_1 \rho_2^2} \tag{11.18}$$

whereas the more satisfactory excluded-volume theories lead to slight modification of this equation

$$A_2 = \frac{F(z)(1/2 - \chi)}{V_1 \rho_2^2} \tag{11.19}$$

where $F(z)$ is a function of the excluded-volume parameter z (Section 10.3.3) and is equal to unity under theta conditions (i.e. when $z=0$). For *small z*, $F(z)$ takes the form

$$F(z) = 1 - b_1 z + b_2 z^2 \tag{11.20}$$

in which b_1 and b_2 are positive constants, the values of which have been calculated from theory and depend upon chain architecture (e.g. for linear chains $b_1 = 2.8654$ and $b_2 = 13.928$).

11.2.2 Measurement of Osmotic Pressure

The membrane osmometers used for measurement of osmotic pressures can be divided into two categories.

1. *Static osmometers* in which the equilibrium osmotic pressure is attained naturally by diffusion of solvent through the membrane.
2. *Dynamic osmometers* in which flow of solvent through the membrane is detected and just prevented by application of pressure to the solution cell.

In order to minimize equilibration times (typically 30–60 min), modern osmometers are designed so that the solution and solvent cells are small (typically 0.3–2 cm^3) and the membrane surface area is high. Additionally, since osmotic pressures are sensitive to temperature, good temperature control is essential and should be ±0.01 °C.

The simplest static osmometers are of the basic form indicated schematically in Figure 11.1 (e.g. the Pinner–Stabin osmometer) and have relatively large cell volumes (typically 3–20 cm^3). The volume of solvent transported across the membrane at equilibrium is reduced by using capillaries for the pressure head. However, these capillary osmometers are cumbersome to use and give relatively long equilibration times (typically several hours). Modern static osmometers have much smaller cell volumes and pressure sensors which monitor, usually indirectly, the pressure in the solution cell or, alternatively, the reduced pressure in the solvent cell. It normally is not necessary to correct the solution concentration for dilution due to movement of solvent into the solution cell because the effect usually is negligible.

In dynamic osmometers, the solvent cell is connected to a solvent reservoir. The vertical position of this reservoir relative to that of the solution cell controls the pressure head and is adjusted by a servo-motor responding to a signal from a sensor that monitors movement of solvent across the membrane. Several designs have emerged. For example, the sensor can be a diaphragm which forms part of the sealed solution side of the cell and is one half of a capacitor. Displacement of the diaphragm due to ingress of solvent into the solution cell causes a change in capacitance which automatically is translated into an increase in the pressure head by servo-motor controlled movement of the solvent reservoir. In a quite different instrument design, solvent flow is monitored by an optical system as movement of an air bubble in a capillary tube connected to the solvent cell, an automatic signal again being sent to the servo-motor. For both of these instrument designs, the solvent reservoir is moved to a position such that the flow of solvent across the membrane is just counteracted by the increase in pressure head (i.e. there is no net flow of solvent across the membrane).

In order to evaluate $\bar{M}_n$ for a particular polymer sample, it is necessary to measure π for each of a series of dilute solutions with different concentrations (usually $c < 20$ g dm^{-3}). A plot of π/c against c gives an intercept at $c = 0$ from which $\bar{M}_n$ can be calculated using Equation 11.15. The slope of the initial linear region of the plot can be used with Equation 11.16 to evaluate the second virial coefficient A_2 from which it is possible to calculate the Flory–Huggins polymer–solvent interaction parameter χ, preferably by use of Equation 11.19 rather than Equation 11.18. In good solvents, A_3 is significant and the plots (see Figure 11.3) show distinct upward curvature at higher c due to the c^2 term in Equation 11.16. The linear region of the plots extends to higher c and A_2 becomes smaller as

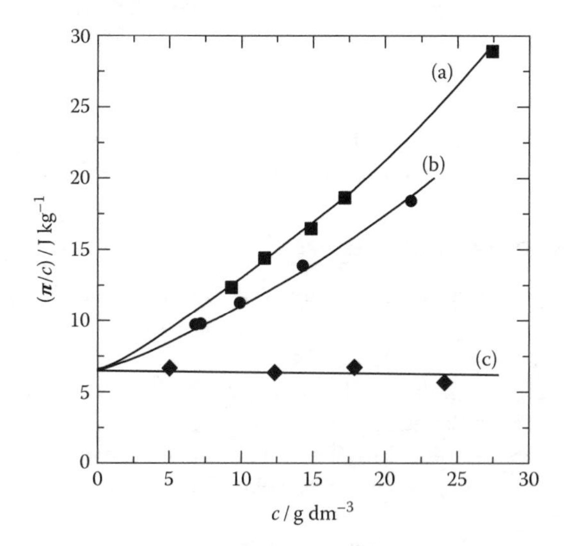

FIGURE 11.3 Plots of π/c versus c for a sample of poly(methyl methacrylate) dissolved in three different solvents at 30 °C: (a) toluene, (b) acetone, and (c) acetonitrile. The common intercept at $c = 0$ gives $\pi/c = 6.48$ J kg^{-1}, which, from Equation 11.15, corresponds to $\bar{M}_n = 388.8$ kg mol^{-1}. (Data taken from Fox, T.G. et al., *Polymer*, 3, 71, 1962.)

the solvency conditions become poorer, as can be seen in Figure 11.3. Under theta conditions $A_2=0$; and so by determining A_2 from measurements at different temperatures and extrapolating the data to $A_2=0$, it is possible to estimate the theta temperature of a polymer–solvent system.

The upper limit for accurate measurement of $\bar{M}_n$ is determined by the sensitivity of the pressure-detecting system and for most osmometers is in the range 5×10^5 to 1×10^6 g mol^{-1}. The lower limit for $\bar{M}_n$ is determined by the extent to which the membrane is permeated by low molar mass polymer molecules. Most membranes are cellulose-based and permeable to molecules with molar masses below about 5000 g mol^{-1}, though more retentive membranes are now available. Thus, low molar mass polymer molecules move through the membrane into the solvent cell and at equilibrium *do not contribute* to the osmotic pressure. Even when these molecules constitute only a small weight fraction of the polymer sample, their number fraction can be significant (because they are of low molar mass) and so their equilibration on either side of the membrane can lead to serious overestimation of $\bar{M}_n$. Therefore, the lower limit for accurate evaluation of $\bar{M}_n$ depends upon the molar mass distribution of the polymer sample and the permeability of the membrane, and typically is in the range 5×10^4 to 1×10^5 g mol^{-1}.

11.3 VAPOUR PRESSURE OSMOMETRY

The technique of *vapour pressure osmometry* (VPO) has its basis in the lowering of solvent vapour pressure by a polymeric solute, and involves measuring the temperature difference between a polymer solution and pure solvent in vapour-phase equilibrium with each other. The principal features of a vapour pressure osmometer are shown schematically in Figure 11.4. Thus, two matched thermistor beads are positioned in a sealed thermostatted (±0.001 °C) chamber, which is saturated with solvent vapour from the solvent contained in the base of the chamber. Using pre-positioned syringes, a drop of pure solvent is placed onto one thermistor bead and a drop of solution onto the other. Due to the lower chemical potential of the solvent in the polymer solution, solvent molecules from the vapour phase condense onto the solution drop giving up their enthalpy of vaporization and causing the temperature of the solution to rise above the temperature T_0 of the chamber and solvent drop. The thermistors form part of a Wheatstone bridge circuit, and the temperature difference ΔT between the solution and solvent drops is measured as a resistance difference ΔR. At equilibrium, the vapour pressure P_{1,T_s} of the solution drop at temperature T_s is exactly equal to the vapour pressure P_{1,T_0} of the pure solvent at temperature T_0. The equilibrium value of $\Delta T_e = T_s - T_0$ is a measure of the vapour pressure lowering caused by the polymeric solute in the solution at T_0 which has vapour pressure P_{1,T_0}.

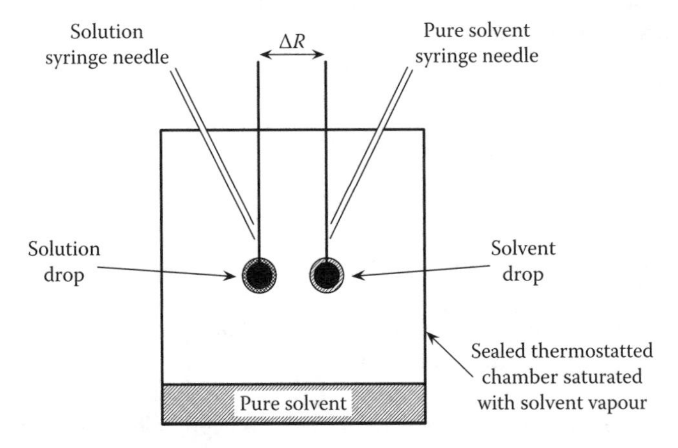

FIGURE 11.4 Schematic illustration of a vapour pressure osmometer showing the solution and solvent drops on matched thermistor beads. The temperature difference between the solution and solvent is measured as the resistance difference ΔR between the thermistor beads.

Assuming that the vapour behaves as an ideal gas, the variation of vapour pressure with temperature is given by the Clausius–Clapeyron Equation

$$\frac{1}{P}\left(\frac{\mathrm{d}P}{\mathrm{d}T}\right)=\frac{\Delta H_{\mathrm{v}}}{\mathbf{R}T^{2}} \tag{11.21}$$

where ΔH_{v} is the molar enthalpy of vaporization of the solvent. Thus,

$$\int_{P_{1,T_0}}^{P_{1,T_s}}\left(\frac{1}{P}\right)\mathrm{d}P=\int_{T_0}^{T_s}\left(\frac{\Delta H_{\mathrm{v}}}{\mathbf{R}T^{2}}\right)\mathrm{d}T$$

Since ΔT is very small, ΔH_{v} can be assumed constant and solution of the integrals gives

$$\ln\left(\frac{P_{1,T_s}}{P_{1,T_0}}\right)=\frac{\Delta H_{\mathrm{v}}\Delta T_{\mathrm{e}}}{\mathbf{R}T_{\mathrm{s}}T_0}$$

Furthermore, at equilibrium $P_{1,T_s}=P^0_{1,T_0}$ and $T_{\mathrm{s}}\approx T_0$ so that the equation simplifies to

$$\ln\left(\frac{P^0_{1,T_0}}{P_{1,T_0}}\right)=\frac{\Delta H_{\mathrm{v}}\Delta T_{\mathrm{e}}}{\mathbf{R}T_0^{2}} \tag{11.22}$$

The activity of the solvent in the solution at T_0 is given by $a_{1,T_0}=P_{1,T_0}/P^0_{1,T_0}$ and so Equation 11.22 may be rewritten as

$$\Delta T_{\mathrm{e}}=-\left(\frac{\mathbf{R}T_0^{2}}{\Delta H_{\mathrm{v}}}\right)\ln a_{1,T_0} \tag{11.23}$$

Substitution of a dilute solution expression for $\ln a_{1,T_0}$ (e.g. Flory–Huggins), followed by a series of approximations similar to those used in deriving the equations for the reduced osmotic pressure, leads to a virial equation of the form

$$\frac{\Delta T_{\mathrm{e}}}{c}=K_{\mathrm{e}}\left[\frac{1}{\bar{M}_{\mathrm{n}}}+A_2c+A_3c^2+\ldots\right] \tag{11.24}$$

where

$K_{\mathrm{e}}=V_1\mathbf{R}T_0^2/\Delta H_{\mathrm{v}}$ and depends only upon the solvent and temperature
V_1 is the molar volume of the solvent
A_2 and A_3 are the second and third virial coefficients defined previously for Equation 11.16
c is the polymer concentration (mass per unit volume).

The dilution effect upon c of the condensed solvent vapour is negligible for most practical values of c.

In practice, ΔT_{e} is not attained due to heat losses from the solution drop to the thermistor bead and its stem, and to the saturated vapour. Instead, a steady-state value ΔT_{ss} is achieved and is given by

$$\frac{\Delta T_{ss}}{c} = K_{ss}\left[\frac{1}{\bar{M}_n} + A_2^{vpo}c + A_3^{vpo}c^2 + \ldots\right] \tag{11.25}$$

where $K_{ss} = K_e/\kappa$ in which κ depends upon the heat transfer processes; A_2^{vpo} and A_3^{vpo} are the VPO second and third virial coefficients and are given by $A_2^{vpo} = A_2 + \kappa_2$ and $A_3^{vpo} = A_3 + \kappa_3$, where κ_2 and κ_3 depend upon the ratio K_{ss}/K_e. Thus, unlike membrane osmometry (MO), VPO is not an absolute method for measurement of $\bar{M}_n$.

Vapour pressure osmometers measure the resistance difference between the two thermistor beads and this is assumed to be proportional to the temperature difference, i.e. $\Delta R_{ss} = k_R \Delta T_{ss}$ where k_R is the proportionality constant. Thus, in order to be of practical use, Equation 11.25 must be modified to the form

$$\frac{\Delta R_{ss}}{c} = K_{Rss}\left[\frac{1}{\bar{M}_n} + A_2^{vpo}c + A_3^{vpo}c^2 + \ldots\right] \tag{11.26}$$

where $K_{Rss} = k_R K_{ss}$. There is experimental evidence which shows that through κ, K_{Rss} is slightly dependent upon the size of the solution drop and upon the molar mass of the solute. However, for most purposes it is reasonable to assume that K_{Rss} is a constant for a given solvent/temperature/osmometer system. Nevertheless, it is good practice to use a consistent procedure and to evaluate K_{Rss} by performing measurements on several samples (rather than one) of accurately known but different $\bar{M}_n$. In this way, a mean value of K_{Rss} can be obtained and any variation of K_{Rss} with $\bar{M}_n$ detected.

Measurements of ΔR_{ss} are made for a series of dilute solutions of the polymer in a solvent which has a relatively high vapour pressure (preferably 100–400 mmHg) at the temperature of analysis. A plot of $\Delta R_{ss}/c$ against c is extrapolated to $c = 0$ to give an intercept from which $\bar{M}_n$ can be calculated

$$\left(\frac{\Delta R_{ss}}{c}\right)_{c\to 0} = \frac{K_{Rss}}{\bar{M}_n} \tag{11.27}$$

The initial linear slopes of such plots give a relative indication of solvency conditions but, unlike MO, cannot yield absolute information.

Despite the non-absolute nature of VPO, it is used extensively to determine $\bar{M}_n$ for polymers of low molar mass. Most commercial instruments are able to detect values of ΔT_{ss} of the order of 10^{-4} °C, giving an upper limit for accurate measurement of $\bar{M}_n$ of about 1.5×10^4 g mol^{-1}. Thus, VPO and MO are complementary techniques.

11.4 EBULLIOMETRY AND CRYOSCOPY

Measurements of the elevation of the solvent boiling point (*ebulliometry*) and of the depression of the solvent freezing point (*cryoscopy*) caused by the presence of a polymeric solute enable $\bar{M}_n$ to be evaluated in a similar way to VPO. Nowadays, however, such measurements are rarely used to evaluate $\bar{M}_n$ for polymers and so will be given only brief consideration here.

Theoretical treatments lead to equations that are identical to Equation 11.24 derived for VPO except for the following.

1. For ebulliometry $\Delta T_e = T_b - T_0$ where T_b and T_0 are the boiling points of the solution and pure solvent, respectively.
2. For cryoscopy $\Delta T_e = T_0 - T_f$ where T_f and T_0 are the freezing points of the solution and pure solvent, respectively, and ΔH_v is replaced by the molar enthalpy of melting of the pure solvent, ΔH_m.

Non-ideality is eliminated in the usual way by extrapolating $\Delta T_e/c$ data to $c=0$ to give an intercept from which $\bar{M}_n$ can be calculated.

Accurate measurements of ΔT_e in ebulliometry and cryoscopy of polymer solutions are made difficult by the small magnitudes of ΔT_e (see Table 11.1), since superheating and supercooling can introduce serious errors. Other practical problems often encountered include separate measurements of T_f and T_0 in cryoscopy, and foaming of the boiling polymer solution in ebulliometry. With current commercial instruments, the upper limit for accurate measurement of $\bar{M}_n$ is about 5×10^3 g mol^{-1}.

11.5 END-GROUP ANALYSIS

The non-thermodynamic methods for evaluation of $\bar{M}_n$ have their basis in the determination of the number of moles of end groups of a particular type in a given mass of polymer and, therefore, are methods of *end-group analysis*. There are four essential requirements that must be satisfied for end-group analysis to be applicable.

1. The end group(s) on the polymer molecules must be amenable to quantitative analysis (usually by titrimetry or spectroscopy).
2. The number of analysable end groups per polymer molecule must be known accurately.
3. Other functional groups which interfere with the end-group analysis either must be absent or their effects must be capable of being corrected for.
4. The concentration of end groups must be sufficient for accurate quantitative analysis.

Thus, end-group analysis is restricted to low molar mass polymers with well-defined structures and distinguishable end-groups. The upper limit for accurate measurement of $\bar{M}_n$ is dependent upon the sensitivity of the technique used to determine the end-group concentration, but typically is 1×10^4 to 1.5×10^4 g mol^{-1}.

End-group analysis is most appropriate to polymers prepared by step polymerization, since they tend to be of relatively low molar mass and have characteristic end-groups suitable for analysis, e.g. polyesters ($-OH$ and $-CO_2H$ end-groups), polyamides ($-NH_2$ and $-CO_2H$ end-groups) and polyurethanes ($-OH$ and $-NCO$ end-groups). Low molar mass polyesters, polyamides and polyethers ($-OH$ end-groups) prepared by ring-opening polymerization similarly are suitable for end-group analysis. The methods most commonly employed are titrimetry (often involving back titrations) and nuclear magnetic resonance spectroscopy (Section 15.6.2). Certain low molar mass polymers prepared from ethylenic monomers by chain polymerization, in particular those with functional end-groups produced by 'living' chain polymerizations (Sections 4.5 and 5.3.2), also are amenable to end-group analysis. However, in most cases polymers prepared by chain polymerization have structures which are insufficiently well-defined and have molar masses which are too high for accurate end-group analysis.

End-group analysis yields the *equivalent mass* M_e (sometimes called the equivalent weight) of the polymer, which is the mass of polymer per mole of end groups. If, for a polymer sample with molar mass dispersity, n_i is the number of moles of polymer molecules of molar mass M_i and f is the number of analysable end-groups per polymer molecule (i.e. its functionality), then

$$M_e = \frac{\sum n_i M_i}{\sum f n_i} \tag{11.28}$$

Since f is a constant and $\bar{M}_n = \sum n_i M_i \Big/ \sum n_i$, Equation 11.28 can be rewritten as

$$M_e = \frac{\bar{M}_n}{f} \tag{11.29}$$

showing that $\bar{M}_n$ can be evaluated from M_e and f.

TABLE 11.2
Theoretical Effects of Low Molar Mass Impurities upon the Value of $\bar{M}_n$ as Calculated using Equation 1.7

Impurity	Weight Fraction of Impurity	$\bar{M}_n$ of Pure Polymer / g mol^{-1}	$\bar{M}_n$ of Impure Polymer / g mol^{-1}
Water (M = 18 g mol^{-1})	0.010	10,000	1,528
	0.001	10,000	6,433
	0.010	200,000	1,784
	0.001	200,000	16,515
Toluene (M = 92 g mol^{-1})	0.010	10,000	4,815
	0.001	10,000	9,028
	0.010	200,000	8,799
	0.001	200,000	63,033

The functionalities f of low molar mass functionalized prepolymers (e.g. polyether polyols used in the preparation of polyurethanes) often are determined from Equation 11.26 by combining the results from VPO (i.e. $\bar{M}_n$) and the functional-group analysis (i.e. M_e). In this respect, the functional groups do not have to be end groups.

11.6 EFFECTS OF LOW MOLAR MASS IMPURITIES UPON $\bar{M}_n$

Purity is of great importance when evaluating $\bar{M}_n$ because low molar mass impurities can give rise to major errors in the measured value of $\bar{M}_n$, as is shown by the data in Table 11.2. These errors are reduced in MO due to equilibration of the impurity on either side of the membrane and in VPO by evaporation of volatile impurities from the solution drop. Thus, the effects of low molar mass impurities upon measurement of $\bar{M}_n$ can be negligible but should never be ignored.

PROBLEMS

11.1 The hydroxyl end-groups of a sample (2.00 g) of linear poly(ethylene oxide) were acetylated by reaction with an excess of acetic anhydride (2.50×10^{-3} mol) in pyridine:

$$\text{R-OH} + \text{CH}_3\text{-}\overset{\text{O}}{\overset{\|}{\text{C}}}\text{-O-}\overset{\text{O}}{\overset{\|}{\text{C}}}\text{-CH}_3 \longrightarrow \text{R-O-}\overset{\text{O}}{\overset{\|}{\text{C}}}\text{-CH}_3 + \text{CH}_3\text{-}\overset{\text{O}}{\overset{\|}{\text{C}}}\text{-OH}$$

After completion of the reaction, water was added to convert the excess acetic anhydride to acetic acid, which together with the acetic acid produced in the acetylation reaction was neutralized by addition of 23.30 cm^3 of a 0.100 mol dm^{-3} solution of sodium hydroxide. Calculate the number-average molar mass for the sample of poly(ethylene oxide) given that each molecule has two hydroxyl end-groups.

11.2 Equilibrium osmotic pressure heads h measured at 22 °C for a series of solutions of a polystyrene sample in toluene are tabulated below:

Polystyrene concentration / g dm^{-3}	2.6	5.2	8.7	11.8	15.1
h / mm toluene	5.85	12.93	24.37	36.34	50.99

Assuming that the densities of toluene and polystyrene at 22 °C are, respectively, 0.867 g cm^{-3} and 1.05 g cm^{-3}, calculate the following.

(a) The number-average molar mass for the sample of polystyrene.
(b) An estimate of the Flory–Huggins polystyrene–toluene interaction parameter.

11.3 Measurements of the equilibrium resistance difference ΔR between the solution and solvent thermistor beads in a vapour pressure osmometer were made under identical conditions for solutions in chloroform of (i) benzil and (ii) a sample of poly(propylene oxide). The results are given in the table below, in which c_B is the concentration of benzil and c_P is the concentration of poly(propylene oxide). Calculate the number-average molar mass for the sample of poly(propylene oxide).

Benzil	c_B / mol dm^{-3}	0.030	0.070	0.100	0.200
	ΔR / ohm	11.541	26.819	38.183	75.481
Poly(propylene oxide)	c_P / g dm^{-3}	6.12	17.52	25.06	50.54
	ΔR / ohm	1.246	4.135	5.920	15.130

11.4 State the methods you would use to determine the number-average molar mass $\bar{M}_n$ for the following polymers.

(a) Samples of poly(ethylene glycol) with $\bar{M}_n$ values in the range 4×10^2 to 5×10^3 g mol^{-1}.

(b) Samples of polyacrylonitrile with $\bar{M}_n$ values in the range 5×10^4 to 2×10^5 g mol^{-1}.

In each case, give the reasons for your choice, name a solvent that would be suitable for the measurements, and discuss briefly possible errors in the determinations.

FURTHER READING

Allen, G. and Bevington, J.C. (eds), *Comprehensive Polymer Science*, Vol. 1 (Booth, C. and Price, C., eds), Pergamon Press, Oxford, U.K., 1989.

Barth, H.G. and Mays, J.W., *Modern Methods of Polymer Characterization*, John Wiley & Sons, New York, 1991.

Billingham, N.C., *Molar Mass Measurements in Polymer Science*, Kogan Page, London, 1977.

Flory, P.J., *Principles of Polymer Chemistry*, Cornell University Press, Ithaca, NY, 1953.

Mark, H.F., Bikales, N.M., Overberger, C.G., and Menges, G. (eds), *Encyclopedia of Polymer Science and Engineering*, Wiley-Interscience, New York, 1985–1989.

Morawetz, H., *Macromolecules in Solution*, 2nd edn., Wiley-Interscience, New York, 1975.

Sun, S.F., *Physical Chemistry of Macromolecules: Basic Principles and Issues*, 2nd edn., Wiley-Interscience, New York, 2004.

Tanford, C., *Physical Chemistry of Macromolecules*, John Wiley & Sons, New York, 1961.

12 Scattering Methods

12.1 INTRODUCTION

There are several possible interactions between radiation and matter, but the most important are absorption, diffraction and scattering. Absorption occurs when the frequency of the radiation corresponds to a specific energy difference for transformation of the constituent molecules or atoms from one energy level to another; this forms the basis for the spectroscopic methods of polymer characterization described in Chapter 15. When the constituent molecules or atoms are organized into regular arrays, as in the crystalline state, radiation with a wavelength similar to the lattice spacings will be diffracted, this forming the basis for crystallography and the well-known Bragg diffraction of X-rays by crystalline solids which is described in Section 17.2.1. When the constituent molecules or atoms are in a disorganized state and the frequency of the radiation does not correspond to energy level differences, the radiation is scattered in all directions by the molecules or atoms.

Scattering of visible light by polymer molecules in solution is the focus of this chapter and provides information on molecular size and shape. In addition, a brief introduction to the use of small-angle X-ray scattering and small-angle neutron scattering for polymer molecular characterization will be given.

12.2 STATIC LIGHT SCATTERING

The phenomenon of light scattering is encountered widely in everyday life. For example, light scattering by airborne dust particles causes a beam of light coming through a window to be seen as a shaft of light, the poor visibility in a fog results from light scattering by airborne water droplets, and laser beams are visible due to scattering of the radiation by atmospheric particles. Also, light scattering by gas molecules in the atmosphere gives rise to the blue colour of the sky and the spectacular colours that can sometimes be seen at sunrise and sunset. These are all examples of *static light scattering* since the time-averaged intensity of scattered light is observed.

Scattering of electromagnetic radiation by molecules results from interaction of the molecules with the oscillating electric field of the radiation, which forces the electrons to move in one direction and the nuclei to move in the opposite direction. Thus, a dipole is induced in the molecules, which for isotropic scatterers is parallel to, and oscillates with, the electric field. Since an oscillating dipole is a source of electromagnetic radiation, the molecules emit light, the *scattered light*, in all directions. Almost all of the scattered radiation has the same wavelength (and hence frequency) as the incident radiation and results from *elastic* (*or Rayleigh*) *scattering* (i.e. with zero energy change). Additionally, a small amount of the scattered radiation has a higher, or lower, wavelength than the incident radiation and arises from *inelastic* (*or Raman*) *scattering* (i.e. with non-zero energy change). Inelastically scattered light carries information relating to bond vibrations and is the basis of Raman spectroscopy (Section 15.5). In this section it is the elastically scattered light from dilute polymer solutions which is of interest since it enables the weight-average molar mass to be determined and can also yield values for the Flory–Huggins interaction parameter and the radius of gyration of the polymer molecules.

12.2.1 Light Scattering by Small Molecules

The theory of light scattering was first developed by Lord Rayleigh in 1871 during his studies on the properties of gases. In his theory, the molecular (or particle) dimensions are assumed to be very

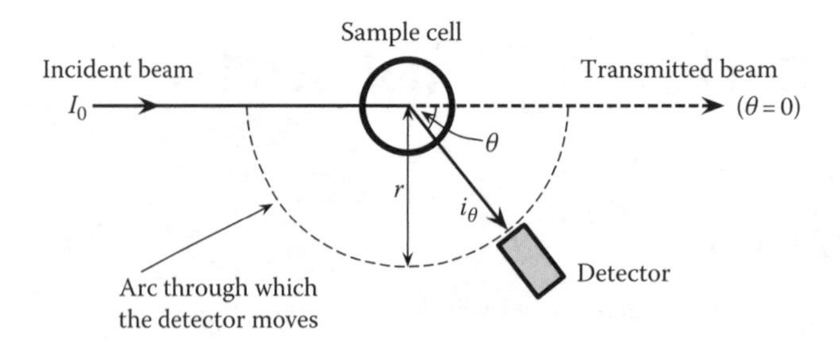

FIGURE 12.1 The main features of apparatus required to measure the light scattered from a polymer solution.

much smaller than the wavelength λ of the incident monochromatic light. Also, in common with all of the theories which will be discussed here, it is assumed implicitly that the scattering is perfectly elastic. The latter assumption is satisfactory because the intensity of the inelastically scattered light is insignificant compared to that of the elastically scattered light. However, the former assumption is not always satisfactory and the Rayleigh theory requires modification when the molecular dimensions approach or are comparable to λ (which is the situation for many polymer molecules when using visible incident radiation).

A schematic diagram which shows the basic principles of a light-scattering experiment is given in Figure 12.1. The basic Rayleigh equation gives the angular variation of the intensity I_θ of the scattered radiation arising from the interaction of a single molecule (or particle) with unpolarized incident radiation of intensity I_0

$$\frac{I_\theta}{I_0} = \frac{8\pi^4\alpha^2(1+\cos^2\theta)}{\lambda^4 r^2} \tag{12.1}$$

where

r and θ are as defined in Figure 12.1

α is the polarizability of the molecule (i.e. the proportionality constant relating the magnitude of the induced dipole to the magnitude of the electric field strength of the incident radiation).

Now consider a volume V of a dilute gas in which there are N gas molecules. Assuming that the *total intensity i_θ of scattered radiation per unit volume* of the dilute gas is the sum of that due to the individual gas molecules present

$$\frac{i_\theta}{I_0} = \left(\frac{N}{V}\right)\frac{8\pi^4\alpha^2(1+\cos^2\theta)}{\lambda^4 r^2} \tag{12.2}$$

Since α is not easily determined, use is made of its relationship to refractive index n according to the equation

$$4\pi\left(\frac{N}{V}\right)\alpha = n^2 - 1 \tag{12.3}$$

The refractive index of a vacuum is unity and so for a dilute gas

$$n = 1 + \left(\frac{\mathrm{d}n}{\mathrm{d}c}\right)c \tag{12.4}$$

where dn/dc is the linear rate of increase of n with gas concentration c (mass per unit volume) and is known as the *refractive index increment*. Since c is small, from Equation 12.4

$$n^2 \approx 1 + 2\left(\frac{\mathrm{d}n}{\mathrm{d}c}\right)c$$

which upon substitution into Equation 12.3 with rearrangement gives

$$\alpha = \frac{(\mathrm{d}n/\mathrm{d}c)c}{2\pi(N/V)}$$

Substitution of this expression into Equation 12.2 yields

$$\frac{i_\theta}{I_0} = \frac{2\pi^2(\mathrm{d}n/\mathrm{d}c)^2 c^2(1+\cos^2\theta)}{\lambda^4 r^2(N/V)} \tag{12.5}$$

Now $N/V = c/(M/\mathbf{N}_\mathrm{A})$ where M is the molar mass of the gas molecules and $\mathbf{N}_\mathrm{A}$ is the Avogadro constant. Thus, Equation 12.5 reduces to

$$\frac{i_\theta}{I_0} = \frac{2\pi^2(\mathrm{d}n/\mathrm{d}c)^2 Mc(1+\cos^2\theta)}{\lambda^4 r^2 \mathbf{N}_\mathrm{A}} \tag{12.6}$$

This is the *Rayleigh Equation for ideal elastic scattering* of unpolarized incident radiation. The scattering envelope (i.e. variation of i_θ with θ) described by this equation is symmetrical (i.e. $i_\theta = i_{180-\theta}$), as is shown in two dimensions in Figure 12.2.

It is convenient at this stage to introduce a quantity called the *Rayleigh ratio R*, which is the reduced relative scattering intensity defined by

$$R = \frac{i_\theta r^2}{I_0(1+\cos^2\theta)} \tag{12.7}$$

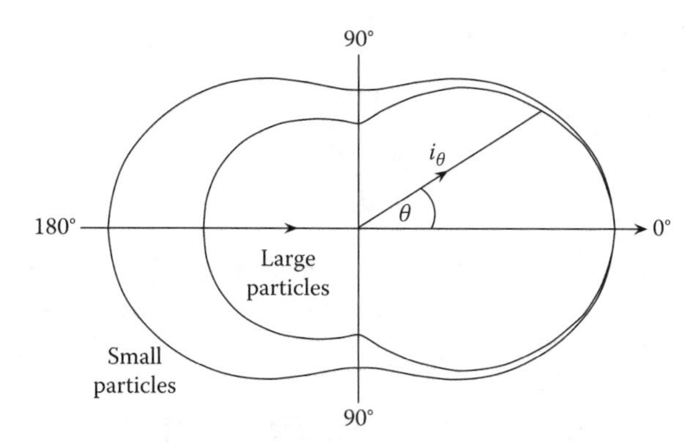

FIGURE 12.2 Two-dimensional scattering envelopes showing the effect of molecular size upon the intensity distribution of light scattered at different angles. The distance from the point of scattering to the perimeter of the scattering envelope represents the relative magnitude of i_θ. The intensity distribution is symmetrical for small particles, but for larger particles is unsymmetrical with the intensities reduced at all angles except zero.

and is independent of both r and θ on the basis of the Rayleigh Equation 12.6. (Note that the Rayleigh ratio sometimes is defined as the quantity $i_\theta r^2/I_0$, which is dependent upon θ; beware!) Hence, the *Rayleigh Equation* may be simplified to

$$R = \frac{2\pi^2 (\mathrm{d}n/\mathrm{d}c)^2 Mc}{\lambda^4 \mathbf{N_A}} \tag{12.8}$$

which shows that M can be determined from measurements of R at a given gas concentration when λ and $\mathrm{d}n/\mathrm{d}c$ are known.

12.2.2 Light Scattering by Liquids and Solutions of Small Molecules

The concentration of molecules in a liquid is very much greater than in an ideal gas. However, the increase in scattering is not proportionate to the increase in concentration, because the molecules have some degree of order with respect to one another and this order gives rise to partial destructive interference of the light scattered from different points within the liquid. It is usual to consider the liquid to be divided into a perfect lattice of volume elements which can contain many molecules, but which are very much smaller than λ and so can be considered as point scatterers. If the liquid were perfectly ordered, then each volume element would contain the same number of molecules and complete destructive interference of the scattered light would take place at all angles except $\theta = 0$. Clearly, this is not the case and light scattering by liquids results from local density fluctuations within the liquid, i.e. at a given instant in time the volume elements do not necessarily contain the same number of molecules (though their time-averaged densities must be equal). Thus, at a given instant in time, there is an excess of scattering from certain volume elements which is not lost by destructive interference.

For dilute solutions, there is an additional contribution resulting from local solute concentration fluctuations. In the study of dilute polymer solutions, only the scattering due to the polymer molecules is required. Thus, the scattering arising from the local solvent density fluctuations is eliminated by taking the difference ΔR between the Rayleigh ratios of the solution and the pure solvent

$$\Delta R = R_{\text{solution}} - R_{\text{solvent}} \tag{12.9}$$

ΔR is commonly known as the *excess Rayleigh ratio* and for static light scattering depends upon the time-averaged mean-square excess polarizability of a volume element arising from the local solute concentration fluctuations, i.e., $\langle (\Delta\alpha)^2 \rangle$. If there are N volume elements each of volume δV, then $N/V = 1/\delta V$ and from Equations 12.2, 12.7 and 12.9

$$\Delta R = \left(\frac{1}{\delta V}\right) \frac{8\pi^4 \langle (\Delta\alpha)^2 \rangle}{\lambda^4} \tag{12.10}$$

At a particular instant in time, $\Delta\alpha = \overline{\Delta\alpha} + \delta\Delta\alpha$ where $\overline{\Delta\alpha}$ is the average excess polarizability (corresponding to the solute concentration c) and $\delta\Delta\alpha$ is the fluctuation about $\overline{\Delta\alpha}$ (due to the solute concentration fluctuation δc). Thus

$$\langle (\Delta\alpha)^2 \rangle = \langle [(\overline{\Delta\alpha})^2 + 2\overline{\Delta\alpha}\delta\Delta\alpha + (\delta\Delta\alpha)^2] \rangle$$

The first term in this relationship is the same for all volume elements at every instant. It therefore makes no overall contribution to scattering because it corresponds to perfect order and complete destructive interference. Additionally, the time-average of the second term is zero because equivalent positive and negative values of $\delta\Delta\alpha$ are equally probable. On this basis, Equation 12.10 can be modified to

$$\Delta R = \left(\frac{1}{\delta V}\right)\frac{8\pi^4\left\langle(\delta\Delta\alpha)^2\right\rangle}{\lambda^4} \tag{12.11}$$

Now from Equation 12.3

$$4\pi\left(\frac{1}{\delta V}\right)\Delta\alpha = n^2 - n_0^2 \tag{12.12}$$

where n and n_0 are the refractive indices of the solution and solvent, respectively. Using the same reasoning as for Equation 12.4

$$n = n_0 + \left(\frac{\mathrm{d}n}{\mathrm{d}c}\right)c$$

and since the solution is dilute

$$n^2 \approx n_0^2 + 2n_0\left(\frac{\mathrm{d}n}{\mathrm{d}c}\right)c$$

Substitution of this relation into Equation 12.12 and differentiating with respect to c gives the relationship between $\delta\Delta\alpha$ and δc

$$\delta\Delta\alpha = \left(\frac{n_0(\mathrm{d}n/\mathrm{d}c)\delta V}{2\pi}\right)\delta c \tag{12.13}$$

Thus, the time-average of $(\delta\Delta\alpha)^2$ is given by

$$\left\langle(\delta\Delta\alpha)^2\right\rangle = \left(\frac{n_0^2(\mathrm{d}n/\mathrm{d}c)^2\delta V^2}{4\pi^2}\right)\left\langle(\delta c)^2\right\rangle$$

which can be substituted into Equation 12.11 to give

$$\Delta R = \frac{2\pi^2 n_0^2(\mathrm{d}n/\mathrm{d}c)^2\delta V\left\langle(\delta c)^2\right\rangle}{\lambda^4} \tag{12.14}$$

The fluctuation theories of Einstein and Smoluchowski lead to the following general expression

$$\left\langle(\delta c)^2\right\rangle = \frac{\mathbf{R}Tc}{\delta V\mathbf{N}_\mathbf{A}(\partial\boldsymbol{\pi}/\partial c)} \tag{12.15}$$

where $\partial\boldsymbol{\pi}/\partial c$ is the first derivative of the osmotic pressure with respect to the solute concentration. From Equation 11.16

$$\frac{\partial\boldsymbol{\pi}}{\partial c} = \mathbf{R}T\left[\frac{1}{M} + 2A_2c + 3A_3c^2 + \ldots\right] \tag{12.16}$$

for a sample of unique molar mass M. Thus, combining Equations 12.14, 12.15 and 12.16 yields

$$\Delta R = \frac{2\pi^2 n_0^2 (\mathrm{d}n/\mathrm{d}c)^2 c}{\mathbf{N_A}\lambda^4[(1/M) + 2A_2c + 3A_3c^2 + \ldots]} \tag{12.17}$$

which shows that as interactions become more significant (i.e. as A_2 and A_3 increase), ΔR decreases. This is because interactions increase the degree of order in the solution.

It is usual to define an *optical constant* K as follows

$$K = \frac{2\pi^2 n_0^2 (\mathrm{d}n/\mathrm{d}c)^2}{\mathbf{N_A}\lambda^4} \tag{12.18}$$

and to rearrange Equation 12.17 into the form

$$\frac{Kc}{\Delta R} = \frac{1}{M} + 2A_2c + 3A_3c^2 + \ldots \tag{12.19}$$

which gives the concentration dependence of the quantity $Kc/\Delta R$ and shows that at $c=0$ it is equal to $1/M$. Thus, the molar mass of the solute can be determined by extrapolation of experimental $Kc/\Delta R$ data to $c=0$.

12.2.3 Light Scattering by Large Molecules in Solution

In deriving Equation 12.19, it was assumed implicitly that the solvent and solute molecules act as point scatterers, i.e. that their dimensions are very much smaller than λ. Whilst in general this assumption is satisfactory for the solvent molecules, it often is inappropriate for polymer solute molecules. The theory for ΔR begins to fail when the solute molecules have dimensions of the order of $\lambda'/20$, where λ' is the wavelength of the light in the medium ($\approx\lambda/n_0$). Thus, the limiting dimensions when using He–Ne laser light ($\lambda=632.8$ nm) are typically about 20 nm. Most polymer molecules of interest have dimensions which either are close to or exceed $\lambda'/20$. The consequence of this is that interference occurs between light scattered from different parts of the same molecule, and so ΔR is reduced. This effect is illustrated schematically in Figure 12.3, which also shows that the path-length (i.e. phase) difference increases with θ and is zero only for $\theta=0$. Thus, the scattering

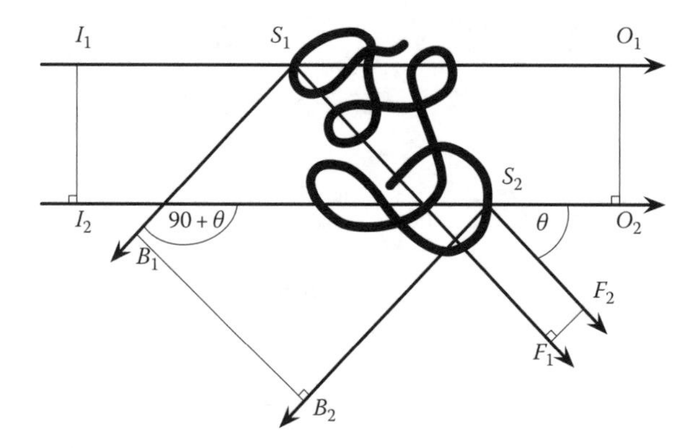

FIGURE 12.3 Schematic illustration of the origin of the interference effects that occur between light scattered by two different segments, S_1 and S_2, in the same molecule. The path-length difference $(I_2S_2B_2-I_1S_1B_1)$ in the backward direction is greater than the path-length difference $(I_1S_1F_1-I_2S_2F_2)$ in the forward direction. The path-length difference is zero at $\theta=0$ $(I_1S_1O_1=I_2S_2O_2)$.

envelope is unsymmetrical, with i_θ greater for forward scattering than for backward scattering, as shown in Figure 12.2. In order to account for such interference effects, a *particle-scattering factor* $P(\theta)$ is introduced and is given by the ratio

$$P(\theta) = \frac{\Delta R_\theta}{\Delta R_{\theta=0}} \tag{12.20}$$

where

ΔR_θ is the measured value of ΔR at the scattering angle θ (i.e. including the effects of interference)
$\Delta R_{\theta=0}$ is the value of ΔR in the absence of interference effects.

For small molecules $P(\theta)=1$ for all values of θ, whereas for large molecules $P(\theta)$ can be very much smaller than unity but increases as θ decreases and at $\theta=0$ becomes equal to unity. By combining Equations 12.19 and 12.20, a more general expression is obtained

$$\frac{Kc}{\Delta R_\theta} = \frac{1}{P(\theta)}\left[\frac{1}{M} + 2A_2c + 3A_3c^2 + \ldots\right] \tag{12.21}$$

Since $\theta=0$ is inaccessible experimentally (i_0 cannot be differentiated from I_0), analytical expressions for $P(\theta)$ are required so that an appropriate extrapolation to $\theta=0$ can be performed in order to eliminate the effects of interference. Debye showed that the functional form of $P(\theta)$ depends upon the shape of the scatterers and derived the following expression for scattering from an isolated Gaussian coil

$$P(\theta) = \left(\frac{2}{u^2}\right)(u - 1 + e^{-u}) \tag{12.22}$$

where $u = q^2\langle s^2\rangle$ in which $\langle s^2\rangle$ is the mean-square radius of gyration of the coil and q is the *scattering vector*, which for dilute solutions is given by

$$q = \frac{4\pi n_0 \sin(\theta/2)}{\lambda} \tag{12.23}$$

Guinier recognized that $P(\theta)$ becomes independent of shape as $\theta \to 0$ and obtained the following general shape-independent expression for $P(\theta)$ when $q\langle s^2\rangle^{1/2} \le 1$

$$P(\theta) = 1 - \left(\frac{q^2\langle s^2\rangle}{3}\right) \tag{12.24}$$

Since $1/(1-x) \approx 1 + x$ when $x \ll 1$, then from Equation 12.24

$$\frac{1}{P(\theta)} \approx 1 + \left(\frac{q^2\langle s^2\rangle}{3}\right) \tag{12.25}$$

Combining Equations 12.21, 12.25 and 12.23 leads to a *general expression for static light scattering*

$$\frac{Kc}{\Delta R_\theta} = \left[\frac{1}{M} + 2A_2c + 3A_3c^2 + \ldots\right]\left[1 + \left(\frac{16\pi^2 n_0^2 \sin^2(\theta/2)}{3\lambda^2}\right)\langle s^2\rangle\right] \quad (12.26)$$

which is applicable at low solute concentrations and scattering angles corresponding to $q\langle s^2\rangle^{1/2} < 1$ for unpolarized incident radiation.

12.2.4 Effect of Molar Mass Dispersity

When using light scattering to characterize polymers, it is necessary to take into account the effect of molar mass dispersity. From Equation 12.21, at infinite dilution

$$\Delta R_\theta = KcMP(\theta) \quad (12.27)$$

and so the contribution to ΔR_θ due to the n_i molecules of molar mass M_i present in the scattering volume V is given by

$$\Delta R_{\theta,i} = K\left(\frac{n_i M_i}{V}\right) M_i P_i(\theta)$$

where $P_i(\theta)$ is the particle-scattering factor for species i. Since $\Delta R_\theta = \Sigma\, \Delta R_{\theta,i}$, then

$$\Delta R_\theta = \left(\frac{K}{V}\right)\sum n_i M_i^2 P_i(\theta)$$

This can be rewritten as

$$\Delta R_\theta = K\left(\frac{\sum n_i M_i}{V}\right)\left(\frac{\sum n_i M_i^2}{\sum n_i M_i}\right)\left(\frac{\sum n_i M_i^2 P_i(\theta)}{\sum n_i M_i^2}\right)$$

which reduces to

$$\Delta R_\theta = Kc\bar{M}_\mathrm{w}\overline{P(\theta)}_\mathrm{z} \quad (12.28)$$

Thus, for a solute with molar mass dispersity, ΔR_θ depends upon the weight-average molar mass and the *z*-average particle-scattering factor. The latter can be examined further by applying Equation 12.24 and leads to

$$\overline{P(\theta)}_\mathrm{z} = 1 - \left(\frac{q^2}{3}\right)\left[\frac{\sum n_i M_i^2 \langle s^2\rangle_i}{\sum n_i M_i^2}\right]$$

which reduces to

$$\overline{P(\theta)}_\mathrm{z} = 1 - \left(\frac{q^2}{3}\right)\langle \overline{s^2}\rangle_\mathrm{z} \quad (12.29)$$

Thus, for a *solute with molar mass dispersity* Equation 12.26 becomes

$$\frac{Kc}{\Delta R_\theta} = \left[\frac{1}{\bar{M}_w} + 2A_2c + 3A_3c^2 + \ldots\right]\left[1 + \left(\frac{16\pi^2 n_0^2 \sin^2(\theta/2)}{3\lambda^2}\right)\left\langle \overline{s^2} \right\rangle_z\right] \quad (12.30)$$

which enables $\bar{M}_w$, A_2 and, for polymer molecules of sufficient size, $\left\langle \overline{s^2} \right\rangle_z$ to be determined from measurements of ΔR_θ at different angles θ for each of several polymer solutions with different concentrations c.

Although Equation 12.30 describes a three-dimensional surface for the variation of $Kc/\Delta R_\theta$ with c and θ, it is usual to plot the experimental data in two dimensions using an elegant technique devised by Zimm. In the *Zimm plot*, $Kc/\Delta R_\theta$ is plotted against the composite quantity $\sin^2(\theta/2) + k'c$, where k' is an arbitrary constant the value of which is chosen so as to give a clear separation of the data points (usually k' is taken as 0.1 when the units of c are g dm^{-3}). A grid-like graph is obtained, such as that shown schematically in Figure 12.4, and consists of two sets of lines, one set joining points of constant c and the other joining points with the same value of θ. In order to evaluate A_2, each of the lines at constant c is extrapolated to $\theta = 0$ (i.e. to $k'c$) and the limiting points thus obtained further extrapolated to $c = 0$. The initial linear slope of this limiting extrapolation then gives A_2 from

$$A_2 = \left(\frac{k'}{2}\right)\left[\frac{\mathrm{d}[Kc/\Delta R_\theta]_{\theta=0}}{\mathrm{d}(k'c)}\right]_{c\to 0} \quad (12.31)$$

Similarly, each of the lines at constant θ are extrapolated to $c = 0$ (i.e. to $\sin^2(\theta/2)$) and the limiting points obtained extrapolated to $\theta = 0$. In this case, the initial linear slope gives $\left\langle \overline{s^2} \right\rangle_z$ in terms of $\bar{M}_w$, λ and n_0

$$\left\langle \overline{s^2} \right\rangle_z = \left(\frac{3\lambda^2 \bar{M}_w}{16\pi^2 n_0^2}\right)\left[\frac{\mathrm{d}[Kc/\Delta R_\theta]_{c=0}}{\mathrm{d}[\sin^2(\theta/2)]}\right]_{\theta\to 0} \quad (12.32)$$

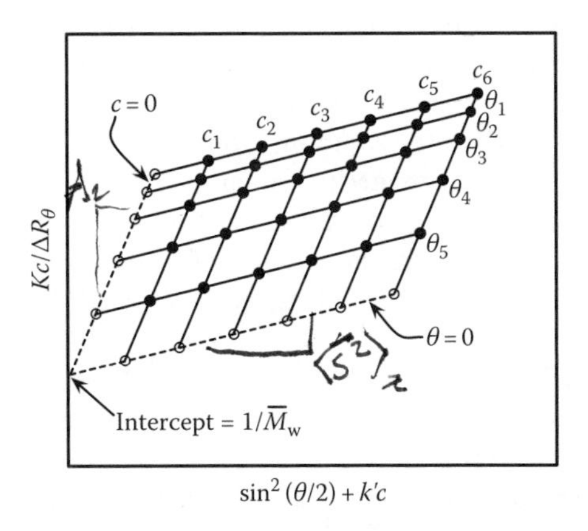

FIGURE 12.4 Schematic illustration of a Zimm plot for analysing light-scattering data. The solid points (•) represent experimental measurements; the open circles (o) are extrapolated points.

The common intercept of the two limiting extrapolations allows $\bar{M}_w$ to be evaluated

$$\left(\frac{Kc}{\Delta R_\theta}\right)_{\substack{c\to 0\\ \theta\to 0}} = \frac{1}{\bar{M}_w} \tag{12.33}$$

and hence $\left\langle \overline{s^2} \right\rangle_z$ from Equation 12.32. Thus, static light scattering measurements yield *absolute values* for $\bar{M}_w$, A_2 (from which χ can be evaluated; see Section 11.2.1) and $\left\langle \overline{s^2} \right\rangle_z$, the latter without reference to molecular shape. Furthermore, the form of the variation of $P(\theta)_{c=0}$ with $q^2\left\langle \overline{s^2} \right\rangle_z$ can be used to deduce the shape of the molecules in solution.

A simpler, though less satisfactory, procedure for evaluation of $\bar{M}_w$ and $\left\langle \overline{s^2} \right\rangle_z$ involves measurement of ΔR_θ at 90° and two other angles symmetrically positioned about $\theta=90°$ (i.e. θ and $180-\theta$, usually 45° and 135°). The *dissymmetry ratio* Z is calculated from

$$Z = \frac{\Delta R_\theta}{\Delta R_{180-\theta}} \tag{12.34}$$

for each polymer solution and extrapolated to $c=0$. This limiting value of Z is used to obtain $P(90)$ and $\left\langle \overline{s^2} \right\rangle_z /\lambda'$ from published tables which give the variation of $P(90)$ with Z and of Z with $\left\langle \overline{s^2} \right\rangle_z /\lambda'$ for scatterers with specific geometries (e.g. Gaussian coils, rigid rods, ellipsoids). Thus, the shape of the polymer molecules must be either assumed or known. It is then a simple matter to calculate $\left\langle \overline{s^2} \right\rangle_z$ from $\left\langle \overline{s^2} \right\rangle_z /\lambda'$ and $\bar{M}_w$ from

$$\left(\frac{Kc}{\Delta R_{90}}\right)_{c\to 0} = \frac{1}{\bar{M}_w P(90)} \tag{12.35}$$

12.2.5 Static Light Scattering Measurements

The basic features of a light-scattering instrument are evident from Figure 12.1, namely, a light source and a thermostatted (±0.01 °C) sample cell positioned on a precision goniometer which enables a photomultiplier detector to be moved accurately to a wide range of scattering angles. Since $i_\theta \propto I_0\lambda^{-4}$, greatest sensitivity is achieved by using high-intensity visible radiation ideally of short wavelength. Modern instruments use monochromatic laser radiation (e.g. He–Ne: $\lambda=632.8$ nm; and Ar ion: $\lambda=488.0$ nm or 514.5 nm) and are able to measure i_θ for θ from about 10° to 160°. It is usual practice to record i_θ as the photomultiplier voltage output, and to evaluate the Rayleigh ratios of the solvent and the solutions from the ratios of their photomultiplier voltage outputs to that of a standard (e.g. benzene), for which the Rayleigh ratio and refractive index are known accurately for the incident radiation at the temperature of measurement. In this way, the need to determine I_0 and to measure absolute values of i_θ are avoided. The Rayleigh ratios thus obtained then are corrected for the effects of refraction and for the variation of the scattering volume with θ.

In addition to the nature of the light source, the sensitivity of the measurements depends strongly upon the refractive index increment dn/dc, which should be as large as possible. Thus, solvents which have refractive indices substantially different from that of the polymer should be chosen (typically to give dn/dc of about 0.0001 $dm^3\ g^{-1}$). The value of dn/dc at the wavelength of the incident radiation must be known accurately in order to perform the calculations of $\bar{M}_w$, A_2 and $\left\langle \overline{s^2} \right\rangle_z$, and if not available in the published literature, can be determined using a differential refractometer.

A further consideration, which is of great practical importance, is the need to ensure that all of the samples analysed are completely dust-free, otherwise the dust particles would contribute to the scattering intensity and lead to erroneous results. Thus, the standard, solvent and solutions analysed must be clarified just prior to use, usually by micropore (e.g. 0.2–0.5 μm porosity) pressure filtration or by high-speed centrifugation.

Light scattering measurements have the potential to be accurate over a broad range of $\bar{M}_w$, typically from 2×10^4 to 5×10^6 g mol^{-1}. The lower limit of $\bar{M}_w$ is determined by the ability to detect small values of i_θ and by the variation of dn/dc with molar mass at very low molar masses. The upper limit corresponds to the molecular dimensions approaching $\lambda'/2$, when complete destructive interference of light scattered from different parts of the same molecule occurs.

12.2.6 Light Scattering by Multicomponent Systems

The treatment of light scattering presented in the previous sections is applicable to a homopolymer dissolved in a single solvent. The modifications required to analyse light scattering by a homopolymer in mixed solvents or by a copolymer in a single solvent are beyond the scope of this book. However, it should be noted that it is necessary to take into account: (i) the fact that a single value of dn/dc is no longer applicable; (ii) the effects of preferential solvent absorption into the molecular coil; and (iii) for copolymers, the variation of repeat unit sequence distribution and of overall composition that occurs from one molecule to another and for different copolymer types (e.g. random, block, graft).

12.3 DYNAMIC LIGHT SCATTERING

In static light scattering experiments, the time-averaged (or 'total') intensity of the scattered light is measured, and for solutions is related to the time-averaged mean-square excess polarizability, which in turn is related to the time-averaged mean-square concentration fluctuation. Whilst such measurements provide a wealth of information, still more can be obtained by considering the real-time random (i.e. Brownian) motion of the solute molecules. This motion gives rise to a *Doppler effect* and so the scattered light possesses a range of frequencies shifted very slightly from the frequency of the incident light (the scattering is said to be *quasi-elastic*). These frequency shifts yield information relating to the movement (i.e. the *dynamics*) of the solute molecules, and can be measured using specialized interferometers and spectrum analysers provided that the incident light has a very narrow frequency band width (i.e. much smaller than the magnitude of the frequency shifts). Thus, the availability of laser light sources has greatly facilitated such measurements.

An alternative, and more popular, means of monitoring the motion of the solute molecules is to record the real-time fluctuations in the intensity of the scattered light. The magnitude and frequency of the intensity fluctuations are at a maximum when light scattered by a single volume element is observed (i.e. corresponding to a specific point in the solution containing relatively few molecules). They are reduced if the overall (i.e. 'mean') intensity of light scattered by several volume elements is measured by the detector. In view of this, highly-sensitive photomultiplier or photodiode detectors with very small apertures are used so that the scattered light entering the detector can be considered to have derived from a single point in the solution. The total numbers of photons of scattered light entering the detector during each of a sequence of time intervals (typically variable from about 50 ns up to about 1 min) are recorded and analysed by a digital correlator interfaced to a computer. The time interval between successive photon countings is known as the *sample time* Δt, and the separation in time between two particular photon countings is known as their *correlation time* τ. It is essential to choose Δt so that it is much smaller than the timescale of the intensity fluctuations, which for polymers is typically in the range 1 μs to 1 ms. Thus, if τ is only a few multiples (e.g. one, two or three) of Δt, the corresponding photon counts will be closely related, and are said to be *correlated*. However, if τ is many multiples of Δt (i.e. much larger than the timescale of the intensity fluctuations), the corresponding photon counts will not be correlated. The *autocorrelation function* $G^{(2)}(\tau)$ of the total intensity i_θ at a particular value of τ is defined by

$$G^{(2)}(\tau) = \lim_{T\to\infty}\left[\frac{1}{T}\int_0^T i_\theta(t)i_\theta(t+\tau)\mathrm{d}t\right] \tag{12.36}$$

and its normalized value $g^{(2)}(\tau)$ by

$$g^{(2)}(\tau) = \frac{G^{(2)}(\tau)}{G^{(2)}(0)} \tag{12.37}$$

where $G^{(2)}(0)$ is the time-averaged value of the square of the intensity at time t, i.e. $\langle i_\theta(t)^2\rangle$. The accuracy with which $g^{(2)}(\tau)$ can be measured depends upon the number N_c of channels in the correlator because each channel records the number of photons hitting the detector in a particular Δt period; for example, a 64-channel correlator can record the photon counts in each of 64 successive Δt periods. Thus, the number of channels controls directly the number of values of τ that can be accessed (64 in the example given, i.e. $\tau=0$, Δt, $2\Delta t$, $3\Delta t$, … $63\Delta t$). In the timescale of the experimental measurements, the correlator evaluates $g^{(2)}(\tau)$ for a series of values of τ from the photon countings as follows

$$g^{(2)}(\tau) = \frac{1}{\langle n^2\rangle}\sum_{i=1}^{N_c} n_i n_{i+j} \tag{12.38}$$

where

$\tau=j\Delta t$ in which j is the number of channels separating the two successive photon counts
n_i and n_{i+j} are the total photon counts recorded in channels i and $i+j$, respectively
$\langle n^2\rangle$ is the time-averaged value of the square of the number of photon counts during a period Δt

(e.g. for $j=3$, $g^{(2)}(3\Delta t)$ is obtained and is the summation for correlation of the photon counts in the first channel with those in the fourth, plus those in the second with the fifth, plus those in the third with the sixth, etc.).

Since the analysis is achieved by photon counting, this technique of dynamic light scattering is known as *photon correlation spectroscopy.*

The decay of $g^{(2)}(\tau)$ with increasing τ carries information relating to the rate of movement of the solute molecules which can be accessed through the Siegert relationship between $g^{(2)}(\tau)$ and the *normalized electric field autocorrelation function* $g^{(1)}(\tau)$

$$g^{(2)}(\tau) = A + B\left|g^{(1)}(\tau)\right|^2 \tag{12.39}$$

where

A and B are constants for a given system
$|g^{(1)}(\tau)|$ is the magnitude of $g^{(1)}(\tau)$.

For independent scattering by solute molecules of unique molar mass undergoing Brownian motion, $g^{(1)}(\tau)$ takes the form of a single exponential decay curve

$$\left|g^{(1)}(\tau)\right| = \exp(-\Gamma\tau) \tag{12.40}$$

where Γ is the *characteristic decay rate*, and so

$$g^{(2)}(\tau) = A + B\exp(-2\Gamma\tau) \tag{12.41}$$

Thus, by fitting the experimental $g^{(2)}(\tau)$ data to an exponential curve, it is possible to evaluate Γ which is related to the *translational diffusion coefficient D* of the solute by

$$\Gamma = q^2 D \tag{12.42}$$

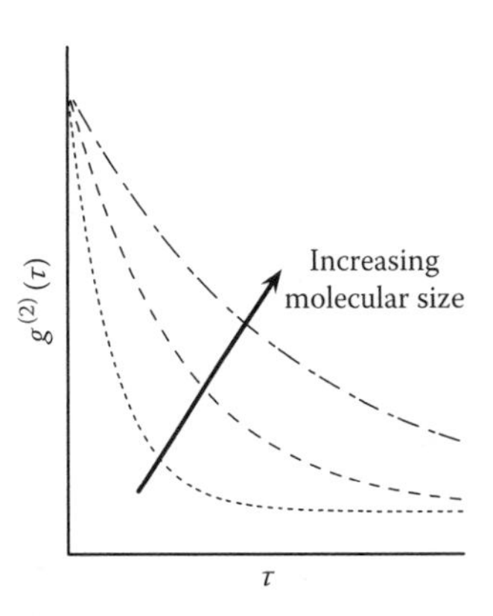

FIGURE 12.5 Schematic illustration of the exponential decay of $g^{(2)}(\tau)$ with τ showing that the rate of decay reduces as molecular size increases.

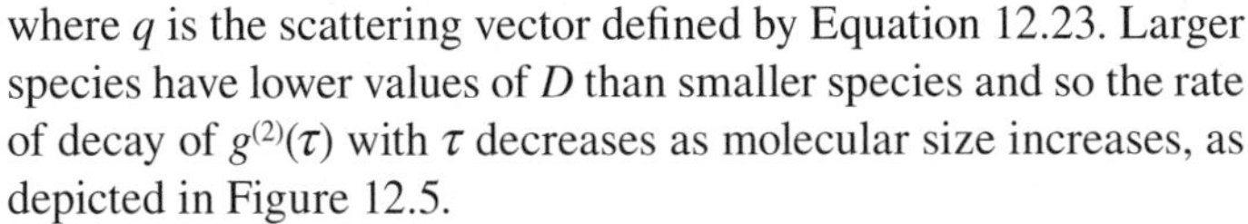

where q is the scattering vector defined by Equation 12.23. Larger species have lower values of D than smaller species and so the rate of decay of $g^{(2)}(\tau)$ with τ decreases as molecular size increases, as depicted in Figure 12.5.

For polymers with molar mass dispersity, $|g^{(1)}(\tau)|$ approximates to a weighted sum of exponentials

$$\left|g^{(1)}(\tau)\right| = \sum C_p \exp(-\Gamma_p \tau) \tag{12.43}$$

where C_p and Γ_p are, respectively, the fractional weighting factor and characteristic decay rate for species p. In practice, sophisticated algorithms are required to fit $g^{(2)}(\tau)$ data accurately to a sum of more than two exponential components and, even then, only the mean value of Γ can be determined with accuracy. Light-scattering theory shows that C_p is proportional to the scattering power of species p in accord with

$$C_p = \frac{z_p I_p}{\sum z_p I_p} \tag{12.44}$$

where

z_p is the number of molecules of species p
I_p is the scattering intensity from a single molecule of species p.

Considering a polymer molecule p as an equivalent solid sphere, I_p is proportional to the square of its molecular volume, V_p^2, and V_p is proportional to its molar mass, M_p, i.e. I_p is proportional to M_p^2, which when combined with Equations 12.42, 12.43 and 12.44 leads to the first cumulant average of Γ

$$\bar{\Gamma} = q^2 \frac{\sum z_p M_p{}^2 D_p}{\sum z_p M_p{}^2} = q^2 \frac{\sum w_p M_p D_p}{\sum w_p M_p} \tag{12.45}$$

where w_p is the total weight of species p. Hence (by analogy with Equation 1.10),

$$\bar{\Gamma} = q^2 \bar{D}_z \tag{12.46}$$

where $\bar{D}_z$ is the *z-average translational diffusion coefficient*. With knowledge of the parameters K_D and a_D in Equation 10.103, i.e. $D = K_D M^{-a_D}$, it is possible to evaluate an average molar mass, though it is not a simple average because

$$\bar{D}_z = K_D \left\langle \overline{\frac{1}{M^{a_D}}} \right\rangle_z \tag{12.47}$$

where $\left\langle \overline{1/M^{a_D}} \right\rangle_z$ is the z-average of $1/M^{a_D}$. The molar mass average obtained is, therefore, given by $\left\langle \overline{1/M^{a_D}} \right\rangle_z^{-1/a_D}$. Additional information relating to the molar mass distribution can be extracted from the distribution of Γ_p, though this is reliable only for relatively narrow distributions. Molecular size in solution also can be determined; for example, from Equation 10.100, for polymers with molar mass dispersity

$$\bar{D}_z = \frac{\mathbf{k}T}{6\pi\eta_0} \left\langle \overline{\frac{1}{R_h}} \right\rangle_z \tag{12.48}$$

where

$\mathbf{k}$ is the Boltzmann constant
T is the temperature
η_0 is the viscosity of the solvent
$\left\langle \overline{1/R_h} \right\rangle_z$ is the z-average of $1/R_h$.

Thus, it is possible to evaluate from D_z the *harmonic z-average hydrodynamic radius* $\left\langle \overline{R_h} \right\rangle_{hz}$ of the polymer molecules (again considered as equivalent solid spheres)

$$\left\langle \overline{R_h} \right\rangle_{hz} = \frac{1}{\left\langle \overline{1/R_h} \right\rangle_z} \tag{12.49}$$

By performing measurements at a series of scattering angles, it is possible to assess the shape of the polymer molecules in solution.

Photon correlation spectroscopy (PCS) is particularly suitable for studies of the hydrodynamic behaviour of polymers, and for studies of the effects of solvency conditions (e.g. solvent, temperature) and skeletal structure (e.g. linear, branched, cyclic) upon chain dimensions. The advances in PCS instrumentation have been greatly assisted by developments in multi-bit processors (which have enabled $g^{(2)}(\tau)$ to be determined with ever greater accuracy) and in the analytical methods by which information is extracted from $g^{(2)}(\tau)$. Nowadays, most light scattering instruments are designed so that they are capable of performing both static (i.e. 'total intensity') and dynamic light scattering measurements, the latter by PCS.

12.4 SMALL-ANGLE X-RAY AND NEUTRON SCATTERING

The basic equations resulting from the theory presented for the scattering of visible radiation can also be applied to the scattering of X-rays and neutrons provided that the optical constant K is redefined to take account of the different origins of the scattering. In each case, K is proportional to the square of the appropriate property difference between the scatterers of interest (e.g. solute molecules) and their environment (e.g. solvent molecules). Scattering of visible radiation results from differences in polarizability which give rise to the $[n_0(\mathrm{d}n/\mathrm{d}c)]^2$ term in Equation 12.18 for K_{LS}. Scattering of X-rays results from *electron density differences* $\Delta\rho_{ed}$ and $K_{XS} \propto (\Delta\rho_{ed})^2$. Finally, scattering of neutrons results from *neutron scattering-length density differences* $\Delta\rho_{nsld}$ and $K_{NS} \propto (\Delta\rho_{nsld})^2$.

The wavelengths of X-rays (typically about 0.1 nm) and neutrons (typically 0.1 nm to 2 nm) are very much smaller than those of visible radiation. Thus, X-ray and neutron scattering measurements are able to provide size information on a much smaller dimensional scale than is possible using light scattering. However, in order to access this information from interference effects (i.e. from $P(\theta)$), it is necessary to make measurements at very small scattering angles ($\theta < 2°$) so that the path-length differences are less than $\lambda'/2$. These measurements are made possible by placing the detector a long distance (typically 1–40 m) from the sample, so that at the detector there is a large separation between the incident beam ($\theta = 0$) and the scattered radiation. Since the measurements are made at small angles (SA), the acronyms SAXS and SANS are used to describe them.

It is possible to evaluate $\bar{M}_w$, A_2 and $\langle \overline{s^2} \rangle_z$ from SAXS and SANS data in essentially the same way as for light scattering. By using the Guinier approximation defined by Equation 12.24 and the corresponding SAXS and SANS equivalents to Equation 12.30, $\langle \overline{s^2} \rangle_z$ can be determined without having to assume a molecular shape. However, the Guinier expression is satisfactory only when $q\langle \overline{s^2} \rangle_z^{1/2} < 1$ and so its use to evaluate $\langle \overline{s^2} \rangle_z^{1/2}$ is valid only when the extrapolation of experimental data is over a range of θ for which $\langle \overline{s^2} \rangle_z^{1/2} < q^{-1}$. In other words, for a given value of λ and range of accessible θ, there is an upper limit for determination of $\langle \overline{s^2} \rangle_z^{1/2}$ using the Guinier approximation. The upper limits for SAXS, SANS and light scattering are, respectively, about 5, 20 and 200 nm, as is indicated by the values of q^{-1} given in Table 12.1 for some typical λ, θ combinations. When $q\langle \overline{s^2} \rangle_z^{1/2} > 1$, complete expressions for $P(\theta)$ are required in order to interpret the experimental data and necessitate the assumption of a particular model for the polymer molecules (e.g. Equation 12.22 for a Gaussian coil).

Synchrotron radiation sources enable X-ray radiation to be selected from wavelengths in the range $0.06 < \lambda < 0.3$ nm, whereas in laboratory instruments the CuK_α line ($\lambda = 0.154$ nm) most

TABLE 12.1
Values of the Reciprocal Scattering Vector q^{-1} Calculated from Equation 12.23 for Some Typical λ, θ Combinations[a]

Scattering Method	λ / nm	θ	q^{-1} / nm
SAXS	0.154	1.0°	1.4
	0.154	0.5°	2.8
	0.154	0.2°	7.0
SANS	1.0	1.0°	9.1
	1.0	0.5°	18.2
	1.0	0.2°	45.6
Light scattering	632.8	60.0°	67.1
	632.8	30.0°	129.7
	632.8	10.0°	385.2

[a] The refractive index n_0 of the medium is assumed arbitrarily to be 1.0 for SAXS and SANS, and 1.5 for light scattering.

commonly is used. Since $\left\langle \overline{s^2} \right\rangle_z^{1/2}$ invariably is greater than 5 nm for polymers, the Guinier region generally is inaccessible by SAXS and model assumptions are required in order to evaluate $\left\langle \overline{s^2} \right\rangle_z^{1/2}$.

The use of thermal neutrons (λ about 0.1 nm) direct from a nuclear reactor for SANS imposes similar restrictions to those encountered with SAXS. However, by reducing the velocities of the thermal neutrons it is possible to produce cold neutrons (λ about 1 nm) which enable $\left\langle \overline{s^2} \right\rangle_z^{1/2}$ up to about 20 nm to be determined by SANS in the Guinier region. Such measurements, therefore, complement similar measurements made by light scattering for which the lower limit of $\left\langle \overline{s^2} \right\rangle_z^{1/2}$ is about 15 nm using an Ar ion laser with $\lambda = 488.0$ nm. In order to gain the necessary contrast (i.e. $\Delta\rho_{\text{nsld}}$) for SANS, use is made of the large difference between the neutron scattering lengths of hydrogen (^{1}H) atoms and deuterium (^{2}H) atoms. Thus, either a deutero-solvent or a deutero-solute is used. Since the ^{1}H and ^{2}H atoms act as individual point scatterers, it is possible to determine $\left\langle \overline{s^2} \right\rangle_z^{1/2}$ for specific blocks in molecules of block copolymers. This is achieved by preparing block copolymers in which only one block is of deutero-polymer and measuring the SANS in an appropriate ^{1}H solvent.

The high penetration of X-rays and neutrons enables SAXS and SANS measurements to be made on solid polymers. SAXS is used widely to measure the size of dispersed phases in polymers and is particularly suitable for studies of polymer morphology (Sections 17.4.3 and 18.3.2.2). SANS measurements on solid polymers are achieved using mixtures of ^{1}H polymer with a small amount (i.e. low c) of the corresponding deutero-polymer (e.g. polystyrene with deutero-polystyrene). Such measurements gave the first irrefutable proof that in a pure amorphous polymer, the polymer molecules adopt their unperturbed dimensions (Section 10.3.3). Similarly, the dimensions of individual blocks in phase-separated block copolymers, and of network chains in network polymers, can be determined using selective deuteration. Thus, SANS is by far the most powerful technique for studies of chain dimensions and has many other applications beyond those mentioned here.

PROBLEMS

12.1 Rayleigh ratios R_θ were obtained at 25 °C for a series of solutions of a polystyrene sample in benzene, with the detector situated at various angles θ to the incident beam of unpolarized monochromatic light of wavelength 546.1 nm. The results are tabulated below. Under the conditions of these measurements, the Rayleigh ratio and refractive index of benzene are 46.5×10^{-4} m^{-1} and 1.502, respectively, and the refractive index increment for the polystyrene solutions is 1.08×10^{-4} dm^3 g^{-1}.

Polystyrene Concentration/g dm^{-3}	$10^4 \times R_\theta$/m^{-1} Measured at θ =			
	30°	60°	90°	120°
0.50	72.3	69.4	66.2	64.3
1.00	89.8	85.7	81.3	78.2
1.50	100.8	97.1	92.0	88.1
2.00	108.7	103.8	99.7	95.9

Using a Zimm plot of the data, determine the following.

(a) The weight-average molar mass $\bar{M}_w$ of the polystyrene sample.

(b) The z-average radius of gyration $\left\langle \overline{s^2} \right\rangle_z^{1/2}$ of the polystyrene molecules in benzene at 25 °C.

(c) The second virial coefficient A_2 for the polystyrene–benzene interaction at 25 °C.

12.2 The z-average diffusion coefficient of a sample of poly(methyl methacrylate) in toluene was measured at 25 °C by dynamic light scattering and found to be 1.63×10^{-11} m^2 s^{-1}. Given that the viscosity of toluene at 25 °C is 5.55×10^{-4} Pa s, calculate the radius of gyration of the poly(methyl methacrylate) chains in toluene. Comment on the type of radius-of-gyration average obtained.

12.3 Discuss the significance of the scattering vector q for measurements of chain dimensions by scattering methods.

FURTHER READING

Allen, G. and Bevington, J.C. (eds), *Comprehensive Polymer Science*, Vol. 1 (Booth, C. and Price, C., eds), Pergamon Press, Oxford, U.K., 1989.

Billingham, N.C., *Molar Mass Measurements in Polymer Science*, Kogan Page, London, U.K., 1977.

Flory, P.J., *Principles of Polymer Chemistry*, Cornell University Press, Ithaca, NY, 1953.

Hiemenz, P.C. and Lodge, T.P., *Polymer Chemistry*, 2nd edn., CRC Press, Boca Raton, FL, 2007.

Huglin, M.B. (ed), *Light Scattering from Polymer Solutions*, Academic Press, London, 1972.

Mark, H.F., Bikales, N.M., Overberger, C.G., and Menges, G. (eds), *Encyclopedia of Polymer Science and Engineering*, Wiley-Interscience, New York, 1985–1989.

Morawetz, H., *Macromolecules in Solution*, 2nd edn., Wiley-Interscience, New York, 1975.

Roe, R.-J., *Methods of X-ray and Neutron Scattering in Polymer Science*, Oxford University Press, New York, 2000.

Sun, S.F., *Physical Chemistry of Macromolecules: Basic Principles and Issues*, 2nd edn., Wiley-Interscience, New York, 2004.

Tanford, C., *Physical Chemistry of Macromolecules*, John Wiley & Sons, New York, 1961.

13 Frictional Properties of Polymers in Solution

13.1 INTRODUCTION

Measurements of the frictional properties of dilute polymer solutions were amongst the earliest methods developed for determination of polymer molar mass and continue to be important today. Solution viscosity is by far the most extensively used frictional property, in part because it is very simple to measure accurately. Thus dilute solution viscometry, which is the principal focus for this chapter, remains a very widely practiced method for evaluation of molar mass and for specification of commercial grades of polymers. Historically, classical measurements of diffusion have been important, but nowadays diffusion mostly is observed using modern techniques, such as dynamic light scattering (Section 12.3) and specialized methods of nuclear magnetic resonance spectroscopy (Section 15.6). Sedimentation methods also have been used throughout the history of polymer science and, although mainly used to characterize biological macromolecules, they are having a renaissance for characterisation of synthetic polymers with the availability of modern analytical ultracentrifuges. Hence, in addition to a more complete description of dilute solution viscometry, a brief introduction will be given to polymer characterisation by ultracentrifugation.

13.2 DILUTE SOLUTION VISCOMETRY

A characteristic feature of a dilute polymer solution is that its viscosity is considerably higher than that of either the pure solvent or similarly dilute solutions of small molecules. This arises because of the large differences in size between polymer and solvent molecules, and the magnitude of the viscosity increase is related to the dimensions of the polymer molecules in solution. Therefore, measurements of the viscosities of dilute polymer solutions can be used to provide information concerning the effects upon chain dimensions of polymer structure (chemical and skeletal), molecular shape, degree of polymerization (hence molar mass) and polymer–solvent interactions. Most commonly, however, such measurements are used to determine the molar mass of a polymer.

13.2.1 Intrinsic Viscosity

The quantities required, and terminology used, in dilute solution viscometry are summarised in Table 13.1 The terminology proposed by IUPAC was an attempt to eliminate the inconsistencies associated with the common names, which define as viscosities, quantities that do not have the dimensions ($\mathrm{M\,L^{-1}T^{-1}}$) of viscosity. However, the common names were well established when the IUPAC recommendations were published, and the new terminology has largely been ignored. The quantity of greatest importance for the purposes of polymer characterisation is the *intrinsic viscosity* $[\eta]$ since it relates to the intrinsic ability of a polymer to increase the viscosity of a particular solvent at a given temperature. The *specific viscosity* of a solution of concentration c is related to $[\eta]$ by a power series in $[\eta]c$

$$\eta_{sp} = k_0[\eta]c + k_1[\eta]^2c^2 + k_2[\eta]^3c^3 + \ldots \tag{13.1}$$

TABLE 13.1
Quantities and Terminology Used in Dilute Solution Viscometry

Common Name	Name Proposed by IUPAC[a]	Symbol and Definition[b]
Relative viscosity	Viscosity ratio	$\eta_r = \eta/\eta_0$
Specific viscosity	—	$\eta_{sp} = \eta_r - 1$
Reduced viscosity	Viscosity number	$\eta_{red} = \eta_{sp}/c$
Inherent viscosity	Logarithmic viscosity number	$\eta_{inh} = \{\ln(\eta_r)\}/c$
Intrinsic viscosity	Limiting viscosity number	$[\eta] = \lim_{c \to 0}(\eta_{red})$

[a] IUPAC is the International Union of Pure and Applied Chemistry.

[b] η_0 is the viscosity of the solvent and η is the viscosity of a polymer solution of concentration c (mass per unit volume).

where k_0, k_1, k_2, are dimensionless constants and $k_0 = 1$. For dilute solutions, Equation 13.1 can be truncated and rearranged to the following expression for the *reduced viscosity*

$$\frac{\eta_{sp}}{c} = [\eta] + k_H[\eta]^2 c \tag{13.2}$$

which is known as the *Huggins equation* and is valid for $[\eta]c \ll 1$. The Huggins constant k_H ($= k_1$) is essentially independent of molar mass and has values which fall in the range 0.3 (for good polymer–solvent pairs) to 0.5 (for poor polymer–solvent pairs).

From the series expansion $\ln(1+x) = x - x^2/2 + x^3/3 - \ldots$ for $|x| < 1$, then for dilute solutions with $\eta_{sp} \ll 1$ (for which η_{sp}^3 and higher terms can be neglected)

$$\ln(\eta_r) = \ln(1+\eta_{sp}) \approx \eta_{sp} - \frac{1}{2}\eta_{sp}^2 \tag{13.3}$$

Rearranging Equation 13.2 for η_{sp} and substituting into Equation 13.3, retaining only the terms in c up to c^2, gives

$$\ln(\eta_r) = [\eta]c + \left(k_H - \frac{1}{2}\right)[\eta]^2 c^2$$

which can be rearranged into the form of the *Kraemer equation* for the *inherent viscosity*

$$\frac{\ln(\eta_r)}{c} = [\eta] + k_K[\eta]^2 c \tag{13.4}$$

where $k_K = k_H - 1/2$ and so k_K is negative (or zero).

Although there are other simplifications of Equation 13.1, the Huggins and Kraemer equations provide the most common procedure for evaluation of $[\eta]$ from experimental data. This involves a dual extrapolation according to these equations and gives $[\eta]$ as the mean intercept (see Figure 13.1).

13.2.2 Interpretation of Intrinsic Viscosity Data

The intrinsic viscosity $[\eta]$ of a polymer is related to its *viscosity-average molar mass* $\bar{M}_v$ by the *Mark–Houwink–Sakurada equation*

$$[\eta] = K\bar{M}_v^a \tag{13.5}$$

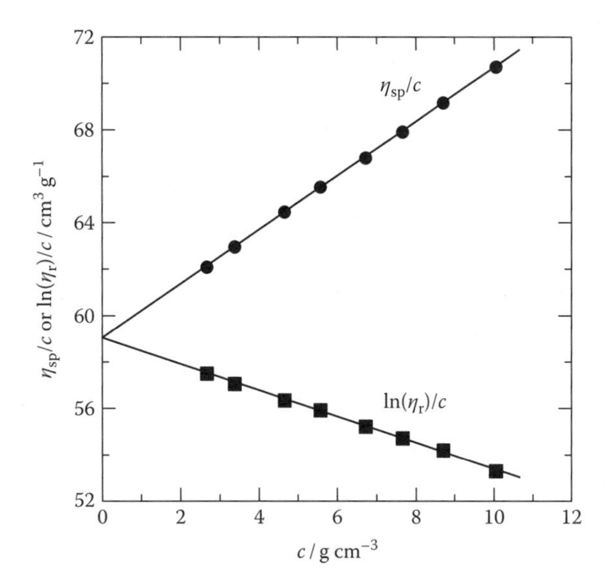

FIGURE 13.1 Evaluation of $[\eta]$ for solutions in toluene of a sample of poly(*tert*-butyl acrylate) at 25 °C using a dual Huggins–Kraemer plot. (Data of Y-F. Wang and P.A. Lovell, The University of Manchester, Manchester, U.K.)

where K and a are characteristic constants for a given polymer/solvent/temperature system and are known as the *Mark–Houwink–Sakurada constants* (or often simply as the Mark–Houwink constants). Theoretical justification for the form of this equation is presented in Section 10.4.2. For Gaussian coils, it was shown that $a=0.5$ under theta conditions, and that a increases to a limiting value of 0.8 with coil expansion (typically $a>0.7$ for polymers in good solvents). The value of K tends to decrease as a increases and for flexible chains it is typically in the range 10^{-3}–10^{-1} cm^3 g^{-1} (g mol^{-1})$^{-a}$.

The form of $\bar{M}_v$ can be deduced on the basis that specific viscosities are additive in the limit of infinite dilution, i.e.

$$\eta_{sp} = \sum (\eta_{sp})_i$$

where $(\eta_{sp})_i$ is the contribution to η_{sp} made by the n_i moles of polymer molecules of molar mass M_i present in unit volume of a dilute polymer solution. Since $c_i = n_i M_i$, $(\eta_{sp})_i = c_i[\eta]_i$ and $[\eta]_i = KM_i^a$ then

$$[\eta] = \left(\frac{\eta_{sp}}{c}\right)_{c\to 0} = K\left(\frac{\sum n_i M_i^{1+a}}{\sum n_i M_i}\right)$$

Comparison of this equation with Equation 13.5 shows that

$$\bar{M}_v = \left(\frac{\sum n_i M_i^{1+a}}{\sum n_i M_i}\right)^{1/a} \tag{13.6}$$

Thus for Gaussian chains $\bar{M}_v$ lies between the number-average and weight-average molar masses, $\bar{M}_n$ and $\bar{M}_w$, but is closer to $\bar{M}_w$. When $a=1$ (as observed for certain inherently stiff or highly extended chains) $\bar{M}_v=\bar{M}_w$.

In order to evaluate $\bar{M}_v$ from $[\eta]$ using the Mark–Houwink–Sakurada equation, it is necessary to know the values of K and a for the system under study. These values most commonly are determined from measurements of $[\eta]$ for a series of polymer samples with known $\bar{M}_n$ or $\bar{M}_w$. Ideally, the samples should have narrow molar mass distributions so that $\bar{M}_n \approx \bar{M}_v \approx \bar{M}_w$; if this is not the case, then provided that their molar mass distributions are of the same functional form (e.g. most probable distribution), the calibration is valid and yields equations that are similar to Equation 13.5, but in which $\bar{M}_v$ is replaced by $\bar{M}_n$ or $\bar{M}_w$. Generally a plot of $\log[\eta]$ against $\log M$ is fitted to a straight line from which K and a are determined. Theoretically, this plot should not be linear over a wide range of M, so that K and a values should not be used for polymers with M outside the range defined by the calibration samples. However, in practice, such plots are essentially linear over wide ranges of M, though curvature at low M often is observed due to the non-Gaussian character of short flexible chains. Some typical values of K and a are given in Table 13.2.

For small expansions of Gaussian polymer chains the expansion parameter α_η for the hydrodynamic chain dimensions is given by a closed expression of the form

$$\alpha_\eta^3 = 1 + bz \tag{13.7}$$

where

z is the excluded volume parameter defined by Equation 10.81

b is a constant, the value of which is uncertain (e.g. values of 1.55 and 1.05 have been obtained from different theories).

Combining this relation with the Flory–Fox Equation 10.96 gives

$$[\eta] = K_\theta M^{1/2} + bK_\theta z M^{1/2}$$

which upon rearrangement, recognizing that from Equation 10.81 z is proportional to $M^{1/2}$, leads to

$$\frac{[\eta]}{M^{1/2}} = K_\theta + BM^{1/2} \tag{13.8}$$

TABLE 13.2
Some Typical Values of the Mark–Houwink–Sakurada Constants *K* and *a*

Polymer	Solvent	Temperature / °C	K [a]	a
Polyethylene	Decalin	135	6.2×10^{-2}	0.70
Polystyrene	Toluene	25	7.5×10^{-3}	0.75
Polystyrene	Cyclohexane	34[b]	8.2×10^{-2}	0.50
Poly(vinyl acetate)	Acetone	30	8.6×10^{-3}	0.74
Cellulose triacetate	Acetone	20	2.38×10^{-3}	1.0

[a] The units of K are $cm^3\ g^{-1}\ (g\ mol^{-1})^{-a}$.

[b] This is the theta temperature for polystyrene in cyclohexane (i.e. $T=\theta$).

where B depends upon the chain structure and polymer–solvent interactions, and is a constant for a given polymer/solvent/temperature system. Thus the corresponding $[\eta]$ and M data obtained for evaluation of Mark–Houwink–Sakurada constants from calibration samples with narrow molar mass distributions, also can be plotted as $[\eta]/M^{1/2}$ against $M^{1/2}$ to give K_θ as the intercept at $[\eta]/M^{1/2}=0$. The value of K_θ can then be used to evaluate for the polymer to which it relates: (i) $\langle s^2 \rangle_0^{1/2}$ (i.e. unperturbed dimensions) for any M by assuming a theoretical value for Φ_0^s in Equation 10.97, and (ii) α_η for a given pair of corresponding $[\eta]$ and M values by using the Flory–Fox Equation 10.96.

The interpretation of $[\eta]$ for branched polymers and copolymers is considerably more complicated, and will not be dealt with here other than to highlight some of the additional complexities. The effect of branching is to increase the segment density within the molecular coil. Thus a branched polymer molecule has a smaller hydrodynamic volume and a lower intrinsic viscosity than a similar linear polymer of the same molar mass. For copolymers of the same molar mass, $[\eta]$ will differ according to the composition, composition distribution, sequence distribution of the different repeat units, interactions between unlike repeat units, and degree of preferential interaction of solvent molecules with one of the different types of repeat unit.

13.2.3 Measurement of Solution Viscosity

The viscosities of dilute polymer solutions most commonly are measured using capillary viscometers of which there are two general classes, namely, U-tube viscometers and suspended-level viscometers (see Figure 13.2). A common feature of these viscometers is that a measuring bulb, with upper and lower etched marks, is attached directly above the capillary tube. The solution is either drawn or forced into the measuring bulb from a reservoir bulb attached to the bottom of the capillary tube, and the time required for it to flow back between the two etched marks is recorded.

In U-tube viscometers, the pressure head giving rise to flow depends upon the volume of solution contained in the viscometer, and so it is essential that this volume is exactly the same for each measurement. This normally is achieved after temperature equilibration by carefully adjusting the liquid level to an etched mark just above the reservoir bulb.

Most suspended-level viscometers are based upon the design due to Ubbelohde, the important feature of which is the additional tube attached just below the capillary tube. This ensures that during measurement, the solution is suspended in the measuring bulb and capillary tube, with atmospheric pressure acting both above and below the flowing column of liquid. Thus, the pressure head depends only upon the volume of solution in and above the capillary, and so is independent of the total volume of solution contained in the viscometer. This feature is particularly useful because it enables solutions to be diluted in the viscometer by adding more solvent. When U-tube viscometers are used, they must be emptied, cleaned, dried and refilled with the new solution each time the concentration is changed.

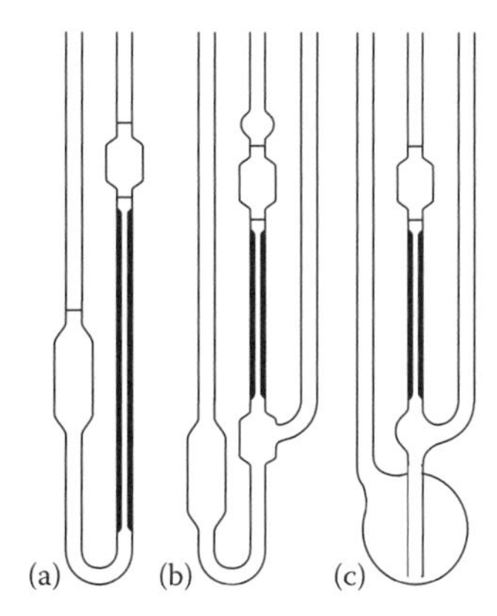

FIGURE 13.2 Schematic illustrations of (a) an Ostwald U-tube viscometer, (b) an Ubbelohde suspended-level viscometer, and (c) a modified Ubbelohde viscometer with a larger reservoir bulb for dilutions.

Before use, it is essential to ensure that the viscometer is thoroughly clean and that the solvent and solutions are freed from dust by filtration, otherwise incorrect and erratic flow times can be anticipated. The viscometer is placed in a thermostatted water (or oil) bath with temperature control of ±0.01 °C or better because viscosity generally changes rapidly with temperature. After allowing sufficient time for temperature equilibration of the solution, several measurements of flow time are made and should be reproducible to ±0.1% when measured visually using a stopwatch. When analysing polyelectrolyte solutions, it is important to suppress the polyelectrolyte effect (see Section 10.4.4) by using an aqueous solution (typically about 0.1 mol dm^{-3}) of an inert electrolyte (e.g. NaCl) as the solvent.

Under conditions of steady laminar Newtonian flow, the volume V of liquid which flows in time t through a capillary of length l and radius r is related to both the pressure difference P across the capillary and the viscosity η of the liquid by *Poiseuille's equation*

$$\frac{V}{t} = \frac{\pi r^4 P}{8\eta l} \tag{13.9}$$

The radial velocity profile corresponding to this equation is parabolic, with maximum velocity along the axis of the capillary tube and zero velocity at the wall. During the measurement of flow time, P continuously decreases and normally is given by

$$P = \langle h \rangle \rho \mathbf{g}$$

where

$\langle h \rangle$ is the average pressure head
ρ is the density of the liquid
$\mathbf{g}$ is the acceleration due to gravity.

Thus Poiseuille's Equation 13.9 can be rearranged to give

$$\eta = \frac{\pi r^4 \langle h \rangle \rho \mathbf{g} t}{8Vl} \tag{13.10}$$

which has the form

$$\eta = A\rho t \tag{13.11}$$

where A is a constant for a given viscometer. Poiseuille's equation does not take into account the energy dissipated in imparting kinetic energy to the liquid, but is satisfactory for most viscometers provided that the flow times exceed about 180 s.

Absolute measurements of viscosity are not required in dilute solution viscometry since it is only necessary to determine the viscosity of a polymer solution relative to that of the pure solvent. Application of Equation 13.11 leads to the following relation for the relative viscosity

$$\eta_r = \frac{\eta_s}{\eta_0} = \frac{\rho_s t_s}{\rho_0 t_0} \tag{13.12}$$

where

η_s and η_0 are the viscosities
ρ_s and ρ_0 are the densities
t_s and t_0 are the flow times
of a polymer solution of concentration c and of the pure solvent, respectively.

Since dilute solutions are used, it is common practice to assume that $\rho = \rho_0$ so that η_r is given simply by the ratio of the flow times t_s/t_0. The other quantities required are then calculated from η_r and c. In more accurate work, kinetic energy and density corrections are applied and, when necessary, the values of $[\eta]$ obtained are extrapolated to zero shear rate. For more thorough analyses of polyelectrolytes, measurements should also be obtained at different added inert electrolyte concentrations and the data extrapolated to infinite ionic strength.

Although dilute solution viscometry is not an absolute method for characterisation of polymers, in comparison to other methods it is simple, fast and uses relatively inexpensive equipment. It also has the advantage that it is applicable over the complete range of attainable molar masses. For these reasons, dilute solution viscometry is widely used for routine measurements of molar mass.

In a more recent development, capillary-bridge differential viscometers have been designed that employ four matched capillary tubes arranged in a manner analogous to a Wheatstone bridge, as shown in Figure 13.3.

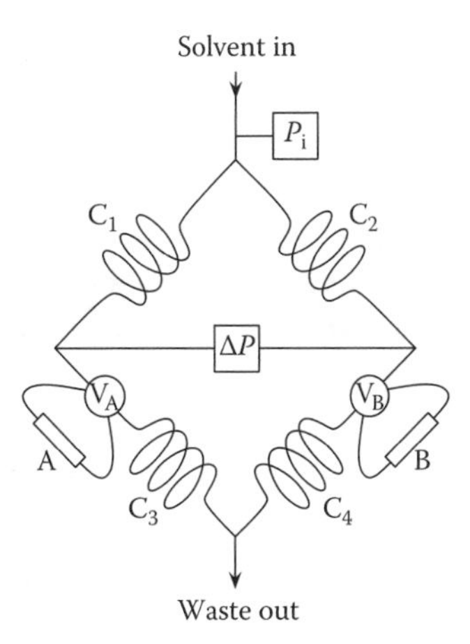

FIGURE 13.3 Schematic illustration of a capillary-bridge differential viscometer in which C_1, C_2, C_3 and C_4 are matched capillary tubes, V_A and V_B are tandem switching valves, and A and B are reservoirs containing polymer solution (A) or solvent (B).

Solvent is fed into the bridge at constant pressure whilst transducers measure the pressure P_i at the inlet and the pressure difference ΔP across the bridge, which should be zero when solvent is flowing through each capillary if the bridge is correctly balanced. To start the analysis, the tandem switching valves are used to switch simultaneously the flow paths to take routes through the two reservoirs A and B. Polymer solution then flows through capillary C_3 until all solvent is purged completely from that capillary. As a consequence of the difference in viscosity between the polymer solution and solvent, ΔP increases to a steady-state value ΔP_s and the overall flow rate Q (= V/t) through capillaries C_1 and C_3 reduces to Q_s. The flow rate Q_0 of solvent through capillaries C_2 and C_4 is unaffected. Applying Equation 13.10 to flow through each of the capillaries with the assumptions that they are of equal flow resistance and that the flow resistance of other components in the bridge is negligible, then

$$\text{for } C_3\text{:} \quad \eta_s = K\left(\frac{P_s}{Q_s}\right) \tag{13.13}$$

$$\text{for } C_1\text{:} \quad \eta_0 = K\left(\frac{P_i - P_s}{Q_s}\right) \tag{13.14}$$

$$\text{for } C_2 \text{ and } C_4\text{:} \quad \eta_0 = K\left(\frac{P_0}{Q_0}\right) \tag{13.15}$$

where

$K = \pi r^4/8l$ (assumed to be the same for each capillary)
P_s is the pressure between capillaries C_1 and C_3
P_0 is the pressure between capillaries C_2 and C_4.

Since the flow rate of solvent through capillaries C_2 and C_4 is the same, it follows that

$$P_0 = \frac{P_i}{2} \tag{13.16}$$

and so

$$P_s = P_0 + \Delta P_s = \frac{P_i}{2} + \Delta P_s \tag{13.17}$$

Taking ratios of Equations 13.13 through 13.15 and Equations 13.14 through 13.15 with rearrangement using the relations defined by Equations 13.16 and 13.17 leads to

$$\frac{\eta_s}{\eta_0} = \left(\frac{P_i + 2\Delta P_s}{P_i}\right)\left(\frac{Q_0}{Q_s}\right) \quad \text{and} \quad \frac{Q_0}{Q_s} = \frac{P_i}{P_i - 2\Delta P_s}$$

which upon combination give an equation for the relative viscosity of the polymer solution

$$\eta_r = \frac{P_i + 2\Delta P_s}{P_i - 2\Delta P_s} \tag{13.18}$$

from which the following equation for the specific viscosity, η_{sp} $(=\eta_r - 1)$, is obtained

$$\eta_{sp} = \frac{4\Delta P_s}{P_i - 2\Delta P_s} \tag{13.19}$$

Kinetic energy and density corrections are neglected, which is not unreasonable given that there is continuous flow through the capillaries and highly dilute polymer solutions are used (e.g. 0.0001 g cm^{-3}). In view of the high dilution it is normal practice to assume that the reduced viscosity is equal to the intrinsic viscosity (i.e. that the term in c in Equation 13.2 is negligible). A more serious assumption is that the effects of shear are negligible, because the shear rates in the capillaries are relatively high (typically 1000–5000 s^{-1}) and can lead to non-Newtonian flow behaviour or, even more seriously, shear degradation (i.e. scission of chains) with consequent errors, especially for very high molar mass polymers. Although rather expensive for use in routine dilute solution viscometry, capillary-bridge differential viscometers are easily adapted for use as detectors in gel permeation chromatography (see Section 14.3) where they provide continuous in-line measurement of intrinsic viscosity, which is by far their most important area of application.

13.3 ULTRACENTRIFUGATION

Measurements of the sedimentation behaviour of polymer molecules in solution can provide a considerable amount of information, e.g. hydrodynamic volume, average molar masses and even some indication of molar mass distribution. Such measurements have been used extensively to characterise biologically active polymers which often exist in solution as compact spheroids or rigid rods and are finding increasing use in the characterisation of synthetic polymers. Hence a brief non-theoretical introduction to sedimentation methods is given here.

The normal gravitational force acting upon a polymer molecule in solution is insufficient to overcome its Brownian motion and does not cause sedimentation. Thus in order to make measurements of sedimentation it is necessary to subject the polymer solutions to much higher gravitational fields. This is achieved using *ultracentrifuges*, which can attain rotational speeds of up to about 70,000 revolutions per minute (rpm) and generate centrifugal fields up to about 400,000 **g** where **g** is the acceleration due to gravity. The polymer solution is placed in a cell which fits into the rotor of the ultracentrifuge and has windows on either side. The cell is in the form of a truncated cone and is positioned so that its peak would be located at the axis of the rotor. This design ensures that sedimentation in radial directions is not restricted. The sedimentation process gives rise to a solvent phase and a concentrated polymer solution phase, which are separated by a boundary layer in which the polymer concentration varies. There is, therefore, a natural tendency for backward diffusion of the molecules in order to equalize the chemical potentials of the components in the different regions of the cell, and this causes broadening of the boundary layer. The breadth of the boundary layer also increases with the degree of molar mass dispersity because molecules of higher molar mass sediment at faster rates. The windows in the cell enable the radial variation in polymer concentration to be measured during ultracentrifugation, typically by monitoring refractive index differences (cf. light scattering; Section 12.2.2). There are two general methods by which sedimentation experiments can be performed and they will now be described briefly.

In a *sedimentation equilibrium* experiment, the cell is rotated at a relatively low speed (typically 5,000–10,000 rpm) until an equilibrium is attained whereby the centrifugal force just balances the tendency of the molecules to diffuse back against the concentration gradient developed. Measurements are made of the equilibrium concentration profiles for a series of solutions with different initial polymer concentrations so that the results can be extrapolated to $c=0$. A rigorous thermodynamic treatment is possible and enables absolute values of $\bar{M}_w$ and z-average molar mass $\bar{M}_z$ to be determined. The principal restriction to the use of sedimentation equilibrium measurements is the long time required to reach equilibrium, since this is at least a few hours and more usually is a few days.

The *sedimentation velocity* method involves rotating the solution cell at very high speeds (typically 60,000–70,000 rpm) and gives results in much shorter timescales than sedimentation equilibrium measurements. The movement of the boundary layer is monitored as a function of time and its steady-state velocity used to calculate the mean sedimentation coefficient S for the polymer in solution. Measurements are made for a series of solution concentrations and enable the limiting sedimentation coefficient S_0 to be obtained by extrapolation to $c=0$. In order to calculate an average molar mass, it is necessary either to know the limiting diffusion coefficient of the polymer in the solvent or to calibrate the system by measuring S_0 for a series of similar polymers, but which have narrow molar mass distributions and known molar masses. The latter procedure is more common and an equation similar in form to the Mark–Houwink–Sakurada Equation 13.5 is used to correlate S_0 data with molar mass for each specific polymer/solvent/temperature system. The resulting average molar mass usually is close to $\bar{M}_w$.

PROBLEMS

13.1 The tables below give the mean flow times t in a suspended-level viscometer recorded for solutions of two of five monodisperse samples of polystyrene at various concentrations c in cyclohexane at 34 °C. Under these conditions, the mean flow time t_0 for cyclohexane is 151.8 s. Determine the intrinsic viscosities $[\eta]$ of samples B and E.

Sample B	
$10^3 \times c$ / g cm^{-3}	t / s
1.586	158.5
5.553	178.6
7.403	189.6
8.884	199.5

Sample E	
$10^3 \times c$ / g cm^{-3}	t / s
1.040	176.1
1.691	195.4
2.255	214.8
2.706	232.4

The intrinsic viscosities of the other three polystyrene samples were evaluated under the same conditions and are given in the following table together with weight-average molar mass $\bar{M}_w$ values determined by light scattering. Using these data together with the calculated values of $[\eta]$ for samples B and E, evaluate the constants of the Mark–Houwink–Sakurada equation for polystyrene in cyclohexane at 34 °C. What can you deduce about the conformation of the polystyrene chains under the conditions of the viscosity determinations?

Polystyrene sample	$\bar{M}_w$ / kg mol^{-1}	$[\eta]$ / cm^3 g^{-1}
A	37.0	15.77
B	102.0	–
C	269.0	42.56
D	690.0	68.12
E	2402.0	–

13.2 The intrinsic viscosity of a sample of poly(methyl methacrylate) is 22.04 cm^3 g^{-1} in toluene at 25 °C and 25.53 cm^3 g^{-1} in ethyl acetate at 20 °C. Given the Mark–Houwink–Sakurada constants tabulated below, calculate the viscosity-average molar mass of the poly(methyl methacrylate) sample.

Solvent	Temperature / °C	K / cm^3 g^{-1} (g mol^{-1})$^{-a}$	a
Toluene	25	7.5×10^{-3}	0.72
Ethyl acetate	20	21.1×10^{-3}	0.64

13.3 The following table gives intrinsic viscosity $[\eta]$ values determined by dilute solution viscometry in toluene at 25 °C and weight-average molar mass $\bar{M}_w$ values determined by light scattering for five samples of poly(vinyl acetate). Using Equation 13.8 derived from the Flory–Fox theory, evaluate the parameter K_θ for poly(vinyl acetate) in toluene at 25 °C. Hence, using Equation 10.96, calculate the expansion parameter α_η for each of the poly(vinyl acetate) samples. Discuss the variation of α_η with molar mass.

$[\eta]$ / cm^3 g^{-1}	$\bar{M}_w$ / kg mol^{-1}
20.91	56.5
37.46	125.6
48.06	176.7
56.54	220.8
70.05	296.1

FURTHER READING

Allen, G. and Bevington, J.C. (eds), *Comprehensive Polymer Science*, Vol. 1 (Booth, C. and Price, C., eds), Pergamon Press, Oxford, U.K., 1989.

Barth, H.G. and Mays, J.W., *Modern Methods of Polymer Characterization*, John Wiley & Sons, New York, 1991.

Billingham, N.C., *Molar Mass Measurements in Polymer Science*, Kogan Page, London, U.K., 1977.

Flory, P.J., *Principles of Polymer Chemistry*, Cornell University Press, Ithaca, NY, 1953.

Hiemenz, P.C. and Lodge, T.P., *Polymer Chemistry*, 2nd edn., CRC Press, Boca Raton, FL, 2007.

Mark, H.F., Bikales, N.M., Overberger, C.G., and Menges, G. (eds), *Encyclopedia of Polymer Science and Engineering*, Wiley-Interscience, New York, 1985–1989.

Morawetz, H., *Macromolecules in Solution*, 2nd edn., Wiley-Interscience, New York, 1975.

Munk, P., *Introduction to Macromolecular Science*, Wiley-Interscience, New York, 1989.

Sun, S.F., *Physical Chemistry of Macromolecules: Basic Principles and Issues*, 2nd edn., Wiley-Interscience, New York, 2004.

Tanford, C., *Physical Chemistry of Macromolecules*, John Wiley & Sons, New York, 1961.

14 Molar Mass Distribution

14.1 INTRODUCTION

In many instances, average molar masses and their ratios (i.e. molar mass dispersities) are insufficient to describe the properties of a polymer, and more complete information on the molar mass distribution (MMD) is required. This is particularly important for polymers that have MMDs which are broad, non-uniform (e.g. having low or high molar mass shoulders) and/or multimodal. Even for polymers with relatively simple MMDs, there is advantage in knowing the complete MMD. Furthermore, any molar mass average can be calculated from the moments of the distribution curve. For example, if $w(M)$ is the weight-fraction MMD function, then the number-average molar mass $\bar{M}_n$ can be defined in integral form (cf. Equation 1.7) as

$$\bar{M}_n = \frac{1}{\int_0^\infty w(M)(1/M)\mathrm{d}M} \tag{14.1}$$

the weight-average molar mass $\bar{M}_w$ (cf. Equation 1.8) as

$$\bar{M}_w = \int_0^\infty w(M)M\mathrm{d}M \tag{14.2}$$

and the z-average molar mass $\bar{M}_z$ (cf. Equation 1.10) as

$$\bar{M}_z = \frac{\int_0^\infty w(M)M^2\mathrm{d}M}{\int_0^\infty w(M)M\mathrm{d}M} \tag{14.3}$$

Similarly from Equation 13.6 (recognizing that $w_i = n_i M_i$ and that $\int_0^\infty w(M)\mathrm{d}M = 1$, i.e. $w(M)$ is normalized), an expression for the viscosity-average molar mass $\bar{M}_v$ is obtained

$$\bar{M}_v = \left[\int_0^\infty w(M)M^a\mathrm{d}M\right]^{1/a} \tag{14.4}$$

where a is the exponent in the Mark–Houwink–Sakurada Equation 13.5.

14.2 FRACTIONATION

During the first few decades of polymer science, there were no techniques available for determining directly complete MMDs of polymers. Hence, it was necessary to separate, i.e. *fractionate*, the polymer into a number of fractions each of which has a narrow distribution of molar mass; the weight and molar mass of each polymer fraction then could be determined (e.g. using the methods described in Chapters 11 through 13), thereby enabling the MMD to be constructed in the form of a histogram. Such procedures are rarely used nowadays because much more rapid and powerful methods are available for determining MMDs, the most important of which is size-exclusion chromatography (Section 14.3). Nevertheless, fractionation itself is still practised, often for purposes of purification, and will be considered here in some detail because it introduces the important topic of phase-separation behaviour of polymers, which will be developed further in Chapter 18.

14.2.1 Phase-Separation Behaviour of Polymer Solutions

The simplest procedure for polymer fractionation is to dissolve the polymer at low concentration in a poor solvent and then to bring about stepwise phase separation (i.e. 'precipitation') of polymer fractions by either changing the temperature or adding a non-solvent. The highest molar mass species phase separate first and so the fractions are obtained in order of decreasing molar mass. Phase separation can be treated theoretically on the basis of Flory–Huggins theory.

The effect of temperature upon phase separation of solutions of non-crystallizing polymers will be considered here since it is easier to analyse. It is usual to deal with molar quantities and so both sides of the Flory–Huggins equation (Equation 10.32) must be divided by (n_1+xn_2) where n_1 and n_2 are the numbers of moles of solvent and polymer present, and x is the *number of segments* in each of the polymer molecules (see Section 10.2.2), which are assumed to have the same molar mass. This gives the following equation for ΔG_m^*, the Gibbs free energy of mixing per mole of lattice sites (which, therefore, can be considered to be the Gibbs free energy of mixing per mol of segments)

$$\Delta G_m^* = \mathbf{R}T\left[\phi_1 \ln\phi_1 + \left(\frac{\phi_2}{x}\right)\ln\phi_2 + \chi\phi_1\phi_2\right] \tag{14.5}$$

where

ϕ_1 $(=n_1/(n_1+xn_2))$ and ϕ_2 $(=xn_2/(n_1+xn_2))$ are the volume fractions of solvent and polymer, respectively

χ is the Flory–Huggins polymer–solvent interaction parameter.

The first two terms in Equation 14.5 are from the combinatorial entropy of mixing and always are negative, thus promoting miscibility, whereas the final term comes from the contact interactions and is positive, thereby reducing miscibility. The balance between these two contributions depends upon temperature and χ (which also is temperature dependent), and determines whether or not a polymer solution will phase separate. Hence, Equation 14.5 describes a series of curves for the variation of ΔG_m^* with ϕ_2, one for each temperature, as depicted in Figure 14.1a, which shows schematically curves that would be consistent with a very low value of x (e.g. 2). As x increases, the curves rapidly become severely asymmetric and, for those which show two minima, the minimum at low ϕ_2 becomes imperceptibly small and moves to extremely low values of ϕ_2, whilst the second minimum at high ϕ_2 becomes more shallow and moves to higher ϕ_2.

Thus, the ΔG_m^* versus ϕ_2 curves can be considered to have one of two general forms, which for the purposes of further analysis are presented in Figure 14.1b. At temperature T_{1b} in Figure 14.1b the polymer and solvent are miscible in all proportions, as is evident from consideration of any point on the curve. For example, if a homogeneous solution with $\phi_2=\phi_{2A}$, corresponding to point A, were to separate into two co-existing phases, conservation of matter demands that one should have $\phi_2<\phi_{2A}$ and the other $\phi_2>\phi_{2A}$, e.g. corresponding to points a′ and a″. It is a relatively simple matter to show that the Gibbs free energy change associated with the phase separation process is given by the difference between (i) the value of ΔG_m^* corresponding to the point of intersection of the *tie-line* (joining points a′ and a″) with the vertical $\phi_2=\phi_{2A}$ line, and (ii) the value of ΔG_m^* on the curve

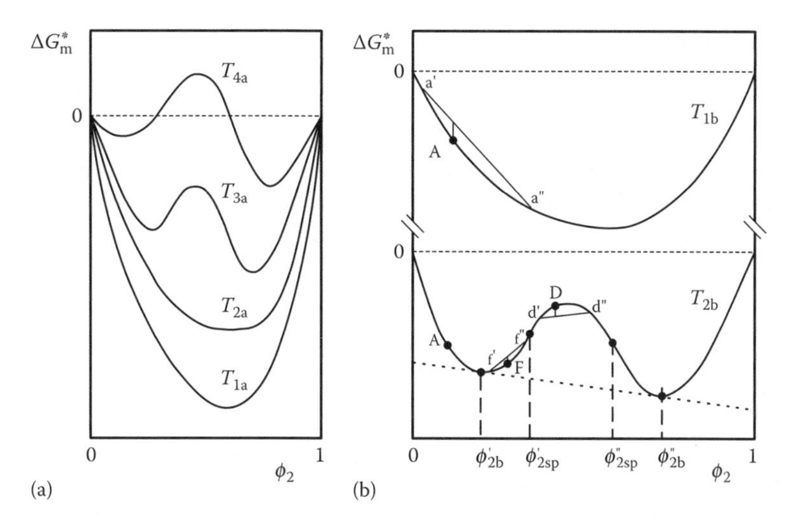

FIGURE 14.1 Schematic illustrations of the variation of ΔG_m^* with ϕ_2 and temperature T: (a) shows how the shape of the curves changes with T, and (b) shows, for the two types of ΔG_m^* versus ϕ_2 curve, the constructions used to identify regions of ϕ_2 where an initially homogeneous solution will be stable, unstable, or metastable towards phase separation.

at A. Clearly, this difference is positive for all points on the curve and so the existence of a single homogeneous phase is favoured for all ϕ_2.

The situation at temperature T_{2b} in Figure 14.1b is rather more complex since two ΔG_m^* minima are present. Consider again phase separation from a homogeneous solution corresponding to ϕ_{2A}. It is easy to see that tie-lines joining any two points on the curve either side of ϕ_{2A} will intersect the vertical $\phi_2 = \phi_{2A}$ line above the curve. This is true for all compositions in the ranges $0 < \phi_2 < \phi'_{2b}$ and $\phi''_{2b} < \phi_2 < 1$ and so homogeneous solutions with ϕ_2 in these ranges are stable at T_{2b}. Now consider phase separation of a homogeneous solution with $\phi_2 = \phi_{2D}$ corresponding to point D. The tie-lines joining two points (such as d′ and d″) immediately on either side of ϕ_{2D} intersect the vertical $\phi_2 = \phi_{2D}$ line *below* the curve. Thus the homogeneous solution is unstable and phase separation takes place until the system becomes stable when the *two co-existing phases* have the *binodal compositions* ϕ'_{2b} and ϕ''_{2b}. All homogeneous solutions with compositions in the range $\phi'_{2sp} < \phi_2 < \phi''_{2sp}$ are similarly unstable and separate into two phases corresponding to ϕ'_{2b} and ϕ''_{2b}. The general condition for equilibrium between two co-existing phases is that for each component, the chemical potential must be the same in both phases, i.e. $\mu'_1 = \mu''_1$ and $\mu'_2 = \mu''_2$. This condition usually is written in terms of chemical potential differences

$$\mu_1 - \mu_1^{\mathrm{o}} = \mu_1'' - \mu_1^{\mathrm{o}} \tag{14.6a}$$

$$\mu_2 - \mu_2^{\mathrm{o}} = \mu_2'' - \mu_2^{\mathrm{o}} \tag{14.6b}$$

since these are more directly related to ΔG_m^*. It can be shown that for any point on the curve, $(\mu_1 - \mu_1^{\mathrm{o}})$ and $(\mu_2 - \mu_2^{\mathrm{o}})/x$ are given by the values of ΔG_m^* which correspond to the intersections of (i) the tangent to the curve at that point, with (ii) the vertical $\phi_2 = 0$ and $\phi_2 = 1$ lines, respectively. Thus Equations 14.6a and 14.6b are satisfied when two points on the curve have a *common tangent,* as for ϕ'_{2b} and ϕ''_{2b} in Figure 14.1b. Since the variation of ΔG_m^* with ϕ_2 is unsymmetrical for polymer solutions, the binodal points do not correspond to the minima in the curve.

Finally, consider a homogeneous solution with $\phi_2 = \phi_{2F}$ corresponding to point F on the curve for T_{2b}. Clearly, phase separation into two phases corresponding to the binodal compositions is favoured thermodynamically. However, in order for this to occur an energy barrier must be overcome because the initial stages of phase separation about ϕ_{2F} (e.g. to points f′ and f″) give rise to an increase in the Gibbs free energy. This is true for all homogeneous solutions with compositions in the ranges $\phi'_{2b} < \phi_2 < \phi'_{2sp}$ and $\phi''_{2sp} < \phi_2 < \phi''_{2b}$, where ϕ'_{2sp} and ϕ''_{2sp} are the *spinodal compositions* corresponding to the points of inflection in the curve. Such solutions are said to be *metastable* and will phase separate to the binodal compositions, but only if the energy barrier can be overcome. Since the spinodal compositions occur at points of inflection, they are located by application of the condition

$$\left(\frac{\partial^2 \Delta G_m^*}{\partial \phi_2^2}\right)' = \left(\frac{\partial^2 \Delta G_m^*}{\partial \phi_2^2}\right)'' = 0 \tag{14.7}$$

The existence of two minima in the variation of ΔG_m^* with ϕ_2 (and hence phase separation) results from the contribution to ΔG_m^* due to contact interactions (i.e. χ and, hence, non-zero enthalpy of mixing, ΔH_m), which cause ΔG_m^* to increase. This contribution reduces as the temperature changes from T_{2b} towards T_{1b} in Figure 14.1b and so the binodal and spinodal points get closer together until at a *critical temperature* T_c they just coincide at a single point corresponding to ϕ_{2c} (T_{2a} in Figure 14.1a would correspond closely to T_c). The curves defined by the binodal points and spinodal points as a function of temperature are known as the *binodal* and *spinodal*, respectively. For most polymer solutions χ reduces as temperature increases (i.e. the contact enthalpy parameter b in Equation 10.28 is positive), so that T_c ($>T_{2b}$) corresponds to the *common maximum* of the binodal and spinodal, and is known as the *upper critical solution temperature* (*UCST*) above which the polymer and solvent are miscible in all proportions (see Figure 14.2a). For the less common situation where χ increases as temperature increases (i.e. the contact enthalpy parameter b in Equation 10.28 is negative), T_c ($<T_{2b}$) corresponds to the *common minimum* of the binodal and spinodal, and is known as the *lower critical solution temperature* (*LCST*) below which the polymer and solvent are completely miscible (see Figure 14.2b). LCST behaviour usually is observed when there are specific favourable polymer–solvent interactions (e.g. hydrogen bonding, charge transfer). The last two examples given in Table 10.1, and poly(ethylene oxide) in water, are systems which show LCST behaviour due to hydrogen bonding interactions, which become weaker as temperature increases. LCST behaviour can also be caused by volume contraction upon mixing because this leads to a reduction in the entropy of mixing. Flory–Huggins theory assumes there to be no volume change and so more advanced theories are required to account quantitatively for the effects of volume changes. Such theories predict that all polymer–solvent systems should show both UCST and LCST behaviour, though obviously in different temperature regimes. However, in all but a few cases (e.g. polystyrene in cyclohexane) only one of these regimes is accessible experimentally.

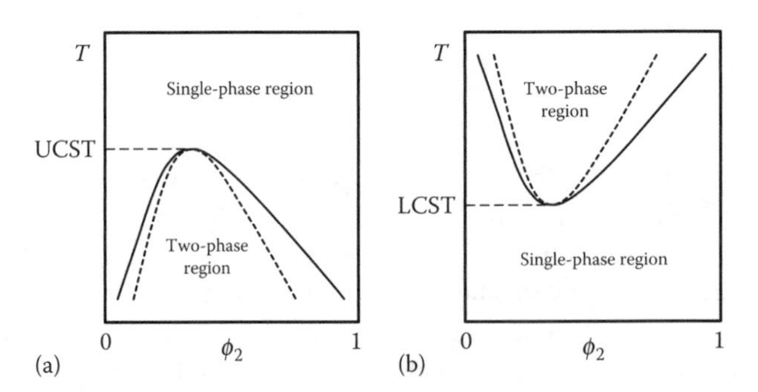

FIGURE 14.2 Schematic illustration of phase diagrams for polymer solutions showing (a) UCST behaviour and (b) LCST behaviour. The solid lines are the binodals and the dashed lines are the spinodals.

The regions outside the binodal correspond to stable homogeneous solutions, whereas the regions within the spinodal correspond to unstable solutions which will spontaneously phase separate. The regions between the binodal and spinodal correspond to metastable solutions which only phase separate if an energy barrier can be overcome.

For both UCST and LCST behaviour, T_c coincides with the turning point in the spinodal and so can be located by application of the condition

$$\left(\frac{\partial^2 \Delta G_m^*}{\partial \phi_2^2}\right)' = \left(\frac{\partial^3 \Delta G_m^*}{\partial \phi_2^3}\right)'' = 0 \tag{14.8}$$

These derivatives can be evaluated easily from Equation 14.5 by recognizing that $\phi_1 = 1 - \phi_2$ and by assuming that χ is independent of ϕ_2. Thus $\left(\partial^2 \Delta G_m^* / \partial \phi_2^2 = 0\right)$ leads to the following equation for the spinodal points

$$\frac{1}{1-\phi_2} + \frac{1}{x\phi_2} - 2\chi = 0 \tag{14.9}$$

and so a theoretical spinodal curve can be constructed if the variation of χ with temperature is known. Application of the condition $\left(\partial^3 \Delta G_m^* / \partial \phi_2^3\right) = 0$ gives the critical composition as

$$\phi_{2c} = \frac{1}{1 + x^{1/2}} \tag{14.10}$$

Hence, ϕ_{2c} is very small for polymers (x is large). The critical value χ_c of the Flory–Huggins interaction parameter is obtained by substituting Equation 14.10 into 14.9

$$\chi_c = \frac{1}{2}\left[1 + \frac{2}{x^{1/2}} + \frac{1}{x}\right] \tag{14.11}$$

Notice that as $x \to \infty$, $\phi_{2c} \to 0$ and $\chi_c \to 1/2$. For most polymers of interest, x is finite but large (>100) and so the critical point corresponds to $0 \le \phi_{2c} < 0.1$ and $0.5 \le \chi_c < 0.7$. For phase separation of a solution of a polymer with molar mass dispersity, the binodal, spinodal and values of ϕ_{2c} and χ_c are obtained by replacing x by its number-average value, and are intermediate to those of the individual polymer species with specific values of x.

14.2.2 Theory of Fractionation by Phase Separation of Dilute Polymer Solutions

It should now be evident that there exist unique pairs of binodal and spinodal curves for each value of x. As x increases, these curves become increasingly skewed towards the $\phi_2 = 0$ axis and T_c moves either to higher temperatures for UCST behaviour or lower temperatures for LCST behaviour (see Figure 14.3). The origin of fractionation by phase separation now is clearly evident, since preferential phase separation of the highest molar mass species can be expected. The next consideration is prediction of how chains with different values of x become distributed between the two co-existing phases in a phase-separated solution. For this purpose, the chemical potential for the n_j chains which have x_j segments is required and can be obtained by partial differentiation of the Flory–Huggins equation (Equation 10.32) with respect to n_j after replacing the second term in the equation by a summation over all chains present in the polymer. Thus, the chemical potential difference $\mu_j - \mu_j^{\circ}$ for chains j in a polymer with molar mass dispersity is given by

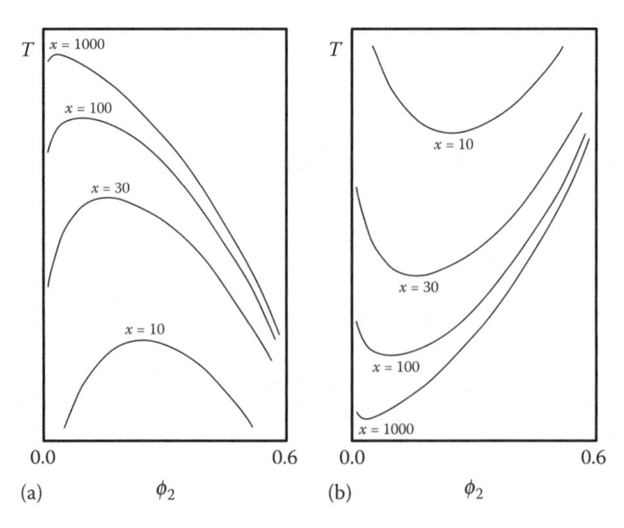

FIGURE 14.3 The effect of the number of chain segments x upon the binodals for (a) UCST behaviour and (b) LCST behaviour.

$$\mu_j - \mu_j^{\mathrm{o}} = \left(\frac{\partial \Delta G_{\mathrm{m}}}{\partial n_j}\right)_{P,T,n_i \neq n_j} = \mathbf{R}T\frac{\partial}{\partial n_j}\left\{n_1 \ln\phi_1 + n_j \ln\phi_j + \sum_{i \neq j} n_i \ln\phi_i + n_1\phi_2\chi\right\}_{P,T,n_i \neq n_j}$$

Using standard mathematical procedures for differentiation and recognizing that ϕ_1 and ϕ_2 are functions of n_j, leads to

$$\mu_j - \mu_j^{\mathrm{o}} = \mathbf{R}T\left\{n_1\frac{\partial \ln\phi_1}{\partial n_j} + \ln\phi_j + n_j\frac{\partial \ln\phi_j}{\partial n_j} + \sum_{i \neq j} n_i\frac{\partial \ln\phi_i}{\partial n_j} + n_1\chi\frac{\partial \phi_2}{\partial n_j}\right\}_{P,T,n_i \neq n_j}$$

which then becomes

$$\mu_j - \mu_j^{\mathrm{o}} = \mathbf{R}T\left\{\left(\frac{n_1}{\phi_1}\right)\frac{\partial \phi_1}{\partial n_j} + \ln\phi_j + \left(\frac{n_j}{\phi_j}\right)\frac{\partial \phi_j}{\partial n_j} + \sum_{i \neq j}\left(\frac{n_i}{\phi_i}\right)\frac{\partial \phi_i}{\partial n_j} + n_1\chi\frac{\partial \phi_2}{\partial n_j}\right\}_{P,T,n_i \neq n_j} \tag{14.12}$$

Considering the definitions of ϕ_1, ϕ_2, ϕ_i and ϕ_j

$$\phi_1 = \frac{n_1}{n_1 + x_j n_j + \sum_{i \neq j} x_i n_i} \text{ and so } \left\{\frac{\partial \phi_1}{\partial n_j}\right\}_{P,T,n_i \neq n_j} = -\frac{x_j n_1}{\left(n_1 + x_j n_j + \sum_{i \neq j} x_i n_i\right)^2} = -\frac{\phi_1 x_j}{n_1 + \sum_i x_i n_i}$$

$$\phi_j = \frac{n_j}{n_1 + x_j n_j + \sum_{i \neq j} x_i n_i} \text{ and so } \left\{\frac{\partial \phi_j}{\partial n_j}\right\}_{P,T,n_i \neq n_j} = \frac{\left(n_1 + x_j n_j + \sum_{i \neq j} x_i n_i\right) - x_j n_j}{\left(n_1 + x_j n_j + \sum_{i \neq j} x_i n_i\right)^2} = \frac{1 - \phi_j x_j}{n_1 + \sum_i x_i n_i}$$

$$\phi_i = \frac{n_i}{n_1 + x_j n_j + \sum_{i \neq j} x_i n_i} \text{ and so } \left\{\frac{\partial \phi_{i \neq j}}{\partial n_j}\right\}_{P,T,n_i \neq n_j} = -\frac{x_j n_i}{\left(n_1 + x_j n_j + \sum_{i \neq j} x_i n_i\right)^2} = -\frac{\phi_i x_j}{n_1 + \sum_i x_i n_i}$$

$$\phi_2 = 1 - \phi_1 \text{ and so } \left\{\frac{\partial \phi_2}{\partial n_1}\right\}_{P,T,n_i \neq n_j} = -\left\{\frac{\partial \phi_1}{\partial n_1}\right\}_{P,T,n_i \neq n_j} = \frac{\phi_1 x_j}{n_1 + \sum_i x_i n_i}$$

which upon substitution into Equation 14.12, with simplification, gives

$$\mu_j - \mu_j^{\circ} = \mathbf{R}T\left[-\phi_1 x_j + \ln\phi_j + 1 - \sum_i \left(\frac{x_j n_i}{n_1 + \sum_i x_i n_i}\right) + \chi \phi_1^2 x_j\right] \tag{14.13}$$

The term containing the summations may be simplified as follows

$$\sum_i \left(\frac{x_j n_i}{n_1 + \sum_i x_i n_i}\right) = x_j \sum_i \left(\frac{n_i}{n_1 + \sum_i x_i n_i}\right)\left(\frac{\sum_i x_i n_i}{\sum_i n_i}\right)\left(\frac{\sum_i n_i}{\sum_i x_i n_i}\right)$$

$$= x_j \sum_i \left(\frac{n_i}{\sum_i n_i}\right)\left(\frac{\phi_2}{\bar{x}_n}\right) = x_j \left(\frac{\phi_2}{\bar{x}_n}\right)$$

where $\bar{x}_n$ is number-average value of x. Hence, with further rearrangement Equation 14.13 becomes

$$\mu_j - \mu_j^{\circ} = \mathbf{R}T\left[\ln\phi_j + 1 - x_j + \phi_2 x_j \left(1 - \frac{1}{\bar{x}_n}\right) + \chi \phi_1^2 x_j\right] \tag{14.14}$$

which reduces to Equation 10.39 for a monodisperse polymer (i.e. when $\bar{x}_n = x_j = x$, and the subscript j is replaced by '2'). Applying the phase equilibrium Equation 14.6b for chains j, i.e. $\mu'_j - \mu_j^{\circ} = \mu''_j - \mu_j^{\circ}$ where the double prime designates the more concentrated phase, leads to the following expression for the partitioning of chains j between the two co-existing phases

$$\frac{\phi''_j}{\phi'_j} = e^{\sigma x_j} \tag{14.15}$$

where the parameter σ is given by

$$\sigma = (\phi'_2 - \phi''_2)\left(1 - \frac{1}{\bar{x}_n}\right) + \chi\left[\phi_1'^2 - \phi_1''^2\right] \tag{14.16}$$

For two particular co-existing phases, the volume fractions in Equation 14.16 and χ have specific values, and σ is a constant. Hence, in principle, Equation 14.16 enables σ to be calculated, but in practice the limitations of Flory–Huggins theory mean that the calculations are not reliable.

However, since phase separation occurs when $\chi > 1/2$ (because $\bar{x}_n < \infty$), such calculations are sufficient to show that values of σ are small but positive. The most important conclusion, therefore, is that the partitioning depends exponentially upon x_j with $\phi_j'' > \phi_j'$ for all values of x_j and the ratio ϕ_j''/ϕ_j' increasing rapidly with x_j.

Assuming that the polymer density is independent of x_j, then on the basis of Equation 14.15 the ratio of the mass fractions f_j'' and f_j' of the x_j-mers in the co-existing phases is given by

$$\frac{f_j''}{f_j'} = Re^{\sigma x_j} \tag{14.17}$$

where $R = V''/V'$, and V'' and V' are the volumes of the concentrated and dilute co-existing phases, respectively. Efficient fractionation requires that $\sigma \ll 1$ and $R \ll 1$. The value of σ is not easily varied, but typically is about 0.01 and so is satisfactory. In contrast, the value of R can be altered and in order to ensure that $V' > V''$ it is necessary to begin with very dilute homogeneous solutions (typically about 2 g dm^{-3}). The data given in Table 14.1 show clearly that whilst all species are present in each phase, the concentrated phase contains almost all of the high molar mass species. The volume fraction ratio ϕ_j''/ϕ_j' is close to unity for the low molar mass species so they are present predominantly in the dilute phase because of its much larger volume. Thus the polymer present in the concentrated phase has a relatively narrow MMD, typically with $\bar{M}_w/\bar{M}_n$ in the range 1.1–1.3.

Flory–Huggins theory gives reasonable predictions for phase separation of dilute polymer solutions because the excluded volume is close to zero under the conditions of phase separation. The limitations of the theory presented here arise principally from the unsatisfactory assumptions that χ is independent of ϕ_2 and that the volume change upon mixing is zero. Whilst the prediction of T_c generally is good, experimentally determined binodals tend to be less sharp with ϕ_{2c} larger than predicted by the theory. Better agreement can be gained by taking into account the dependence of χ upon ϕ_2 and the effects of volume changes, but such theories are beyond the scope of this book.

Measurements of T_c can be used to determine theta temperatures. Comparison of the Flory–Huggins dilute solution Equations 3.50 and 3.57 leads to

$$\chi - \frac{1}{2} = \psi\left[\left(\frac{\theta}{T}\right) - 1\right] \tag{14.18}$$

At $T = T_c$, χ can be substituted by χ_c from Equation 14.11 to give after rearrangement

$$\frac{1}{T_c} = \frac{1}{\theta} + \frac{1}{\psi\theta}\left[\frac{1}{x^{1/2}} + \frac{1}{2x}\right] \tag{14.19}$$

TABLE 14.1
Some Values of ϕ_j''/ϕ_j' and f_j''/f_j' Calculated from Equations 14.15 and 14.17 with $\sigma = 0.01$ and $R = 0.005$

x_j	10	100	500	1000	5000
ϕ_j''/ϕ_j'	1.1	2.7	148.4	2.2×10^4	5.2×10^{21}
f_j''/f_j'	0.006	0.014	0.74	110	2.6×10^{19}

in which x is replaced by $\bar{x}_n$ for a polymer with molar mass dispersity. Thus a plot of $1/T_c$ against $(1/\bar{x}_n^{1/2} + 1/2\bar{x}_n)$ gives a straight line with intercept $1/\theta$. It is found that theta temperatures obtained in this way are in good agreement with those obtained from osmotic pressure measurements (Section 11.2.2). Inspection of Equation 14.19 reveals that $T_c \rightarrow \theta$ as $x \rightarrow \infty$, thus providing another alternative definition of θ as the critical temperature for miscibility in the limit of infinite molar mass.

14.2.3 Procedures for Fractionation

The basic requirements for fractionation were established in the preceding section. Thus phase separation of a very dilute polymer solution is brought about by causing the solvency conditions to deteriorate (i.e. causing χ to increase). This can be achieved either by adding a non-solvent to the solution or by changing the temperature. The former procedure is preferred and involves addition of non-solvent to the polymer solution until phase separation is clearly evident. The addition of non-solvent is stopped at this point and the solution heated (assuming UCST behaviour) to redissolve the concentrated phase (i.e. χ is decreased). This solution then is cooled *slowly* to achieve equilibrium phase separation. After allowing the concentrated phase to settle at the bottom of the fractionation flask, it is removed and the remaining dilute solution phase subjected to further similar fractionations. The MMDs of the fractions obtained are shown schematically in Figure 14.4 in relation to the MMD of the polymer before fractionation. Thus there is considerable overlap of the distribution curves of the individual fractions and this has to be taken into account when constructing the distribution curve for the original polymer.

The procedure just described is rather time-consuming and so attempts have been made to automate fractionation by phase separation using methods of chromatography. One such technique involves depositing the polymer at the top of a column of glass beads and then passing a mixture of solvent and non-solvent through the column which usually has a gradient of temperature along its length (decreasing temperature for UCST behaviour). The concentration of solvent in the eluant is gradually increased so that the low molar mass species dissolve first, the eluant stream being collected in successive portions and the different polymer fractions recovered by removal of the solvent from each fraction. Nowadays, continuous liquid–liquid extraction procedures are employed whereby the polymer initially is dissolved in one solvent and is extracted continuously from this solution by another solvent which is only partially miscible with the first solvent.

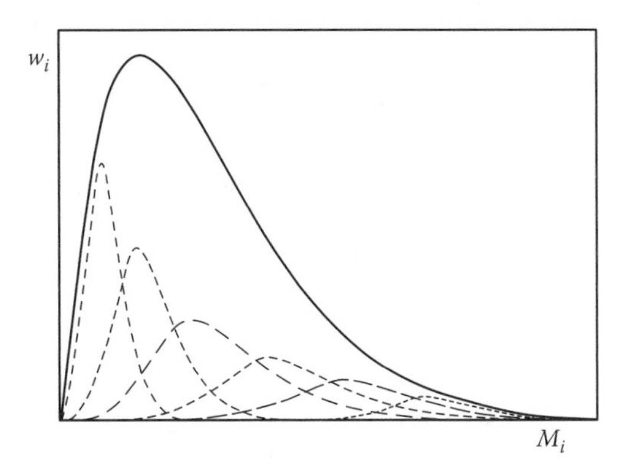

FIGURE 14.4 Schematic representation of typical weight fraction w_i versus molar mass M_i distribution curves for the fractions obtained from successive phase separations of a dilute solution of a polymer sample with molar mass dispersity, the full MMD of which is shown as the solid line.

14.3 GEL PERMEATION CHROMATOGRAPHY

The technique of *gel permeation chromatography* (GPC) was developed during the mid-1960s and is an extremely powerful method for determining the complete MMD of a polymer. In GPC a dilute polymer solution is injected into a solvent stream, which then flows through a column packed with beads of a porous gel. The porosity of the gel is of critical importance and typically is in the range 5–10^5 nm. The small solvent molecules pass both through and around the beads, carrying the polymer molecules with them where possible. The smallest polymer molecules are able to pass through most of the pores in the beads and so have a relatively long flow-path through the column. However, the largest polymer molecules are excluded from all but the largest of the pores because of their greater molecular size and consequently have a much shorter flow-path. Thus GPC is a form of *size-exclusion chromatography* in which the polymer molecules elute from the chromatography column in order of decreasing molecular size in solution. The concentration of polymer in the eluate is monitored continuously and the chromatogram obtained is a plot of concentration against elution volume, which provides a qualitative indication of the MMD. The chromatogram also can reveal the presence of low molar mass additives (e.g. plasticizers), since they appear as separate peaks at large elution volumes.

14.3.1 SEPARATION BY SIZE EXCLUSION

The volume of solvent contained in a GPC system from the point of solution injection, through the column to the point of concentration detection can be considered as the sum of a void volume V_0 (i.e. the volume of solvent in the system outside the porous beads) and an internal volume V_i (i.e. the volume of solvent inside the beads). The volume of solvent required to elute a particular polymer species from the point of injection to the detector is known as its *elution volume* V_e and on the basis of *separation by size exclusion* is given by

$$V_e = V_0 + K_{se}V_i \tag{14.20}$$

where K_{se} is the fraction of the internal pore volume penetrated by those particular polymer molecules. For small polymer molecules that can penetrate all of the internal volume $K_{se}=1$ and $V_e=V_0+V_i$, whereas for very large polymer molecules that are unable to penetrate the pores $K_{se}=0$ and $V_e=V_0$. Clearly for polymer molecules of intermediate size K_{se} and V_e lie between these limits.

For the flow rates typically used in GPC (about 1 cm^3 min^{-1}) it is found that V_e is independent of flow rate. This means that there is sufficient time for the molecules to diffuse into and out of the pores such that equilibrium concentrations c_i and c_0 of the molecules are attained inside and outside the pores, respectively. Since the separation process takes place under equilibrium conditions, K_{se} can be considered as the *equilibrium constant* defined by $K_{se}=c_i/c_0$ (note that when the polymer molecules penetrate the pores to the same extent as the solvent $c_i=c_0$ and $K_{se}=1$). From thermodynamics the relation between K_{se} and the Gibbs free energy change ΔG_p^o for permeation of the pores is given by

$$\Delta G_p^o = -\mathbf{R}T \ln K_{se} \tag{14.21}$$

For separation exclusively by size exclusion (i.e. without contributions from interactions), the enthalpy change associated with transfer of solute species into the pores *must be zero*. Thus ΔG_p^o is controlled by the corresponding entropy change ΔS_p^o and is given by $\Delta G_p^o = -T\Delta S_p^o$. Hence from Equation 14.21, K_{se} is given by

$$K_{se} = \exp\left(\frac{\Delta S_p^o}{\mathbf{R}}\right) \tag{14.22}$$

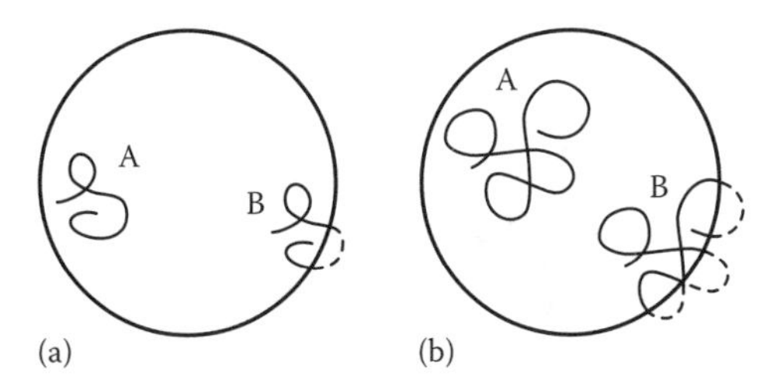

FIGURE 14.5 Schematic illustration of (a) a small and (b) a large polymer molecule inside a cylindrical pore. At point A, the molecule is positioned with its centre of mass at its distance of closest approach to the wall, whereas at point B the same molecular conformation is excluded because the molecule is positioned very close to the wall.

The number of conformations available to a polymer molecule is reduced inside a pore because of its close proximity to the impenetrable walls of the pore, and so $\Delta S_{\mathrm{p}}^{\mathrm{o}}$ is negative. This loss of conformational entropy is equivalent to exclusion of the centre of mass of the molecule from regions closer to the walls of the pores than the hydrodynamic radius of the molecule (see Figure 14.5). Methods of statistical thermodynamics can be used to obtain expressions for $\Delta S_{\mathrm{p}}^{\mathrm{o}}$ on the basis of simplifying models (e.g. Gaussian coils entering cylindrical pores). The details will not be considered here, but the general result is of the form

$$\Delta S_{\mathrm{p}}^{\mathrm{o}} = -\mathbf{R}A_{\mathrm{s}}\left(\frac{\bar{L}}{2}\right) \tag{14.23}$$

where

A_{s} is the surface area per unit pore volume

$\bar{L}$ is the mean molecular projection of the molecule when free in solution (e.g. for a spherical molecule, $\bar{L}$ is its mean diameter).

Substituting Equations 14.22 and 14.23 into 14.20 gives

$$V_{\mathrm{e}} = V_0 + V_{\mathrm{i}} \exp\left(\frac{-A_{\mathrm{s}}\bar{L}}{2}\right) \tag{14.24}$$

Thus V_{e} is predicted to vary exponentially with $\bar{L}$ (i.e. with molecular size) and $V_{\mathrm{e}} \rightarrow (V_0 + V_{\mathrm{i}})$ as $\bar{L} \rightarrow 0$, with $V_{\mathrm{e}} \rightarrow V_0$ as $\bar{L} \rightarrow \infty$. An alternative interpretation of Equation 14.24 is that V_{e} can be expected to decrease approximately *linearly* with $\log \bar{L}$ (i.e. with the logarithm of molecular size). This interpretation is particularly useful for the purposes of calibration.

14.3.2 Calibration and Evaluation of Molar Mass Distributions

In order to convert a GPC chromatogram into a MMD curve it is necessary to know the relationship between molar mass M and V_{e}. This relationship results from Equation 14.24 due to the dependence of molecular size upon M. The molecular size of a polymer molecule in solution can be taken as its hydrodynamic volume (see Section 10.4.2), which from Equation 10.94 is proportional to $[\eta]M$ where $[\eta]$ is the intrinsic viscosity. Hence it can be predicted that $\log([\eta]M)$ will decrease approximately linearly with V_{e}. From the Mark–Houwink–Sakurada equation (Equation 10.98)

$$\log([\eta]M) = \log K + (1+a)\log M \tag{14.25}$$

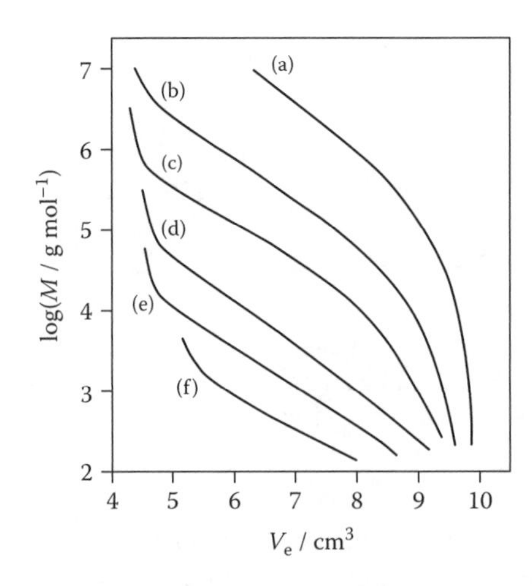

FIGURE 14.6 Some typical GPC calibration curves for polystyrene in tetrahydrofuran obtained using a GPC system comprising a series of six different GPC columns packed with gels of porosity: (a) 10^5 nm, (b) 10^4 nm, (c) 10^3 nm, (d) 10^2 nm, (e) 50 nm, (f) 5 nm. In each case, the calibration curve is approximately linear over the range of resolvable molar masses. (Data provided by Polymer Laboratories Ltd., Shropshire, U.K.)

and so $\log([\eta]M)$ increases linearly with $\log M$ as long as the Mark–Houwink–Sakurada constants K and a remain truly constant. Thus a calibration plot of $\log M$ against V_e should be approximately linear, as is found in practice. The calibration plot is specific to the polymer under study (through the values of K and a) and is obtained by measuring V_e for a series of narrow MMD samples of the polymer with different M (see Figure 14.6). The value of V_e usually is taken as that at the peak of the chromatogram for this purpose.

Conversion of the GPC chromatogram of detector response $f(V_e)$ against V_e to the *weight-fraction MMD* (i.e. $w(M)$ against M) can now be considered (see Figure 14.7). Assuming that $f(V_e)$ is directly proportional to the mass concentration of the polymer in the eluate and that the proportionality constant is independent of M, then the weight fraction dw of polymer which elutes between V_e and $(V_e + dV_e)$ is given by

$$dw = \frac{f(V_e)dV_e}{\int_0^\infty f(V_e)dV_e} \tag{14.26}$$

where the integral in the denominator is simply the area A under the GPC chromatogram and serves to normalize $f(V_e)dV_e$ to give a weight fraction. Thus Equation 14.26 can be rewritten as

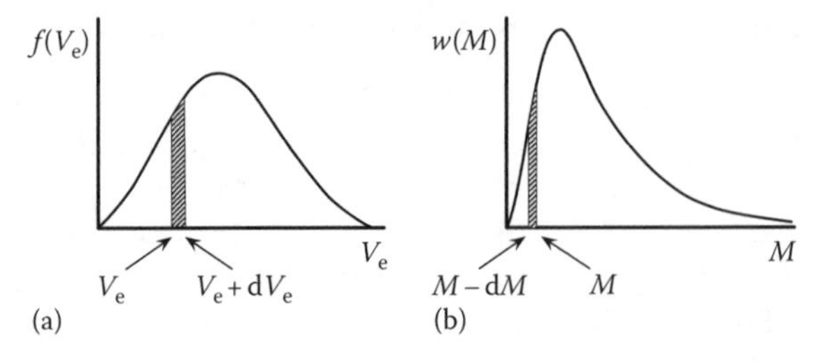

FIGURE 14.7 Schematic illustrations of (a) a GPC chromatogram and (b) the weight-fraction MMD into which it transposes.

$$\frac{\mathrm{d}w}{\mathrm{d}V_e} = \frac{f(V_e)}{A} \tag{14.27}$$

Corresponding to the elution volume interval V_e to $(V_e + \mathrm{d}V_e)$, the same weight fraction $\mathrm{d}w$ of polymer exists between M and $(M - \mathrm{d}M)$ in the weight-fraction MMD (because M decreases as V_e increases). The distribution function $w(M)$ is normalized and so $\mathrm{d}w = -w(M)\mathrm{d}M$ from which

$$w(M) = -\frac{\mathrm{d}w}{\mathrm{d}M} \tag{14.28}$$

Applying the chain rule for derivatives

$$-\frac{\mathrm{d}w}{\mathrm{d}M} = \left(\frac{\mathrm{d}w}{\mathrm{d}V_e}\right)\left(-\frac{\mathrm{d}V_e}{\mathrm{d}\ln M}\right)\left(\frac{\mathrm{d}\ln M}{\mathrm{d}M}\right) \tag{14.29}$$

Now $-\mathrm{d}V_e/\mathrm{d}\ln M = 1/(2.303|S(V_e)|)$ where $|S(V_e)|$ is the magnitude of the (negative) slope of the calibration plot (of $\log M$ against V_e) at V_e, and $\mathrm{d}\ln M/\mathrm{d}M = 1/M(V_e)$ where $M(V_e)$ is the value of M corresponding to V_e. Combining these relationships with Equations 14.27, 14.28 and 14.29 leads to

$$w(M) = \left(\frac{f(V_e)}{A}\right)\left(\frac{1}{2.303\,|S(V_e)|}\right)\left(\frac{1}{M(V_e)}\right) \tag{14.30}$$

The first term in this equation is obtained from the chromatogram and the remaining two terms from the calibration plot. Thus $w(M)$ can be evaluated point by point to give the weight-fraction MMD. Finally, with knowledge of $w(M)$, any molar mass average can be calculated from moments of the distribution (see Equations 14.1 through 14.4).

Due to the dependence of V_e upon the *logarithm* of M, accurate conversion of the GPC chromatogram to the weight-fraction MMD necessitates precise knowledge of the form of the calibration curve. Whilst the curve nominally is linear, in detail it is not, and use of a linear fit to the calibration data can lead to serious errors in $w(M)$. Nowadays it is common practice to fit $\log M$ to a power series in V_e (i.e. letting $\log M = \sum_{x=1}^{n} k_x V_e^x$ and using the lowest value of n which affords a good fit, often $n = 4$ or 5). The first differential of the power series enables $S(V_e)$ $(= \mathrm{d}\log M/\mathrm{d}V_e)$ to be calculated for each value of V_e. In this way small, but significant, local variations in $S(V_e)$ are taken into account to give $w(M)$ more accurately. Obviously, a wide range of calibration standards are required to define accurately the calibration curve over the full range of M.

A further improvement in the accuracy of $w(M)$ can be achieved by first correcting the GPC chromatogram for diffusional broadening of the peak. Such corrections are difficult to perform and are negligible for most polymers analysed using modern gels with small beads diameters. However, they are important in the analysis of narrow MMD polymers which generally give diffusion-broadened peaks that are skewed towards low V_e.

14.3.3 Universal Calibration

The need to determine the precise form of the calibration curve for each polymer to be analysed using a given solvent/temperature/GPC column system is a formidable task. Furthermore, with the exception of the more common polymers, standard samples suitable for calibration are not available for most polymers. Thus the concept of a universal calibration that applies to all polymers analysed using a given GPC system is very attractive.

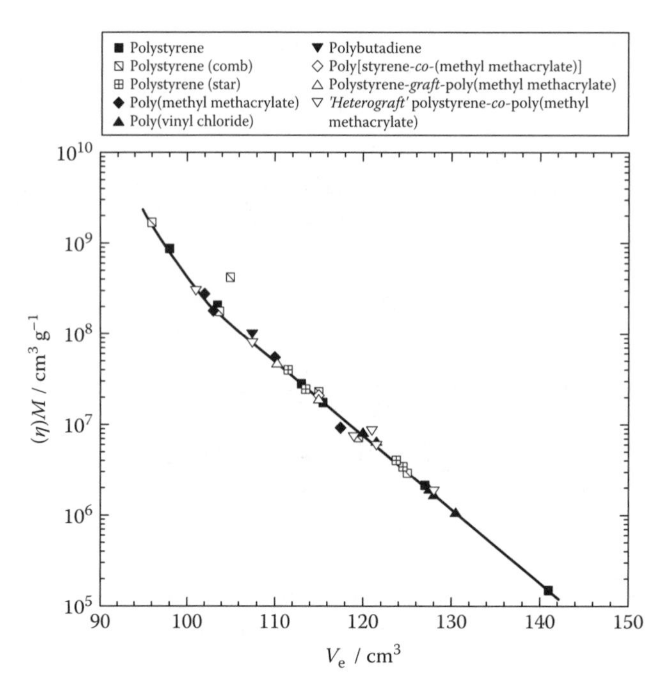

FIGURE 14.8 Universal calibration curve for a GPC system comprising crosslinked polystyrene gels with tetrahydrofuran as solvent. (Data taken from Grubisic, B. et al., *J. Polym. Sci. B Polym. Lett.*, 5, 753, 1967.)

In Section 14.3.2, it was stated that molecular size in solution can be represented by the *hydrodynamic volume* of the molecule, which is proportional to $[\eta]M$ (see Equation 10.94). Thus a plot of $\log([\eta]M)$ against V_e should be approximately linear and *independent of the polymer* under consideration. As shown by the data presented in Figure 14.8, there is ample experimental evidence confirming that such a plot does indeed provide a *universal calibration* for a given solvent/temperature/GPC column system. From the Mark–Houwink–Sakurada equation (Equation 10.98), $[\eta]M = KM^{1+a}$ and so the molar mass $M(V_e)$ corresponding to V_e for a polymer for which the calibration curve is not known, is given by

$$M(V_e) = \left[\frac{([\eta]M)_{V_e}}{K}\right]^{1/(1+a)} \tag{14.31}$$

where

$([\eta]M)_{V_e}$ is the value of $[\eta]$M taken from the universal calibration curve at V_e

K and a are the Mark–Houwink–Sakurada constants for the polymer under the conditions of the GPC analysis.

Thus the system is calibrated using readily available standards, such as narrow MMD polystyrenes, for which the Mark–Houwink–Sakurada constants, K_S and a_S, are known. This enables a universal calibration curve of $\log([\eta]_S M_S)$ $(= \log(K_S M_S^{1+a_S})$ against V_e to be obtained from which the $\log M$ against V_e calibration curve can be evaluated for any other polymer using Equation 14.31 provided that K and a are known.

14.3.4 Porous Gels and Eluants for GPC

The GPC column packings most commonly used with organic solvents are rigid porous beads of either crosslinked polystyrene or surface-treated silica gel. For aqueous GPC separations, porous beads of water-swellable crosslinked polymers (e.g. crosslinked polyacrylamide gels), glass or silica are employed.

The ability to resolve the different molar mass species present in a polymer sample depends upon a number of factors. As is evident from Figure 14.6, an appropriate range of gel porosities is required to obtain the necessary resolution and is obtained by using either a connected series of shorter columns each of which is packed with a gel of different porosity, or one long column packed with mixed gels. Resolution increases approximately with $l^{1/2}$ where l is the total column length, and also with $(1/d^2)$ where d is the bead diameter. The early GPC gels had d of about 50 μm and large column lengths were required for the separations. Nowadays, however, most GPC gels have d of either 5 or 10 μm, and much shorter column lengths can be used. This means that the elution volumes are much smaller and the separation process is much faster, typically 30–45 min. Initially, GPC performed using gels of small bead diameter was termed *high-performance GPC*, but that term is redundant now that the large size gels are no longer used. The most recent developments have been in *high-throughput* GPC where some resolution is sacrificed in the drive to achieve very fast analyses (10–15 min) for which low-volume, monolithic porous columns have been developed.

The choice of eluant is of considerable importance because it can have a significant influence upon the contribution of secondary modes of separation. For example, adsorption of polymer molecules to the walls of the pores will retard their movement through the column, causing V_e to increase beyond its value for size exclusion alone. Such effects are more probable if the solvency conditions are poor, and so a good solvent for the polymer should be used as the eluant. The most common eluants are toluene and tetrahydrofuran for non-polar and moderately-polar polymers that are soluble at room temperature, dimethylacetamide for more polar polymers that do not dissolve in tetrahydrofuran, 1,2-dichlorobenzene at about 130 °C for crystalline polyolefins, and 2-chlorophenol at about 90 °C for crystalline polyesters and polyamides.

14.3.5 Practical Aspects of GPC

The essential components of a GPC apparatus are: (i) a solvent pump, (ii) an injection valve, (iii) a column (or series of columns) packed with beads of porous gel, (iv) a detector and (v) a computer for data analysis (see Figure 14.9). Since most size-exclusion separations are performed using 5 or 10 μm diameter high performance gels, only these GPC systems will be considered here.

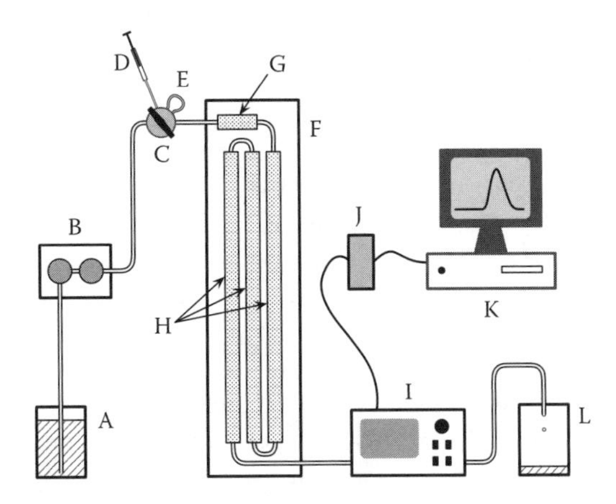

FIGURE 14.9 Schematic illustration of a GPC apparatus: A, solvent reservoir; B, solvent pump; C, injection valve; D, syringe for injecting the polymer solution; E, injection loop; F, column oven; G, guard column (which protects the more expensive GPC columns from adventitious blockages); H, GPC columns; I, detector; J, analogue-to-digital converter; K, computer; and L, waste solvent/solution reservoir.

High-quality solvent pumps which give pulse-free constant volumetric flow rates Q are essential for GPC because it is more accurate to record elution times t_e rather than measure elution volumes ($=Qt_e$) due to the relatively small values of V_0+V_i (typically 30–40 cm^3). Also, the pump must generate high pressures (typically 4–14 MPa) to force the solvent through the columns (of tightly packed small-diameter beads of gel) at the usual flow rate of about 1 cm^3 min^{-1}. To facilitate adjustment for any slight fluctuations in flow rate from one run to another it is normal to include a flow marker in the polymer solution, typically a small inert organic compound (e.g. decalin) that elutes at the full permeation limit (i.e. corresponding to V_0+V_i).

A dilute solution (e.g. 2 g dm^{-3}) of the polymer to be analysed is transferred by syringe to the injection loop of the injection valve; the loop typically has a volume of about 0.05 cm^3. In the normal valve position, the solvent stream passes directly through the valve, but when it is switched to the injection position, the flow path is changed to pass around the injection loop thus carrying the polymer solution from the loop towards the columns. The injection valve is then moved back to its normal position so that another polymer solution can be placed in the injection loop ready for the next analysis. The GPC columns usually are placed inside an oven so that their temperature can be kept constant; this also enables higher temperatures to be used for solvents that are too viscous to pump at room temperature or for polymers that dissolve only at higher temperatures.

Since only about 0.1 mg of polymer is injected, highly sensitive stable detectors are required. In this respect, detectors which measure absorption of ultraviolet (UV) light are best, though they are restricted to systems in which the polymer absorbs UV light at wavelengths at which the solvent does not absorb. Many polymers do not absorb UV light and detectors which measure absorption of infrared (IR) radiation are more appropriate, but much more expensive. Evaporative light scattering (ELS) detectors atomize the eluate into a heated chamber, which causes the solvent to evaporate rapidly, leaving tiny suspended particles of solute, the amount of which is detected by turbidity (i.e. light scattering). The most commonly used detectors, however, are differential refractometers (RI) which measure continuously the difference between the refractive index of the eluate and that of the pure solvent. The responses from UV, IR, RI and ELS detectors are proportional to the concentration of polymer in the eluate.

More specialized detectors which enable the molar mass of the polymer in the eluate to be measured continuously also are available. The most important are on-line capillary-bridge differential viscometers (IV; see Section 13.2.3) and low-angle laser light scattering (LALLS) detectors that measure, respectively, the specific viscosity η_{sp} and the limiting excess Rayleigh ratio $\Delta R_{\theta\to 0}$ of the eluate continuously. Thus, by using a concentration detector (usually RI), an IV detector and an LALLS detector in series, so-called *triple-detector GPC*, it is possible to apply the principles of universal calibration directly for each sample analysed since c, $[\eta]$, and M are measured continuously. The combined signals enable $M(V_e)$ to be determined continuously on the basis that η_{sp} and $\Delta R_{\theta\to 0}$ are equal to their limiting values at $c=0$ (because c is very small). The attraction for such triple-detector systems is that they yield the weight-fraction MMD directly, thus obviating the need for calibration. Multi-angle laser light scattering (MALLS) detectors that measure light scattering at several angles simultaneously also are available and facilitate continuous in-line measurement of both molar mass and molecular dimensions (see Section 12.2.3 and Equation 12.26).

The signal from the detector(s) is analogue and so is converted to digital before being stored in the memory of a computer which then is used to perform the necessary data manipulation. In this way, the MMD and the molar mass averages can be obtained soon after the sample has eluted. Where more than one detector is used in series, the detectors need to be synchronized using the signals from the flow marker. When analysing very high molar mass polymers by GPC care must be taken to check for evidence of polymer degradation because the shear rates in the columns are high and the shear forces can cause scission of the largest chains (typically those above 10^6 g mol^{-1}). Taken together with the difficulty in producing gels that can resolve very high molar mass chains, GPC is not so suitable when a significant proportion of the polymer has molar mass above 10^6 g mol^{-1}.

Other than the issues in analysing very high molar mass polymers, it should now be obvious why GPC is a very important technique for polymer characterization and why its use is widespread throughout academy and industry. A natural progression from the analytical systems described here, has been the development of much larger volume *preparative* GPC systems for fractionation of polymers. Typically, the volume of preparative systems is about 10× that of analytical systems and enables 10–25 mg of polymer to be fractionated in a single run requiring about the same time as an analytical GPC analysis. The eluate is collected in successive portions corresponding to specific ranges of molar mass in the MMD, and the different polymer fractions are recovered from these portions by removal of the solvent. If larger quantities of each fraction are required, the procedure can be repeated several times. Thus preparative GPC has significant advantages over classical methods for fractionation (Section 14.2.3), the most important being that it is much faster and can yield fractions with molar masses in pre-defined ranges.

GPC of copolymers is complicated by superposition of the molecular size and copolymer composition distributions. Thus when analysing copolymers, it is good practice to use complementary concentration detectors in series (e.g. RI+UV or RI+IR), which makes it possible to observe the complete MMD from the RI signal and the average composition of the copolymer molecules at each point in the distribution from the UV or IR signal. For example, this facilitates determination of the uniformity with which the UV- or IR-active repeat units are distributed across copolymer chains of different molar masses for correlation with the conditions under which the copolymer was synthesized. However, a more recent development known as *chromatographic cross-fractionation* is far more elegant for characterization of copolymers and gives both the molecular size distribution and the distribution of copolymer composition. In its most efficient form, it involves connecting the outlet from a GPC system directly to another high-performance column chromatography system in which the molecules are separated according to composition via differential solubility and/or adsorption effects. Thus the copolymer molecules are separated first according to their size in solution and then according to their composition.

14.4 FIELD-FLOW FRACTIONATION

The principle of *field-flow fractionation* (*FFF*) is to pass a polymer solution through a narrow channel whilst applying simultaneously a field which generates (*cross-flow*) forces perpendicular to the forward flow direction, as depicted in Figure 14.10.

The flow rates are low enough to ensure Newtonian fluid behaviour and so the velocity profile is parabolic with maximum velocity at the centre of the channel and zero velocity at the walls, in accord with Poiseuille's equation. The cross-flow is strong and causes the chains to move into the region of flow close to the *accumulation wall*, but this movement is countered by the natural tendency of the chains to diffuse towards regions where their concentration is lower thereby redistributing their positions within the channel. These opposing forces are illustrated on the left side of Figure 14.11 where (for clarity) the chain is shown massively oversized compared to the dimensions of the channel, which is typically 50–300 μm deep. The system is dynamic with individual chains able to move in either direction across the flow at different points as they pass down the channel. After only a short distance along the channel, a steady state is attained in which there is a well-defined (statistical) concentration gradient of chains away from the accumulation wall. The distribution can be represented by a characteristic layer thickness which varies with molecular size because the largest chains are most affected by the cross flow and also diffuse more slowly than smaller chains. Thus the layer thickness is smaller for larger chains, which, therefore, experience (on average) lower velocities than smaller chains for which the layer thickness is larger and extends into regions of higher velocity. The differences in concentration distribution for chains of different size are exemplified on the right-hand side of Figure 4.11. Hence, chains emerge from the channel in order of increasing size, with the smallest chains emerging first and the largest last.

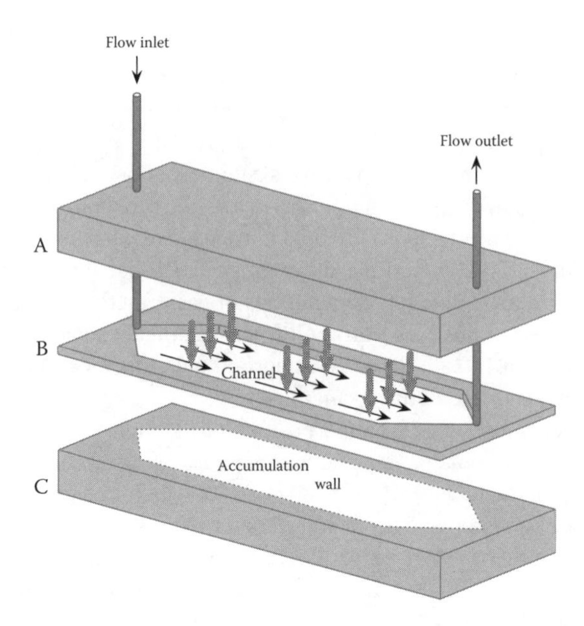

FIGURE 14.10 Schematic illustration of an FFF apparatus showing the three components of the flow channel, which are bolted together in operation: A, the upper block that delivers the cross-field forces (shown as broad arrows), which operate perpendicular to the flow direction (shown as thin arrows); B, a thin sheet of spacer material from which the channel is cut; and C, the lower block, which acts as the accumulation wall. The outflow passes into a detector that is interfaced to a computer for data collection and analysis.

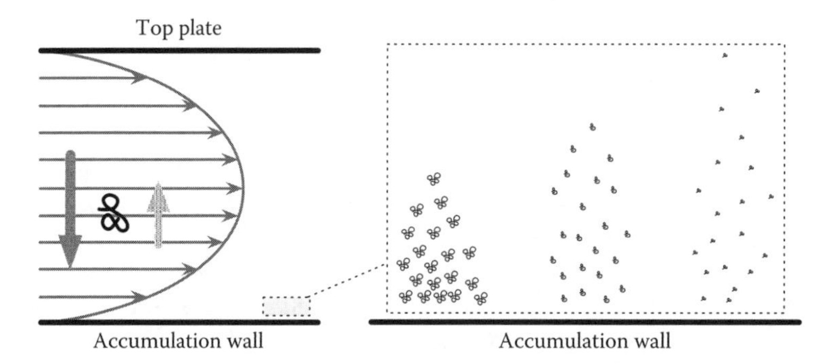

FIGURE 14.11 Schematic diagram illustrating the molecular separation mechanism operating in FFF. The diagram on the left shows the three forces exerted on chains: (i) the horizontal arrows are vectors representing the flow velocities carrying the solution forwards at different points in the cross-section of the flow channel; (ii) the broad, dark arrow represents the cross-flow field that forces chains towards the accumulation wall; and (iii) the broad, light arrow represents the diffusional forces that carry chains back towards regions further away from the accumulation wall where their concentration is depleted. The diagram on the right shows how under steady-state conditions chains are distributed across the channel within a small region close to the accumulation wall; the smaller chains diffuse further into the channel and so (on average) experience larger forward flow velocities, which carry them through the channel faster than larger chains.

14.4.1 FFF Techniques

A wide range of cross-flow fields have been utilized and give rise to a range of FFF techniques. Only those most commonly used, and for which commercial instruments are available, will be described here. The most widely applicable technique is *flow-FFF* (Fl-FFF) in which an additional perpendicular flow of solvent is used. In *symmetrical-Fl-FFF* (S-Fl-FFF) both the upper and lower blocks are separated from the channel by semi-permeable membranes that allow solvent molecules, but not polymer chains, to pass through. In addition to the normal inlet flow, solvent is pumped into the channel through the upper block and can exit through the lower block, thus creating the cross-flow field. *Asymmetrical-Fl-FFF* (A-Fl-FFF) is an alternative (but more complex) design in which only the lower plate is permeable and only the normal inlet flow is used during elution; since solvent passes out through the lower block as it moves along the channel, a trapezoidal channel profile is used (with the channel width narrowest near the exit) to compensate for the loss of solvent and maintain a uniform flow velocity along the length of the channel (though, of course, this pertains only for rather specific flow conditions). The mode of operation of A-Fl-FFF also is more complex than for S-Fl-FFF, requiring a separate inlet for injection of the polymer solution directly into the channel and an initial focusing step in which solvent enters the channel through both the inlet and outlet tubes in order to *focus* the sample at its point of entry into the channel prior to allowing normal flow down the channel for separation and elution of the sample. A-Fl-FFF does, however, have advantages of simpler construction and higher resolution. *Thermal-FFF* (Th-FFF) employs a temperature gradient across the channel that creates a strong thermal diffusion cross-flow field; typically the upper plate is 20–80 °C above the temperature of the lower plate, giving temperature gradients of about 100–1000 °C mm^{-1} depending on the channel depth. The other common FFF technique is *sedimentation-FFF* (Sd-FFF) in which the channel is placed around the circumference of a centrifuge rotor such that the centrifugal forces act perpendicular to the flow along the channel. Electrical, magnetic and gravitational cross-flow fields also have been used, but are less important. In addition, there are several different modes of operation for FFF instruments. Recent reviews should be consulted for a more thorough appreciation of the different FFF techniques and their underlying theory.

Other than the FFF unit itself, the apparatus for FFF resembles that for GPC. Solvent is supplied from a reservoir to a pump that must deliver highly stable flow rates (but at much lower pressures than for modern GPC). A dilute solution of the polymer to be analysed (typically ~0.01 cm^3 of a ~10 g dm^{-3} solution) is injected between the pump and the entry point into the FFF unit either directly into the flow stream or via an injection valve. As for GPC, a low molar mass compound normally is included in the polymer solution to act as a flow-rate marker. The eluate emerging from the FFF channel can be monitored using any of the types of detectors described for GPC. Although RI detectors are common, if they are used alone, the FFF instrument must be calibrated with polymers of known molar mass. Thus, in common with GPC, there is a major advantage in using a concentration detector in series with a more advanced detector that is capable of measuring molar mass. The series combination of RI and MALLS detectors is particularly powerful for FFF of high molar mass polymers, providing accurate molar mass and molecular size distributions.

14.4.2 Theory of Solute Separation by FFF

Since the same basic principle of separation is common to all FFF techniques, the conditions for separation can be considered generically. It is usual to consider the separation in terms of an elution ratio R which is the ratio of the elution time t_0 for solvent molecules to that for a solute species t_s. Since the time for elution of a given molecule is inversely proportional to its mean velocity along the column, R also is given by the ratio of the mean migration velocity $\langle \bar{v}_s \rangle_{XS}$ of the solute species along the channel to the linear average velocity $\langle \bar{v} \rangle_{XS}$ of flow, where $\langle \ \rangle_{XS}$ indicates an average over the cross-section of the channel

$$R = \frac{t_0}{t_s} = \frac{\langle \bar{v}_s \rangle_{XS}}{\langle \bar{v} \rangle_{XS}} \tag{14.32}$$

In order to examine how molecules are separated it is necessary to understand the factors which affect $\langle \bar{v}_s \rangle_{XS}$. However, an expression for $\langle \bar{v} \rangle_{XS}$ will be considered first and can be obtained from Poiseuille's equation for the variation of flow velocity $v(x)$ with distance x from a wall for flow down a channel that is extremely shallow compared to its width (as in FFF)

$$v(x) = \left(\frac{P}{2\eta l} \right)(w - x)x \tag{14.33}$$

where

- P is the pressure difference along the channel
- η is the viscosity of the fluid (which can be taken to be that of the solvent)
- l is the length of the channel
- w its depth (i.e. the maximum value of x).

Thus

$$\langle \bar{v} \rangle_{XS} = \frac{\int_0^w v(x)\mathrm{d}x}{w}$$

which upon substitution of Equation 14.33 is solved easily to give

$$\langle \bar{v} \rangle_{XS} = \frac{Pw^2}{12\eta l} \tag{14.34}$$

Analysis for $\langle \bar{v}_s \rangle_{XS}$ is less straightforward, though not difficult. It is necessary to know both $v(x)$ and the concentration $c(x)$ of the solute species at a distance x from the accumulation wall, in which case

$$\langle \bar{v}_s \rangle_{XS} = \frac{\int_0^w c(x)v(x)\mathrm{d}x}{\int_0^w c(x)\mathrm{d}x} \tag{14.35}$$

Thus an expression for $c(x)$ is required and can be obtained by considering the opposing cross-flow field and diffusional forces acting on the solute species, which may be expressed as follows under steady-state conditions when they are exactly equal

$$Uc(x) = D\frac{\mathrm{d}c(x)}{\mathrm{d}x} \tag{14.36}$$

where

- U is the drift velocity across the channel caused by the cross-flow field
- D is the translational diffusion coefficient of the solute species in the solvent.

Note that U is a vector quantity and, in relation to this analysis, has a negative value because it is directed towards the accumulation wall whereas distances in the channel are measured from the accumulation wall (i.e. in the opposite direction). Hence, it is usual to use $-|U|$ rather than U where $|U|$ is the magnitude of U. Rearranging Equation 14.36 and applying integration limits gives

$$\int_{c_0}^{c(x)} \frac{\mathrm{d}c(x)}{c(x)} = -\left(\frac{|U|}{D}\right)\int_0^x \mathrm{d}x \tag{14.37}$$

where c_0 is the concentration of the solute species at the accumulation wall. The integration is simple and gives

$$c(x) = c_0 \exp\left(-\frac{|U|x}{D}\right) \tag{14.38}$$

The ratio $D/|U|$ is a measure of the mean distance of the solute species from the accumulation wall during its path along the channel and often is referred to as the mean thickness of the solute layer; more significantly, it relates to the important dimensionless retention parameter λ which is given by

$$\lambda = \frac{D}{|U|w} \tag{14.39}$$

Hence Equation 14.38 can be rewritten as

$$c(x) = c_0 \exp\left(-\frac{x}{w\lambda}\right) \tag{14.40}$$

which is helpful for the integrations required in solving Equation 14.35. Substituting Equations 14.33 and 14.40 into Equation 14.35 and simplifying gives

$$\langle \bar{v}_\mathrm{s} \rangle_\mathrm{XS} = \left(\frac{P}{2\eta l}\right)\left\{\frac{\int_0^w \left[wx\exp\left(-\frac{x}{w\lambda}\right) - x^2\exp\left(-\frac{x}{w\lambda}\right)\right]\mathrm{d}x}{\int_0^w \exp\left(-\frac{x}{w\lambda}\right)\mathrm{d}x}\right\} \tag{14.41}$$

The two terms in the numerator are integrated by parts, leading to

$$\langle \bar{v}_\mathrm{s} \rangle_\mathrm{XS} = \left(\frac{P}{2\eta l}\right)\left\{\frac{\left(\left[w^3\lambda^2 + 2w^3\lambda^3\right]\left[1+\exp\left(-\frac{1}{\lambda}\right)\right]\right) - 4w^3\lambda^3}{w\lambda\left[1-\exp\left(-\frac{1}{\lambda}\right)\right]}\right\} \tag{14.42}$$

Recognising that $\exp(-1/\lambda)\cdot\exp(1/2\lambda)=\exp(-1/2\lambda)$, by multiplying both the numerator and denominator in the right-hand parentheses by $\exp(1/2\lambda)$ and simplifying, Equation 14.42 can be reduced to the following form

$$\langle \overline{v}_s \rangle_{xs} = \left(\frac{Pw^2\lambda}{2\eta l} \right) \left\{ \frac{\left[\exp\left(\frac{1}{2\lambda}\right) + \exp\left(-\frac{1}{2\lambda}\right) \right] - 2\lambda \left[\exp\left(\frac{1}{2\lambda}\right) - \exp\left(-\frac{1}{2\lambda}\right) \right]}{\left[\exp\left(\frac{1}{2\lambda}\right) - \exp\left(-\frac{1}{2\lambda}\right) \right]} \right\}$$

which may be written more simply as

$$\langle \overline{v}_s \rangle_{xs} = \left(\frac{Pw^2\lambda}{2\eta l} \right) \left[\coth\left(\frac{1}{2\lambda}\right) - 2\lambda \right] \tag{14.43}$$

Substituting Equations 14.34 and 14.43 into Equation 14.32 gives

$$R = 6\lambda \left[\coth\left(\frac{1}{2\lambda}\right) - 2\lambda \right] \tag{14.44}$$

from which the importance of λ is now fully evident, as it is the factor which controls the elution time of the solute species. Equation 14.44 simplifies further under conditions where $\lambda \leq 0.2$ since $\coth(1/2\lambda) \approx 1$ for which $R = 6(\lambda - 2\lambda^2)$ and in the limit where λ is very small (≤ 0.01), $R = 6\lambda$. Inspection of Equation 14.39 shows that such small values of λ correspond to use of strong cross-flow fields, which is normal.

Although for several FFF techniques λ is amenable to thorough theoretical treatment and it is possible to predict R, the precise mathematical analysis is unique to each type of cross-flow field used. In each case, it is necessary only to derive an equation for $|U|$ because D is measurable and w is known from the design of the FFF unit. To illustrate the principles, S-Fl-FFF will be considered because the analysis is simple. The volumetric cross-flow rate Q_{cf} in S-Fl-FFF is measurable and given by $|U|A_{wall}$ where A_{wall} is the cross-sectional area of the accumulation wall which is given by $A_{wall} = V_{channel}/w$ where $V_{channel}$ is the volume of the channel. Hence,

$$|U| = \frac{Q_{cf} w}{V_{channel}} \tag{14.45}$$

and so, from Equation 14.39, for S-Fl-FFF

$$\lambda = \frac{DV_{channel}}{Q_{cf} w^2} \tag{14.46}$$

from which it is now clear that differences in D are responsible for separation of different solute species because $V_{channel}$, Q_{cf} and w are fixed by the experimental conditions for all species in the channel. Thus solute species are separated according to their hydrodynamic size (see Section 10.4.3). Given that the relationship between D and molar mass M is given by Equation 10.103, $D = K_D M^{-a_D}$, where K_D and a_D are parameters with positive values unique to the polymer-solvent-temperature system, it is possible to predict the variation of retention time t_s with M. By combining Equations 14.46 and 10.103 together with Equation 14.32 and the limiting form of Equation 14.44 for very small λ, i.e. $R = 6\lambda$, the following equation for t_s is obtained

$$t_s = \left(\frac{t_0 Q_{cf} w^2}{6K_D V_{channel}} \right) M^{a_D} \tag{14.47}$$

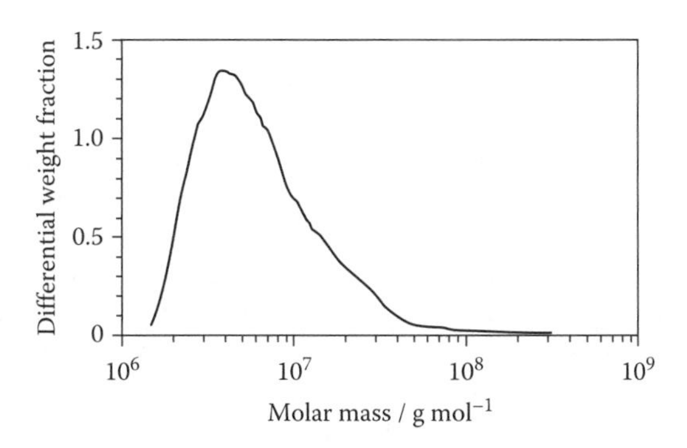

FIGURE 14.12 Differential weight fraction MMD of a sample of xanthan gum obtained at 50 °C by asymmetrical flow-FFF with RI and MALLS dual detectors. (Data taken from Viebke, C. and Williams, P.A., *Food Hydrocoll.*, 14, 265, 2000.)

which indicates that calibration can be achieved using a logarithmic plot

$$\log t_s = A + a_D \log M \tag{14.48}$$

where $A = \log(t_0 Q_{cf} w^2 / 6K_D V_{channel})$. As for GPC, the assumptions inherent in Equation 14.48 are such that the logarithmic calibration plots are only approximately linear and more usually are fitted to a low-order polynomial. Of course, if a molar mass detector (such as MALLS) is employed, calibration is not necessary.

Sedimentation-FFF also can be treated rigorously, but thermal-FFF is rather more complex because D changes with temperature and so varies across the channel due to the thermal gradient. In all cases, including flow-FFF, there are subtleties which complicate the mathematical analysis compared to the simple approach presented here (e.g. correction for diffusional broadening is important in FFF). Thus, original papers and reviews should be consulted for more complete treatments of the different FFF techniques and modes of operation.

14.4.3 Applications of FFF

Flow-FFF and thermal-FFF are not capable of resolving oligomeric chains with molar masses below about 10^3 g mol^{-1}, but for most polymers, this is not a serious limitation. Sedimentation-FFF, however, is suited only to very high molar mass polymers because the centrifugal forces needed for separation of low–medium molar mass chains are not attainable with current instrumentation. The major advantages of flow-FFF and thermal-FFF are that they can provide MMDs across an extremely wide range of molar mass from around 10^3 g mol^{-1} upwards (with no upper limit) and that, because the shear forces are low, shear scission of very large chains is rarely a problem (unlike in GPC). Hence, FFF techniques are important for molar mass characterization of naturally-occurring polymers, many of which (e.g. polysaccharides) have non-uniform broad MMDs that extend to extremely high molar masses. The ability of FFF to determine MMDs of such high molar mass polymers is illustrated by the MMD of a xanthan gum shown in Figure 14.12, which extends up to about 4×10^8 g mol^{-1}. Thus FFF techniques are far better than GPC (and other techniques) for analysis of polymers of very high molar mass. Analysis of broad MMDs is best achieved by varying the cross-flow field strength in a programmed manner during the run.

14.5 MASS SPECTROSCOPY

Mass spectroscopy (MS) involves rapid conversion of a sample into vapour-phase molecular ions, which are characterized by their mass:charge ratio m/z, where m is the molecular mass and z is the charge given in multiples of **e**, the elementary electronic charge ($=1.602\times10^{-19}$ C). The molecular ions generated are transported along an evacuated tube under the influence of electrical and magnetic fields in a manner such that individual molecular ions of specific mass:charge ratio m/z can be separated and detected with great precision, e.g. species with m/z values differing by as little as 0.001 Dalton $\mathbf{e}^{-1}$ can be resolved in favourable cases (1 Dalton $\equiv$ 1 Da = mass of a single atom of $^{12}C = 1.66\times10^{-24}$ g and is equivalent to a molar mass of 1 g mol^{-1}). Thus the output from a mass spectrometer is effectively a plot of abundance against molar mass, which has the form of a number-based MMD.

MS evolved through analysis of small molecules which, in early instruments, were vaporized and then ionized by exposure to a beam of high-energy electrons. Under these conditions, the molecules usually undergo significant fragmentation into smaller species (of lower m/z), which is not an issue for analysis of small molecules of unique molar mass, but for polymers is a serious issue because, even when the MMD is narrow, they comprise chains with different molar masses (e.g. if fragmentation occurred, it would not possible to know whether signals at lower m/z were from whole polymer chains or from chain fragments). Less harsh (*softer*) methods of ionization, such as chemical ionization or field ionization were developed to reduce fragmentation in MS of small molecules and are more appropriate to polymers, but there still was the major issue of getting macromolecules into the vapour phase without degradation. For this reason, for many years, MS of polymers was used mainly in studies of polymer degradation, i.e. for analysis of low molar mass degradation products (see Section 15.7.2).

The problem of analysing non-volatile substances by MS is not unique to polymers and received considerable attention over several decades. As for many of the developments in polymer molar mass characterization, the breakthroughs were driven by work on biopolymers. During the 1970s, several new approaches to soft ionization were developed, but it was not until the 1980s, with the advent of *electrospray ionization* (ESI) and then *matrix-assisted laser desorption/ionization* (*MALDI*), that it became possible to generate and vaporize molecular ions from polymers without causing fragmentation. Although both methods are used for MS analysis of polymers, ESI tends to be used for biopolymers (e.g. proteins) whereas MALDI has become by far the most important method for synthetic polymers. Techniques and instrumentation developed significantly during the 1990s to the point where, nowadays, MALDI MS is used widely in support of research on polymers for evaluation of MMD and for elucidating fine details of chemical structure. This section focuses on the principles of MS and its use for measurement of MMD. Use of MALDI MS for probing the structure of polymers is deferred to Section 15.7, which is concerned with determination of molecular structure.

14.5.1 Mass Spectra of Polymers

Polymer chains have molar masses that are multiples of the repeat unit molar mass (neglecting end groups) and, unlike any other method for analysis of MMD, the mass resolution of MS is high enough to reveal this discrete nature of the distribution, as is evident from the MALDI mass spectra of two low molar mass polystyrene samples shown in Figure 14.13. The strength of the signal depends upon how many molecular ions with that particular m/z value hit the detector and so represents the number of such species. The insets in Figure 14.13 show the m/z resolution more clearly and reveal that (within experimental error) the separation of the molecular ion peaks in each spectrum correspond to the repeat unit mass for polystyrene chains (104 Da), as is expected. Thus the regular separation of the main peaks in the mass spectrum of a polymer also can be used to determine the repeat unit molar mass and, in some cases, this is helpful in identifying or confirming the basic structure of a polymer.

The style of presentation of the mass spectra shown in Figure 14.13 is common practice, i.e. it is normal to convert the detector signal directly into percentage abundance and to plot this against m/z, showing the units of m/z (incorrectly) as Dalton (when strictly they should be Dalton $\mathbf{e}^{-1}$). Thus, polymer MMDs obtained by MALDI MS most commonly are shown as abundance (which is proportional to number fraction) against m/z (which is equivalent to molar mass in g mol^{-1} for

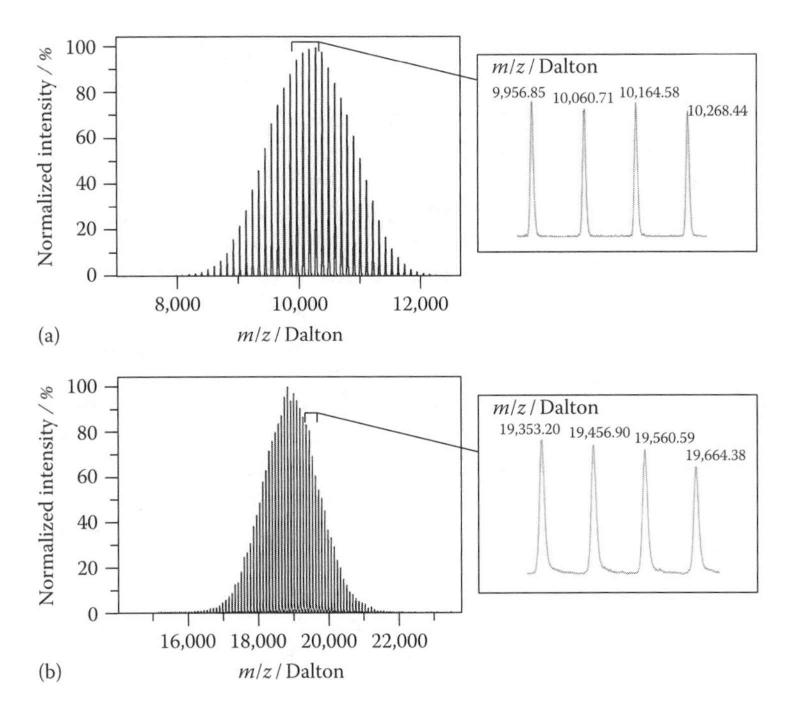

FIGURE 14.13 MALDI time-of-flight mass spectra of narrow MMD polystyrene samples with number-average molar masses of: (a) 10,210 g mol^{-1}; and (b) 19,880 g mol^{-1}. The insets show expansions of regions of the spectra near to the maxima in the MMDs. The samples were desorbed/ionized using Ag^+ ions with *trans*-2-[3-(4-*t*-butylphenyl)-2-methyl-2-propenylidene]malononitrile as the matrix. (Spectra provided by D. Haddleton, University of Warwick, Warwick, U.K.)

m/z species with $z = +1$ **e**); i.e. as number-based distributions. However, the normal practice when considering polymer MMDs is to plot the weight fraction of species against molar mass, i.e. the weight-fraction distribution, which has a quite different appearance. This needs to be borne in mind when comparing MMDs from MALDI MS with those measured by the other methods described in this chapter.

The inset in Figure 14.12 reveals that the m/z peaks are not sharp spikes as might be expected for molecular ions from species of unique mass. Peak broadening arises in part from the limited m/z resolution of the mass spectrometer, but also due to the presence of different atomic isotopes. Of the most common elements found in polymer structures (C, H, N, O), carbon has an isotope (^{13}C) with an abundance (1.1%) that is significant in relation to the degree of polymerization. For example, in a C–C backbone of 100 carbon atoms, one of those atoms is likely to be ^{13}C and there is an approximately 1% chance that a chain has two ^{13}C atoms, etc. Thus in any sample of polymer, there will be polymer molecules with the same degree of polymerization, but different masses depending on the number of ^{13}C atoms present in the molecule (including side-groups), each molecule giving a molecular ion with a unique m/z differing from that of the equivalent all-^{12}C molecule in steps of 1 m/z unit. These weak signals are not resolved and lead to peak broadening, and naturally will increase in number as the degree of polymerization increases due to the increasing number of finite probabilities for the presence of different numbers of ^{13}C atoms.

14.5.2 Methods of Soft Ionization for Polymers

Creation of molecular ions from polymers requires methods of soft ionization in order to prevent (or at least minimize) fragmentation of the polymer molecules. The two most important methods are described here and their optimization for each polymer analysed is absolutely crucial for reliable analysis.

14.5.2.1 Electrospray Ionization

Although the origins of ESI can be traced to the 1960s, it was not until the mid-1980s that it became reliable for analysis of polymers. ESI involves spraying a polymer solution into a chamber through a needle, the tip of which is subjected to a high voltage. The resulting tiny, highly charged droplets are heated by a stream of nitrogen gas and divide rapidly into smaller and smaller droplets due to Coulombic repulsions as the solvent evaporates, ultimately producing individual charged molecular species that pass into the mass separator/detector system. A feature of ESI is that species with multiple charges are generated easily, but not necessarily reproducibly; i.e. not all macromolecules of the same molecular mass are vaporized/ionized into molecular ions with the same charge (i.e. giving molecular ions with different m/z values from species of identical molecular mass). This is less of a complication for biopolymers which have a unique molar mass since it gives m/z peaks with separations equal to well-defined fractions of the molar mass; in fact, it can be an advantage because if the molar mass is very high and the molecular ion has multiple charges, m/z is reduced and the molecular ions are more easily detected. However, in analysis of polymers with molar mass dispersity, even for synthetic polymers with narrow molar distributions, the generation of molecular ions species with different numbers of charges creates considerable uncertainty in the interpretation of ESI mass spectra.

14.5.2.2 Matrix-Assisted Laser Desorption/Ionization

MALDI involves vaporization of a solid solution of polymer dispersed at low concentration (0.01–1 wt%) within a low molar mass crystalline matrix material in which metal ions also are present. The sample is vaporized by a short pulse (a few nanoseconds) of high-energy laser radiation (usually UV light of 337 nm wavelength), during which each polymer molecule associates with a metal ion and is carried into the vapour phase. After a short delay (typically 100–800 ns), the molecular ions are moved into the mass separator/detector system. MALDI invariably produces singly-charged species and so is much more useful than ESI for analysis of polymers with molar mass dispersity. This is one of the main reasons why MALDI has become dominant in MS characterization of synthetic polymers. The mechanism by which macromolecules are vaporized and ionized in MALDI is uncertain, but it is well-established that selection of the matrix material is critical for reliable measurements and that the matrix material needs to be optimized for each type of polymer analysed. The matrix material must be capable of forming uniform solid solutions with the polymer and metal salt, able to absorb strongly at the wavelength of the laser radiation and give rapid vaporization. Hence a suitable matrix material is likely to have a solubility parameter (Section 10.2.5) similar to that of the polymer to be analysed and should be a strong absorber of ultraviolet light. Extensive lists of matrix materials suitable for particular polymers are now available (see Further Reading). The metal salt used for *cationization* of the polymer molecules also needs to be selected in relation to polymer structure. For polymers that contain heteroatoms (e.g. polyethers, poly[(meth)acrylates], polyesters and polyamides), alkali metal salts (e.g. Na^+ and K^+ salts) can be used because their ions readily associate with the heteroatoms present in the polymer; Na^+ is particularly good and widely used because it exists as a single isotope, which simplifies interpretation of the mass spectrum. Non-polar polymers do not take up alkali metal ions and, in the absence of sites of unsaturation, are very difficult to cationize; hence polyethylene and polypropylene are not amenable to analysis by MALDI MS. However, polystyrene and (with more difficulty) poly(1,3-dienes) can be analysed using transition metal ions (usually the Ag^+ salt of trifluoroacetic acid), which are believed to associate with the electrons in π-bonds. There are many different methods of sample preparation. The simplest involves mixing solutions of the matrix material, metal salt and polymer, then placing a drop of the mixed solution onto a MALDI plate and evaporating the solvent rapidly in order to minimize the chance for segregation of the components and ensure that the polymer and metal salt are distributed intimately and uniformly throughout the matrix material as it crystallizes during solvent evaporation.

14.5.3 Time-of-Flight Mass Spectroscopy

Mass spectrometers have four main components: (i) a system for vaporizing/ionizing the sample (ESI or MALDI for polymers); (ii) an ion accelerator, which forces the individual molecular ions to move at high speed; (iii) a system of magnetic and electric fields, which focus the molecular ions and allow their separation according to the value of m/z; and (iv) a detector, which captures and counts the number of molecular ions.

Analysis of polymers requires separation and detection of molecular ions with a wide range of m/z that extends to very high values. Conventional mass spectrometers operate on the principle of detecting molecular ions with different m/z values in sequence by sweeping the power to the focusing fields; only molecular ions with a particular value of m/z can enter the detector at each point in the field sweep. Hence, for each m/z, only a small fraction of the total number of molecular ions generated reach the detector (i.e. detection is wasteful). Additionally, detection of molecular ions with very high values of m/z is not easily accessible. In time-of-flight (ToF) detection, all molecular ions follow exactly the same path to the detector, irrespective of m/z, and are separated simply by differences in the time taken to reach the detector. Thus, in ToF MS, all the molecular ions generated are captured by the detector, greatly improving sensitivity and facilitating easy detection of molecular ions with high m/z values. For this reason, most ESI and MALDI mass spectrometers available nowadays employ ToF detection.

14.5.3.1 Principles of MALDI Time-of-Flight Mass Spectrometry

In MALDI ToF mass spectrometers, the plume of molecular ions generated from the MALDI plate are accelerated by applying a fixed potential difference (typically 15–35 kV) between the plate and a grounded extractor plate (which together constitute the *ion accelerator* chamber). The accelerated molecular ions pass through an opening in the extractor plate into a linear drift tube (1–2 m in length) which is held at constant electric field. All the molecular ions produced follow a single path down the drift tube to the detector, ensuring that they all are detected, but they proceed at velocities which depend on the value of m/z and so molecular ions with different m/z values arrive in sequence at the detector. Individual spectra are acquired in a very short timescale (e.g. <100 μs). Further improvement in resolution and detection of higher m/z species can be achieved using *reflectron ToF* in which an electric field is used to reflect the molecular ions at an angle back from the end of their linear path down the drift tube and along a new linear path to another detector. Most modern MALDI ToF mass spectrometers have this facility, which can be activated when analysing higher molar mass polymers and when greater m/z resolution is required.

The potential energy gained by each molecular ion in moving through the electric field in the ion accelerator is converted into kinetic energy according to the following simple energy balance equation (in which the left-hand side = potential energy, and the right-hand side = kinetic energy)

$$zV_x = \frac{1}{2}mv_x^2 \tag{14.49}$$

where

V_x is the potential, and

v_x the velocity of the molecular ion, a distance x from the MALDI plate.

Since $V_x = E_V x$, where E_V is the linear electric field gradient along the ion accelerator chamber, then

$$v_x = \left(\frac{2E_V}{m/z}\right)^{1/2} x^{1/2} \tag{14.50}$$

and so the velocity v_A of a molecular ion at the point when it passes through the opening in the extractor plate is

$$v_A = \left(\frac{2V_A}{m/z}\right)^{1/2} \tag{14.51}$$

where
V_A ($=E_V x_A$) is the potential difference, and
x_A the distance

between the two plates in the ion accelerator chamber.

Thus v_A reduces as m/z increases, showing that the ToF to the detector will increase as m/z increases, i.e. the lowest molar mass species are detected first.

The time t_A taken for a molecular ion to exit the accelerator can be evaluated by recognizing that $v_x = (\mathrm{d}x/\mathrm{d}t)_x$, which upon substitution into Equation 14.50, followed by rearrangement and application of integration limits, gives

$$\int_0^{t_A} \mathrm{d}t = \left(\frac{m/z}{2E_V}\right)^{1/2} \int_0^{x_A} x^{-1/2}\mathrm{d}x$$

Solution of the integrals and substituting for $E_V = V_A/x_A$ leads to

$$t_A = 2x_A\left(\frac{m/z}{2V_A}\right)^{1/2} \tag{14.52}$$

On exit from the accelerator, each molecular ion travels at constant velocity v_A down the drift tube to the detector surface, which is placed a linear distance x_D from the extractor plate and reaches the detector after a further time t_D, which is given simply by x_D/v_A, i.e. from Equation 14.51

$$t_D = x_D\left(\frac{m/z}{2V_A}\right)^{1/2} \tag{14.53}$$

Thus the total ToF t_f of a molecular ion from the point at which the accelerating potential is applied to the point when it hits the detector is $t_A + t_D$, which, from Equations 14.52 and 14.53, gives

$$t_f = (2x_A + x_D)\left(\frac{m/z}{2V_A}\right)^{1/2} \tag{14.54}$$

For a particular instrument and fixed value of V_A, Equation 14.54 takes the form

$$t_f = K(m/z)^{1/2} \tag{14.55}$$

where K is a constant ($=\{2x_A + x_D\}/\{2V_A\}^{1/2}$). It is instructive to put some numbers into this equation. For example, if $2x_A + x_D = 1$ m, $V_A = 20$ kV and $m/z = 10{,}000$ Da $\mathbf{e}^{-1}$ ($=10{,}000$ Da $\times 1.66 \times 10^{-27}$ kg Da$^{-1} \times (1/1.602 \times 10^{-19}$ C$) = 1.04 \times 10^{-4}$ kg C^{-1}), recognizing that 1 C V = 1 J = 1 N m and that

1 N = 1 kg m s^{-2}, then substitution into Equation 14.54 gives $t_f = 51$ μs. Thus MALDI ToF MS analysis times are rapid and it is normal to accumulate many hundreds of spectra recorded from a rapid sequence of pulses of the laser (typically with a pulse interval of 0.02–0.1 s) in order to increase the signal-to-noise ratio.

In order to ensure that all molecular ions have been vaporized, it is usual to delay applying the accelerator voltage momentarily (typically by 100–800 ns) following the pulse of MALDI-inducing laser radiation, but the pulse is still used to trigger the recording of t_f. This is known as *delayed extraction* and greatly improves the quality of MALDI ToF mass spectra. The measured ToF with delayed extraction is given by

$$t_f = t_{delay} + K(m/z)^{1/2} \tag{14.56}$$

where t_{delay} is the delay time. Because the value of t_{delay} is imprecisely known, it is usual to calibrate MALDI ToF mass spectrometers with compounds of well-defined molar mass that span the range of interest. Achieving calibration with low molar mass calibrants is relatively easy, but calibration at higher molar masses requires care and usually is achieved by analysis of proteins which have known, unique molar masses.

14.5.3.2 Evaluation of Molar Mass Distribution from Time-of-Flight Mass Spectra

Equation 15.33 makes it clear that, because the molecular ions are detected in linear intervals of t_f, the mass spectrum obtained by MALDI ToF MS has peaks collected in intervals of $(m/z)^{1/2}$, i.e. the individual m/z peaks are spaced in unequal intervals of m/z with the separation growing as m/z increases. Thus the discrete MMD obtained is non-linear in molar mass with more peaks at lower molar mass than at higher molar mass on a linear scale of molar mass. As a consequence, molar mass averages cannot be calculated directly from the discrete abundance versus molar mass data from MALDI ToF mass spectra (because this will give values that are significantly low). This is similar to the situation with GPC, where the molar mass intervals obtained from the chromatogram are linear in log$M$ (see Section 14.3.2), and so MALDI ToF MS data must be handled in an analogous way. The number fraction dn of polymer molecules which are present in the ToF mass spectrum between t_f and $(t_f + dt_f)$ is given by

$$dn = \frac{f(t_f)dt_f}{\int_{t_{delay}}^{\infty} f(t_f)dt_f} \tag{14.57}$$

where $f(t_f)$ is the function describing how detector signal varies with t_f (i.e. the mass spectrum) and the integral in the denominator is the area A_n under the mass spectrum, which simply normalizes $f(t_f)$. Hence

$$dn = \frac{f(t_f)dt_f}{A_n} \tag{14.58}$$

Corresponding to this ToF interval, the same number fraction dn of polymer molecules exist between M and $(M + dM)$ in the normalized number-fraction MMD $n(M)$ and so

$$dn = n(M)dM \tag{14.59}$$

Equating Equations 14.58 and 14.59, with rearrangement, gives

$$n(M) = \left\{ \frac{f(t_\mathrm{f})}{A_\mathrm{n}} \right\} \frac{\mathrm{d}t_\mathrm{f}}{\mathrm{d}M} \tag{14.60}$$

Recognizing that, invariably, $z = +1$ **e**, then $m/z = M$ and Equation 14.56 can be rewritten as

$$t_\mathrm{f} = t_\mathrm{delay} + KM^{1/2} \tag{14.61}$$

which upon differentiation gives

$$\frac{\mathrm{d}t_\mathrm{f}}{\mathrm{d}M} = \left(\frac{K}{2} \right) M^{-1/2} \tag{14.62}$$

Substitution of Equation 14.62 into Equation 14.60 leads to

$$n(M) = \left(\frac{K}{2A_\mathrm{n}} \right) f(t_\mathrm{f}) M^{-1/2} \tag{14.63}$$

which enables the mass spectrum to be converted into the number-fraction MMD. Molar mass averages can be determined from moments of $n(M)$, e.g. $\bar{M}_\mathrm{n}$ and $\bar{M}_\mathrm{w}$ are given by

$$\bar{M}_\mathrm{n} = \int_0^\infty n(M) M \mathrm{d}M$$

and

$$\bar{M}_\mathrm{w} = \frac{\int_0^\infty n(M) M^2 \mathrm{d}M}{\int_0^\infty n(M) M \mathrm{d}M}$$

Alternatively, though less satisfactorily, $\bar{M}_\mathrm{n}$ and $\bar{M}_\mathrm{w}$ can be determined using discrete values of $n(M)$ calculated by applying Equation 14.63 to each individual peak present in the mass spectrum

$$\bar{M}_\mathrm{n} = \sum_{M_{t\mathrm{min}}}^{M_{t\mathrm{max}}} n(M_{t_\mathrm{f}}) M_{t_\mathrm{f}}$$

and

$$\bar{M}_\mathrm{w} = \frac{\sum_{M_{t\mathrm{min}}}^{M_{t\mathrm{max}}} n(M_{t_\mathrm{f}}) M_{t_\mathrm{f}}^2}{\sum_{M_{t\mathrm{min}}}^{M_{t\mathrm{max}}} n(M_{t_\mathrm{f}}) M_{t_\mathrm{f}}}$$

where M_{t_min} and M_{t_max} are the molar masses corresponding to the first and final m/z species detected.

For comparison with MMDs obtained by other methods, the normalized weight-fraction MMD $w(M)$ is needed. Following a similar analysis to that for evaluation of $n(M)$, the weight fraction dw of polymer molecules which are present in the ToF mass spectrum between t_f and (t_f+dt_f) must be equal to the corresponding weight fraction between M and $(M+dM)$ in the weight-fraction MMD, i.e.

$$dw = \frac{f(t_f)M dt_f}{A_w} = w(M)dM \quad (14.64)$$

where A_w normalizes $f(t_f)M$ and is given by $\int_{t_{delay}}^{\infty} f(t_f)M dt_f$. Combining Equations 14.62 and 14.64, with rearrangement, gives

$$w(M) = \left(\frac{K}{2A_w}\right) f(t_f)M^{3/2} \quad (14.65)$$

which enables the mass spectrum to be converted into the weight-fraction MMD. Combining Equations 14.63 and 14.65 reveals the relationship between $w(M)$ and $n(M)$

$$w(M) = n(M)\left(\frac{A_n}{A_w}\right)M^2$$

14.5.4 Molar Mass Distributions Obtained by MALDI Mass Spectroscopy

As MALDI MS evolved during the 1990s, there was great hope that it would provide absolute MMDs. However, as will be evident from the preceding sections, there are significant uncertainties in the MMDs obtained from MALDI mass spectra. Incomplete desorption/ionization of the sample, especially the highest molar mass species, is one source of uncertainty. Another arises from the (usually unknown) extent to which chains fragment (degrade) during the desorption/ionization process. Also, most modern instruments are based on ToF separation of the molecular ions, for which the detector response at higher m/z may be diminished due to partial saturation of the detector by the molecular ions with lower m/z values which reach the detector first. In addition to these effects, the signal-to-noise ratio increases and the ability to resolve polymer chains with different degrees of polymerization reduces as chain length increases because the signal depends on the number of species hitting the detector, which for molecules of high molar mass is small even if they constitute a significant proportion of the total polymer by weight. Each of these uncertainties increases as molar mass increases and so MALDI MS often yields MMDs that, particularly for moderate–higher molar mass polymers, are skewed to lower molar mass species compared to the true MMD.

Another factor specific to ToF mass spectra is that they are recorded with a non-linear m/z axis, the peaks being spaced in increasingly large intervals of m/z as m/z increases (Section 14.5.3.2). Since 'MMDs' from MALDI ToF MS often are reported as mass spectra of abundance against m/z, the true appearance of the number-fraction MMD is masked.

Several comparisons have been made of MMDs obtained for the same polymer sample by GPC (Section 14.3) and MALDI MS. In general, MALDI MS provides accurate measurement of molar mass but has uncertainty associated with the number (and weight) fraction of each species (particularly for the higher molar mass species), whereas in GPC, there is greater uncertainty in the molar masses obtained (through calibration or use of multiple detectors), but high certainty in the weight fraction of each species detected. For low molar mass polymers with narrow MMDs, MALDI MS has been shown to give accurate data if carried out using appropriate sample preparation and spectrometer settings (Sections 14.5.2 and 14.5.3). However, for moderate–high molar mass polymers and polymers with broad MMDs, MALDI MS is less reliable and the data need to be scrutinized carefully, especially before use in calculating molar mass averages.

For all the reasons outlined above, although modern instruments now are capable of analysing polymers with molar masses up to around 1000 kg mol^{-1}, MALDI MS is most suited to determining the MMDs of polymers which have MMDs that extend up to about 50,000 g mol^{-1}. MALDI MS is particularly powerful, and can provide unique information, when it used for combined determination of MMD and molecular structure, as is discussed in Section 15.7.1.

PROBLEMS

14.1 A polymer solution was cooled very slowly until phase separation took place to give two phases in equilibrium. Subsequent analysis of the phases showed that the volume fractions of polymer in the phases were $\phi_2'' = 0.89$ and $\phi_2' = 0.01$, respectively. Using the equation for $\overline{\Delta G_1}$ given in Problem 10.1, together with the equilibrium condition $\overline{\Delta G_1}'' = \overline{\Delta G_1}'$, calculate an estimate (two significant figures) of the Flory–Huggins polymer–solvent interaction parameter for the conditions of phase separation.

14.2 Polystyrene and poly(methyl methacrylate) samples of narrow MMD were analysed by under identical conditions using the same GPC system. The elution time for one of the poly(methyl methacrylate) samples was found to be the same as that for the polystyrene sample of molar mass 97,000 g mol^{-1}. Evaluate the molar mass of this poly(methyl methacrylate) sample given that for the solvent and temperature used, the Mark–Houwink–Sakurada constants are $K = 1.03 \times 10^{-2}$ cm^3 g^{-1} (g mol^{-1})$^{-a}$ with $a = 0.74$ for polystyrene and $K = 5.7 \times 10^{-3}$ cm^3 g^{-1} (g mol^{-1})$^{-a}$ with $a = 0.76$ for poly(methyl methacrylate). State clearly any assumptions you make in your calculation.

14.3 A flow FFF experiment was performed at 25 °C on a narrow MMD sample of polystyrene in tetrahydrofuran using a cross-flow rate of 1.00 cm^3 min^{-1} and a channel flow rate of 1.00 cm^3 min^{-1} on a system with a total channel volume of 0.79 cm^3 and a channel depth of 220 μm. The elution time t_s corresponding to the peak in the plot of detector signal versus time was 15.43 min. Given that the elution time t_0 for solvent molecules under the same analysis conditions is 0.81 min, calculate the value of the retention parameter λ and determine the diffusion coefficient of the polystyrene sample.

14.4 In a ToF matrix-assisted laser desorption/ionization mass spectrometry measurement, the total ToF for one of the species detected was 77 μs. Given that the total distance traversed by the species in the instrument (from ionization to detection) is 0.9 m and that a 15 kV accelerating voltage was used, calculate the molar mass (in units of g mol^{-1}) of the species detected. State any assumptions that you make.

FURTHER READING

General Reading

Allen, G. and Bevington, J.C. (eds), *Comprehensive Polymer Science*, Vol. 1 (Booth, C. and Price, C., eds), Pergamon Press, Oxford, U.K., 1989.

Barth, H.G. and Mays, J.W., *Modern Methods of Polymer Characterization*, John Wiley & Sons, New York, 1991.

Billingham, N.C., *Molar Mass Measurements in Polymer Science*, Kogan Page, London, 1977.

Flory, P.J., *Principles of Polymer Chemistry*, Cornell University Press, Ithaca, NY, 1953.

Mark, H.F., Bikales, N.M., Overberger, C.G., and Menges, G. (eds), *Encyclopedia of Polymer Science and Engineering*, Wiley-Interscience, New York, 1985–1989.

Morawetz, H., *Macromolecules in Solution*, 2nd edn., Wiley-Interscience, New York, 1975.

Sun, S.F., *Physical Chemistry of Macromolecules: Basic Principles and Issues*, 2nd edn., Wiley-Interscience, New York, 2004.

Tager, A., *Physical Chemistry of Polymers*, 2nd edn, Mir Publishers, Moscow, Russia, 1978.
Tanford, C., *Physical Chemistry of Macromolecules*, John Wiley & Sons, New York, 1961.

Field-Flow Fractionation

Cölfen, H. and Antonietti, M., Field-flow fractionation techniques for polymer and colloid analysis, *Advances in Polymer Science*, **150**, 67 (2000).
Giddings, J.C., Field-flow fractionation of macromolecules, *Journal of Chromatography*, **470**, 327 (1989).
Kowalkowski, T., Buszewski, B., Cantado, C., and Dondi, F., Field-flow fractionation: theory, techniques, applications and the challenges, *Critical Reviews in Analytical Chemistry*, **36**,129 (2006).
Schimpf, M., Polymer analysis by thermal field-flow fractionation, *Journal of Liquid Chromatography and Related Technologies*, **25**, 2101 (2002).

Mass Spectroscopy

Batoy, S.M.A.B., Akhmetova, E., Miladinovic, S., Smeal, J., and Wilkins, C.L., Developments in MALDI mass spectrometry: the quest for the perfect matrix, *Applied Spectroscopy Reviews*, **43**, 485 (2008).
de Hoffman, E. and Stroobant, V., *Mass Spectrometry: Principles and Applications*, John Wiley & Sons, Chichester, U.K., 2007.
El-Aneed, A., Cohen, A., and Banoub, J., Mass spectrometry, review of the basics: electrospray, MALDI, and commonly used mass analyzers, *Applied Spectroscopy Reviews*, **44**, 210 (2009).
Hanton, S.D., Mass spectrometry of polymers and polymer surfaces, *Chemical Reviews*, **101**, 527 (2001).
Montaudo, G. and Lattimer, R.P. (eds), *Mass Spectrometry of Polymers*, CRC Press, Boca Raton, FL, 2002.
Montaudo, G., Samperi, F., and Montaudo, M.S., Characterization of synthetic polymers by MALDI-MS, *Progress in Polymer Science*, **31**, 277 (2006).
Nielen, M.W.F., MALDI time-of-flight mass spectrometry of synthetic polymers, *Mass Spectrometry Reviews*, **18**, 309 (1999).

15 Chemical Composition and Molecular Microstructure

15.1 INTRODUCTION

The previous chapters on polymer characterization have been concerned with methods for determination of molar mass averages, molar mass distributions and molecular dimensions. In many instances, this information is all that is necessary to characterize a homopolymer when its method of preparation is known. However, for certain homopolymers (e.g. polypropylene, polyisoprene), knowledge of molecular microstructure is of crucial importance. Additionally, for a copolymer it is necessary to determine the chemical composition in terms of the mole or weight fractions of the different repeat units present. It is also desirable to determine the distribution of chemical composition amongst the different copolymer molecules that constitute the copolymer (e.g. using methods described in Section 14.3.5), and to determine the sequence distribution of the different repeat units in these molecules. Furthermore, when characterizing a sample of an unknown polymer, the first requirement is to identify the repeat unit(s) present. Thus, methods for determination of the chemical composition and molecular microstructure are of great importance.

Simple methods of chemical analysis are useful in certain cases. Qualitative chemical tests such as the Beilstein and Lassaigne tests for the presence of particular elements (e.g. halogens, nitrogen, sulphur) are of use in the identification of an unknown polymer. *Combustion analysis* for quantitative determination of elemental composition can be used to confirm the purity of a homopolymer and to determine the average chemical composition of a copolymer for which the repeat units are known and have significantly different elemental compositions (e.g. the acrylonitrile contents of different nitrile rubbers, which are poly(butadiene-*co*-acrylonitrile)s, can be evaluated from their percentage carbon and/or percentage nitrogen contents). The usefulness of these analyses, however, is restricted by their inability to give structural information. *Spectroscopic methods* are by far the most powerful for this purpose and there now are available an enormous number of fully interpreted spectra of low molar mass compounds and polymers which provide a firm foundation for structural characterization. In addition to giving overall structural information (e.g. identification of repeat units, determination of the average chemical composition of a copolymer), spectroscopic methods can be used to investigate molecular microstructure, repeat unit sequence distributions and structural defects. Hence, they will be the focus for this chapter.

15.2 PRINCIPLES OF SPECTROSCOPY

Spectroscopy is concerned with the measurement and interpretation of the absorption and emission of radiation. This arises from transitions between quantized energy states of the constituent atoms, groups of atoms or molecules (depending on the energy of the radiation) and so corresponds to specific energy differences ΔE. The crucial link between the frequency of radiation ν and the energy E associated with one photon is provided by *Planck's equation*

$$E = \mathbf{h}\nu \tag{15.1}$$

where **h** is the Planck constant ($=6.626\times10^{-34}$ J s). Often the wavelength λ of radiation is reported and is related to ν through the velocity of light **c** ($=2.998\times10^{8}$ m s^{-1})

$$\lambda = \frac{\mathbf{c}}{\nu} \tag{15.2}$$

Thus a transition from a lower to a higher energy state will result from absorption of radiation if its frequency (or wavelength) corresponds, through Planck's equation (Equation 15.1), exactly to the difference in energy ΔE between the two quantized states, i.e. when

$$\nu = \frac{\Delta E}{\mathbf{h}} \quad \text{or} \quad \lambda = \frac{\mathbf{hc}}{\Delta E} \tag{15.3}$$

Similarly, Equation 15.3 defines the frequency (or wavelength) of radiation emitted as a consequence of the reverse transition from the higher to the lower energy state.

15.2.1 Uses of Electromagnetic Radiation in Polymer Science

The principal regions of the electromagnetic spectrum are shown in Figure 15.1 together with indications of how the different regions of radiation are used in analysis.

Scattering and diffraction of high-energy (short wavelength, high frequency) X-rays, electrons and neutrons find use in determining the size and shape of polymer molecules in solution and at interfaces, studying polymer morphology and elucidating the arrangement of atoms in ordered phases, aspects of which are described in Sections 12.4, 17.2, 17.4 and 18.3.2.2.

Although microwave radiation nowadays is used to promote chemical reactions, including polymerizations, *microwave spectroscopy* is only useful for analysing small molecules, so is not important in polymer science.

In *electron spin resonance* (ESR) spectroscopy, absorption of microwave radiation arises from transitions between different alignments, in a strong applied magnetic field, of the magnetic moments resulting from the spins of *unpaired electrons* (i.e. from paramagnetism). Thus ESR spectroscopy can be used to study species such as radicals, radical-anions and radical-cations. The splitting of an ESR absorption due to spin–spin coupling with magnetic nuclei (e.g. ^{1}H) in the molecular species containing the unpaired electron provides details of the molecular structure of that species. Thus, ESR spectroscopy finds use in the identification and study of reactive intermediates in polymerizations that have radical or single-electron transfer mechanisms and species generated during oxidation, high-energy irradiation and thermal and mechanical degradation of polymers. Additionally, ESR spectroscopy is used for characterizing polymers that possess unpaired electrons (e.g. doped conducting polymers; Section 25.3.5), and also has been used to probe molecular dynamics.

The types of spectroscopy most important for determining the chemical structure of polymers are, however, *ultraviolet and visible light absorption spectroscopy*, *infrared spectroscopy*, *Raman spectroscopy*, *nuclear magnetic resonance spectroscopy* and *mass spectroscopy*. These methods of spectroscopy are reviewed in this chapter with emphasis on their use for analysis of polymers. Fundamental aspects, and descriptions of the instrumentation and experimental procedures employed, will be treated only briefly because a full account would require a disproportionate amount of space and there already exist many excellent texts on spectroscopy which deal with these aspects. Also due to the limitations of space, only a few spectra and short surveys of the potential of these spectroscopic methods for characterization of polymers are given. The reader is referred to specialist texts on spectroscopy of polymers for a more complete appreciation of their uses.

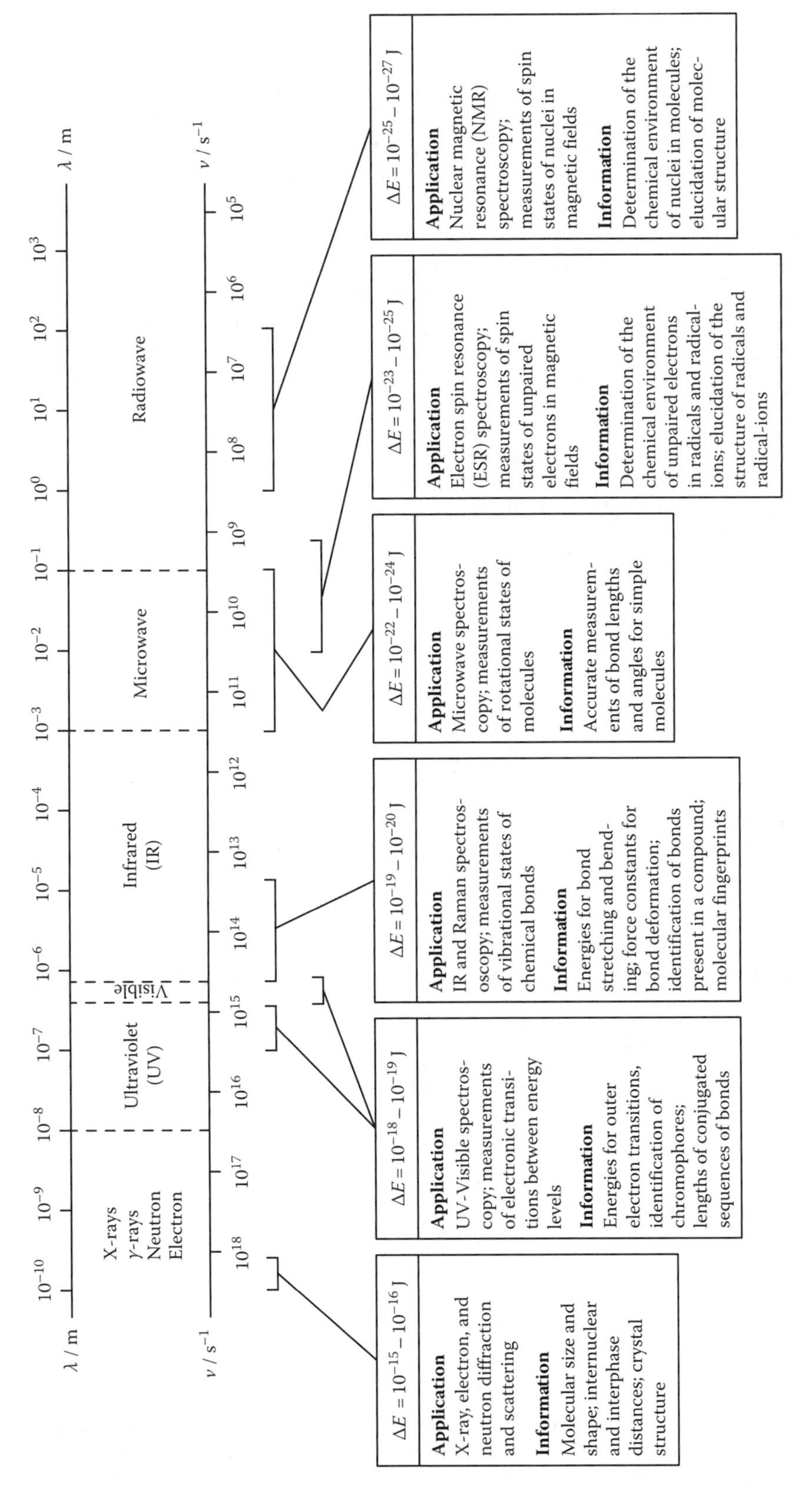

FIGURE 15.1 The spectrum of electromagnetic radiation and some of its uses in analysis. ΔE is the range of energy changes corresponding to one photon of radiation for the typical range of radiation used in practice.

15.2.2 The Beer–Lambert Law for Absorption of Electromagnetic Radiation

Equation 15.3 gives the frequency (or wavelength) of radiation that will be absorbed by a single species undergoing a transition between energy states, but provides no information on the amount of light absorbed by a substance, which, of course, will comprise an extremely large number of those species. If it is assumed that each individual species absorbs one photon with an efficiency φ and that the total amount of light absorbed is a simple sum of all those photons of radiation absorbed, then analysis for the total intensity of radiation absorbed by a substance is quite straightforward. Consider the situation shown in Figure 15.2 where incident radiation of intensity I_0 passes through a thickness ℓ of a substance that contains N_{ab} absorbing species per unit volume.

At a distance z into the substance, the intensity of the radiation has reduced to I_z and as the radiation passes through the infinitesimally thin layer of cross-sectional area a and thickness dz the intensity is reduced by a further amount dI due to its interaction with the $N_{ab}a\mathrm{d}z$ absorbing species present in that thin disc of the substance. Thus the intensity $I_{z+\mathrm{d}z}$ of radiation at $z+\mathrm{d}z$ is given by

$$I_{z+\mathrm{d}z} = I_z - I_z \varphi N_{ab} a\mathrm{d}z$$

Since $\mathrm{d}I = I_{z+\mathrm{d}z} - I_z$, then the intensity I_T of radiation emerging from the far side of the substance can be obtained by a simple integration over the whole thickness ℓ

$$\int_{I_0}^{I_T} \frac{\mathrm{d}I_z}{I_z} = -\varphi N_{ab} a \int_0^{\ell} \mathrm{d}z$$

and so

$$\ln\left(\frac{I_0}{I_T}\right) = \varphi a N_{ab} \ell \quad (15.4)$$

in which φa often is referred to as the *absorption cross-section*.

It is more convenient to write Equation 15.4 in the form of the *Beer–Lambert equation*

$$A = \varepsilon c \ell \quad (15.5)$$

where

ε is the *molar absorptivity* ($=\varphi a\mathbf{N}_A/2.303$ in which $\mathbf{N}_A$ is the Avogadro constant)
c is the molar concentration of absorbing species ($=N_{ab}/\mathbf{N}_A$)
A is the *absorbance*, which is related to I_0, I_T and the percentage transmission, $\%T$ ($=100I_T/I_0$)

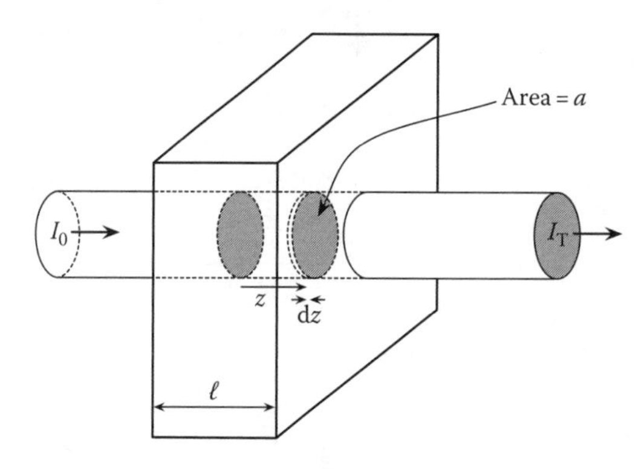

FIGURE 15.2 A schematic diagram showing a beam of radiation with incident intensity I_0 and cross-sectional area a passing through a thickness ℓ of a substance and emerging with a lower intensity I_T due to the absorption of the radiation by absorbing species in the substance.

by

$$A=\log_{10}\left(\frac{I_0}{I_T}\right)=\log_{10}\left(\frac{100}{\%T}\right) \tag{15.6}$$

Note that it is standard practice to have ε in units of $dm^3\ mol^{-1}\ cm^{-1}$, with c in units of $mol\ dm^{-3}$ and ℓ in cm.

Extensive use is made of the Beer–Lambert equation (Equation 15.5) for measuring concentrations of functional groups, particularly in UV-visible and IR spectroscopy. However, to ensure that meaningful results are obtained, the key assumptions inherent in derivation of the equation must be borne in mind when making such quantitative measurements: (i) loss of intensity due to light scattering must be negligible (i.e. samples must be free from dust and other light scatterers) and (ii) the concentration of absorbing species must be quite low (typically no more than 10 mmol dm^{-3}) because ε becomes concentration dependent (tending to reduce) at higher concentrations due to interactions between molecules. The effect of concentration on the value of ε is particularly significant when analysing solutions of ion-containing molecules (e.g. polyelectrolytes and ionomers) because concentration has a very significant effect on the extent of ion separation across a very wide range of concentration from very dilute to moderately concentrated. Fluorescence and phosphorescence also can lead to errors in the application of Equation 15.5.

15.3 ULTRAVIOLET AND VISIBLE LIGHT ABSORPTION SPECTROSCOPY

Absorption of ultraviolet and visible (UV-vis) radiation results from electronic transitions of bonding electrons, most commonly from the *highest occupied bonding or non-bonding molecular orbital (HOMO)* to the *lowest unoccupied molecular orbital (LUMO)*. The relative energies for electronic orbitals are shown in Figure 15.3, from which it can be seen that electronic transitions for electrons in σ-bonds (principally $\sigma \rightarrow \sigma^*$ transitions) are of relatively high energy, so occur in the less accessible far-UV region (100 nm $<\lambda<$ 200 nm) and are of little practical interest. In contrast, electronic transitions of electrons in π-bonds (principally $\pi \rightarrow \pi^*$ transitions) tend to fall in the much more accessible near/middle-UV regions (spanning 200 nm $<\lambda<$ 400 nm) or, if the π-bonds are in long conjugated sequences, in the visible region (400 nm $<\lambda<$ 700 nm) and are the basis of most UV-vis spectroscopy. Weak *forbidden* electronic $n \rightarrow \pi^*$ transitions of non-bonding lone-pair electrons on heteroatoms in π-bonds also are observed in the near/middle-UV regions. Thus UV-vis spectroscopy is most useful for studying molecules that contain π-bonds. The groups which give rise to the absorptions are termed *chromophores*.

Although ΔE for a given electronic transition might be considered to have a unique value, this is not the case because associated with each electronic energy state are (lower energy) vibrational

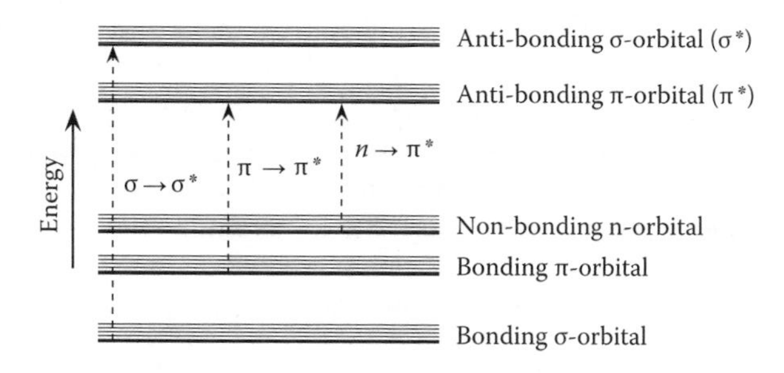

FIGURE 15.3 Schematic diagram showing the relative energies (bold lines) of the different electronic orbitals and the more common types of transition between them. The additional fine lines above each electronic energy level represent the corresponding bond vibrational energy levels.

and rotational energy states. Hence, the same electronic transition can be accompanied by different transitions between bond vibrational and rotational energy states, giving rise to a range of ΔE values for the same electronic transition. The effect is dominated by different transitions between bond vibrational energy states because they correspond to much larger energy differences than transitions between bond rotational energy states. However, because the differences between the vibrational energy states are small compared to ΔE (see Figure 15.3), the different values of ΔE often are not resolved experimentally and, in many cases, UV-vis absorptions appear as quite broad bands rather than as clusters of discrete absorptions. It is normal practice to measure the maximum absorbance A_{max} in a broad absorption band and to report the location of an absorption as the corresponding wavelength λ_{max}. Similarly, values of ε reported in the literature are those corresponding to the absorbance maximum, i.e. ε_{max}.

One very significant feature of UV-vis spectroscopy is that the values of ε_{max} for *allowed* electronic transitions of chromophores tend to be very large and facilitate measurement of very low concentrations of such chromophores with great accuracy. Some values of λ_{max} and ε_{max} for electronic transitions of common chromophores are given in Table 15.1.

15.3.1 Applications of UV-vis Spectroscopy in Polymer Science

UV-vis spectroscopy is not capable of providing detailed information on chemical structure beyond establishing the presence of a particular type of chromophore and so its use is limited to situations where there is good knowledge of the system to be analysed. Furthermore, the most experimentally accessible region ($200 \text{ nm} < \lambda < 700 \text{ nm}$) is restricted mainly to analysis of molecules that contain conjugated π-bonds either in aromatic groups or as extended sequences of double bonds. Unfortunately, analysis of non-conjugated C=O groups and C=C groups is more difficult because the strong $\sigma \rightarrow \sigma^*$ C=O absorption (180–200 nm) and the strong $\pi \rightarrow \pi^*$ C=C absorption (160–200 nm) occur in wavelength ranges where molecular oxygen absorbs strongly (requiring the analysis cell to be flushed with nitrogen or under vacuum). The $n \rightarrow \pi^*$ transition of C=O groups is accessible (see Table 15.1) and is widely used for analysis, but it is very weak, which limits accuracy. In view of all these considerations, it is easy to understand why UV-vis spectroscopy has a relatively narrow range of uses in polymer science.

As is evident from Table 15.1, monomers with conjugated π-bonds (e.g. acrylic and methacrylic acids and their esters and amides, styrenic monomers and other aromatic monomers) have strong $\pi \rightarrow \pi^*$ absorptions and so are detected easily by UV spectroscopy, even when present at low levels. For such monomers, UV spectroscopy can be used to determine the monomer concentration in a formulated resin, for monitoring conversion during polymerization and for quantifying levels of residual monomer at the end of a polymerization or in a product. Similarly, polymers that contain aromatic groups (e.g. from styrene repeat units or phthaloyl-containing structural units) can be analysed using their $\pi \rightarrow \pi^*$ transitions (see Figure 15.4), an approach that is most useful for determining copolymer composition when the repeat unit structures are known. For example, the composition of poly(butadiene-*co*-styrene) can be determined from the $\pi \rightarrow \pi^*$ B-band absorption of the styrene repeat units because the other parts of the copolymer chemical structure do not contribute to that region of the UV spectrum. This is particularly suitable for gel permeation chromatography analysis of copolymers (Section 14.3.5); by using two detectors (e.g. a differential refractometer to give the total copolymer concentration in the eluant and a UV-vis detector set to record only the absorption from aromatic repeat units) it is possible to determine the distribution of copolymer composition within the molar mass distribution.

As the length of a conjugated sequence of π-bonds increases, the $\pi \rightarrow \pi^*$ transition moves to even lower energies (longer wavelengths). This effect is illustrated by the data for the polyene series shown as the first six entries in Table 15.1. The value of ΔE for the transition can be predicted from simple one-dimensional solution of the Schrödinger equation for an electron constrained to move within a rectangular box, the length of which corresponds to the length of the conjugated sequence

TABLE 15.1
Values of λ_{max} and ε_{max} for Electronic Transitions of Common Chromophores in the Near/Middle UV and Visible Regions

Chromophore [a]	Transition Type	Band Type [b]	λ_{max} / nm	ε_{max} / dm^3 mol^{-1} cm^{-1}
$H\!-\!(CH{=}CH)_2\!-\!H$	$\pi \rightarrow \pi^*$	K	217	21,000
$H\!-\!(CH{=}CH)_3\!-\!H$	$\pi \rightarrow \pi^*$	K	258	35,000
$H\!-\!(CH{=}CH)_4\!-\!H$	$\pi \rightarrow \pi^*$	K	296	52,000
$H\!-\!(CH{=}CH)_5\!-\!H$	$\pi \rightarrow \pi^*$	K	335	118,000
$H\!-\!(CH{=}CH)_8\!-\!H$	$\pi \rightarrow \pi^*$	K	415	210,000
$H\!-\!(CH{=}CH)_{11}\!-\!H$	$\pi \rightarrow \pi^*$	K	470	185,000
R(R')C=O	$n \rightarrow \pi^*$	R	270–300	10–20
R'(R')C=C(R')–C(R')=O	$\pi \rightarrow \pi^*$ $n \rightarrow \pi^*$	K R	215–250 310–330	10,000–20,000 20–30
R–C(=O)–O–R'	$n \rightarrow \pi^*$	R	~200	~60
R'(R')C=C(R')–C(=O)–O–R'	$\pi \rightarrow \pi^*$	K	200–220	10,000–15,000
substituted benzene ring	$\pi \rightarrow \pi^*$ $\pi \rightarrow \pi^*$	K B	200–240 250–280	6,000–15,000 200–2,000

Sources: Data taken from Lambert, J.B. et al., *Introduction to Organic Spectroscopy*, Macmillan, New York, 1987; Silverstein, R.M. et al., *Spectrometric Identification of Organic Compounds*, 5th edn., John Wiley & Sons, New York, 1991.

[a] R = alkyl and R′ = alkyl or H.

[b] Burawoy classification.

of π-bonds, thus enabling the conjugation length to be estimated from λ_{max} for the absorption. Hence, UV-vis spectroscopy can be used to monitor changes in the length of conjugated sequences of π-bonds, for example the growth of polyene sequences in the first stage of converting polyacrylonitrile fibres into carbon fibres and in the dehydrochlorination degradation of poly(vinyl chloride). The measurements of conjugation length also are useful in studies of dyes, most of which are aromatic compounds with extended sequences of π-bonds that move the $\pi \rightarrow \pi^*$ transition into the visible region; by varying the conjugation length, the colour of the dye is controlled. Hence, visible light spectroscopy can be used to analyse dyes present in a polymer product. Similarly,

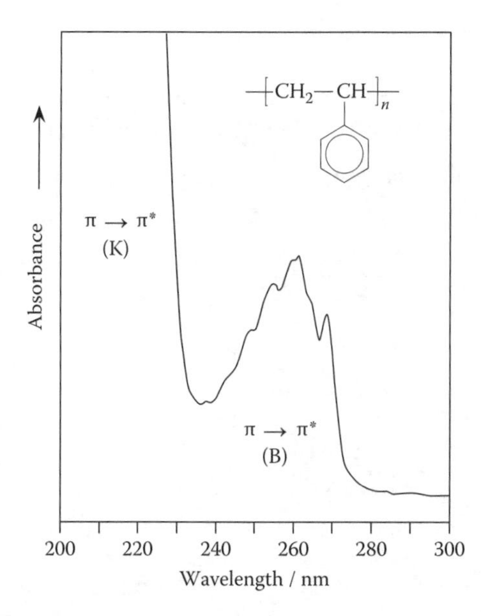

FIGURE 15.4 UV spectrum of polystyrene showing the B $\pi \rightarrow \pi^*$ transition band with vibrational energy fine structure; the K $\pi \rightarrow \pi^*$ transition band is present but off-scale due to its much higher ε_{max} (see Table 15.1). (Data taken from Meal, L., *J. Appl. Polym. Sci.*, 41, 2521, 1990.)

UV-vis spectroscopy can be used to analyse polymers for the presence of other additives that contain chromophores, such as antioxidants.

Fluorescence and *phosphorescence* techniques are closely related to UV-vis spectroscopy and are important in studies of polymer photophysics. *Confocal laser scanning microscopy* is now an important technique for providing two-dimensional (2D) and three-dimensional (3D) images of fluorescent species within a material and is most important in biomedical applications of polymers.

15.3.2 Practical Aspects of UV-vis Spectroscopy

One attraction of UV-vis spectroscopy is that simple instruments are available at very low cost and often are sufficient for most measurements. The instruments operate on the *double-beam* principle in which the difference in absorbance between the sample and a reference (the open atmosphere or a cell containing just solvent) is recorded one wavelength at a time (in the most simple instruments) or, more commonly, over a range of wavelengths at a defined scanning rate. In most cases, the absorptions under investigation have high values of ε and so it usually is necessary to make a dilute solution of the sample in a solvent that does not absorb in the wavelength range of interest. The solvents must be of high purity (free from trace impurities of UV-vis absorbers) and need to be chosen with care because dipolar and hydrogen-bonding interactions with the chromophoric group can give rise to changes in λ_{max} and/or ε_{max}. The analysis cells are precision-made and of well-defined path length (typically 2–20 mm). Thin films must be used to analyse polymers in the solid state so that the absorbance is reduced to values in the range where the Beer–Lambert equation (Equation 15.5) is obeyed. Additionally, since contributions from overlapping absorptions normally cannot be resolved, a prerequisite for analysis by UV-vis spectroscopy is that only one component in the sample gives a UV-vis absorption in the wavelength range of interest. Another factor that needs to be considered in the analysis of copolymers is whether interactions between repeat units lead to effects on λ_{max} and/or ε_{max}.

As indicated earlier, the speed and sensitivity of UV-vis spectroscopy makes it ideal for real-time analyses. This is not used very often to monitor reactions, but is used extensively in chromatography for detection of solutes in eluants, the most important application in polymer science being the use of UV-vis detectors in gel permeation chromatography.

15.4 INFRARED SPECTROSCOPY

The energies associated with vibrations of atoms in a molecule with respect to one another are quantized (as depicted by the fine lines in Figure 15.3) and absorption of electromagnetic radiation in the infrared (IR) region (approximately $1\ \mu m > \lambda > 50\ \mu m$) gives rise to transitions between these different vibrational states. Absorption results from coupling of a vibration with the oscillating electric field of the IR radiation, and this interaction can occur only when the vibration produces an oscillating dipole moment (cf. Raman spectroscopy, Section 15.5). Since vibrating atoms are linked together by chemical bonds, it is usual to refer to the vibrations as bond deformations, of which the simplest types are stretching and bending. For a particular IR-active bond deformation (e.g. C=O stretching), absorption will occur when the frequency ν of the radiation is that defined by Equation 15.3 in which ΔE is the energy difference between the upper and lower vibrational energy levels for the particular bond deformation. Usually, the only significant absorptions correspond to promotion of bond deformations from their ground states to their next highest energy levels. The basis of *IR spectroscopy* is that for a particular type of bond deformation ΔE, and hence ν, depend upon the atoms involved (e.g. ΔE and ν for N–H stretching are higher than for C–H stretching). Thus by measuring the absorption of IR radiation over a range of ν a spectrum is obtained which contains a series of absorptions at different ν. Each absorption corresponds to a specific type of deformation of a particular bond (or sequence of bonds) and is characteristic of that bond deformation with ν not greatly affected by the other atoms present in the molecule. Thus the absorption due to a particular bond deformation occurs at approximately the same ν for all molecules (including polymers) which contain that bond.

The *characteristic absorption regions* for some of the most commonly observed bond deformations are shown in Figure 15.5. Note that IR radiation is defined either by its wavelength or, more usually nowadays, by its *wavenumber* $\bar{\nu}$, which is the reciprocal of its wavelength (i.e. $\bar{\nu} = 1/\lambda$) and is related to its frequency by $\nu = \bar{\nu}\mathbf{c}$. Thus the most common absorptions occur in the wavenumber range 4000–650 cm^{-1} and for this reason IR spectra usually are recorded over this range. It is convenient to divide IR spectra into three regions for interpretation. In the 3500–1300 cm^{-1} (functional group) region, absorptions can easily be assigned to particular bond deformations on the basis of

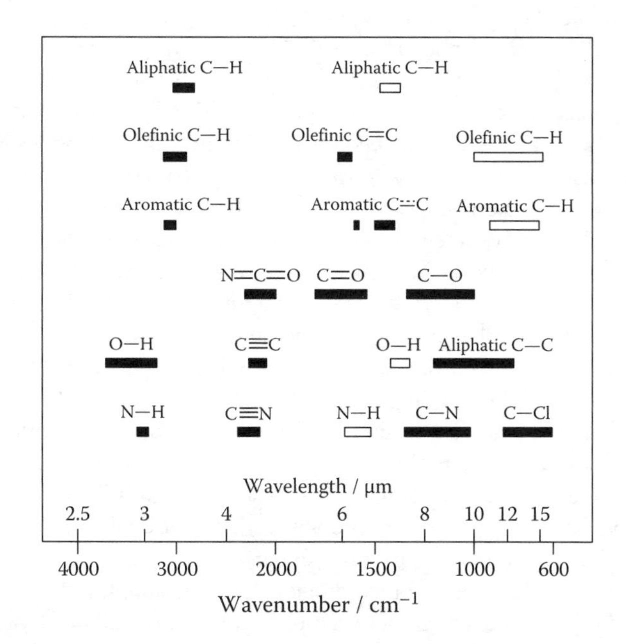

FIGURE 15.5 Characteristic IR absorption regions for some commonly observed bond stretching (■) and bending (▭) deformations.

their position, strength and nature (e.g. O–H stretching gives a very strong, broad absorption centred around 3300 cm^{-1} whereas N–H stretching, which appears in the same wavenumber range, gives a much weaker and sharper absorption that if present in a primary amine group ($-NH_2$) appears as a doublet due to symmetric and asymmetric simultaneous deformation of the two N–H bonds). Absorptions in the 1300–1000 cm^{-1} region arise from C–O stretching and more complex skeletal C–C stretching vibrations, often of several bonds together, and are difficult to assign with any certainty; hence, the number, position and strength of the absorptions in this region tend to be unique to a particular material, which is why it is known as the 'fingerprint' region. Finally, the 1000–650 cm^{-1} (olefinic/aromatic) region shows absorption patterns that are characteristic of specific types of substituted C=C groups (*vinyl*, *cis*- or *trans*-) or benzene rings (mono-substituted or *ortho*-, *meta*- or *para*-disubstituted) and are due to C–H bending vibrations which are out of the plane defined by the sp^2-hybridized carbon atoms. Full details of the nature of each of the characteristic IR absorptions indicated in Figure 15.5 (plus absorptions due to other bond vibrations) can be found in the many textbooks on spectroscopic analysis of organic compounds.

15.4.1 Applications of IR Spectroscopy in Polymer Science

The simplest application of IR spectroscopy is for *polymer identification*. Comparison of the positions of absorptions in the IR spectrum of a polymer sample with the characteristic absorption regions, leads to identification of the bonds and functional groups present in the polymer. In many cases, this information is sufficient to identify the polymer. However, for confirmation, the spectrum can be compared in detail with that of an authentic sample since the two spectra should be identical if the inference is correct, i.e. the IR spectrum of a polymer can be considered as a 'fingerprint' for this purpose. For example, see the IR spectrum of polystyrene shown in Figure 15.6a. The interpretation of the spectra can become more complex when the polymer contains additives which contribute to the absorptions in the spectrum. For example, Figure 15.6b shows the IR spectrum of a sample of flexible poly(vinyl chloride) (PVC) which contains absorptions from C=O stretching and out-of-plane aromatic C–H bending deformations that are due to the phthalate ester plasticizer, i.e. the IR spectrum is the sum of the spectra of the plasticizer and PVC. A similar situation obtains in the identification of blends of two or more polymers.

IR spectroscopy also finds application in the characterization of branched polymers and copolymers. The *degree of branching* in a branched polymer can be determined provided that characteristic absorptions due to the branches can be identified. For example, branching in low-density polyethylene (Section 4.2.4.2) has been determined using absorptions due to the terminal CH_3 groups on the branches. *Average compositions of copolymers* can be determined if the different repeat units give characteristic IR absorptions which are not coincident. For example, the composition of a sample of poly[styrene-*co*-(*n*-butyl acrylate)] can be determined from the out-of-plane C–H bending absorptions of the benzene rings in styrene repeat units and the C=O stretching absorption of the ester groups in *n*-butyl acrylate repeat units. For such quantitative determinations, the spectra are recorded as absorbance A against wavenumber (or wavelength) (see Equation 15.6). The difference ΔA between the absorbance at the absorption maximum and the background absorbance at that position (obtained by linking the baselines before and after the absorption) is directly related by the Beer–Lambert equation (Equation 15.5) to the molar concentration c of the bond(s) or group(s) giving rise to the absorption. Since a single sample is being analysed, the path length ℓ of the radiation through the sample is the same for all absorptions in the spectrum. Hence, by taking the ratio of the ΔA values for the characteristic absorptions, the molar ratio of the concentrations of the bonds or groups of interest can be calculated (e.g. $c_1/c_2 = \varepsilon_2\Delta A_1/\varepsilon_1\Delta A_2$) as long as the ratio of the molar absorptivities has been determined previously using samples of known composition. As for UV-vis spectroscopy, when carrying out such quantitative measurements, it is important to be aware that their accuracy can suffer if interactions between groups on the different repeat units in the copolymer and the repeat unit sequencing along the chain influences the wavenumber and/or the

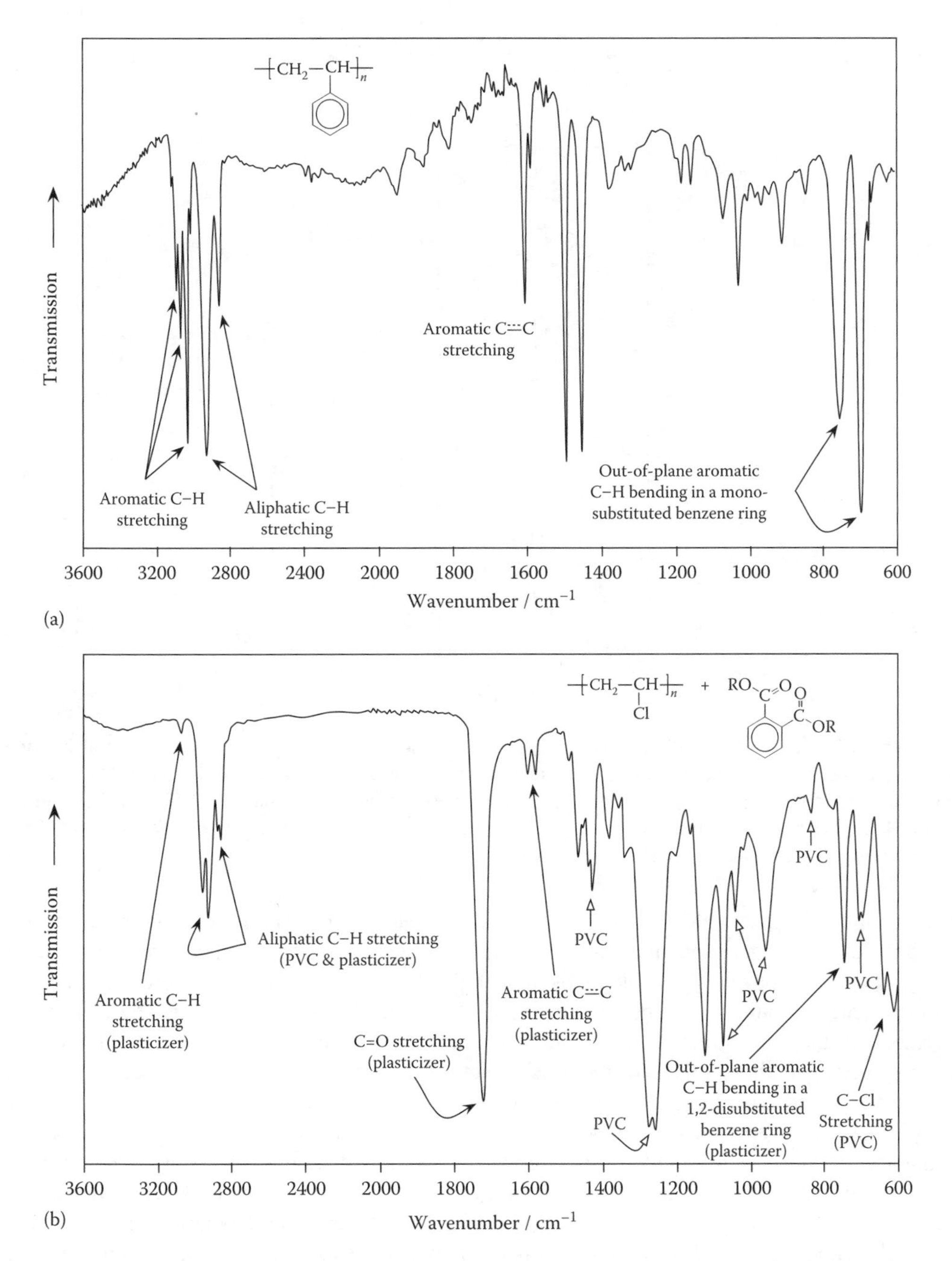

FIGURE 15.6 IR spectra of (a) polystyrene and (b) poly(vinyl chloride) (PVC) plasticized with a phthalate (R is typically a 2-ethylhexyl group). Assignments of the main absorptions are shown; note that the hollow-headed arrows in (b) identify the locations of PVC fingerprint absorptions.

molar absorptivity for an absorption (i.e. when these properties also are dependent on the copolymer composition).

Bond deformations, particularly bending deformations, are not completely insensitive to the presence of other atoms and bonds in the molecular structure or to molecular conformation and interactions with other molecules. Hence, the precise location of an IR absorption carries more information than simply the existence of a particular bond or group. It is for this reason, and also

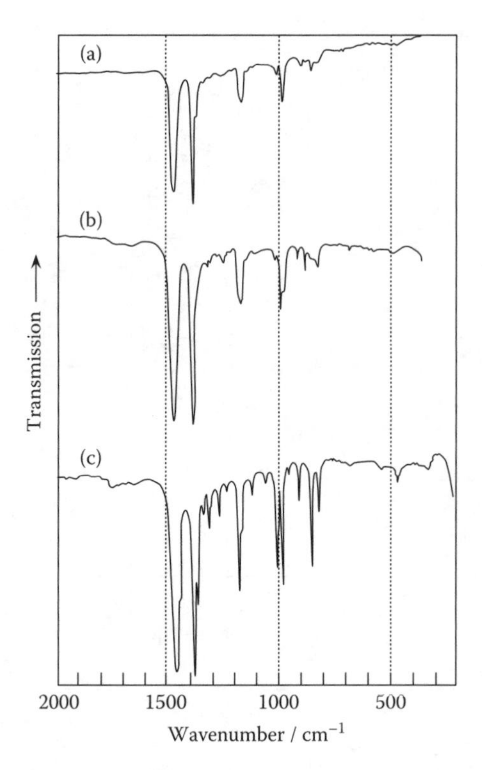

FIGURE 15.7 IR spectra of (a) atactic, (b) syndiotactic and (c) isotactic polypropylene. The absorptions at 970 and 1460 cm^{-1} do not depend upon tacticity, whereas the absorptions at 840, 1000 and 1170 cm^{-1} are characteristic of the 2*3/1 helix of isotactic polypropylene, and the absorption at 870 cm^{-1} is characteristic of the 4*2/1 helix of syndiotactic polypropylene. (Data taken from Klöpffer, W., *Introduction to Polymer Spectroscopy*, Springer-Verlag, Berlin, Germany, 1984.)

because more complex deformations are specific to a particular structure, that IR spectra can be used as polymer fingerprints and can provide information on molecular microstructure. Thus by using IR spectroscopy, it is possible to distinguish the different repeat unit structures that can arise from polymerization of 1,3-diene monomers (see Table 6.1). Similarly, the IR spectra of isotactic, syndiotactic and atactic forms of a polymer show characteristic differences, as shown in Figure 15.7 for polypropylene, though these differences do not result directly from the differences in configuration but instead arise from the effects of configuration upon local chain conformation (Section 10.3.2). Such differences, for example, have been used to estimate the fractions of isotactic and syndiotactic sequences in samples of polypropylene and poly(methyl methacrylate). The effects of chain conformation upon IR absorptions are more general and also apply to polymers which are not tactic. For example, the IR spectrum of semi-crystalline poly(ethylene terephthalate) shows additional absorptions to that of the purely amorphous polymer due to differences in the conformations of chains in the crystalline and amorphous phases.

Other uses of IR spectroscopy include monitoring reactions through the appearance or loss of absorption and measurements of orientation by use of polarized IR radiation.

15.4.2 Practical Aspects of IR Spectroscopy

There are two general types of instrument for recording IR spectra, namely *double-beam* and *Fourier transform (FT) IR spectrometers*. Modern instruments are of the latter type which have a number of advantages over the former, including better signal-to-noise ratios and the ability to record complete spectra in very much shorter timescales (typically a few seconds rather than several minutes). Thus using FTIR spectrometers it also is possible to make *time-resolved measurements* such as monitoring fast reactions, continuous monitoring of the composition of the eluant in gel permeation chromatography, and, with more specialized instrumentation, it is even possible to examine molecular dynamics (e.g. segmental motion) in timescales of less than a second.

Sample handling is the same for both types of spectrometer and use is made of salts, such as potassium bromide and sodium chloride, which do not absorb in the IR region. Solid polymers usually are analysed in the form of either (i) discs pressed from finely powdered dilute (1–2 wt%) dispersions of polymer in potassium bromide or (ii) melt-pressed or solution-cast thin films. Liquid polymers are analysed as thin films between the polished faces of two blocks (known as plates) of sodium chloride. Analysis of polymers in solution tends to be avoided where possible because a significant proportion of the IR spectrum of the polymer is obscured by the IR absorptions of the solvent. IR spectra of polymer surfaces can be recorded using techniques such as *attenuated total reflectance* and *specular reflectance*, and their use has grown with the increasing importance of polymer surface chemistry.

IR spectroscopy is the most widely used method for characterizing the molecular structures of polymers, principally because it provides a lot of information and is relatively inexpensive and easy to perform. However, it is not simple to interpret absolutely the more subtle features of IR spectra, such as those due to differences in tacticity. Such interpretations are usually made on the basis of information obtained from other techniques, in particular *nuclear magnetic resonance spectroscopy* (Section 15.6), which is by far the most powerful method for determining the detailed molecular microstructures of polymers.

FTIR microscopy is a special type of IR spectroscopy in which samples are imaged by recording IR absorptions in focused regions of a sample, typically a few microns in size. Most commonly, FTIR microscopy is used to scan across an area of a sample to give a 2D IR spectrum of the surface, thereby enabling surface inhomogeneities to be detected easily. Modern FTIR microscopes have array detectors that can quickly scan areas of up to several square centimetres at a range of spatial resolutions from around 50 μm down to a few microns. Use of FTIR imaging is growing in importance, particularly in biomedical and forensic applications.

15.5 RAMAN SPECTROSCOPY

Raman spectroscopy is another form of vibrational spectroscopy, but unlike absorptions of IR radiation, Raman spectra arise from inelastic scattering of visible light. This difference is important because the bond vibration must produce a change in polarizability to be Raman active, but a change in dipole moment for IR absorption. Hence, Raman spectroscopy complements IR spectroscopy in that bond vibration modes which are Raman active tend to be weak or inactive in IR spectroscopy and vice versa.

The energy changes that take place during absorption of IR radiation, elastic (Rayleigh) light scattering (Section 12.2) and inelastic (Raman) light scattering are depicted in Figure 15.8. Absorption of IR radiation results in promotion of a bond vibration from the ground state to the next higher-energy vibrational state. In Rayleigh (elastic) light scattering there is no change in vibrational state and the scattered radiation has the same frequency ν_0 as the incident visible radiation, but is of much lower intensity than the incident light (see Section 12.2.1). In Raman (inelastic) light scattering there is a change in vibrational state and the frequency of the scattered radiation differs from ν_0 by $\Delta\nu$, which is known as the *Raman shift* (see Figure 15.8). The lines resulting from Stokes scattering have frequency $\nu_0 - \Delta\nu$ and are stronger than the anti-Stokes lines, which have frequency $\nu_0 + \Delta\nu$ (because the population of the ground vibrational state is greater than the higher-energy vibrational state). Thus the Raman scattering lines are symmetrically placed on either side of the Rayleigh scattering line and separated from it by $\Delta\nu$. Since the Raman lines represent only a very small fraction of the total scattered radiation (the Stokes lines are about 1/1000 the intensity of the Rayleigh line) and $\nu_0 \gg \Delta\nu$, highly monochromatic incident radiation is required for their detection. In view of this, it

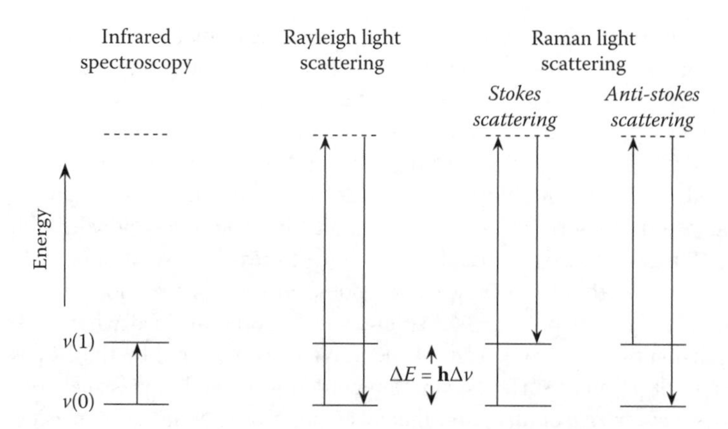

FIGURE 15.8 Schematic diagram showing the energy changes that take place during absorption of IR radiation, Rayleigh (elastic) light scattering and Raman (inelastic) light scattering. Note that $\nu(0)$ and $\nu(1)$ are, respectively, the ground state and lowest excited-state energy levels for a particular bond vibration. In Raman spectroscopy $\Delta\nu$ values, known as *Raman shifts*, are measured as the difference between the frequency of the incident radiation and that of the Stokes scattering lines.

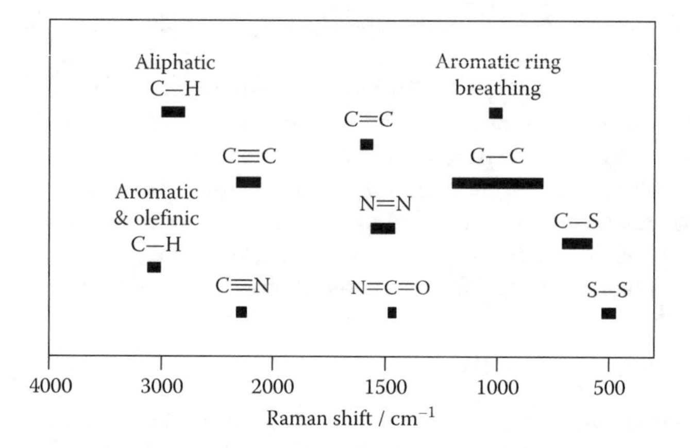

FIGURE 15.9 Wavenumber regions for bond stretching vibrations that give strong Raman lines.

is not surprising that even though Raman and Krishnan first reported the effect in 1928 and Raman was awarded the 1930 Nobel Prize for Physics for the discovery, use of Raman spectroscopy did not become widespread until highly efficient monochromators and laser light sources became available. By measuring each of the Raman lines (i.e. unique values of $\Delta\nu$ corresponding to different bond vibrations) a spectrum is obtained which can be interpreted in terms of *characteristic group frequencies*, just as in IR spectroscopy. Symmetrical bond stretching modes tend to give strong Raman lines, the most important of which for polymers are highlighted in Figure 15.9.

15.5.1 Applications of Raman Spectroscopy in Polymer Science

Raman spectroscopy can be used to identify particular bonds and functional groups in the structure of a polymer and for fingerprint identification in the same way as IR spectroscopy. For example, the Raman spectrum of polystyrene is shown in Figure 15.10; comparison with its IR spectrum shown in Figure 15.6a highlights the complementary nature of Raman and IR spectroscopy. However, many applications of Raman spectroscopy in polymer science arise because it involves measurement of scattered radiation (rather than transmitted radiation) and so the system or material to be

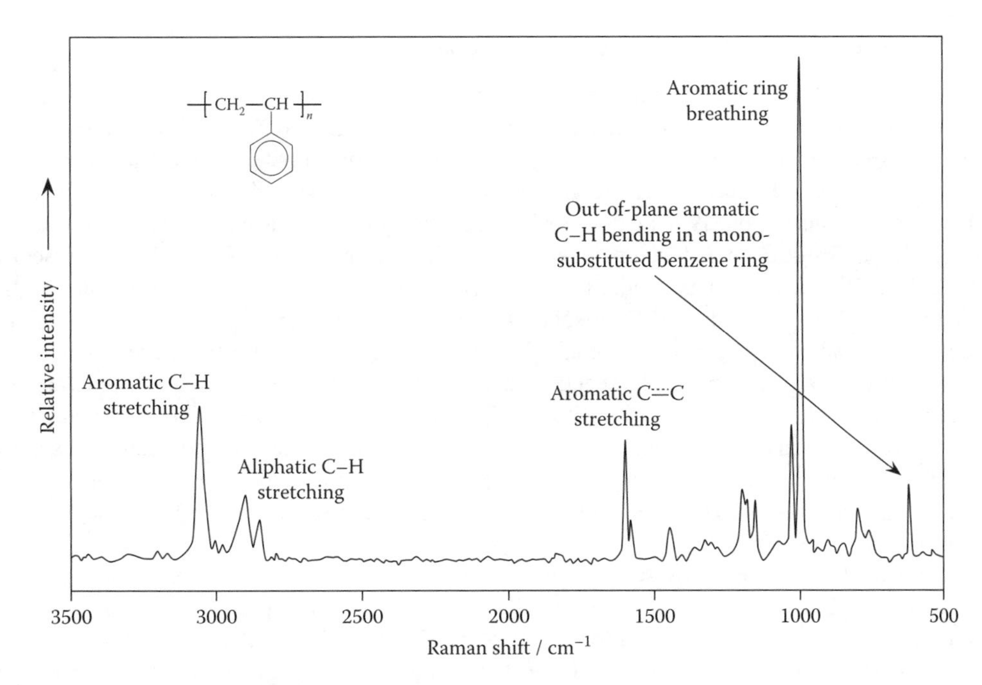

FIGURE 15.10 Raman spectrum of polystyrene. Note the characteristically strong breathing mode of the benzene ring.

studied can be analysed directly without need for sample preparation. For example, Raman spectroscopy is used to monitor polymerizations of olefinic monomers directly (on both laboratory and industrial production scales) through the intensity of the line for C=C stretching, the exact position of which varies for different monomers thus enabling conversion of individual monomers in a copolymerization to be followed simultaneously. Raman spectroscopy also has been used to study crosslinking reactions in coatings and in the vulcanization of natural rubber.

Due to the symmetry of particular modes of bond vibration, Raman spectra are sensitive to the local environment and orientation of the bonds. Hence, Raman spectroscopy can be used to investigate morphological features of bulk polymers. For example, Raman spectroscopy can be used to study crystallization of polymers, for which purpose it is common to use the longitudinal acoustical mode (LAM), i.e. accordion-like stretching, of the relatively short chain sequence that forms the stem in a chain-folded crystal (see Sections 17.3 and 17.4.1). Similarly, chain conformation, orientation and anisotropy in polymer materials can be studied by measuring polarized Raman spectra. Another important application of Raman spectroscopy is in the study of polymer deformation micromechanics, which is made possible because Raman lines move to lower wavenumbers in proportion to the stress in the bond(s). Thus it is possible to monitor the change in $\Delta\nu$ with stress for particular bond vibrations (double and triple bonds are most sensitive to stress) and to then use those data as a calibration with which to monitor the stress (or strain) in artefacts made from that polymer. For example, model composites have been prepared that contain a single poly(*p*-phenylene benzoxazole) (PBO) fibre incorporated centrally aligned within the constricted region of an epoxy resin tensile testing specimen; the model composite specimens then were subjected to tensile testing with the overall stress-strain curve being measured conventionally whilst measuring simultaneously by Raman microscopy the strain in the PBO fibre at several points along its length from the shift in the position of the Raman line for aromatic C=C stretching, analysis of the fibre being possible because it was within the optically transparent epoxy resin (see Section 24.6.3.2 and Figure 24.16). Such measurements led to the first experimental demonstration of the Cox shear lag model for composites (see Section 24.6.3.1). Nowadays, Raman microscopy is used widely for studying the micromechanics of deformation in polymer fibres, composites and blends.

15.5.2 Practical Aspects of Raman Spectroscopy

Separation of the Rayleigh scattering line from the Raman lines is critical to the measurement of Raman spectra and requires highly efficient monochromators. Early commercial Raman spectrometers used large (about 1 m in length) double or triple monochromators to achieve separation of the Raman lines from each other and the Rayleigh line. The development of efficient, compact single monochromators during the 1990s led to further growth in the use of Raman spectroscopy and to portable Raman spectrometers that employ diode lasers as the light source. Unlike IR spectroscopy, no special sample preparation is required because scattered light is measured. Thus Raman spectra can be recorded from liquids and from solids in different forms (e.g. powders, fibres, films and sheets) with equal ease. Background signals due to fluorescence often are the biggest issue since they may obscure large regions of the spectrum and can be particularly serious with aromatic polymers, though use of different lasers (i.e. different values of v_0) can mitigate or even eliminate these problems. Raman signals can be enhanced substantially if v_0 of the incident radiation corresponds to visible light absorptions of the polymer, this technique being referred to as *resonance Raman spectroscopy*, which is especially useful when recording Raman spectra of highly conjugated and highly aromatic polymers.

As indicated in the previous section, techniques of Raman microscopy have become very important in probing molecular-level deformation micromechanics of polymers and most Raman spectrometers nowadays are designed for microscopy. The laser light is transmitted to the sample through the lens of the microscope and the scattered light returns back through the microscope, part of it being reflected into the monochromator for detection of the Raman lines and the other part being used to view (and locate) the point on the sample from which the light was scattered. By rastering across the sample, 2D surface maps of Raman spectra can be obtained. Additionally, probes that record light scattered from a short distance of penetration into a liquid have been designed specifically for measurements from within liquids and, with light piping, allow for location of the detection system remote from the vessel containing the liquid (e.g. a polymerization reactor). Such light-piping systems also facilitate remote analysis of solids.

15.6 NUCLEAR MAGNETIC RESONANCE SPECTROSCOPY

All atomic nuclei have a positive charge due to the proton(s) they contain. In the nuclei of isotopes which contain an odd number of protons and/or an odd number of neutrons, this charge spins about the nuclear axis generating a *magnetic moment* along the axis. Such nuclei have *nuclear spin quantum numbers* I that are either integral (1, 2, ...) or half integral (1/2, 3/2, ...) and in the presence of an external magnetic field align themselves in $2I+1$ different ways, each having a characteristic energy. This is the origin of *nuclear magnetic resonance (NMR) spectroscopy*, which is a form of spectroscopy in which the absorption of electromagnetic radiation from the radiofrequency range (typically 1–500 MHz) promotes nuclei from low-energy to high-energy alignments in an external magnetic field. The simplest situation is that for nuclei with $I=1/2$ (e.g. ^{1}H, ^{13}C, ^{19}F) since their nuclear magnetic moments align either parallel with (low-energy state) or parallel against (high-energy state) the magnetic field. The energy difference between these two states depends upon the *magnetogyric ratio* γ of the nucleus and the *strength* B of the *external magnetic field* ***at the nucleus***. The frequency v of radiation required to flip the nuclear magnetic moment from being aligned with, to being aligned against, the external magnetic field is known as the *resonance frequency* and is given by

$$v = \frac{\gamma B}{2\pi} \tag{15.7}$$

showing that v is proportional to both γ and B. The properties of some important nuclei are given in Table 15.2. The energy differences corresponding to the values of v are very small (about

TABLE 15.2
Properties of Some Important Nuclei[a]

Isotope	a / % [b]	I	γ / rad $T^{-1}s^{-1}$	ν_0 / MHz [c]
^{1}H	99.98	1/2	2.674×10^{8}	100.0
^{2}H	0.016	1	4.106×10^{7}	15.3
^{13}C	1.108	1/2	6.724×10^{7}	25.1
^{14}N	99.63	1	1.932×10^{7}	7.2
^{19}F	100.00	1/2	2.516×10^{8}	94.1

[a] ^{12}C and ^{16}O have no spin ($I=0$) and do not show an NMR effect.
[b] a is the natural abundance of the isotope.
[c] ν_0 is the fundamental resonance frequency calculated from Equation 15.7 using $B_0=2.3487$ T where B_0 is the strength of the applied magnetic field.

0.01–0.1 J mol^{-1}) and so even prior to absorption the population of a low-energy state is only very slightly greater than that of the high-energy state.

For each nucleus in a molecule, the effects of its electronic environment and neighbouring nuclei cause B to be slightly different from the strength B_0 of the applied magnetic field. The largest effects are those due to the electrons surrounding a nucleus. These electrons circulate under the influence of B_0 and in so doing they generate a local magnetic field which opposes B_0, thus *shielding* the nucleus from the external field. As the electron density at a nucleus increases, B (and hence ν) decreases. Inductive effects influence local electron density and so are important in determining the magnitude of the shielding. When a group bonded to a nucleus donates electrons, it is said to *shield* the nucleus; alternatively, if it withdraws electrons it is said to *deshield* the nucleus. Additionally, circulation of π-electrons can have very significant effects upon B at adjacent nuclei. For example, ^{1}H nuclei of benzene rings (which have C–C π-bonds) and of aldehyde groups (which have a C–O π-bond) are deshielded strongly by the magnetic fields arising from circulation of the π-electrons. It should now be evident that if nuclei of the same type (e.g. ^{1}H) in the same molecule do not have exactly identical local environments in the molecule, then they will have different B and will absorb (i.e. *magnetic resonance* will occur) at different frequencies. Although these differences in B are only of the order of parts per million of B_0, they are measurable and form the basis of NMR spectroscopy.

NMR spectra can be obtained either (i) by fixing B_0 and measuring the absorption of radiation over a range of ν close to the fundamental resonance frequency ν_0 ($=\gamma B_0/2\pi$) of the type of nuclei under study or (ii) by fixing ν at ν_0 and measuring the absorption of this radiation as B_0 is varied over a small range. These methods are equivalent, though modern spectrometers invariably use method (i). A strongly shielded nucleus will absorb at low ν (corresponding to low B) using method (i), or equivalently, at high B_0 using method (ii). In both cases, the absorption is said to occur *upfield*. Similarly, an absorption which due to deshielding effects occurs at relatively high ν, or equivalently, at relatively low B_0, is said to be *downfield*. For both methods, the absorptions of the sample are recorded as chemical shifts from an absorption of a reference compound. The dimensionless δ-scale of chemical shifts is used universally nowadays and does not depend upon the method used to record the spectrum. It is defined by

$$\delta = \frac{10^6(\nu_0^{\text{sample}} - \nu_0^{\text{ref}})}{\nu_0} \tag{15.8a}$$

for measurements made using method (i), or by

$$\delta = \frac{10^6(B_0^{ref} - B_0^{sample})}{B_0^{ref}} \quad (15.8b)$$

for those made using method (ii). (Note that the older dimensionless τ-scale for ^{1}H NMR spectroscopy is related to the δ-scale by $\tau = 10 - \delta$.) The δ-scale is used because it is independent of B_0 and ν_0, whereas the differences $\nu_0^{sample} - \nu_0^{ref}$ and $B_0^{ref} - B_0^{sample}$ increase as B_0 and ν_0 increase. Thus NMR spectra with greatly improved resolution (i.e. separation) of the absorptions, recorded by using much stronger applied magnetic fields, can easily be compared with spectra recorded on older, lower-resolution instruments because the δ-scale is unaffected.

In NMR studies of polymers, ^{1}H and ^{13}C NMR spectroscopy are most widely used. Both employ tetramethylsilane (TMS), $Si(CH_3)_4$, as the reference compound. Since silicon is more electropositive than carbon, the hydrogen and carbon atoms in TMS give single ^{1}H and ^{13}C absorptions (which, by convention, correspond to $\delta = 0$) that almost invariably are upfield of ^{1}H and ^{13}C absorptions of organic compounds and polymers. Thus, ^{1}H and ^{13}C δ-values normally are positive and increase in the downfield (i.e. deshielding) direction. Some generalized correlations between chemical environment and chemical shift for ^{1}H and ^{13}C nuclei are shown in Figures 15.11 and 15.12, respectively. These correlations illustrate the potential of NMR spectroscopy for evaluation of molecular structure and show that while most ^{1}H absorptions occur in the range $0 < \delta < 13$, the corresponding ^{13}C absorptions occur over the much wider range $0 < \delta < 220$. Thus ^{13}C NMR spectroscopy has the distinct advantage of having far greater resolution then ^{1}H NMR spectroscopy.

For quantitative studies, the areas under each of the absorptions are required. These areas are evaluated by an integrator and so are called *integrations*. Modern NMR spectrometers record spectra by summing individual spectra obtained from a succession of pulses of radiation. Provided that the procedures for recording an NMR spectrum are chosen properly (see Section 15.6.5), the ratios of the integrations for the different absorptions are equal to the ratios of the numbers of the respective nuclei present in the molecules, and can be used, for example, to evaluate relative molar compositions of copolymers. This is the normal situation for ^{1}H NMR spectra, even when short intervals between pulses are used. However, ^{13}C nuclei relax more slowly between pulses and can quickly become saturated, so under the usual conditions of fast pulsing, the areas under the absorptions are not normally proportional to the number of ^{13}C nuclei. Special conditions are required for quantitative ^{13}C NMR spectroscopy (see Section 15.6.5).

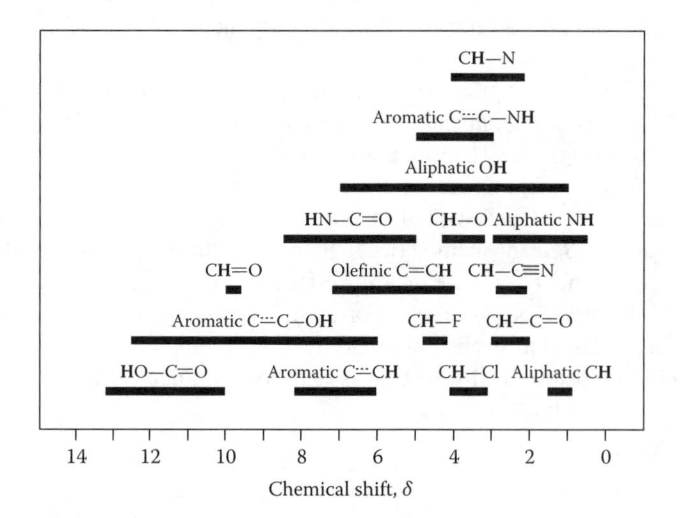

FIGURE 15.11 Characteristic ^{1}H NMR absorption regions.

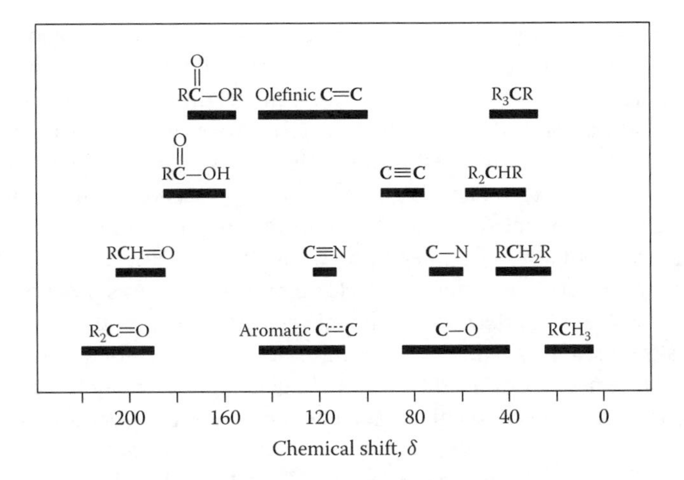

FIGURE 15.12 Characteristic ^{13}C NMR absorption regions.

The effects, intimated earlier, of neighbouring nuclei upon NMR absorptions arise because nuclear spins interact through the intervening bonding electrons. In this interaction, known as *scalar spin–spin coupling* (referred to as *J-coupling*), the magnetic moment of one nucleus is sensed by other nuclei in the same molecule and vice versa, though the interaction usually is significant only when the coupled nuclei are separated by no more than three bonds. This J-coupling causes an NMR absorption to be split into essentially symmetric multiple absorptions, the number of absorptions in most cases of interest being one greater than the number of neighbouring nuclei to which the nucleus is coupled and their separation being determined by the strength of the coupling. Hence, the splitting of an NMR absorption due to J-coupling provides valuable structural information.

The low natural abundance of ^{13}C means that in ^1H NMR, the effects of ^1H–^{13}C J-coupling are insignificant, but J-coupling of ^1H nuclei on the same (H–C–H) or adjacent (H–C–C–H) carbon atoms is significant and of great importance for the elucidation of molecular structure. Note, however, that splitting of an absorption due to J-coupling can be observed only if the ^1H nuclei have different environments (i.e. different chemical shifts in the spectrum). For example, consider the molecular fragment and its associated ^1H NMR spectrum shown in Figure 15.13. In this molecular

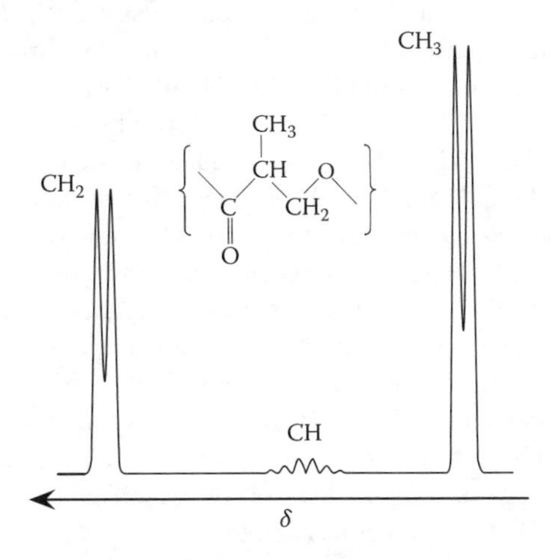

FIGURE 15.13 Schematic representation of the ^1H NMR spectrum resulting from the molecular fragment shown. The area under each multiplet peak is proportional to the number of ^1H nuclei in that chemical environment and the splitting of the absorptions due to J-coupling provides important information about the total number of ^1H nuclei on adjacent carbon atoms.

fragment (due to fast bond rotation), the three ^1H nuclei in the CH_3 group are identical and the two ^1H nuclei in the CH_2 group are identical, but the CH, CH_2 and CH_3 groups have unique chemical environments. The ^1H nuclei of the CH_3 group have a single, distinct neighbouring ^1H atom (from the CH group) and so the ^1H NMR absorption for CH_3 group appears as a doublet in the range $1<\delta<2$ (since it is remote from the deshielding effects of the ether oxygen atom and carbonyl group) with a relative integral of 3. The CH group has five distinct neighbouring ^1H atoms (from the CH_3 and CH_2 groups) and so appears as a sextet in the range $2<\delta<3$ (due to the deshielding effect of the carbonyl group) with a relative integral of 1. Finally, the ^1H nuclei of the CH_2 group have one distinct neighbouring ^1H atom (from the CH group) and appear as a doublet around $3<\delta<4$ (due to the stronger deshielding effect of the ether oxygen atom) with a relative integral of 2. Notice how weak the peaks due to the single CH group have become as a consequence of splitting due to J-coupling with a large number of neighbouring ^1H nuclei. By inversing the analysis just carried out, the precise structure of the molecular fragment could have been deduced from the ^1H NMR spectrum.

In most cases, as indicated in this example, the ^1H–^1H J-coupling is small compared with the chemical shift differences between the different ^1H nuclei environments, thus enabling ^1H NMR spectra to be interpreted without difficulty. When overlap does occur, the J-splitting information (and hence detailed structural information) still can be obtained using more advanced NMR techniques, which enable 2D NMR spectra to be recorded in which the chemical shifts of the different ^1H absorptions are shown on one axis and their splitting by ^1H–^1H J-coupling is shown on a second perpendicular axis with intensities represented as contour lines.

In ^{13}C NMR spectroscopy the effects of ^{13}C–^{13}C J-coupling are completely negligible due to the low abundance of ^{13}C nuclei. However, ^1H–^{13}C J-coupling is significant but makes the interpretation of ^{13}C NMR spectra difficult because the splitting is strong and, in many cases, is greater than the chemical shift differences between the absorptions of the different ^{13}C nuclei. Thus when recording ^{13}C NMR spectra, it is usual to employ techniques which *decouple* ^{13}C spins from ^1H spins so that the different ^{13}C nuclei each give rise to only a single absorption in the spectrum. Although this facilitates interpretation of the spectrum, the structural information available from ^1H–^{13}C J-coupling patterns is lost. This information can, however, be obtained without prejudicing interpretation by using specialized NMR techniques that make it possible to obtain a ^1H-decoupled ^{13}C NMR spectrum from which it is easy to identify the numbers of ^1H atoms bonded directly to each carbon atom giving rise to an absorption in the spectrum. For example, *distortionless enhancement by polarization transfer (DEPT)* is a widely used multi-pulse ^{13}C NMR technique that addresses ^{13}C and ^1H nuclei with the final pulse set to give a 45°, 90° or 135° flip of the ^1H nuclei spins, these pulse sequences being referred to as DEPT45, DEPT90 and DEPT135. In all DEPT spectra, each ^{13}C NMR absorption appears as a single peak (as in a normal ^1H-decoupled ^{13}C NMR spectrum), but ^{13}C nuclei with no directly attached ^1H atoms (i.e. quaternary carbon atoms, C_q) are absent. The DEPT45 spectrum shows CH, CH_2 and CH_3 absorptions positive, whereas the DEPT90 (in theory) shows only absorptions from ^{13}C nuclei with one attached ^1H atom (i.e. CH groups), though it often is difficult to find pulse sequence parameters that eliminate completely CH_2 and CH_3 absorptions. In DEPT135 spectra, ^{13}C nuclei with odd numbers of directly attached ^1H atoms (i.e. those in CH and CH_3 groups) appear positive and those with an even number of directly attached ^1H atoms (i.e. those in CH_2 groups) appear negative. Figure 15.14 shows schematically the nature of the normal, DEPT45 and DEPT135 ^1H-decoupled ^{13}C NMR spectra that would be expected from the molecular fragment shown in Figure 15.13. As can be seen, in combination with a normal ^1H-decoupled ^{13}C NMR spectrum and knowledge of δ for each absorption, the DEPT spectra make it quite easy to assign each absorption to a C_q, CH, CH_2 or CH_3 group.

In this relatively brief introduction it has been possible only to give a general indication of the immense power of NMR spectroscopy for elucidation of molecular structures. In the following sections, this is demonstrated by considering some important applications of NMR spectroscopy in probing details of the molecular structure of polymers.

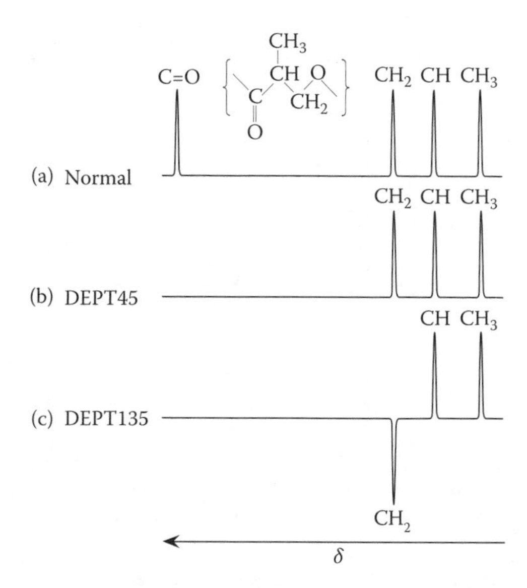

FIGURE 15.14 Schematic representations of (a) the normal ^{1}H-decoupled ^{13}C NMR spectrum, (b) the DEPT45 and (c) the DEPT135 ^{13}C NMR spectra of the molecular fragment shown in Figure 15.13. Note that the C=O absorption is absent in the DEPT spectra because the carbon atom has no attached ^{1}H atoms and that in the DEPT135 spectrum the absorption due the CH_2 group is negative. (Note that the absorptions are shown with equal areas, i.e. assuming that the spectra have been recorded under quantitative conditions.)

15.6.1 Analysis of Molecular Structure and Composition by NMR Spectroscopy

Many of the applications of NMR spectroscopy are the same as those of IR spectroscopy (see Section 15.4.1). Thus both ^{1}H and ^{13}C NMR are used widely for routine purposes such as polymer identification, confirmation of molecular structure and evaluation of average copolymer composition. NMR spectroscopy, however, offers much greater scope than IR spectroscopy for elucidating detailed features of molecular microstructure. This is especially true of ^{13}C NMR spectroscopy, which nowadays is by far the most important technique for detailed structural characterization of polymers.

15.6.2 Analysis of End Groups and Branch Points by NMR Spectroscopy

NMR spectra of polymers of low molar mass often show unique absorptions due to the end groups. By referencing these absorptions to those of nuclei in the repeat units, it is possible to obtain the ratio of the number of end groups to the number of repeat units and thereby evaluate the number-average molar masses of such polymers provided that the number of those end groups per polymer molecule is known from the method of synthesis. An example is shown in Figure 15.15. Alternatively, NMR spectra of low molar mass polymers can be analysed in order to determine the structure of end groups, thereby providing evidence of the chain initiation and/or termination events that occurred during polymerization.

For branched polymers, NMR absorptions due to the branch points can be identified and reveal the chemical structure of those branch points, thus leading to a better understanding of the mechanism(s) by which the branches form. The degree of branching can be determined by taking the ratio of the total integral for branch point absorptions to that for nuclei present in all the repeat units. The chemistry of chain transfer to polymer in radical polymerization of *n*-butyl acrylate is identical to that shown for methyl acrylate in Figure 4.3. This mechanism was first revealed by ^{13}C NMR analysis of poly(*n*-butyl acrylate) prepared by emulsion polymerization, a typical spectrum of which is shown in Figure 15.16. The distinct absorptions from the poly(*n*-butyl acrylate) branch point were identified and assigned with the help of DEPT spectra.

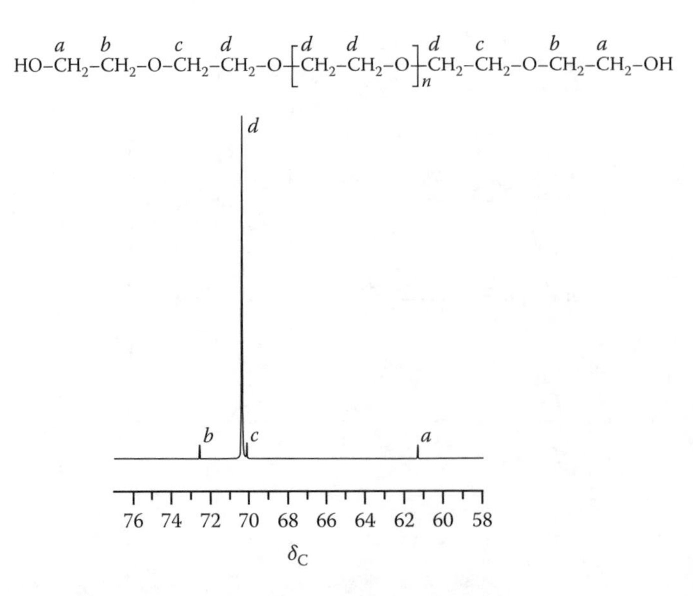

FIGURE 15.15 ^{1}H-decoupled ^{13}C NMR spectrum of a sample of linear poly(ethylene oxide) with hydroxyl end-groups. The (weak) absorptions at δ 72.5, 70.1 and 61.2 are due to the ^{13}C atoms nearest the chain-ends, and are shifted from that of the main-chain ^{13}C atoms at δ 70.3 because of effects arising from the hydroxyl end-groups. From the ratio of the integrations for the end-group absorptions to that for the main-chain absorption, the number-average molar mass of the poly(ethylene oxide) sample is calculated to be 1500 g mol^{-1}. (Spectrum and analysis courtesy of F. Heatley, The University of Manchester, Manchester, U.K.)

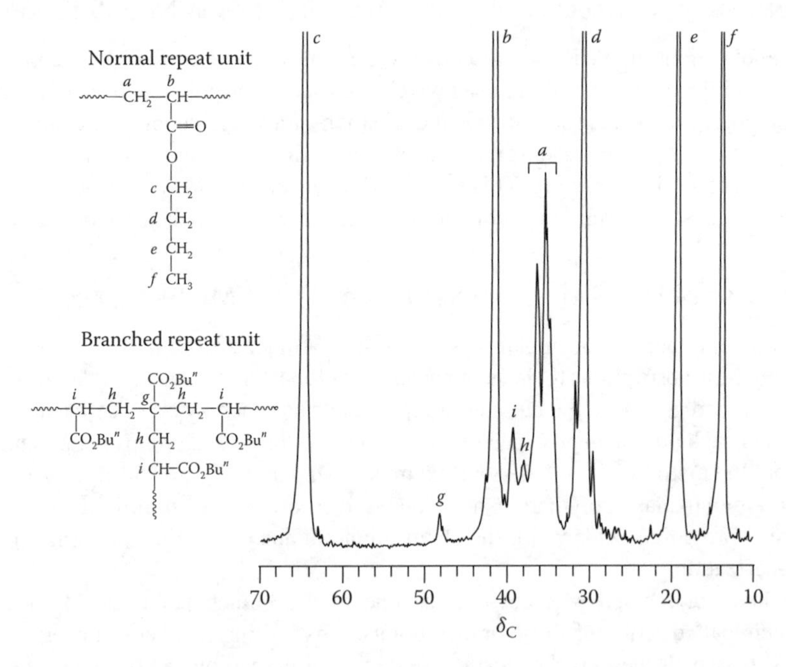

FIGURE 15.16 ^{1}H-decoupled ^{13}C NMR spectrum in the aliphatic region $70>\delta>10$ for a sample of poly(n-butyl acrylate) prepared by emulsion polymerization at 80 °C. The branch points are characterized by distinct resonances from the branch C_q and the three CH_2 and three CH carbon atoms adjacent to the branch point. Quantitative analysis of the spectrum by taking the ratio of one-third of the branch CH_2+CH integral (for resonances $h+i$) to the total backbone CH_2+CH integral (for resonances $a+b+h+i$) shows that the sample comprises 4.1 mol% branched repeat units. (Spectrum taken from the research of P.A. Lovell and F. Heatley, The University of Manchester, Manchester, U.K.)

15.6.3 Determination of Molecular Microstructure by NMR Spectroscopy

The importance of NMR spectroscopy (especially ^{13}C NMR) is most clearly demonstrated by its ability to yield quantitative information on features of molecular microstructure not accessible by other techniques. These features include (i) head-to-tail and head-to-head repeat unit linkages, (ii) the different types of repeat unit structures arising from 1,3-diene monomers, (iii) isotactic and syndiotactic sequences of repeat units, and (iv) the sequence distribution of the different repeat units in a copolymer chain. The latter two applications of NMR spectroscopy will be considered here in detail because they provide an insight into the statistics of sequence distributions as well as demonstrating the ability of NMR spectroscopy to reveal fine detail about molecular microstructure.

15.6.3.1 Determination of Tacticity

Tacticity was introduced in Section 6.2 and is very important because it has a major affect on the physical properties of polymers, in particular the glass transition temperature and the ability to crystallize. Accurate measurement of polymer tacticity was made possible by NMR spectroscopy, which can discriminate between different sequences of repeat units through differences in the chemical shifts of similar nuclei in slightly different environments. Before considering an example of how NMR spectroscopy can be used to reveal tacticity, general sequence distribution statistics will be considered.

As described in Section 6.2, during chain growth, the creation of each repeat unit can yield either an isotactic relationship with the previous repeat unit, known as a *meso* (m) placement, or a syndiotactic relationship, known as a *racemic* (r) placement. When considering sequence distributions it is usual to describe them using the m and r symbols. Hence an isotactic polymer will have a continuous sequence of *meso* links whereas a syndiotactic polymer will have a continuous sequence of racemic links. Sequence distributions of isotactic polymers are rarely absolutely perfect and a real section of an isotactic polymer might look something like

~*mmmmmmmmmmmmmmmmmrmmmmmmmmmmmmmmmmmmmmmmmmm*~

where the single r link is significant because it changes the nature of stereochemistry either side of it (the two all-m sequences have repeat units with opposite configurations). NMR spectroscopy is capable of revealing such detailed information about sequence distributions both qualitatively and quantitatively. From the perspective of NMR spectroscopy it is normal to represent the chains using simple 2D projections, as shown in Figure 15.17. Such projections make it is easy to write down longer sequences and to see the different possible environments for the carbon and hydrogen nuclei.

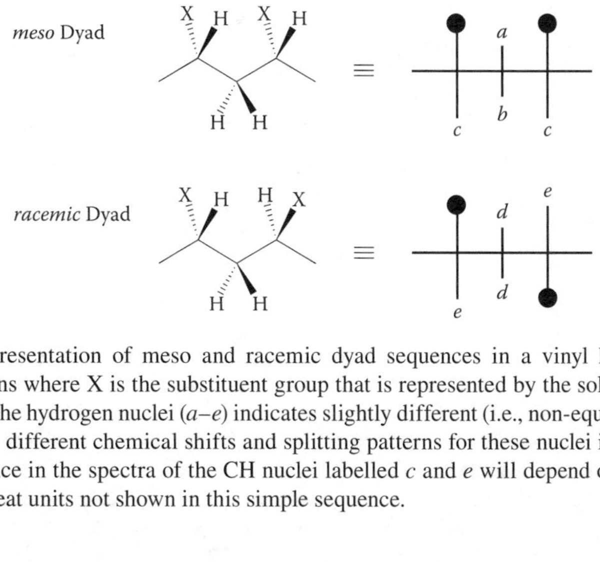

meso Dyad

X H X H

H H

≡

a b c c

racemic Dyad

X H H X

H H

≡

e d d e

FIGURE 15.17 Representation of meso and racemic dyad sequences in a vinyl homopolymer as two-dimensional projections where X is the substituent group that is represented by the solid circle in the projection. The labelling of the hydrogen nuclei (*a*–*e*) indicates slightly different (i.e., non-equivalent) environments, which can give rise to different chemical shifts and splitting patterns for these nuclei in ^{1}H NMR spectra. In practice, the appearance in the spectra of the CH nuclei labelled *c* and *e* will depend on the configuration of the other adjacent repeat units not shown in this simple sequence.

If the probability of a meso sequence being created is P_m and that for a racemic sequence is P_r and the probabilities are not influenced by the preceding chain structure, then $P_m + P_r = 1$ and only P_m (or P_r) needs to be known in order to predict the probability of any particular sequence. For example, the probability of an *mrm* tetrad is $P_m P_r P_m = P_m^2(1 - P_m)$. The probabilities for these simple conditions (where the preceding chain structure has no effect) conform to *Bernoullian statistics*. Using NMR spectroscopy it is possible to measure the mole fraction of a particular sequence and, since this must be equal to its probability of formation, the value of P_m can be determined. Table 15.3 shows dyad, triad and tetrad sequences and gives the equations for their mole fractions in terms of P_m. The dyad and tetrad projections highlight different environments for the central CH_2 nuclei and the triad projection does so for the central CH nuclei. Provided that NMR absorptions can be assigned to particular sequences, the integrals for those absorptions can be used to measure the corresponding mole fractions and by fitting to the probability equations in Table 15.3, P_m can be evaluated. Although P_m is the overall mole fraction of meso placements in the homopolymer and

TABLE 15.3
Sequences in Vinyl Homopolymers and Their Mole Fractions according to Bernoullian Statistics

Sequence Type	Designation [a]	Projection	Mole Fraction of Sequence in Terms of P_m [a]
Dyad	*m*		P_m
	r		$1 - P_m$
Triad	*mm*		P_m^2
	mr (and *rm*)		$2P_m(1 - P_m)$
	rr		$(1 - P_m)^2$
Tetrad	*mmm*		P_m^3
	mmr (and *rmm*)		$2P_m^2(1 - P_m)$
	rmr		$P_m(1 - P_m)^2$
	mrm		$P_m^2(1 - P_m)$
	rrm (and *mrr*)		$2P_m(1 - P_m)^2$
	rrr		$(1 - P_m)^3$

[a] Where there are two equivalent sequences (e.g. *mr* is indistinguishable from *rm*), this is accounted for by the factor of 2 in the mole fraction equation.

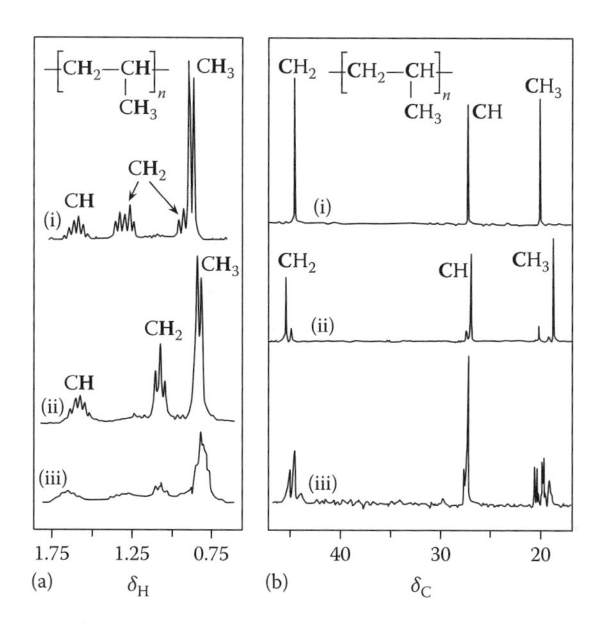

FIGURE 15.18 (a) ^1H NMR spectra and (b) ^1H-decoupled ^{13}C NMR spectra of (i) isotactic, (ii) syndiotactic and (iii) atactic polypropylene. (Data and assignments taken from Tonelli, A.E., *NMR Spectroscopy and Polymer Microstructure: The Conformational Connection*, VCH, New York, 1989.)

may be taken as the mole fraction isotacticity, the mole fractions of longer isotactic sequences (e.g. of *mm* triads or *mmm* tetrads) provide a better measure of tacticity.

Figure 15.18 shows ^1H NMR spectra and ^1H-decoupled ^{13}C NMR spectra for samples of polypropylene with different tacticities. The ^1H NMR spectra in Figure 15.18a show multiplet absorptions due to J-coupling, the effects of which are most clearly evident in the spectrum for isotactic polypropylene. The CH_3 ^1H absorption is split by J-coupling with the adjacent CH ^1H nucleus and so is observed as a doublet. The CH ^1H absorption is split by J-coupling with adjacent CH_2 and CH_3 ^1H nuclei and so is observed as an octet, though the outermost peaks of the octet are not evident at the scale shown in Figure 15.18a. As is evident from the consideration of Figure 15.17, the two ^1H nuclei in the CH_2 of isotactic polypropylene are not equivalent and so give ^1H absorptions with quite different chemical shifts, each being split by J-coupling, the full extent of which is masked by the CH_3 ^1H absorption for the CH_2 ^1H absorption at $\delta < 1$. However, the two ^1H nuclei in the CH_2 of syndiotactic polypropylene are equivalent and give rise to a ^1H absorption with an intermediate chemical shift that appears as a triplet due to J-coupling with the two adjacent CH ^1H nuclei. Although the CH_2 ^1H absorption is sensitive to the sequencing, the splittings due to J-coupling complicate determination of P_m. This is generally true for ^1H NMR spectra of vinyl homopolymers. In contrast, the ^1H NMR spectra of homopolymers with two non-H substituents, such as poly(methyl methacrylate), are not affected by J-coupling and the integrals for the different CH_2 ^1H absorptions can be used to determine P_m.

The ^{13}C absorptions for polypropylene in Figure 15.18b appear only as singlets because the ^{13}C NMR spectra were recorded under conditions of ^1H decoupling. Thus each ^{13}C absorption corresponds to a specific sequence, which is why ^1H-decoupled ^{13}C NMR spectra are most widely used for measurements of P_m and tacticity. Inspection of the spectra shows that the CH_2 and CH_3 ^{13}C absorptions are most sensitive to the sequencing. Comparison of the spectra for isotactic and syndiotactic polypropylene reveals that the syndiotactic sample has some *mm* sequences. The spectrum for atactic polypropylene is the most interesting because it highlights the ability of ^1H-decoupled ^{13}C NMR to see fine structural detail. This is revealed in Figure 15.19, which shows an expansion of the CH_3 ^{13}C absorption region for another sample of atactic polypropylene together with assignments

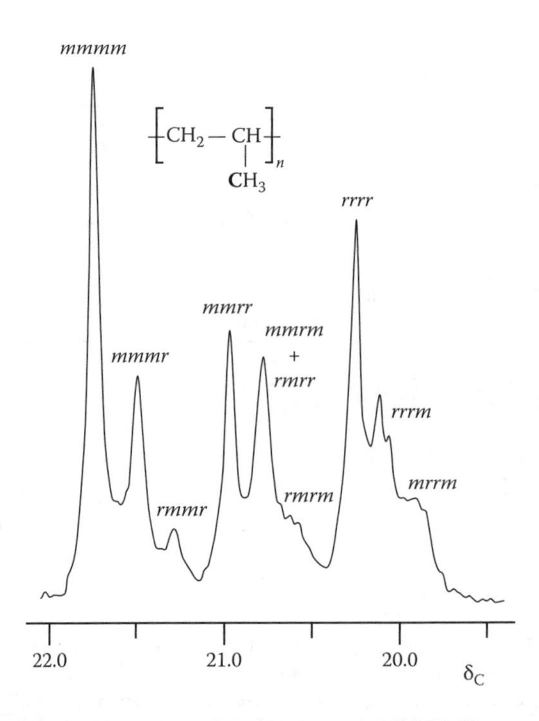

FIGURE 15.19 Expansion of the CH_3 region of the ^1H-decoupled ^{13}C NMR spectrum of a sample of atactic polypropylene showing the assignment of individual absorptions to specific pentad sequences. (Data and assignments taken from Randall, J.C., *Polymer Sequence Determination: Carbon-13 NMR Method*, Academic Press, New York, 1977.)

of the individual absorptions to specific pentad sequences, these being established by comparison with spectra of simple hydrocarbons (model compounds) that contain the same sequences and with theoretical predictions of chemical shifts. Through use of even higher-resolution NMR spectrometers it is possible to distinguish much longer sequences, such as hexads and heptads.

Careful comparison of measured probabilities with the predictions of Table 15.3 for homopolymers prepared by radical polymerization shows that they conform to Bernoullian statistics, which is because there is no control of the stereochemistry of propagation. However, polypropylene and other homopolymers prepared under conditions of polymerization where stereochemical control is exerted (e.g. coordination polymerization and anionic polymerization performed under conditions of strong coordination, see Chapter 6) do not conform to Bernoullian statistics because the previously created repeat unit does influence the nature of the next placement. Thus the statistics of their sequence distributions is more complicated and described using four probabilities: P_{mm}, P_{mr}, P_{rm} and P_{rr}, where the first subscript represents the nature of the previously created repeat unit and the second represents the nature of the newly created repeat unit (e.g. P_{mr} is the probability for creation of a racemic placement following a previously created meso placement). Since a propagating chain must create either a meso or a racemic placement

$$P_{mm} + P_{mr} = 1 \tag{15.9}$$

$$P_{rm} + P_{rr} = 1 \tag{15.10}$$

and only two probabilities (e.g. P_{mr} and P_{rm}) are needed to define the probability of any particular sequence. In order to develop equations relating mole fractions of particular sequences to the probabilities P_{mr} and P_{rm}, the equivalence and indistinguishability of the *mr* and *rm* triad sequences is used. The mole fractions of these triads must be equal, a condition that can be expressed mathematically by

$$x_m P_{mr} = x_r P_{rm} \tag{15.11}$$

where x_m and x_r are the mole fractions of the meso and racemic dyads, respectively, i.e. the overall mole fractions of the two types of placement. Recognizing that $x_m + x_r = 1$ and substituting $x_r = 1 - x_m$ with rearrangement gives

$$x_m = \frac{P_{rm}}{P_{mr} + P_{rm}} \tag{15.12}$$

Similarly,

$$x_r = \frac{P_{mr}}{P_{mr} + P_{rm}} \tag{15.13}$$

The mole fractions of longer sequences can now be related to P_{mr} and P_{rm}. For example, using Equations 15.9 through 15.13, equations for the mole fractions of the triad sequences are obtained

$$x_{mm} = x_m P_{mm} = x_m(1 - P_{mr}) = \frac{P_{rm}(1 - P_{mr})}{P_{mr} + P_{rm}}$$

$$x_{mr} = x_m P_{mr} + x_r P_{rm} = \frac{2P_{mr}P_{rm}}{P_{mr} + P_{rm}}$$

$$x_{rr} = x_r P_{rr} = x_r(1 - P_{rm}) = \frac{P_{mr}(1 - P_{rm})}{P_{mr} + P_{rm}}$$

where the definition of x_{mr} again recognizes that the *mr* and *rm* triads are equivalent and indistinguishable (i.e. x_{mr} includes both sequences). A similar treatment leads to equations for mole fractions of tetrads, pentads, etc., in terms of P_{mr} and P_{rm}. The results of these analyses conform to *first-order Markov statistics* and are presented in Table 15.4, including the equations for the mole fractions of tetrads. Sequence distributions in polymers prepared under conditions of stereochemical control often fit well to first-order Markov statistics, but in some situations the pre-penultimate repeat unit also can influence the sequencing and second-order Markov statistics apply.

The mole fractions of triads can be used to check whether the sequence distribution for a homopolymer obeys Bernoullian statistics, because $P_{mm} = P_{rm}$ and $P_{rr} = P_{rm}$. The equations for the triad mole fractions given in Table 15.3 can be combined to give the following equality

$$\frac{4x_{mm}x_{rr}}{x_{mr}^2} = 1 \tag{15.14}$$

If Equation 15.14 is satisfied, then the homopolymer sequence distribution is Bernoullian; if not, then higher-order statistics must be operating. Conformity with first-order Markov statistics can be established using tetrad mole fractions, which, by combining the equations in Table 15.4, must conform to the equalities

$$\frac{4x_{mmm}x_{mrm}}{x_{mmr}^2} = 1 \tag{15.15}$$

TABLE 15.4
Mole Fractions of Sequences in Vinyl Homopolymers according to First-Order Markov Statistics

Sequence Type	Mole Fraction of[a]	Relation to P_{mr} and P_{rm} [a]
Dyad	*m*	$x_m = \frac{P_{rm}}{P_{mr} + P_{rm}}$
	r	$x_r = \frac{P_{mr}}{P_{mr} + P_{rm}}$
Triad	*mm*	$x_{mm} = \frac{P_{rm}(1 - P_{mr})}{P_{mr} + P_{rm}}$
	mr (includes *rm*)	$x_{mr} = \frac{2P_{mr}P_{rm}}{P_{mr} + P_{rm}}$
	rr	$x_{rr} = \frac{P_{mr}(1 - P_{rm})}{P_{mr} + P_{rm}}$
Tetrad	*mmm*	$x_{mmm} = \frac{P_{rm}(1 - P_{mr})^2}{P_{mr} + P_{rm}}$
	mmr (includes *rmm*)	$x_{mmr} = \frac{2P_{mr}P_{rm}(1 - P_{mr})}{P_{mr} + P_{rm}}$
	rmr	$x_{rmr} = \frac{P_{mr}P_{rm}^2}{P_{mr} + P_{rm}}$
	mrm	$x_{mrm} = \frac{P_{mr}^2 P_{rm}}{P_{mr} + P_{rm}}$
	rrm (includes *mrr*)	$x_{rrm} = \frac{2P_{mr}P_{rm}(1 - P_{rm})}{P_{mr} + P_{rm}}$
	rrr	$x_{rrr} = \frac{P_{mr}(1 - P_{rm})^2}{P_{mr} + P_{rm}}$

[a] Where there are two equivalent sequences (e.g. *mr* is indistinguishable from *rm*), this is accounted for by the factor of 2 in the mole fraction equation.

and

$$\frac{4x_{rrr}x_{rmr}}{x_{rrm}^2} = 1 \tag{15.16}$$

If both equalities are satisfied, then by taking the ratio x_{mrm}/x_{rmr} using the equations in Table 15.4, the following relationship between the first-order Markov probabilities P_{mr} and P_{rm} is obtained

$$\frac{P_{mr}}{P_{rm}} = \frac{x_{mrm}}{x_{rmr}} \tag{15.17}$$

The values of P_{mr} and P_{rm} can be obtained by an iterative process of fitting the measured mole fractions to the equations given in Table 15.4. A simpler, though less rigorous procedure, is to establish simple relations between sequence mole fractions and P_{mr} and P_{rm}. For example, the ratios x_{mmm}/x_{mm} and x_{rrr}/x_{rr} can be used, which from Table 15.4 lead to

$$P_{mr} = 1 - \frac{x_{mmm}}{x_{mm}} \tag{15.18}$$

$$P_{rm} = 1 - \frac{x_{rrr}}{x_{rr}} \tag{15.19}$$

These equations are informative because it is clear that when propagation is either perfectly isotactic or perfectly syndiotactic, $P_{mr}=0$ and $P_{rm}=0$.

15.6.3.2 Determination of Repeat Unit Sequence Distributions in Copolymers

From a statistical perspective, sequence distributions in copolymers comprising two types of repeat unit (represented by A and B) may be considered in exactly the same way as for tactic homopolymers – the two stereochemical configurations in a tactic homopolymer are simply replaced by the two different types of repeat unit in the copolymer with no change in the underlying mathematics. However, unlike tacticity, for which the Bernoullian statistics set out in Table 15.3 are satisfied when there is no stereochemical control, the probability for addition of a particular monomer to a propagating chain invariably depends on the nature of the active end unit, i.e. whether it is of type A or B (see Sections 9.3.1 and 9.3.2). Thus first-order Markov statistics must be used to analyse sequence distributions in copolymers, redefining the four probabilities as P_{AA}, P_{AB}, P_{BA} and P_{BB}, where P_{AA} is the probability for addition of monomer A to a propagating chain ending in an A-type repeat unit, i.e. the first subscript represents the active end unit and the second the monomer that adds to it (as for the four rate coefficients for propagation in binary copolymerization; see Section 9.3.1). Since a propagating chain must add either monomer A or monomer B

$$P_{AA} + P_{AB} = 1$$

$$P_{BA} + P_{BB} = 1$$

and so, as for tacticity, only two probabilities (P_{AB} and P_{BA}) are needed to define the probability of any particular sequence. (This is exactly analogous to the use of two monomer reactivity ratios in the terminal model for binary copolymerization; see Section 9.3.1.) Thus the equations given in Table 15.4 can be used for binary copolymers, simply replacing each m by A and each r by B. The same is true for all the first-order Markov equations presented in the preceding section. Thus, provided that NMR absorptions can be assigned to particular repeat unit sequences, P_{AB} and P_{BA} can be evaluated from the integrals for those absorptions. These probabilities also can be defined in terms of the kinetics parameters used in the terminal model for copolymerization

$$P_{AB} = \frac{k_{AB}[B]}{k_{AA}[A] + k_{AB}[B]} \tag{15.20}$$

$$P_{BA} = \frac{k_{BA}[A]}{k_{BB}[B] + k_{BA}[A]} \tag{15.21}$$

where the terms are as defined in Section 9.3.1. Dividing both the numerator and denominator of Equation 15.20 by $k_{AB}/([A]+[B])$ and those of Equation 15.21 by $k_{BA}/([A]+[B])$ and taking into account the definitions of the monomer reactivity ratios r_A and r_B given in Equation 9.6, the following relations are obtained

$$P_{AB} = \frac{f_B}{r_A f_A + f_B} \tag{15.22}$$

$$P_{BA} = \frac{f_A}{r_B f_B + f_A} \tag{15.23}$$

which can be rearranged to give r_A and r_B in terms of P_{AB} and P_{BA} determined from NMR measurements of copolymer repeat unit sequence distribution and the mole fractions (f_A and f_B) of the two monomers used in the copolymerization

$$r_A = \frac{f_B(1 - P_{AB})}{P_{AB} f_A} \tag{15.24}$$

$$r_B = \frac{f_A(1 - P_{BA})}{P_{BA} f_B} \tag{15.25}$$

Thus it is possible to determine monomer reactivity ratios from the copolymer repeat unit sequence distribution probabilities P_{AB} and P_{BA} as well as from the overall copolymer composition (see Section 9.3.4), with both sets of data coming from the same NMR spectra.

As for measurement of tacticity, ^{1}H-decoupled ^{13}C NMR spectra are most useful for the analysis of copolymer repeat unit sequences because of the greater separation of the absorptions and the absence of splitting. The assignment of absorptions to particular repeat unit sequences is done using advanced NMR techniques (DEPT being particularly useful for copolymers) and theoretical predictions of chemical shifts. To illustrate the ability of ^{13}C NMR to reveal fine detail of copolymer repeat unit sequence distributions, ^{1}H-decoupled ^{13}C NMR spectra of four copolymers of ethylene and vinyl acetate are shown in Figure 15.20 together with the assignment of the absorptions to triad sequences. As would be expected, the relative areas of the absorptions vary in accord with the overall copolymer composition, i.e. as the mol% vinyl acetate increases, triads containing ethylene repeat units become less probable and their absorptions become weaker, while the triads containing vinyl acetate repeat units become stronger.

The analysis of copolymers considered so far neglects effects of repeat unit stereochemistry. Inclusion of such effects makes the analysis far more complex. For example, taking account of stereochemistry in a binary copolymer prepared from monomers that yield tactic homopolymers increases the number of distinguishable dyads from 2 to 6 (see Figure 15.21) and the number of distinguishable triads from 3 to 20. Hence, complete analysis and assignment of absorptions in the NMR spectra of such copolymers is difficult unless there are some simplifying effects that make many of the sequences insignificant (e.g. if the copolymerization has a strong alternating tendency).

15.6.4 Other Uses of NMR Spectroscopy in Polymer Science

Another important application of NMR spectroscopy is the study of *chain dynamics* by monitoring the magnetic relaxation behaviour of the excited nuclei. It is not possible here to give details of such measurements, but it should be noted that the motion of side groups and chain segments can be characterized, and that diffusion coefficients can be determined.

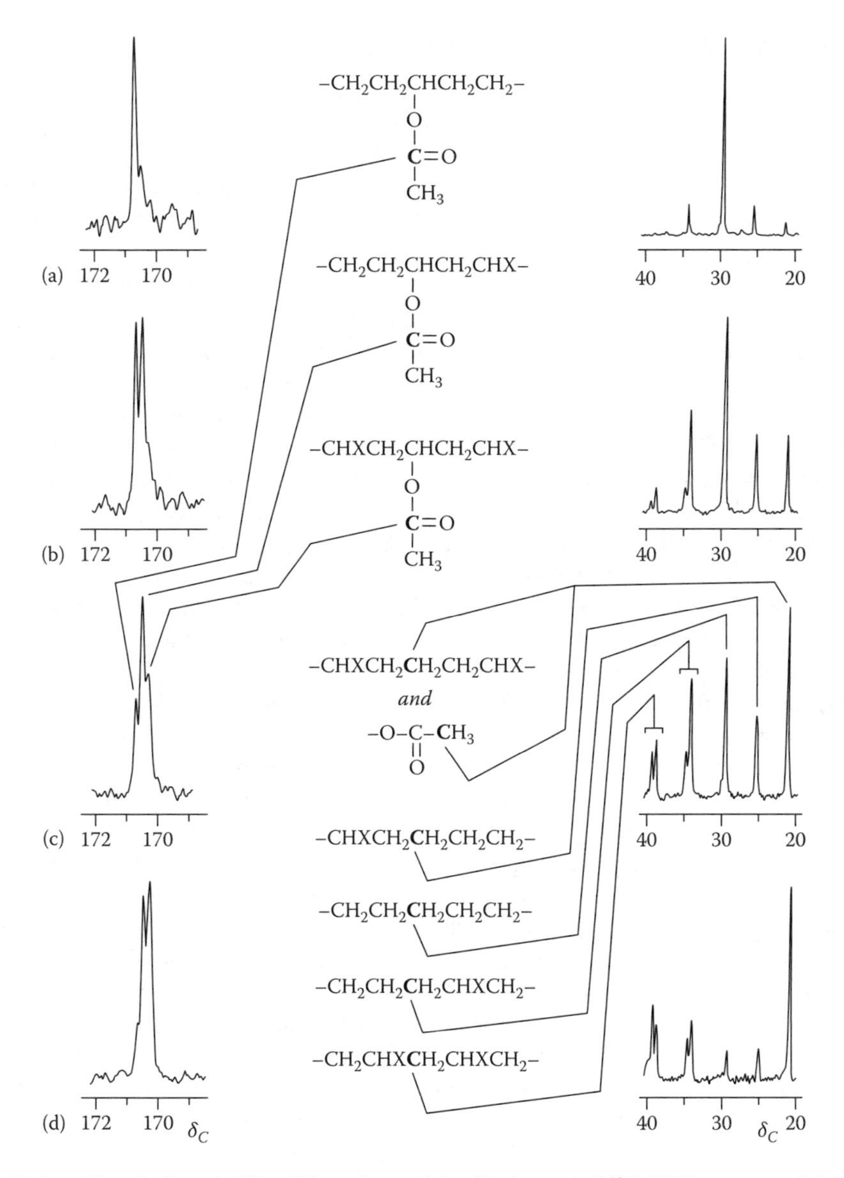

FIGURE 15.20 The C=O and CH_2+CH_3 regions of the ^{1}H-decoupled ^{13}C NMR spectra of four samples of poly[ethylene-*co*-(vinyl acetate)] comprising: (a) 10 mol%, (b) 34 mol%, (c) 48 mol% and (d) 74.5 mol% vinyl acetate repeat units. Assignments of the absorptions to triad sequences (where X = $OCOCH_3$) are shown against the spectrum for (c) and were established by comparison with spectra of model compounds and with theoretical predictions of chemical shifts. (Data and assignments were taken from Wu, K. et al., *J. Polym. Sci. Polym. Phys. Edn.*, 12, 901, 1974.)

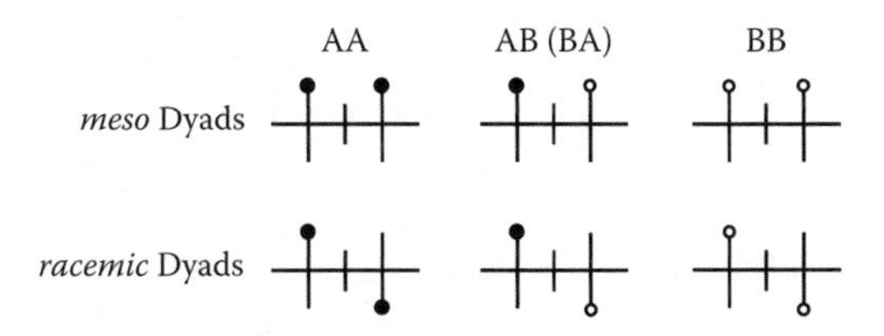

FIGURE 15.21 The six distinguishable dyad sequences in a copolymer where both A and B repeat units have two distinct stereochemical configurations.

15.6.5 Practical Aspects of NMR Spectroscopy

All NMR spectrometers have large magnets which give highly homogeneous magnetic fields and commonly are specified in terms of the fundamental ^{1}H resonance frequency corresponding to the strength of the magnet, for example 100 MHz if $B_0=2.3$ T (see Table 15.2). Early NMR spectrometers used either permanent magnets or electromagnets (e.g. 60 and 100 MHz) and were of the *continuous-wave (CW) type* in which a single radiowave frequency was applied continuously whilst a small sweep coil slowly swept B_0 over an appropriate range. Modern NMR spectrometers commonly have superconducting magnets (e.g. 300 and 500 MHz) and are of the *pulsed Fourier transform (PFT) type*. In PFT NMR, B_0 is fixed and a short pulse (e.g. a few microseconds) of radiowave frequencies covering the full range of interest is applied to promote all the nuclei to their high-energy states. The simultaneous relaxation of these nuclei back to their low-energy states is detected as an interferogram, known as a *free induction decay* (FID), which is converted to the NMR spectrum by Fourier transformation. Usually, this procedure is repeated many times allowing a suitable time interval (e.g. a few seconds) between successive pulses for acquisition of the FIDs. The data from each of these scans is accumulated to give the final NMR spectrum. The number of scans required to obtain a good spectrum depends upon the sensitivity of the NMR effect. For example, even a single scan may be adequate for ^{1}H NMR whereas several thousand scans often are required for ^{13}C NMR due to the low natural abundance of ^{13}C. In quantitative studies, it is essential to ensure that the time interval between successive pulses is sufficient to ensure that all of the excited nuclei have relaxed fully back to their low-energy states before each subsequent pulse is applied; typically, a pulse interval of 0.5–1 s is sufficient for recording ^{1}H NMR spectra, whereas for ^{13}C NMR spectra pulse intervals of ~10 s are required. For ^{13}C NMR spectroscopy, ^{1}H-decoupling enhances the intensity of the ^{13}C absorptions due to an effect known as the *nuclear Overhauser effect* (NOE) and so, in quantitative ^{13}C NMR studies, the NOE enhancement must either be taken into account or suppressed by use of specialized techniques (e.g. by *gated decoupling*).

Polymer samples can be analysed in the form of dilute, semi-dilute or concentrated solutions, as solvent-swollen gels and also as solids. Polymer solutions and gels for ^{1}H and ^{13}C NMR studies are prepared using deuterated solvents such as deuterochloroform ($CDCl_3$) and deuterobenzene (C_6D_6). With the general exception of dilute solutions, as the polymer concentration increases the absorptions become broader leading ultimately to loss of information such as that available from the splitting due to J-coupling. This can be particularly severe in spectra recorded from solid polymers. However, techniques have been developed which greatly suppress this broadening (e.g. *dipolar decoupling* with *magic-angle spinning*) and enable useful spectra to be recorded from solid polymers. Thus solid-state NMR spectroscopy is finding increasing use in the study of bulk polymers.

15.7 MASS SPECTROSCOPY

Mass spectroscopy (MS) is fundamentally different to the other forms of spectroscopy in that it is not based on the absorption or emission of electromagnetic radiation, but instead involves vaporizing and ionizing the molecules within a sample and then separating the resulting molecular ions according to their mass:charge ratio m/z. The word 'spectroscopy' is used by analogy to true forms of spectroscopy, since MS provides information about molecular structure and the output from a mass spectrometer appears like a spectrum, but it actually is a plot of abundance against m/z. The principles of MS for analysis of polymers were described in the context of determining molar mass distribution in Section 14.5, which should be consulted to gain an appreciation of the theory, practice and limitations of MS.

As described in Section 14.5.2, MS of polymers requires use of soft ionization techniques, the most important being matrix-assisted laser desorption/ionization (MALDI), which enables mass spectra to be obtained from high molar mass polymers. In addition to the measurement of molar mass distribution, MALDI MS has proven to be especially powerful for examining the chemical structure of polymers in fine detail. In this section, the use of MALDI MS for probing molecular structure will be exemplified and some long-established general uses of MS in polymer characterization will be outlined.

15.7.1 Elucidation of Structural Features by Mass Spectroscopy

Although MALDI MS cannot resolve structural isomers (NMR spectroscopy is needed for this), the mass resolution and precise molecular mass it provides have proven to be particularly powerful for evaluating other details of polymer structure. The molar masses and molar mass separations of the peaks in the mass spectrum of a polymer can be used to determine the repeat unit molar mass, the chemical structure of end groups and the presence of other repeat units or modified side-groups, even if they are present at low levels because a single end-group or non-standard repeat unit in a polymer molecule will change its molecular mass by a fixed amount, irrespective of its chain length. If the repeat unit structure, method of polymer synthesis and nature of any subsequent reaction carried out on the polymer are known, then values for *m*/*z* obtained by MALDI MS can be compared with values predicted from this knowledge with the aim of confirming that the target polymer structure has been achieved.

The use of MALDI MS for elucidation of structural features will be exemplified here by results from a study of the relative contributions of termination by disproportionation and combination (Section 4.2.3) in free-radical polymerization of methyl methacrylate (MMA) initiated using azobisisobutyronitrile (Section 4.2.1) at 90 °C. Three types of poly(methyl methacrylate) (PMMA) chains can be expected, distinguished by the end groups arising from the two modes of termination

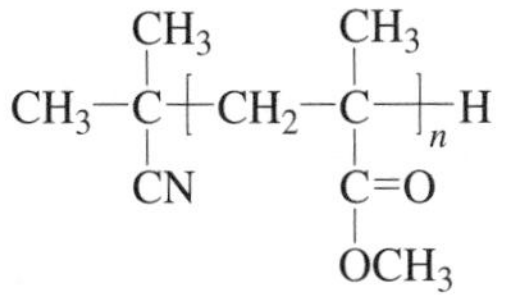

$$CH_3-\overset{CH_3}{\underset{CN}{C}}\!-\!\left[CH_2-\overset{CH_3}{\underset{\underset{OCH_3}{C=O}}{C}}\right]_n\!-H \qquad M_{exact} = 68.05 + \{100.05 \times n\} + 1.01 \quad (g\ mol^{-1})$$

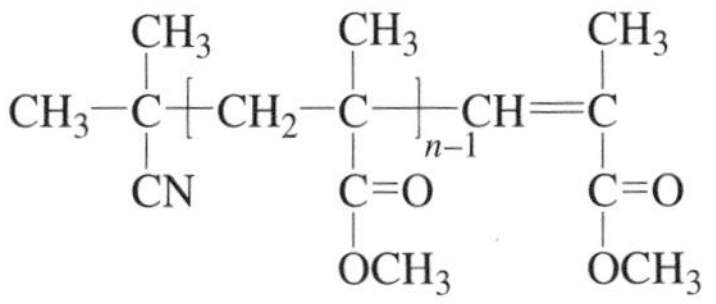

$$CH_3-\overset{CH_3}{\underset{CN}{C}}\!-\!\left[CH_2-\overset{CH_3}{\underset{\underset{OCH_3}{C=O}}{C}}\right]_{n-1}\!-CH{=}\overset{CH_3}{\underset{\underset{OCH_3}{C=O}}{C}} \qquad M_{exact} = 68.05 + \{100.05 \times (n-1)\} + 99.04 \quad (g\ mol^{-1})$$

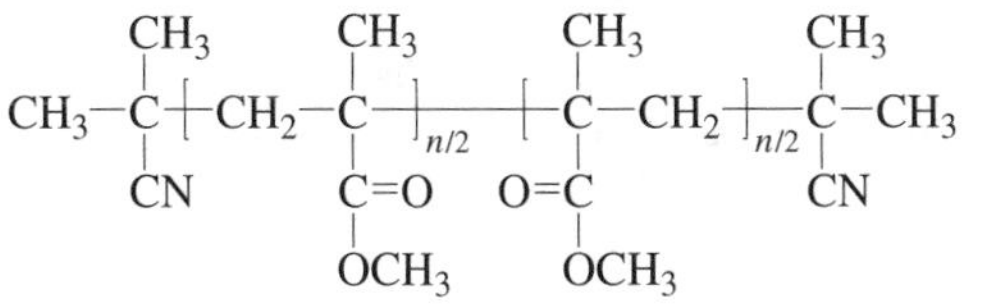

$$CH_3-\overset{CH_3}{\underset{CN}{C}}\!-\!\left[CH_2-\overset{CH_3}{\underset{\underset{OCH_3}{C=O}}{C}}\right]_{n/2}\!\!-\!\left[\overset{CH_3}{\underset{\underset{OCH_3}{O=C}}{C}}-CH_2\right]_{n/2}\!-\overset{CH_3}{\underset{CN}{C}}-CH_3 \qquad M_{exact} = \{2 \times 68.05\} + \{100.05 \times n\} \quad (g\ mol^{-1})$$

Equations for exact molar masses M_{exact} are given here because MS separates individual isotopic species and so comparison of *m*/*z* values must be with M_{exact} values. In the equations, n is the number of MMA-derived units in the chain. Figure 15.22 shows the MMD determined by MALDI MS for a sample of PMMA prepared as described above (note that, as is common practice in MS, molar masses were equated to *m*/*z* values and so were not corrected for the mass contributed by the Na^+ ions associated with the vaporized chains). The MMD reveals the apparent superposition of two slightly displaced MMDs, which is very clear in the expanded MMD where they are labelled as series (i) and (ii). As would be expected, the peaks in each series are separated by the molar mass of an MMA repeat unit (100 g mol^{-1}) and the two distributions are for chains resulting from the two different modes of termination. Table 15.5 provides a comparison of the measured molar mass data with values of $M_{exact}+M_{Na^+}$ calculated using the equations given above (where M_{Na^+} is the exact molar mass of $^{23}Na^+$). This shows, unambiguously, that the series (i) peaks are for chains

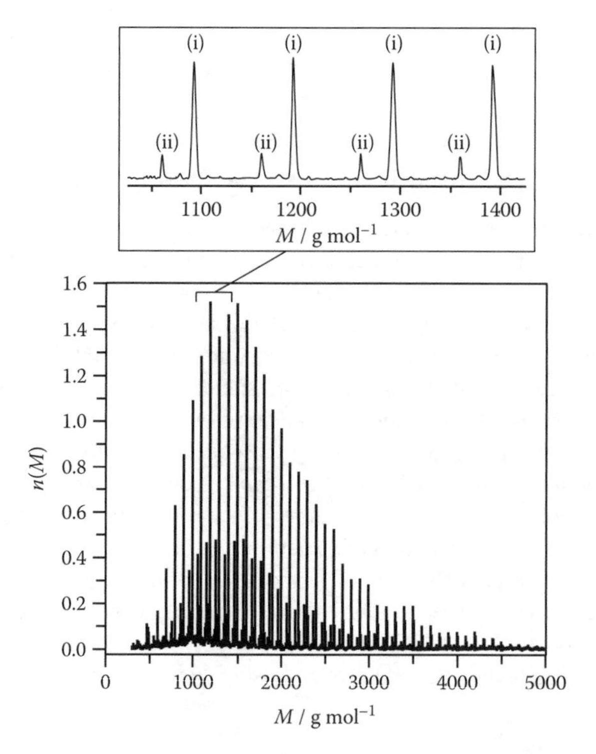

FIGURE 15.22 Number fraction $n(M)$ molar mass distribution obtained by MALDI ToF MS of a low molar mass poly(methyl methacrylate) sample prepared by free-radical polymerization at 90 °C. The inset shows expansion of a region close to the maximum in the distribution and highlights the two major series of peaks, labelled (i) and (ii). The sample was desorbed/ionized using Na^+ ions with 2,5-dihydroxybenzoic acid as the matrix. (Data taken from Zammit, M.D. et al., *Macromolecules*, 30, 1915, 1997.)

TABLE 15.5
Analysis of Selected Molar Mass M_{MALDI} Data Taken from the MALDI MS Molar Mass Distribution Shown in Figure 15.22

Peak Series	M_{MALDI}[a] / g mol^{-1}	n	$M_{exact}+M_{Na^+}$ / g mol^{-1} [b]		
			Terminated by disproportionation [c]		Terminated by combination
			Saturated chain end	Unsaturated chain end	
(i)	1092.555	10	1092.55	1090.53	–
	1192.620	11	1192.55	1190.58	–
	1292.676	12	1292.65	1290.63	–
(ii)	1060.257	9	–	–	1059.54
	1160.164	10	–	–	1159.59
	1260.231	11	–	–	1259.64

[a] Data taken from Zammit et al., *Macromolecules*, 30, 1915, 1997.

[b] The M_{MALDI} data are m/z values taken from the MS spectrum and include the Na^+ ion associated with the vaporized PMMA chains, hence the exact molar mass of $^{23}Na^+$ ($M_{Na^+}=22.90$ g mol^{-1}) has been added to M_{exact} to enable comparison to M_{MALDI}.

[c] The peaks for the saturated and unsaturated chain ends resulting from disproportionation are not resolved in the mass spectrum because the m/z resolution of the spectrometer was 3–4 Dalton e^{-1}.

terminated by disproportionation and the series (ii) peaks are for chains terminated by combination. The results confirm that termination in free-radical polymerization of MMA is dominated by disproportionation (Section 4.2.3). Analysis of the peak areas revealed that the ratio of rate coefficients for disproportionation and combination (k_{td}/k_{tc}) is about 4.4 at 90 °C, which shows that about 81% of termination events proceed by disproportionation (i.e. $q \approx 0.81$; see Section 4.3.2).

15.7.2 Other Uses of Mass Spectroscopy in Polymer Science

Prior to the development of MALDI, MS was primarily used in polymer science for analysis of low molar mass molecules obtained by solvent extraction from commercial polymers (e.g. analysis for plasticizers, antioxidants and UV stabilizers), for investigation of polymer composition (e.g. in a competitor company's product) and for studies of polymer degradation (where it assists in identifying the degradation products, knowledge of which enables the mechanism of degradation to be discerned). These applications of MS continue to be important today; the latter two are essentially the same because the use of MS to evaluate composition involves analysis of the degradation products from pyrolysis of the polymer.

For all these applications, it is common to couple MS with gas–liquid chromatography (GC) such that the output stream from the latter is subjected to continuous MS analysis, i.e. the mass spectrometer is used as a final-stage detector for the products separated in the GC column. When pyrolysed under specific conditions directly into the GC column, homopolymers often give characteristic GC-MS 'fingerprints', which give rise to the use of this technique for investigating the composition of polymers of unknown structure.

PROBLEMS

15.1 The table below gives the results from the measurement of the percentage transmission %T of ultraviolet light of wavelength 254 nm for a series of dilute solutions of polystyrene in tetrahydrofuran that were analysed using a cuvette with a 2.0 mm path length. Using an appropriate plot of the data, determine the molar absorptivity of polystyrene repeat units at 254 nm, giving the value in units of $dm^3\ mol^{-1}\ cm^{-1}$.

Polystyrene solution concentration / g dm^{-3}	0.081	0.167	0.339	0.550	0.689
%T at 254 nm	91.1	82.6	67.9	53.3	45.5

15.2 Explain how IR spectroscopy could be used to distinguish between polymers **I–V** shown below. In your answer, you should specify the differences between the spectra of the polymers that would lead to unambiguous identifications.

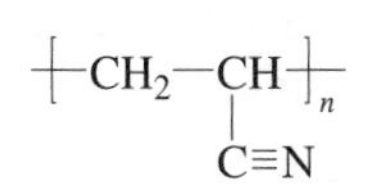

$-[CH_2-CH(C\equiv N)]_n-$

I: Polyacrylonitrile

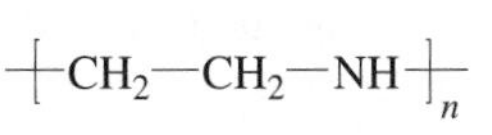

$-[CH_2-CH_2-NH]_n-$

II: Poly(ethylene imine)

$-[CH_2-CH(C(=O)-O-CH_2-CH_2OH)]_n-$

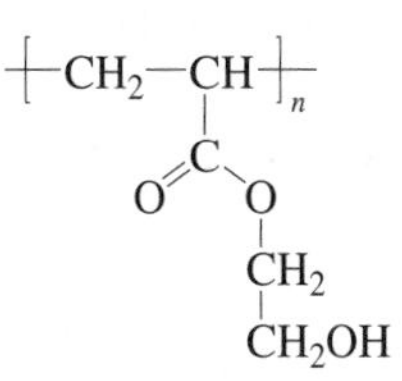

III: Poly(2-hydroxyethyl acrylate)

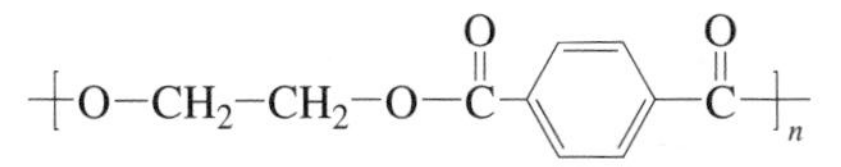

$-[O-CH_2-CH_2-O-C(=O)-C_6H_4-C(=O)]_n-$

IV: Poly(ethylene terephthalate)

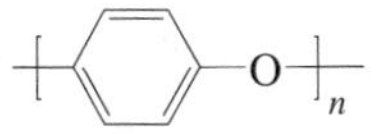

$-[C_6H_4-O]_n-$

V: Poly(1,4-phenylene oxide)

15.3 Sketch: (a) the ^{1}H NMR spectrum, (b) the ^{1}H-decoupled ^{13}C NMR spectrum and (c) the DEPT135 ^{13}C NMR spectrum you would expect to observe for each of the following polymers.

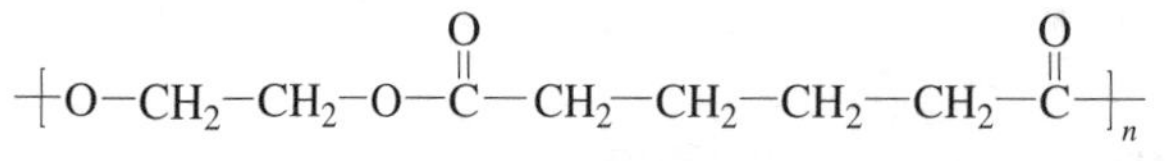

$$\left[O-CH_2-CH_2-O-\overset{O}{\overset{\|}{C}}-CH_2-CH_2-CH_2-CH_2-\overset{O}{\overset{\|}{C}} \right]_n$$

VI: Poly(ethylene adipate)

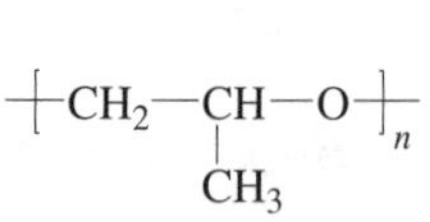

$$\left[CH_2-\underset{CH_3}{\underset{|}{CH}}-O \right]_n$$

VII: *Isotactic* poly(propylene oxide)

15.4 ^{1}H-decoupled ^{13}C NMR spectra of two samples of poly(methyl methacrylate) were recorded under quantitative conditions. The α-methyl carbon resonances in the region $15<\delta<25$ showed clear separation of the *mm*, *mr* and *rr* triads for which the peak integrals are given in the table below. For each sample:

(a) Calculate the mole fraction of each triad sequence.
(b) Determine the Bernoullian probability P_m for a *meso* placement.
(c) Test whether the data conform to Bernoullian statistics.
(d) Comment on the tacticity of the polymer and thereby deduce the method of polymerization most likely to have been used for synthesis of the sample.

Poly(Methyl Methacrylate) sample	Peak Integrals for the Triad α-Methyl Carbon Resonances		
	mm	*mr*	*rr*
A	7.8	64.5	123.2
B	218.2	9.2	2.3

15.5 Discuss how MALDI mass spectroscopy could be used to perform the following evaluations.

(a) Establish whether a sample of low molar mass linear poly(ethylene oxide) has hydroxyl groups at both chain ends or one methoxy end-group and one hydroxyl end-group.

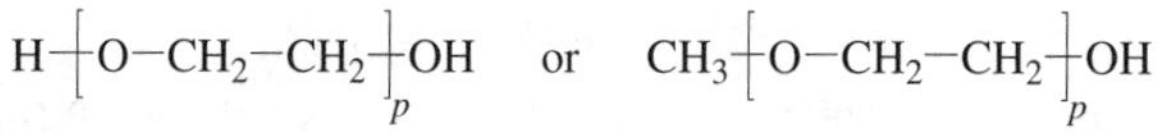

$$H\left[O-CH_2-CH_2 \right]_p OH \quad \text{or} \quad CH_3\left[O-CH_2-CH_2 \right]_p OH$$

(b) Show that the conversion of a low molar mass sample of poly(*t*-butyl acrylate) into poly(acrylic acid) has gone to completion according to the following reaction scheme.

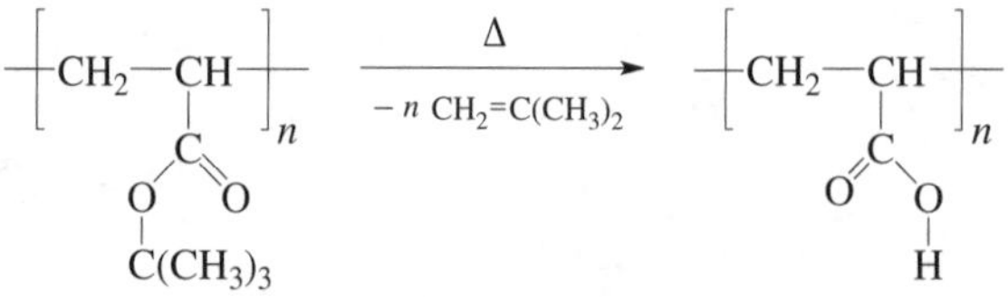

$$\left[CH_2-\underset{\underset{O=C-O-C(CH_3)_3}{|}}{CH} \right]_n \xrightarrow[-\,n\,CH_2=C(CH_3)_2]{\Delta} \left[CH_2-\underset{\underset{O=C-O-H}{|}}{CH} \right]_n$$

(c) Provide confirmation that a sample of poly(ethyl methacrylate) with a number-average molar mass of about 3 kg mol^{-1}, prepared by free-radical catalytic chain transfer polymerization, has the following structure.

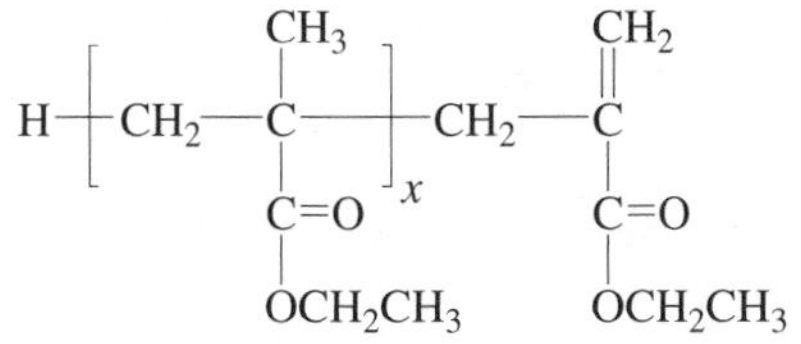

$$H{-}\!\left[CH_2{-}\underset{\underset{OCH_2CH_3}{|}}{\underset{C=O}{|}}\!\!\overset{CH_3}{\overset{|}{C}}\right]_x\!{-}CH_2{-}\underset{\underset{OCH_2CH_3}{|}}{\underset{C=O}{|}}\!\!\overset{CH_2}{\overset{\|}{C}}$$

FURTHER READING

General Reading

Allen, G. and Bevington, J.C. (eds), *Comprehensive Polymer Science*, Vol. 1 (Booth, C. and Price, C., eds), Pergamon Press, Oxford, 1989.

Klöpffer, W., *Introduction to Polymer Spectroscopy*, Springer-Verlag, Berlin, Germany, 1984.

Koenig, J.L., *Spectroscopy of Polymers*, 2nd edn., Elsevier, Amsterdam, The Netherlands, 1999.

Lambert, J.B., Shurvell, H.F., Lightner, D.A., and Cooks, R.G., *Introduction to Organic Spectroscopy*, Macmillan, New York, 1987.

Mark, H.F., Bikales, N.M., Overberger, C.G., and Menges, G. (eds), *Encyclopedia of Polymer Science and Engineering*, Wiley-Interscience, New York, 1985–1989.

Silverstein, R.M., Bassler, G.C., and Morrill, T.C., *Spectrometric Identification of Organic Compounds*, 5th edn., John Wiley & Sons, New York, 1991 (last edition to include UV-visible spectroscopy).

Silverstein, R.M., Webster, F.X., and Kiemle, D.J., *Spectrometric Identification of Organic Compounds*, 7th edn., John Wiley & Sons, Hoboken, NJ, 2005.

Vibrational Spectroscopy

Bower, D.I. and Maddams, W.F., *The Vibrational Spectroscopy of Polymers*, Cambridge University Press, Cambridge, MA, 1989.

Colthrup, N.B., Daly, L.H., and Wiberley, S.E., *Introduction to Infrared and Raman Spectroscopy*, 2nd edn., Academic Press, New York, 1975.

Nuclear Magnetic Resonance Spectroscopy

Abraham, R.J., Fisher, J., and Loftus, P., *Introduction to NMR Spectroscopy*, John Wiley & Sons, New York, 1988.

Bovey, F.A., *Nuclear Magnetic Resonance Spectroscopy*, 2nd edn., Academic Press, San Diego, CA, 1988.

Bovey, F.A. and Mirau, P.A., *NMR of Polymers*, Academic Press, New York, 1996.

Ibbett, R.N., *NMR Spectroscopy of Polymers*, Blackie, Glasgow, U.K., 1993.

Mirau, P.A., *A Practical Guide to Understanding the NMR of Polymers*, Wiley-Interscience, New York, 2005.

Matsuzaki, K., Uryu, T., and Asakura, T., *NMR Spectroscopy and Steroregularity of Polymers*, Japan Scientific Societies Press, Tokyo, Japan, 1996.

Randall, J.C., *Polymer Sequence Determination: Carbon-13 NMR Method*, Academic Press, New York, 1977.

Tonelli, A.E., *NMR Spectroscopy and Polymer Microstructure: The Conformational Connection*, Wiley-VCH, New York, 1989.

Wehrli, F.W., Marchand, A.P., and Wehrli, S., *Interpretation of Carbon-13 NMR Spectra*, 2nd edn., John Wiley & Sons, New York, 1988.

Mass Spectroscopy

Batoy, S.M.A.B., Akhmetova, E., Miladinovic, S., Smeal, J., and Wilkins, C.L., Developments in MALDI mass spectrometry: the quest for the perfect matrix, *Applied Spectroscopy Reviews*, **43**, 485 (2008).

de Hoffman, E. and Stroobant, V., *Mass Spectrometry: Principles and Applications*, John Wiley & Sons, Chichester, U.K., 2007.

El-Aneed, A., Cohen, A., and Banoub, J., Mass spectrometry, review of the basics: electrospray, MALDI, and commonly used mass analyzers, *Applied Spectroscopy Reviews*, **44**, 210 (2009).

Hanton, S.D., Mass spectrometry of polymers and polymer surfaces, *Chemical Reviews*, **101**, 527 (2001).
Montaudo, G. and Lattimer, R.P. (eds), *Mass Spectrometry of Polymers*, CRC Press, Boca Raton, FL, 2002.
Montaudo, G., Samperi, F., and Montaudo, M.S., Characterization of synthetic polymers by MALDI-MS, *Progress in Polymer Science*, **31**, 277 (2006).
Nielen, M.W.F., MALDI time-of-flight mass spectrometry of synthetic polymers, *Mass Spectrometry Reviews*, **18**, 309 (1999).

Part III

Phase Structure and Morphology of Bulk Polymers

16 The Amorphous State

16.1 INTRODUCTION

The term *amorphous* is used to describe materials that are completely non-crystalline, such as polymer glasses and rubbers. Perhaps the most widely-known amorphous glassy polymer is polystyrene which, although relatively brittle, is used widely because it can be produced relatively cheaply and has good processing properties. The use of poly(methyl methacrylate) (PMMA), with its excellent optical properties, has now been extended to general thermoforming and moulding applications. Polycarbonate also has been introduced, and its clarity and toughness have been exploited widely. Although we will see in subsequent chapters that the properties of these materials have been studied at length, it is difficult to characterize their detailed morphology in simple terms, as there is no well-defined molecular order in amorphous polymers.

16.1.1 Structure in Amorphous Polymers

Amorphous polymers in the solid state can be thought of simply as frozen polymer liquids. Over the years, there have been many attempts to analyse the structure of liquids of small molecules, and they have met with a varied success. As may be expected, the necessity of having long chains makes the problem more difficult to solve in the case of polymer melts and amorphous polymers. Investigation of the morphology of amorphous polymers has yielded only a limited amount of information. Figure 16.1a is an X-ray diffraction pattern of an amorphous polymer, showing only one diffuse ring. This is confirmed by Figure 16.1b which shows the intensity of scattering across a radial section on the diffraction pattern. In crystalline materials the discrete rings or spots on X-ray diffraction patterns correspond to Bragg reflections from particular crystallographic planes within the crystals (Section 17.2.1). The lack of such discrete reflections in non-crystalline materials is taken as an indication of the lack of crystalline order. The very diffuse scattering observed in Figure 16.1 corresponds simply to the regular spacing of the atoms along the polymer chain or in side groups, and the variable separation of the atoms in adjacent molecules. It was pointed out in Section 12.4 that small-angle neutron scattering has been used to show that, in a pure amorphous polymer, the polymer molecules adopt their unperturbed dimensions (Section 10.3).

There have been many suggestions over the years that both polymer melts and polymer glasses have domains of order. These are microscopic regions in which there is a certain amount of molecular order. This idea is not unique for polymers, and similar ideas have been put forward to explain the structure of liquids and inorganic glasses. The main evidence for ordered regions in glassy polymers has come from examination of these materials by electron microscopy either by taking replicas of fracture surfaces or looking directly at thin films where a nodular structure on a scale of ~100 nm can sometimes be seen. Similar features have been seen in many glassy polymers, and the nodules have been interpreted probably incorrectly as domains of order. The presence of such order, however, has not been confirmed by X-ray diffraction (e.g. Figure 16.1).

There are some polymers, such as poly(ethylene terephthalate) and isotactic polypropylene which normally crystallize rather slowly, that can be obtained in an apparently amorphous state by rapidly quenching the melt, which does not allow sufficient time for crystals to develop by the normal processes of nucleation and growth. Crystallization can be induced by annealing the quenched polymer at an elevated temperature. The X-ray diffraction patterns that are obtained from the quenched polymer tend to be diffuse and ill-defined. This could be either because the samples are

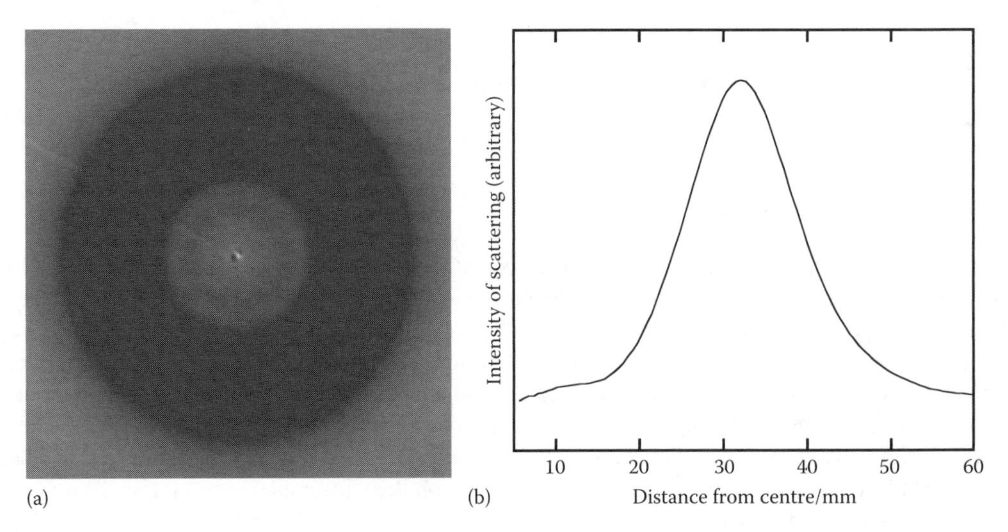

FIGURE 16.1 Wide-angle X-ray diffraction from an epoxy resin. (a) Flat-plate diffraction pattern. (b) Intensity of scattering as a function of distance from the centre of the pattern.

truly non-crystalline or due to the presence of very small and imperfect crystals. Crystallization is manifest during annealing (heating at elevated temperature) by the diffraction patterns becoming more well-defined. There clearly must be a certain amount of molecular rearrangement during annealing, but this cannot take place through large-scale molecular diffusion as the process occurs in the solid state. The mechanism of crystallization is, therefore, unclear, and it is very difficult to decide whether the quenched polymers are truly amorphous or just contain very small crystals.

16.2 THE GLASS TRANSITION

When the melt of a polymer is cooled, it becomes more viscous and flows less readily. If the polymer is not able to crystallize and the temperature is reduced low enough, it becomes rubbery and then as the temperature is reduced further, it becomes a relatively hard and elastic polymer glass. The temperature at which the polymer undergoes the transformation from a rubber to a glass is known as the *glass transition temperature,* T_g. The ability to form glasses is not confined to non-crystallisable polymers. Any material that can be cooled sufficiently below its melting temperature without crystallizing will undergo a glass transition.

There is a dramatic change in the properties of a polymer at the glass transition temperature. For example, there is a sharp increase in the stiffness of an amorphous polymer when its temperature is reduced below T_g. This will be dealt with further in the section on transitions (Section 20.6). There also are abrupt changes in other physical properties, such as heat capacity and thermal expansion coefficient. One of the most widely used methods of demonstrating the glass transition and determining T_g is by measuring the specific volume of a polymer sample as a function of the temperature, as shown in Figure 16.2. In the regimes above and below the glass transition temperature, there is an approximately linear variation in specific volume with temperature, but in the vicinity of the T_g, there is change in slope of the curve that occurs over several degrees. The T_g is normally taken as the point at which the extrapolations of the two lines meet. Another characteristic of the T_g is that its value depends upon the rate at which the temperature is changed during the measurements. It is found that the lower the cooling rate, the lower the value of T_g that is obtained. It is still a matter of some debate as to whether a limiting value of T_g would eventually be reached if the cooling rate were low enough. It also is possible to detect a glass transition in the amorphous regions of a semi-crystalline polymer, but the change in properties at T_g is usually less marked than for a fully amorphous polymer because the amorphous regions comprise only a small fraction of the total polymer.

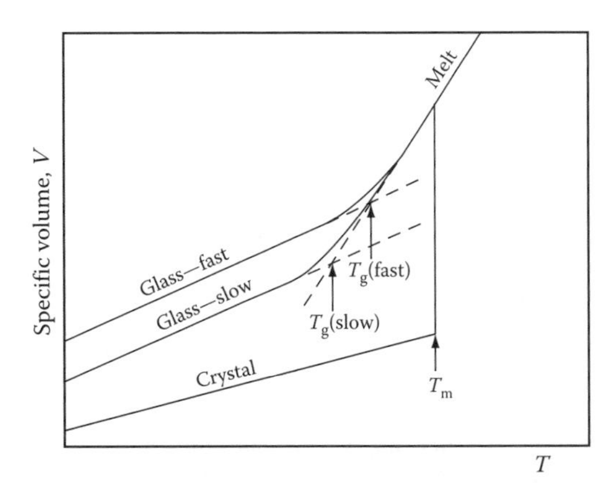

FIGURE 16.2 Variation of the specific volume with temperature for different materials. Examples are shown for the formation of a glassy polymer, by either slow or fast cooling, and the melting of a crystalline material at a temperature T_m.

16.2.1 Thermodynamics of the Glass Transition

There have been attempts to analyse the glass transition from a thermodynamic viewpoint. Thermodynamic transitions are classified as being either first order or second order. In a first-order transition, the free energy as a function of any given state variable (V, P, T) is continuous, but the first partial derivatives of the free energy with respect to the relevant state variables are discontinuous. Hence, for a first-order transition, the Gibbs free energy, G, at the transition temperature is continuous, but $(\partial G/\partial T)_P$ and $(\partial G/\partial P)_T$ are discontinuous. For a first-order transition such as melting or vaporization there are discontinuities in volume V, enthalpy H and entropy S, since

$$\left(\frac{\partial G}{\partial P}\right)_T = V \tag{16.1}$$

$$\left(\frac{\partial (G/T)}{\partial (1/T)}\right)_P = H \tag{16.2}$$

$$\left(\frac{\partial G}{\partial T}\right)_P = -S \tag{16.3}$$

For example, Figure 16.2 shows that there is a step change in V at the melting temperature, T_m.

In the case of a second-order transition, there is a discontinuity in the second partial derivatives of the free-energy function, but continuity of both the free energy and its first partial derivatives. Hence, there are no discontinuities in V, H or S at the temperature of the transition, but there are discontinuities in the variations of thermal expansion coefficient α, compressibility κ and heat capacity C_p with temperature since

$$\left(\frac{\partial}{\partial T}\left(\frac{\partial G}{\partial P}\right)_T\right)_P = \left(\frac{\partial V}{\partial T}\right)_P = \alpha V \tag{16.4}$$

$$\left(\frac{\partial^2 G}{\partial P^2}\right)_T = \left(\frac{\partial V}{\partial P}\right)_T = -\kappa V \tag{16.5}$$

$$\frac{\partial}{\partial T}\left(\left(\frac{\partial(G/T)}{\partial(1/T)}\right)_P\right)_P = \left(\frac{\partial H}{\partial T}\right)_P = C_P \tag{16.6}$$

$$-\left(\frac{\partial^2 G}{\partial T^2}\right)_P = \left(\frac{\partial S}{\partial T}\right) = \frac{C_P}{T} \tag{16.7}$$

Although we are most familiar with first-order transitions such as melting, there are a number of phenomena that may be termed second-order transitions found in materials, such as the onset of ferromagnetism.

It now is possible to consider whether the glass transition can be considered as a thermodynamic transition. Figure 16.3 shows the temperature dependence of the thermodynamic quantities, G, V, H and S, and the derivatives α, κ and C_p for first- and second-order transitions and for the glass transition. It can be seen that the glass transition appears to be have the characteristics of a second-order transition such as a change in slope of the plot of V, H or S against T rather than a step change as is found at the T_m. However, Figure 16.2 shows that T_g shifts to higher temperature as the cooling rate increases which could not happen for a true second-order transition.

The features described above indicate that the process of glass formation cannot be regarded as a true transition in the thermodynamic sense, but rather as an inhibition of kinetic processes. The

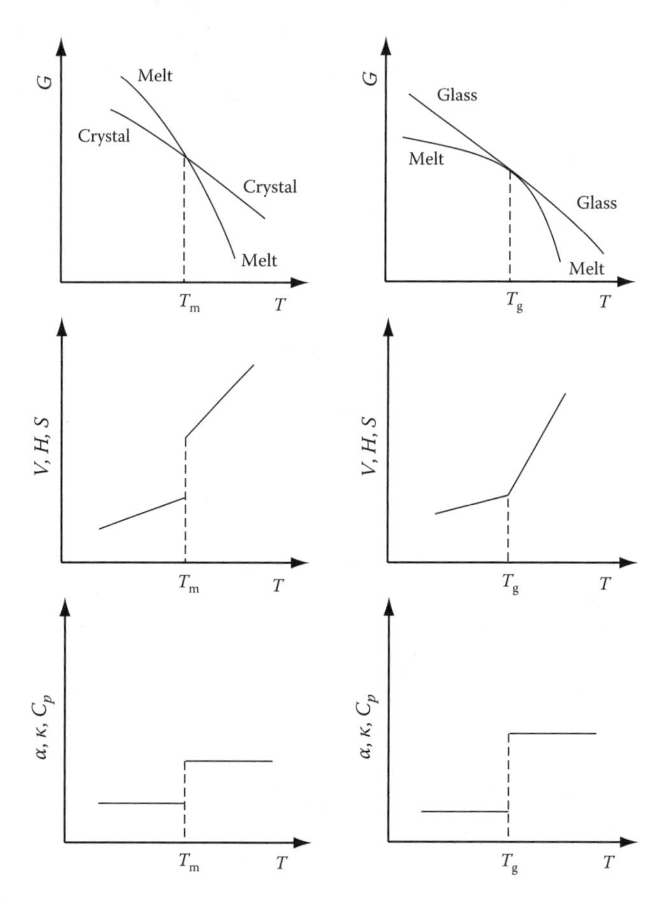

FIGURE 16.3 Schematic illustration of the difference between melting and glass transitions in terms of the effect upon the different thermodynamic parameters. (Adapted from Rehage, G. and Borchard, W., *Physics of Glassy Polymers*, Applied Science Publishers, London, U.K., 1973.)

fundamental reason for this is that there does not exist an internal thermodynamic equilibrium on both sides of the T_g, as would be the case for a second-order transition. Because of this, it has been suggested that it might be better to use the term 'glassy solidification' to describe the process. Although there is much to commend this suggestion, such terminology is not in general use, and the term 'glass transition' will be used throughout this chapter.

16.2.2 Free Volume

One of the most useful approaches to analysing the glass transition is to use the concept of *free volume*. This concept has been used in the analysis of liquids, and it can be readily extended to the consideration of the glass transition in polymers. The free volume is the space in a solid or liquid sample that is not occupied by polymer molecules, i.e. the 'empty-space' between molecules. In the liquid state, it is supposed that the free volume is high, and so molecular motion is able to take place relatively easily because the unoccupied volume allows the molecules space to move and so change their conformations freely. A reduction in temperature will reduce the amount of thermal energy available for molecular motion. It also is envisaged that the free volume will be sensitive to the change in temperature, and that most of the thermal expansion of the polymer rubber or melt can be accounted for by a change in the free volume. As the temperature of the melt is lowered, the free volume will be reduced until eventually there will not be enough free volume to allow molecular rotation or translation to take place. The temperature at which this happens corresponds to T_g as below this temperature the polymer glass is effectively 'frozen'. The situation is represented schematically in Figure 16.4. The free volume is represented as a shaded area. It is assumed to be constant at V_f^* below T_g and to increase as the temperature is raised above T_g. The total sample volume V therefore consists of volume occupied by molecules V_o and free volume V_f such that

$$V = V_o + V_f \tag{16.8}$$

It is more convenient to talk in terms of fractional free volume f_V, which is defined as $f_V = V_f/V$. At and below the T_g, f_V is given by $f_g = V_f^*/V$ and can be considered as being effectively constant. Above the T_g, there will be an important contribution to V_f from the expansion of the melt. The free volume above T_g is then given by

$$V_f = V_f^* + (T - T_g)\left(\frac{\partial V}{\partial T}\right) \tag{16.9}$$

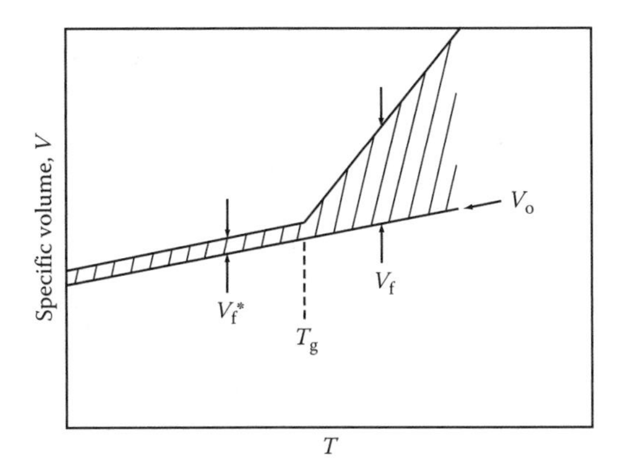

FIGURE 16.4 Schematic illustration of the variation of the specific volume V of a polymer with temperature T. The free volume is represented by the shaded area.

Dividing through by V gives

$$f_V = f_g + (T - T_g)\alpha_f \tag{16.10}$$

where α_f is the thermal expansion coefficient of the free volume which will be given, close to T_g, by the difference between the thermal expansion coefficients of the rubbery and glassy polymer. The equation for the fractional free volume will only be strictly valid over small increments of temperature above T_g. It is found for a whole range of different glassy polymers that f_g is remarkably constant, and this concept of free volume has found important use in the analysis of the rate and temperature dependence of the viscoelastic behaviour of polymers between T_g and $T_g + 100$ K (Section 20.7).

16.3 FACTORS CONTROLLING THE T_g

Different polymers have different values of glass transition temperature. This affects many of their physical properties and the variation of these properties with temperature, and so it is important at this stage to consider the factors that affect the value of T_g of polymers.

16.3.1 CHEMICAL STRUCTURE

The onset of the glass transition marks a significant change in the physical properties of the polymer. A glassy polymer will lose its stiffness and have a tendency to flow above T_g. As many polymers are used in practical applications in the glassy state, it is vital to know what factors control the T_g. At the glass transition, the molecules that are effectively frozen in position in the polymer glass become free to rotate and translate, and so it is not surprising that the value of the T_g will depend upon the physical and chemical nature of the polymer molecules. The effect of factors such as molar mass and branching will be considered in Section 16.3.3. Here the way in which the chemical nature of the polymer chain, as shown in Table 16.1, affects T_g will be examined.

TABLE 16.1
Approximate Values of Glass Transition Temperature, T_g, for Various Polymers

Repeat Unit	Polymer Name	T_g/°C
$-CH_2-CH_2-$	Polyethylene	−130 to −10
$-CH_2-CH_2-O-$	poly(ethylene oxide)	−67
$-H_2C-C_6H_4-CH_2-$	poly(*p*-xylylene)	80
Vinyl polymers	Side group (X)	
$-CH_2-CHX-$	$-CH_3$	−23
	$-CH_2-CH_3$	−24
	$-CH_2-CH_2-CH_3$	−40
	$-CH_2-CH(CH_3)_2$	30
	$-C_6H_5$	100
	−Cl	81
	−OH	85
	−CN	97

The most important factor in controlling the T_g of a polymer is chain flexibility, which is controlled by the nature of the chemical groups that constitute the main chain. Polymers, such as polyethylene $(-CH_2-CH_2-)_n$ and poly(ethylene oxide) $(-CH_2-CH_2-O-)_n$ that have relatively flexible chains because of easy rotation about main chain bonds, tend to have low values of T_g. The value of the T_g of polyethylene is a matter of some dispute because samples are normally highly crystalline which makes measurement of T_g rather difficult. Values have been quoted between −130°C and −30°C. However, the incorporation of units that impede rotation and stiffen the chain clearly causes a large increase in T_g. For example, the presence of a *p*-phenylene ring in the polyethylene chain causes poly(*p*-xylylene) to have a T_g of the order of 80°C.

In vinyl polymers of the type $(-CH_2-CHX-)_n$ the nature of the side group has a profound effect upon T_g as can be seen in Table 16.1. The presence of side groups on the main chain has the effect of increasing T_g through a restriction of bond rotation. Large and bulky side groups tend to cause the greatest stiffening, and if the side group itself is flexible, the effect is not so marked. The presence of polar groups such as −Cl, −OH, or −CN tend to raise T_g more than non-polar groups of equivalent size. This is because the polar interactions restrict rotation further, and so poly(vinyl chloride) $(-CH_2-CHCl-)_n$ has a higher T_g than polypropylene $(-CH_2-CHCH_3-)_n$ (Table 16.1).

16.3.2 Copolymerisation

It is possible to analyse the glass transition behaviour of copolymers using the free-volume concept outlined in the previous section. If a single-phase copolymer is produced from two monomers by varying their relative proportions and T_g is plotted against composition as shown in Figure 16.5, then it is found that the T_g of the copolymer lies on, or more usually, below a straight line joining the T_gs of the two homopolymers.

The dependence of the T_g of a copolymer upon composition can be predicted if it is assumed that there is a characteristic amount of free volume associated with each type of repeat unit and that it is the same in both a copolymer or homopolymer. If the two monomers are A and B, then Equation 16.10 can be written as

$$f_V^A = f_g^A + (T - T_g^A)\alpha_f^A \tag{16.11}$$

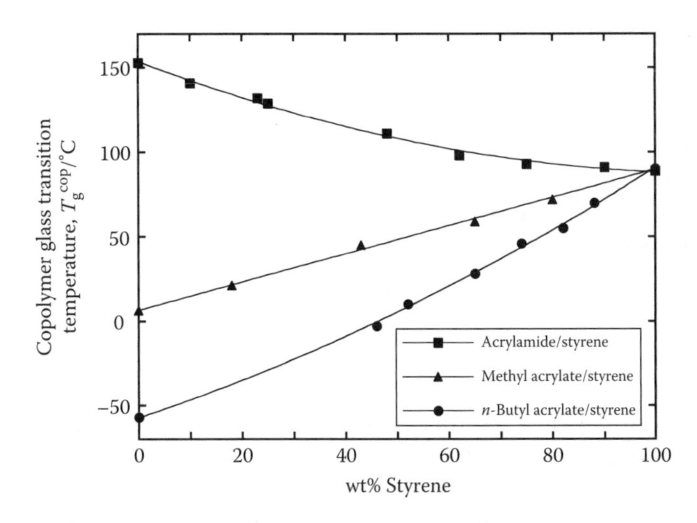

FIGURE 16.5 Variation of copolymer glass transition temperature with the weight fraction of styrene for various styrene-based statistical copolymers. (Data taken from Illers, K.H., *Kolloid-Zeitschrift.*, 190, 16, 1963.)

$$f_V^B = f_g^B + (T - T_g^B)\alpha_f^B \tag{16.12}$$

and for the copolymer

$$f_V^{cop} = f_g^{cop} + (T - T_g^{cop})\alpha_f^{cop} \tag{16.13}$$

where

the left-hand terms are the fractional free volume at T ($>T_g$)

f_g is the fractional free volume at the glass transition temperature

α_f is the thermal expansion coefficient of the free volume.

If it is assumed that in the copolymer the free volumes add in proportion to the weight fractions w_A and w_B of comonomers A and B, then

$$f_g^{cop} = w_A f_g^A + w_B f_g^B \tag{16.14}$$

and in general

$$f^{cop} = w_A f^A + w_B f^B \tag{16.15}$$

If the thermal expansion coefficients of the free volume add in a similar way, then

$$\alpha_f^{cop} = w_A \alpha_f^A + w_B \alpha_f^B \tag{16.16}$$

Combining Equations 16.11 through 16.16 and rearranging leads to

$$T_g^{cop}(w_A + Cw_B) = T_g^A w_A + CT_g^B w_B \tag{16.17}$$

where $C = \alpha_f^B / \alpha_f^A$. If $C \sim 1$, then

$$T_g^{cop} = T_g^A w_A + T_g^B w_B \tag{16.18}$$

which gives a straight line relationship between T_g^{cop} and the copolymer composition. It is found that Equation 16.18 generally overestimates T_g^{cop} and other theoretical calculations have led to a relationship of the form

$$\frac{1}{T_g^{cop}} = \frac{w_A}{T_g^A} + \frac{w_B}{T_g^B} \tag{16.19}$$

This gives a curve below the straight line relationship and actual T_g^{cop} values tend to lie between values given by Equations 16.18 and 16.19.

Similar behaviour with a T_g between that of the two components is found in the case of miscible polymer blends. In multiphase systems such as phase-separated block copolymers and polymer blends, a separate T_g for each phase, similar to those of the constituent homopolymers, is found.

16.3.3 Molecular Architecture

We have, so far, only been concerned with the way in which the chemical nature of the polymer molecules affects T_g. It also is known that the physical characteristics of the molecules such as molar mass, branching and crosslinking affect the temperature of the glass transition. The value of the T_g is found to increase as the molar mass of the polymer is increased, and the behaviour can be approximated to an equation of the form

$$T_g^{\infty} = T_g + \frac{K}{M} \tag{16.20}$$

where
T_g^{∞} is the value of T_g for a polymer sample of infinite molar mass
K is a constant.

The form of this equation can be justified using the concept of free volume if it is assumed that there is more free volume involved with each chain end than a corresponding segment in the middle part of the chain. In a polymer sample of density ρ and number-average molar mass $\bar{M}_n$, the number of chains per unit volume is given by $\rho \mathbf{N}_A/\bar{M}_n$, and so the number of chain ends per unit volume is $2\rho \mathbf{N}_A/\bar{M}_n$, where $\mathbf{N}_A$ is the Avogadro number. If θ is the contribution of one chain end to the free volume, then the total fractional free volume due to chain ends f_c is given by

$$f_c = \frac{2\rho \mathbf{N}_A \theta}{\bar{M}_n} \tag{16.21}$$

It can be reasoned that if a polymer with this value of f_c has a glass transition temperature of T_g, then f_c will be equivalent to the increase in the free volume on expanding the polymer thermally between T_g and T_g^{∞}. This means that

$$f_c = \alpha_f (T_g^{\infty} - T_g) \tag{16.22}$$

where α_f is the thermal expansion coefficient of the free volume (Section 16.2.2). Combining Equations 16.21 and 16.22 and rearranging leads to the final result

$$T_g = T_g^{\infty} - \frac{2\rho \mathbf{N}_A \theta}{\alpha_f \bar{M}_n} \tag{16.23}$$

which is of exactly the same form as Equation 16.20 with $K = 2\rho \mathbf{N}_A \theta/\alpha_f$. The observed dependence of the T_g upon molar mass is, therefore, predicted from considerations of free volume.

A small number of branches on a polymer chain are found to reduce the value of T_g. This again can be analysed using the free-volume concept. It follows from the above analysis that the glass transition temperature of a chain possessing a total number of y ends per chain is given by an equation of the form

$$T_g = T_g^{\infty} - \frac{\rho y \mathbf{N}_A \theta}{\alpha_f \bar{M}_n} \tag{16.24}$$

where T_g^{∞} is again the glass transition temperature of a linear chain of infinite molar mass. Since a linear chain has two ends, the number of branches per chain is $(y - 2)$. Equation 16.24 is only valid

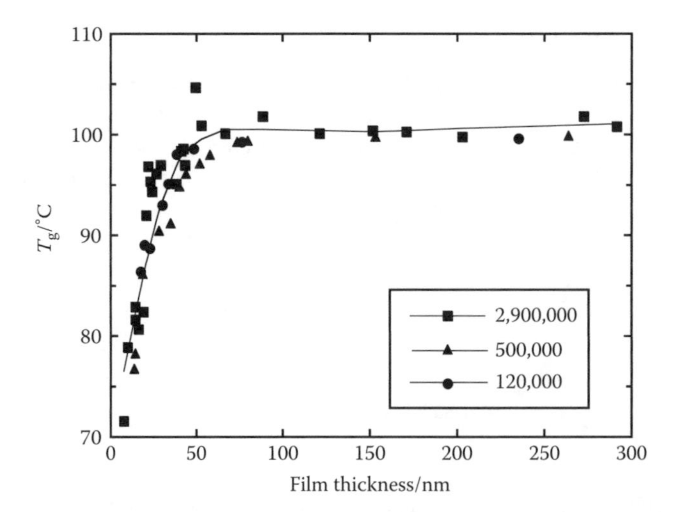

FIGURE 16.6 Measured glass transition temperatures for thin films of polystyrene of different indicated molar mass in g mol^{-1}. (Data taken from Keddie, J.L. et al., *Europhys. Lett.*, 27, 59, 1994.)

when the number of branches is low. A high density of branching will have the same effect as side groups in restricting chain mobility and hence raising T_g.

The presence of chemical *crosslinks* in a polymer sample has the effect of increasing T_g, although when the density of crosslinks is high the range of the transition region is broadened and the glass transition may not occur below the polymer degradation temperature in highly crosslinked materials. Crosslinking tends to reduce the specific volume of the polymer which means that the free volume is reduced, and so the T_g is raised because molecular motion is made more difficult.

16.3.4 Film Thickness

Recent studies have found that the glass transition temperature in thin polymer films can differ significantly from the bulk value. This is demonstrated in Figure 16.6 for thin films of polystyrene coated on single-crystal silicon wafers. The measurements were undertaken using ellipsometry, which is sensitive to the refractive index of the film, allowing accurate determination of the T_g in very thin films. It can be seen that for a 10 nm thick film T_g falls by up to 30 K from the bulk value. Even larger depression of the T_g has been reported for freestanding thin polymer films. This behaviour has been interpreted in terms of the onset of molecular mobility at the T_g. It is thought that there is considerably more mobility possible in the surface layers of polymers due to the lack of surrounding constraints. A 10 nm film has dimensions on the scale of the polymer coils and will have a high proportion of molecules near the surface. Hence the T_g, corresponding to the onset of general molecular mobility, will be reduced. It should be noted that there is no strong molar mass dependence found in the data shown in Figure 16.6 for the range of molar mass employed (10^5–10^6 g mol^{-1}), since the molar masses are relatively high. Although the behaviour shown here is typical of that of many polymers, in some cases the T_g may increase for thin films, especially where there is a strong interaction with the supporting surface.

16.4 MACROMOLECULAR DYNAMICS

Macromolecular dynamics is concerned with the way in which polymer chains are able to move, and this concept is essential for understanding of the onset of molecular motion at the T_g as well as a variety of other phenomena such as crystallization (Chapter 17) and viscoelasticity (Chapter 20).

In a gas, the small molecules move by translation in space, colliding with the containing walls and other molecules. In a liquid, small molecules also move by translation, but the distance between collisions is confined to molecular dimensions. The motion of long polymer molecules can be broken down into one of two different processes. Either the chain can change its overall conformation without moving its centre of mass or its centre of mass can move relative to that of the neighbouring molecules. Both types of motion occur though Brownian motion induced by thermal activation processes and can be considered as types of self diffusion. The diffusion of polymer chains is further constrained by their inability to move in a sideways direction as this is blocked by neighbouring chains; such motion is very slow and can only be achieved through cooperative processes between different chains.

16.4.1 The Rouse–Bueche Theory

One of the first theories on the motion of polymer chains was developed by Rouse and Beuche. It assumes that the molecule is long enough to be represented by a number of molecular subunits each of which obeys the Gaussian distribution function (Section 10.3.1). It is envisaged that these subunits can be represented by a series of beads connected by springs as shown in Figure 16.7. The springs are assumed to obey Hooke's law and also act as universal joints along the chain. It is thought that the dynamics of the polymer chains depends upon the interplay of *three* different forces acting on the repeat units.

- A frictional force, which is proportional to the relative velocity of the repeat unit and the surrounding medium.
- The force between adjacent repeat units along the chain maintaining connectivity.
- A random force on the chain from collisions with the surrounding medium leading to Brownian motion.

The Rouse–Bueche model allows the contribution of these forces upon chain dynamics to be determined and enables the prediction of behaviour such as the self diffusion of polymer molecules, stress-relaxation and creep (Chapter 20). It has been found to be particularly useful for the behaviour of dilute solutions (<1%) but is less successful for bulk polymers.

One of the important predictions of this model is that the viscosity of a polymer melt η should be proportional to the number of repeat units along the chain and hence the molar mass of the polymer M, i.e.

$$\eta \propto M \tag{16.25}$$

Such behaviour is found at low values of molar mass, as shown in Figure 16.8, but then for all molten polymers that are above some critical value of molar mass, M_{cr}, there is a much stronger dependence of the viscosity upon molar mass. Typically for high molar mass polymers

$$\eta \propto M^{3.5} \tag{16.26}$$

This change in behaviour has been accounted for in terms of the concept of chain *entanglements*, which lead to a slowdown in the chain dynamics. Polymers will only be in the form of isolated coils in dilute solution. As the concentration of the solution is increased, and in the melt, long polymer chains will become entangled due

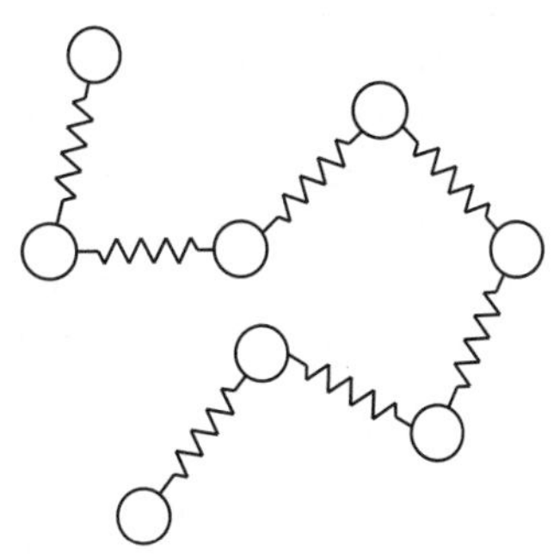

FIGURE 16.7 Schematic diagram of the Rouse–Bueche bead and spring model of a polymer chain. (Adapted from Sperling, L.H., *Introduction to Physical Polymer Science*, 4th edn., Wiley Interscience, Hoboken, NJ, 2006.)

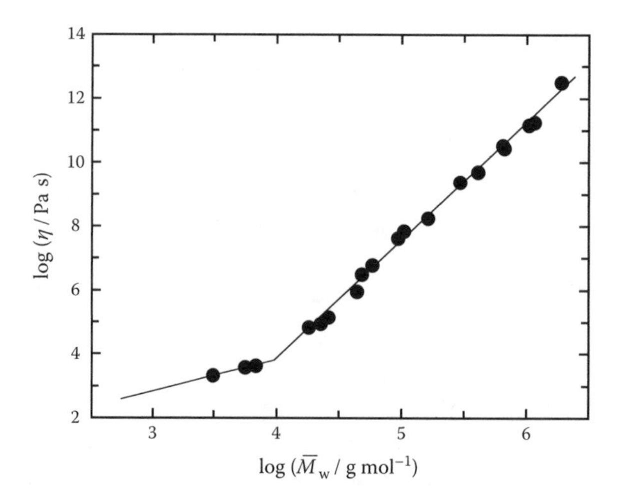

FIGURE 16.8 Viscosity of a series of monodisperse polyisoprene samples as a function of molar mass. (Data taken from Nemoto N. et al., *Macromolecules*, 5, 531, 1972.)

to the mutual uncrossability of the covalently-bonded backbone chains. The Rouse–Bueche model does not take into account the inability of chains to cross each other.

Entanglements manifest themselves in other situations in polymer science. For example, high molar mass elastomers may exhibit rubber elasticity (Chapter 21) in the uncrosslinked state, even if the polymer is completely soluble, due to the entanglements acting as physical crosslinks The effective molar mass between entanglements, M_e, can be determined from the plateau modulus (Figure 20.12) of these uncrosslinked elastomers using rubber elasticity theory. The strength of glassy thermoplastics also can decrease markedly for polymer samples below the critical molar mass for entanglements, M_e. The change in slope observed at a critical molar mass M_{cr} in Figure 16.8 is due to entanglements, but M_{cr} is not exactly the same as M_e. The relationship between the two parameters is a little complicated and thought to be $M_{cr} \cong 2M_e$.

16.4.2 The de Gennes Reptation Theory

Although the Rouse–Bueche theory was useful in that it established the relationship between chain motion and phenomena such as creep, relaxation and viscous flow, it was less successful in giving quantitative agreement of its predictions with experimental measurements. Over recent years attention has shifted to the approach of de Gennes who introduced the idea of the motion of polymer chains taking place by a process he termed *reptation*, shown schematically in Figure 16.9a. In this case, motion of the chain is impeded by a series of fixed obstacles that it is unable to cross, and the chain has to move through them in a snake-like manner—hence the term 'reptation' from the word reptile. The motion of the chain through the set of obstacles is therefore thought to be equivalent to it, being constrained to a tube as shown in Figure 16.9b. The molecule itself is allowed to adopt different conformations within the tube, and the length of the tube is considerably less than the contour length of the molecule.

Development of this model to predict different polymer properties was first achieved by Doi and Edwards, but the theory is complex. It is, however, relatively easy to predict the dependence of the melt viscosity upon molar mass. It is assumed that if a steady force f is applied to the chain, it will move along the tube at a velocity v. The mobility of the molecule in the tube is then defined as μ_{tube}, where

$$v = \mu_{tube} f \tag{16.27}$$

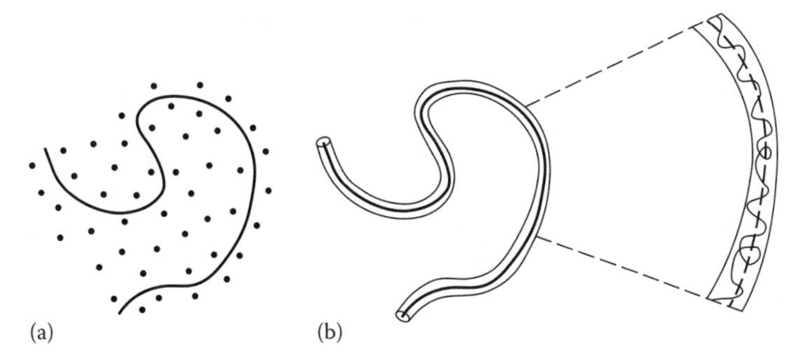

FIGURE 16.9 Reptation model of de Gennes. (a) Polymer chain surrounded by obstacles (b) The same chain in an equivalent tube, showing details of the conformation of the molecule in the tube (enlargement). (Adapted from de Gennes, P.G., *Scaling Concepts in Polymer Physics*, Cornell University Press, Ithaca, NY, 1979.)

If the number of repeat units in the chain is x, then the force must be proportional to x, so that

$$f = \frac{v}{\mu_{\text{tube}}} \propto x \tag{16.28}$$

and so

$$\mu_{\text{tube}} = \frac{\mu_1}{x} \tag{16.29}$$

where the parameter μ_1 is independent of chain length. The concept of a tube diffusion coefficient D_{tube} can now be introduced which is related to μ_{tube} by the Einstein relationship (c.f. Equation 10.99)

$$D_{\text{tube}} = \mu_{\text{tube}}\mathbf{k}T = \frac{\mu_1\mathbf{k}T}{x} = \frac{D_1}{x} \tag{16.30}$$

where

$\mathbf{k}$ is the Boltzmann constant
T is the temperature
D_1 is also independent of chain length.

In order to eliminate the original tube, the molecule must diffuse over a length comparable to its overall length L and using standard diffusion theory, the time τ taken for this to happen will be given by

$$\tau \cong \frac{L^2}{D_{\text{tube}}} \cong \frac{xL^2}{D_1} \propto x^3 \tag{16.31}$$

since L is a linear function of x.

The viscosity of the system related to the reptation of the chain is given by the relation $\eta = \tau E$, where E is a modulus that depends upon the distance between the obstacles and not the chain length. Hence, since $x \propto M$, the reptation model leads to the relationship

$$\eta \propto M^3 \tag{16.32}$$

This is a much stronger dependence of the viscosity upon molar mass than was predicted by the Rouse–Bueche model and, even though this derivation is relatively simple, it has been heralded as a great success of the reptation theory. It does not quite predict the measured exponent of ~3.5 found experimentally (Figure 16.8 and Equation 16.26), but it serves to demonstrate the nature of the dynamics of entangled polymer molecules. A further consequence of this approach is that it predicts that the reptation of branched chains will be very difficult and consequently they will have much higher viscosities in the melt than linear chains of the same molar mass, as is found in practice. Since many commercial polymers are non-linear, this finding is of considerable practical importance.

PROBLEMS

16.1 The refractive index, $\bar{n}$, of a polymer was determined as a function of temperature. The results are given below. Determine the glass transition temperature of the polymer.

T/°C	$\bar{n}$
20	1.5913
30	1.5898
40	1.5883
50	1.5868
60	1.5853
70	1.5838
80	1.5822
90	1.5801
100	1.5766
110	1.5725
120	1.5684
130	1.5643

16.2 It is intended to toughen poly(methyl methacrylate) (PMMA) by the addition of particles of rubbery poly[(*n*-butyl acrylate)-*co*-styrene] made by emulsion polymerization. In order to retain the good optical clarity of the PMMA, it is necessary to match the refractive indices $\bar{n}$ of the PMMA and copolymer particles. Estimate the relative proportions of the two comonomers required in the particles given that at room temperature the refractive indices of the homopolymers are PMMA, 1.489; poly(*n*-butyl acrylate), 1.466; and polystyrene, 1.591. Would you expect the material to retain its clarity if the temperature were changed from ambient?

16.3 Discuss the reasons for the differences in glass transition temperature for the following pairs of polymers with similar chemical structures.

(a) Polyethylene (~−120 °C) and polypropylene (~−20 °C).
(b) Poly(methyl acrylate) (10 °C) and poly(vinyl acetate) (32 °C).
(c) Poly(but-1-ene) (−24 °C) and poly(but-2-ene) (−73 °C).
(d) Poly(ethylene oxide) (−41 °C) and poly(vinyl alcohol) (85 °C).
(e) Poly(ethyl acrylate) (−24 °C) and poly(methyl methacrylate) (105 °C).

16.4 The following data were obtained for the values of glass transition temperature T_g of poly[(vinylidene fluoride)-*co*-chlorotrifluoroethylene] as a function of the weight fraction w_A of vinylidene fluoride. By fitting these data to a suitable curve estimate, the value of T_g for a copolymer with $w_A = 0.75$.

W_A	0	0.14	0.35	0.4	0.54	1.0
T_g/K	319	292	270	265	258	235

16.5 The glass transition temperature T_g of a linear polymer with number-average molar mass $\bar{M}_n = 2300$ was found to be 121 °C. The value of T_g increased to 153 °C for a sample of the same linear polymer with $\bar{M}_n = 9000$. A branched grade of the same polymer with $\bar{M}_n = 5200$ was found to have a T_g of 115 °C. Determine the average number of branches on the molecules of the branched polymer.

FURTHER READING

Booth, C. and Price, C. (eds), *Comprehensive Polymer Science*, Vol. 2, Pergamon Press, Oxford, U.K., 1989.

de Gennes, P.-G., *Scaling Concepts in Polymer Physics*, Cornell University Press, Ithaca, NY, 1979.

Doi, M., *Introduction to Polymer Physics*, Clarendon Press, Oxford, 1996.

Elias, H.-G., *Macromolecules 1: Structure and Properties*, 2nd edn., Plenum Press, New York, 1984.

Haward, R.N. (ed.), *The Physics of Glassy Polymers*, Applied Science Publishers, London, U.K., 1973.

Haward, R.N. and Young R.J. (eds.), *The Physics of Glassy Polymers*, 2nd edn., Chapman & Hall, London, 1997.

Jones R.A.L., *Soft Condensed Matter*, Oxford University Press, Oxford, U.K., 2002.

Shaw, M.T. and MacKnight, W.J., *Introduction to Viscoelasticity*, 3rd edn., John Wiley & Sons, Hoboken, NJ, 2005.

Sperling, L.H., *Introduction to Physical Polymer Science*, 4th edn., Wiley-Interscience, Hoboken, NJ, 2006.

17 The Crystalline State

17.1 INTRODUCTION

From the inception of polymer science, it has been recognized that polymer molecules possess the ability to crystallize. The extent to which this occurs varies with the type of polymer and its molecular microstructure. The main characteristic of crystalline polymers that distinguishes them from most other crystalline solids is that they are normally only *semi-crystalline*. This is self-evident from the fact that the density of a crystalline polymer is normally between that expected for fully crystalline polymer and that of amorphous polymer. The degree of crystallinity and the size and arrangement of the crystallites in a semi-crystalline polymer have a profound effect upon the physical and mechanical properties of the polymer, and in order to explain these properties fully, it is essential to have a clear understanding of the nature of these crystalline regions.

17.1.1 CRYSTALLINITY IN POLYMERS

The crystallization of polymers is of enormous technological importance. Many thermoplastic polymers will crystallize to some extent when the molten polymer is cooled below the melting point of the crystalline phase. This is a procedure that is done repeatedly during polymer processing and the presence of the crystals has an important effect upon polymer properties. There are many factors that can affect the rate and extent to which crystallization occurs for a particular polymer. They can be processing variables such as the rate of cooling, the presence of orientation in the melt and the melt temperature. Other factors include the tacticity and molar mass of the polymer, the amount of chain branching and the presence of any additives such as nucleating agents.

The obvious questions to ask concerning crystallinity in polymers are why and how do polymer molecules crystallize? The answer to the question why is given by consideration of the thermodynamics of the crystallization process. The Gibbs free energy G of any system is related to the enthalpy H and the entropy S by the equation, $G = H - TS$, where T is the thermodynamic temperature. The system is in equilibrium when G is a minimum. A polymer melt consists of randomly coiled and entangled chains. This gives a much higher entropy than if the molecules are in the form of extended chains since there are many more conformations available to a coil than for a fully extended chain. The higher value of S leads to a lower value of G. Now, if the melt is cooled to a temperature below the melting point of the polymer T_m, crystallization may occur. There is clearly a high degree of order in polymer crystals, as in any other crystalline material, and this ordering leads to a considerable reduction in S. This entropy penalty is, however, more than offset by the large reduction in enthalpy that occurs during crystallization. If the magnitude of the enthalpy change ΔH_m (latent heat) is greater than that of the product of the melting temperature and the entropy change ($T_m \Delta S_m$), crystallization will be favoured thermodynamically since a lower value of G will result.

As with all applications of thermodynamics, it can only be strictly applied to processes that occur quasi-statistically, i.e. very slowly. Polymers often are cooled rapidly from the melt, particularly when they are being processed industrially. In this situation, crystallization is controlled by kinetics and the rate at which the crystals nucleate and grow becomes important. With many crystallizable polymers, it is possible to cool the melt so rapidly that crystallization is completely absent and an amorphous glassy polymer results. With these systems, crystallization normally can be induced by annealing the amorphous polymer at a temperature between the glass transition temperature and the melting point T_m of the crystals.

The second question that was raised is how do polymer molecules crystallize? A useful way of imagining a molten polymer is as an enormous bowl of tangled, very long wriggling spaghetti. The actual strands of polymer spaghetti would have a length to diameter ratio typically of 10,000 to 1 rather than about 100 to 1 for spaghetti. It is quite remarkable that such a random mass could crystallize to give an ordered structure and clearly significant molecular rearrangement and considerable cooperative movement is required for crystallization to occur.

Well-defined examples of polymer crystals can be found on crystallization from dilute polymer solutions. Small, isolated lamellar crystals are obtained, although detailed measurements have shown that they are still only semi-crystalline. It is thought that in this case, the molecules are folded and the fold regions give rise to the non-crystalline component in the crystals. There is a higher degree of perfection in solution-grown polymer crystals than in the melt-crystallized counterparts. The polymer molecules are known to be in the form of isolated coils in dilute solution and crystallization is not hindered by factors such as entanglements.

Single crystals of many materials occur naturally (e.g. quartz and diamond). For other materials such as metals and semi-conductors, the growth of single crystals from the melt is now a routine matter. With polymers, it is only recently that true single crystals have been prepared. This can only be done by using the process of solid-state polymerization outlined in Section 8.2. The most perfect crystals are obtained with certain substituted diacetylene monomers that polymerize topochemically to give polymer crystals of macroscopic dimensions that can be 100% crystalline.

17.1.2 Crystal Structure and Unit Cell

Crystalline solids consist of regular three-dimensional arrays of atoms. In polymers, the atoms are joined together by covalent bonds along the macromolecular chains. These chains pack together side-by-side and lie along one particular direction in the crystals. It is possible to specify the structure of any crystalline solid by defining a regular pattern of atoms that is repeated in the structure. This repeating unit is known as the *unit cell* and the crystals are made up of stacks of the cells. In atomic solids such as metals, the atoms lie in simple close-packed arrays (like billiard or pool balls), and the structures can be specified by defining simple cubic or hexagonal unit cells containing only a few atoms. The packing units in molecular solids such as polymer crystals are more complicated than in atomic solids. In this case, the unit cells are made up of the repeating segments of the polymer chains packed together, often with several segments in each unit cell. Depending upon the complexity of the polymer chain, there can be tens or even hundreds of atoms in the unit cell. The spatial arrangement of the atoms is controlled by the covalent bonding within a particular molecular segment, with the polymer segments held together in the crystals by secondary van der Waals forces or hydrogen bonding. Since the polymer chains lie along one particular direction in the cell and there is only relatively weak secondary bonding between the molecules, the crystals have very anisotropic physical properties.

17.2 DETERMINATION OF CRYSTAL STRUCTURE

17.2.1 X-Ray Diffraction

When the crystal structure of a molecular compound is analysed, both the relative positions of the atoms on the molecular repeat units and the arrangement of these segments in the unit cell must be determined. This three-dimensional structure is normally determined using *X-ray diffraction*, involving measurement of the positions and intensities of all the diffraction maxima from a single crystal sample. Four-circle diffractometers allow the relative positions of all the atoms in crystals with quite complex structures to be determined as a matter of routine in a period of a few days in laboratory apparatus or in a few hours using synchrotron X-rays, but such techniques require good quality, relatively large crystals.

17.2.1.1 Polymer Single Crystals

For most conventional polymers, large single crystals are not available, but the complete crystal structures of several polydiacetylenes polymerized as macroscopic single crystals in the solid state (Section 8.2) have been determined. The degree of perfection shown by these polydiacetylene single crystals can be seen in Figure 17.1. This shows a single crystal rotation photograph obtained by

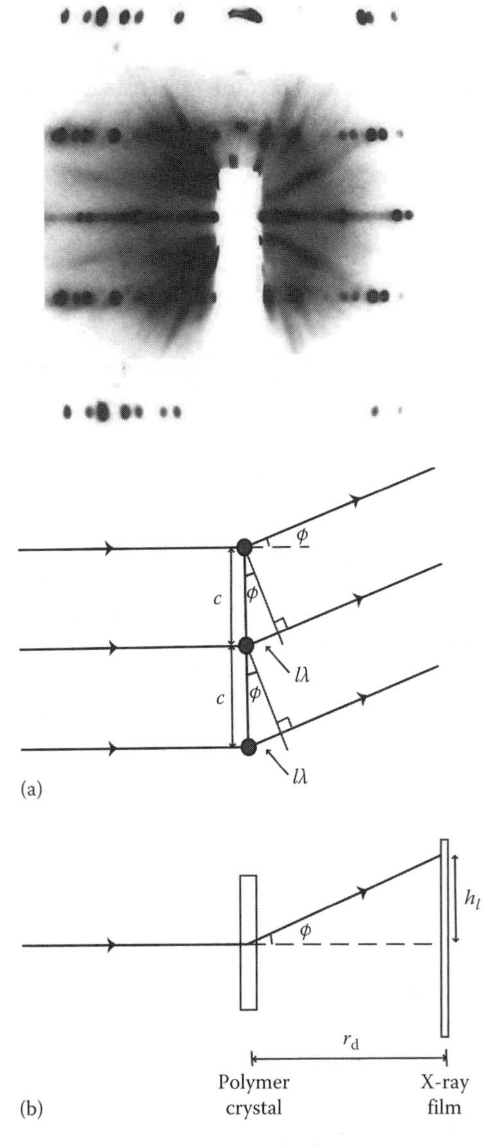

FIGURE 17.1 Single-crystal X-ray oscillation photograph obtained from a polydiacetylene single crystal. The chain direction is vertical. Schematic illustrations are shown of (a) the formation of the layer lines on an oscillation photograph for diffraction by a lattice in one dimension and (b) the experimental arrangement.

using a cylindrical film with the chain direction as the oscillation axis. The diffraction pattern is obtained by rotating the crystal about one particular axis and is in the form of spots lying along parallel layer lines. The lattice parameter in the direction of the rotation axis of the crystal can be determined directly from the layer line spacing. Since the rotation axis coincides with the chain direction, this is, of course, the chain direction repeat conventionally indexed for polymers as c in the unit cell. If the layer lines are indexed $l=0$ for the equatorial line and $l=1$ for the first layer etc., then c is simply related to the height h_l of the lth layer line above the equatorial line. The crystal lattice can be considered as acting as a diffraction grating, as shown in Figure 17.1a. The layer lines are then due to discrete differences in path length $l\lambda$, where λ is the wavelength of the radiation. The angle of deviation of the beam ϕ is therefore given as

$$\sin\phi = \frac{l\lambda}{c} \tag{17.1}$$

The angle ϕ is related to h_l and the radius of the film r_d (the specimen-to-film distance) through the simple geometrical construction shown in Figure 17.1b. This leads to the relation

$$\tan\phi = \frac{h_l}{r_d} \tag{17.2}$$

Combining Equations 17.1 and 17.2 gives the lattice parameter c as

$$c = \frac{l\lambda}{\sin\tan^{-1}(h_l/r_d)} \tag{17.3}$$

17.2.1.2 Semi-Crystalline Polymers

Macroscopic single crystals are a special case and large single crystals of conventional polymers are not available. Crystal-structure determinations for semi-crystalline polymers are less precise and the sophisticated methods of crystal-structure determination cannot be used without modification. Instead of using single crystals, samples normally are prepared in the form of highly oriented, drawn or rolled fibres. They can also be analysed by obtaining rotation photographs, but flat-plate X-ray patterns are more usually obtained. Nowadays, two-dimensional CCD cameras often are employed to obtain the diffraction patterns digitally, as shown schematically in Figure 17.2. The diameters of the rings in the diffraction pattern in Figure 17.2b can be related to the spacing of the crystal planes d_{hkl} through the scattering geometry shown in Figure 17.2a and Bragg's law:

$$n\lambda = 2d_{hkl}\sin\theta \tag{17.4}$$

where

n is the order of the ring
θ is the scattering angle.

X-ray diffraction patterns from melt-crystallized polymers are usually in the form of rings superimposed on a diffuse background, as shown in Figure 17.3a. This background indicates the presence of a non-crystalline (amorphous) phase and the rings indicate that there is a second phase consisting of small randomly-oriented crystallites also present. A flat film pattern of oriented polypropylene is shown in Figure 17.3b. This type of X-ray photograph often is called a fibre pattern and they are characteristic of oriented fibres of semi-crystalline polymers. They are analogous to rotation photographs, but the layer lines are in the form of hyperbolae because the film is flat rather than cylindrical. It is not necessary to rotate the polymer fibres because they are polycrystalline with a spread of crystal orientation about the fibre axis. In polymer fibres, the polymer molecules are oriented approximately

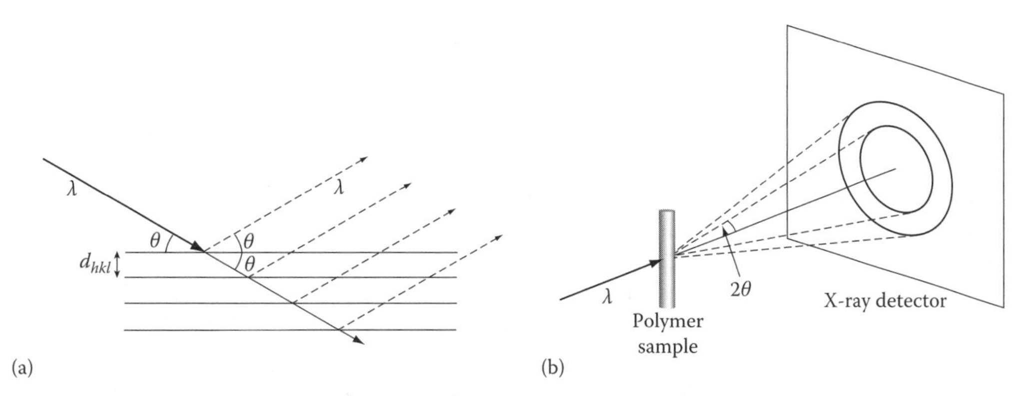

FIGURE 17.2 Schematic illustration of (a) the diffraction of X-rays by a crystal lattice and (b) X-ray diffraction from a polycrystalline polymer sample.

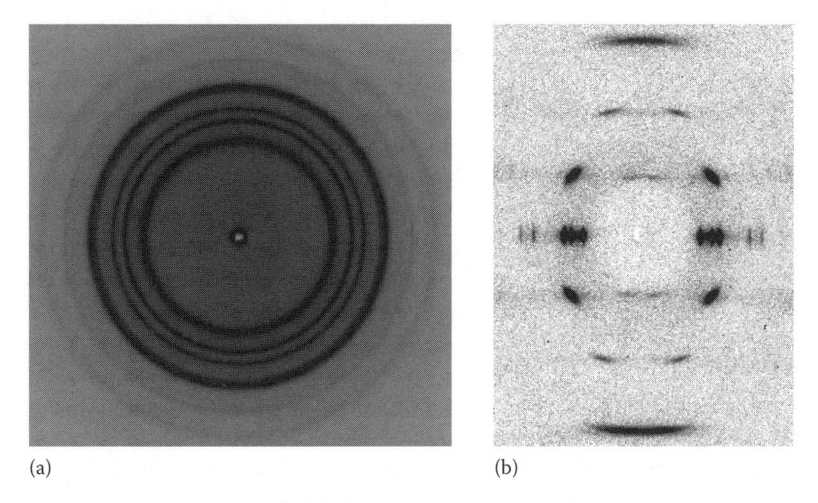

FIGURE 17.3 Flat-plate X-ray diffraction patterns obtained from polypropylene samples: (a) isotropic melt-crystallized material (Courtesy of Y.-T. Shyng, University of Manchester, Manchester, U.K.) and (b) an oriented polypropylene fibre. (Courtesy of C. Meakin, University of Manchester, Manchester, U.K.)

parallel to the fibre axis and since this is parallel to a crystal axis (normally defined as *c*), the fibre pattern is equivalent to a rotation photograph about this crystallographic axis for a single crystal sample. It is possible to analyse the layer lines in fibre patterns in a similar way to rotation photographs if certain precautions are taken. Equation 17.3 can only be applied when h_l is measured directly above and below the central spot and if r_d is taken as the specimen-to-film distance rather than the film radius. Since the chain direction normally coincides with the fibre axis in oriented polymer samples, the dimensions of the *c*-axis in the unit cell can be found directly from the height of the *l*th layer line.

In order to determine other unit cell dimensions it is necessary to assign the crystallographic indices (*hkl*) to each spot in the fibre pattern. This is often difficult to do for single crystal patterns and there are even greater problems with semi-crystalline polymers. The reflections tend to be in the form of arcs rather than spots because the orientation of the crystals is somewhat imperfect. Also they tend to be rather diffuse because the dimensions of the crystallites are usually small (a few tens of nanometres) and they contain many lattice imperfections. It is sometimes possible to improve the resolution of the diffraction patterns by using doubly oriented samples. These are tapes that have been rolled as well as stretched to induce orientation in perpendicular directions within the plane of the tape. They then have a three-dimensional texture rather than the normal cylindrical symmetry possessed by fibres.

Once the spacings and intensities of all the (*hkl*) reflections or rings have been measured as accurately as sample perfection allows, the polymer crystallographer does not normally have sufficient information to determine the crystal structure of the polymer and so has to resort to a series of educated guesses as to what this structure may be. There is now a considerable amount of experience in this area and some empirical rules have been noted to help with crystal structure determination. The polymer chains are normally in their lowest energy conformations within the crystal as the bonding in the crystal is not sufficiently strong to perturb the nature of the chain. The chains pack into the crystal so as to fill space as efficiently as possible to maximize the intermolecular forces. The van der Waals' radii of the pendant atoms on the chain limit the distance of closest approach of each chain.

It also is possible to call upon other information that is generally available. The structures of chemically similar polymers often are known and these can be useful as a starting-point for crystal structure determination. The stereochemical nature of the polymer chains depends upon the method of synthesis (Chapter 6) and knowledge of this is important. Spectroscopic techniques (Chapter 15) give the detailed microstructure of the polymer molecules and information on the conformation and packing of the chains within the crystals. Also, it is possible to determine the lowest-energy crystal structure by taking into account inter- and intra-molecular interactions for different structural models.

Once a promising structure has been found, it is necessary to compare the measured positions and intensities of the (*hkl*) reflections with those predicted for the structure that is proposed. The agreement that is obtained is never perfect and normally one has to refine the proposed structure until a structure with the best fit to the measured data is obtained. This means that the crystal structures that are quoted in the following sections tend to be somewhat idealized and in many cases better fits almost certainly are possible.

17.2.2 Polymer Crystal Structures

The crystal structures of several hundred polymer molecules have now been determined. Those of a few common polymers are listed in Table 17.1. The name and chemical repeat unit of the polymer are given in column one. The unit cell of any crystal structure can be assigned to one of the seven basic crystal systems (triclinic, monoclinic, etc.) and this is given as the first entry in column two. The packing and symmetry of the polymer segments in the unit cell allows the structure to be assigned to one of the 230 space groups that are given as the second entry in column two. Polymer molecules pack into crystals either in the form of zig-zags or helices and this also is indicated in column two. The nomenclature used to describe the zig-zag or helix is $A*u/t$, where A represents the class of the helix given by the number of skeletal atoms in the asymmetric unit of the chain. The parameter u represents the number of these units on the helix corresponding to the crystallographic (chain direction) repeat and t is the number of turns of the helix in this crystallographic repeat. This nomenclature can be used to describe both helices and zig-zags. For example, if polyethylene is considered as polymethylene $(-CH_2-)_n$, the 1*2/1 helix represents a planar zig-zag. Alternatively, it could be considered as $(-CH_2-CH_2-)_n$ and the zig-zag chain conformation would be designated 2*1/1. The unit cell dimensions are listed in column three in the order *a*, *b*, *c* and the chain axis is indicated in italics. The angles between the axes of the unit cell α, β and γ are given in order in column four. The number of chain repeat units per unit cell is given in column five, and the densities ρ_c of the crystals determined from the crystal structure are given in the final column.

17.2.3 Factors Determining Crystal Structure

There are certain structural requirements that are essential before a polymer molecule can crystallize. It is necessary for the polymer chains to be linear, although a limited number of branches on a polymer chain tend to limit the extent of crystallization but do not stop it completely. In a similar way, copolymerization affects the symmetry perfection of a polymer chain, but other repeat units can be tolerated to a limited extent. One of the most important factors that determines the extent of

TABLE 17.1
Details of the Crystal Structures of Various Common Polymers

Polymer Repeat Unit	Crystal System, Space Group, Mol. Helix	Unit Cell Axes/nm	Unit Cell Angles	No. Units	ρ_c/Mg m^{-3}
Polyethylene I	Orthorhombic	0.7418	90°	4	0.9972
$-CH_2-$	Pnam	0.4946	90°		
	1*2/1	*0.2546*	90°		
Polyethylene II	Monoclinic	0.809	90°	4	0.998
$-CH_2-$	C2/m	*0.253*	107.9°		
	1*2/1	0.479	90°		
Polytetrafluoroethylene I	Triclinic	0.559	90°	13	2.347
$-CF_2-$	P1	0.559	90°		
	1*13/6	*1.688*	119.3°		
Polytetrafluoroethylene II	Trigonal	0.566	90°	15	2.302
$-CF_2-$	$P3_1$ or $P3_2$	0.566	90°		
	1*15/7	*1.950*	120°		
Polypropylene (iso)	Monoclinic	0.666	90°	12	0.946
$-CH_2-CHCH_3-$	$P2_1/c$	2.078	99.62°		
	2*3/1	*0.6495*	90°		
Polystyrene (iso)	Trigonal	2.19	90°	18	1.127
$-CH_2-CHC_6H_5-$	$R\bar{3}c$	2.19	90°		
	2*3/1	*0.665*	120°		
Polypropylene (Syndio)	Orthorhombic	1.450	90°	8	0.930
$-CH_2-CHCH_3-$	$C222_1$	0.560	90°		
	4*2/1	*0.740*	90°		
Poly(vinyl chloride) (Syndio)	Orthorhombic	1.040	90°	4	1.477
$-CH_2-CHCl-$	Pbcm	0.530	90°		
	4*1/1	*0.510*	90°		
Poly(vinyl alcohol) (atac)	Monoclinic	0.781	90°	2	1.350
$-CH_2-CHOH-$	P2/m	*0.251*	91.7°		
	2*1/1	0.551	90°		
Poly(vinyl fluoride) (atac)	Orthorhombic	0.857	90°	2	1.430
$-CH_2-CHF-$	Cm2m	0.495	90°		
	2*1/1	*0.252*	90°		
Poly(4-methyl-1-pentene) (iso)	Tetragonal	2.03	90°	28	0.822
$-CH_2-CH(CH_2-CH(CH_3)_2)-$	$P\bar{4}$	2.03	90°		
	2*7/2	*1.38*	90°		
Poly(vinylidene chloride)	Monoclinic	0.673	90°	4	1.957
$-CH_2-CCl_2-$	$P2_1$	*0.468*	123.6°		
	4*1/1	1.254	90°		
1,4-Polyisoprene (*cis*)	Orthorhombic	1.246	90°	8	1.009
$-CH_2-CCH_3=CH-CH_2-$	Pbac	0.886	90°		
	8*1/1	*0.81*	90°		
1,4-Polyisoprene (*trans*)	Orthorhombic	0.783	90°	4	1.025
$-CH_2-CCH_3=CH-CH_2-$	$P2_12_12_1$	1.187	90°		
	4*1/1	*0.475*	90°		
Polyoxymethylene I	Trigonal	0.4471	90°	9	1.491
$-CH_2-O-$	$P3_1$ or $P3_2$	0.4471	90°		
	2*9/5	*1.739*	120°		

(*continued*)

TABLE 17.1 (continued)
Details of the Crystal Structures of Various Common Polymers

Polymer Repeat Unit	Crystal System, Space Group, Mol. Helix	Unit Cell Axes/nm	Unit Cell Angles	No. Units	ρ_c/Mg m^{-3}
Polyoxymethylene II	Orthorhombic	0.4767	90°	4	1.533
$-CH_2-O-$	$P2_12_12_1$	0.7660	90°		
	2*2/1	*0.3563*	90°		
Poly(ethylene terephthalate)	Triclinic	0.456	98.5°	1	1.457
$-(CH_2)_2-O-CO-C_6H_4-CO-O-$	$P\bar{1}$	0.596	118°		
	12*1/1	*1.075*	112°		
Nylon 6, α	Monoclinic	0.956	90°	8	1.235
$-(CH_2)_5-CO-NH-$	$P2_1$	*1.724*	67.5°		
	7*2/1	0.801	90°		
Nylon 6, γ	Monoclinic	0.933	90°	4	1.173
$-(CH_2)_5-CO-NH-$	$P2_1/a$	*1.788*	121°		
	7*2/1	0.478	90°		
Nylon 6.6, α	Triclinic	0.49	48.5°	1	1.24
$-(CH_2)_6NH-CO-$	$P\bar{1}$	0.54	77°		
$(CH_2)_4-CO-NH-$	14*1/1	*1.72*	63.5°		
Nylon 6.6, β	Triclinic	0.49	90°	2	1.25
$-(CH_2)_6NH-CO-$	$P\bar{1}$	0.80	77°		
$(CH_2)_4-CO-NH-$	14*1/1	*1.72*	67°		

Source: Adapted from Wunderlich, B., *Macromolecular Physics*, Vols. 1–3, Academic Press, London, 1973, 1976 and 1980.

crystallization and crystal structure is *tacticity* (Section 6.2). In general, isotactic and syndiotactic polymers will crystallize whereas atactic polymers are non-crystalline, although there are some exceptions to this rule. The most useful way of considering the structural factors controlling the crystallization of polymers is to look at specific examples.

17.2.3.1 Polyethylene

The most widely studied crystalline polymer is *polyethylene*. The lowest-energy chain conformation of this macromolecule can be evaluated from considerations of the conformations of the *n*-butane molecule (Figure 10.6). A series of all-*trans* bonds produce the lowest-energy conformation in *n*-butane because of steric interactions and this is reflected in polyethylene where the most stable molecular conformation is the crystal with all-*trans* bonds, i.e. the planar zig-zag. These molecules then pack into the orthorhombic unit cell given in Table 17.1. The arrangement of the molecules in the polyethylene crystal structure is illustrated in Figure 17.4 with the polymer molecules lying parallel to the *c*-axis. The molecules are held in position in the cell by the secondary van der Waals bonds between the chain segments. The interactions between the H atoms on the polymer chain determine the setting angle of the molecules in the cell. This is the angle that the molecular zig-zags make with the *a*- or *b*-axes. Orthorhombic polyethylene (I) is the most stable crystal structure but a monoclinic modification (II) can be formed by mechanical deformation of orthorhombic polyethylene. The molecules are still in the form of a planar zig-zag and the chain direction repeat and crystal density are virtually unchanged. The two structures differ just in the way the molecules are packed in the unit cells. *Polymorphism*, i.e. the existence of more than one type of crystal structure, due to deformation or caused by different crystallization conditions, is not uncommon in polymers and other examples will be discussed below.

The copolymerization of ethylene with small amounts of propylene produces a branched form of polyethylene that is still capable of crystallization. The orthorhombic crystal structure is maintained

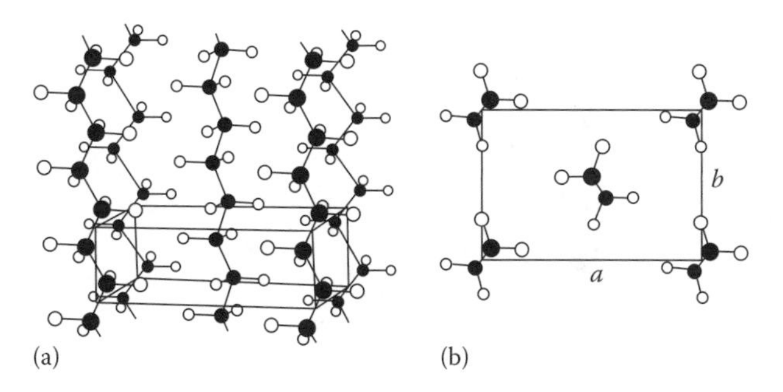

FIGURE 17.4 Crystal structure of orthorhombic polyethylene. (a) General view of the unit cell. (b) Projection of the unit cell parallel to the chain direction, *c*. (• = carbons atoms, o = hydrogen atoms).

and the *b*- and *c*-axes are almost unchanged. On the other hand, the *a* crystal axis undergoes an expansion which depends upon the proportion of $-CH_3$ groups on the polymer chain.

17.2.3.2 Polytetrafluoroethylene

Polytetrafluoroethylene exists in at least two crystal modifications. The low-temperature form I is stable below 19°C and above this temperature the trigonal II form is found. The polymer can be considered as an analogue of polyethylene and it would undoubtedly have a similar crystal structure if the F atoms were not just too large to allow a planar zig-zag to be stable. The F atoms are significantly larger than H atoms, and there is a considerable amount of steric repulsion when $(-CF_2-)_n$ is in the form of a planar zig-zag. Because of this, the molecule adopts a helical conformation that allows the larger F atoms to be accommodated. Below 19°C, the molecules are in the form of a 13/6 helix and at higher temperatures, they untwist slightly into a 15/7 helix. The two types of polytetrafluoroethylene helices have very smooth molecular profiles, as molecular models confirm and can be considered as cylinders. This smooth profile has been suggested as being responsible for the good low-friction properties of this polymer. The molecular cylinders pack very efficiently into the crystal structures in a hexagonal array, rather like a stack of pencils, and give rise to unit cells which have hexagonal dimensions (Table 17.1). They are not, however, assigned hexagonal crystal structures because the twisted molecules do not have a sufficiently high degree of symmetry. *Poly(vinyl fluoride)* is a related polymer which only has 25% of the F atoms possessed by polytetrafluoroethylene. Consequently, it does not suffer from the steric repulsion of the pendant atoms and even the atactic form adopts a crystal structure similar to polyethylene I (Table 17.1). In fact, the *b* and *c* unit cell dimensions are virtually identical to those of polyethylene, but the *a* repeat is considerably larger, to accommodate the bigger F atoms, as for the ethylene–propylene copolymers.

17.2.3.3 Vinyl Polymers

It is found that atactic vinyl polymers $(-CH_2-CHX-)_n$ will not crystallize when X is large. We have just seen that when X = F, crystallization can still take place. This also is the case for atactic *poly(vinyl alcohol)*. Since the –OH group is relatively small, the molecules crystallize in the form of a planar zig-zag into a distorted monoclinic version of the polyethylene crystal structure. In general, it is found that vinyl polymers must be either isotactic or syndiotactic before crystallization can occur. Isotactic vinyl polymers invariably crystallize with the polymer molecules in a helical conformation to accommodate the relatively large side groups. For example, the molecules in the isotactic forms of both *polystyrene* and *polypropylene* are in the form of a 3/1 helix, as illustrated schematically in Figure 17.5. This helix is formed readily by having alternating *trans* and *gauche* positions (Section 10.3.2) along the polymer chain. If the side groups are large and bulky, they require more space than is available in the 3/1 helix and the molecules form looser helices.

The packing of the helices in isotactic vinyl polymers depends upon the type of helix present and the nature and size of the side groups. The packing is dictated by the intermeshing of the side

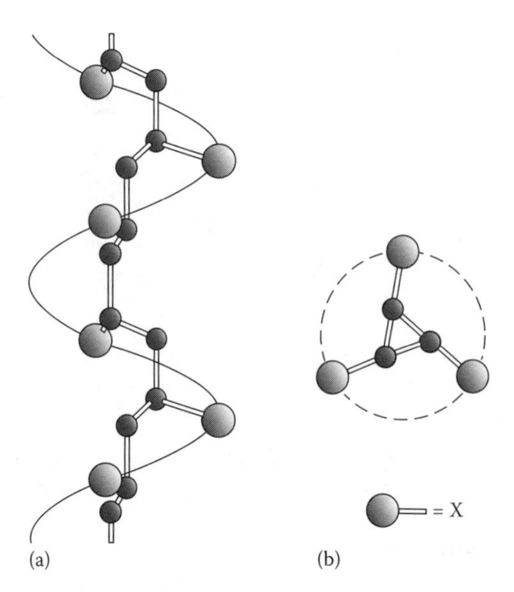

FIGURE 17.5 Schematic representation of the 3/1 helix found for isotactic vinyl polymers of the type $(-CH_2-CHX-)_n$ where, for example, X=$-CH_3$, $-C_6H_5$, $-CH{=}CH_2$, etc. (a) Side view of helix. (b) View along helix.

groups, but also is complicated by the possibility of having two different kinds of helix, both of which can be either right- or left-handed. The trigonal crystal structure of isotactic polystyrene (Table 17.1) is typical of many 2*3/1 helices whereas the monoclinic structure of polypropylene is somewhat exceptional. In *syndiotactic* vinyl polymers the conformation of the molecules again is controlled by the size of the side groups. For example, the Cl atoms in syndiotactic *poly(vinyl chloride)* are sufficiently small to allow the backbone to remain in a planar zig-zag.

17.2.3.4 Polyamides

The final types of crystal structure that will be considered in detail are those of *nylons* (aliphatic polyamides) where hydrogen bonding, which can be many times stronger than van der Waals bonding, controls chain packing. In both forms of *nylon 6* and *nylon 6.6*, the molecules are in the form of extended planar zig-zags joined together in hydrogen-bonded sheets. An example of this is given for nylon 6.6 in Figure 17.6a where the hydrogen bonds between oxygen atoms and >NH groups in adjacent molecules are indicated by dashed lines. The α and β forms of this polymer differ in the way in which the hydrogen-bonded sheets are stacked, as illustrated in Figure 17.6b and c. In the α form, successive planes are displaced in the *c* direction, whereas in the β form, the hydrogen-bonded sheets have alternating up and down displacements. The situation is slightly different in nylon 6 where the α and γ forms differ in the orientation of the molecules in the hydrogen-bonded sheets. They are anti-parallel in the α form and parallel to each other in the γ form. This difference in orientation leads to a slightly shorter chain direction repeat in the γ form.

17.3 POLYMER SINGLE CRYSTALS

17.3.1 Solution-Grown Single Crystals

Although it has been known for many years that crystallization can occur in polymers, it was not until 1957 that the preparation of isolated polymer single crystals was first reported, having been grown by precipitation from dilute solution. The first single crystals prepared were of polyethylene but, in the years since, single crystals of nearly every known soluble crystalline polymer have been prepared. Polyethylene crystals are, by far, the most widely studied examples of polymer single crystals and most of the discussion here will be concerned with polyethylene crystals.

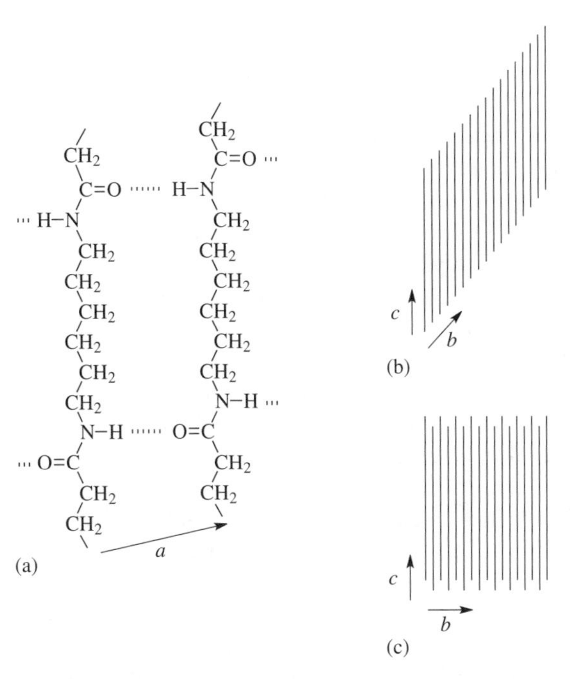

FIGURE 17.6 Schematic representation of the α and β forms of nylon 6.6. (a) Molecular arrangement of hydrogen-bonded sheets common to both forms. (b) Stacking of sheets in the α form, viewed parallel to the sheets. (c) Stacking of the sheets in the β form. (Adapted from Wunderlich, B., *Macromolecular Physics*, Vols. 1–3, Academic Press, London, 1973, 1976 and 1980.)

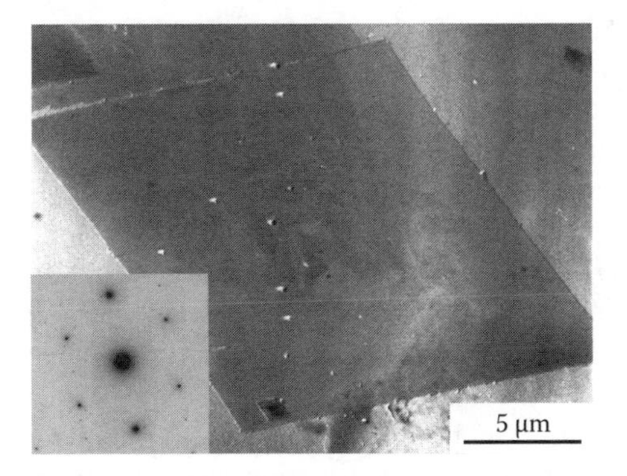

FIGURE 17.7 Solution-grown polyethylene single crystal. Electron micrograph of a lamellar crystal (Courtesy of P. Allan, Brunel University, Middlesex, U.K.) with a selected-area diffraction pattern. (Courtesy of I.G. Voigt–Martin, University of Mainz, Mainz, Germany.)

A characteristic feature of solution-grown polymer crystals is that they are small. They can usually just about be resolved in an optical microscope, but they are most readily examined in an electron microscope. Figure 17.7 shows an electron micrograph of a polyethylene single crystal grown by precipitation from dilute solution. The crystals normally are precipitated either by cooling a hot solution, which is the most widely used method, or by the addition of a non-solvent. The polyethylene crystal in Figure 17.7 is plate-like (lamellar) and of the order of 10 μm across. The crystal thickness can be measured by shadowing the crystal under vacuum with a heavy metal such as gold, at a known angle. A simple geometrical construction allows the crystal thickness to be calculated from the length of the shadow. For most solution-grown crystals, the thickness is found to be of the order of 10 nm. It also can be seen from Figure 17.7 that the lamellar crystals have straight edges that make angles that are characteristic for a given polymer. This regular shape reflects the underlying crystalline nature of the lamellar crystals.

The individual single crystals are too small to be studied with X-rays and the relative orientation of the polymer molecules within the lamellar crystals can only be determined by electron diffraction. This can conveniently be done in an electron microscope set up for selected-area electron diffraction. An electron diffraction pattern from a polyethylene single crystal with the electron beam perpendicular to the crystal surface is shown in the inset of Figure 17.7. The diffraction pattern consists of a regular array of discrete spots and there is no evidence of an amorphous halo. This means that the entity in Figure 17.7 diffracts in the same way as single crystals of non-polymeric solids such as metals. Because of this, the diffraction pattern can be analysed in the standard way and so it can be considered as corresponding to a section of the reciprocal lattice perpendicular to the crystallographic direction parallel to the electron beam. The diffraction pattern in Figure 17.7 can be indexed as that expected for conventional orthorhombic polyethylene (Table 17.1) and it corresponds to a beam direction of [001]. Since the chain direction in orthorhombic polyethylene is c, it means that the molecular axis in polyethylene single crystals is *perpendicular to the crystal surface*. This observation is somewhat surprising since, as the polymer molecules are many hundreds of nanometers long and the crystals only ~10 nm thick, it must mean that the polymer molecules *fold back and forth* between the top and bottom surfaces of the lamellae. Although the example given in Figure 17.7 is for polyethylene, it has been found that solution-grown crystals of polymers are normally lamellar and that the molecules always are folded such that they are perpendicular (or nearly perpendicular) to the crystal surface.

A distinctive feature of polymer single crystals is that the crystal edges are normally straight and make well-defined angles with each other. It is also found from electron diffraction that the edge faces correspond to particular crystallographic planes. The crystal shown in Figure 17.7 has four {110} type edge faces. Sometimes, polyethylene crystals are truncated and can have {100} edges as well. The different types of polyethylene crystals are shown schematically in Figure 17.8. It also is found that the crystals are divided into sectors. The crystal illustrated in Figure 17.8a shows four {110} sectors and that in Figure 17.8b has two additional {100} sectors. This sectorization gives a clue concerning the nature of the folding that will be discussed later.

Measurements of the density and other properties of solution-grown polymer crystals have shown that they are not perfect single crystals. The density of the crystals is always less than the theoretical density, which means that non-crystalline material must also be present in the crystals. It is generally thought that the bulk of this non-crystalline component resides in the *fold surfaces* of the crystals. Examination of molecular models of the folding molecule shows that if bonds are to remain unstrained, the folding can be achieved in polyethylene within the space of not less than five bonds with three of them in *gauche* conformations and it is obvious that the packing of folded

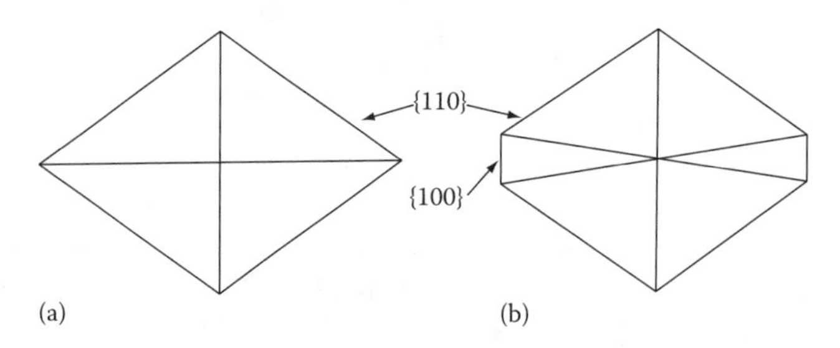

FIGURE 17.8 Schematic diagrams of two forms of lamellar single crystals of polyethylene. (a) Crystal with {110} sectors. (b) Crystal with {100} sectors as well.

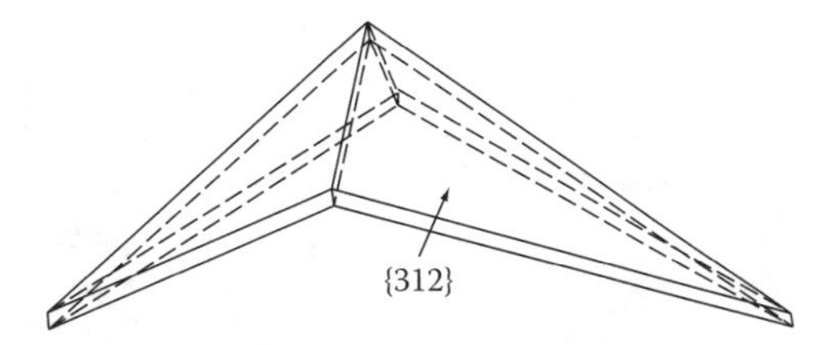

FIGURE 17.9 Schematic representation of a pyramidal polyethylene single crystal. (Adapted from Schultz, J.M., *Polymer Materials Science*, Prentice Hall, Englewood Cliffs, NJ, 1974.)

molecules must be less efficient than that of extended molecules. The non-crystalline nature of the fold surface is also reflected in the observation that single crystals of polyethylene and other polymers are not always flat. Such crystals can often be seen in a collapsed form in the electron microscope containing pleats or corrugations. The true form of the crystals was only found by direct observation of single crystals suspended in a liquid using optical microscopy. It was shown that some polyethylene crystals were in the form of a hollow pyramid, as shown schematically in Figure 17.9. The fold surface in pyramidal polyethylene single crystals has been shown to be close to the {312} plane of the polyethylene unit cell. The exact form of the single crystals depends upon several factors, such as the temperature at which the crystals were grown. Pyramidal crystals tend to grow at high temperatures whereas corrugated crystals form at lower crystallization temperatures.

There has been considerable argument over the years concerning the way in which folding occurs and the nature of the fold plane. The different models that have been suggested to account for folding in polymer crystals are illustrated schematically in Figure 17.10. The models range from random re-entry ones, where a molecule leaves and re-enters a crystal randomly, to adjacent re-entry models, whereby molecules leave and re-enter the crystals in adjacent positions. Two particular adjacent re-entry models have been suggested where the folds are either regular and tight, or irregular and of variable length.

The main consensus of opinion appears to be that in single crystals grown from dilute solution the molecules undergo adjacent re-entry. This leads to there being a crystallographic plane on which the molecules fold, normally known as a *fold plane*. In the polyethylene crystals drawn in Figure 17.8a, it is thought that the fold planes are parallel to the crystal edges. This means that the fold planes must be of the type {110} and in truncated crystals (Figure 17.8b) there are thought to be sectors of {100} folding as well. The fold surface itself is thought to be fairly irregular, consisting of a mixture of loose and tight folds and chain ends.

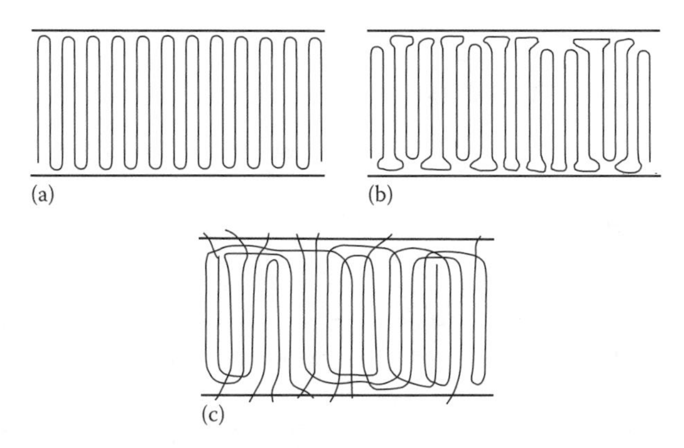

FIGURE 17.10 Schematic illustrations of the different types of folding suggested for polymer single crystals. (a) Adjacent re-entry with sharp folds. (b) Adjacent re-entry with loose folds. (c) Random re-entry or 'switchboard' model.

(a)

(b)

FIGURE 17.11 (a) Photograph of macroscopic polydiacetylene single crystals fibres on graph paper with a millimetre spacing. (b) High-resolution transmission electron micrograph showing the lattice image of a polydiacetylene single crystal (the polymer chain direction *c* is vertical).

17.3.2 Solid-State Polymerized Single Crystals

It is possible to prepare true 100% crystalline polymer crystals by the process of solid-state polymerization described in Section 8.2. An example of polydiacetylene single crystal fibres formed by solid-state polymerization is shown in Figure 17.11a. The crystals can also be obtained in the form of lozenges. Both types have regular facets, are of macroscopic dimensions and so are similar to single crystals that can be prepared for non-polymeric materials such as metals, ceramics or semiconductors. A typical X-ray rotation photograph of a similar polydiacetylene crystal was given in Figure 17.1, and it indicates that in such crystals, the molecules are in a chain-extended conformation. The polymer molecules are parallel to the fibre axis in the fibrous crystals and examination of the fibres using high-resolution transmission electron microscopy confirms this, as shown in Figure 17.11b. This is an image of the crystal lattice showing perfectly aligned polymer chains, with an inter-chain spacing of the order of 1.2 nm. Density measurements have shown that the crystals have their theoretical densities calculated from knowledge of the crystal structure and so it can be concluded that the crystals are virtually defect-free with chain-folding being completely absent. Such crystals have allowed considerable advances to be made in the understanding of the fundamental physical properties of crystalline polymers. In particular, important new aspects of the deformation and fracture of polymer crystals have been revealed.

17.4 SEMI-CRYSTALLINE POLYMERS

The isolated chain-folded lamellar single crystals described in Section 17.3.1 are obtained only by crystallization from dilute polymer solutions. As the solutions are made more concentrated, the crystal morphologies change and more complex crystal forms are obtained. In dilute solution, the polymer coils are isolated from each other, but as it becomes more concentrated the molecules become entangled. Particular chains, therefore, have the probability of being incorporated in more than one crystal, giving rise to inter-crystalline links. Crystals grown from concentrated solutions are often found either as lamellae with spiral overgrowth, as shown in Figure 17.12 or as aggregates of lamellar crystals. Under certain conditions, the crystals can be even more complex showing twinned and/or dendritic structures similar to the morphologies found in non-polymeric crystalline solids.

This particular section is concerned with the structures that are formed principally by crystallization from the melt. This is not fundamentally different from solution crystallization, especially

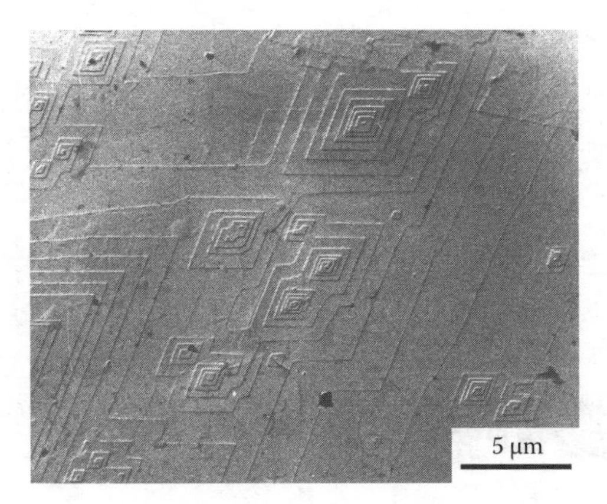

FIGURE 17.12 Electron micrograph of polyethylene crystals grown from CCl_4 at 80°C showing large growth spirals. The micrograph was obtained by replication and the replica has been shadowed with a heavy metal. (Courtesy of G. Lieser, MPI, Mainz, Germany.)

when the solutions are concentrated. In the melt, chain entanglements are of extreme importance and consequently the crystals that form are more irregular than those obtained from dilute solution. A feature common to both melt and solution crystallization is that the solid polymer is only *semi-crystalline* and contains both crystalline and amorphous components.

17.4.1 Spherulites

If a melt-crystallized polymer is prepared in the form of a thin film, either by sectioning a bulk sample or casting the film directly, and then viewed in an optical microscope using polarized light, a characteristic structure is normally obtained. It consists of entities known as *spherulites* that show a characteristic Maltese cross pattern in polarized light. A typical spherulite is shown in Figure 17.13. They form by nucleation at different points in the sample and then grow as spherical entities; the growth of the spherulites stops when impingement of adjacent spherulites occurs. At first sight, they look very similar to grains in a metal being roughly of the same dimensions and growing in a similar way, as shown in Figure 17.14a. However, the grains in a metal are individual

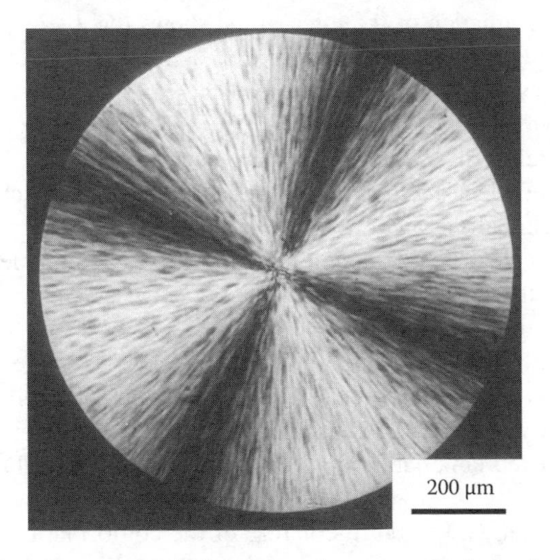

FIGURE 17.13 Single spherulite growing in isotactic polystyrene. Optical micrograph taken in polarized light. (Courtesy of D. Tod, Queen Mary College, University of London, London, U.K.)

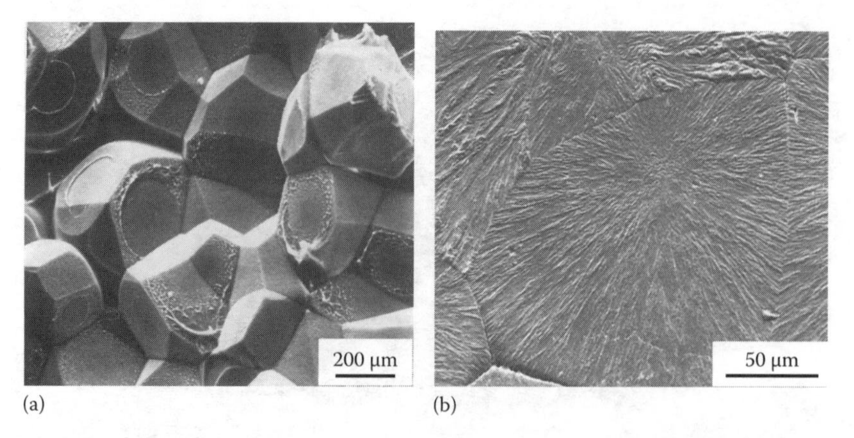

FIGURE 17.14 Scanning electron micrographs of polypropylene spherulites. (a) Fracture surfaces of a brittle sample of low molar mass polymer showing interspherulitic fracture. (Courtesy of K. Friedrich, University of Kaiserslautern, Kaiserslautern, Germany.) (b) An etched surface of polypropylene crystallized at 128°C showing the structure of the individual spherulites and the spherulite boundaries. (Courtesy of M. Burke, University of Manchester, Manchester, U.K.)

single crystals whereas detailed investigation of polymeric spherulites shows that they consist of numerous crystals radiating from a central nucleus.

There are many examples of spherulites in other crystalline solids, particularly in naturally occurring mineral rocks. The typical extinction patterns seen in polarized light are due to the orientation of the crystals within the spherulites. Analysis of the Maltese cross patterns has indicated that the molecules normally are aligned tangentially in polymer spherulites. It has been shown that the *b* crystals axis is radial in polyethylene spherulites and the *a* and *c* crystallographic directions are tangential. In some polymer samples, the spherulites are seen to be ringed, which has been attributed to a regular twist in the crystals.

The detailed structure of polymer spherulites can best be studied by electron microscopy because of the small size of the individual crystals. Figure 17.14b shows a scanning electron micrograph of an etched surface of polypropylene crystallized at 128°C. The surface was microtomed and then etched with a strongly oxidizing solution, that preferentially attacks the amorphous regions of the polymer. It exposes the crystals radiating from the central nucleus and terminating at the spherulite boundaries. Close examination of the micrograph shows that the crystals are lamellar. When viewed edge-on, they appear as fine lines of the order of 10 nm thick and as flat areas when they are parallel to the sample surface. The dimensions of these crystal lamellae are, of course, similar to those of solution-grown lamellar single crystals. Further evidence for the lamellar nature of the crystals in polymer spherulites has been obtained by examination of the debris left after the degradation of spherulitic polyethylene and polypropylene in nitric acid. The acid selectively oxidizes the inter-crystalline material and the crystal fragments that are left are found by electron microscopy to be lamellar. Selected-area electron diffraction has shown that the polymer molecules are oriented approximately normal to the lamellar surfaces. This is further evidence of the similarity between melt-crystallized and solution-grown lamellae.

Electron microscopy also has revealed the twist of the crystals in the spherulites that was suspected from optical microscopy. Surface replicas taken from spherulitic polymers have shown clearly that the lamellae twist but the reason for this twisting is not yet understood. A possible structure of the twisted lamellae in polyethylene spherulites is illustrated schematically in Figure 17.15.

An obvious question that arises is of the nature of the conformation of the molecules within the crystals. There is accumulated evidence that inter-crystalline links must exist between the lamellae in spherulitic polymers. The similarity between the morphologies of and molecular

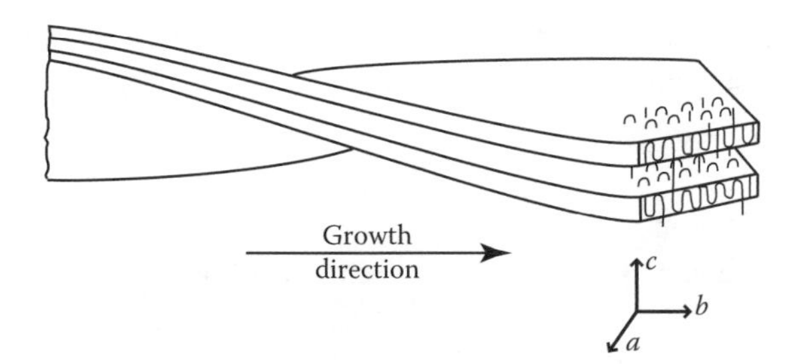

FIGURE 17.15 Schematic representation of a possible model for twisted lamellae in spherulitic polyethylene showing chain-folds and intercrystalline links.

orientation in these lamellae and solution-grown crystals clearly implies that a certain degree of chain folding must also take place. However, definitive experiments in this area have been extremely difficult to perform. The best evidence of the conformation of polymer molecules in melt-crystallized polymers has come from neutron-scattering studies of mixtures of a homopolymer and deuterated version of the same polymer. The results of such experiments are open to different interpretations but the consensus of opinion appears to be that the structure is similar to that shown schematically in Figure 17.16. There are thought to be small segments of fairly sharp chain folding and a large number of inter-crystalline links with a given molecule shared between at least two crystals.

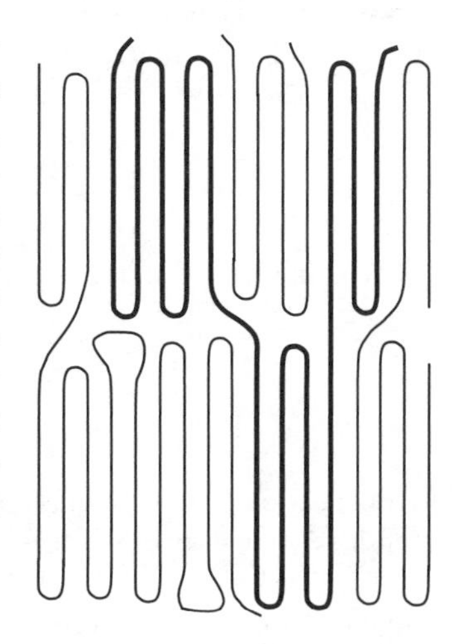

FIGURE 17.16 Schematic representation of one possible model for the conformation of a deuterated polymer molecule (heavy line) in a matrix of protonated molecules (light lines).

17.4.2 Degree of Crystallinity

Melt-crystallized polymers are never completely crystalline. This is because there are an enormous number of chain entanglements in the melt and it is impossible for the amount of organization required to form a 100% crystalline polymer to take place during crystallization. The degree of crystallinity is of great technological and practical importance and several methods have been devised to measure it, although they do not always produce precisely the same results.

The crystallization of a polymer from the melt is accompanied by a reduction in specimen volume due to an increase in density. This is because the crystals have a higher density than the molten or non-crystalline polymer and this effect provides the basis of the *density method* for the determination of the degree of crystallinity. The technique relies upon the observation that there is a relatively large and measurable difference (up to 20%) between the densities of the crystalline and amorphous regions of the polymer. This method can yield both the volume fraction of crystals ϕ_c and the mass fraction x_c from measurement of sample density ρ.

If V_c is the volume of crystals and V_a the volume of amorphous material, then the total specimen volume V is given by

$$V = V_c + V_a \tag{17.5}$$

Similarly the mass of the specimen W is given by

$$W = W_c + W_a \tag{17.6}$$

where W_c and W_a are the masses of crystalline and amorphous materials in the sample, respectively. Since density ρ is mass per volume, it follows from Equation 17.6 that

$$\rho V = \rho_c V_c + \rho_a V_a \tag{17.7}$$

Substituting for V_a from Equation 17.5 into Equation 17.7 and rearranging leads to

$$\frac{V_c}{V} = \left(\frac{\rho - \rho_a}{\rho_c - \rho_a}\right) = \phi_c \tag{17.8}$$

since ϕ_c is equal to the volume of crystals divided by the total specimen volume. The mass fraction x_c of crystals is similarly defined as

$$x_c = \frac{W_c}{W} = \frac{\rho_c V_c}{\rho V} \tag{17.9}$$

and combining Equations 17.8 and 17.9 gives

$$x_c = \frac{\rho_c}{\rho}\left(\frac{\rho - \rho_a}{\rho_c - \rho_a}\right) \tag{17.10}$$

This equation relates the mass fraction degree of crystallinity x_c to the specimen density ρ and the densities, ρ_c and ρ_a, of the crystalline and amorphous components.

The density of the polymer sample can readily be determined by *flotation* in a density-gradient column. This is a long vertical tube containing a mixture of liquids with different densities. The column is set up with a light liquid at the top and a steady increase in density towards the bottom. It is calibrated with a series of floats of known density. The density of a small piece of the polymer is determined from the position it adopts when it is dropped into the column. The density of the crystalline regions ρ_c can be calculated from knowledge of the crystal structure (Table 17.1). The term ρ_a can sometimes be measured directly if the polymer can be obtained in a completely amorphous form, for example by rapid cooling of a polymer melt. Otherwise, it can be determined by extrapolating either the density of the melt to the temperature of interest or that of a series of semi-crystalline samples to zero crystallinity. Equations 17.8 and 17.10 are valid only if the sample contains none of the holes or voids that are often present in moulded samples, and if ρ_a remains constant. In practice, since the packing of the molecules in the amorphous areas is random, it is likely that ρ_a will be different in specimens that have had different thermal treatments.

A powerful method of determining the degree of crystallinity is *wide-angle X-ray scattering* (WAXS). A typical WAXS curve for a semi-crystalline polymer is given in Figure 17.17 where the intensity of X-ray scattering is plotted against diffraction angle, 2θ. Such a scan can be obtained by measuring the intensity of scattering outwards from the centre of the diffraction pattern in Figure 17.3a. The relatively sharp peaks (rings in Figure 17.3a) are due to scattering from the crystalline regions and the broad underlying 'halo' is due to scattering from non-crystalline areas. In principle, it should be possible to determine the degree of crystallinity from the relative areas under the crystalline peaks and the amorphous halo in Figure 17.17. In practice, it is often difficult to resolve the curve into areas due to each phase. The shape of the amorphous halo can be determined from the WAXS curve for a completely amorphous sample, obtained by rapidly cooling a molten sample. For certain polymers, such as polyethylene, this can be difficult or even impossible to do and the amorphous scattering can only be estimated. Also, corrections should be made for disorder in the crystalline regions that can give rise to a reduction in area of the sharp peaks. Nevertheless,

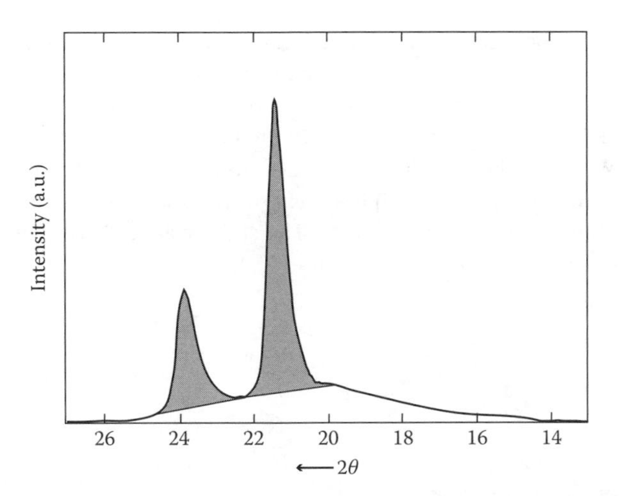

FIGURE 17.17 WAXS curves for a medium-density polyethylene. The intensity of scattering is plotted as a function of 2θ. The crystalline peaks are shaded and the amorphous halo is white.

an approximate idea of the crystallinity can be obtained from the simple construction shown in Figure 17.17. A horizontal base-line is drawn between the extremities of the scattering curve to remove the background scattering. The amorphous halo is then traced in either from knowledge of the scattering from a purely amorphous sample or by estimation using experience with other polymers. The mass fraction of crystals x_c is then given to a first approximation by

$$x_c = \frac{A_c}{(A_a + A_c)} \tag{17.11}$$

where

A_a is the area under the amorphous halo
A_c is the area remaining under the crystalline peaks.

A more sophisticated analysis involving resolution of the curve in Figure 17.17 into three distinct components, two crystalline peaks and one amorphous halo, lying on the same baseline, can produce slightly more accurate results.

Although the density and WAXS methods of determining the degree of crystallinity of semi-crystalline polymers are widely used techniques several others have been employed with varying degrees of success. One technique often used is differential scanning calorimetry (DSC; Section 17.8.1). This involves determining the change in enthalpy ΔH_m during the melting of a semi-crystalline polymer. The degree of crystallinity can then be calculated by comparing with ΔH_m for a 100% crystalline sample. Spectroscopic methods such as NMR and infrared spectroscopy (Sections 15.6 and 15.4) also have been employed. In general, there is found to be an approximate correlation between the different methods for measurement of the degree of crystallinity, although the results often differ in detail.

17.4.3 Crystal Thickness and Chain Extension

It is known that lamellar crystalline entities are obtained by the crystallization of polymers from both the melt and dilute solution. The most characteristic dimension of these lamellae is their thickness and this is known to vary with the crystallization conditions. The most important variable that controls the crystal thickness is the *crystallization temperature* and the most detailed measurements of this have been made for solution-crystallized lamellae. Several methods of crystal thickness determination have been used. They include shadowing lamellae deposited upon a substrate and measuring the length of the shadow by electron microscopy, measuring the thickness by interference microscopy and determining it by using small-angle X-ray scattering (SAXS) from a stack of deposited crystals. When the lamella single crystals are allowed to settle out from solution, they form a solid mat of parallel crystals. The periodicity of the stack in the mat is related to both the crystal thickness and the degree of crystallinity. The stack scatters X-rays in a similar way to the atoms in a crystal lattice and Bragg's law (Equation 17.4) is obeyed:

$$n\lambda = 2d\sin\theta \tag{17.12}$$

where
- n is an integer
- λ the wavelength of the radiation
- d is now the periodicity of the array
- θ the diffraction angle.

Since d is of the order of 10 nm, by using X-rays ($\lambda \sim 0.1$ nm) θ will be very small and the diffraction maxima are very close to the main beam.

The variation of lamellar thickness with crystallization temperature for polyoxymethylene crystallized from a variety of solvents is shown in Figure 17.18a. For a given solvent, the lamellar thickness is found to increase with increasing crystallization temperature. This behaviour is typical of many crystalline polymers, such as polyethylene or polystyrene, for which the length of the fold period increases as the crystallization temperature is raised. The different behaviour in the various solvents displayed in Figure 17.18a again is typical of solution crystallization, and Figure 17.18b is a plot of the lamellar thickness against the reciprocal of the supercooling ΔT ($=T_s - T_c$) which gives a master curve of all the data in Figure 17.18a. This indicates that the crystallization process is controlled by the difference between the crystallization temperature T_c and solution temperature T_s, rather than the actual temperature of crystallization. This has important implications for the theories of polymer crystallization that will be discussed later (Section 17.7.3).

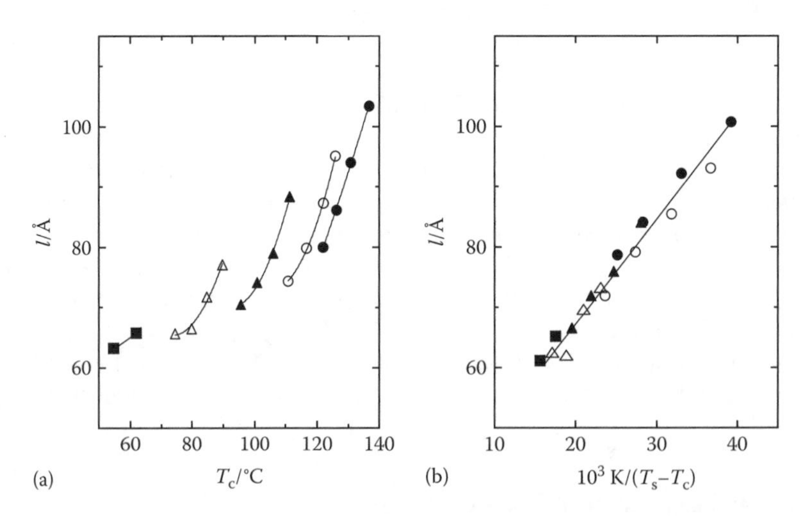

FIGURE 17.18 (a) Dependence of lamellar thickness l upon crystallization temperature T_c for polyoxymethylene crystallized from different solvents (1 Å = 0.1 nm). (b) Master curve for all the data in (a) plotted against the reciprocal of the supercooling. Solvents: ■, phenol; Δ, *m*-cresol; ▲, furfuryl alcohol; o, benzyl alcohol; •, acetophenone. (Data taken from Magill, J., *Treatise on Materials Science and Technology*, Vol. 10A, J.M. Schultz (ed), Academic Press, New York, 1977.)

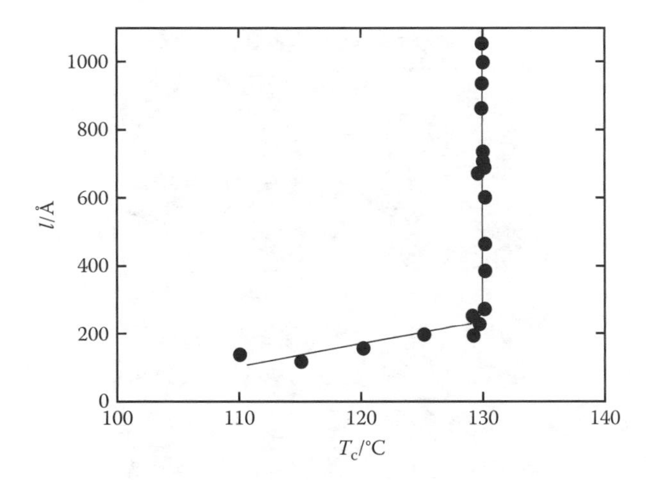

FIGURE 17.19 Lamellar thickness l (1 Å = 0.1 nm) as a function of crystallization temperature for isothermally melt-crystallized polyethylene. (Data taken from Wunderlich, B., *Macromolecular Physics*, Vols. 1–3, Academic Press, London, 1973, 1976 and 1980.)

The lamellar thickness for melt-crystallized polymers is rather more difficult to determine to any great accuracy. SAXS curves can be obtained from such samples but they tend to be much weaker and less well-defined than their counterparts from single-crystal mats. If the lamellar thickness is sufficiently large (>50 nm), it is possible to measure it by electron microscopic examination of sections of bulk samples. The variation of lamellar thickness with crystallization temperature for melt-crystallized polyethylene is shown in Figure 17.19. The behaviour is broadly similar to that of solution-crystallized polymers in that the lamellar thickness increases as the crystallization temperature is raised. In fact, it is found that at high supercoolings the thickness of solution and melt-crystallized lamellae are comparable. At low supercoolings, however, there is a rapid rise in lamellar thickness with crystallization temperature that is not encountered with solution crystallization. It is thought that this is due to an isothermal thickening process whereby the crystals become thicker with time when held at a constant temperature close to the melting temperature. On the other hand, there is very little effect of crystallization time upon the thickness of solution-grown lamellae.

As might be expected, the *molar mass* of the polymer has a strong effect upon lamellar thickness only when the length of the molecules is comparable to the crystal thickness. Normally, the molecular length is many times greater than the fold length and so even large increases in molar mass produce only correspondingly small increases in lamellar thickness. Another variable that has a profound effect upon the crystallization of polyethylene and a few other polymers from the melt is *pressure*. The melting temperature of polymers increases rapidly with pressure and it is found that crystallization of polyethylene at low supercoolings at pressures above about 3 kbar (300 MPa) can produce crystal lamellae in which some of the molecules are in fully extended conformations. Crystals of up to 10 μm thick have been reported for high-molar-mass-polymers. This chain-extended type of morphology can also be found in polytetrafluoroethylene crystallized slowly at ambient pressure and a micrograph of a replica of a fracture surface of this polymer is given in Figure 17.20. The micrograph shows long tapering lamellae containing striations approximately perpendicular to the lamellar surfaces. The striations are caused by steps on the fracture surface and define the chain direction within each lamella. In the chain-extended lamellae seen in Figure 17.20, it is thought that the molecules fold backwards and forwards several times as it is known that the molecular length is somewhat greater than the average crystal thickness. Chain-extended morphologies can have extremely high degrees of crystallinity, sometimes in excess of 95%. However, their mechanical properties are extremely disappointing. Chain-extended polyethylene is relatively

FIGURE 17.20 Electron micrograph of a replica of a fracture surface of melt-crystallized polytetrafluoroethylene showing chain-extended crystals. The striations define the chain-direction in each crystal.

TABLE 17.2
Effect of Changing Supercooling ΔT, Pressure p, Molar Mass M and Time t upon the Lamellar Thickness of Polyethylene

	ΔT	p	M	t
Solution crystallized	--	0	0	0
Melt crystallized (High T)	--	+++	0	+
Melt crystallized (Low T)	---	++++	+	++
Annealed	--	++	(+)	++

Source: Adapted from Wunderlich, B., *Macromolecular Physics*, Vols. 1–3, Academic Press, London, 1973, 1976 and 1980.)

stiff but absence of inter-crystalline link molecules causes it to be very brittle and it crumbles when deformed mechanically.

The effect of the different variables upon the fold length and lamellar thickness of polyethylene is summarized in Table 17.2. The number of positive or negative signs indicates the magnitude of the effect in each case.

17.4.4 Crystallization with Orientation

If polymer melts or solutions are crystallized under stress, morphologies are found that are strikingly different from those obtained in the absence of stress. This observation has important technological consequences as many polymers are processed in the form of fibres, injection moulded or extruded and all of these processes involve the application of stress to the material during crystallization.

Some of the earliest observations on the effect of stress upon crystallization were from studies of thin stretched films of natural rubber. It was found that crystals formed transverse to the direction of the applied stress. The crystals were found to nucleate on a central backbone and grow in perpendicular directions. Similar row-nucleated structures have since been found on crystallizing oriented melts of many other polymers. A related structure is found on cooling rapidly stirred polymer solutions where structures are found that have been termed *shish-kebab morphologies*. They consist of a central backbone with lamellar crystal overgrowth. It has been found by electron diffraction that in both the backbone and the lamellar overgrowths, the polymer molecules are parallel to the shish-kebab axis. It is thought that the molecules are folded in the overgrowth, but that in the backbone there is a considerable degree of chain extension and alignment, although some folds may also be present.

17.4.5 Polymer Fibres

Polymer fibres can be produced through two general routes, *melt* spinning and *solution* spinning, and they have microstructures in which the polymer molecules generally are aligned in the direction of the fibre axis. The degree of alignment or orientation depends upon a number of factors that include the processing method, the processing conditions and the properties of the polymer molecules. Polymer fibres are extremely important as commercial materials and are used widely in numerous applications, such as in textiles, ropes and reinforcements for polymer-based composites (Chapter 24). The fibres consist of aligned molecules, which allow the high unidirectional stiffness and strength of the polymer molecules to be exploited.

In the melt spinning process, a molten polymer is extruded though a small hole or die and the extruded stream of molten polymer is cooled rapidly to quench in the molecular orientation. The degree of orientation depends upon the processing conditions, high levels of orientation being obtained with high extrusion rates and fast cooling. The degree of crystallinity in the polymer fibre depends upon the crystallisation kinetics of the polymer. The rapid cooling of a slowly crystallising polymer can lead to an amorphous fibre whereas for other polymers, such as polyethylene or polypropylene, the fibres are invariably crystalline. The polypropylene fibre used in Figure 17.3b was processed by melt spinning, and the diffraction pattern shows a high degree of both crystallinity and orientation.

17.5 LIQUID CRYSTALLINE POLYMERS

Liquid crystallinity is a state of matter that shows properties of both liquids and crystals. For instance, a liquid crystalline material may be able to flow like a liquid but it also has molecular order similar to that found in crystalline solids. Liquid crystalline materials were first identified through observation in an optical microscope using a polarised light source. The liquid crystalline phases were observed to display regions of different texture in the microscope which correspond to areas where the molecules are aligned in particular directions. Although much of the original work upon liquid crystalline materials was confined to low molar mass molecules and indeed, such materials are now widely used in liquid crystalline displays (LCDs), it has long been recognised that there also are polymers that show liquid crystallinity. The development of liquid crystalline polymers (LCPs) has led to a class of new polymeric materials with both novel processing characteristics and a unique set of properties.

17.5.1 Classes of Liquid Crystals

The general classification of liquid crystalline materials goes back to the work of Friedel who identified three general types of liquid crystals: *nematic*, *cholesteric* and *smectic*. This classification also is used to classify liquid crystalline polymers. Another terminology that needs to be introduced is the word *mesogenic* that describes a molecule or part of a molecule capable of forming liquid

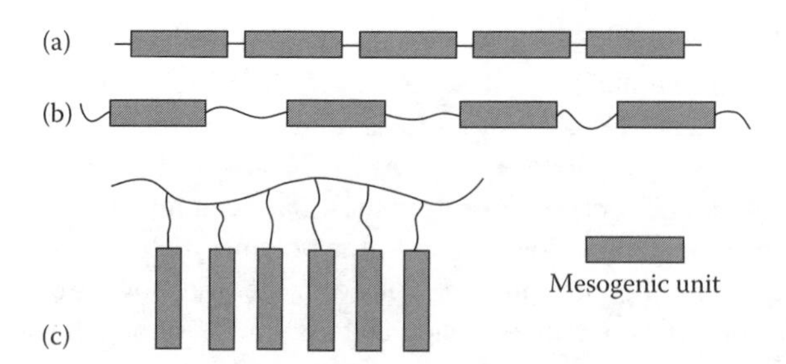

FIGURE 17.21 Schematic illustration of different types of polymer liquid crystals. (a) Rigid rod polymer. (b) Polymer with flexible links between the meosogenic units to give thermotropic behaviour. (c) Side-chain liquid crystalline polymer.

crystalline phases. In the case of small molecules, the whole molecule may be a mesogen and polymers may have mesogenic units if parts of the molecule are sufficiently rigid and straight that they can form liquid crystalline domains; the mesogenic units may be either in the main chain or side chain of the polymer, as shown schematically in Figure 17.21.

Figure 17.22 shows schematic diagrams of the arrangement of the mesogens in the three general classifications of liquid crystals. In the *nematic* phase, the mesogens are aligned imperfectly in one direction and a vector known as the director **n** describes locally the average orientation of the mesogens (Figure 17.22a). The nematic phase possesses order in terms of the orientation of the mesogenic units but it differs from a solid crystal in that there is no long-range order in the centres-of-mass of the mesogenic units.

The *cholesteric* phase is illustrated in Figure 17.22b and is named as such since it was first identified for derivatives of cholesterol. This is essentially a nematic phase in which there is a periodic twist about an axis perpendicular to the direction of the director. Cholesteric phases are formed by molecules that are chiral in nature and associated with the presence of an asymmetric carbon atom, such that left- and right-handed isomers can exist. The asymmetric packing of these chiral molecules causes a twist about the axis perpendicular to the director and leads to the long-range chiral order found in the cholesteric phase.

The *smectic* phase has a layered arrangement of mesogens, as shown in Figure 17.22c. There is no long-range positional order of the mesogens within the layers nor is there any correlation in the positions of the mesogens between the layers.

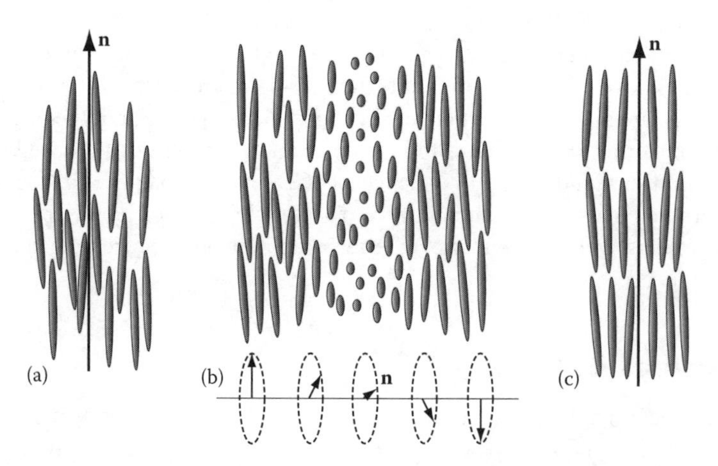

FIGURE 17.22 Three general classes of liquid crystals: (a) nematic, (b) cholesteric and (c) smectic. The director **n** is indicated in each case.

17.5.2 Polymer Liquid Crystals

Polymer liquid crystals contain mesogenic units that are rigid and usually aromatic in nature, similar to the mesogenic groups found in low molar mass liquid crystal compounds. The liquid crystalline (LC) phase is only found between the melting temperature of the polymer and a higher so-called upper transition temperature at which point the LC polymer melt becomes isotropic. Aromatic polymers generally have high melting temperatures (Section 17.8.3.1) and so the challenge in terms of the design of LCPs is therefore to synthesize materials that can be processed in the liquid crystalline state under conditions that do not cause degradation of the material at high temperature.

17.5.2.1 Thermotropic Systems

Thermotropic liquid crystalline systems are ones in which the transition to liquid crystalline behaviour occurs on reaching the melting temperature T_m and persists in the melt to the upper transition temperature at which point it becomes an isotropic liquid. The main considerations in the design of thermotropic LCPs have been concerned with reducing the transition temperatures to a useful working range whilst maintaining the desirable liquid crystalline characteristics of the material.

Based upon the experience with low molar mass organic compounds, *para*-substituted aromatic rigid-rod polymer molecules such as poly(*p*-phenylene terephthalamide) (PPTA) or poly(hydroxybenzoic acid) (PHBA) might be expected to form liquid crystalline phases above their melting temperatures. The problem is, however, that such polymers have very high melting temperatures (>500°C) and decompose at or near their T_m. One way around this problem is to produce random copolymers containing the addition of different groups that interrupt the structure and reduce T_m whilst maintaining the mesogenic characteristics. Aromatic copolyesters serve as a good example of how this can be achieved.

One aromatic copolyester system is based upon copolymers of hydroxybenzoic acid (HBA) and hydroxynaphthoic acid (HNA), that has been commercialized under the name Vectra®, as shown in Table 3.2. Figure 17.23 shows the variation of the melting temperature of this copolymer with composition, from which it can be seen that the T_m has a minimum of about 250°C at a composition containing around 40% of HNA repeat units, with the copolymer being liquid crystalline above T_m for all compositions. The presence of the HNA units has the beneficial effect of reducing T_m by destroying the periodicity of the chain and inhibiting crystallization, whilst maintaining good

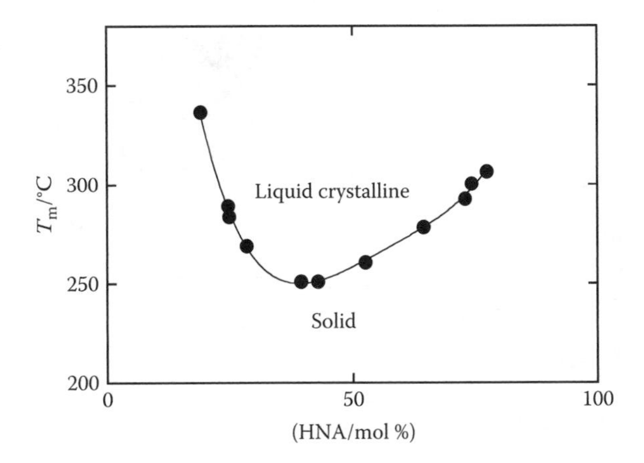

FIGURE 17.23 Dependence of the melting temperature T_m upon composition for random copolymers of hydroxybezoic acid (HBA) and hydroxynaphthoic acid (HNA). (Data taken from Donald, A.M. and Windle, A.H., *Liquid Crystalline Polymers*, Cambridge University Press, Cambridge, U.K., 1992.)

mechanical properties and without detracting from the good thermal and chemical resistance of the material. The liquid crystalline nature of the molten polymer also allows LCP fibres to be produced in which the molecules are highly aligned. The fibres are produced by melt spinning followed by post-spin drawing and have high levels of stiffness and strength (see Chapter 19). These liquid crystalline copolyester fibres have been commercialised under the trade name of Vectran®.

17.5.2.2 Lyotropic Systems

Another approach to facilitate the processing of rigid rod molecules is by the addition of low molar mass solvent to form a *lyotropic* system. This enables the processing of polymers, such as PPTA, that would degrade during any attempt at melt processing of the polymer. The presence of the solvent has the effect of reducing the melting temperature of the polymer whilst maintaining the liquid crystalline nature of the material in the liquid state.

Solution spinning is used widely in the production of polymer fibres and a recent development has been for the processing of fibres from polymers such as PPTA or poly(*p*-phenylene benzobisoxazole) (PBO) that do not exhibit thermotropic behaviour. The molecules of these polymers are inherently stiff and will form lyotropic solutions in suitable solvents. PPTA typically is spun from a lyotropic dope consisting of 20% PPTA in 99.8% sulphuric acid which is solid at room temperature and so is spun between 70°C and 90°C. The molecules in the lyotropic solution line up parallel to each other forming nematic liquid-crystalline domains and when such solutions are extruded through holes of small diameter in the die (*spinneret*), the molecules become aligned in the fibres. The strands of the fibres emerge from the spinneret and pass into a coagulation bath (e.g. water at 5°C) that removes the solvent and then are subjected to post-spin drawing. The fibres produced have a high degree of molecular alignment of the constituent rigid molecules and so have exceptional mechanical properties. The process known as *dry-jet wet-spinning* is shown schematically in Figure 17.24 and is used commercially to produce PPTA fibres such as Kevlar® and Twaron®. A similar process is used to produce fibres of PBO Zylon® with the lyotropic dope in this case being a 5% solution of the polymer in poly(phosphoric acid).

Fibres of PPTA and PBO are highly crystalline and have a high degree of molecular orientation parallel to the fibre axis. Figure 17.25 shows a scanning electron micrograph of a single

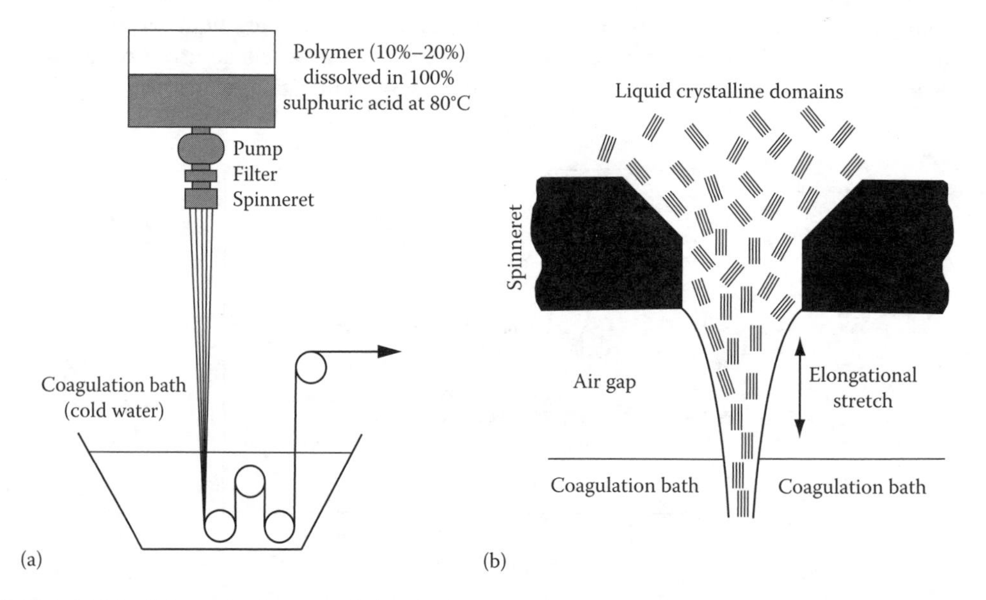

FIGURE 17.24 Schematic diagram of the dry-jet wet-spinning process used to produce PPTA fibres. (a) Overview of the process. (b) Alignment of the liquid crystalline domains in the spinneret. (Adapted from Rebouillat, S., High Performance Fibres, J.W.S. Hearle (ed) Woodhead Publishing Ltd., Cambridge, U.K., 2000.)

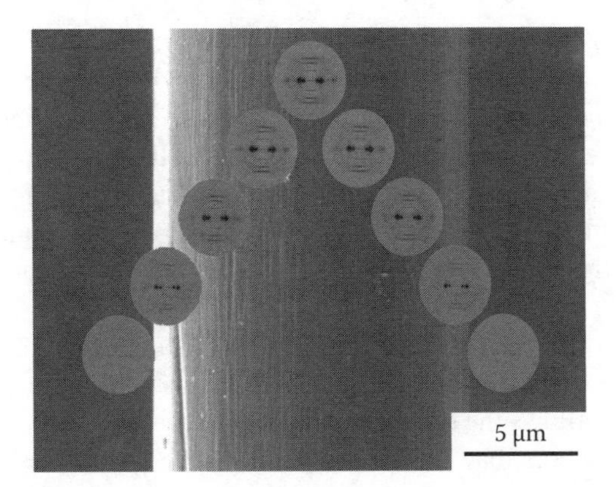

FIGURE 17.25 Scanning electron micrograph of a PBO fibre with superimposed synchrotron microfocus X-ray diffraction patterns taken at different positions across the fibre. (Courtesy of R.J. Davies, ESRF, Grenoble, France.)

12 μm diameter PBO fibre. Superimposed upon the fibre are a series of microfocus X-ray diffraction patterns obtained using synchrotron radiation with a 2 μm diameter beam at different positions between the fibre edge and centre. Although the intensity of the patterns varies with the distance through which the beam has to pass through the fibre, it can be seen that they all show a high degree of molecular orientation throughout the fibre. As the PPTA and PBO molecules are inherently stiff, these rigid-rod polymer fibres have high levels of strength and stiffness (see Chapter 19).

17.6 DEFECTS IN CRYSTALLINE POLYMERS

Although they have well-defined crystal structure and morphologies, crystalline polymers are by no means perfect from a structural viewpoint. They contain crystalline and amorphous regions and also areas that are partially disordered. It has been recognized for many years that crystals of any material contain imperfections such as dislocations or point defects and there is no fundamental reason why even the relatively well-ordered crystalline regions of crystalline polymers should not also contain such defects. In fact, it is now known that polymer crystals contain defects that are similar to those found in other crystalline solids, but in considering such imperfections it is essential to take into account the macromolecular nature of the molecules which form the crystals.

The presence of defects in a crystal will give rise to broadening of any diffraction maxima, such as the rings seen in the X-ray diffraction pattern in Figure 17.3a. In principle, the degree of broadening can be used to determine the amount of disorder in the crystal caused by the presence of defects. However, broadening of the diffraction maxima also can be caused by the crystals having a finite size and, since polymer crystals normally are relatively small, most of the broadening observed usually is caused by the size effect.

In considering the types of defects that may be present in polymer crystals, it is best to look at the different types separately and defects that have been found or postulated to exist in polymer crystals are discussed next.

17.6.1 Point Defects

The presence of point defects such as vacancies or interstitial atoms or ions is well established in atomic and ionic crystals. The situation is somewhat different in macromolecular crystals where the types of point defects are restricted by the long-chain nature of the polymer molecules. It is relatively easy to envisage the types of defects that may occur. They could include chain ends, short branches, folds or copolymer units. There is accumulated evidence that the majority of this chain disorder is excluded from the crystals and incorporated in non-crystalline regions. It also is clear, however, that at least some of it must be present in the crystalline regions.

Another type of point defect that could occur and is unique for macromolecules is a molecular kink. For example, a kink could be generated in a planar zig-zag chain consisting of all-*trans* bonds by making two of the bonds *gauche*. The sequence of five consecutive bonds would be ... tg^+ tg^- t ... rather than ... *ttttt* ... Such a kinked chain could be incorporated into a crystal with only a relatively slight local distortion of the structure. One important aspect of such a kink defect is that in polyethylene it enables an extra $-CH_2-$ group to be incorporated in the crystal and motion of the defect along the chain allows the transport of material across the crystal. It has been suggested that the motion of such defects may be the mechanism whereby crystals thicken during crystallization (Section 17.4.3).

17.6.2 Dislocations

It is found that in metal crystals, slip can take place at stresses well below the theoretically calculated shear stress. The movement of line defects known as dislocations was invoked to account for this observation and with the advent of electron microscopy the presence of such defects was finally proved. The two basic types of dislocations found in crystals, screw and edge, are shown in Figure 17.26. A dislocation is characterized by its line and Burgers vector. If the Burgers vector is parallel to the line, it is termed a *screw dislocation* and if it is perpendicular it is called an *edge dislocation*. In general, the Burgers vector and dislocation line may be at any angle, as shown in Figure 17.26c. When this is the case, the dislocation has both edge and screw components and it is said to be *mixed*. The dislocations are drawn in Figure 17.26 for an atomic crystal but there is no a priori reason why

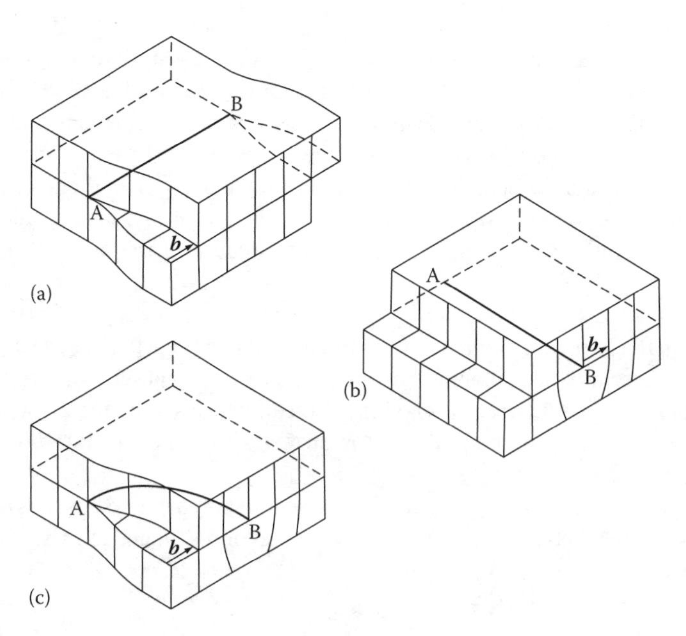

FIGURE 17.26 Schematic representation of dislocations in crystals. (a) Screw dislocation. (b) Edge dislocation. (c) Mixed screw and edge dislocation. The line of dislocation AB and the Burgers vector (***b***) are indicated in each case. (Adapted from Kelly, A. and Groves, G.W., *Crystallography and Crystal Defects*, Longman, London, U.K., 1970.)

they cannot exist in polymer crystals. There are, however, several important differences between polymeric and atomic crystals that lead to restrictions upon the type of dislocations that may occur in polymer crystals. The differences are as follows.

- There is a high degree of anisotropy in the bonding in polymer crystals. The covalent bonding in the chain direction is very much stronger than the relatively weak transverse secondary bonding.
- Polymer molecules strongly resist molecular fracture and so dislocations that tend to do this are not favoured.
- The crystal morphology can also affect the stability and motion of the dislocations. The occurrence of chain folding and the fact that polymer crystals often are relatively thin may affect matters.

The possibility of dislocations being involved in the deformation of polymer crystals has received considerable attention over recent years. This will be considered later (Section 22.4.3), and this section will be concerned with dislocations that may be present in undeformed crystals.

The most obvious example of the occurrence of dislocations in polymer crystals is seen in Figure 17.12 where the crystals contain growth spirals. It is thought that there is a screw dislocation with a Burgers vector of the size of the fold length (~10 nm) at the centre of each spiral. Such a dislocation is illustrated schematically in Figure 17.27, and the Burgers vector and dislocation line are clearly parallel to each other and the chain direction.

Dislocations with Burgers vectors of a similar size to the unit cell dimensions have been deduced from Moiré patterns obtained from overlapping lamellar single crystals observed in the electron microscope. The Moiré pattern is caused by double diffraction when the two crystals are slightly misaligned and they allow the presence of defects in the crystals to be established. This technique has allowed edge dislocations with Burgers vectors perpendicular to the chain direction to be identified in several different types of polymer crystals. Such dislocations are thought to be due to the presence of a terminated fold plane in crystals containing molecules folding by adjacent re-entry. If this is so, it seems likely that the dislocation will not be able to move without breaking bonds at the fold surface.

With recent advances and improvements in the instrumentation of electron microscopy, there have been numerous reports of the observation of dislocations in atomic, ionic and molecular crystals through direct imaging of the crystal lattice. Although this technique is difficult to apply to polymer crystals because of radiation damage, recently, dislocations have been imaged directly in polymer crystals. An example of this is shown in Figure 17.28 for a polydiacetylene single crystal. Edge dislocations due to a row of chain ends can be seen in the high-resolution micrograph. Such dislocations may have been present in the monomer crystal and then locked in by formation of polymer chains during the solid-state polymerization reaction (Figure 8.1)

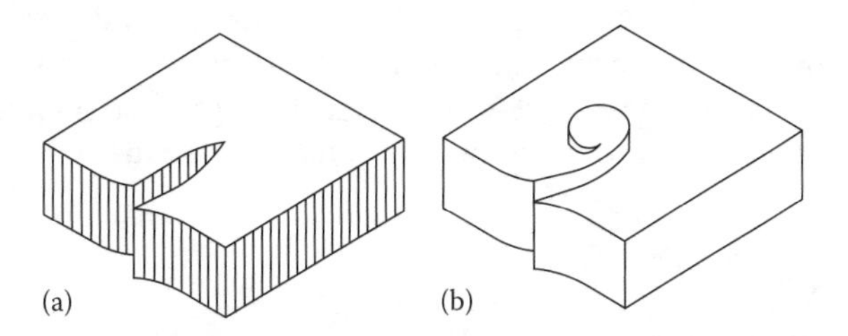

FIGURE 17.27 Schematic representation of screw dislocations with Burgers vectors parallel to the chain direction in lamellar polymer crystals. (a) Illustration of relation between the dislocation line and the chain direction (indicated by striations). (b) Screw dislocation leading to growth spiral.

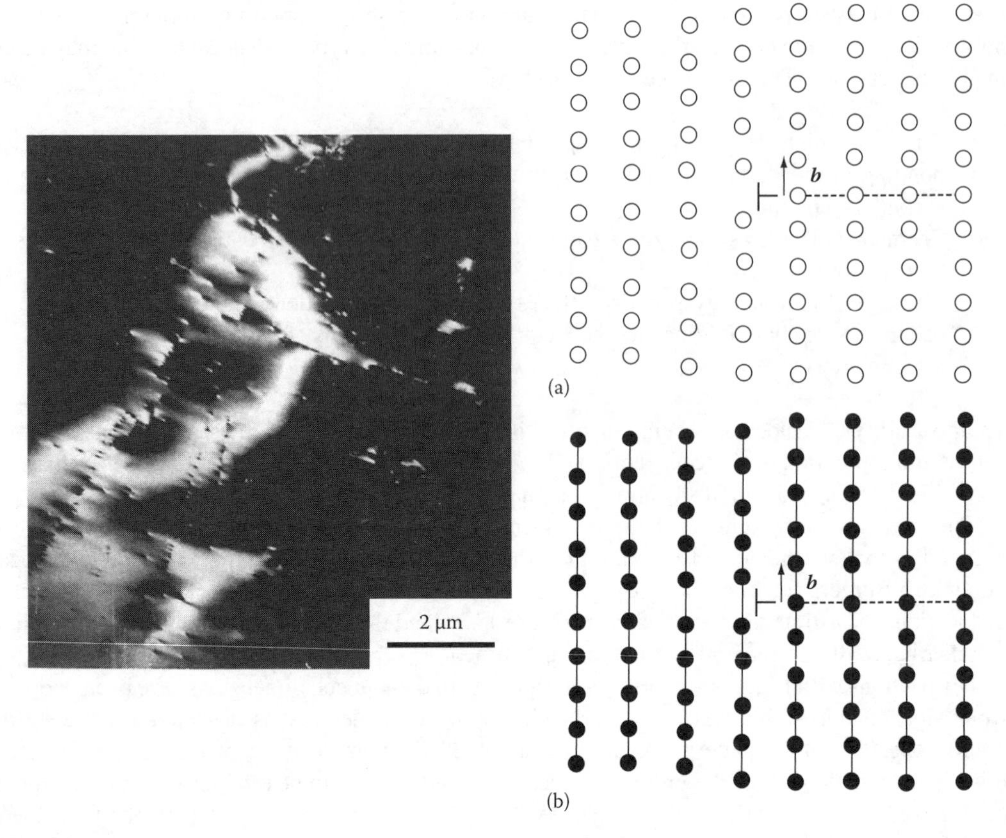

FIGURE 17.28 Transmission electron micrograph showing edge dislocations, seen as rows of spots with a black-white contrast, in a polydiacetylene single crystal. Schematic diagrams of the dislocations in (a) the monomer crystal and (b) the polymer crystal.

17.6.3 Other Defects

There are many other possible types of imperfections in crystalline polymers as well as point-defects and dislocations. The fold surface and chain folds can be considered as defects. It is also possible to consider the boundaries between crystals as areas containing defects.

The effect of the many types of types of defect discussed above is to introduce disorder into the polymer crystals. The result of their presence is to deform and distort the crystal lattice and produce broadening of the X-ray diffraction maxima (Figure 17.3). This type of disorder in a crystalline structure has been termed *paracrystallinity.* The polymer paracrystal is modelled by allowing the unit cell dimensions to vary from one cell to another and this allows the broadening of the X-ray patterns, over and above that expected from crystal size effects, to be explained. The concept of paracrystallinity has proved extremely useful in characterizing the structures of semi-crystalline polymers, but there is more clearly to be learned about the detailed structure of these materials.

17.7 CRYSTALLIZATION

17.7.1 General Considerations

Crystallization is the process whereby an ordered structure is produced from a disordered phase, usually a melt or dilute solution, and melting can be thought of as being essentially the opposite of this process. When the temperature of a polymer melt is reduced to the melting temperature, there is a tendency for the random tangled molecules in the melt to become aligned and form small ordered regions. This process is known as *nucleation* and the ordered regions are called *nuclei*. These nuclei are stable only below the melting temperature of the polymer since they are disrupted by thermal motion above this temperature. The second step in the crystallization process is *growth* whereby the crystal nuclei grow by the addition of further chains. Crystallization is, therefore, a process that takes place in two distinct steps, nucleation and growth, that may be considered separately.

Nucleation is classified as being either homogeneous or heterogeneous. During homogeneous nucleation in a polymer melt or solution, it is envisaged that small nuclei form randomly throughout the melt. Although this process has been analysed in detail from a theoretical viewpoint, it is thought that, in the majority of cases of crystallization from polymer melts and solutions, nucleation takes place heterogeneously on foreign bodies such as dust particles or the walls of the containing vessel. When all other factors are kept constant, the number of nuclei formed depends upon the temperature of crystallization. At low undercooling, nucleation tends to be sporadic and during melt crystallization, a relatively small number of large spherulites form. On the other hand, when the undercooling is increased many more nuclei form and a large number of small spherulites are obtained.

The growth of a crystal nucleus can, in general, take place in one, two or three dimensions with the crystals in the form of rods, discs or spheres, respectively. The growth of polymer crystals takes place by the incorporation of the macromolecular chains within crystals which normally are lamellar. Crystal growth will then be manifest as the change in the lateral dimensions of lamellae during crystallization from solution or the change of spherulite radius during melt crystallization. An important experimental observation that simplifies the theoretical analysis is that the change in linear dimensions of the growing entities at a given temperature of crystallization usually is linear with time. This means that the spherulite radius r will be related to time t through an equation of the form

$$r = vt \tag{17.13}$$

where v is known as the growth rate. This equation usually is valid until the spherulites become so large that they touch each other. The change in the linear dimensions of lamellae tends to obey an equation of a similar form for solution crystallization. The growth rate v is strongly dependent upon the crystallization temperature, as shown in Figure 17.29. It is found that the growth rate is relatively low at crystallization temperatures just below the melting temperature of the polymer, but

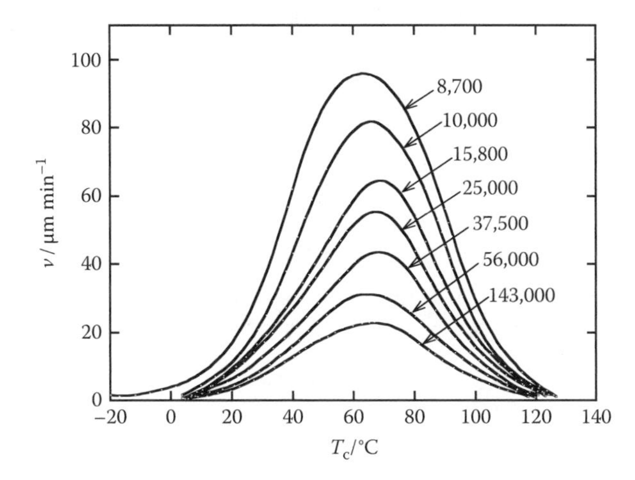

FIGURE 17.29 Dependence of crystal growth rate v upon crystallization temperature T_c for different fractions of poly(tetramethyl-*p*-sil-phenylene-siloxane) (molar mass given in g mol^{-1}). (Data taken from Magill, J., *Treatise on Materials Science and Technology*, Vol. 10A, J.M. Schultz (ed), Academic Press, New York, 1977.)

as the supercooling is increased there is a rapid increase in v. However, it is found that eventually there is a peak in v and further lowering of the crystallization temperature produces a reduction in v. Also, for a given polymer it is found that the growth rate at a particular temperature depends upon the molar mass with v increasing as M is reduced (Figure 17.29). The reason for the peak in v is thought to be due to two competing effects. The thermodynamic driving force for crystallization will increase as the crystallization temperature is lowered. But as the temperature is reduced, there will be an increase in viscosity, and transport of material to the growth point will be more difficult and so v peaks and eventually decreases as the temperature is reduced even though the driving force continues to increase.

17.7.2 Overall Crystallization Kinetics

The degree of crystallinity has an important effect upon the physical properties of a polymer. If a polymer melt is allowed to crystallize, the crystallinity clearly increases as the spherulites form and grow and so the analysis of the crystallization of a polymer is of profound importance in understanding structure–property relationships in polymers. The nucleation and growth of spherulites in a polymer liquid can be readily analysed if a series of assumptions are made.

If a polymer melt of mass W_0 is cooled below the crystallization temperature, then spherulites will nucleate and grow over a period of time. Assuming that nucleation is homogeneous and that at a given temperature the number of nuclei formed per unit time per unit volume (i.e. the rate of nucleation) N is a constant, then the total number of nuclei formed in time interval dt will be $NW_0\mathrm{d}t/\rho_L$ where ρ_L is the density of the liquid polymer. After a length of time t, these nuclei will have grown into spherulites of radius r. The volume of each spherulite will be $4\pi r^3/3$ or using Equation 17.13, $4\pi v^3 t^3/3$. If the density of spherulitic material is ρ_S, then the mass of each spherulite will be $4\pi v^3 t^3 \rho_S/3$. The total mass of spherulitic material, dW_S, present at time t, and grown from the nuclei formed in the time interval, dt, will be given by

$$\mathrm{d}W_S = \frac{4}{3}\pi v^3 t^3 \rho_S N W_0 \frac{\mathrm{d}t}{\rho_L} \tag{17.14}$$

The total mass of spherulitic material formed after time t from all nuclei is then given by

$$W_S = \int_0^t \frac{4\pi v^3 \rho_S N W_0 t^3}{3\rho_L}\,\mathrm{d}t \tag{17.15}$$

which can be integrated to give

$$\frac{W_S}{W_0} = \frac{\pi N v^3 \rho_S t^4}{3\rho_L} \tag{17.16}$$

Alternatively, this can be expressed in terms of the mass of liquid, W_L, remaining after time t, since $W_S + W_L = W_0$

$$\frac{W_L}{W_0} = 1 - \frac{\pi N v^3 \rho_S t^4}{3\rho_L} \tag{17.17}$$

This analysis is highly simplified and the equations are valid only for the early stages of crystallization. However, the essential features of spherulitic crystallization are predicted. It is expected that the mass fraction of the crystals should depend initially upon t^4. It also follows that if the nuclei are formed instantaneously, then a t^3 dependence would be expected as only the change in spherulite volume is time dependent.

Equation 17.17 is valid only in the initial stages of crystallization and must be modified to account for impingement of the spherulites. Also, the analysis is slightly incorrect because, during crystallization, there is a reduction in the overall volume of the system and the centre of the spherulites move closer to each other. When this impingement is taken into account, it is found that W_L/W_0 is related to t through an equation of the form

$$W_L/W_0 = \exp(-zt^4) \tag{17.18}$$

This type of equation is generally known as an Avrami equation and when t is small Equations 17.17 and 17.18 have the same form. If types of nucleation and growth other than those considered here are found, the Avrami equation can be expressed as

$$W_L/W_0 = \exp(-zt^n) \tag{17.19}$$

where n is called the Avrami exponent.

From an experimental viewpoint, it is much easier to follow the crystallization process by measuring the change in specimen volume rather than the mass of spherulitic material. If the initial and final specimen volumes are defined as V_o and V_∞, respectively, and the specimen volume at time t is given by V_t, then it follows that

$$V_t = \frac{W_L}{\rho_L} + \frac{W_S}{\rho_S} = \frac{W_0}{\rho_S} + W_L\left(\frac{1}{\rho_L} - \frac{1}{\rho_S}\right) \tag{17.20}$$

and since

$$V_0 = \frac{W_0}{\rho_L} \quad \text{and} \quad V_\infty = \frac{W_0}{\rho_S}$$

then Equation 17.20 becomes

$$V_t = V_\infty + W_L\left(\frac{V_0}{W_0} - \frac{V_\infty}{W_0}\right) \tag{17.21}$$

Rearranging and combining Equations 17.19 and 17.21 gives

$$\frac{W_L}{W_0} = \frac{V_t - V_\infty}{V_0 - V_\infty} = \exp(-zt^n) \tag{17.22}$$

This equation then allows the crystallization process to be monitored by measuring how the specimen volume changes with time. In practice, this normally is done by dilatometry. The crystallizing polymer sample is enclosed in a dilatometer and the change in volume is monitored from the change in height of a liquid in the dilatometer, which follows the specimen volume. In terms of heights measured in the dilatometer, Equation 17.22 can be expressed as

$$\left(\frac{V_t - V_\infty}{V_0 - V_\infty}\right) = \left(\frac{h_t - h_\infty}{h_0 - h_\infty}\right) = \exp(-zt^n) \tag{17.23}$$

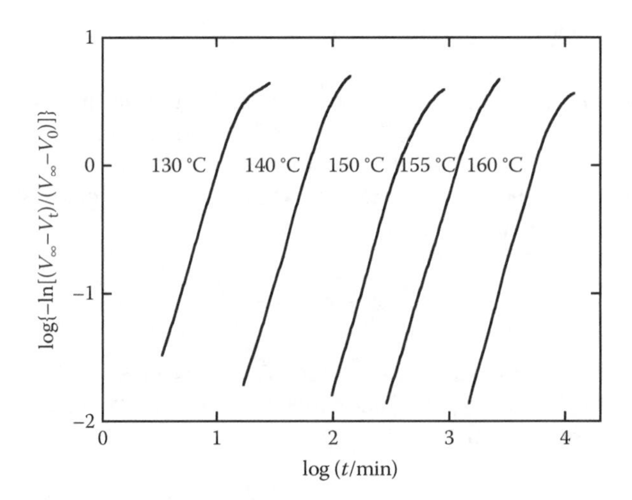

FIGURE 17.30 Avrami plot for polypropylene crystallizing from the melt at different indicated temperatures. The data points have been left off for clarity. (Data taken from Parrini, P. and Corrieri, G., *Makromol. Chem.*, 62, 83, 1963.)

The Avrami exponent n can be determined from the slope of a plot of log $\{\ln[(h_t - h_\infty)/(h_0 - h_\infty)]\}$ against log t. Figure 17.30 shows an Avrami plot for polypropylene crystallizing at different temperatures. It often is difficult to estimate n from such plots because its value can vary with time. Also, non-integral values can be obtained and care must be exercised in using the Avrami analysis, as interpretation of the value of n in terms of specific nucleation and growth mechanisms can sometimes be ambiguous.

Serious deviations from the Avrami expression can be found, particularly towards the later stages of crystallization, because secondary crystallization often occurs and there is usually an increase in crystal perfection with time. These processes can take place relatively slowly and make the estimation of V_∞ rather difficult.

17.7.3 Molecular Mechanisms of Crystallization

Although the Avrami analysis is fairly successful in explaining the phenomenology of crystallization, it does not give any insight into the molecular process involved in the nucleation and growth of polymer crystals. There have been many attempts to develop theories to explain the important aspects of crystallization. Some of the theories are highly sophisticated and involve lengthy mathematical treatments. To date, none have been completely successful in explaining all the aspects of polymer crystallization. The important features that any theory must explain are the following characteristic experimental observations.

- Polymer crystals usually are thin and lamellar when crystallized from both dilute solution and the melt.
- A unique dependence is found between the lamellar thickness and crystallization temperature and, in particular, the lamellar thickness is found to be proportional to $1/\Delta T$ (cf. Figure 17.18b).
- Chain folding is known to occur during crystallization from dilute solution and occurs to a certain extent during melt crystallization.
- The growth rates of polymer crystals are found to be highly dependent upon the crystallization temperature and molar mass of the polymer.

There are major difficulties in testing the various theories, such as the lack of reliable experimental data on well-characterized materials and an absence of some important thermodynamic data on the crystallization process. Nevertheless, it is worth considering the predictions of the various theories.

The most widely accepted approach is the kinetic description due to Hoffman, Lauritzen and others that has been used to explain effects observed during polymer crystallization. It is essentially an extension of the approach used to explain the kinetics of crystallization of small molecules. The process is divided into two stages, nucleation and growth and the main parameter that is used to characterize the process is the Gibbs free energy G, which has been described in Section 17.1.1. It follows that the change in free energy ΔG on crystallization at a constant temperature T is given by

$$\Delta G = \Delta H - T\Delta S \tag{17.24}$$

where
ΔH is the enthalpy change
ΔS is the change in entropy.

It is envisaged that in the primary nucleation step, a few molecules pack side by side to form a small cylindrical crystalline embryo. This process involves a change in free energy since the creation of a crystal surface, that has a surface energy, will tend to cause G to increase whereas incorporation of molecules in a crystal causes a reduction in G (Section 17.1.1) that will depend upon the crystal volume. The result of these two competing effects upon ΔG is illustrated schematically as a function of crystal size in Figure 17.31. When the embryo is small, the surface-to-volume ratio is high and so the overall value of ΔG increases because of the rapid increase in surface energy. However, as the embryo becomes larger, the surface-to-volume ratio decreases and there will be a critical size above which ΔG starts to decrease and eventually the free energy will be less than that of the original melt. This concept of primary nucleation has been widely applied to many crystallizing systems. The main difference between polymers and small molecules or atoms is in the geometry of the embryos. It is expected that the form of the free-energy change in all cases will be as shown in Figure 17.31. The peak in the curve may be regarded as an energy barrier and it is envisaged that at the crystallization temperature there will be sufficient thermal fluctuations to allow it to be overcome. Once the nucleus is greater than the critical size, it will grow spontaneously as this will cause ΔG to decrease.

The theories of crystallization envisage the growth of the polymer crystals as taking place by a process of secondary nucleation on a pre-existing crystal surface. This process is illustrated schematically in Figure 17.32, whereby molecules are added to a molecularly smooth crystal surface. This process is similar to primary nucleation but differs somewhat because less new surface per unit volume of crystal is created than in primary nucleation and so the activation energy barrier is lower. The first step in the secondary nucleation process is the laying down of a molecular strand on an otherwise smooth crystal surface. This is followed by the subsequent addition of further

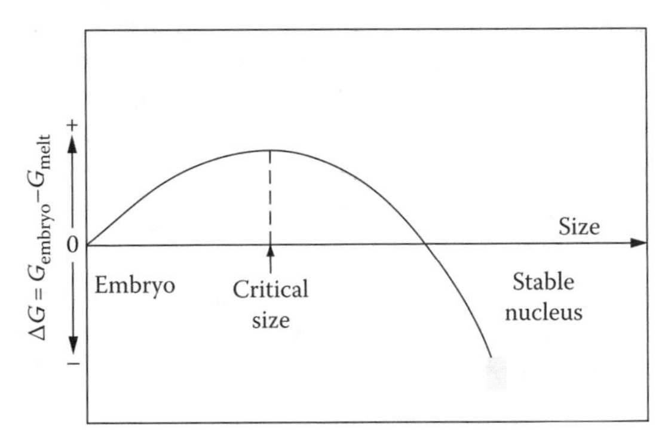

FIGURE 17.31 Schematic representation of change in free energy ΔG for the nucleation process during polymer crystallization.

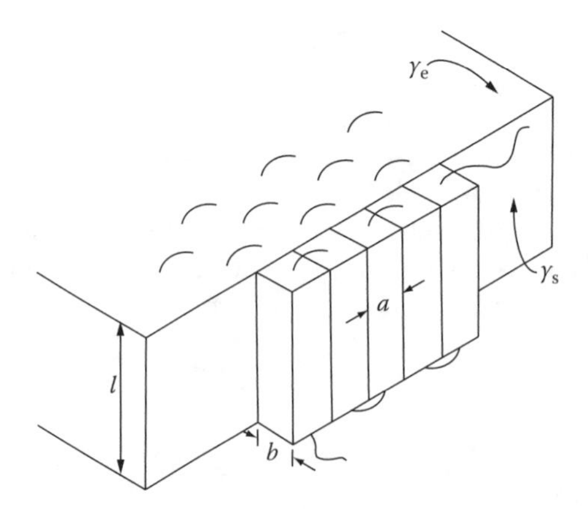

FIGURE 17.32 Model of the growth of a lamellar polymer crystal through the successive laying down of adjacent molecular strands. The different parameters are defined in the text. (Adapted from Magill, J., *Treatise on Materials Science and Technology*, Vol. 10A, J.M. Schultz (ed.), Academic Press, New York, 1977.)

segments through a chain-folding process. It is found that chain folding occurs only for flexible polymer molecules. Extended-chain crystals are obtained from more rigid molecules. It is thought that the folding only takes place when there is relatively free rotation about the polymer backbone. The basic energetics of the crystallization process can be developed if it is assumed that the polymer lamellae have a fold surface energy of γ_e, a lateral surface energy of γ_s and that the free-energy change on crystallization is ΔG_V per unit volume. If the secondary nucleation process illustrated in Figure 17.32 is considered on this basis, then the increase in surface free energy ΔG_n involved in laying down n adjacent molecular strands of length l will be

$$\Delta G_n(\text{surface}) = 2bl\gamma_s + 2nab\gamma_e$$

when each strand has a cross-sectional area of ab. Because of the incorporation of molecular strands, there will be a reduction in free energy in a crystal that will be given by

$$\Delta G_n(\text{crystal}) = -nabl\,\Delta G_V$$

The overall change in free energy when n strands are laid down will then be

$$\Delta G_n = 2bl\gamma_s + 2nab\gamma_e - nabl\,\Delta G_V \tag{17.25}$$

The value of ΔG_V can be estimated from a simple calculation. It is assumed that polymer crystals have an *equilibrium melting temperature* T_m^o, which is the temperature at which a crystal without any surface would melt. If unit volume of crystal is considered, then at this temperature, it follows from Equation 17.24 that

$$\Delta G_V = \Delta H_V - T_m^o \Delta S_V \tag{17.26}$$

where ΔH_V and ΔS_V are the enthalpies and entropies of fusion per unit volume, respectively. However, at T_m^o, there is no change in free energy for the idealized boundaryless crystal since melting and crystallization are equally probable and so $\Delta G_V = 0$ at this temperature, which means that

$$\Delta S_V = \frac{\Delta H_V}{T_m^o} \tag{17.27}$$

For crystallization below this temperature, ΔG_V will be finite and it is envisaged that ΔS_V will not be very temperature dependent and so at a temperature $T(< T_m^o), \Delta G_V$ can be approximated to

$$\Delta G_V = \Delta H_V - \frac{T\Delta H_V}{T_m^o} \tag{17.28}$$

Since the degree of undercooling ΔT is given by $(T_m^o - T)$, rearranging Equation 17.28 gives

$$\Delta G_V = \frac{\Delta H_V \Delta T}{T_m^o} \tag{17.29}$$

Inspection of Equation 17.25 shows that for a given value of n, ΔG_n has a maximum value when l is small and decreases as l increases. Eventually, a critical length of strand l^* will be achieved and the secondary nucleus will be stable ($\Delta G_n = 0$). Also, normally, n is large and so the term $2bl\gamma_s$ is negligible. A final approximate equation relating l^* to ΔT can be obtained by combining Equations 17.25 and 17.29 to give

$$l^* \sim \frac{2\gamma_e T_m^o}{\Delta H_V \Delta T} \tag{17.30}$$

Thus, the inverse proportionality between l and ΔT observed experimentally (Figure 17.18b) is predicted theoretically. The analysis outlined above is highly simplified, but it serves to illustrate important aspects of the kinetic approach to polymer crystallization.

The kinetic approach also can be used to explain why chain folding is found for solution-grown crystals. The separation between individual molecules is relatively high in dilute solutions and so growth can take place most rapidly by the successive deposition of chain-folded strands of the same molecule once it starts to be incorporated in a particular crystal.

17.8 MELTING

The melting of polymer crystals is essentially the reverse of crystallization, but it is more complicated than the melting of low molar mass crystals. Melting also controls the upper-use temperature of semi-crystalline polymers. Significant information can be obtained upon the morphology of polymers from their melting behaviour but care has to be taken in following the melting behaviour in the laboratory. Experimental aspects of melting are discussed first of all, followed by a description of the effects of structure upon the melting behaviour of a polymer.

17.8.1 Differential Scanning Calorimetry

Differential scanning calorimetry (DSC) is a method of thermal analysis that has become very important in polymer science for investigation of processes that involve a change of heat capacity (so-called second order transformations) or a change of enthalpy (so-called first order transformations). Here, the focus will be on use of DSC to study melting, but it is important to be aware that DSC has a much wider range of applications, including studies of crystallization, polymerization (particularly for monitoring the exotherm associated with curing/crosslinking), and measurements of the glass transition temperature(s) for homopolymers, copolymers and polymer blends.

There are two classes of DSC instrument which differ fundamentally in their design and operation. In *power-compensation* DSC, a sample and inert reference material are independently heated (or cooled) at a controlled rate in adjacent, separate cells whilst recording simultaneously their

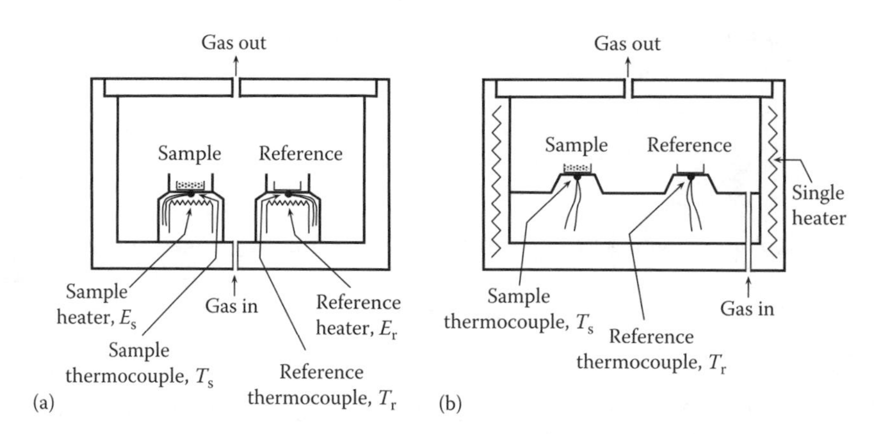

FIGURE 17.33 Schematic diagrams showing the key features of cell design for (a) power-compensation DSC and (b) heat-flux DSC. The sample is shown contained in an aluminium pan with an empty aluminium pan as the reference. T_s and T_r are, respectively, the temperatures of the sample and reference. In power-compensation DSC, the separate energy inputs to the sample and reference, E_s and E_r respectively, are adjusted such that $T_s = T_r$ as both cells are heated or cooled. Heat-flux DSC employs a single heater and ΔT (= $T_s - T_r$) is monitored as the cell is heated or cooled. The atmosphere can be controlled by flowing a gas through the cell enclosure.

temperature and monitoring the differential heat input to the sample and reference required to keep them at exactly the same temperature. In *heat-flux* DSC, the sample and inert reference material are heated (or cooled) at a controlled rate side by side in a single cell whilst recording simultaneously the temperature difference between them. Both types of DSC use the same sample containers, which for most work on polymers are small aluminium pans (ca. 6 mm diameter × 1–2 mm deep) onto which an aluminium lid can be crimped if desired. The mass of sample used depends on its form and density (since this determines how much can be fitted into the pan), but typically is 5–20 mg. Special hermetically sealable aluminium pans/lids can be used to prevent loss of volatiles if this is important. The key features that distinguish the designs for power-compensation and heat-flux DSC instruments are shown in Figure 17.33.

Data from power-compensation DSC are plotted as the differential energy input ΔE (= $E_s - E_r$) against temperature. Because the sample and reference have slightly different heat capacities, ΔE is never zero, but it attains a steady-state value which provides a baseline, displacements from which occur when the sample undergoes a first- or second-order transformation. For example, if the sample undergoes an endothermic event (such as melting), more energy needs to be input to the sample cell (i.e. $E_s > E_r$) to keep it at the same temperature as the reference ($\Delta T = T_s - T_r = 0$), and there is an upward displacement from the baseline because ΔE is positive. The opposite occurs if the sample undergoes an exothermic event (such as crystallization). At the glass transition temperature T_g in amorphous phases, there is a step change in specific heat capacity (Section 16.2), which causes an abrupt change in the steady-state value of ΔE; hence, the glass transition is manifested as an abrupt change in the position of the baseline, which moves to a higher value of ΔE above T_g (because specfic heat capacity in the rubbery state is higher than in the glassy state). Each of these thermal events is shown in the schematic power-compensation DSC heating trace in Figure 17.34a, which is typical of a sample of a crystallizable polymer that has been quench-cooled from the molten state to a temperature well below its T_g; the rapid cooling does not allow time for crystallization and because the amorphous polymer obtained is below T_g, it remains unable to crystallize. Hence, the first thermal event in the DSC trace is the glass–rubber transition, above which the polymer chains have mobility and eventually reach a temperature (the *cold crystallization temperature* T_{cc}) at which they are able organize into crystals; the process of crystallization is exothermic and so ΔE is negative. Once crystallization is complete, a new baseline is established which relates to the specific heat capacity of the semi-crystalline

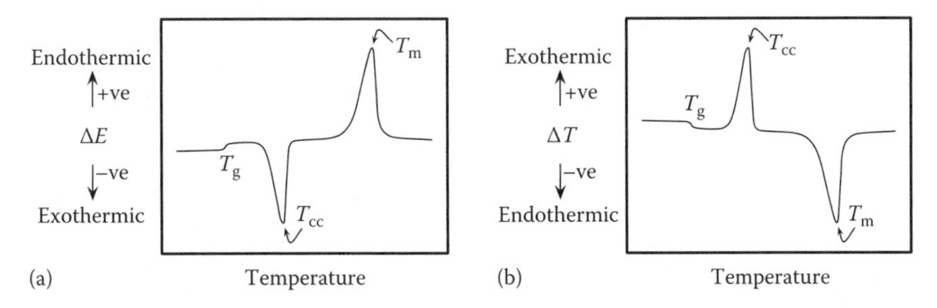

FIGURE 17.34 Schematic diagrams showing (a) power-compensation and (b) heat-flux DSC traces expected for an amorphous sample of a crystallizable polymer, where T_g is the glass transition temperature, T_{cc} is the cold crystallization temperature (i.e. for crystallization in the solid state) and T_m is the melting temperature.

polymer. As temperature increases further, there is little change until the melting temperature T_m is reached, at which point the polymer crystals begin to melt, a process that is endothermic and signified by a positive change in ΔE.

Analysis of the same quench-cooled polymer by heat-flux DSC will show the same thermal events, but the appearance of the trace will be inverted about the temperature axis compared to the power-compensation DSC trace, as is depicted in Figure 17.34b. This is because in heat-flux DSC, sample and reference have a common heater and the plot is of the temperature difference ΔT (= $T_s - T_r \neq 0$) against temperature; hence as a sample undergoes an exothermic event, T_s goes slightly above T_r (ΔT is positive), whereas during an endothermic event T_s goes slightly below T_r (ΔT is negative). Thus power-compensation and heat-flux DSC traces of any particular sample always appear inverted about the temperature axis when compared to each other, as summarized in Table 17.3.

DSC is used routinely for measurements of polymer transition temperatures and for evaluation of enthalpies of crystallization and melting during heating of a solid polymer or cooling of a liquid polymer from the melt. Figure 17.35 shows how quantitative data should be obtained from DSC traces. The value of T_g normally is taken as the onset temperature, as shown in Figure 17.35a, but in some cases the temperature at the midpoint in the baseline step is used. Careful integration of exotherms and endotherms is necessary to obtain accurate areas A because the change in baseline position during the thermal event (due to the smooth change in polymer specific heat capacity) should be taken into account, as shown in Figure 17.35b. DSC instruments record the value of ΔE or ΔT as electrical power (i.e. *heat flow*) and so the values of A are obtained in units of W K (i.e. J K s^{-1}). Hence, the energy associated with the exotherm/endotherm is given by the ratio A/k, where

TABLE 17.3
Comparison of the Different Appearances of Power-Compensation and Heat-Flux DSC Traces

			Baseline Change in DSC Trace When the Sample Undergoes an			
Type of DSC	**ΔT[a]**	**Ordinate Axis[a]**	**Increase in C_P[b]**	**Decrease in C_P[b]**	**Exotherm**	**Endotherm**
Power-compensation	=0	ΔE	+ve shift	−ve shift	−ve peak	+ve peak
Heat-flux	≠0	ΔT	−ve shift	+ve shift	+ve peak	−ve peak

[a] ΔT (=$T_s - T_r$) and ΔE (=$E_s - E_r$); see Figure 17.33.
[b] C_P = specific heat capacity.

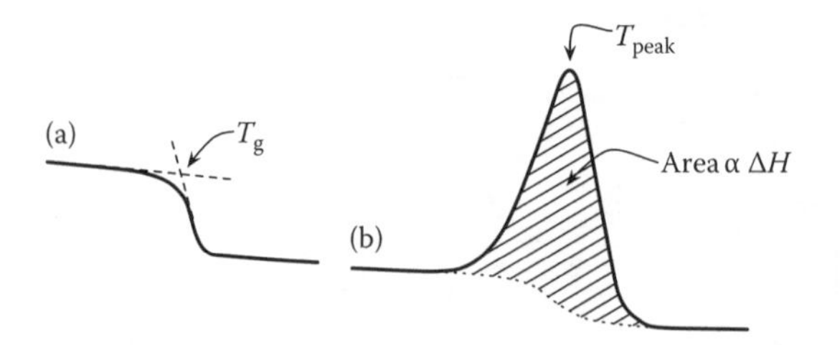

FIGURE 17.35 Schematic diagrams (of heat-flux DSC traces) showing: (a) how the glass transition temperature T_g is located and (b) how, for an exotherm, the peak temperature T_{peak} (e.g. T_{cc}), and peak area should be determined from DSC traces (the same principles apply to analysis of an endotherm).

k is the heating/cooling rate. The value of the enthalpy change ΔH normally is quoted as energy per unit mass and so is given by

$$\Delta H = K_{DSC}\left(\frac{A}{kW}\right) \tag{17.31}$$

where

K_{DSC} is the instrument calibration factor (the ratio of the true heat flow to the measured heat flow ≈1)
W is the mass of material analysed.

In practice, K_{DSC} varies slightly with k as a consequence of the time-dependency of thermal lags in data acquisition and so, for accurate measurements of ΔH it is necessary to determine K_{DSC} for each heating rate used. For both types of DSC, calibration is achieved by analysing materials which show an exotherm or endotherm with a known value of ΔH; most commonly this is done by analysing melting of metals, such as indium, that have low melting points and known enthalpies of melting. For power-compensation DSC instruments, K_{DSC} is independent of temperature (because the power supplied to the sample and reference are measured directly), but for heat-flux DSC, K_{DSC} varies with temperature (because values of ΔT must be converted to electrical power using the heat capacity, which varies with temperature). Modern heat-flux DSC instruments correct for the variation of K_{DSC} with temperature in the data analysis software.

Modulated-temperature DSC (MTDSC) is a more recent development that is applicable to both power-compensation and heat-flux DSC instrument designs. The only change in performing MTDSC is that a modulation is placed on the nominally linear heating (or cooling) rate such that (for a sinusoidal modulation) the temperature T_t at time t is given by

$$T_t = T_0 + kt + T_{mod}\sin\omega t \tag{17.32}$$

where

T_0 is the initial temperature
k is the underlying linear heating/cooling rate
T_{mod} and ω are respectively the amplitude and frequency of the temperature modulation.

For normal DSC, the last term in Equation 17.32 is absent. This apparently small change has major significance because, unlike normal DSC, the MTDSC trace enables thermal events that are reversing and non-reversing to be separated and also facilitates easy measurement of heat capacity. Thus, by using MTDSC, simultaneous reversing and non-reversing thermal events can be detected independently. For example, MTDSC analysis of quench-cooled poly(ethylene terephthalate) (PET) samples reveals that re-crystallization occurs throughout melting, i.e. crystals of lower T_m melt and the polymer then re-crystallizes to produce new, more perfect crystals with a higher T_m. Normal

DSC gives only the total heat change and so the larger thermal event dominates; in the case of quench-cooled PET, only the melting endotherm is seen and its area is much smaller than it should be due to the compensating effect of the heat released by the simultaneous re-crystallization. Thus MTDSC should be used when making measurements of enthalpies of melting.

Nowadays, most differential scanning calorimeters are capable of both normal DSC and MTDSC. MTDSC typically is carried out using $k = 1°C–3°C\ min^{-1}$ (with $T_{mod} = 1°C–2°C$ and $\omega^{-1} = 30–60$ s) and so is much slower than normal DSC for which $k = 10°C–20°C\ min^{-1}$ is typical. The results can be influenced by procedural variables and so it is important to report the instrument used, the sample size and form, the heating/cooling rate (including T_{mod} and ω^{-1} for MTDSC), the gaseous atmosphere and flow rate, the calibrant(s) used and the method(s) used to locate transition temperatures and determine enthalpy changes.

17.8.2 Melting of Polymer Crystals

There are several characteristics of the melting behaviour of polymers that distinguishes them from other materials. They can be summarized as follows.

- It is not possible to define a single melting temperature for a polymer sample as the melting generally takes place over a range of temperature.
- The melting behaviour depends upon the specimen history and in particular upon the temperature of crystallization.
- The melting behaviour also depends upon the rate at which the specimen is heated.

These observations are a reflection of the peculiar morphologies that polymer crystals can possess and, in particular, the fact that polymer crystals normally are thin. The concept of an equilibrium melting temperature T_m^o (Section 17.7.3) is introduced because of the variability in the melting behaviour. This corresponds to the melting temperature of an infinitely large crystal. However, there is still a good deal of disagreement over the values of T_m^o even for widely studied polymers such as polyethylene. This is particularly disturbing because accurate values of T_m^o are required in order to test the theories of crystallization and melting quantitatively.

The value of T_m^o can be estimated by a simple extrapolation procedure. It is found that the observed melting temperature, T_m, for a polymer sample is always greater than the crystallization temperature, T_c and a plot of T_m versus T_c is usually linear, as shown in Figure 17.36. Since T_m can

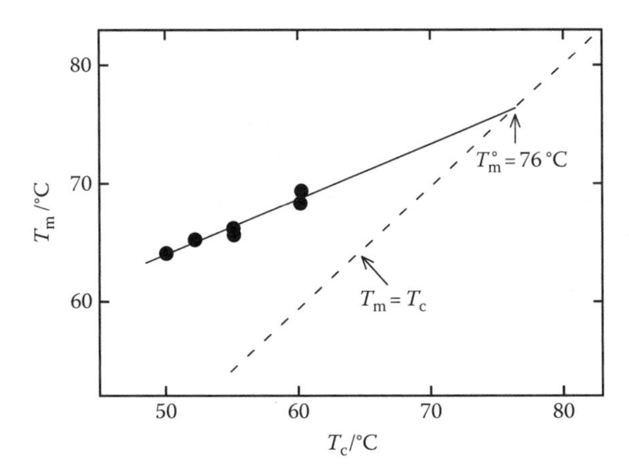

FIGURE 17.36 Plot of melting temperature T_m against crystallization temperature T_c for poly(*dl*-propylene oxide). (Data taken from Magill, J., *Treatise on Materials Science and Technology*, Vol. 10A, J.M. Schultz (ed), Academic Press, New York, 1977.)

never be lower than T_c, the line $T_m = T_c$ will represent the lower limit of the melting behaviour. The point at which the extrapolation of the upper line meets the $T_m = T_c$ line then represents the melting temperature of a polymer crystallized infinitely slowly and for which crystallization and melting would take place at the same temperature. This intercept therefore gives T_m^o.

There is found to be a strong dependence of the observed melting temperature T_m of a polymer crystal upon the crystal thickness l. This can be explained by considering the thermodynamics of melting a rectangular lamellar crystal with lateral dimensions x and y. If it is assumed that the crystal has side surface energy of γ_s and top and bottom surface energies of γ_e, then melting causes a decrease in surface free energy of $(2xl\gamma_s + 2yl\gamma_s + 2xy\gamma_e)$. This is compensated by an increase in free energy of ΔG_V per unit volume due to molecules being incorporated in the melt rather than in a crystal. The overall change in free energy on melting the lamellar crystal is given by

$$\Delta G = xyl\Delta G_V - 2l(x+y)\gamma_s - 2xy\gamma_e \quad (17.33)$$

The value of ΔG_V is given by Equation 17.29. For lamellar crystals, the area of the top and bottom surfaces will be very much larger than the sides and so the term $2l(x + y)\gamma_s$ can be neglected. At the melting point of the crystal, $\Delta G = 0$ and so it follows from Equations 17.29 and 17.33 that

$$T_m = T_m^o - \frac{2\gamma_e T_m^o}{l\Delta H_V} \quad (17.34)$$

where ΔH_V is the enthalpy of fusion per unit volume of the crystals. Inspection of this equation shows that for finite size crystals T_m will always be less than T_m^o. Also it allows T_m^o and γ_e to be calculated if T_m is determined as a function of l. A plot of T_m against $1/l$ is predicted to have a slope of $-2\gamma_e T_m^o/\Delta H_V$ with T_m^o as the intercept. The value of ΔH_V can be measured by DSC in a separate experiment.

A process that affects the melting behaviour of crystalline polymers and is of interest in its own right is *annealing*. This is a term normally used to describe the heat treatment of metals and the annealing of polymers bears a similarity to that of metals. It is found that when crystalline polymers are heated to temperatures just below the melting temperature there is an increase in lamellar crystal thickness. The driving force is the reduction in free energy gained by lowering the surface area of a lamellar crystal when it becomes thicker and less wide. The lamellar thickening only happens at relatively high temperatures when there is sufficient thermal energy available to allow the necessary molecular motion to take place.

A certain amount of annealing usually takes place when a crystalline polymer sample is heated and melted. The increase in lamellar thickness l causes an increase in T_m (Equation 17.34). This means that the measured melting temperature will depend upon the heating rate because annealing effects will be lower for more rapid rates of heating. However, high heating rates can give rise to other problems, such as those arising from thermal conductivity limitations and so great care is necessary when measuring the melting temperatures of polymer samples, especially when the melting behaviour is being related to the as-crystallized structure which could change during heating.

17.8.3 Factors Affecting T_m

The use of polymers in many practical applications often is limited by their relatively low melting temperatures. Because of this, there has been considerable interest in determining the factors which control the value of T_m and in synthesizing polymers which have high melting temperatures.

17.8.3.1 Chemical Structure

The over-riding factor that determines the melting points of different polymers is their chemical structure. The melting points of several polymers are listed in Table 17.4. It is most convenient to

TABLE 17.4
Approximate Values of Melting Temperature, T_m, for Various Polymers

Repeat Unit		T_m/°C
$-CH_2-CH_2-$		137–146
$-CH_2-CH_2-O-$		67
$-CH_2-CH_2-CO-O-$		122
$-H_2C-C_6H_4-CH_2-$ (p-phenylene)		397
$-CH_2-CH_2-CO-NH-$		330
$-CH_2-CH_2-CH_2-CO-NH-$		260
$-CH_2-CH_2-CH_2-CH_2-CO-NH-$		258
Side group (X)		
$-CH_2-CHX-$	$-CH_3$	187
	$-CH_2-CH_3$	125
	$-CH_2-CH_2-CH_3$	78
	$-CH_2-CH(CH_3)_2$	235
	$-C_6H_5$ (phenyl)	240

consider the melting points of different polymers using polyethylene $(-CH_2-CH_2-)_n$ as a reference. The first factor that must be considered is the stiffness of the main polymer chain. This is controlled by the ease at which rotation can take place about the chemical bonds along the chain. In general, incorporation of linking groups such as –O– or –CO–O– in the main chain increases flexibility and so lowers T_m (Table 17.4). On the other hand, the presence of a *p*-phenylene group in the main chain increases the stiffness and causes a large increase in T_m.

Another important factor that causes an increase in T_m is the presence of polar groups such as the amide linkage –CONH– that allows intermolecular hydrogen bonding to take place within the crystals. The presence of the hydrogen bonding tends to make the melting of the crystals more difficult and so raise their T_m. This is because there will be an enthalpy of fusion $\Delta H_{\text{H-bond}}$ associated with each hydrogen bond that adds to the overall enthalpy of fusion. The melting points of different polyamides are very sensitive to the degree of intermolecular bonding and the value of the T_m is reduced as the number of $-CH_2-$ groups between the amide linkages is increased (Table 17.4). The effect of the chemical structure of nylon polymers upon their melting points is shown in Figure 17.37. It can be seen that the density of hydrogen bonds decreases going from nylon 6.6 (Figure 17.37a) to nylon 6.10 (Figure 17.37b) and there is a consequent decrease in T_m from 265°C to 222°C. There is a more subtle effect when the nylons contain an odd number of carbon atoms in the chemical repeat, as shown for nylon 7.7 in Figure 17.37c. In this case, half of the >N–H groups are not lined up adjacent to –C=O groups and so are unable to form hydrogen bonds. The value of T_m for nylon 7.7 is consequently only 210°C and much lower than that of nylon 6.6.

A third factor that governs the value of T_m in different polymers is the type and size of any side groups present on the polymer backbone. This is most easily shown when the effect of having different hydrocarbon side groups in vinyl polymers of the type $(-CH_2-CHX-)_n$ is considered. The presence of a $-CH_3$ side group regularly placed on each alternate carbon atom along the polyethylene chain leads to a reduction in chain flexibility and means that polypropylene has a higher melting point than polyethylene. However, if the side group is long and flexible,

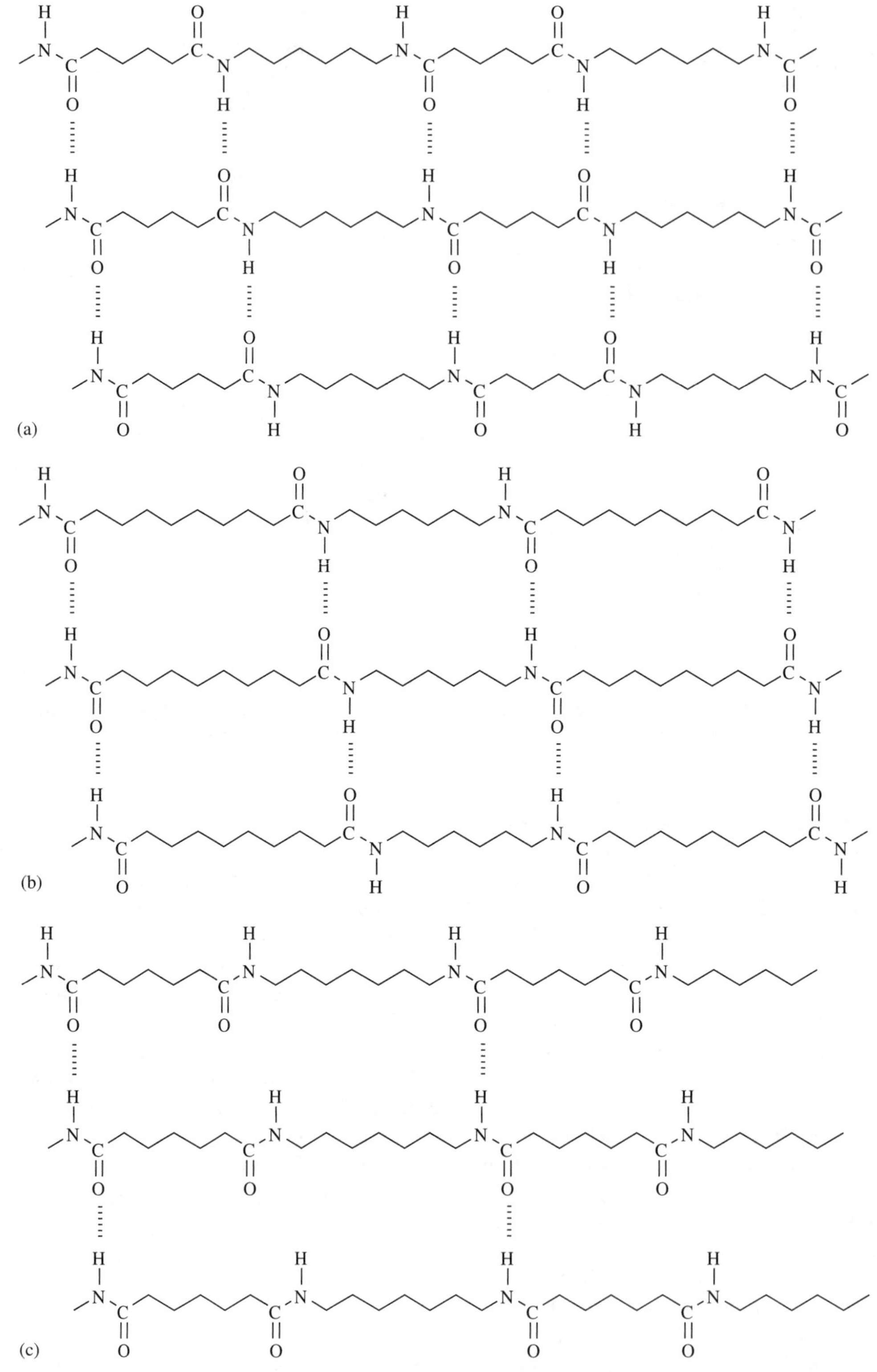

(a)

(b)

(c)

FIGURE 17.37 Differences in patterns of hydrogen bonding in crystals of various nylon polymers: (a) nylon 6.6, $T_m = 265$ °C; (b) nylon 6.10, $T_m = 222$ °C; (c) nylon 7.7, $T_m = 210$ °C.

the T_m is lowered as its length is increased. On the other hand, an increase in the bulkiness of the side group restricts rotation about bonds in the main chain and so has the effect of raising T_m (Table 17.4).

It can be seen from the considerations outlined above that it is possible to exert a good deal of control upon the melting temperatures of polymers. It must be borne in mind that, generally, factors that affect the T_m of a polymer also change the glass transition temperature T_g and in general these two parameters cannot be varied independently of each other. Also, changing the structure of the polymer may affect the ease of crystallization and although the potential melting point of a crystalline phase may be high, the amount of this phase that may form could be low.

17.8.3.2 Molar Mass and Branching

For a particular type of polymer, the value of T_m depends upon the molar mass and degree of chain branching. This is because of the effects of chain ends, which are in low molar mass polymers and the branches in non-linear polymers, both have the effect of introducing defects into the crystals and so lower their T_m. If the molar mass of the polymer is sufficiently high for the polymer to have useful mechanical properties, the effect of varying M upon T_m is not strong. In contrast, the presence of branches in a high-molar-mass sample of polyethylene can reduce T_m by 30 K.

Chains ends and branches can be thought of as impurities that depress the melting temperatures of polymer crystals. The behaviour can be analysed in terms of the chemical potentials (Section 10.2.3) per mole of the polymer repeat units in the crystalline state μ_u^c and in the pure liquid μ_u^o (the standard state). For the *pure* polymer, which is a single-component system,

$$\mu_u^c - \mu_u^o = -\Delta G_u = -(\Delta H_u - T\Delta S_u) \tag{17.35}$$

where

ΔG_u is the free energy of fusion per mole of repeat units

ΔH_u and ΔS_u are the enthalpy and entropy of fusion per mole of repeat units.

It was shown in Section 17.7.3 that ΔH and ΔS would not be expected to be very temperature dependent between T and T_m^o and so from Equation 17.27, Equation 17.35 becomes

$$\mu_u^c - \mu_u^o = -\Delta H_u\left(1 - \frac{T}{T_m^o}\right) \tag{17.36}$$

For a multicomponent system in equilibrium, the chemical potential of a component is given by Equation 10.35. Thus, the difference in chemical potential in Equation 17.36 is given by

$$\mu_u^c - \mu_u^o = \mathbf{R}T\ln a$$

where a is the activity of the crystalline phase, which is less than unity due to the presence of the 'impurity,' that depresses the melting point to T_m. Combining this equation with Equation 17.36 gives

$$\frac{1}{T_m} - \frac{1}{T_m^o} = -\frac{\mathbf{R}}{\Delta H_u}\ln a \tag{17.37}$$

If the mole fraction of crystallizable polymer is X_A, and that of the impurity X_B, then to a first approximation Equation 17.37 becomes, assuming ideal behaviour

$$\frac{1}{T_m} - \frac{1}{T_m^o} = -\frac{\mathbf{R}}{\Delta H_u} \ln X_A \tag{17.38}$$

and for small values of X_B it is easy to show that $-\ln X_A \approx X_B$. Hence it follows that

$$\frac{1}{T_m} - \frac{1}{T_m^o} = \frac{\mathbf{R}}{\Delta H_u} X_B \tag{17.39}$$

Linear polymers have two chain ends and so the mole fraction of chain ends is given approximately by $2/\bar{x}_n$. If these chain ends are considered to be the 'impurity,' then Equation 17.39 becomes

$$\frac{1}{T_m} - \frac{1}{T_m^o} = \frac{\mathbf{R}}{\Delta H_u} \frac{2}{\bar{x}_n} \tag{17.40}$$

This equation shows clearly that $T_m \rightarrow T_m^o$ as $\bar{x}_n \rightarrow \infty$ and is found to give a reasonable prediction of the dependence of melting temperature upon molar mass. It is easy to modify Equation 17.40 to account for branches where the term $2/\bar{x}_n$ becomes $y/\bar{x}_n$ if there are y ends per chain.

17.8.3.3 Copolymers

The melting point of a polymer also will be affected by *copolymerization*. In the case of random or statistical copolymers (Section 1.2.3), the structure is very irregular and so crystallization normally is suppressed and the copolymers usually are amorphous. In contrast, in block and graft copolymers, crystallization of one or more of the blocks may take place. It is possible to analyse the melting behaviour for a copolymer system in which there are a small number of non-crystallizable comonomer units incorporated in the chain using Equation 17.39. These units will act as 'impurities' (cf. chain ends) and so the melting point of the copolymer will again be given by Equation 17.39 with X_B defined as the mole fraction of non-crystallizable comonomer units. However, it must be emphasized that this equation only holds for low values of X_B.

17.8.4 Relationship between T_m and T_g

It is clear that the chemical factors outlined above as controlling T_m are also ones that have an effect upon T_g (Section 16.3). In fact, the same factors tend to raise or lower both T_m and T_g as both are controlled principally by main-chain stiffness. It is not surprising, therefore, that a correlation is found between T_m and T_g for polymers which exhibit both types of transition. It is found that when they are expressed in Kelvins, the value of T_g is generally between 0.5 T_m and $0.8T_m$. This behaviour is shown schematically in Figure 17.38. It demonstrates that it is not possible to control T_g and T_m independently for hompolymers. There are possibilities of having more control over T_g and T_m separately by making copolymers. For example, random copolymers of nylon 6.6 and nylon 6.10 can be made for which the T_g is very little different from that of the homopolymers, because the stiffness of the main chain will be virtually unchanged. On the other hand, the irregularity introduced into the main chain reduces the ability of the molecules to crystallize and so lowers T_m.

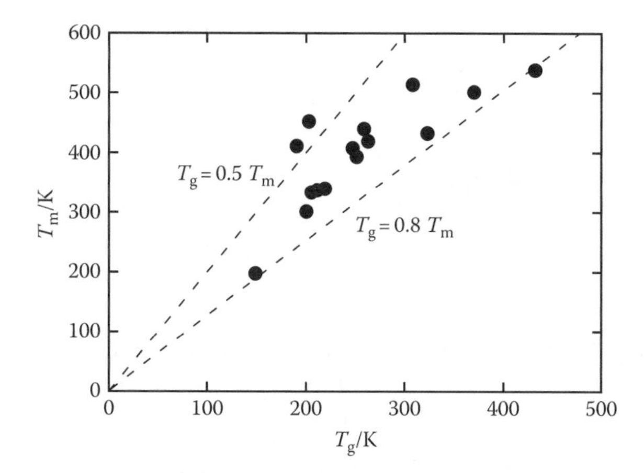

FIGURE 17.38 Plot of T_m against T_g for a variety of common polymers such as polyethylene, polypropylene, polystyrene, poly(ethylene oxide) etc. (Data taken from Boyer, R.F., *Rubb. Chem. Technol.*, 36, 1303, 1963.)

PROBLEMS

Unit cell dimensions for polyethylene may be assumed to be as follows:

$$a = 0.741 \text{ nm}, b = 0.496 \text{ nm}, c = 0.254 \text{ nm}, \alpha = \beta = \gamma = 90°$$

The wavelength of CuK_α radiation is 0.1542 nm.

17.1 A flat-plate X-ray diffraction pattern was obtained from an unoriented sample of polyethylene using CuK_α radiation. It consisted of three rings of radius 22.2, 36.6 and 19.7 mm. The specimen-to-film distance was 50 mm. Calculate the *d*-spacing of the planes giving rise to these reflections. It is thought that these reflections are from (*hk*0) type planes. Draw a scale diagram of the unit cell of polyethylene projected along the *c*-axis. Measure the spacing of the low-index (*hk*0) type planes and hence identify the planes giving rise to the observed reflections.

17.2 A flat-plate X-ray diffraction pattern was obtained from an oriented polymer fibre using CuK_α radiation. It consisted of a series of spots arranged along a horizontal line containing the central spot and first-order layer lines above and below the central line. It was found that with a specimen-to-film distance of 30 mm the first order lines were 22.9 mm above and below the central spot.

(a) Calculate the spacing of the chain repeat *c* of the polymer crystals.
(b) Explain why only first-order layer lines are observed for this polymer and suggest how higher order lines may be obtained.
(c) Determine the angle through which the specimen would have to be tilted in order to obtain the (001) reflection on the diffraction pattern.

17.3 Isotactic polypropylene $[-CH_2-CHCH_3-]_n$ has a monoclinic unit cell dimensions $a = 0.666$ nm, $b = 2.078$ nm and $c = 0.649$ nm (the polymer chain direction) and with $\gamma = 99.6°$ (the angle between *a* and *b*).

(a) Given that there are 12 polymer repeat units in the unit cell, determine the density of isotactic polypropylene crystals.
(b) If the density of amorphous polypropylene is 886 kg m^{-3}, use the information in (a) to determine the volume fraction of crystals in a sample of polypropylene with a density of 905 kg m^{-3}.
(c) Account for the observation that solution-grown polymer single crystals are not 100% crystalline.

[The Avogadro constant is 6.022×10^{23} mol^{-1} and the molar mass of C and H atoms may be taken as 12 g mol^{-1} and 1 g mol^{-1}, respectively.]

17.4 The following data were determined during the calibration of a density gradient column:

Marker Float Density/kg m^{-3}	Position in Column/cm
921.1	59.23
929.1	50.74
939.1	40.85
949.1	30.96
959.2	20.84

The density of polyethylene crystals is 998 kg/m^3 and that of amorphous polyethylene is 886 kg/m^3. Use the above information to determine both the *volume fraction* and *weight fraction of crystals* in a sample of polyethylene that settles in the column at a height of 44.41 cm.

17.5 A sample of poly(1,4-phenylene sulphide) (PPS) is known to have a glass transition temperature of 89°C, a cold-crystallization temperature of 137°C and a melting temperature of 278°C. Sketch the *heat-flux* differential scanning calorimetry (DSC) trace you would expect to observe on heating a completely amorphous specimen of this PPS from 25°C–325°C, labelling the axes and each of the thermal events you show.

A sample of this PPS (15.3 mg) that had been annealed at 150°C for 1 h was analysed by DSC using a heating rate of 20°C min^{-1} and the area of the melting endotherm found to be 192.55 mW K. Melting of indium (23.2 mg) was used for calibration and gave a melting endotherm of area 216.87 mW K on heating at 20°C min^{-1}. Given that the enthalpy for melting of indium is 28.6 J g^{-1} and that the enthalpy of melting for 100% crystalline PPS is 55.9 J g^{-1}, calculate

(a) The DSC instrument calibration factor, K_{DSC}

(b) The enthalpy of melting and degree of crystallinity of the annealed sample of PSS

17.6 A linear homopolymer was crystallized from the melt at crystallization temperatures (T_c) within the range 270–330 K. Following complete crystallization, the melting temperatures (T_m) were measured by differential scanning calorimetry. Determine the equilibrium melting temperature, T_m°.

T_c/K	T_m/K
270	300.0
280	306.5
290	312.5
300	319.0
310	325.0
320	331.0
330	337.5

Small-angle X-ray scattering experiments using CuK_α radiation gave the positions of the first maxima shown in the table below. Using the Bragg equation calculate the values of the long period for each crystallization temperature.

T_c/K	θ/°
270	0.44
280	0.39
290	0.33
300	0.27
310	0.22
320	0.17
330	0.11

The degree of crystallinity was measured to be 45% for all samples. Calculate the lamellar thickness in each case and hence determine graphically the fold surface energy, γ_e, if the enthalpy of fusion of the polymer per unit volume ΔH_V is 1.5×10^8 J m^{-3}.

17.7 The melting temperature (T_m) of a sample of poly(decamethylene adipate) with a number-average degree of polymerization ($\bar{x}_n$) of 3 was found to be 65°C. T_m was found to be 75°C for a sample of the same polymer with $\bar{x}_n = 10$. Estimate the equilibrium melting temperature T_m° for poly(decamethylene adipate).

17.8 The following melting temperature data were obtained for a series of polyamides of the type $-[-HN-(CH_2)_x-NH-CO(CH_2)_8CO-]-_n$:

x	T_m/°C
2	258
3	221
4	240
5	196
6	220

(a) Giving your reasons, account for the dependence of T_m upon x.
(b) What would be the effect on T_m of replacing the $-(CH_2)_x-$ groups in the polyamides with a *p*-phenylene group?

FURTHER READING

Alexander, L.E., *X-Ray Diffraction Methods in Polymer Science*, Wiley-Interscience, New York, 1969.
Bassett, D.C., *Principles of Polymer Morphology*, Cambridge University Press, Cambridge, U.K., 1981.
Bassett, D.C., *Developments in Crystalline Polymers-1*, Applied Science Publishers, London, U.K., 1982.
Bassett, D.C., *Developments in Crystalline Polymers-2*, Elsevier Applied Science, London, U.K., 1988.
Donald, A.M., Windle, A.H., and Hanna, S., *Liquid Crystalline Polymers*, 2nd edn., Cambridge University Press, Cambridge, U.K., 2005.
Hall, I.H., *Structure of Crystalline Polymers*, Elsevier Applied Science, London, U.K., 1984.
Hatakeyama, T. and Quinn, F.X., *Thermal Analysis: Fundamentals and Applications to Polymer Science*, 2nd edn., John Wiley & Sons, Chichester, U.K., 1999.
Hearle, J.W.S. (ed), *High Performance Fibres*, Woodhead Publishing Ltd., Cambridge, U.K., 2000.
Höhne, G.W.H., Hemminger, W.F., and Flammersheim, H.-J., *Differential Scanning Calorimetry*, 2nd edn., Springer-Verlag, Berlin, Germany, 2003.
Kelly, A. and Groves, G.W., *Crystallography and Crystal Defects*, Longman, London, U.K., 1970.
Magill, J.H., Morphogenesis of solid polymers, in *Treatise on Materials Science and Technology*, Vol. 10A (ed. J.M. Schultz), Academic Press, New York, 1977.
Menczel, J.D. and Prime, R.B. (eds), *Thermal Analysis of Polymers – Fundamentals and Applications*, John Wiley & Sons, Hoboken, NJ, 2009.
Reading, M. and Hourston, D.J. (eds), *Modulated Temperature Differential Scanning Calorimetry*, Springer, New York, 2006.
Sawyer L.C., Grubb D.T., and Meyers G.F., *Polymer Microscopy*, 3rd edn., Springer, New York, 2008.
Schultz, J.M., *Polymer Materials Science*, Prentice Hall, Englewood Cliffs, NJ, 1974.
Turi, E.A., *Thermal Characterization of Polymeric Materials*, 2nd edn., Academic Press, San Diego, CA, 1997.
Wunderlich, B., *Macromolecular Physics*, Vol. 1, Academic Press, London, 1973.
Wunderlich, B., *Macromolecular Physics*, Vol. 2, Academic Press, London, 1976.
Wunderlich, B., *Macromolecular Physics*, Vol. 3, Academic Press, London, 1980.

18 Multicomponent Polymer Systems

18.1 INTRODUCTION

The preceding two chapters were concerned principally with the structure of amorphous and crystalline single-component polymer systems. Polymers, however, often are used in practical applications in the form of multicomponent systems. This terminology covers a very wide range of structural forms that includes mixtures of polymers as blends, a variety of copolymer types such as block or graft, and different kinds of polymer-based composites in which the polymer is usually the matrix material. The term 'composite' can cover a variety of structural forms that include particulate- or fibre-reinforced polymers, nanocomposites and even foams (mixtures of a polymer and a gas). The properties of polymer-based composites reinforced with particles or fibres are covered in detail in Chapter 24. The present chapter will be restricted to two types of multiphase polymer systems, namely blends and block copolymers.

The reasons for employing multicomponent systems are manyfold. One of the main drivers may be the simple matter of a reduction in cost – the addition of a relatively cheap second component may reduce the cost of an expensive polymer without affecting its performance significantly. On the other hand, the addition of a second component may impart an improvement in properties that can be either mechanical or functional (e.g. electrical and optical). In other cases, it may be possible to produce unique phase structures, particularly at the nanometre level, in multicomponent polymer systems that lead to materials with special properties not attainable with a single-component polymer system.

18.2 POLYMER BLENDS

If two chemically-different polymers are mixed together, it often is found that they undergo almost complete phase separation and form two separate phases. Figure 18.1 shows a transmission electron micrograph of a thin section of a blend of an engineering polymer with a high temperature elastomer. It can be seen that the blend consists of rubber particles of variable size, less than 0.5 μm in diameter, dispersed uniformly throughout the polymer matrix. Historically, it was thought that polymer–polymer miscibility would take place only relatively rarely, but these days it is recognized that different polymer–polymer combinations can show a variety of different types of behaviour ranging from complete phase separation to partial miscibility and even complete miscibility in some cases. Table 18.1 lists some examples of particular polymer pairs that are found to be either miscible or immiscible.

18.2.1 THERMODYNAMICS OF POLYMER BLENDS

The basic thermodynamic relationship governing the mixing of dissimilar polymers is

$$\Delta G_{\mathrm{m}} = \Delta H_{\mathrm{m}} - T\Delta S_{\mathrm{m}} \tag{10.2}$$

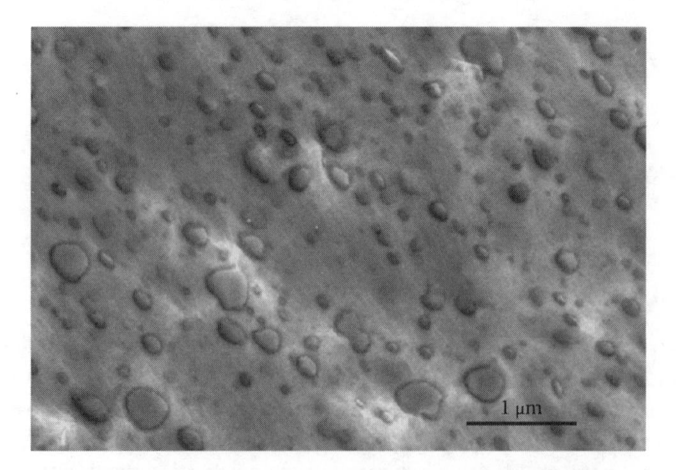

FIGURE 18.1 Transmission electron micrograph of a section from an immiscible blend of an engineering polymer with a high temperature elastomer which is stained and present as dispersed particles. (Courtesy of S. Rannou, University of Manchester, Manchester, U.K.)

TABLE 18.1
Examples of Some Miscible and Immiscible Polymer Pairs

Polymer 1	Polymer 2	Miscible
Polystyrene	Poly(2,6-dimethyl-1,4-phenylene oxide)	✓
Polystyrene	Poly(vinyl methyl ether)	✓
Poly(methyl methacrylate)	Poly(vinylidene fluoride)	✓
Poly(vinyl chloride)	Poly(butylene terephthalate)	✓
Poly(ethylene oxide)	Poly(acrylic acid)	✓
Polystyrene	Polybutadiene	×
Polystyrene	Poly(methyl methacrylate)	×
Polystyrene	Poly(dimethyl siloxane)	×
Nylon 6	Poly(ethylene terephthalate)	×

Source: Adapted from Sperling, L.H., *Introduction to Physical Polymer Science*, 4th edn., Wiley Interscience, Hoboken, NJ, 2006.

where
ΔG_m is the free energy of mixing
ΔH_m is the enthalpy of mixing
ΔS_m is the entropy of mixing.

For the two polymers to be miscible, ΔG_m must be negative. In the case of low molar mass materials, an increase in temperature usually leads to an increase in the $T\Delta S_m$ term thus making ΔG_m more negative and leading to better miscibility as the temperature is increased. This is not always the case with polymer–polymer mixtures which usually are found to become less miscible as the temperature is increased.

It is possible to employ the Flory–Huggins theory (Section 10.2.2), developed initially for polymer–solvent mixtures, to model also the thermodynamic behaviour of polymer blends. In this case, the two types of polymer chains are considered to be made up of segments that sit on a lattice of identical cells as shown in Figure 18.2. Polymer 1 occupies a continuous sequence of x_1

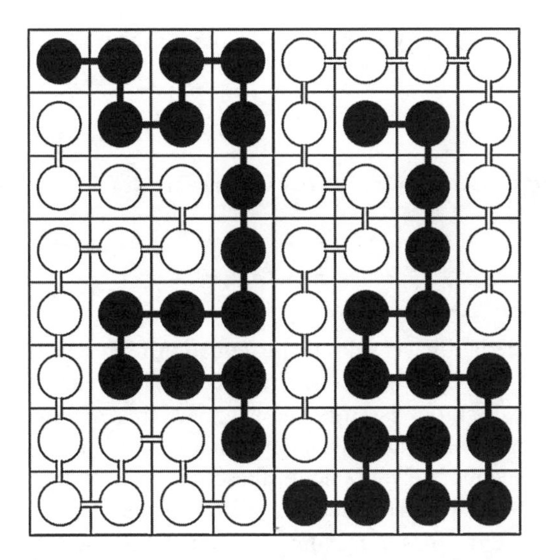

FIGURE 18.2 Flory–Huggins lattice model for a polymer blend. The black segments represent one type of polymer and white ones the other type.

segments and polymer 2 occupies a continuous sequence of x_2 segments. The volume fractions of the two polymers are given by

$$\phi_1 = \frac{N_1 x_1}{(N_1 x_1 + N_2 x_2)} \quad \text{and} \quad \phi_2 = \frac{N_2 x_2}{(N_1 x_1 + N_2 x_2)} \tag{18.1}$$

where N_1 and N_2 are the number of molecules of each type of polymer. In blends, the interactions that have to be considered are either between segments of the same polymer or between segments of the different polymers. It is possible to use the analysis for a polymer solution in Section 10.2.2 with a slight modification taking into account that there are now two polymers rather than a single polymer and solvent. Equation 18.1 leads to Equation 10.26 becoming in the case of a blend

$$\Delta G_{\mathrm{m}}^{\mathrm{contact}} = (z-2)x_1 N_1 \phi_2 \Delta g_{12} \tag{18.2}$$

and so to Equation 10.31 becoming

$$\Delta G_{\mathrm{m}}^{\mathrm{contact}} = \mathbf{R}T x_1 n_1 \phi_2 \chi \tag{18.3}$$

where

n_1 represent the number of moles of polymer 1

χ is now the Flory–Huggins polymer–polymer interaction parameter.

The combinational entropy Equation 10.23 is generic for any mixture and so is applicable to polymer–polymer mixtures

$$\Delta S_{\mathrm{m}}^{\mathrm{comb}} = -\mathbf{R}[n_1 \ln \phi_1 + n_2 \ln \phi_2] \tag{10.23}$$

The free energy of mixing can again be determined from the relation

$$\Delta G_{\rm m} = \Delta G_{\rm m}^{\rm contact} - T\Delta S_{\rm m}^{\rm comb}$$

to give the *Flory–Huggins equation for the Gibbs free energy of mixing* for the polymer–polymer blend

$$\Delta G_{\rm m} = \mathbf{R}T[n_1 \ln \phi_1 + n_2 \ln \phi_2 + x_1 n_1 \phi_2 \chi] \quad (18.4)$$

The first two terms on the right-hand side of this equation represent the combinational entropy of mixing and the third is the enthalpy term. In the case of solvent–solvent or polymer–solvent mixtures, the combinational entropy is large and negative (since ϕ_1 and ϕ_2 always are fractions, their logarithms will be negative). Hence as the temperature is raised, $\Delta G_{\rm m}$ will become more negative and mixing will become more favourable for these mixtures. In the case of polymer–polymer blends, the combinational entropy term becomes less important because both n_1 and n_2 are relatively small and the thermodynamics of mixing is dominated by the enthalpy term, $\mathbf{R}Tx_1n_1\phi_2\chi$. It was shown in Section 10.2.2 that the interaction parameter χ has a temperature dependence of the form

$$\chi = a + \frac{b}{T} \quad (10.28)$$

This temperature dependence of χ can lead to the enthalpy term and hence $\Delta G_{\rm m}$ in Equation 18.4 becomes less negative as the temperature is raised resulting in decreasing miscibility for the polymer blend.

Although the Flory–Huggins approach is useful in analysing the characteristics of polymer blends, it is only able to show trends and is not capable of predicting the behaviour accurately as it ignores a number of factors such as any volume changes that may take place upon mixing. The solubility parameter approach (Section 10.2.5) also is a useful guide to predicting miscibility in polymer blends. For example, poly(2,6-dimethyl-1,4-phenylene oxide) has a solubility parameter δ of around 19 $J^{1/2}$ $cm^{-3/2}$, and Table 18.1 shows that it is miscible with polystyrene which has a similar solubility parameter (see Table 10.2).

18.2.2 Phase Behaviour

The phase behaviour of polymer blends has similarities with that of polymer solutions (Section 14.2.1) and again can be treated theoretically using the Flory–Huggins theory. It is normal to deal with molar quantities so both sides of Equation 18.4 have to be divided by $(x_1n_1 + x_2n_2)$, where n_1 and n_2 are the numbers of moles of the two polymers present in the blend and x_1 and x_2 are the numbers of segments of each type of polymer. The Gibbs free energy of mixing per mole of segments is then given by

$$\Delta G_{\rm m}^* = \mathbf{R}T\left[\left(\frac{\phi_1}{x_1}\right)\ln \phi_1 + \left(\frac{\phi_2}{x_2}\right)\ln \phi_2 + \chi\phi_1\phi_2\right] \quad (18.5)$$

A schematic *phase diagram* of a polymer blend is shown in Figure 18.3 (c.f. Figure 14.2). This plot captures the characteristics of the processes, mapping out areas where different phases are present as a function of both the composition of the mixture and temperature. The diagram shows three regions with different degrees of miscibility: a single-phase region between the two binodals,

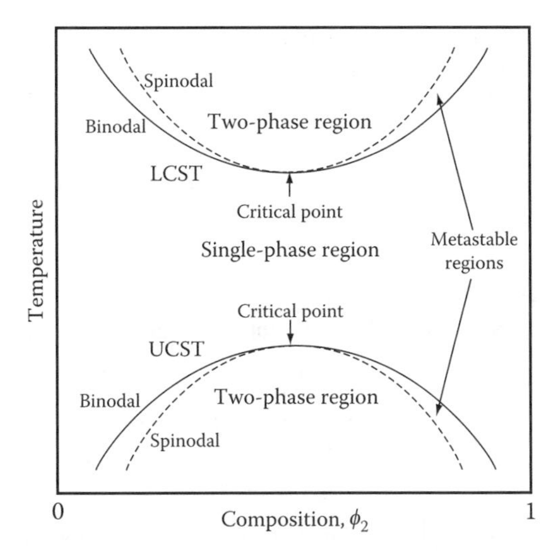

FIGURE 18.3 Phase diagram of a polymer blend showing the LCST and UCST and the binodals and spinodals.

metastable regions between the binodals and spinodals, and two-phase regions of immiscibility that are bordered by spinodals. The diagram also shows two critical solution temperatures; the lower critical solution temperature (LCST) at high temperature and the upper critical solution temperature (UCST) at lower temperature. The miscible (one-phase) and metastable regions are separated by the binodals and the metastable and two-phase regions are separated by the spinodals. Phase separation will take place when a single-phase system undergoes either a change in composition or, more commonly, a change in temperature that forces it to cross the binodals or spinodals. The two-phase regions within the spinodals correspond to areas where phase separation occurs by a process akin to crystallization, where by there is nucleation and growth of the phase-separated domains. In contrast, in the metastable regions between the binodals and spinodals, if the energy barrier can be overcome, phase separation will occur by a spontaneous process known as *spinodal decomposition*.

It was shown in Section 14.2.1 that the condition for phase separation is given in general for the spinodal as

$$\left(\frac{\partial^2 \Delta G_{\mathrm{m}}^*}{\partial \phi^2}\right) = 0 \tag{18.6}$$

and the critical temperature T_c by application of the general condition

$$\left(\frac{\partial^2 \Delta G_{\mathrm{m}}^*}{\partial \phi^2}\right) = \left(\frac{\partial^3 \Delta G_{\mathrm{m}}^*}{\partial \phi^3}\right) = 0 \tag{18.7}$$

These derivatives can be determined readily from Equation 18.5 by assuming that χ is independent of ϕ_1 and remembering that $\phi_2 = 1 - \phi_1$. Hence, $(\partial^2 \Delta G_{\mathrm{m}}^* / \partial \phi_1^2) = 0$ leads to the following equation for the spinodal points

$$\frac{1}{x_1 \phi_1} + \frac{1}{x_2 (1 - \phi_1)} - 2\chi = 0 \tag{18.8}$$

and so a theoretical spinodal can be constructed if the variation of χ with temperature is known (e.g. Equation 10.28). The critical composition can be determined from application of the condition $(\partial^3\Delta G_m^*/\partial\phi_1^3)=0$ to give

$$\phi_{1c}=\frac{1}{1+(x_1/x_2)^{1/2}} \tag{18.9}$$

Hence, it follows that if $x_1=x_2$, $\phi_{1c}=0.5$. The critical value of the Flory–Huggins polymer–polymer interaction parameter χ_c can be obtained by substituting Equation 18.9 into 18.8

$$\chi_c=\frac{1}{2}\left[\frac{1}{x_1^{1/2}}+\frac{1}{x_2^{1/2}}\right]^2 \tag{18.10}$$

Phase diagrams are employed widely in Materials Science and often are used in analysis of the composition and microstructures of metal alloys. They can be used in a similar way with polymer blends. The phase diagram for a polymer blend can be determined from the above analysis, although numerical methods have to be employed unless the situation is simplified, such as for the case of a mixture of two polymers which have an equal number of segments, where $x_1=x_2=x$. In this case, the critical point is located at $\phi_{1c}=0.5$ and Equation 18.10 gives

$$\chi_c=\frac{2}{x} \tag{18.11}$$

The analysis outlined above yields an analytical solution for the phase diagram of a polymer blend if the variation of χ with temperature is known.

A variety of different types of phase behaviour have been found for binary polymer blends. Some blends undergo phase separation on heating whereas others phase separate on cooling, depending upon whether χ increases or decreases with temperature. In general, solvent–solvent and polymer–solvent mixtures usually show an upper critical solution temperature (UCST) and become more miscible as the temperature is increased, whereas polymer–polymer mixtures, which normally become less miscible with increasing temperature, exhibit a lower critical solution temperature (LCST). Experimental phase diagrams for polymer blends are invariably asymmetrical (unlike Figure 18.3), unless the molar mass of the two polymers is similar. Where there is a large difference in molar mass they can be very asymmetric. The phase diagram illustrated in Figure 18.3 is for liquid–liquid phase separation. If either of the components is capable of crystallizing, the phase behaviour can be complicated further.

In the case of highly miscible polymers, it may be difficult in practice to determine the LCST or UCST since at low temperatures the blend may undergo a glass transition that restricts molecular motion and does not allow phase separation to occur. On the other hand, degradation of the polymers may occur at high temperature to make this region also inaccessible experimentally. For highly immiscible polymer blends, the phase diagram will normally show a two-phase region over the entire temperature and composition range.

Figure 18.4 is a schematic phase diagram for a polymer blend showing LCST behaviour. For a given temperature above the LCST, the points of intersection with the phase boundary lines denote the limits of miscibility of the two polymers, ϕ_{2a} and ϕ_{2b} at this temperature. For a given blend composition ϕ_2, the volume fractions of the phase rich in component 1, $\phi_2^{1\text{rich}}$ and that rich in component 2, $\phi_2^{2\text{rich}}$ in the two-phase region can be determined by undertaking the tie-line calculation illustrated in Figure 18.4 such that

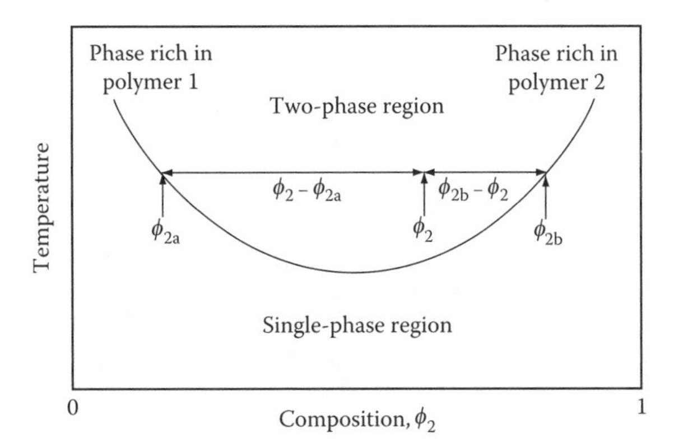

FIGURE 18.4 The tie-line calculation to determine the phase compositions in a polymer blend. (Adapted from Robeson, L.M., *Polymer Blends: A Comprehensive Review*, Hanser, Cincinnati, OH, 2007.)

$$\frac{\phi_2^{1\,\text{rich}}}{\phi_2^{2\,\text{rich}}} = \frac{\phi_{2b} - \phi_2}{\phi_2 - \phi_{2a}} \tag{18.12}$$

18.2.3 Glass Transition Behaviour

The glass transition behaviour of a polymer blend depends very much on its phase behaviour (Figure 18.3). If the two polymers are totally immiscible, then the blend will exhibit two separate glass transition temperatures, similar to those of the two-component polymers. On the other hand, for completely miscible polymers the glass transition temperature of the blend T_g^{blend} can be determined using a relationship such as Equation 16.18 or 16.19 derived for single-phase copolymers, for example

$$\frac{1}{T_g^{\text{blend}}} = \frac{w_1}{T_g^1} + \frac{w_2}{T_g^2} \tag{18.13}$$

where the superscripts 1 and 2 refer to the weight fractions w of the two polymer components.

The behaviour becomes more complicated when there is partial miscibility, such as that shown in Figure 18.4. The blend will again show two glass transition temperatures, but in this case one will be for the phase rich in component 1 and the other for the phase rich in component 2. These two glass transition temperatures will be between those of the two-component polymers. The values of each T_g can be determined using Equation 18.13 knowing the values of w_1 and w_2 at the phase boundaries, determined from the volume fractions ϕ_1 and ϕ_2 at the phase boundaries and densities of each component polymer.

18.2.4 Compatibilization of Polymer Blends

It has been pointed out already that most polymer mixtures are immiscible. Moreover, they also are incompatible, which means that if an attempt is made to mix the polymers together in the melt, the particles of the dispersed phase invariably tend to coalesce and grow larger to reduce the surface area of the interface. Hence, in order to obtain good dispersions of small particles of the second phase, such as that shown in Figure 18.1, action has to be taken to modify the interface between the two phases.

There are a number of ways in which compatibilization can be achieved. The process of *in situ polymerization* can be used to produce covalent bonding between the constituent polymers resulting

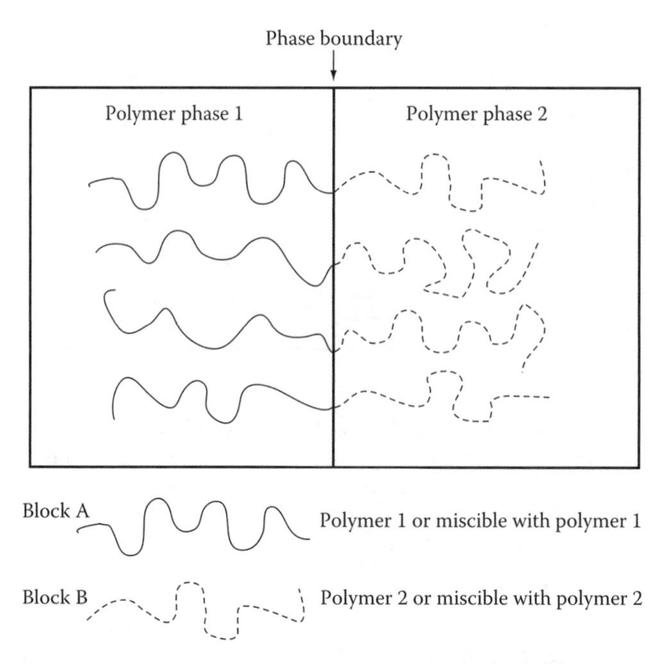

FIGURE 18.5 Schematic diagram of the compatibilization of a polymer blend using a di-block copolymer. (Adapted from Robeson, L.M., *Polymer Blends: A Comprehensive Review*, Hanser, Cincinnati, OH, 2007.)

in graft or block copolymers that stabilizes the interface between the two phases. Another method is to add a *ternary polymeric component* to the blend of immiscible polymers. This normally is a polymer that has good interfacial adhesion with both phases and so concentrates in the interfacial regions leading to a good dispersion and facilitating efficient stress transfer across the interface between the two phases. There are a number of methods whereby *reactive compatibilization* can be used. A third polymer can be added which is miscible with one of the component polymers and capable of reacting with the other. This leads to the formation of a graft copolymer that will bridge the interface. *Reactive extrusion* can be used in which chemical reactions take place between the components of the blend in the extreme conditions of the extruder, again leading to the formation of graft copolymers. *Polymer–polymer reactions* also can be used to improve compatibility. One example is the ester interchange reaction (Section 3.2.5), that can take place between two immiscible polyesters leading to the formation of some 'blocky' copolymer. In the extreme, this process can be used to promote the formation of a miscible blend from two polymers that were initially immiscible.

One of the neatest methods of compatibilization is simply to add a *block copolymer* in which each of the blocks has the same or similar composition to that of the component polymers of the blend. This system will have its lowest free energy when the block copolymer is concentrated at the interface between the phases, as shown in Figure 18.5. This is an example of an AB di-block copolymer in a two-component system where the A-block of the block copolymer resides in one phase and the B-block in the other phase, which places covalent bonds across the interface. This concept is used widely in incompatible blend systems and leads to good control of the morphology combined with an improvement in interfacial adhesion and hence mechanical properties.

18.3 BLOCK COPOLYMERS

Copolymers are polymers derived from more than one species of monomer, as described in Section 1.2.3. *Block copolymers* are made up of sequences, or blocks, of chemically different repeat units (A and B). Methods for their synthesis were described in Section 9.4 and some typical block

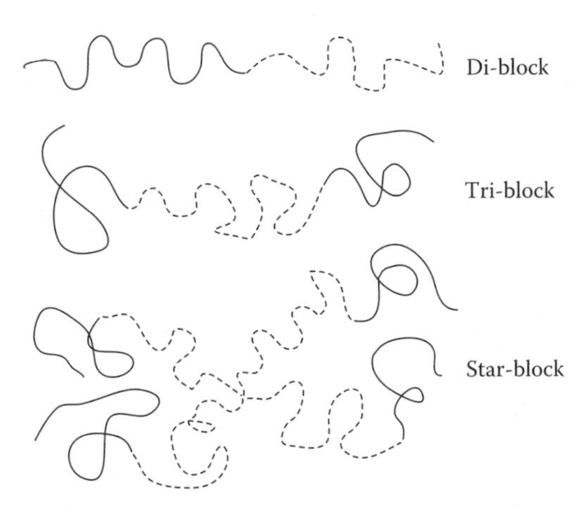

FIGURE 18.6 Schematic illustration of different block copolymer architectures. The solid lines represent block A and the dashed lines represent block B.

copolymer architectures are shown in Figure 18.6. The simplest example is a di-block copolymer consisting of single blocks of the two different repeat units, A and B. Tri-block copolymers can have either ABA or BAB block architecture or may be of the type ABC if a third monomer is employed in the synthesis. Another example that is given in Figure 18.6 is that of a star-block copolymer, and although there is an infinite number of possibilities of novel and different architectures for block copolymers, only di-block and tri-block copolymers will be considered in detail here.

18.3.1 Thermodynamics of Block Copolymer Phase Behaviour

One interesting and useful aspect of the behaviour of block copolymers is that they are able to self assemble on the nanometre scale into ordered structures through the process of *phase separation*. The driving force for this is the enthalpy of de-mixing of the blocks which leads to a tendency for blocks to phase separate in a similar way to polymer blends. In block copolymers, however, there is the extra constraint in that the constituent blocks are connected through covalent bonds that prevent the macroscopic phase separation found in polymer blends (Figure 18.1). This also means that the dimensions of the ordered structures are relatively small compared with blends, typically no more than a few times the radius of gyration $\langle s^2 \rangle^{1/2}$ of the copolymer (Section 10.3). The overall phase separation process in block copolymers is, therefore, controlled by two factors: the system is trying to minimize the area of the interface between the two micro-phases whilst trying to maximize the entropy of the polymer chains.

It is found that the phase structure of block copolymers is controlled by only three parameters: the composition of the copolymer characterized by f_A (the overall volume fraction of component A), the overall degree of polymerization x and the Flory–Huggins A–B segment interaction parameter χ. The transformation of a homogeneous molten di-block copolymer into heterogeneous micro-phases is known as an *order–disorder transition*. This is shown schematically in Figure 18.7 and for a given overall block copolymer composition, the transition is found to occur at a critical value of the product χx. It can be predicted using the mean-field theory that, for a symmetric di-block copolymer ($f_A = f_B = 0.5$), the critical value of this product for the order–disorder transition will be $(\chi x)_{crit} \sim 10.5$. The product χx characterizes the balance between enthalpic and entropic factors for the di-block copolymer and it can be contrasted with a critical value of $(\chi x)_{crit} = 2$ for phase separation in a symmetric polymer blend given by Equation 18.11. The difference between the symmetric di-block copolymer and the symmetric polymer blend is that, in addition to the loss of mixing entropy with a one-component system, due to the connectivity in the di-block copolymer there also is a loss in entropy due to chain stretching in the two separated phases, shown in Figure 18.7, not encountered

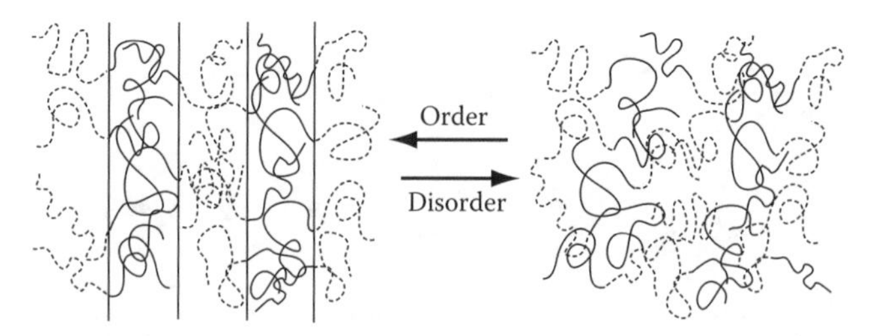

FIGURE 18.7 Schematic illustration of the order–disorder transformation in a di-block copolymer.

in a phase-separated blend. This entropy penalty is proportional to the overall degree of polymerization x. For most di-block copolymers, the interaction parameter has a temperature dependence given by Equation 10.28, $\chi = a + b/T$, where a and b (>0) are constants. Increasing the interaction parameter χ of a disordered block copolymer melt by, for example, reducing the temperature will drive the system toward local ordering as shown in Figure 18.7. On the other hand if χ (or x) is decreased sufficiently for an ordered di-block copolymer, it will transform to a disordered phase.

It is possible to extend considerations of the thermodynamics of block copolymers using the mean-field approach to predict the dependence of the order–disorder transition and other transitions upon χx and f_A. The predicted *morphology diagram* is shown in Figure 18.8, and although it is not a phase diagram in the traditional sense (a phase diagram is normally a map of phase behaviour as a function of temperature and composition as shown in Figure 18.3) it can be treated in essentially the same way (remembering that χ decreases as T increases). It can be seen from Figure 18.8 that when the product χx exceeds ~10.5, the molecular segments in symmetric di-block copolymers ($f_A = f_B = 0.5$) self assemble to form a lamellar micro-phase structure (LAM) which consists of alternating parallel layers. In this region, just above the order–disorder transition, the compositional variation across the micro-domain interface is not very abrupt and it is more like a local sinusoidal

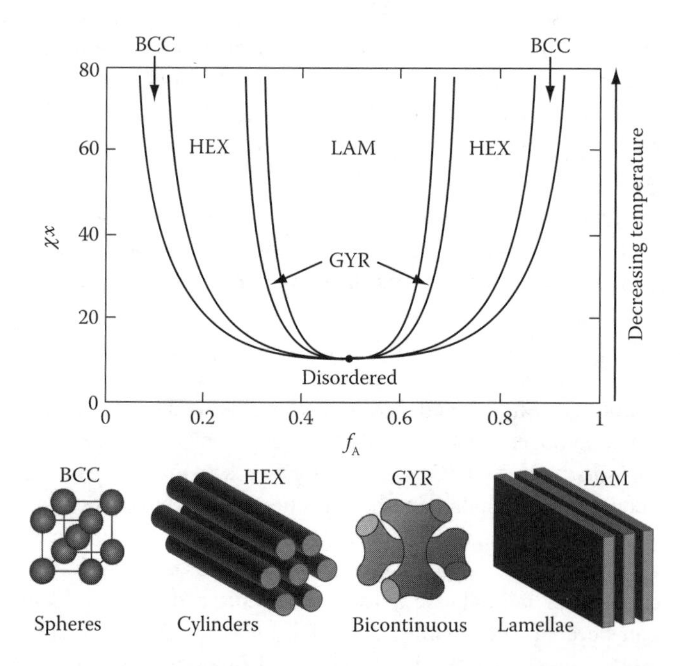

FIGURE 18.8 Theoretical block copolymer morphology diagram for a di-block copolymer showing regions where different phase structures are predicted: LAM, Lamellar; GYR, bicontinuous gyroid; HEX, hexagonal close-packed cylinders; BCC, body-centred close-packing of spheres. Schematic diagrams of the different structures are also presented. (Adapted from Matsen, M.W. and Bates, F.S., *Macromolecules*, 29, 1091, 1996.)

variation in composition rather than the formation of nearly pure phases. This generally is termed the *weak segregation* regime. It is not until $\chi x \sim 100$ that the components of the two phases are relatively pure and there is only a very narrow interphase between them with *strong segregation*.

In the case of asymmetric di-block copolymers ($f_A \neq f_B$), it is more energetically favourable for the phases to have curved interfaces and the molecular segments self assemble into different structures. The formation of the different phase structures also is constrained by the entropy penalty of ensuring that the polymer chains are not stretched too much. Hence, the microstructures obtained involve keeping a balance between interface curvature and chain stretching. When there is strong segregation, then as the copolymer composition moves away from $f_A = 1/2$ increasing interfacial curvature leads to firstly a bicontinuous gyroid phase (GYR), then a phase of hexagonal close-packed cylinders (HEX) and finally a body-centred cubic phase (BCC) as shown in Figure 18.8. If the length of the segments A and B is the same in the copolymer, this behaviour is symmetrical either side of $f_A = f_B = 0.5$. A phase rich in component B becomes the continuous phase in the microstructure as f_A decreases, and similarly a phase rich in component A becomes the continuous phase in the microstructure as f_A increases.

18.3.2 Morphology of Block Copolymers

Since block copolymers have phase domains on the scale of tens of nanometres, it is not possible to observe their morphology using optical microscopy and so higher-resolution specialized techniques such as electron microscopy or small-angle X-ray scattering have to be employed.

18.3.2.1 Transmission Electron Microscopy

The most direct method of visualizing the phase domains in block copolymers is transmission electron microscopy. Specimens generally are prepared by casting films from solutions that are then annealed to produce the equilibrium morphology, which is then ultra-microtomed to produce ultrathin sections. The sectioning is undertaken ideally at cryogenic temperatures (typically −100 °C) to avoid damage to or distortion of the morphology during sectioning. Contrast between the two phases often can be enhanced by staining using a variety of different staining agents depending upon the constituent polymers. For example, if one of the phases is a polydiene rubber, then OsO_4 can be employed as it selectively stains the rubbery component dark. The morphologies of a series of block copolymers based upon different compositions of polystyrene and polyisoprene or polybutadiene are described next.

Figure 18.9a shows that the morphology of a polystyrene-*block*-polybutadiene-*block*-polystyrene ABA tri-block copolymer with similar volume fractions of the two components ($f_{PS} \sim 0.45$) is lamellar in form (LAM) as predicted in Figure 18.8. The lamellae are visualized edge-on in this section

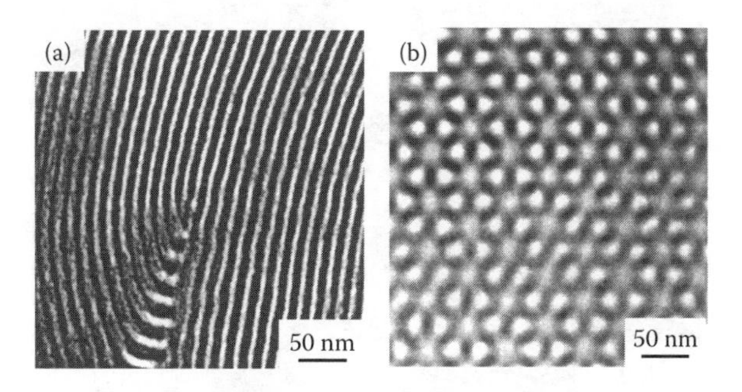

FIGURE 18.9 Transmission electron micrographs of stained sections of block copolymers. (a) A polystyrene-*block*-polybutadiene-*block*-polystyrene ABA tri-block copolymer with similar volume fractions of the two components ($f_{PS} \sim 0.45$). (b) A polystyrene-*block*-polyisoprene-*block*-polystyrene ABA tri-block copolymer with $f_{PS} \sim 0.33$. (Courtesy of B.J. Dair and E.L. Thomas, MIT, Cambridge, MA.)

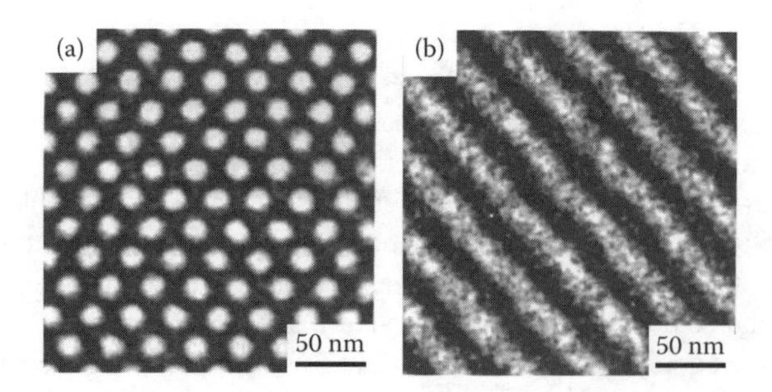

FIGURE 18.10 Transmission electron micrographs of stained sections of a polystyrene-polyisoprene star-block copolymer with the polyisoprene blocks at the core of the star (Figure 18.6) and $f_{PS} \sim 0.30$. (a) Transverse section showing hexagonal packing of the polystyrene cylinders. (b) Section parallel to the cylinders. (Courtesy of D.J. Kinning and E.L. Thomas, MIT, Cambridge, MA.)

although there is a region where they are twisted and form a grain boundary. When the volume fraction of the minority component is changed to around 0.33, a completely different morphology is obtained as shown in Figure 18.9b for a polystyrene-*block*-polyisoprene-*block*-polystyrene ABA tri-block copolymer with $f_{PS} \sim 0.33$. In this case, a bicontinuous gyroid (GYR) morphology is obtained which has a characteristic wagon-wheel appearance in this section (along a [111] projection, i.e. the diagonal of the cubic unit cell) with threefold symmetry. This again is consistent with the prediction in Figure 18.8 for this value of f_{PS}.

The effect of reducing the volume fraction of polystyrene further is shown in Figure 18.10 where a hexagonal close-packed structure (HEX) is found in a polystyrene-polyisoprene star-block copolymer with the polyisoprene blocks at the core of the star (Figure 18.6) and $f_{PS} \sim 0.30$. A HEX structure is predicted in Figure 18.8 for this composition and the hexagonal packing of cylinders in the morphology can be clearly visualized by taking transverse and parallel sections.

Figure 18.11 shows the microstructure of a polystyrene-*block*-polybutadiene copolymer containing a low volume fraction of polystyrene ($f_{PS}=0.074$) with a body-centred cubic (BCC) structure. The continuous phase is polybutadiene (stained dark) and the most stable structure is that of small polystyrene spheres on a BCC lattice (see Figure 18.8), which appears as a square lattice in this section of the morphology.

One of the disadvantages of TEM is that there can often be misidentification of the morphology as only 2D projections are investigated and they are often only from small regions of the samples.

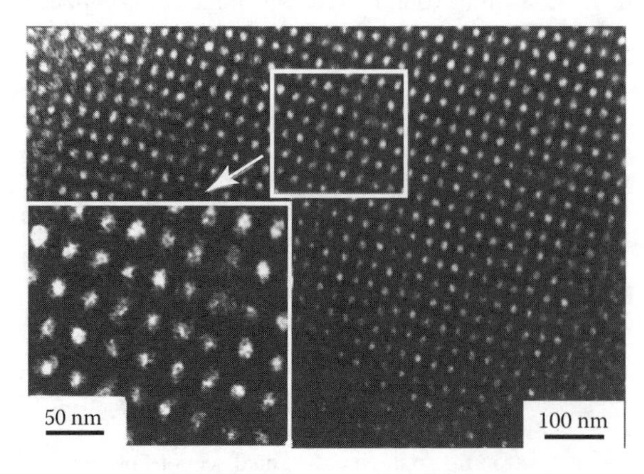

FIGURE 18.11 Transmission electron micrograph of a stained section of a polystyrene-*block*-polybutadiene copolymer ($f_{PS}=0.074$) with a BCC structure. Details of the microstructure are shown inset. (Courtesy of D.J. Kinning and E.L. Thomas MIT, Cambridge, MA.)

Hence, a complete 3D analysis of the morphology needs to be confirmed by a technique that probes much larger and more representative volumes of the material (~1 mm^3), such as small-angle X-ray scattering.

18.3.2.2 Small-Angle X-Ray Scattering

Block copolymers have phase domains with dimensions typically on the length scale of 1–100 nm and so small-angle X-ray and neutron scattering (Section 12.4) are ideal techniques for the characterization of their morphology. Figure 18.12 shows a small-angle X-ray scattering (SAXS) plot of the scattered intensity $I(q)$ as a function of the scattering vector q for a polystyrene-*block*-polyisoprene copolymer with a lamellar microstructure. The scattering vector in the case of SAXS is defined conventionally as

$$q = \frac{4\pi \sin\theta}{\lambda} \tag{18.14}$$

where

λ is the wavelength of the radiation
θ is half the scattering angle 2θ
q has dimensions of reciprocal length.

(It should be pointed out that in the case of light scattering [Equation 12.23], the scattering angle is defined as θ whereas, to be consistent with the Bragg equation, the scattering angle for SAXS is 2θ. Also, the value of n_0 in Equation 12.23 is 1 for X-rays.) It should be recalled that Bragg's law for X-ray scattering (Equation 17.4) is given by

$$n\lambda = 2d\sin\theta \tag{18.15}$$

where

d is the spacing of the planes (lamellae in this case)
n is the order of the reflection.

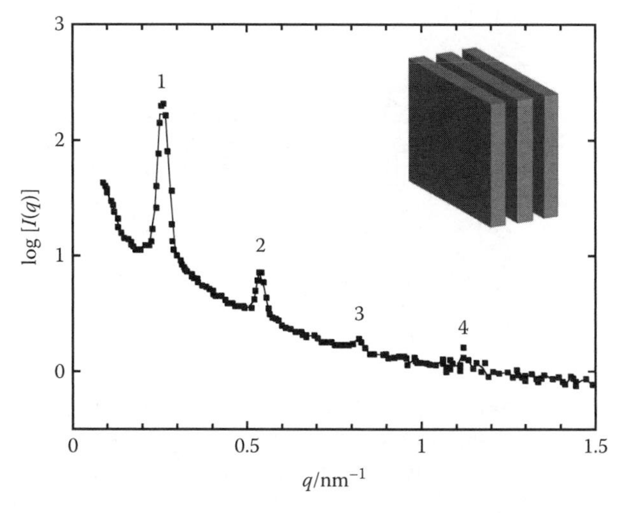

FIGURE 18.12 Small-angle X-ray scattering data for a polystyrene-*block*-polyisoprene copolymer (f_{PS} = 0.64) with a lamellar morphology showing a plot of $I(q)$ versus q. The different orders of reflection n are indicated. (Data taken from Khandpur, A.K. et al., *Macromolecules*, 28, 8796, 1995.)

By combining Equations 18.14 and 18.15, the lamellar plane spacing d can be related to the scattering vector by

$$d = \frac{2\pi n}{q} \quad (18.16)$$

The advantage of using the scattering vector q is that data from a variety of sources and using different wavelengths can be easily compared. The scattering pattern in Figure 18.12 was obtained with the X-ray beam parallel to the layers of an aligned lamellar structure. It consists of a series of equally spaced peaks for different indicated orders of reflection n and the lamellar plane spacing d is calculated to be 23.3 nm.

18.3.3 Thermoplastic Elastomers

Chemical crosslinks normally are required for a polymer to display elastomeric behaviour but a novel category of polymers, known as *thermoplastic elastomers*, are capable of behaving as elastomers without the necessity of having chemical crosslinks. The materials can be processed as thermoplastics at elevated temperatures, but when they are cooled to ambient temperature they become elastomers. The 'crosslinks' in thermoplastic elastomers are physical rather than chemical in nature and these physical crosslinks anchor a network of flexible molecules in the material. This transition in behaviour from a thermoplastic melt to an elastomer is completely reversible and so, unlike conventional elastomers, the thermoplastic elastomers can be reprocessed without difficulty. Several types of polymers have been developed for use as thermoplastic elastomers and the materials are best described by considering specific examples.

Thermoplastic polyurethane elastomers are segmented copolymers prepared by the reaction of a diisocyanate with a prepolymer polyol and a short-chain diol as described in Section 9.4.1. The prepolymer polyol blocks form the *soft segments* and diisocyanate reacts with the diol to form polyurethane *hard segments*. The hard segments tend to aggregate to give ordered crystalline domains aided by the possibility of forming hydrogen bonds between adjacent hard segment units. In contrast, the soft segments are above their T_g and remain amorphous at ambient temperature. The structure is shown schematically in Figure 18.13 where it can be seen that the aggregates of hard segments act as

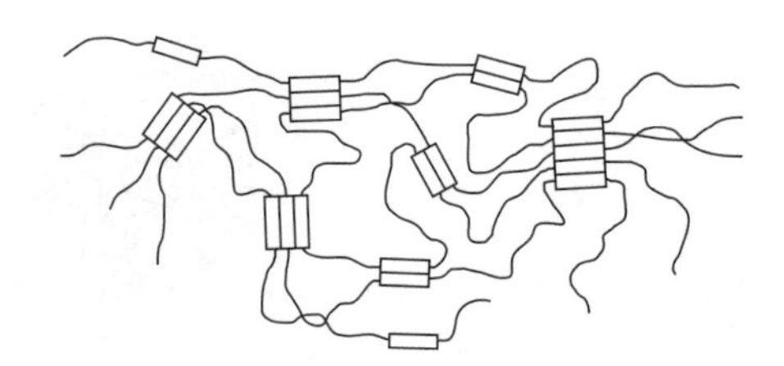

FIGURE 18.13 Schematic illustration of the formation of hard segments in a thermoplastic polyurethane elastomer. (Adapted from Schollenberger, C.S. and Dinbergs, K., *J. Elast. Plast.*, 1, 65, 1975.)

FIGURE 18.14 Schematic representation of the structure of an ABA tri-block copolymer of the polystyrene-*block*–polydiene-*block*–polystyrene type with $f_{PS} \sim 0.2$. The thicker lines represent the polystyrene blocks and the thinner lines the polydiene blocks.

junction points in a network of flexible molecules. Although the hard segment domains are relatively stable, they can be disrupted by heating the material to its processing temperature or by the action of a solvent. This means that the materials can be readily moulded into solid elastomeric artefacts or, by using a solvent, they can be applied as thin elastomeric coatings. Analogous behaviour is found for segmented copolyesters (Section 9.4.1).

As described in Section 9.4.2, methods such as 'living' polymerization can be used to prepare ABA tri-block copolymers suitable for use as thermoplastic elastomers. In such copolymers, the A blocks are normally of a homopolymer which is glassy (e.g. polystyrene) and the B-block is of a rubbery homopolymer (e.g. a polydiene such as polybutadiene or polyisoprene). The characteristic properties of these materials stem from the fact that two polymers that contain repeat units of a different chemical type tend to be incompatible on the molecular level. Thus, the block copolymers phase separate into domains that are rich in one or the other type of repeat unit. In the case of the polystyrene-*block*-polydiene-*block*-polystyrene ABA tri-block copolymers used as thermoplastic elastomers (with about 20% by volume polystyrene blocks), the structure is phase separated at ambient temperature into approximately spherical polystyrene-rich domains which are dispersed in a matrix of the polydiene chains (Figure 18.8). This type of structure is shown schematically in Figure 18.14, where it can be seen that the polystyrene blocks are anchored in the spherical domains. At ambient temperature, the polystyrene is below its T_g whereas the polydiene is above its T_g. Hence, the material consists of a rubbery matrix containing a rigid dispersed phase.

However, since the flexible polydiene blocks are linked by covalent bonds to the polystyrene blocks at the boundaries of the rigid domains, the structure is effectively 'physically crosslinked' and the materials display elastomeric properties. On heating the material to elevated temperatures well above the T_g of polystyrene, the polystyrene-rich domains become disrupted and the material can be processed as a normal thermoplastic.

PROBLEMS

18.1 The Gibbs free energy of mixing per mole of segments in a two-component polymer blend is given by

$$\Delta G_m^* = \mathbf{R}T\left[\left(\frac{\phi_1}{x_1}\right)\ln\phi_1 + \left(\frac{\phi_2}{x_2}\right)\ln\phi_2 + \chi\phi_1\phi_2\right]$$

where

ϕ_1 and x_1 are the volume fraction and number of segments of polymer 1
ϕ_2 and x_2 are the volume fraction and number of segments of polymer 2
χ is the Flory–Huggins polymer–polymer interaction parameter.

(a) If the condition for phase separation is given in general for the spinodal as

$$\left(\frac{\partial^2 \Delta G_m^*}{\partial \phi^2}\right) = 0$$

show that the equation defining the spinodal points is

$$\frac{1}{x_1\phi_1} + \frac{1}{x_2(1-\phi_1)} - 2\chi = 0$$

(b) If at the critical temperature, T_c, the general condition for the spinodal is

$$\left(\frac{\partial^2 \Delta G_m^*}{\partial \phi^2}\right) = \left(\frac{\partial^3 \Delta G_m^*}{\partial \phi^3}\right) = 0$$

(i) show that the critical composition is given by

$$\phi_{1c} = \frac{1}{1+(x_1/x_2)^{1/2}} \text{ and}$$

(ii) that the critical value of χ_c, the Flory–Huggins interaction parameter is

$$\chi_c = \frac{1}{2}\left[\frac{1}{x_1^{1/2}} + \frac{1}{x_2^{1/2}}\right]^2$$

(c) What is the value of ϕ_{1c} at the critical point when the two polymers have the same number of segments and the corresponding critical value of the Flory–Huggins interaction parameter χ_c?

18.2 The following data were obtained for the glass transition temperature, T_g, of a blend of poly(methyl methacrylate) and poly(vinylidene fluoride) as a function of the volume fraction of poly(vinylidene fluoride), ϕ_{PVDF}. Determine the T_g of a blend with $\phi_{PVDF}=0.8$ and comment upon its expected morphology.

T_g/°C	ϕ_{PVDF}
94	0.0
74	0.1
62	0.3
40	0.4
5	0.6
–45	1.0

18.3 The following data were obtained upon the glass transition temperatures T_g of various blends of natural rubber (NR) with different elastomers. For each pair of elastomers, explain the differences between the values of T_g for the 50/50 blends and the polymers separately.

Elastomer			T_g/°C	
I	II	Mixing Ratio I/II	I	II
NR	NBR	—	−69	−30
NR	NBR	50/50	−67	−33
NR	CR	—	−69	−43
NR	CR	50/50	−65	−44
NR	BR	—	−69	−48
NR	BR	50/50	−60	−60

NBR, poly(butadiene-*co*-acrylonitrile); CR, polychloroprene; BR, polybutadiene.

18.4 The small-angle X-ray scattering data for a polystyrene-block-polyisoprene copolymer with a lamellar morphology in Figure 18.12 show maxima at q values of approximately 0.264, 0.542 and 0.819 nm^{-1}. Determine the lamellar plane spacing of the domains and the thickness of the polyisoprene layers, given that the volume fraction of polystyrene f_{PS} is 0.64. Justify any assumptions you might make.

FURTHER READING

Abetz, V. and Simon, P.F.W., Phase behaviour and morphologies of block copolymers, *Advances in Polymer Science*, **189**, 125 (2005).
Blackley, D.C., *Synthetic Rubbers*, Applied Science Publishers, London, U.K., 1983.
Hamley I.W., *The Physics of Block Copolymers*, Oxford Science Publications, Oxford, 1998.
Hamley I.W. (ed.), *Developments in Block Copolymer Science and Technology*, John Wiley & Sons, Chichester, NY, 2004.
Haward, R.N. and Young R.J. (eds.), *The Physics of Glassy Polymers*, Chapman & Hall, London, U.K., 1997.
Jones R.A.L., *Soft Condensed Matter*, Oxford University Press, Oxford, U.K., 2002.
Robeson, L.M., *Polymer Blends – A Comprehensive Review*, Hanser, Cincinnati, OH, 2007.
Sperling, L.H., *Introduction to Physical Polymer Science*, 4th edn., Wiley-Interscience, Hoboken, NJ, 2006.

Part IV

Properties of Bulk Polymers

19 Elastic Deformation

19.1 INTRODUCTION

Polymers are being used increasingly in structural engineering applications where they are subjected to appreciable stress. Their increase in use is due to a number of factors. One of the most important is that although on an absolute basis their mechanical strength and stiffness may be relatively low compared with other materials, when the low density of polymers is taken into account, their specific strength and stiffness become comparable with those of metals and ceramics. Also, the fabrication costs of polymeric components usually are considerably lower than for other materials. Polymers melt at relatively low temperature and can be readily moulded into quite intricate components using a single moulding process.

Perhaps the most over-riding reason for using polymers is that they can display unique mechanical properties. One polymer that has been used for many years is natural rubber, which is elastomeric. It can be stretched to very high extensions and will snap back as soon as the stress is removed. Polyolefins are used widely in domestic containers and in packaging because of their remarkable toughness and flexibility and ease of fabrication. Other polymers are employed as adhesives. Epoxy resins can produce bonds between metals which can be as strong as those made by conventional joining methods such as welding or riveting.

With the now increasingly more widespread use of polymers, an understanding of their mechanical properties is essential. The mechanical properties will be considered here both from a phenomenological viewpoint and in terms of the molecular deformation processes that occur. This latter approach has been facilitated by the better understanding of polymer morphology that was outlined in earlier chapters.

19.2 ELASTIC DEFORMATION

In order to understand the elastic deformation behaviour of polymers, it is necessary, first of all, to revise basic definitions of parameters such as stress and strain and the relationships between them.

19.2.1 STRESS

Force can be exerted externally on a body in two particular ways. Gravity and inertia can be thought of as *body forces* since they act directly on all the individual atoms and molecules in the body. The other types are *surface* or *contact forces* that act only on the surface of the body, but their effect is transmitted to the atoms and molecules inside the body through the atomic and molecular bonds. In consideration of the mechanical properties of polymers, we are mainly interested in the effect of applying surface forces such as stress or pressure to the material, but it must be remembered that certain polymers, such as non-crosslinked elastomers, will flow under their own weight, a simple response to the body force of gravity.

Although the response to these forces reflects the displacement of the individual atoms and molecules within the body, the system normally is considered from a macroscopic viewpoint and is regarded as a continuum. In order to define the state of stress at a point within the body, we consider the surface forces acting on a small cube of material around that point as shown in Figure 19.1. Each of these surface forces is divided by the area upon which it acts and then it is resolved into

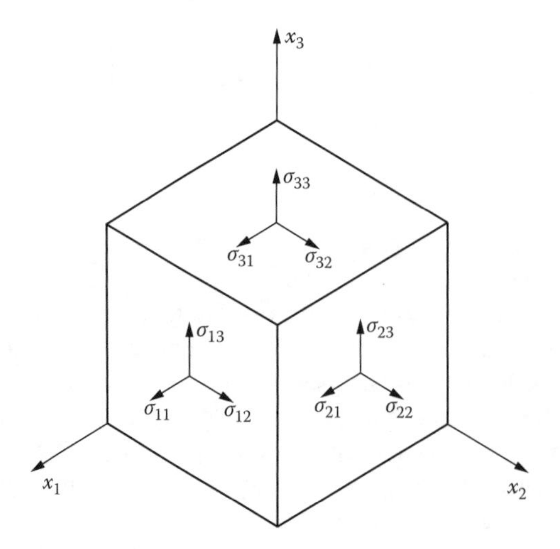

FIGURE 19.1 Schematic illustration of the nine stress components for a rectangular coordinate system.

components that are parallel to the three coordinate axes. In total, there will be nine stress components given as

$$\begin{matrix} \sigma_{11} & \sigma_{12} & \sigma_{13} \\ \sigma_{21} & \sigma_{22} & \sigma_{23} \\ \sigma_{31} & \sigma_{32} & \sigma_{33} \end{matrix}$$

where the first subscript gives the normal to the plane on which the stress acts and the second subscript defines the direction of the stress. The components σ_{11}, σ_{22} and σ_{33} are known as normal stresses as they are perpendicular to the plane on which they act and the others are shear stresses. The nine stress components are not independent of each other. If the cube is in equilibrium and not rotating, it is necessary that for the shear stresses

$$\sigma_{12} = \sigma_{21}, \sigma_{23} = \sigma_{32}, \sigma_{31} = \sigma_{13}$$

and so the state of stress at a point in a body in equilibrium can be defined by a stress tensor σ_{ij} that has six independent components

$$\sigma_{ij} = \begin{pmatrix} \sigma_{11} & \sigma_{12} & \sigma_{13} \\ \sigma_{21} & \sigma_{22} & \sigma_{23} \\ \sigma_{31} & \sigma_{32} & \sigma_{33} \end{pmatrix} \tag{19.1}$$

The knowledge of the state of stress at a point provides enough information to calculate the stress acting on any plane within the body.

It is possible to express the stress acting at a point in terms of three *principal stresses* acting along principal axes. In this case, the shear stresses (σ_{ij} where $i \neq j$) are all zero and the only terms remaining in the stress tensor are σ_{11}, σ_{22} and σ_{33}. It often is possible to determine the principal axes from simple inspection of the body and the state of stress. For example, two principal axes always lie in the plane of a free surface. When considering the mechanical properties of polymers, it often is useful to divide the stress tensor into its hydrostatic and deviatoric components. The hydrostatic pressure p is given by

$$p = \frac{1}{3}(\sigma_{11} + \sigma_{22} + \sigma_{33}) \tag{19.2}$$

and the deviatoric stress tensor σ'_{ij} is found by subtracting the hydrostatic stress components from the overall tensor such that

$$\sigma'_{ij} = \begin{pmatrix} (\sigma_{11} - p) & \sigma_{12} & \sigma_{13} \\ \sigma_{21} & (\sigma_{22} - p) & \sigma_{23} \\ \sigma_{31} & \sigma_{32} & (\sigma_{33} - p) \end{pmatrix} \tag{19.3}$$

19.2.2 Strain

When forces are applied to a material, the atoms change position in response to the force and this change is known as *strain*. A simple example is that of a thin rod of material of length l, that is extended a small amount δl by an externally applied stress. In this case, the strain e can be represented by

$$e = \frac{\delta l}{l} \tag{19.4}$$

For a general type of deformation consisting of extension and compression in different directions and shear, the situation is very much more complicated. However, the analysis can be simplified greatly by assuming that any strains are small and terms in e^2 can be neglected. In practice, the analysis can only normally be strictly applied to the elastic deformation of crystals and a rather different approach has to be applied in the case of, for example, elastomer deformation where very high strains are involved.

In the analysis of *infinitesimal strains*, the displacements of particular points in a body are considered. It is most convenient to use a rectangular coordinate system with axes x_1, x_2 and x_3 meeting at the origin O. The analysis is most easily explained by looking at the two-dimensional case shown in Figure 19.2. If P is a point in the body with coordinates (x_1, x_2), then when the body is deformed it will move to a point P′ given by coordinates (x_1+u_1, x_2+u_2). However, another point Q that is infinitesimally close to P will not be displaced by the same amount. If Q originally has coordinates of (x_1+dx_1, x_2+dx_2), its displacement to Q′ will have components (u_1+du_1, u_2+du_2). Since dx_1 and dx_2 are infinitesimal, we can write

$$du_1 = \frac{\partial u_1}{\partial x_1}dx_1 + \frac{\partial u_1}{\partial x_2}dx_2 \tag{19.5}$$

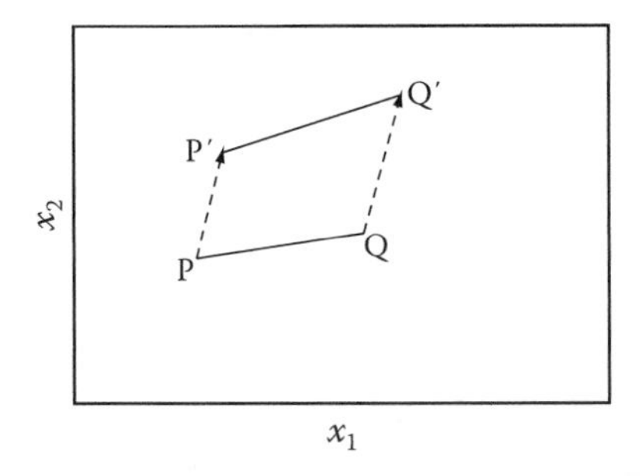

FIGURE 19.2 General displacement in two dimensions of points P and Q in a body. (Adapted from Kelly, A. and Groves, G.W., *Crystallography and Crystal Defects*, Longman, London, U.K., 1970.)

and

$$du_2 = \frac{\partial u_2}{\partial x_1}dx_1 + \frac{\partial u_2}{\partial x_2}dx_2 \tag{19.6}$$

It is necessary to define the partial differentials and this can be done by writing them as follows:

$$e_{11} = \frac{\partial u_1}{\partial x_1} \quad e_{12} = \frac{\partial u_1}{\partial x_2}$$

$$e_{21} = \frac{\partial u_2}{\partial x_1} \quad e_{22} = \frac{\partial u_2}{\partial x_2}$$

Equations 19.5 and 19.6 can be written more neatly as

$$du_i = e_{ij}dx_j \quad (j = 1, 2)$$

Since du_i and dx_j are vectors, e_{ij} must be a tensor and the significance of the various e_{ij} can be seen by taking particular examples of states of strain.

The simplest case is where the strains are small and the vector PQ can be taken as first parallel to the x_1 axis and then parallel to the x_2 axis as shown in Figure 19.3a. This enables the distortion of a rectangular element to be determined by considering the displacement of each vector separately. Since PQ_1 is parallel to x_1, $dx_2 = 0$ and so

$$du_1 = \frac{\partial u_1}{\partial x_1}dx_1 = e_{11}dx_1 \tag{19.7}$$

$$du_2 = \frac{\partial u_2}{\partial x_1}dx_1 = e_{21}dx_1 \tag{19.8}$$

The significance of these equations can be seen from Figure 19.3a. When e_{11} and e_{21} are small, the term e_{11} corresponds to the extensional strain of PQ_1 resolved along x_1 and e_{21} gives the anticlockwise rotation of PQ_1. It also follows that e_{22} will correspond to the extensional strain of PQ_2 resolved along x_2 and that e_{12} is the clockwise rotation of PQ_2.

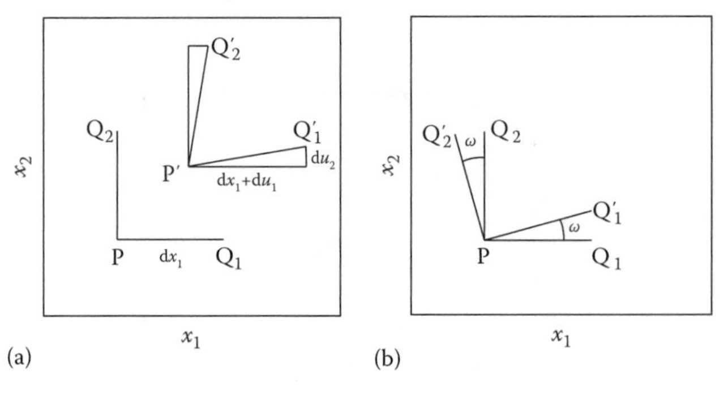

FIGURE 19.3 Two-dimensional displacements used in the definition of strain. (a) General displacements. (b) Rotation when $e_{11} = e_{22} = 0$ but $e_{12} = -\omega$ and $e_{21} = \omega$. (Adapted from Kelly, A. and Groves, G.W., *Crystallography and Crystal Defects*, Longman, London, U.K., 1970.)

The simple tensor notation of strain outlined above is not an entirely satisfactory representation of the strain in a body, as it is possible to have certain components of $e_{ij} \neq 0$ without the body being distorted. This situation is illustrated in Figure 19.3b. In this case $e_{11} = e_{22} = 0$, but a small rotation ω of the body occurs and is given by $e_{12} = -\omega$ and $e_{21} = \omega$. This problem can be overcome by expressing the general strain tensor e_{ij} as a sum of an anti-symmetrical tensor and a symmetrical tensor such that

$$e_{ij} = \frac{1}{2}(e_{ij} - e_{ji}) + \frac{1}{2}(e_{ij} + e_{ji})$$

or

$$e_{ij} = \omega_{ij} + \varepsilon_{ij} \tag{19.9}$$

The term ω_{ij} then gives the rotation and ε_{ij} represents the pure strain.

This analysis can be readily extended to three dimensions, and the strain tensor e_{ij} has nine rather than four components. The tensor ε_{ij} that represents the pure strain is again given by

$$\varepsilon_{ij} = \frac{1}{2}(e_{ij} + e_{ji})$$

and it can be written more fully as

$$\begin{pmatrix} \varepsilon_{11} & \varepsilon_{12} & \varepsilon_{13} \\ \varepsilon_{21} & \varepsilon_{22} & \varepsilon_{23} \\ \varepsilon_{31} & \varepsilon_{32} & \varepsilon_{33} \end{pmatrix} = \begin{pmatrix} e_{11} & \frac{1}{2}(e_{12} + e_{21}) & \frac{1}{2}(e_{13} + e_{31}) \\ \frac{1}{2}(e_{21} + e_{12}) & e_{22} & \frac{1}{2}(e_{23} + e_{32}) \\ \frac{1}{2}(e_{31} + e_{13}) & \frac{1}{2}(e_{32} + e_{23}) & e_{33} \end{pmatrix} \tag{19.10}$$

Inspection of this equation shows that $\varepsilon_{12} = \varepsilon_{21}$, $\varepsilon_{23} = \varepsilon_{32}$, $\varepsilon_{31} = \varepsilon_{13}$, which means that the strain tensor ε_{ij} has only six independent components (cf. Stress tensor σ_{ij} in the last section). The diagonal components of ε_{ij} correspond to the tensile strains parallel to the axes and the off-diagonal ones are then shear strains.

As with stresses, it is possible to choose a set of axes such that all the shear strains are zero and the only terms remaining are the *principal strains*, ε_{11}, ε_{22} and ε_{33}. In fact, for an isotropic body the axes of principal stress and principal strain coincide. This type of deformation can be readily demonstrated by considering what happens to a unit cube with the principal axes parallel to its edges. On deformation, it transforms into a rectangular brick of dimensions $(1 + \varepsilon_{11})$, $(1 + \varepsilon_{22})$ and $(1 + \varepsilon_{33})$.

19.2.3 Relationship between Stress and Strain

For many materials the relationship between stress and strain can be expressed, at least at low strains, by *Hooke's law* that states that stress is proportional to strain. It must be borne in mind that this is not a fundamental law of nature. It was discovered by empirical observation and certain materials, particularly polymers, tend not to obey it except at low strains. Hooke's law enables us to define Young's modulus E of a material that for simple uniaxial extension or compression is given by E = stress/strain.

It will be readily appreciated that this simple relationship cannot be applied without modification to complex systems of stress. In order to do this, Hooke's law must be generalized and the relationship between σ_{ij} and ε_{ij} is obtained in its most general form by assuming that every stress component is a linear function of every strain component.

$$\text{i.e. } \sigma_{11} = A\varepsilon_{11} + B\varepsilon_{22} + C\varepsilon_{33} + D\varepsilon_{12} \ldots$$

$$\text{and } \sigma_{22} = A'\varepsilon_{11} + B'\varepsilon_{22} + C'\varepsilon_{33} + D'\varepsilon_{12} \ldots$$

This can be expressed most neatly using tensor notation as

$$\sigma_{ij} = c_{ijkl}\varepsilon_{ij} \tag{19.11}$$

where c_{ijkl} is a fourth-rank tensor containing all the stiffness constants. In principle, there are $9\times9=81$ stiffness constants, but the symmetry of σ_{ij} and ε_{ij} reduces them to $6\times6=36$.

The notation used often is simplified to a contracted form

$$\begin{aligned}
&\sigma_{11}\to\sigma_1, \quad \sigma_{22}\to\sigma_2, \quad \sigma_{33}\to\sigma_3\\
&\sigma_{23}\to\sigma_4, \quad \sigma_{13}\to\sigma_5, \quad \sigma_{12}\to\sigma_6\\
&\varepsilon_{11}\to\varepsilon_1, \quad \varepsilon_{22}\to\varepsilon_2, \quad \varepsilon_{33}\to\varepsilon_3\\
&2\varepsilon_{23}\to\gamma_4, \quad 2\varepsilon_{13}\to\gamma_5, \quad 2\varepsilon_{12}\to\gamma_6
\end{aligned}$$

It is conventional to replace the tensor shear strains by the angles of shear γ that are twice the corresponding shear component in the strain tensor. There also is a corresponding contraction in c_{ijkl} that can then be written as

$$c_{ij} = \begin{pmatrix} c_{11} & c_{12} & c_{13} & c_{14} & c_{15} & c_{16}\\ c_{21} & c_{22} & c_{23} & c_{24} & c_{25} & c_{26}\\ c_{31} & c_{32} & c_{33} & c_{34} & c_{35} & c_{36}\\ c_{41} & c_{42} & c_{43} & c_{44} & c_{45} & c_{46}\\ c_{51} & c_{52} & c_{53} & c_{54} & c_{55} & c_{56}\\ c_{61} & c_{62} & c_{63} & c_{64} & c_{65} & c_{66} \end{pmatrix} \tag{19.12}$$

Only 21 of the 36 components of the matrix are independent since strain energy considerations show that $c_{ij}=c_{ji}$. Also, if the strain axes are chosen to coincide with the symmetry axes in the material, there may be a further reduction in the number of independent components. For example, in a cubic crystal the x_1, x_2 and x_3 axes are equivalent and so

$$c_{11} = c_{22} = c_{33}, \quad c_{12} = c_{23} = c_{31}, \quad c_{44} = c_{55} = c_{66}$$

and it can be proved further that all the other stiffness constants are zero. This means that there are only three independent elastic constants for cubic crystals, c_{11}, c_{12} and c_{44}. Polymer crystals are always of lower symmetry, and consequently there are more independent constants. Orthorhombic polyethylene (Table 17.1) requires nine stiffness constants to characterize fully its elastic behaviour.

In the case of elastically isotropic solids, there are only two independent elastic constants, c_{11} and c_{12}, because $2c_{44}=c_{11}-c_{12}$. Since glassy polymers and randomly oriented semi-crystalline polymers fall into this category, it is worth considering how the stiffness constants can be related to quantities such as Young's modulus E, Poisson's ratio v, shear modulus G and bulk modulus K that are measured directly. The *shear modulus* G relates the shear stress σ_4 to the angle of shear γ_4 through the equation

$$\sigma_4 = G\gamma_4 \tag{19.13}$$

and since $c_{44}=\sigma_4/\gamma_4$, we have the simple result that the shear modulus is given by

$$G = c_{44} \tag{19.14}$$

The situation is not so simple in the case of *uniaxial tension*, where all stresses other than σ_1 are equal to zero and $\varepsilon_2 = \varepsilon_3$. Generalized Hooke's law can be expressed as

$$\sigma_1 = c_{11}\varepsilon_1 + c_{12}\varepsilon_2 + c_{13}\varepsilon_3 = c_{12}(\varepsilon_1 + \varepsilon_2 + \varepsilon_3) + (c_{11} - c_{12})\varepsilon_1 = c_{12}\Delta + 2c_{44}\varepsilon_1 \tag{19.15}$$

where $\Delta = \varepsilon_1 + \varepsilon_2 + \varepsilon_3$ represents, at low strains, the fractional volume change or dilatation. For triaxial deformation (σ_1, σ_2, $\sigma_3 \neq 0$), the *bulk modulus K* relates the hydrostatic pressure p to Δ through

$$p = K\Delta \tag{19.16}$$

Since $p = \frac{1}{3}(\sigma_1 + \sigma_2 + \sigma_3)$, using variants of Equation 19.15

$$p = \left(c_{12} + \frac{2}{3}c_{44} \right)\Delta \tag{19.17}$$

Comparing Equations 19.16 and 19.17 enables K to be expressed as

$$K = c_{12} + \frac{2}{3}c_{44} \tag{19.18}$$

This equation then allows an expression for Young's modulus E to be obtained. Young's modulus relates the stress and strain in uniaxial tension through

$$\sigma_1 = E\varepsilon_1 \tag{19.19}$$

In uniaxial tension, $\sigma_2 = \sigma_3 = 0$ and so

$$\sigma_1 = 3p = (3c_{12} + 2c_{44})\Delta \tag{19.20}$$

Substituting for σ_1 from Equation 19.15 and rearranging gives

$$\varepsilon_1 = \frac{(c_{12} + c_{44})\Delta}{c_{44}} \tag{19.21}$$

and so the modulus is given by

$$E = \frac{\sigma_1}{\varepsilon_1} = \frac{c_{44}(3c_{12} + 2c_{44})}{(c_{12} + c_{44})} \tag{19.22}$$

The final constant that can be derived is *Poisson's ratio* v, which is the ratio of the lateral strain to the longitudinal extension measured in a uniaxial tensile test. It then is given by

$$v = -\frac{\varepsilon_2}{\varepsilon_1} = -\frac{\varepsilon_3}{\varepsilon_1} \tag{19.23}$$

This can be related to Δ through

$$v = \frac{1}{2}\left(1 - \frac{\Delta}{\varepsilon_1} \right) \tag{19.24}$$

which using Equation 19.21 gives

$$\nu = \frac{c_{12}}{2(c_{12} + c_{44})} \tag{19.25}$$

It is clear that since the four constants G, K, E and ν can be expressed in terms of three stiffness constants, only two of which are independent, G, K, E and ν must all be related to each other. It is a matter of simple algebraic manipulation to show that, for example

$$G = \frac{E}{2(1+\nu)} \quad \text{and} \quad K = \frac{E}{3(1-2\nu)} \quad \text{etc.}$$

We are now in a position to show how an isotropic body will respond to a general system of stresses. A stress of σ_1 will cause a strain of σ_1/E along the x_1 direction, but will also lead to strains of $-\nu\sigma_1/E$ along the x_2 and x_3 directions. Similarly, a stress of σ_2 will cause strains of σ_2/E along x_2 and of $-\nu\sigma_2/E$ in the other directions. If it is assumed that the strains due to the different stresses are additive, then using the original notation it follows that

$$\varepsilon_{11} = \frac{\sigma_{11}}{E} - \frac{\nu}{E}(\sigma_{22} + \sigma_{33}) = e_{11} \tag{19.26}$$

$$\varepsilon_{22} = \frac{\sigma_{22}}{E} - \frac{\nu}{E}(\sigma_{33} + \sigma_{11}) = e_{22} \tag{19.27}$$

$$\varepsilon_{33} = \frac{\sigma_{33}}{E} - \frac{\nu}{E}(\sigma_{11} + \sigma_{22}) = e_{33} \tag{19.28}$$

The shear strains can be obtained directly from Equation 19.13 and are $\varepsilon_{12} = \sigma_{12}/2G$, $\varepsilon_{23} = \sigma_{23}/2G$, $\varepsilon_{31} = \sigma_{31}/2G$.

19.3 ELASTIC DEFORMATION OF POLYMERS

The stress–strain curves of several different types of polymers tested to failure are shown schematically in Figure 19.4. Young's modulus E can be determined from the initial slope of these curves. One of the most useful aspects of polymers is the wide range of moduli that can be obtained. Typical values of E for the materials indicated are given in Table 19.1. *Elastomers* tend to have very low Young's moduli of the order of 10^6 Pa. This is because their deformation consists essentially of uncoiling the molecules in a crosslinked network. It is possible to calculate a value for the modulus of an elastomer using the statistical theory and this gives very good agreement with the experimentally determined values (see Chapter 21). *Semi-crystalline polymers* can be considered as a composite consisting of polymer crystals distributed in an amorphous matrix and have Young's modulus of the order of 10^8–10^9 Pa depending on whether the amorphous phase polymer is above or below its glass transition temperature. The moduli of *polymer glasses* are typically of the order of 10^9 Pa. The deformation of a polymer glass is somewhat more difficult to model theoretically. The structure is thought to be completely random and so elastic deformation will involve the bending and stretching of the strong covalent bonds on the polymer backbone, as well as the displacement of adjacent molecules which is opposed by relatively weak secondary (e.g. van der Waals) bonding.

The most striking feature of the elastic properties of *polymer fibres* is that they are very anisotropic. The moduli parallel to the chain direction are ~10^{11} Pa, which is similar to the values of

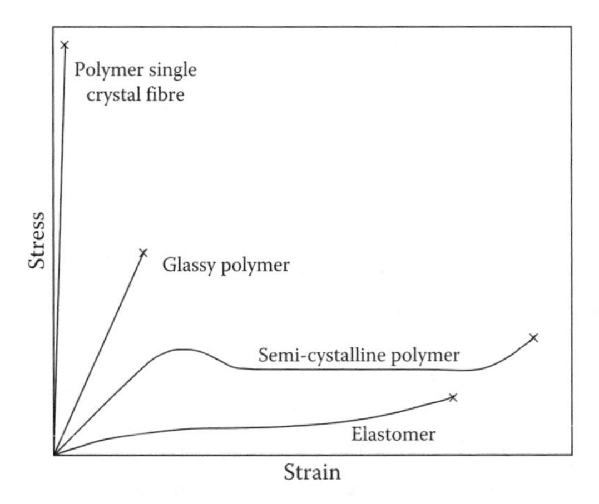

FIGURE 19.4 Schematic stress–strain curves of different types of polymers, drawn approximately to scale. The point at which the material fractures is marked as '×'.

TABLE 19.1
Typical Values of Young's Modulus for Different Types of Polymer

Type of Polymer	Young's Modulus/Pa
Elastomer	$\sim 10^6$–10^7
Semi-crystalline polymer	$\sim 10^8$–10^9
Glassy polymer	$\sim 10^9$
Polymer fibre	
($\perp$ to axis)	$\sim 10^9$
($\parallel$ to axis)	$\sim 10^{11}$

moduli found for metals (e.g. steel, $E \approx 2.1 \times 10^{11}$ Pa). The moduli of polymer fibres deformed in directions perpendicular to the chain axes are, however, much lower at about 10^9 Pa. This large difference in modulus reflects the molecular orientation (Section 17.4.5) and anisotropy of the bonding in the fibres. Deformation parallel to the chain direction involves stretching of the strong covalent chain backbone bonds and changing the bond angles, or if the molecule is in a helical conformation distorting the molecular helix. Deformation in the transverse direction is opposed by only relatively weak secondary van der Waals forces or hydrogen bonds.

19.3.1 Deformation of a Polymer Chain

Because of the high anisotropy of the bonding in polymer crystals, it is possible with knowledge of the crystal structure (Section 17.2) and the force constants of the chemical bonds to calculate the theoretical moduli of polymer crystals form the deformation of individual polymer chains. An exact calculation is rather complex as it requires detailed knowledge of inter- and intramolecular interactions, but an estimate can be made of the chain-direction modulus of a polymer that crystallizes with its backbone in the form of a planar zig-zag. This involves a relatively simple calculation, undertaken originally by Treloar, who assumed that deformation involves only the bending and stretching of the bonds along the molecular backbone.

The model of the polymer chain used in the calculation is shown schematically in Figure 19.5. It is treated as consisting of n rods of length l, each rod being capable of stretching along its length but not of bending, and joined together by torsional springs. The bond angles are taken initially to be Θ and

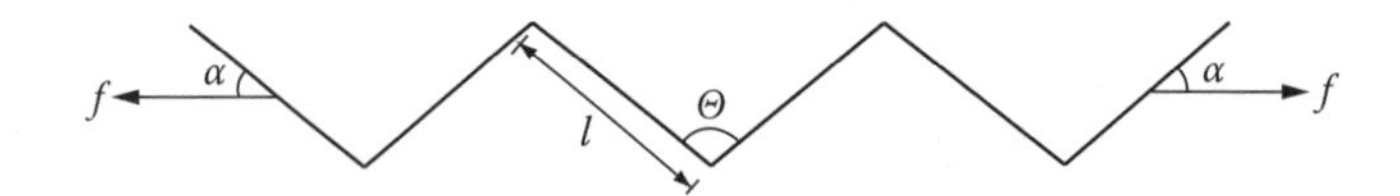

FIGURE 19.5 Model of a polymer chain undergoing deformation. (Adapted from Treloar, L.G.R., *Polymer,* 1, 95, 1960.)

the angle between the applied force f and the individual bonds as initially α. The original length of the chain L is then $nl\cos\alpha$ and so the change in length δL on deforming the chain is given by

$$\delta L = n\delta(l\cos\alpha) = n(\cos\alpha\ \delta l - l\sin\alpha\ \delta\alpha) \tag{19.29}$$

It is possible to obtain expressions for δl and $\delta\Theta$ in terms of the applied force f. Consideration of the bond stretching allows δl to be determined. The component of the force acting along the bond direction is $f\cos\alpha$. This can be related to the extension through the force constant k_l for bond stretching that can be determined from infrared or Raman spectroscopy (Sections 15.4 and 15.5) and is the constant of proportionality relating the force along the bond to its extension. It follows, therefore, that

$$\delta l = \frac{f\cos\alpha}{k_l} \tag{19.30}$$

The force required to cause valence angle deformation can be determined by using the angular deformation force constant k_Θ that relates the change in bond angle $\delta\Theta$ to the torque acting around each of the bond angles. This torque is equal to the moment of the applied force about the angular vertices, $\frac{1}{2}fl\sin\alpha$ and so the change in the angle between the bonds is given by

$$\delta\Theta = \frac{(fl\sin\alpha)}{2k_\Theta} \tag{19.31}$$

But since it can be shown by a simple geometrical construction that $\alpha = 90° - \Theta/2$, it follows that

$$\delta\alpha = -\frac{\delta\Theta}{2}$$

and Equation 19.31 becomes

$$\delta\alpha = -\frac{(fl\sin\alpha)}{4k_\Theta} \tag{19.32}$$

Putting Equations 19.30 and 19.32 into 19.29 gives

$$\frac{\delta L}{f} = n\left[\frac{\cos^2\alpha}{k_l} + \frac{l^2\sin^2\alpha}{4k_\Theta}\right] \tag{19.33}$$

Young's modulus of the polymer is given by

$$E = \frac{(f/A)}{(\delta L/L)} \tag{19.34}$$

where A is the cross-sectional area supported by each chain and so

$$E = \frac{l\cos\alpha}{A}\left[\frac{\cos^2\alpha}{k_l} + \frac{l^2\sin^2\alpha}{4k_\Theta}\right]^{-1}$$

or in terms of the bond angle Θ

$$E = \frac{l\sin(\Theta/2)}{A}\left[\frac{\sin^2(\Theta/2)}{k_l} + \frac{l^2\cos^2(\Theta/2)}{4k_\Theta}\right]^{-1} \tag{19.35}$$

This equation allows the modulus to be calculated with knowledge of the crystal structure, which gives A, l and Θ, and the spectroscopically measured values of the force constants k_l and k_Θ. It predicts a value of E for polyethylene of about 180 Pa which is generally rather lower than values obtained from more sophisticated calculations. This is because the analysis in the preceding text does not allow for intramolecular interactions that tend to increase the modulus further. The chain-direction modulus does not completely describe the elastic behaviour of a polymer crystal. A complete set of elastic constants c_{ij} (Equation 19.12) are required in order to do this. The number of constants depends upon the crystal symmetry and for an orthorhombic crystal, such as polyethylene, nine constants are necessary. The measurement of all the elastic constants is very difficult even for good single crystals like those of polydiacetylenes (Section 17.3.2). A set of elastic constants has been calculated for orthorhombic polyethylene and they are given in Table 19.2. They were obtained by Odajima and Maeda who calculated the forces required to both stretch the bonds along the polymer backbone and to open up the angles between the bonds, as in the above model. In addition, inter- and intramolecular interactions were taken into account.

19.3.2 Polymer Crystal Moduli

Single crystals of polymers are not normally available with dimensions that are suitable to allow them to be tested. The production of 100% crystalline macroscopic polydiacetylene single crystals by solid-state polymerization (Section 8.2) has made it possible to measure the moduli of these crystals directly using simple mechanical methods. A stress–strain curve for a polydiacetylene single-crystal fibre is given in Figure 19.6. The modulus calculated from the initial slope of the curve is 61 GPa, which is very close to the value (~65 GPa) calculated using the simple bond-stretching and bond-bending method of Treloar.

Estimates of chain-direction moduli can be made experimentally using an X-ray diffraction method. This involves measuring the relative change under an applied stress in the position of any (*hkl*) Bragg reflection that has a component of the chain-direction repeat (usually indexed *c*). In order to do this, it is necessary either to use fibres in which the molecules are aligned parallel to the fibre axis or to mechanically draw an isotropic polymer into a filament or a tape. A bundle of the fibres, or the filament or tape, are then deformed in an X-ray beam, either from a laboratory X-ray generator or a synchrotron radiation source, employing a geometry similar to that shown in Figure 17.1b.

TABLE 19.2
Calculated Values of the Nine Elastic Constants for a Polyethylene Crystal at 25°C

Stiffness Component	c_{ij}/GPa
c_{11}	4.83
c_{22}	8.71
c_{33}	257.1
c_{12}	1.16
c_{13}	2.55
c_{23}	5.84
c_{44}	2.83
c_{55}	0.78
c_{66}	2.06

Source: Data taken from Odajima, A. and Maeda, T., *J. Polym. Sci.*, C15, 55, 1966.

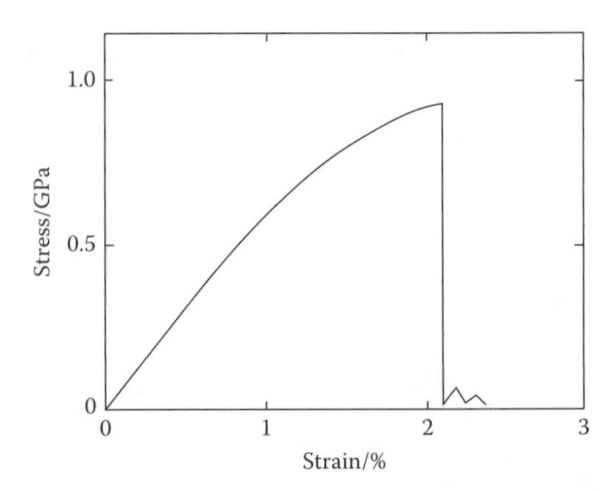

FIGURE 19.6 A stress–strain curve obtained for a polydiacetylene single-crystal fibre.

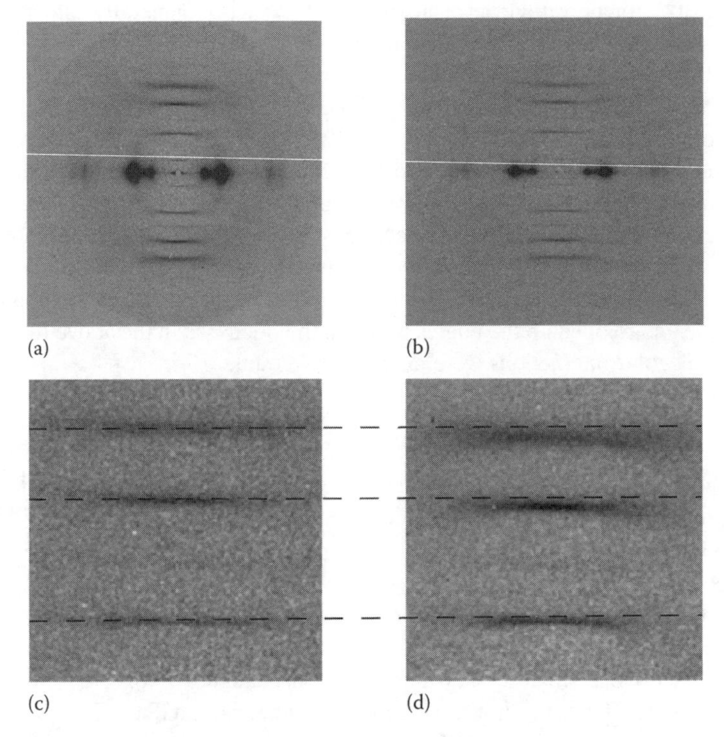

FIGURE 19.7 X-ray diffraction patterns of a PBO fibre with the chain direction and fibre axis vertical (a) before deformation and (b) under stress. (c) Enlargement of the layer lines above the central spot before deformation and (d) under stress. (Courtesy of R.J. Davies, ESRF, Grenoble, France.)

It is found that under tensile stress, the position of the Bragg reflections on the layer lines in the diffraction patterns, such as those shown in Figure 17.25, move closer to the beam centre indicating that the crystal lattice is subjected to stress. This is shown for a single poly(*p*-phenylene benzobisoxazole) (PBO) fibre in Figure 19.7. The lattice parameter c can be determined using Equation 17.3 and the axial crystal strain e_c can be determined from

$$e_c = \frac{c_\sigma - c_0}{c_0} \tag{19.36}$$

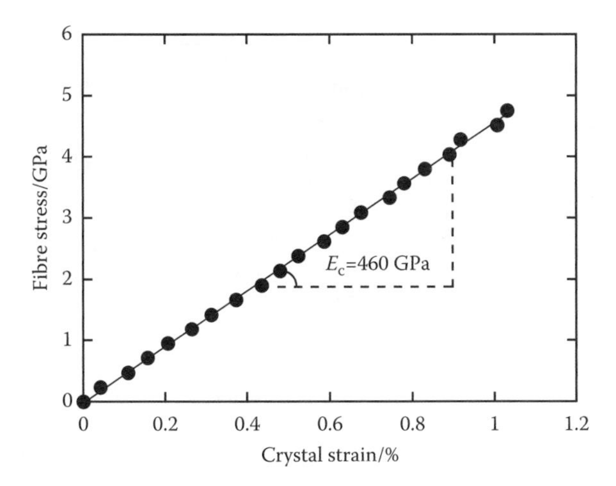

FIGURE 19.8 Variation of crystal strain with fibre stress for a single PBO fibre.

where

c_σ is the value of lattice parameter under stress
c_0 is its value before deformation.

The chain-direction crystal modulus can be determined by making the assumption that the stress on the crystal is equal to that on the whole fibre, filament or tape. If the applied stress is σ_f, then the chain-direction crystal modulus is given by

$$E_c = \frac{\sigma_f}{e_c} \quad (19.37)$$

Normally a series of measurements are made of the variation of axial crystal strain with applied stress, and E_c is determined from the slope of a plot such as that shown in Figure 19.8.

It should be noted that E_c approximates to c_{11} in Equation 19.12. Other components of c_{ij}, such as transverse values, are more difficult to determine as it is difficult to apply stress in a controlled manner in directions perpendicular to the fibre axis. The chain-direction modulus E_c always is greater than Young's modulus E_f of the fibre, filament or tape. This is because the crystals and lower-modulus amorphous material are lined up in parallel in the fibre morphology. Although they are subjected to the same level of stress, the strain in the amorphous material is higher than that in the crystals leading to $E_f < E_c$.

Table 19.3 gives some calculated and experimentally measured values of chain-direction crystal moduli, E_c, of several different polymer crystals. Highly oriented semi-crystalline polymers tend to have moduli of no more than 50% of the theoretical values. This is because even though they may possess good crystal orientation, the presence of an amorphous phase tends to cause a significant reduction in modulus. The directly measured and calculated values of moduli for some polydiacetylene single crystals also are given in Table 19.3.

A general picture emerges concerning the values of chain direction moduli of polymer crystals. They tend to be higher than the theoretical prediction if the molecule is in the form of a planar zig-zag rather than a helix. For example, polyethylene is stiffer than polyoxymethylene or polytetrafluoroethylene and both have molecules in helical conformations (Table 17.1). The helices can be extended more easily than the polyethylene planar zig-zag. Also, the presence of large side groups tends to reduce the modulus because they increase the separation of molecules in the crystal. This causes an increase in the area supported by each chain (A in Equation 19.35).

TABLE 19.3
Calculated and Measured Values of Chain-Direction Moduli for Several Different Polymer Crystals

Polymer		Calculated		Measured	
		E/GPa	Method[a]	*E*/GPa	Method
Polyethylene		182	VFF	240	X-ray
		340	UBFF	358	Raman
		257	UBFF	329	Neutron (CD_2)
Polypropylene		49	UBFF	42	X-ray
Polyoxymethylene		150	UBFF	54	X-ray
Polytetrafluoroethylene		160	UBFF	156	X-ray
				222	Neutron
Substituted	(i)	49	VFF	45	Mechanical
Polydiacetylenes[b]	(ii)	65	VFF	61	Mechanical
PPTA[c]				224	X-ray
PBO[c]				460	X-ray

Source: Data taken from Young, R.J., in *Developments in Polymer Fracture*, E.H. Andrews (ed.), Applied Science, London, U.K., 1979.

[a] VFF, valence force field (e.g. Treloar's method); UBFF, Urey–Bradley force field (e.g. used to determine the values in Table 19.2).

[b] (i) Phenyl urethane derivative; (ii) ethyl urethane derivative (see Figure 8.1).

[c] PPTA, poly(*p*-phenylene terephthalamide); PBO, poly(*p*-phenylene benzobisoxazole).

Although polymers generally are considered to be relatively low modulus flexible materials, it is clear from Table 19.3 that their potential unidirectional moduli are relatively high. This feature has been utilized to a limited extent for many years in conventional polymer fibres. The development of stiff and inflexible polymer molecules such as PPTA and PBO (Table 3.2) has led to very high modulus polymer fibres such as 'Kevlar' and 'Zylon' being produced by the spinning of liquid crystalline solutions. For many purposes it is the *specific modulus* (Young's modulus divided by specific gravity) that is important, especially when light high-stiffness materials are required. Polymers tend to have low specific gravities, typically 0.8–1.5 compared with 7.9 for steel, and so their specific stiffnesses compare even more favourably with metals.

19.3.3 Elastic Deformation of Semi-Crystalline Polymers

The presence of crystals in a polymer has a profound effect upon its mechanical behaviour. This effect has been known for many years from experience with natural rubber. Figure 19.9a shows the increase in Young's modulus of natural rubber with crystallinity, that can occur due to crystallization that can occur during prolonged storage of the material at ambient temperature. There is a 100-fold increase in modulus when the degree of crystallinity increases from 0 to 0.25. A similar effect is found in other semi-crystalline polymers. Figure 19.9b shows how the modulus of a linear polyethylene increases as the crystallinity is increased through use of different heat treatments. The degree of crystallinity in linear polyethylene is usually quite high (>60%) and the data in Figure 19.9b can be extrapolated to give a modulus of the order of 5 GPa for non-oriented 100% crystalline material.

It is apparent from considerations of the structure in Section 17.4 that semi-crystalline polymers are essentially two-phase materials and that the increase in modulus is due to the presence of the crystals. Traditional ideas of the stiffening effect due to the presence of crystals were based upon the statistical theory of elastomer deformation (Section 21.3). It was thought that the crystals in the amorphous

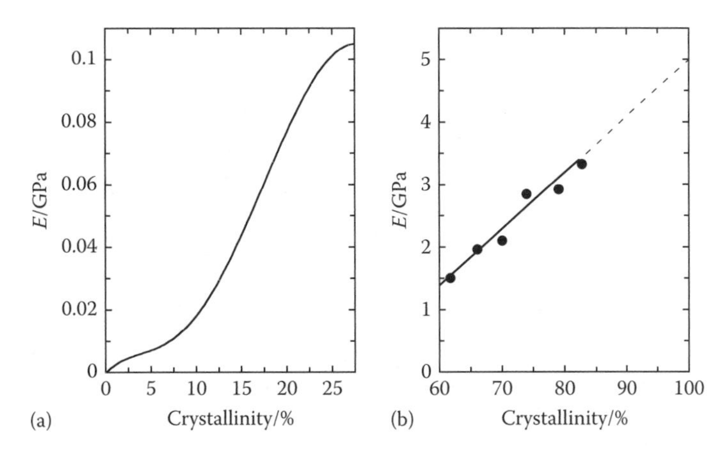

FIGURE 19.9 Variation of Young's modulus E with the degree of crystallinity for different polymers. (a) Increase in modulus of natural rubber with crystallization. (Data taken from Leitner, M., *Trans. Farad. Soc.*, 51, 1015, 1955.) (b) Dependence of the modulus of polyethylene upon crystallinity. (Data taken from Wang, T.T., *J. Appl. Phys.*, 44, 4052, 1973.)

'rubber' behaved like crosslinks and produced the stiffening through an increase in crosslink density rather than through their own inherent stiffness. Although this mechanism may be relevant at very low degrees of crystallinity, it is clear that most semi-crystalline polymers behave as composites (Chapter 24) and the observed modulus is a reflection of the combined modulus of the amorphous and crystalline regions. The exact way in which this combination of moduli can be represented mathematically is not a trivial problem. For example, spherulitic polymers are elastically isotropic, but the individual crystals within the spherulites are themselves highly anisotropic. It is, therefore, necessary to calculate an effective modulus for the particular type of morphology under consideration. The effective crystal modulus depends not only upon the proportion of crystalline material present, given by the degree of crystallinity, but also upon the size, shape and distribution of the crystals within the polymer sample. It is clear, therefore, that reliable estimates of the properties of the semi-crystalline polymer composite can be made only for specific well-defined morphologies.

A model system that can be used is the polydiacetylene for which the stress–strain curve was given in Figure 19.6. By controlling the polymerization conditions, it is possible to prepare single-crystal fibres which contain both monomer and polymer molecules. The monomer has a modulus of only 9 GPa along the fibre axis compared with 61 GPa for the polymer; and the partly polymerized fibres which contain both monomer and polymer molecules are found to have values of modulus between these two extremes as shown in Figure 19.10. The variation of the modulus with the proportion of polymer (approximately equal to the conversion) can be predicted by two simple composite models. The first one due to Reuss assumes that the elements in the structure (i.e. the monomer and polymer molecules) are lined up in series and experience the same stress. In this case, the modulus E_R will be given by an equation of the form

$$\frac{1}{E_R} = \frac{V_p}{E_p} + \frac{(1 - V_p)}{E_m} \tag{19.38}$$

where

V_p is the volume fraction of polymer
E_p and E_m are the moduli of the polymer and matrix (monomer), respectively.

The second model due to Voigt assumes that all the elements are lined up in parallel and so experience the same strain. The modulus for the Voigt model E_V is given by the rule of mixtures as

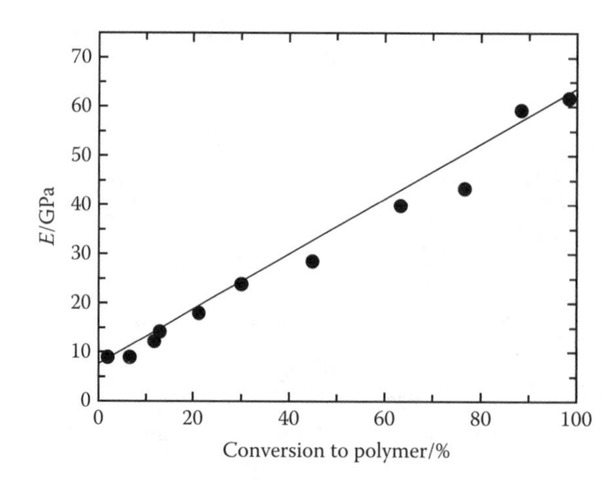

FIGURE 19.10 Dependence of the modulus of polydiacetylene single-crystal fibres upon the percentage conversion of monomer into polymer. The line corresponds to the prediction of Equation 19.39.

$$E_V = E_p V_p + E_m (1 - V_p) \tag{19.39}$$

The line corresponding to the prediction of Equation 19.39 is also given in Figure 19.10. It can be seen that the experimental points lie close to the Voigt line indicating that the components are experiencing approximately the same strain in this particular structure. It will be shown in Chapter 24 that equations similar to Equations 19.38 and 19.39 can be used to predict the moduli of fibre-reinforced composites, and it is necessary that in all similar composite structures the measured moduli fall either between or on the Voigt and Reuss lines. The agreement that is found between the measured moduli of the polydiacetylene fibres and the Voigt prediction is due to the structure of the fibres which consist of long polymer molecules lying parallel to the fibre axis and embedded in a monomer matrix. This morphology ensures that the strain will be uniform in the deformed structure.

The situation is considerably more complicated with more random structures such as spherulitic polymers. The measured modulus of such a structure is an average of that of the crystalline and amorphous regions. Even when it is assumed that the amorphous material is isotropic, there are problems in determining how to estimate the effective average modulus of the crystals and how to combine this with the modulus of the amorphous material to give an overall modulus for a polymer sample. This latter problem can be overcome by extrapolating measurements of modulus as a function of crystallinity to 100% crystallinity as shown in Figure 19.9b. The structure of such a material can be thought of as a randomly oriented polycrystalline structure not unlike a polycrystalline metal. The Reuss and Voigt models can again be applied with their respective assumption of uniform stress and uniform strain. Using the nine crystal moduli from Table 19.2, the Reuss model predicts a modulus of 4.9 GPa compared with 15.6 GPa for the Voigt model. Since the extrapolated value is ~5 GPa, it can be seen that in the case of spherulitic polyethylene, the Reuss model gives a better prediction of the observed behaviour.

PROBLEMS

19.1 In this question on the deformation of a polymer assume that it behaves as a homogeneous isotropic elastic material.

(a) Consider a cube of the polymer with the stresses σ_1, σ_2 and σ_3 acting on the three mutually perpendicular faces. If the cube deformed elastically under the influence of the three stresses, use the principle of the additive nature of elastic strains to derive the three-dimensional elastic stress–strain equations:

$$e_1 = \frac{1}{E}[\sigma_1 - v(\sigma_2 + \sigma_3)]$$

$$e_2 = \frac{1}{E}[\sigma_2 - v(\sigma_3 + \sigma_1)]$$

$$e_3 = \frac{1}{E}[\sigma_3 - v(\sigma_1 + \sigma_2)]$$

where
- E is Young's modulus
- v is Poisson's ratio

(b) For the cube of polymer subjected to hydrostatic stress, $p = \sigma_1 = \sigma_2 = \sigma_3$ use the equations in (a) to derive the following equation:

$$K = \frac{E}{3(1-2v)}$$

where K is the bulk modulus of the material.

(c) What would be Poisson's ratio, v, of an isotropic polymer that was deformed uniaxially elastically at constant volume? Elastomers are normally assumed to deform at approximately constant volume and so what implications would this have about the bulk modulus, K, of elastomers?

(d) For an isotropic polymer subjected to three-dimensional stressing, what relationship exists between elastic volume change and the algebraic sum of the principal stresses $(\sigma_1 + \sigma_2 + \sigma_3)$?

(e) What uniaxial stress σ_1 would result in the same volume change for the polymer as that resulting from the three-dimensional stresses $\sigma_1 = +30$ MPa, $\sigma_2 = -40$ MPa, $\sigma_3 = +10$ MPa?

19.2 Treloar showed that the modulus of a zig-zag polymer chain, E, can be given by:

$$E = \frac{l \sin(\Theta/2)}{A}\left[\frac{\sin^2(\Theta/2)}{k_l} + \frac{l^2 \cos^2(\Theta/2)}{4k_\Theta}\right]^{-1}$$

where
- A is the cross-sectional area supported by each chain
- l is the link length
- k_l is the bond force constant
- k_Θ is the angular deformation constant
- Θ is the initial bond angle

(a) Calculate E for a polyethylene single crystal by using the values given below.

(b) Comment on the accuracy of your calculated value for the modulus of the polyethylene crystal.

(For polyethylene, $A = 1.824 \times 10^{-19}$ m^2, $l = 1.53 \times 10^{-10}$ m, $\Theta = 109.5°$, $k_l = 436$ Pa m^{-1}, $k_\Theta = 8.19 \times 10^{-19}$ Pa m).

19.3 A sample of an oriented semi-crystalline polymer is deformed in tension in an X-ray diffractometer. The position of the (002) peak is found to change as the stress on the sample is increased as indicated below.

Stress/MPa	(Bragg Angle)/degrees
0	37.483
40	37.477
80	37.471
120	37.466
160	37.460
200	37.454

Determine Young's modulus of the crystals in the polymer in the chain direction assuming that the stress on the crystals is equal to the stress applied to the specimen as a whole. (The radiation used was CuK_{α}, $\lambda = 0.1542$ nm.)

FURTHER READING

Hearle, J.W.S. (ed), *High Performance Fibres*, Woodhead Publishing Ltd, Cambridge, U.K., 2004.
Hibbeler, R.C., *Mechanics of Materials*, 8th edn., Pearson, Upper Saddle River, NJ, 2010.
Kelly, A., Groves, G.W., and Kidd, P., *Crystallography and Crystal Defects*, John Wiley & Sons, New York, 2000.
Ward, I.M. and Sweeney, J., *An Introduction to the Mechanical Properties of Solid Polymers*, 2nd edn., Wiley, New York, 2004.
Williams, J.G., *Stress Analysis of Polymers*, Longman, London, 1973.
Young, R.J., in *Developments in Polymer Fracture* (E.H. Andrews, ed.), Applied Science, London, U.K., 1979.

20 Viscoelasticity

20.1 INTRODUCTION

A distinctive feature of the mechanical behaviour of polymers is the way in which their response to an applied stress or strain depends upon the rate or time period of loading. This dependence upon rate and time is in marked contrast to the behaviour of elastic solids such as metals and ceramics that, at least at low strains, obey *Hooke's law* such that the stress is proportional to the strain and independent of loading rate. On the other hand, the mechanical behaviour of viscous liquids is time dependent. It is possible to represent their behaviour at low rates of strain by *Newton's law* whereby the stress is proportional to the strain-rate and independent of the strain. The behaviour of most polymers can be thought of as being somewhere between that of elastic solids and viscous liquids. At low temperatures and high rates of strain they display elastic behaviour, whereas at high temperatures and low rates of strain they behave in a viscous manner, flowing like a liquid. Polymers are, therefore, termed *viscoelastic* as they display aspects of both viscous and elastic types of behaviour.

Polymers used in engineering applications often are subjected to stress for prolonged periods of time, but it is not possible to know how a polymer will respond to a particular load without a detailed knowledge of its viscoelastic properties. There have been many studies of the viscoelastic properties of polymers and several textbooks written specifically about the subject of viscoelasticity. The intention here is to give a brief introduction to what is a vast subject and to leave the reader to consult more advanced texts for further details.

The exact nature of the time dependence of the mechanical properties of a polymer sample depends upon the type of stress or straining cycle employed. The variation of the stress σ and strain e with time t is illustrated schematically in Figure 20.1 for a simple polymer tensile specimen subjected to four different deformation histories. In each case, the behaviour of an elastic material also is given as a dashed line for comparison. During *creep* loading, a constant stress is applied to the specimen at $t=0$ and the strain increases rapidly at first, slowing down over longer periods of time. In an elastic solid, the strain stays constant with time. The behaviour of a viscoelastic material during *stress relaxation* is shown in Figure 20.1b. In this case, the strain is held constant and the stress decays slowly with time, whereas in an elastic solid it would remain constant. The effect of deforming a viscoelastic material at a *constant stress-rate* is shown in Figure 20.1c. The increase in strain is not linear and the curve becomes steeper with time and also as the stress-rate is increased. If different *constant strain-rates* are used the variation of stress with time is not linear. The slope of the curve tends to decrease with time, but it is steeper for higher strain-rates. The variation of both strain and stress with time is linear for constant stress- and strain-rate tests upon elastic materials.

20.2 VISCOELASTIC MECHANICAL MODELS

It is clear that any theoretical explanations of the above phenomena should be able to account for the dependence of the stress and strain upon time. Ideally, it should be possible to predict, for example, the stress relaxation behaviour from knowledge of the creep curve, but in practice, with real polymers this is somewhat difficult to do. Hence, the situation often is simplified by assuming that the polymer behaves as a *linear viscoelastic* material, i.e. it is assumed that the deformation of the polymer comprises independent elastic and viscous components so that the deformation of the

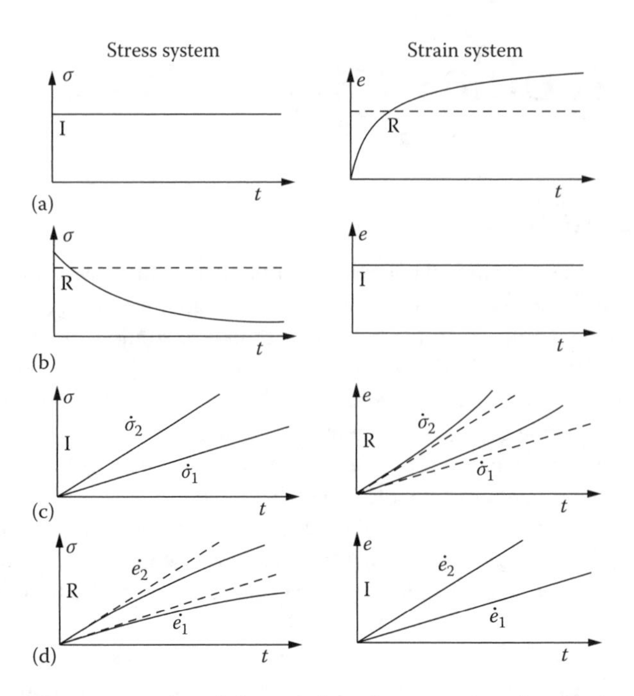

FIGURE 20.1 Schematic representation of the variation of stress and strain with time indicating the input (I) and responses (R) for different types of loading: (a) creep, (b) relaxation, (c) constant stressing-rate and (d) constant straining-rate. (Adapted from Williams, J.G., *Stress Analysis of Polymers*, Longman, London, U.K., 1973.)

polymer can be described by a combination of Hooke's law and Newton's law. The linear elastic behaviour is given by Hooke's law as

$$\sigma = Ee \tag{20.1}$$

or

$$\frac{\mathrm{d}\sigma}{\mathrm{d}t} = \frac{E\mathrm{d}e}{\mathrm{d}t} \tag{20.2}$$

where E is the elastic modulus. Newton's law is used to describe the linear viscous behaviour through the equation

$$\sigma = \eta\frac{\mathrm{d}e}{\mathrm{d}t} \tag{20.3}$$

where
- η is the viscosity
- de/dt the strain rate.

It should be noted that these equations only apply at small strains. A particularly useful method of formulating the combination of elastic and viscous behaviour is through the use of mechanical models. The two basic components used in the models are an elastic spring of modulus E, which obeys Hooke's law (Equation 20.1) and a viscous dashpot containing an incompressible liquid of viscosity η, which obeys Newton's law (Equation 20.3). The various models that have been proposed involve different combinations of these two basic elements.

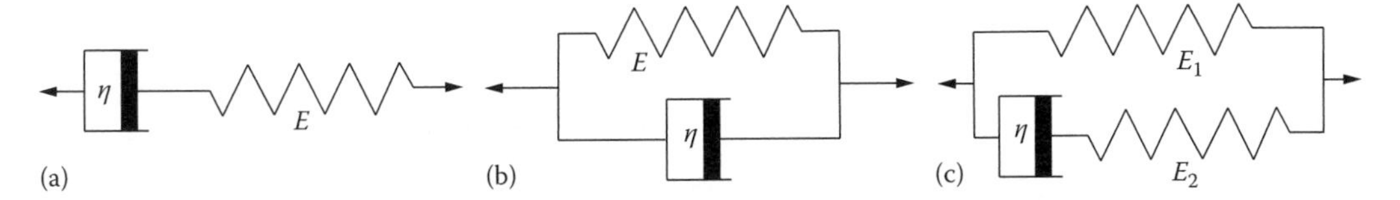

FIGURE 20.2 Mechanical models used to represent the viscoelastic behaviour of polymers: (a) Maxwell model, (b) Voigt model and (c) standard linear solid.

20.2.1 Maxwell Model

This model was proposed in the nineteenth century by Maxwell to explain the time-dependent mechanical behaviour of viscous materials, such as tar or pitch. It consists of a spring and dashpot in series, as shown in Figure 20.2a. Under the action of an overall stress σ, there will be an overall strain e in the system, which is given by

$$e = e_1 + e_2 \tag{20.4}$$

where

e_1 is the strain in the spring
e_2 the strain in the dashpot.

Since the elements are in series, the stress will be identical in each one and so

$$\sigma_1 = \sigma_2 = \sigma \tag{20.5}$$

Equations 20.2 and 20.3 can be written, therefore, as

$$\frac{\mathrm{d}\sigma}{\mathrm{d}t} = E\frac{\mathrm{d}e_1}{\mathrm{d}t} \quad \text{and} \quad \sigma = \eta\frac{\mathrm{d}e_2}{\mathrm{d}t}$$

for the spring and dashpot, respectively. Differentiation of Equation 20.4 gives

$$\frac{\mathrm{d}e}{\mathrm{d}t} = \frac{\mathrm{d}e_1}{\mathrm{d}t} + \frac{\mathrm{d}e_2}{\mathrm{d}t} \tag{20.6}$$

and so for the Maxwell model

$$\frac{\mathrm{d}e}{\mathrm{d}t} = \frac{1}{E}\frac{\mathrm{d}\sigma}{\mathrm{d}t} + \frac{\sigma}{\eta} \tag{20.7}$$

At this stage, it is worth considering how closely the Maxwell model predicts the mechanical behaviour of a polymer. In the case of creep, the stress is held constant at $\sigma = \sigma_0$ and so $\mathrm{d}\sigma/\mathrm{d}t = 0$. Equation 20.7 can, therefore, be written as

$$\frac{\mathrm{d}e}{\mathrm{d}t} = \frac{\sigma_0}{\eta} \tag{20.8}$$

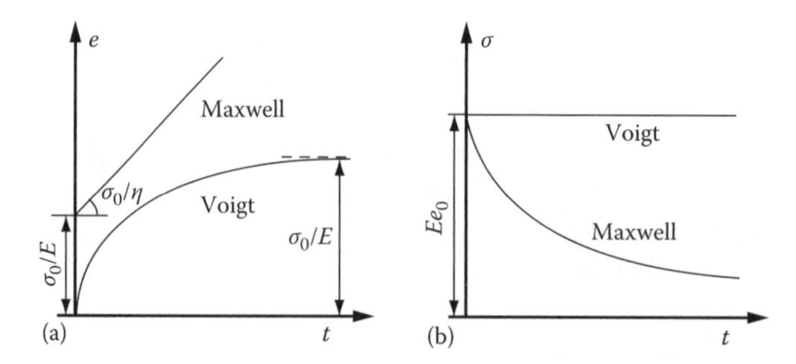

FIGURE 20.3 Behaviour of the Maxwell and Voigt models during different types of loading: (a) creep (constant stress σ_0) and (b) relaxation (constant strain e_0). (Adapted from Williams, J.G., *Stress Analysis of Polymers*, Longman, London, U.K., 1973.)

Thus, the Maxwell model predicts Newtonian flow, as shown in Figure 20.3a. The strain is expected to increase linearly with time, which is clearly not the case for a viscoelastic polymer (Figure 20.1) where $\mathrm{d}e/\mathrm{d}t$ decreases with time.

The model is of more use in predicting the response of a polymer during stress relaxation. In this case, a constant strain $e = e_0$ is imposed on the system and so $\mathrm{d}e/\mathrm{d}t = 0$. Equation 20.7 then becomes

$$0 = \frac{1}{E}\frac{\mathrm{d}\sigma}{\mathrm{d}t} + \frac{\sigma}{\eta} \tag{20.9}$$

and, hence,

$$\frac{\mathrm{d}\sigma}{\sigma} = -\frac{E}{\eta}\mathrm{d}t \tag{20.10}$$

This can be readily integrated if at $t = 0$, $\sigma = \sigma_0$ to give

$$\sigma = \sigma_0 \exp\left(-\frac{Et}{\eta}\right) \tag{20.11}$$

where σ_0 is the initial stress. The term η/E will be constant for a given Maxwell model and it sometimes is referred to as a *relaxation time* τ_0. The equation can then be written as

$$\sigma = \sigma_0 \exp\left(\frac{-t}{\tau_0}\right) \tag{20.12}$$

which predicts an exponential decay of stress, as shown in Figure 20.3b. This is a rather better representation of polymer behaviour than the Maxwell model for creep. However, Equation 20.12 predicts that the stress relaxes completely over a long period of time, which is not normally the case for a real polymer.

20.2.2 Voigt Model

This also is known as the Kelvin model. It consists of the same elements as are in the Maxwell model but in this case they are in parallel (Figure 20.2b) rather than in series. The parallel arrangement of the spring and the dashpot means that the strains are uniform, i.e.

$$e = e_1 = e_2 \tag{20.13}$$

and the stresses in each component will add to make an overall stress of σ such that

$$\sigma = \sigma_1 + \sigma_2 \tag{20.14}$$

The individual stresses σ_1 and σ_2 can be obtained from Equations 20.1 and 20.3 and are

$$\sigma_1 = Ee \quad \text{and} \quad \sigma_2 = \eta\frac{\mathrm{d}e}{\mathrm{d}t}$$

for the spring and the dashpot, respectively. Putting these expressions into Equation 20.14 and rearranging gives

$$\frac{\mathrm{d}e}{\mathrm{d}t} = \frac{\sigma}{\eta} - \frac{Ee}{\eta} \tag{20.15}$$

The usefulness or otherwise of the model can again be determined by examining its response to particular types of loading. The Voigt or Kelvin model is particularly useful in describing the behaviour during creep where the stress is held constant at $\sigma = \sigma_0$. Equation 20.15 then becomes

$$\frac{\mathrm{d}e}{\mathrm{d}t} + \frac{Ee}{\eta} = \frac{\sigma_0}{\eta} \tag{20.16}$$

This simple differential equation has the solution

$$e = \frac{\sigma_0}{E}\left[1 - \exp\left(-\frac{Et}{\eta}\right)\right] \tag{20.17}$$

The constant ratio η/E can again be replaced by τ_0, the relaxation time and so the variation of strain with time for a Voigt model undergoing creep loading is given by

$$e = \frac{\sigma_0}{E}\left[1 - \exp\left(\frac{-t}{\tau_0}\right)\right] \tag{20.18}$$

This behaviour is shown in Figure 20.3a, and it clearly represents the correct form of behaviour for a polymer undergoing creep. The strain rate decreases with time and $e \to \sigma_0/E$ as $t \to \infty$. On the other hand, the Voigt model is unsuccessful in predicting the stress relaxation behaviour of a polymer. In this case, the strain is held constant at e_0 and since $\mathrm{d}e/\mathrm{d}t = 0$, then Equation 20.15 becomes

$$\frac{\sigma}{\eta} = \frac{Ee_0}{\eta} \tag{20.19}$$

or

$$\sigma = Ee_0 \tag{20.20}$$

which is the linear elastic response indicated in Figure 20.3b.

20.2.3 Standard Linear Solid

It has been demonstrated that the Maxwell model describes the stress relaxation of a polymer to a first approximation, whereas the Voigt model similarly describes creep. A logical step forward is, therefore, to find some combination of these two basic models that can account for both phenomena. A simple model that does this is known as the standard linear solid, one example of which is shown in Figure 20.2c. In this case, a Maxwell element and spring are in parallel. The presence of the second spring will stop the tendency of the Maxwell element to undergo simple viscous flow during creep loading, but will still allow the stress relaxation to occur.

There have been many attempts at devising more complex models that can give a better representation of the viscoelastic behaviour of polymers. As the number of elements increases, the mathematics becomes more complex. It must be stressed that the mechanical models only give a mathematical representation of the mechanical behaviour and as such do not give much help in interpreting the viscoelastic properties on a molecular level.

20.3 BOLTZMANN SUPERPOSITION PRINCIPLE

A cornerstone of the theory of linear viscoelasticity is the *Boltzmann superposition principle.* It allows the state of stress or strain in a viscoelastic body to be determined from knowledge of its entire deformation history. The basic assumption is that during viscoelastic deformation in which the applied stress is varied, the overall deformation can be determined from the algebraic sum of strains due to each loading step. Before use of the principle can be demonstrated, it is necessary to define a parameter known as the *creep compliance* $J(t)$, which is a function only of time. It allows the strain after a given time $e(t)$ to be related to the applied stress σ for a linear viscoelastic material since

$$e(t) = J(t)\sigma \tag{20.21}$$

The use of the superposition principle can be demonstrated by considering the creep deformation caused by a series of step loads, as shown in Figure 20.4. If there is an increment of stress $\Delta\sigma_1$ applied at a time τ_1 then the strain due this increment at time t is given by

$$e_1(t) = \Delta\sigma_1 J(t - \tau_1) \tag{20.22}$$

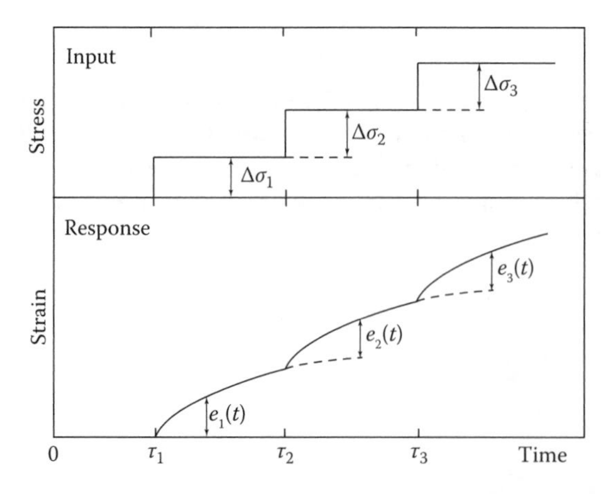

FIGURE 20.4 Response of a linear viscoelastic material to a series of loading steps. (Adapted from Ward, I.M., *Mechanical Properties of Solid Polymers*, 2nd edn., Wiley-Interscience, New York, 1983.)

If further increments of stress, which may be either positive or negative, are applied then the principle assumes that the contributions to the overall strain $e(t)$, from each increment are additive so that

$$
\begin{aligned}
e(t) &= e_1(t) + e_2(t) + \ldots \\
&= \Delta\sigma_1 J(t-\tau_1) + \Delta\sigma_2 J(t-\tau_2) + \ldots \\
&= \sum_{n=0}^{n} J(t-\tau_n)\Delta\sigma_n
\end{aligned} \tag{20.23}
$$

It is possible to represent this summation in an integral form such that

$$e(t) = \int_{-\infty}^{t} J(t-\tau)\mathrm{d}\sigma(t) \tag{20.24}$$

The integral is taken from $-\infty$ to t because it is necessary to take into account the entire deformation history of the viscoelastic sample as this will affect the subsequent behaviour. This equation can be used to determine the strain after any general loading history. It normally is expressed as a function of τ, which gives

$$e(t) = \int_{-\infty}^{t} J(t-\tau)\frac{\mathrm{d}\sigma(\tau)}{\mathrm{d}\tau}\mathrm{d}\tau \tag{20.25}$$

The usefulness of the Boltzmann superposition principle can be seen by considering specific simple examples. A trivial case is for a single loading step of σ_0 at a time $\tau=0$. In this case, $J(t-\tau)=J(t)$ and so $e(t)=\sigma_0 J(t)$. A more useful application is demonstrated in Figure 20.5 whereby a stress of σ_0 is applied at time $\tau=0$ and taken off at time $\tau=t_1$. The strain e_1 due to loading is given by

$$e_1 = \sigma_0 J(t) \tag{20.26}$$

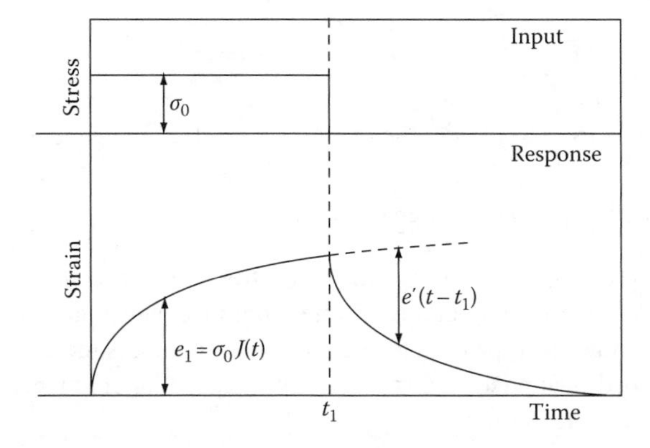

FIGURE 20.5 Response of a linear viscoelastic material to loading followed by unloading, illustrating the Boltzmann superposition principle.

and that due to unloading is similarly

$$e_2 = -\sigma_0 J(t - t_1) \tag{20.27}$$

The total strain after time $t(>t_1)$, $e(t)$ is given by

$$e(t) = \sigma_0 J(t) - \sigma_0 J(t - t_1) \tag{20.28}$$

In this region, $e(t)$ is decreasing and this process is known as recovery. If the recovered strain $e'(t-t_1)$ is defined as the difference between the strain that would have occurred if the initial stress had been maintained and the actual strain then

$$e'(t - t_1) = \sigma_0 J(t) - e(t) \tag{20.29}$$

Combining Equations 20.28 and 20.29 gives the recovered strain as

$$e'(t - t_1) = \sigma_0 J(t - t_1) \tag{20.30}$$

which is identical to the creep strain expected for a stress of $+\sigma_0$ applied after a time t_1. It is clear, therefore, that the extra extension or recovery that results from each loading or unloading event is independent of the previous loading history and is considered by the superposition principle as a series of separate events that add up to give the total specimen strain.

It is possible to analyse stress relaxation in a similar way using the Boltzmann superposition principle. In this case, it is necessary to define a stress relaxation modulus $G(t)$, which relates the time-dependent stress $\sigma(t)$ to the strain e through the relationship

$$\sigma(t) = G(t)e \tag{20.31}$$

If a series of incremental strains Δe_1, Δe_2 etc., are applied to the specimen at times τ_1, τ_2 etc., the total stress will be given by

$$\sigma(t) = \Delta e_1 G(t - \tau_1) + \Delta e_2 G(t - \tau_2) + \dots \tag{20.32}$$

after time t. This equation is analogous to Equation 20.23 and can be given in a similar way as an integral (cf. Equation 20.25)

$$\sigma(t) = \int_{-\infty}^{t} G(t - \tau)\frac{\mathrm{d}e(\tau)}{\mathrm{d}\tau}\mathrm{d}\tau \tag{20.33}$$

This equation can then be used to predict the overall stress after any general straining programme.

20.4 DYNAMIC MECHANICAL TESTING

So far, only the response of polymers during creep and stress relaxation has been considered. Often, they are subjected to variable loading at a moderately high frequency and so it is pertinent to consider their behaviour during this type of deformation. The situation is most easily analysed when an oscillating sinusoidal load is applied to a specimen at a particular frequency. If the applied stress varies as a function of time according to

$$\sigma = \sigma_0 \sin \omega t \tag{20.34}$$

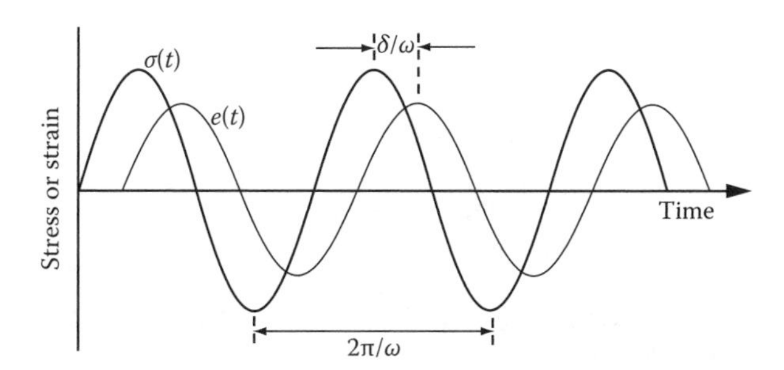

FIGURE 20.6 Variation of stress and strain with time for a viscoelastic material. The parameters are defined in the text.

where ω is the angular frequency ($2\pi f$ where f is the frequency), the strain for an elastic material obeying Hooke's law would vary in a similar manner as

$$e = e_0 \sin \omega t \tag{20.35}$$

However, for a viscoelastic material, the strain lags somewhat behind the stress (e.g. during creep). This can be considered as a damping process and the result is that when a stress defined by Equation 20.34 is applied to the sample the strain varies in a similar sinusoidal manner, but out of phase with the applied stress. Thus, the variation of stress and strain with time can be given by expressions of the type

$$\left.\begin{aligned} e &= e_0 \sin \omega t \\ \sigma &= \sigma_0 \sin(\omega t + \delta) \end{aligned}\right\} \tag{20.36}$$

where δ is the 'phase angle' or 'phase lag', the relative angular displacement of the stress and strain (Figure 20.6). The equation for stress can be expanded to give

$$\sigma = \sigma_0 \sin \omega t \cos \delta + \sigma_0 \cos \omega t \sin \delta \tag{20.37}$$

The stress, therefore, can be considered as being resolved into two components: one of $\sigma_0 \cos \delta$ which is in phase with the strain and another $\sigma_0 \sin \delta$ which is 90° ($\pi/2$ rad) out of phase with the strain. Hence, it is possible to define two dynamic moduli: E_1 which is in phase with the strain and E_2, which is $\pi/2$ rad out of phase with the strain. Since $E_1 = (\sigma_0/e_0)\cos \delta$ and $E_2 = (\sigma_0/e_0)\sin \delta$, Equation 20.37 becomes

$$\sigma = e_0 E_1 \sin \omega t + e_0 E_2 \cos \omega t \tag{20.38}$$

The phase angle δ is then given by

$$\tan \delta = \frac{E_2}{E_1} \tag{20.39}$$

A complex notation often is favoured for representation of the dynamic mechanical properties of viscoelastic materials. The stress and strain are given as

$$e = e_0 \exp \mathrm{i} \omega t$$

and

$$\sigma = \sigma_0 \exp \mathrm{i}(\omega t + \delta) \tag{20.40}$$

where $\mathrm{i} = (-1)^{1/2}$. The overall complex modulus $E^* = \sigma/e$ is then given by

$$E^* = \frac{\sigma_0}{e_0} \exp \mathrm{i}\delta = \frac{\sigma_0}{e_0}(\cos\delta + \mathrm{i}\sin\delta) \tag{20.41}$$

From the definitions of E_1 and E_2, it follows that

$$E^* = E_1 + \mathrm{i}E_2 \tag{20.42}$$

and because of this, E_1 and E_2 are sometimes called the *real* and *imaginary* parts of the modulus, respectively.

At this stage, it is worth considering how the dynamic mechanical properties of a polymer can be measured experimentally. This can be done quite simply by using a *torsion pendulum* similar to the one illustrated schematically in Figure 20.7. It consists of a cylindrical rod of polymer, which is fixed at its upper end and connected to a large inertial disc at its bottom end. The apparatus normally is arranged such that the weight of the disc is counter-balanced so that the specimen is not under any uniaxial tension or compression. The system can be set into oscillation by twisting and releasing the inertial disc and it undergoes sinusoidal oscillations at a frequency ω that depends upon the dimensions of the system and the properties of the rod. If the system is completely frictionless, it will oscillate indefinitely when the rod is perfectly elastic. For a viscoelastic material, such as a polymer, the oscillations are damped and their amplitude decreases with time, as shown in Figure 20.7. The apparatus is sufficiently compact to allow measurements to be made over a wide range of temperatures (typically −200°C to +200°C) but since the dimensions of the specimen cannot be varied greatly, ω is limited to a relatively narrow range (typically 0.01–10 Hz).

Where a large range of frequency is required, measurements of viscoelastic properties usually are made by employing forced-vibration techniques. Dynamic mechanical testing instruments are now available that subject a specimen to sinusoidal cycles of deformation in tension or bending, and measure directly the stress, strain and $\tan\delta$.

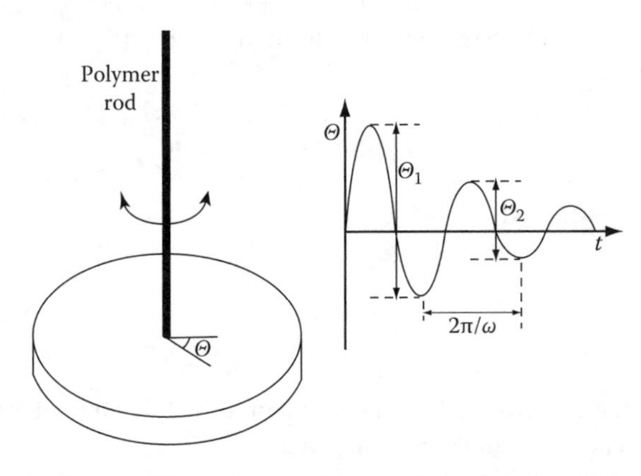

FIGURE 20.7 Schematic diagram of a torsion pendulum and the damping of the oscillations for a viscoelastic material.

The motion of a torsion pendulum can be analysed relatively simply. If a circular cross-sectioned rod of radius a and length l is used, then the torque T in the rod when one end is twisted by an angle Θ relative to the other is given by

$$T = \frac{G\pi a^4}{2l}\Theta \tag{20.43}$$

where G is the shear modulus of the rod. If the material is perfectly elastic, then the equation of motion of the oscillating system is

$$I\ddot{\Theta} + T = 0 \tag{20.44}$$

where

I is the moment of inertia of the disc

$\ddot{\Theta}$ is the second derivative of Θ with respect to time

Hence,

$$I\ddot{\Theta} + \frac{G\pi a^4}{2l}\Theta = 0 \tag{20.45}$$

describes the motion of the system. An equation of this form has the standard solution

$$\Theta = \Theta_0 \cos(\omega t + \alpha)$$

where

$$\omega = \left(\frac{G\pi a^4}{2lI}\right)^{1/2} \tag{20.46}$$

The shear modulus for an elastic rod is therefore given by

$$G = \frac{2lI\omega^2}{\pi a^4} \tag{20.47}$$

It was shown earlier that the Young's modulus of a viscoelastic material can be considered as a complex quantity E^*. It is possible to define the complex shear modulus in a similar way as

$$G^* = G_1 + \mathrm{i}G_2 \tag{20.48}$$

where G_1 and G_2 are the real and imaginary parts of the shear modulus. If the rod is made from a viscoelastic material, the equation of motion (Equation 20.45) then becomes

$$I\ddot{\Theta} + \frac{\pi a^4}{2l}(G_1 + \mathrm{i}G_2)\Theta = 0 \tag{20.49}$$

If this equation is assumed to have a solution

$$\Theta = \Theta_0 e^{-\lambda t} e^{\mathrm{i}\omega t} \tag{20.50}$$

then equating real and imaginary parts gives

$$G_1 = \frac{2lI}{\pi a^4}(\omega^2 - \lambda^2) \tag{20.51}$$

and

$$G_2 = \frac{2lI}{\pi a^4} 2\omega\lambda \tag{20.52}$$

Also the phase angle δ is given by

$$\tan\delta = \frac{E_2}{E_1} = \frac{G_2}{G_1} = \frac{2\omega\lambda}{\left(\omega^2 - \lambda^2\right)} \tag{20.53}$$

A quantity that normally is measured for a viscoelastic material in the torsion pendulum is the logarithmic decrement Λ. This is defined as the natural logarithm of the ratio of the amplitude of two successive cycles, i.e.

$$\Lambda = \ln\left(\frac{\Theta_n}{\Theta_{n+1}}\right) \tag{20.54}$$

The determination of Λ is shown in Figure 20.7. Since the time period between two successive cycles is $2\pi/\omega$, then combining Equations 20.50 and 20.54 gives

$$\Lambda = \frac{2\pi\lambda}{\omega} \tag{20.55}$$

Hence Equation 20.51 becomes

$$G_1 = \frac{2lI\omega^2}{\pi a^4}\left(1 - \frac{\Lambda^2}{4\pi^2}\right) \tag{20.56}$$

and Equation 20.52

$$G_2 = \frac{2lI\omega^2}{\pi a^4}\frac{\Lambda}{\pi} \tag{20.57}$$

and δ is given by

$$\tan\delta = \frac{\Lambda/\pi}{1 - \Lambda^2/4\pi^2} \tag{20.58}$$

Usually Λ is small ($\ll 1$) and so the term $\Lambda^2/4\pi^2$ normally is neglected so that Equation 20.56 becomes

$$G_1 \simeq \frac{2lI\omega^2}{\pi a^4} = G \tag{20.59}$$

and Equation 20.58 approximates to

$$\tan\delta \simeq \frac{\Lambda}{\pi} \tag{20.60}$$

when the damping is small.

The real part of the complex modulus G_1 often is called the *storage modulus* because it can be identified with the in-phase elastic component of the deformation. Elastic materials 'store' energy during deformation and release it on unloading. The imaginary part G_2 sometimes is called the *loss modulus* since it gives a measure of the energy dissipated during each cycle through its relation with Λ (Equation 20.57) and $\tan\delta$. The real and imaginary parts of E^* can be treated in an analogous way.

20.5 FREQUENCY DEPENDENCE OF VISCOELASTIC BEHAVIOUR

It is found experimentally that when the values of storage modulus E_1, loss modulus E_2 and $\tan\delta$ are measured for a polymer at a fixed temperature, their values depend upon the testing frequency or rate. Generally, it is found than $\tan\delta$ and E_2 usually are small at very low and very high frequencies and their values peak at some intermediate frequency. On the other hand, E_1 is high at high frequencies when the polymer is displaying glassy behaviour and low at low frequencies when the polymer is rubbery. The value of E_1 changes rapidly at intermediate frequencies in the viscoelastic region where the damping is high and $\tan\delta$ and E_2 peak (normally at slightly different frequencies).

The frequency dependence of the viscoelastic properties of a polymer can be demonstrated very simply using the Maxwell model (Section 20.2.1). The mechanical behaviour of the model is represented by Equation 20.7 as

$$\frac{\mathrm{d}e}{\mathrm{d}t} = \frac{1}{E}\frac{\mathrm{d}\sigma}{\mathrm{d}t} + \frac{\sigma}{\eta} \tag{20.7}$$

where
E is the modulus of the spring
η is the viscosity of the dashpot (Figure 20.2a).

It was shown earlier that the model has a characteristic relaxation time $\tau_0 = \eta/E$ and so Equation 20.7 becomes

$$E\tau_0\frac{\mathrm{d}e}{\mathrm{d}t} = \tau_0\frac{\mathrm{d}\sigma}{\mathrm{d}t} + \sigma \tag{20.61}$$

It was shown in Section 20.4 that if the stress on a viscoelastic material is varied sinusoidally at a frequency ω then the variation of stress and strain with time can be represented by complex equations of the form

$$e = e_0 \exp \mathrm{i}\omega t$$

and

$$\sigma = \sigma_0 \exp \mathrm{i}(\omega t + \delta) \tag{20.40}$$

where δ is the 'phase lag'. Substituting the relationships for σ and e given in Equation 20.40 into Equation 20.61 gives

$$\frac{E\tau_0 \mathrm{i}\omega}{(\tau_0 \mathrm{i}\omega + 1)} = \frac{\sigma_0 \exp \mathrm{i}(\omega t + \delta)}{e_0 \exp \mathrm{i}\omega t} = \frac{\sigma}{e} \tag{20.62}$$

The ratio σ/e is the complex modulus E^* and so

$$E_1 + \mathrm{i}E_2 = \frac{E\tau_0 \mathrm{i}\omega}{(\tau_0 \mathrm{i}\omega + 1)} \tag{20.63}$$

Equating the real and imaginary parts then gives

$$E_1 = \frac{E\tau_0^2\omega^2}{(\tau_0^2\omega^2 + 1)} \quad \text{and} \quad E_2 = \frac{E\tau_0\omega}{(\tau_0^2\omega^2 + 1)} \tag{20.64}$$

and since $\tan\delta = E_2/E_1$, it follows that

$$\tan\delta = \frac{1}{\tau_0\omega} \tag{20.65}$$

The variation of E_1, E_2 and $\tan\delta$ with frequency for a Maxwell model with a fixed relaxation time is sketched in Figure 20.8. The behaviour of E_1 and E_2 is predicted quite well by the analysis with a peak in E_2 at $\tau_0\omega = 1$, but $\tan\delta$ drops continuously as ω increases and the peak that is found experimentally does not occur. Better predictions can be obtained by using more complex models, such as the standard linear solid, and combinations of several models that have a range of relaxation times.

One of the main reasons for studying the frequency dependence of the dynamic mechanical properties is that it is often possible to relate peaks in E_2 and $\tan\delta$ to particular types of molecular motion in the polymer. The peaks can be regarded as a 'damping' effect and occur at the frequency of a particular molecular motion in the polymer structure. There is particularly strong damping at the *glass transition*. We have, so far, considered this a purely static phenomenon (Chapter 16) being observed through changes in properties such as specific volume or heat capacity. The large increase in free volume that occurs above the T_g allows space for molecular motion to occur more easily. Significant damping occurs at T_g when the applied test frequency matches the natural frequency

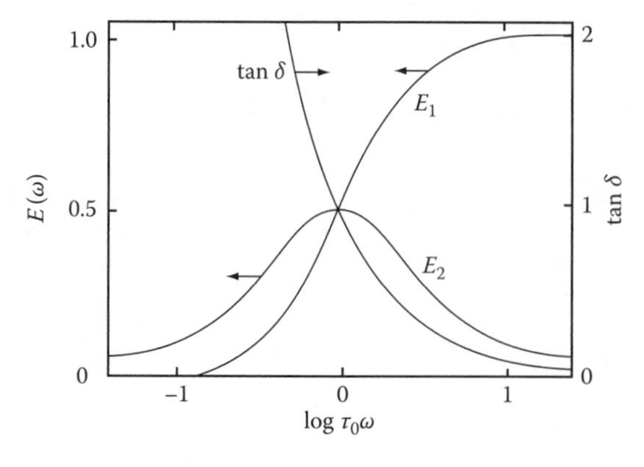

FIGURE 20.8 Variation of E_1, E_2 and $\tan\delta$ with frequency ω for a Maxwell model with a spring modulus $E = 1$ and a fixed relaxation time τ_0.

for main-chain rotation. At higher frequencies, there will be insufficient time for chain uncoiling to occur and the material will be relatively stiff. On the other hand, at lower frequencies, the chains will have more time to move and the polymer will appear to be soft and rubbery.

Peaks in E_2 also can be found at the *melting temperature* T_m in semi-crystalline polymers. This is due to the greater freedom of molecular motion possible when the molecules are no longer packed regularly into crystals.

It is possible that other types of molecular motion can occur such as the rotation of side groups, as these are often reflected in other damping occurring at somewhat different frequencies. The magnitude of the peaks in E_2 and $\tan\delta$ in this case are usually smaller than those obtained at T_g and such phenomena are known as 'secondary transitions'. The fact that dynamic mechanical testing can be used to follow main-chain and side-group motion in polymers makes it a powerful technique for the characterization of polymer structure.

The exact frequency at which the T_g and secondary transitions occur depends upon the temperature of testing as well as the structure of the polymer. In general, the frequency at which the transition takes place increases as the testing temperature is increased, and for a given testing frequency it is possible to induce a transition by changing the testing temperature. From a practical viewpoint, it is normally much easier to keep the frequency fixed and vary the testing temperature rather than do measurements at a variable frequency and so most of the data available on the dynamic mechanical behaviour of polymers have been obtained at a fixed frequency and over a range of temperature. For a complete picture of the dynamic mechanical behaviour of a polymer, it is desirable that measurements should be made over as wide a range of frequency and temperature as possible and it sometimes is possible to relate data obtained at different frequencies and temperatures through a procedure known as time–temperature superposition (see Section 20.7).

20.6 TRANSITIONS AND POLYMER STRUCTURE

The transition behaviour of a very large number of polymers has been studied widely as a function of testing temperature using dynamic mechanical testing methods. The types of behaviour observed are found to depend principally upon whether the polymers are amorphous or crystalline. Figure 20.9 shows the variation of the shear modulus G_1 and $\tan\delta$ with temperature for atactic polystyrene, which is typical for amorphous polymers. The shear modulus decreases as the testing temperature is increased and drops sharply at the glass transition where there is a corresponding large peak in $\tan\delta$. Figure 20.9b shows the variation of $\tan\delta$ with temperature and it is possible to see minor peaks at low temperatures. These correspond to secondary transitions. The peak that occurs at the highest temperature normally is labelled α and the subsequent ones are called β, γ etc. In an amorphous polymer, the α-transition or α-relaxation corresponds to the glass transition, which can be detected by other physical testing methods (Section 16.2) and so the temperature of the α-relaxation depends upon the chemical structure of the polymer. In general, it is found that the transition temperature is increased as the main chain is made stiffer (see Section 16.3). Side groups raise the T_g or α-relaxation temperature if they are polar or large and bulky and reduce it if they are long and flexible.

The assignment of particular mechanisms to the secondary β, γ etc. transitions can sometimes be difficult since their position and occurrence depends upon which polymer is being studied. At least three secondary transitions have been reported for atactic polystyrene (Figure 20.9b). The position of the β-peak at about 50°C is rather sensitive to testing frequency and it merges with the α-relaxation at high frequencies. It has been assigned to rotation of phenyl groups around the main chain or alternatively the co-operative motion of segments of the main chain containing several atoms. It is thought that the γ-peak may be due to the occurrence of head-to-head rather than head-to-tail repeat sequences (Section 4.2.3) and the δ-peak has been assigned to rotation of the phenyl group around its link with the main chain.

The interpretation of the relaxation behaviour of semi-crystalline polymers can be extremely difficult because of their complex two-phase structure (Section 17.4). It sometimes is possible to

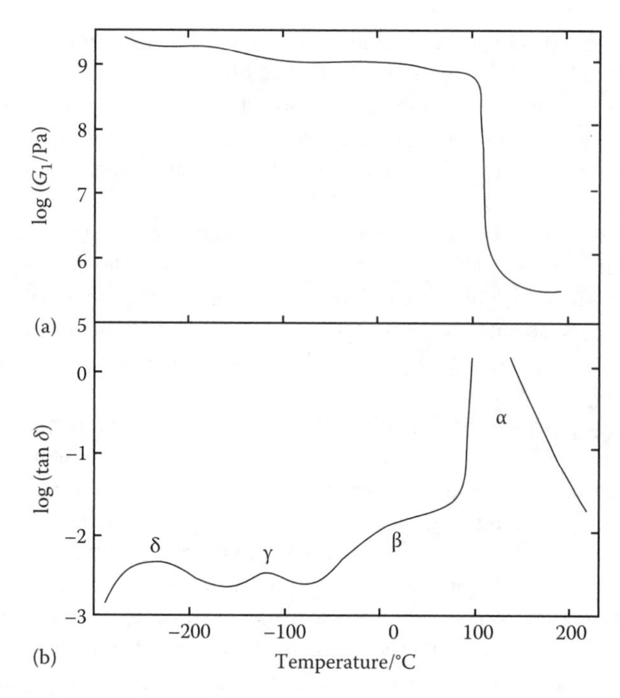

FIGURE 20.9 Variation of (a) shear modulus G_1 and (b) tan δ with temperature for polystyrene. (Data taken from Arridge, R.G.C., *Mechanics of Polymers*, Clarendon Press, Oxford, U.K., 1975.)

consider the crystalline and amorphous regions as separate entities with the amorphous domains having a glass transition. It is clear, however, that as particular molecules can transverse both the amorphous and crystalline regions, this picture may be too simplistic. The relaxation behaviour of polyethylene has been more widely studied than that of any other semi-crystalline polymer and so this will be taken as an example, even though it is not completely understood. Figure 20.10 shows the variation of tan δ with temperature for two samples of polyethylene, one of high density (linear) and the other of low density (branched). In polyethylene, there are four transition regions designated α', α, β, and γ. The γ transition is very similar in both the high- and low-density samples, whereas the β-relaxation is virtually absent in the high-density polymer. The α and α' peaks also are somewhat different in the pre-melting region.

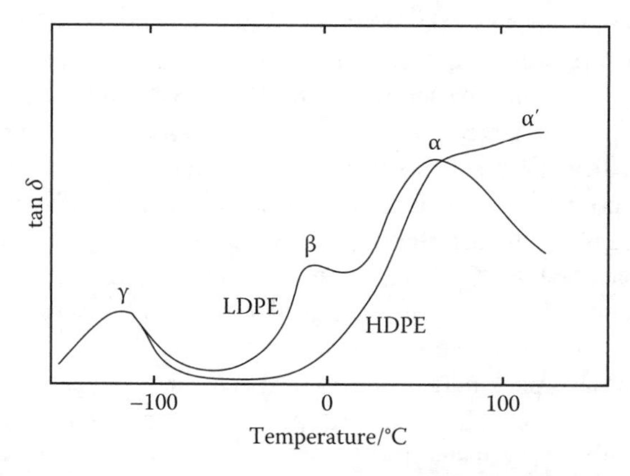

FIGURE 20.10 Variation of tan δ with temperature for high- and low-density polyethylene. (Data taken from Ward, I.M., *Mechanical Properties of Solid Polymers*, 2nd edn., Wiley-Interscience, New York, 1983.)

The presence of branches in the polyethylene molecules produces a difference in polymer morphology. The degree of crystallinity, crystal size and crystal perfection all are reduced by the branching and this can lead to complications in interpretation of the relaxation behaviour. The intensities of the α′- and α-relaxation decrease as the degree of crystallinity is reduced, implying that they are associated with motion within the crystalline regions. On the other hand, the intensity of the γ-relaxation increases with a reduction in crystallinity indicating that it is associated with the amorphous material and it has been tentatively assigned to a glass transition in the amorphous domains. The disappearance of the β-transition with the absence of branching has been taken as a strong indication that it is associated with relaxations at the branch points. In general, the assignment of the peaks in crystalline polymers to particular types of molecular motion is sometimes difficult and often a matter for some debate.

20.7 TIME–TEMPERATURE SUPERPOSITION

It has been suggested earlier that it may be possible to interrelate the time and temperature dependence of the viscoelastic properties of polymers. It is thought that there is a general equivalence between time and temperature. For instance, a polymer that displays rubbery characteristics under a given set of testing conditions can be induced to show glassy behaviour by either reducing the temperature or increasing the testing rate or frequency. This type of behaviour is shown in Figure 20.11 for the variation of the shear compliance J of a polymer with testing frequency measured at different temperatures in the region of the T_g. The material is rubber-like with a high compliance at high temperatures and low frequencies, and becomes glassy with a low compliance as the temperature is reduced and the frequency increased.

It was found empirically that all the curves in Figure 20.11 could be superposed by keeping one fixed and shifting all the others by different amounts, horizontally parallel to the logarithmic frequency axis. If an arbitrary reference temperature T_s is taken to fix one curve then if ω_s is the frequency of a point on the curve at T_s with a particular compliance and ω is the frequency of a point with the same compliance on a curve at a different temperature, then the shift required to superpose the two curves is a displacement of $(\log \omega_s - \log \omega)$ along the log frequency axis. The 'shift factor' a_T is defined by

$$\log a_T = \log \omega_s - \log \omega = \log\left(\frac{\omega_s}{\omega}\right) \tag{20.66}$$

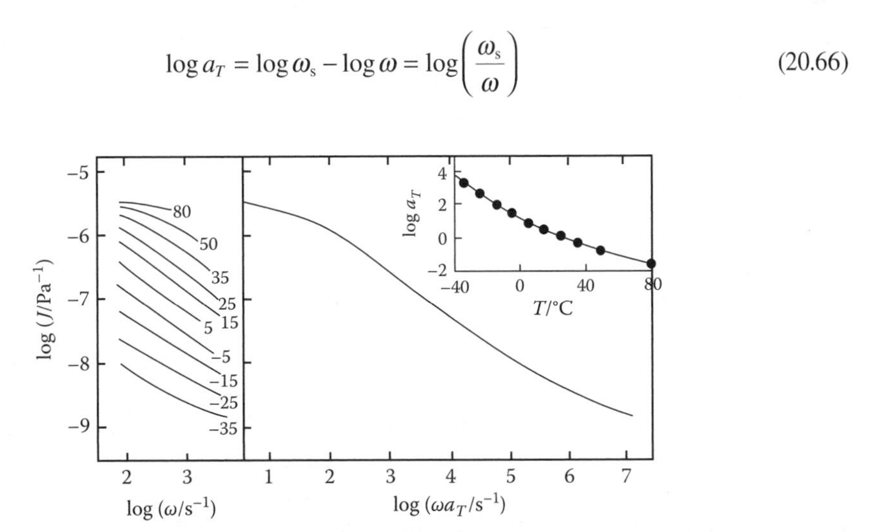

FIGURE 20.11 Example of the time–temperature superposition principle using shear compliance J data for polyisobutylene. The curves of J as a function of frequency obtained at different temperatures (in °C) can all be shifted to lie on a master curve using the shift factor a_T. (Data taken from Fitzgerald, E.R. et al., *J. Appl. Phys.*, 24, 650, 1953.)

and this parameter is a function only of temperature. The values of $\log a_T$, which must be used to superpose the curves in Figure 20.11, are given in the inset. The density of the polymer changes with temperature and so this will also affect the compliance of the polymer. It normally is unnecessary to take this into account unless a high degree of accuracy is required. Much of the early work upon time–temperature superposition was done by Williams, Landel and Ferry and they proposed that a_T could be given by an equation of the form

$$\log a_T = \frac{-C_1(T - T_s)}{C_2 + (T - T_s)} \tag{20.67}$$

where
C_1 and C_2 are constants
T_s is the reference temperature.

This normally is called the *WLF equation* and was originally developed empirically. It holds extremely well for a wide range of polymers in the vicinity of the glass transition and if T_s is taken as T_g, measured by a static method such as dilatometry, then

$$\log a_T = \frac{-C_1^g(T - T_g)}{C_2^g + (T - T_g)} \tag{20.68}$$

and the new constants C_1^g and C_2^g becomes 'universal' with values of 17.4 and 51.6 K, respectively. In fact, the constants vary somewhat from polymer to polymer, but it often is quite safe to assume the universal values as they usually give shift factors that are close to measured values.

The WLF master curve produced by superposing data obtained at different temperatures is a very useful way of presenting the mechanical behaviour of a polymer. The WLF equation also is useful in predicting the mechanical behaviour of a polymer outside the range of temperature and frequency (or time) for which experimental data are available.

Although the WLF equation was developed originally by curve fitting, it is possible to justify it theoretically from considerations of free volume. The concept of free volume was introduced in Section 16.2 in the context of the variation of the specific volume of a polymer with temperature and the abrupt change in slope of the $V-T$ curve at T_g. The fractional free volume f_V in a polymer is given as (Equation 16.10):

$$f_V = f_g + (T - T_g)\alpha_f \tag{20.69}$$

where
f_g is the fractional free volume at T_g
α_f is the thermal expansion coefficient of the free volume.

There are several ways of deriving a relationship of the form of the WLF equation, but perhaps the simplest is by assuming that the polymer behaves according to a viscoelastic model having a relaxation time τ_0 (Section 20.2). For the Maxwell model $\tau_0 = \eta/E$ where η is the viscosity of the dashpot and E is modulus of the spring. It can be assumed that E is independent of temperature and that only η varies with temperature. If the shifts in the time–temperature superposition are thought of as a process of matching the relaxation times, then the shift factor is given by

$$a_T = \frac{\tau_0(T)}{\tau_0(T_g)} = \frac{\eta(T)}{\eta(T_g)} \tag{20.70}$$

when the T_g is used as the reference temperature. It is possible to relate the viscosity to the free volume through a semi-empirical equation developed by Doolittle from the study of the viscosities of liquids. He was able to show that for a liquid, η is related to the free volume V_f through an equation of the form

$$\ln \eta = \ln A + \frac{B(V - V_f)}{V_f} \quad (20.71)$$

where
V is the total volume
A and B are constants.

This equation can be rearranged to give

$$\ln \eta(T) = \ln A + B\left(\frac{1}{f_V} - 1\right) \quad (20.72)$$

and at the T_g

$$\ln \eta(T_g) = \ln A + B\left(\frac{1}{f_g} - 1\right) \quad (20.73)$$

Substitution for f_V from Equation 20.69 in Equation 20.72 and subtracting Equation 20.73 from 20.72 gives

$$\ln \frac{\eta(T)}{\eta(T_g)} = B\left(\frac{1}{f_g + \alpha_f(T - T_g)} - \frac{1}{f_g}\right) \quad (20.74)$$

This equation can be rearranged to give

$$\log \frac{\eta(T)}{\eta(T_g)} = \log a_T = \frac{-(B/2.303 f_g)(T - T_g)}{f_g/\alpha_f + (T - T_g)} \quad (20.75)$$

which has an identical form as the WLF Equation 20.67. The universal nature of C_1^g and C_2^g in the WLF equation implies that f_g and α_f should also be the same for different polymers. It is found that f_g is of the order of 0.025 and α_f approximately 4.8×10^{-4} K^{-1} for most amorphous polymers.

20.8 EFFECT OF ENTANGLEMENTS

When polymers are undergoing the transition between the rubbery and glassy states, the viscoelastic behaviour is dominated by local rearrangement of the polymer conformations so that entanglements or crosslinks play only a minor role. Because of this, the molar mass of the polymer does not affect the behaviour in this region significantly unless the molar mass is very low. The properties of the polymer in the rubbery region, however, are strongly dependent upon molar mass, as shown in Figure 20.12. This shows, for a series of polystyrene fractions with narrow molar mass distributions, a WLF log–log plot of the dependence of G_1 at 160°C upon the frequency ω of testing multiplied by the WLF shift factor a_T. In the case of the low molar mass fractions, there is a steady increase in $\log G_1$ with an increase in the testing frequency. The curves then shift progressively to lower frequencies as the molar mass M is increased. In the case of the higher molar mass fractions, $\log G_1$ first increases as the frequency is increased until it reaches a plateau level and the length of

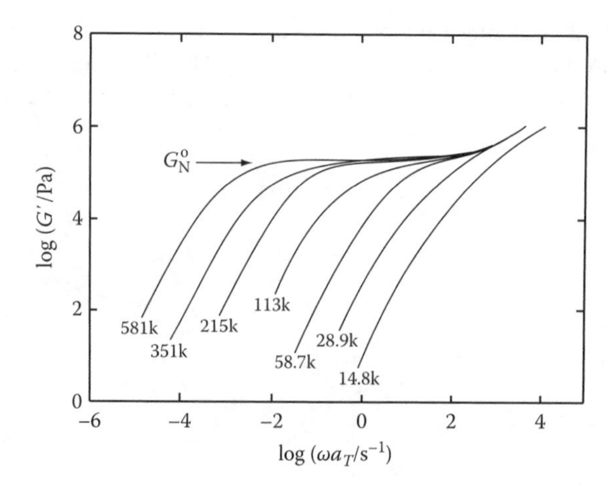

FIGURE 20.12 WLF plot of storage modulus of a series of polystyrene samples with narrow distribution of molar mass (values of $\bar{M}_w$ are indicated in g mol^{-1}) measured at 160°C. (Data taken from Onogi, T. et al., *Macromolecules*, 3, 109, 1970.)

the plateau increases as M increases. The development of this plateau region with increasing molar mass can be correlated with the onset of formation of an entanglement network. The modulus G_N^o of the polymer in this region is often termed the *plateau modulus*, essentially the pseudo equilibrium modulus of the entanglement network.

The parameter, G_N^o is different for different polymers, but essentially is independent of molar mass, once the polymer chains are long enough to form an entangled network. Doi and Edwards, using the de Gennes reptation theory for the concept of a polymer molecule confined to a tube (Figure 16.9), predicted that

$$G_N^o \propto M^0 \tag{20.76}$$

i.e. G_N^o is independent of M, which is precisely what is observed. This result, along with the prediction of the dependence of the viscosity η of a polymer upon M^3 (Equation 16.32), helped to confirm the success of the reptation theory, culminating with the award of the 1991 Nobel Prize to de Gennes for his work on both polymers and liquid crystals.

The reason why G_N^o is different for different polymers was investigated by Graessley and Edwards. They used the analogy with rubber elasticity (Section 21.3.3), which led them to define $G_N^o/\mathbf{k}T$ as an interaction density, or the number of elastic polymer chains per unit volume ($\mathbf{k}$ is the Boltzmann constant and T is the temperature). In this case, differences in G_N^o for different polymers will arise from differences in polymer structure and chain conformation. These chain properties can be described using the *Kuhn length b* (Section 10.3.2), which can be thought of as the length of a freely jointed link required to form a theoretical chain that is characteristic of the particular polymer and gives chain dimensions equal to those of the real chain. The number of elastic chains per equivalent link is then proportional to $G_N^o b^3/\mathbf{k}T$ since b^3 is proportional to the volume per equivalent link. This quantity can be related to the volume fraction of equivalent links given by the product of the number of polymer molecules per unit volume v and the chain volume. The chain volume can be taken as the product of the *contour length* of the chain L (section 10.3.1) and its cross-sectional area, which should be proportional to b^2. Hence, the volume fraction of links must be proportional to vLb^2. Graessley and Edwards postulated that $G_N^o b^3/\mathbf{k}T$ could be written explicitly as a function of vLb^2, i.e.

$$\frac{G_N^o b^3}{\mathbf{k}T} = \mathrm{F}(vLb^2) \tag{20.77}$$

Through the careful analysis of experimental data for 15 different polymers, Graessley and Edwards concluded that the function $F(\nu Lb^2)$ must have the form of a power law, such that

$$\frac{G_N^o b^3}{\mathbf{k}T} = K_1(\nu Lb^2)^a \tag{20.78}$$

where the constant of proportionality K_1 and exponent a are both universal constants.

Figure 20.13 shows a plot of the dimensionless interaction density $G_N^o b^3/\mathbf{k}T$ versus dimensionless contour length concentration νLb^2 for a number of polymer melts, and it can be seen that the data fall close to single straight line of slope 2 (=a). The best fit to the data for this value of a leads to $K_1 = 0.0108 \pm 0.0033$. It is possible to rationalise the power law dependence in Equation 20.78, since for binary interactions in which two links participate in each interaction, the number of entanglements per unit volume is proportional theoretically to the square of the volume fraction of equivalent links, in which case $a = 2$.

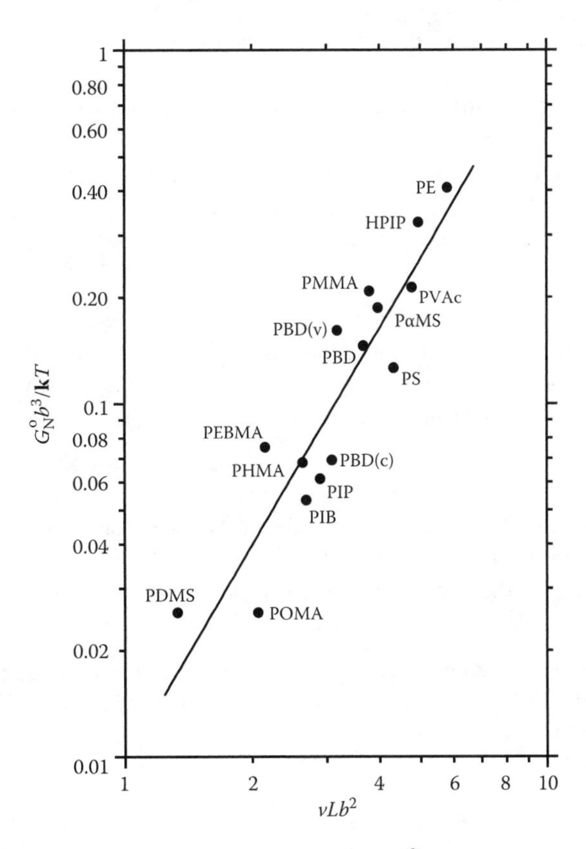

FIGURE 20.13 Plot of dimensionless interaction parameter $G_N^o b^3/\mathbf{k}T$ versus dimensionless contour length νLb^2 for a range of polymer melts. The abbreviations for the polymers are as follows: PE, polyethylene; HPIP, hydrogenated polyisoprene; PMMA, poly(methyl methacrylate); PVAc, poly(vinyl acetate); PαMS, poly(α-methyl styrene); PBD(v), high vinyl-content polybutadiene; PBD, polybutadiene; PS, polystyrene; PEBMA, poly(2-ethylbutyl methacrylate); PHMA, poly(*n*-hexyl methacylate); PBD(c), high *cis*-content polybutadiene; PIP, high *cis*-content polyisoprene; PIB, polyisobutylene; PDMS, polydimethylsiloxane; POMA, poly(*n*-octyl methacrylate). (Data taken from Graessley, W.W. and Edwards, S.F., *Polymer*, 22, 1329, 1981.)

20.9 NON-LINEAR VISCOELASTICITY

An unfortunate aspect of the analysis of the viscoelastic behaviour of polymers is that for most polymers deformed in practical situations, the theory of linear viscoelasticity does not apply. It was emphasized earlier (Section 20.2) that Hooke's law and Newton's law are obeyed only at low strains. However, many polymers and especially semi-crystalline ones do not obey the Boltzmann superposition principle even at low strains and it is found that the exact loading route affects the final state of stress and strain. In this case, empirical approaches normally are used and the mechanical properties are evaluated for the range of conditions of service expected (e.g. stress, strain, time and temperature) and are tabulated or displayed graphically for use in practical applications.

PROBLEMS

20.1 Mechanical models are often used to describe the viscoelastic behaviour of polymers. The Maxwell model, which consists of a spring and dashpot in series, predicts that the variation of strain, e, with time, t, for a viscoelastic material can be described by an equation of the form

$$\frac{\mathrm{d}e}{\mathrm{d}t} = \frac{1}{E}\frac{\mathrm{d}\sigma}{\mathrm{d}t} + \frac{\sigma}{\eta} \qquad \begin{array}{l} E = \text{modulus of spring} \\ \eta = \text{viscosity of dashpot} \end{array}$$

where σ is the applied stress. The Voigt model, which consists of a spring and dashpot in parallel, leads to an equation of the form

$$\frac{\mathrm{d}e}{\mathrm{d}t} = \frac{\sigma}{\eta} - \frac{Ee}{\eta}$$

(a) Derive equations describing the variation of strain with time for both models subjected to a constant stressing-rate (i.e. $\mathrm{d}\sigma/\mathrm{d}t = \text{constant}$).

(b) Derive equations describing the variation of stress with time when both models are subjected to a constant strain-rate ($\mathrm{d}e/\mathrm{d}t = \text{constant}$).

20.2 A ring-shaped seal, made from a viscoelastic material, is used to seal a joint between two rigid pipes. When incorporated in the joint, the seal is held at a fixed compressive strain of 0.2. Assuming that the seal can be treated as a Maxwell model, determine the time before the seal begins to leak under an internal fluid pressure of 0.3 MPa. It can be assumed that the relaxation time τ_0 of the material is 300 days and the short-term (instantaneous) modulus of the material is 3 MPa.

20.3 A viscoelastic polymer, which can be assumed to obey the Boltzmann superposition principle, is loaded by uniform tensile stress of 15 MPa applied at $t=0$ and removed at $t=100$ s. If the creep compliance is

$$J(t) = J_0\left(1 + \left\{\frac{t}{500}\right\}\right)$$

what is the total strain at $t=200$ s if $J_0 = 2\ \mathrm{m^2/GN}$?

20.4 A viscoelastic polymer that can be assumed to obey the Boltzmann superposition principle is subjected to the following deformation cycle. At a time, $t=0$, a tensile stress of 20 MPa is applied instantaneously and maintained for 100 s. The stress is then removed at a rate of 0.2 MPa $\mathrm{s^{-1}}$ until the polymer is unloaded. If the creep compliance of the material is given by

$$J(t) = J_0\left(1 - \exp\left\{\frac{-t}{\tau_0}\right\}\right)$$

where
$J_0 = 3\ \mathrm{m^2/GPa}$
$\tau_0 = 200\ \mathrm{s}$

determine
(a) The strain after 100s (before the stress is reversed)
(b) The residual strain when the stress falls to zero.

20.5 A viscoelastic polymer is subjected to an oscillating sinusoidal load applied at an angular frequency, ω. Assuming that the variation of the strain e and stress σ with time t can be represented by equations of the form

$$e = e_0 \sin \omega t$$

$$\sigma = \sigma_0 \sin(\omega t + \delta)$$

where δ is the phase lag between the stress and strain, show that the energy dissipated per cycle of the deformation ΔU can be given by

$$\Delta U = \sigma_0 e_0 \pi \sin \delta$$

FURTHER READING

Arridge, R.G.C., *Mechanics of Polymers*, Clarendon Press, Oxford, U.K., 1975.
Doi, M. and Edwards, S.F., *The Theory of Polymer Dynamics*, Clarendon Press, Oxford, 1986.
Ferry, J.D., *Viscoelastic Properties of Polymers*, Wiley, New York, 1970.
Lakes, R.S., *Viscoelastic Solids*, CRC Press, Boca Raton, FL, 1999.
Lin Y.-H., *Polymer Viscoelasticity: Basics, Molecular Theories, and Experiments*, World Scientific Publishing Co. Pte. Ltd, Singapore, 2003.
Mark, J.E. (ed), *Physical Properties of Polymers Handbook*, Springer, New York, 2007.
Shaw, M.T. and MacKnight, W.J., *Introduction to Viscoelasticity*, 3rd edn., John Wiley & Sons, Hoboken, NJ, 2005.
Sperling, L.H., *Introduction to Physical Polymer Science*, 4th edn., Wiley-Interscience, New York, 2006.
Ward, I.M., *Mechanical Properties of Solid Polymers*, 2nd edn., Wiley-Interscience, New York, 1983.
Williams, J.G., *Stress Analysis of Polymers*, Longman, London, U.K., 1973.

21 Elastomers

21.1 INTRODUCTION

Crosslinked rubbers, *elastomers*, possess the remarkable ability of being able to stretch to 5–10 times their original length and then retract rapidly to near their original dimensions when the stress is removed. This behaviour sets them apart from materials such as metals and ceramics where the maximum reversible strains that can be tolerated usually are less than 1%. The mechanical behaviour of elastomers will be considered from both phenomenological and molecular viewpoints and it will be demonstrated that it is possible to predict their mechanical properties from first principles.

21.1.1 General Considerations

It is worthwhile to consider in detail the full molecular structural requirements that are needed of a polymer before it can have elastomeric properties. The main *three* requirements are as follows.

- The polymer must be above its glass transition temperature T_g.
- The polymer must have a very low degree of crystallinity ($x_c \rightarrow 0$).
- The polymer should be lightly crosslinked.

Although many polymers have some of these characteristics, there are only relatively few that display true elastomeric behaviour. For example, polyethylene is above its T_g at room temperature but it also is highly crystalline and so it is not an elastomer. The copolymerization of ethylene with propylene destroys the crystallinity of the polymer and can lead to materials commonly known as ethylene/propylene rubbers. On crosslinking such polymers display elastomeric properties. The elastomeric properties of crosslinked *natural rubber*, which is a naturally occurring form of *cis*-1, 4-polyisoprene, have been known for many years. Although it is rather tacky and tends to flow in its natural state, on crosslinking it exhibits useful elastomeric behaviour.

The importance of crystallization again can be demonstrated by considering the physical properties of the two naturally occurring isomers of polyisoprene, the structure of which was described in Section 6.3. When the chains have a *cis* configuration crystallization is relatively difficult. The degree of crystallinity in natural rubber is fairly low and the melting point (~35 °C) is only just above ambient temperature. The chains in *cis*-1,4-polyisoprene also can coil relatively easily by rotation about single bonds and when the polymer is crosslinked it has good elastomeric properties. On the other hand, in gutta percha and balata, which are two forms of *trans*-1,4-polyisoprene, the chain is more regular and the degree of crystallinity is higher because crystallization takes place more readily to give crystals that are more stable (T_m ~ 75 °C). The chains in the *trans* form also are less flexible. These factors make the material relatively hard and rigid with no useful elastomeric properties. Although natural rubber is used widely in a large number of applications, synthetic elastomers are often also employed. Natural rubber is rather prone to chemical degradation by ozone and tends to swell in the presence of solvents. Synthetic elastomers are able to overcome some of these problems. Also, elastomers are often used with a variety of additives, such as carbon black, that reinforce the materials and improve stiffness, strength and abrasion resistance.

21.1.2 Vulcanization

The process of crosslinking rubbers to produce elastomers is known, for historical reasons, as vulcanization (Section 1.1). Sulphur vulcanization was discovered first, specifically for crosslinking natural rubber, but can be used to crosslink most rubbers that possess sites of unsaturation in their structure. Sulphur is used at a level of 0.5–5 parts per hundred of rubber and the crosslinking reaction takes place upon heating during the shaping process, typically at temperatures in the range 120–180 °C. Some examples of the types of crosslinks (i.e. junction points) formed in this highly complex reaction are shown below for a polymer with repeat units derived from butadiene. In the absence of other components, the reaction is relatively slow and inefficient (m is large). Thus, it is usual to include an accelerator (e.g. a mercaptobenzothiazole) and an activator (e.g. zinc stearate) to increase the rate of reaction and improve its efficiency ($m = 1$ or 2).

$$n \sim\sim CH_2-CH=CH-CH_2 \sim\sim + \text{sulphur}$$

$n = 2$:

$$\begin{array}{l} \sim\sim CH-CH=CH-CH_2\sim\sim \\ \quad | \\ \quad S_m \\ \quad | \\ \sim\sim CH_2-CH-CH_2-CH_2\sim\sim \end{array}$$

$n = 3$:

$$\begin{array}{l} \sim\sim CH-CH=CH-CH_2\sim\sim \\ \quad | \\ \quad S_m \\ \quad \quad \backslash \\ \sim\sim CH_2-CH-CH-CH_2\sim\sim \\ \qquad\qquad | \\ \qquad\qquad S_m \\ \qquad\qquad | \\ \sim\sim CH_2-CH-CH_2-CH_2\sim\sim \end{array}$$

Methods of non-sulphur vulcanization were first developed for crosslinking rubbers that do not possess sites of unsaturation (e.g. poly(ethylene-*co*-propylene)s and siloxanes), but now also are used to crosslink unsaturated rubbers. Crosslinking is effected using free-radical initiators, in particular peroxides (e.g. cumyl peroxide and benzoyl peroxide), at temperatures similar to those for sulphur vulcanization. In saturated rubbers, the free radicals ($R^\bullet$) generated from the initiator, abstract hydrogen atoms from C–H bonds in the rubber to give polymeric radicals, which couple to produce the crosslinks

$$2\,R^\bullet + 2 \sim\sim CH_2-CH_2\sim\sim \longrightarrow 2 \sim\sim \dot{C}H-CH_2\sim\sim + 2\,R-H$$

$$\downarrow \text{coupling}$$

$$\begin{array}{l} \sim\sim CH-CH_2\sim\sim \\ \quad | \\ \sim\sim CH-CH_2\sim\sim \end{array}$$

A similar reaction involving the formation and coupling of allylic radicals occurs with unsaturated rubbers. Additionally, the allylic radicals can add to C=C bonds in other molecules to produce crosslinks

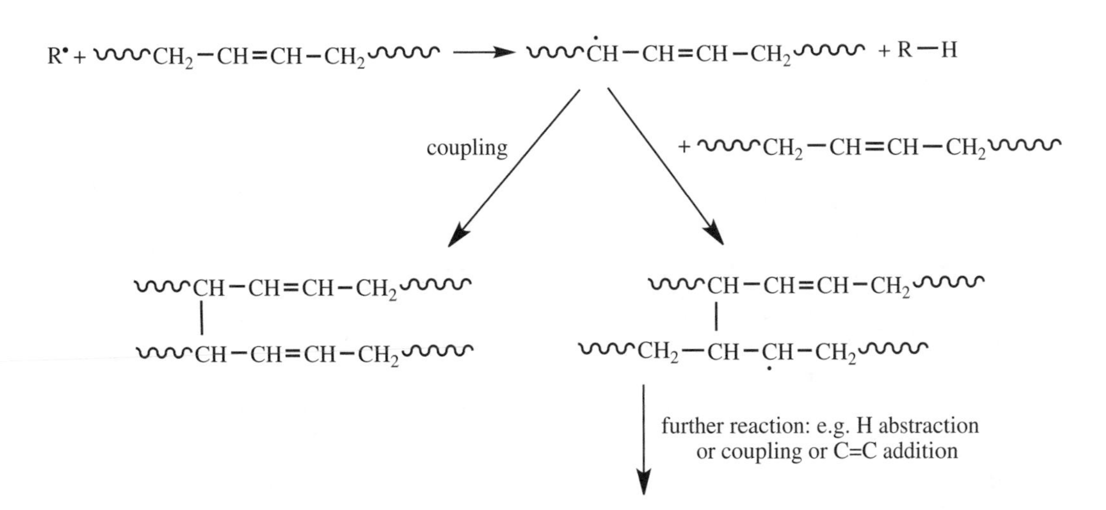

Other methods of non-sulphur vulcanization have been developed (e.g. for polychloroprene, and for room temperature vulcanization of siloxanes) but these are beyond the scope of this brief survey.

The properties of a particular elastomer are controlled by the nature of the crosslinked network. Ideally, this can be envisaged as an amorphous network of polymer chains with junction points from which at least three chains emanate. Every length of polymer chain can be thought of as being anchored at two separate sites. In reality, this is a great oversimplification and the likely situation is sketched in Figure 21.1. In a polymer of finite molar mass, there will be some loose chain ends. Also it is possible that intramolecular crosslinking can take place leading to the presence of loops. Neither of these two structural irregularities will contribute to the stiffness of the network. On the other hand, entanglements can act as effective crosslinks, even in an unvulcanized rubber, at least during short-term loading.

Crystallization takes place only very slowly in natural rubber at ambient temperature since ΔT is small, but when a sample is stretched crystallization takes place relatively rapidly. The rate and degree of crystallization depend upon the extension of the sample and the length of time the extension is maintained. It is very rapid at strains of above about 300%–400% and degrees of crystallinity of over 30% have been reported at the highest extensions. Crystallization is manifested by the appearance of arcs in the X-ray diffraction patterns. The unstretched material is non-crystalline and the diffraction pattern shows only diffuse rings. On deforming natural rubber the molecules tend to become aligned parallel to the stretching direction. Since polymer molecules always pack

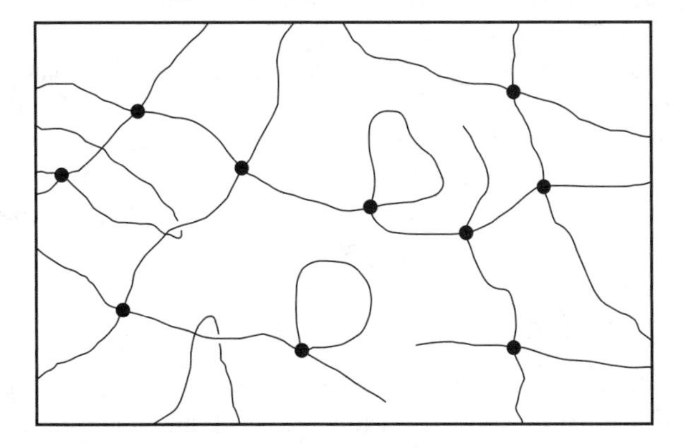

FIGURE 21.1 Schematic diagram of the molecules in an elastomer showing normal crosslinking, chain ends, intramolecular loops and entanglements.

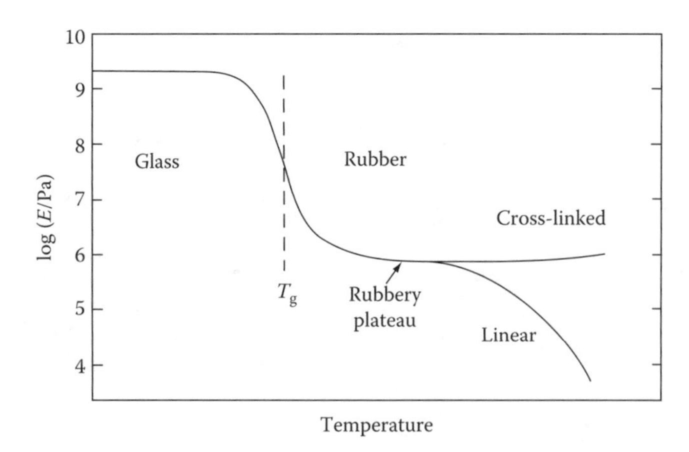

FIGURE 21.2 Typical variation of Young's modulus E with temperature for a polymer showing the rubbery plateau and the effect of crosslinking upon E in the rubbery state.

side-by-side into crystals this alignment reduces the entropy change on crystallization (Section 17.1.1). The stretching process, therefore, aids the crystallization process and the observed arcing in the diffraction patterns follows from the molecular alignment that causes the crystallites to form with their c-axes approximately parallel to the stretching direction. On the other hand, when crystallization takes place in an unstretched sample, the diffraction pattern contains only rings showing that the crystallites have random orientations.

21.1.3 Mechanical Behaviour

Most non-crystallizable polymers are capable of being obtained in the rubber-like state. Figure 21.2 shows the variation of Young's modulus E of an amorphous polymer with temperature for a fixed rate or frequency testing. At low temperatures, the polymer is glassy with a high modulus (~10^9 Pa). The modulus then falls through the glass transition region where the polymer is viscoelastic and the modulus becomes very rate- and temperature-dependent. If the polymer has a sufficiently high molar mass such that entanglements are able to form (i.e. $M > M_e$, where M_e is the entanglement molar mass (Section 16.4.1)), then as the temperature is increased, the modulus will become constant in the rubbery plateau region. Depending upon the value of M, the modulus of a linear polymer drops effectively to zero at a sufficiently high temperature when it behaves like a liquid (Section 20.8), but if the polymer is crosslinked it will remain approximately constant at ~10^6 Pa or increase slightly as the temperature is increased further.

The unique deformation behaviour of elastomeric materials has fascinated scientists for many years and there are even reports of investigations into the deformation of natural rubber from the beginning of the nineteenth century. Elastomer deformation is particularly amenable to analysis using thermodynamics, as an elastomer behaves essentially as an 'entropy spring'. It is even possible to derive the form of the basic stress–strain relationship from first principles by considering the statistical thermodynamic behaviour of the molecular network.

21.2 THERMODYNAMICS OF ELASTOMER DEFORMATION

As well as having the ability to sustain large reversible amounts of extension there are several other more subtle aspects of the deformation of elastomers, which are rather unusual and merit further consideration. A well-known property is that when an elastomer is stretched rapidly, it becomes warmer and conversely when it is allowed to equilibriate for a while at constant extension and then allowed to contract rapidly, it cools down. This can be readily demonstrated by putting an elastic band to ones lips, which are quite sensitive to changes in temperature, and stretching it quickly. Another property is that the length of an elastomer sample held at constant stress decreases as the temperature is increased whereas an unstressed specimen exhibits conventional thermal expansion behaviour. It is possible to rationalize these aspects of the thermomechanical behaviour by considering the deformation of elastomers in thermodynamic terms. This allows the basic relationship between force, length and temperature to be established for an elastomeric specimen in terms of parameters such as internal energy and entropy. One important experimental observation that allows the analysis to be simplified is that the deformation of elastomers takes place approximately at constant volume (Poisson's ratio ~0.5).

The first law of thermodynamics establishes the relationship between the change in internal energy of a system $\mathrm{d}U$ and the heat $đQ$ absorbed by the system and work $đW$ done by the system as

$$\mathrm{d}U = đQ - đW \tag{21.1}$$

The bars indicate that $đQ$ and $đW$ are inexact differentials because Q and W, unlike U, are not macroscopic functions of the system. If the length of an elastic specimen is increased, a small amount $\mathrm{d}l$ by a tensile force f then an amount of work $f\,\mathrm{d}l$ will be done *on* the system. There is also a change in volume $\mathrm{d}V$ during elastic deformation and work $p\,\mathrm{d}V$ is done against the pressure p. Since elastomers deform at roughly constant volume, the contribution of $p\,\mathrm{d}V$ to $đW$ at ambient pressure will be small and so the work done *by* the system (i.e. the specimen) when it is extended is given by

$$đW = -f\mathrm{d}l \tag{21.2}$$

The deformation of elastomers can be considered as a reversible process and so $đQ$ can be evaluated from the second law of thermodynamics that states that for a reversible process

$$đQ = T\mathrm{d}S \tag{21.3}$$

where

T is the thermodynamic temperature
$\mathrm{d}S$ is the change in entropy of the system.

Combining Equations 21.1 through 21.3 gives

$$f\mathrm{d}l = \mathrm{d}U - T\mathrm{d}S \tag{21.4}$$

Most of the experimental investigations on elastomers have been done under conditions of constant pressure. The thermodynamic function that can be used to describe equilibrium under these conditions is the Gibbs free energy, but, since elastomers tend to deform at constant volume, it is possible to use the Helmholtz free energy A in the consideration of equilibrium. It is defined as

$$A = U - TS \tag{21.5}$$

and for a change taking place under conditions of constant temperature

$$\mathrm{d}A = \mathrm{d}U - T\mathrm{d}S \tag{21.6}$$

Comparing Equations 21.4 and 21.6 gives

$$f\,\mathrm{d}l = \mathrm{d}A \text{ (at constant } T)$$

and so

$$f = \left(\frac{\partial A}{\partial l}\right)_T \tag{21.7}$$

Combining Equations 21.6 and 21.7 enables the force to be expressed as

$$f = \left(\frac{\partial U}{\partial l}\right)_T - T\left(\frac{\partial S}{\partial l}\right)_T \tag{21.8}$$

The first term refers to the change in internal energy with extension and the second to the change in entropy with extension. An expression of this form could be used approximately to describe the response of any solid to an applied force. For most materials, the internal energy term is dominant, but in the case of elastomers the change in entropy gives the largest contribution to the force. This can be demonstrated after modifying the entropy term in Equation 21.8. The Helmholtz free energy can be expressed in a differential form for any general change as

$$\mathrm{d}A = \mathrm{d}U - T\mathrm{d}S - S\mathrm{d}T \tag{21.9}$$

and combining this equation with Equation 21.4 gives

$$\mathrm{d}A = f\mathrm{d}l - S\mathrm{d}T \tag{21.10}$$

and partial differentiation under conditions first of all of constant temperature and then of constant length gives

$$\left.\begin{aligned} (\partial A/\partial l)_T &= f \\ \text{and}\quad (\partial A/\partial T)_l &= -S \end{aligned}\right\} \tag{21.11}$$

But we have the standard relation for partial differentiation

$$\frac{\partial}{\partial l}\left(\frac{\partial A}{\partial T}\right)_l = \frac{\partial}{\partial T}\left(\frac{\partial A}{\partial l}\right)_T \tag{21.12}$$

and applying this to Equations 21.11 gives

$$\left(\frac{\partial S}{\partial l}\right)_T = -\left(\frac{\partial f}{\partial T}\right)_l \tag{21.13}$$

Equation 21.8 then becomes

$$f = \left(\frac{\partial U}{\partial l}\right)_T + T\left(\frac{\partial f}{\partial T}\right)_l \tag{21.14}$$

These last two equations contain parameters that can be measured experimentally and so allow the entropy and internal energy terms to be calculated. This can be done by measuring how the force required to hold an elastomer at constant length varies with the absolute temperature. Figure 21.3

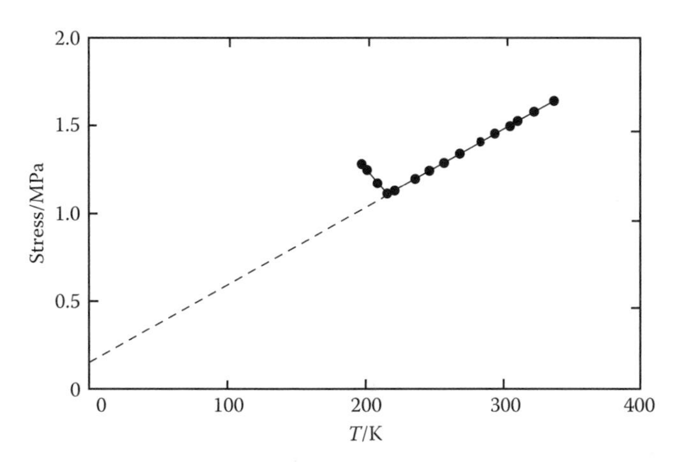

FIGURE 21.3 Variation of the stress in an elastomer (vulcanized natural rubber) as a function of temperature when the specimen was held at a fixed extension of 350%. (Data taken from Meyer, K.H. and Ferri, C., *Helv. Chim. Acta*, 18, 570, 1935.)

shows the results of an early experiment of this type. The stress in the elastomer drops as the temperature is reduced until about 220 K (−53 °C) at which point it starts to rise again. This change in slope corresponds to the transition to a glassy state when the material is no longer elastomeric. The data in the rubbery region are virtually linear and can be extrapolated approximately to zero tension at absolute zero. The curve in Figure 21.3 is of great significance and can be related directly to Equation 21.14. The slope of the curve at any point $(\partial f/\partial T)_l$ gives the variation of entropy with extension at that temperature. Since the curve is linear, it suggests that $(\partial S/\partial l)_T$ is fairly temperature independent and as the intercept is approximately zero it also implies that $(\partial U/\partial l)_T$ is small. This means that there is very little change in internal energy during the extension of a rubber and that the deformation is dominated by the change in entropy.

The data in Figure 21.3 were obtained at a relatively high extension (350%). When the experiment is repeated at lower extensions, the slope of the curve decreases and below about 10% extension it becomes negative. This is caused by a reduction in stress due to thermal expansion of the material as the temperature is increased and it is known as the *thermo-elastic inversion effect*. If the effective change in extension due to thermal expansion is allowed for then the thermo-elastic inversion effect disappears and the stress increases proportionately with temperature at low extensions as well as high extensions.

All the considerations of thermo-elastic effects have so far been concerned with changes in stress at constant extension, but they can be extended directly to experiments done at constant tension. In this case, the tendency of the stress to increase as the temperature is raised will cause a specimen subjected to a constant stress to reduce its length as the temperature is increased. This can be demonstrated readily by hanging a weight on a length of an elastomer and changing the temperature. The weight moves upwards when the elastomer is heated and it goes downwards when the elastomer is cooled.

Another thermo-elastic effect that was mentioned earlier is the tendency of an elastomer to become warm when stretched rapidly. This type of behaviour is illustrated in Figure 21.4 for the adiabatic extension of an elastomer. Some of the heating can be explained by crystallization that also can occur when natural rubber is stretched but a temperature rise is found even in non-crystallizable elastomers. This can be explained by the consideration of the thermodynamic equilibrium. The rise in temperature as the length is increased adiabatically is characterized by $(\partial T/\partial l)_S$, where the subscript refers to a change at constant entropy, which is given by the standard relation

$$\left(\frac{\partial T}{\partial l}\right)_S = -\left(\frac{\partial T}{\partial S}\right)_l \left(\frac{\partial S}{\partial l}\right)_T \tag{21.15}$$

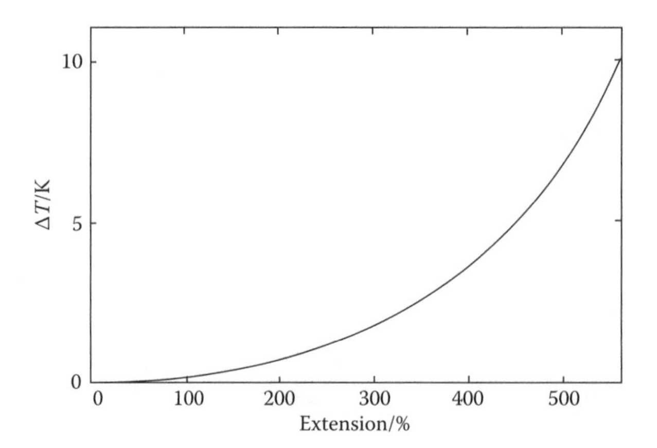

FIGURE 21.4 Increase in temperature ΔT upon adiabatic extension of an elastomer (vulcanized natural rubber). (Data taken from Dart, S.L. et al., *Ind. Engng. Chem.*, 34, 1340, 1942.)

The first factor on the right-hand side can be expressed as

$$\left(\frac{\partial T}{\partial S}\right)_l = \left(\frac{\partial T}{\partial H}\right)_l \left(\frac{\partial H}{\partial S}\right)_l \tag{21.16}$$

but

$$\left(\frac{\partial T}{\partial H}\right)_l = \frac{1}{C_l} \quad \text{and} \quad \left(\frac{\partial H}{\partial S}\right)_l = T$$

where C_l is the heat capacity of the elastomer held at constant length and so Equation 21.15 becomes

$$\left(\frac{\partial T}{\partial l}\right)_S = -\frac{T}{C_l}\left(\frac{\partial S}{\partial l}\right)_T \tag{21.17}$$

Combining Equations 21.13 and 21.17 gives

$$\left(\frac{\partial T}{\partial l}\right)_S = \frac{T}{C_l}\left(\frac{\partial f}{\partial T}\right)_l \tag{21.18}$$

This equation shows clearly that the temperature of an elastomer will rise when it is stretched adiabatically as long as $(\partial f/\partial T)_l$ is positive as is normally the case (Figure 21.3). It is a direct consequence of the fact that when the elastomer is stretched, work is transformed reversibly into heat.

21.3 STATISTICAL THEORY OF ELASTOMER DEFORMATION

Although the thermodynamic approach shows clearly that when an elastomer is deformed it behaves like an entropy spring, no specific deformation mechanisms are implied. In fact, none of the thermodynamic equations depend upon the macromolecular nature of the material, although the behaviour obviously stems from the presence of polymer molecules in the elastomer. The statistical approach looks directly at how the molecular structure changes during deformation, allowing the change in entropy during deformation to be calculated and the stress–strain curve to be derived from first principles. It is similar to the calculation of the pressure of an ideal gas using statistical thermodynamics and it makes use of the calculations of the conformations of freely-jointed chains outlined in Section 10.3.

21.3.1 Entropy of an Individual Chain

The material is assumed to be made up of a network of crosslinked polymer chains with the individual lengths of chain adopting random conformations. When the network is extended, the molecules become uncoiled and their entropy is reduced. The force required to deform the elastomer can, therefore, be related directly to this change in entropy. The first step in the analysis involves calculation of the entropy of an individual polymer molecule. This can be done with the help of Equation 10.64 which gives the probability per unit volume $W(x,y,z)$ of finding one end of a freely jointed chain at a point (x,y,z) a distance r from the other end, that is fixed at the origin. This probability can be expressed as

$$W(x,y,z) = \left(\frac{\beta}{\pi^{1/2}}\right)^3 \exp\left(-\beta^2 r^2\right) \tag{10.64}$$

and the function has been plotted graphically in Figure 10.4(a). The parameter β is defined as $\beta = (3/(2nl^2))^{1/2}$ where n is the number of links of length l in the chain and β can be considered to be characteristic for a particular chain. Equation 10.64 allows the entropy of a single chain to be calculated using the Boltzmann relationship from statistical thermodynamics, which is

$$S = \mathbf{k} \ln \Omega \tag{21.19}$$

where

k the Boltzmann constant

Ω is the number of possible conformations the chain can adopt.

It can be assumed that Ω will be proportional to $W(x,y,z)$. This is clearly a reasonable assumption since when r is small $W(x,y,z)$ is large (Figure 10.4a) and there will be many conformations available to the chain. $W(x,y,z)$ drops as r is increased and the number of possible conformations is reduced as the chain becomes extended. The entropy S of a single chain can then be expressed by an equation of the form

$$S = c - \mathbf{k}\beta^2 r^2 \tag{21.20}$$

where c is a constant.

21.3.2 Deformation of the Polymer Network

Having obtained an expression for the entropy of a single chain it is a relatively simple matter to determine the change in entropy on deforming an elastomeric polymer network. It is assumed that when the elastomer is in either the strained or unstrained state, the junction points can be regarded as being fixed at their mean positions and that the lengths of chain between the points behave as freely-jointed chains so that Equation 21.20 can be applied. It also is assumed that when the material is deformed the change in the components of the displacement vector r of each chain is proportional to the corresponding change in specimen dimensions, i.e. with an assumption of *affine* deformation. Since large strains are encountered in the deformation of elastomers, it is more convenient to use *extension ratios* λ_1, λ_2, λ_3 rather than infinitesimal strains. The extension ratio λ in a particular

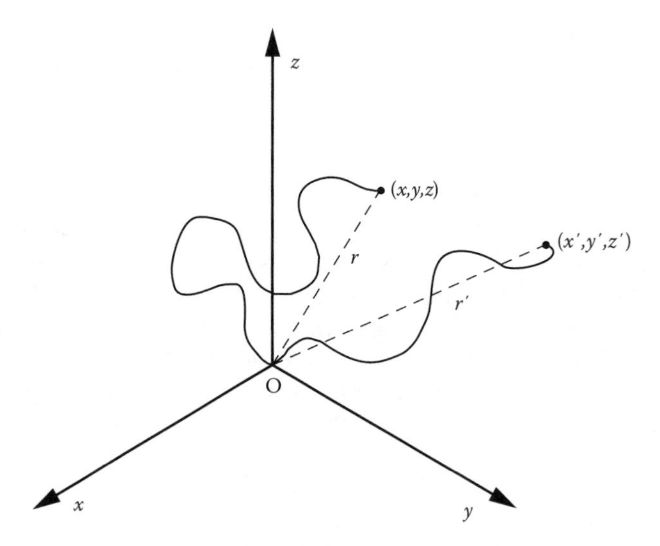

FIGURE 21.5 Schematic representation of the displacement of a junction point in an elastomer network from (x,y,z) to (x′,y′,z′) during a general deformation. (Adapted from Treloar, L.R.G., *The Physics of Rubber Elasticity*, 3rd edn., Clarendon Press, Oxford, 1975.)

direction is defined as the deformed length of the specimen in that direction divided by the original length. The axes of these ratios are chosen to coincide with the rectangular coordinate system used to characterize the conformations of the original lengths of chain. If an individual chain is considered, first of all, then if one junction point is fixed at the origin. O, a general deformation of the elastomer will displace the other junction point from (x,y,z) to (x′,y′,z′), as shown in Figure 21.5. The affine deformation assumption means that the change in the components of the displacement vector will be proportional to the overall change in specimen dimensions means, i.e.

$$x' = \lambda_1 x, \quad y' = \lambda_2 y, \quad z' = \lambda_3 z \tag{21.21}$$

Since r^2 is equal to $(x^2 + y^2 + z^2)$, it follows from Equation 21.20 that the entropies of the individual chain will be

$$S = c - \mathbf{k}\beta^2(x^2 + y^2 + z^2) \quad \text{(before deformation)}$$

$$S' = c - \mathbf{k}\beta^2(\lambda_1^2 x^2 + \lambda_2^2 y^2 + \lambda_3^2 z^2) \quad \text{(after deformation)}$$

and so the change in entropy of an individual chain which has its end displaced from (x,y,z) to (x′,y′,z′) will be given by

$$\Delta S_i = S' - S = -\mathbf{k}\beta^2[(\lambda_1^2 - 1)x^2 + (\lambda_2^2 - 1)y^2 + (\lambda_3^2 - 1)z^2] \tag{21.22}$$

The polymer consists of a network of many such chains with a range of displacement vectors. If the number of these chains per unit volume is defined as N, the number dN, which have ends initially in the small volume element dxdydz at the point (x,y,z), can be determined from the Gaussian distribution function as

$$\mathrm{d}N = NW(x, y, z)\mathrm{d}x\mathrm{d}y\mathrm{d}z$$

or

$$\mathrm{d}N = N\left(\frac{\beta}{\pi^{1/2}}\right)^3 \exp[-\beta^2(x^2+y^2+z^2)]\mathrm{d}x\mathrm{d}y\mathrm{d}z \quad (21.23)$$

On deformation, the change in entropy of these chains is given by $\Delta S_i \mathrm{d}N$ and the total entropy change ΔS per unit volume of the sample during deformation is given by the sum of the entropy changes for all chains. This can be expressed in an integral form as

$$\Delta S = \int \Delta S_i \mathrm{d}N$$

or

$$\Delta S = \int_{-\infty}^{\infty}\int_{-\infty}^{\infty}\int_{-\infty}^{\infty} -\frac{N\mathbf{k}\beta^5}{\pi^{3/2}}[(\lambda_1^2-1)x^2+(\lambda_2^2-1)y^2+(\lambda_3^2-1)z^2]\exp[-\beta^2(x^2+y^2+z^2)]\mathrm{d}x\mathrm{d}y\mathrm{d}z$$

which when evaluated using the standard relationships

$$\int_{-\infty}^{\infty} \exp(-\beta^2x^2)x^2\mathrm{d}x = \frac{\pi^{1/2}}{2\beta^3} \quad \text{and} \quad \int_{-\infty}^{\infty} \exp(-\beta^2x^2)\mathrm{d}x = \frac{\pi^{1/2}}{\beta}$$

gives the simple result

$$\Delta S = -\frac{1}{2}N\mathbf{k}(\lambda_1^2+\lambda_2^2+\lambda_3^2-3) \quad (21.24)$$

This equation relates the change in entropy to the extension ratios and the number of chains between crosslinks per unit volume. It was shown earlier (Section 21.2) that ideally there is no change in internal energy U when an elastomer is deformed and, since the deformation takes place at constant volume, the change in the Helmholtz free energy per unit volume ΔA is

$$\Delta A = -T\Delta S = \frac{1}{2}N\mathbf{k}T(\lambda_1^2+\lambda_2^2+\lambda_3^2-3) \quad (21.25)$$

for isothermal deformation. This will be identical with the isothermal reversible work of deformation w per unit volume and hence

$$w = \frac{1}{2}N\mathbf{k}T(\lambda_1^2+\lambda_2^2+\lambda_3^2-3) \quad (21.26)$$

21.3.3 Limitations and Use of the Theory

The statistical theory is remarkable in that it enables the macroscopic deformation behaviour of an elastomer to be predicted from considerations of how the molecular structure responds to an applied strain. However, it is important to realize that it is only an approximation to the actual behaviour and has significant limitations. Perhaps the most obvious problem is with the assumption that end-to-end distances of the chains can be described by the Gaussian distribution. This problem has been highlighted earlier in connection with solution properties (Section 10.3) where it was shown that the distribution cannot be applied when the chains become extended. It can be overcome to a certain extent with the use of more sophisticated distribution functions, but the use of such functions is beyond the scope of this present discussion. Another problem concerns the value of N. This will be governed by the number of junction points in the polymer network that can be either chemical (crosslinks) or physical (entanglements) in nature. The structure of the chain network in an elastomer has been discussed earlier (Section 21.1). There will be chain ends and loops that do not contribute to the strength of the network, but, if their presence is ignored, it follows that if all network chains are anchored at two crosslinks then the density ρ of the polymer can be expressed as

$$\rho = \frac{NM_c}{\mathbf{N}_A} \tag{21.27}$$

where
M_c is the number-average molar mass of the chain lengths between crosslinks
$\mathbf{N_A}$ is the Avogadro constant.

It follows that N is given by

$$N = \frac{\rho \mathbf{N}_A}{M_c} = \frac{\rho \mathbf{R}}{M_c \mathbf{k}} \tag{21.28}$$

and Equation 21.26 then becomes

$$w = \frac{\rho \mathbf{R} T}{2M_c}(\lambda_1^2 + \lambda_2^2 + \lambda_3^2 - 3) \tag{21.29}$$

This normally is written as

$$w = \frac{1}{2} G(\lambda_1^2 + \lambda_2^2 + \lambda_3^2 - 3) \tag{21.30}$$

where G is given by

$$G = \frac{\rho \mathbf{R} T}{M_c} \tag{21.31}$$

The parameter G relates w, the work of deformation per unit volume (which has dimensions of stress) to the extension ratios and so G also has dimensions of stress. It is, therefore, often referred to as the shear modulus of the elastomer. Inspection of Equation 21.31 shows that G has some interesting properties. As might be expected, G increases as M_c is reduced. This means that the material becomes stiffer as the crosslink density increases and the network becomes tighter. A rather more surprising prediction of the equation, that has been substantiated experimentally, is that, unlike almost every other material, the modulus of an elastomer increases as the temperature is increased. This is yet another consequence of the deformation of elastomers being dominated by changes in entropy rather than of internal energy.

Network defects, such as the *entanglements, loops* and *chain ends* shown in Figure 21.1, will have an important effect upon the mechanical behaviour of elastomers. Entanglements will act as physical crosslinks and tend to increase the modulus. Loops will not contribute to the network elasticity and so do not affect the modulus. Chain ends also will have no significant effect upon the network elasticity.

21.3.3.1 Entanglements

It was pointed out in Chapter 16 that it is possible for high molar mass polymers above their T_g to display elastomeric behaviour even if they have not been crosslinked, since the entanglements can act as physical crosslinks. Figure 20.12 shows that this leads to a rubbery plateau over which the plateau modulus G_N^o is remarkably constant and independent of the polymer molar mass. Since the entanglements are responsible for this behaviour, Equation 21.31 can be rearranged to give

$$M_e = \frac{\rho \mathbf{R} T}{G_N^o} \tag{21.32}$$

where M_e is now the molar mass between entanglements – often called the *entanglement molar mass*. The form of this equation enabled Graessley and Edwards to relate the term $G_N^o/\mathbf{k}T$ to the interaction density or the number of elastic chains per unit volume (Section 20.8). As explained in Chapter 16, M_e also can be related to the parameter M_{cr} determined from the variation of melt viscosity with molar mass (Figure 16.8).

21.3.3.2 Chain Ends

The effect of chain ends has been treated quantitatively by Flory. If the number of original molecules before crosslinking is N_p, then $N_p - 1$ intermolecular linkages will be needed to form a continuous molecular structure. A further additional linkage will then give rise to either a closed loop or two network chains, as shown in Figure 21.6. It is only the additional linkages forming network

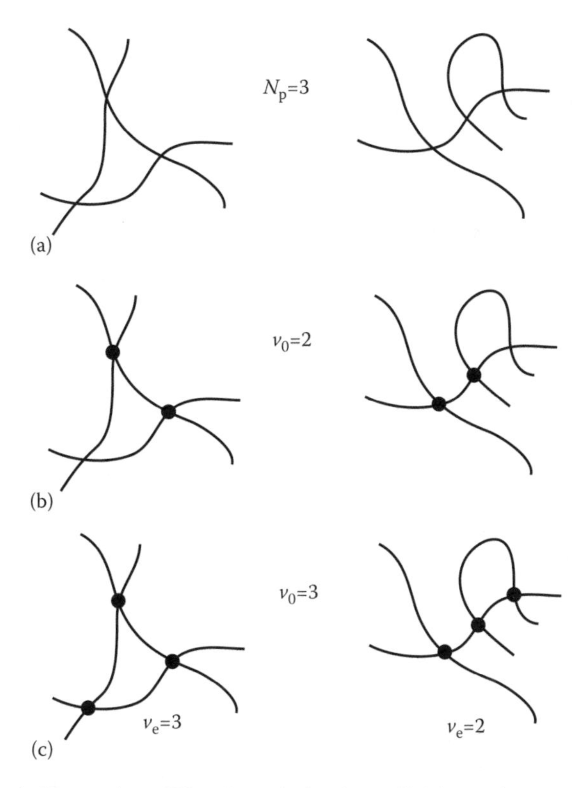

FIGURE 21.6 Schematic illustration of Flory's analysis of crosslinking to form a polymer network for three original molecules ($N_p = 3$): (a) two different sets of three uncrosslinked molecules, (b) continuous molecular structures formed when $\nu_0 = 2$ ($= N_p - 1$) and (c) effect of introducing three crosslinks into the structures ($\nu_0 = 3$), demonstrating how loops give rise to fewer effective crosslinks.

chains that lead to an elastomeric network. If ν_0 is taken as the total number of crosslinks that have been introduced, then the number ν_e of these that act as effective linkages will be

$$\nu_e = \nu_0 - (N_p - 1) \tag{21.33}$$

The number N_e of effective network chains will be twice the number of effective crosslinks and since $N_p \gg 1$ then

$$N_e = 2\nu_e \approx 2\nu_0\left(1 - \frac{N_p}{\nu_0}\right) \tag{21.34}$$

With no defects present, then $N_e = 2\nu_0$ and so the term $(1 - N_p/\nu_0)$ represents the fractional reduction in N_e due to loose ends. It is possible to put this in terms of the molar mass M of the original molecules before crosslinking by modifying the relationship in Equation 21.27 to give $N_p = \rho\mathbf{N}_A/M$. In a similar way, $2\nu_0 = \rho\mathbf{N}_A/M_c$ for an ideal network with no chain ends and so Equation 21.34 becomes

$$N_e = 2\nu_0\left(1 - \frac{2M_c}{M}\right) \tag{21.35}$$

Since the number of effective chains controlling the network's mechanical properties is reduced, Equation 21.31 for the shear modulus will be modified to become

$$G = \frac{\rho \mathrm{R} T}{M_c}\left(1 - \frac{2M_c}{M}\right) \tag{21.36}$$

The inspection of this equation shows that as the initial molecules become longer and $M \to \infty$, chain ends are less important and the equation becomes the same as Equation 21.31.

21.4 STRESS–STRAIN BEHAVIOUR OF ELASTOMERS

The statistical theory allows the stress–strain behaviour of an elastomer to be predicted. The calculation is simplified greatly when the observation that elastomers tend to deform at constant volume is taken into account. This means that the product of the extension ratios must be unity, i.e.

$$\lambda_1\lambda_2\lambda_3 = 1 \tag{21.37}$$

The form of the stress–strain behaviour depends upon the loading geometry employed. The present analysis will be restricted to simple uniaxial tension or compression when specimens are deformed to an extension ratio λ in the direction of the applied stress. It is possible to replace λ_1 by λ but Equation 21.37 requires that $\lambda_2\lambda_3 = 1/\lambda_1$ and so $\lambda_2 = \lambda_3 = 1/\lambda^{1/2}$. Equation 21.30 can, therefore, be written as

$$w = \frac{1}{2}G\left(\lambda^2 + \frac{2}{\lambda} - 3\right) \tag{21.38}$$

for uniaxial tension or compression. If the specimen has an initial cross-sectional area of A_0 and length l_0 and it is extended δl by a uniaxial tensile force f, then the amount of work δW done *on* the specimen is given by

$$\delta W = f\delta l$$

Equation 21.37 gives an expression for w, which is the work per unit volume. The volume of the specimen is A_0l_0 and so

$$\delta w = \frac{\delta W}{A_0 l_0} = \left(\frac{f}{A_0}\right)\left(\frac{\delta l}{l_0}\right) \tag{21.39}$$

The quantity f/A_0 is the force per unit initial cross-sectional area, i.e. the nominal stress σ_n and $\delta l/l_0 = \delta e \approx \delta\lambda$. A relationship between σ_n and λ can then be derived from Equation 21.38 since rearranging Equation 21.39 and expressing it in a differential form gives

$$\sigma_n = \frac{dw}{d\lambda} = G\left(\lambda - \frac{1}{\lambda^2}\right) \tag{21.40}$$

It is possible to compare the relationship between σ_n and λ predicted by this equation with experimental results by fitting the theoretical curve to experimental data with a suitably-chosen value of G. Figure 21.7a shows the results of such an exercise for a sample of vulcanized natural rubber deformed in tension. The value of G chosen was ~0.4 MPa and it can be seen that the agreement between experiment and theory is good at low elongations ($\lambda < 1.5$). However, the theoretical curve falls below the experimental one as the strain is increased further, but eventually rises above it at elongations of $\lambda > 6$. It can be seen that the theory predicts the basic form of the relationship between σ_n and λ, especially at low strains when the Gaussian approximation may be expected to hold well. The lack of agreement at high strains is thought to be due to at least two factors. The Gaussian function will no longer apply and crystallization also can occur in the material. The theory predicts the deformation behaviour rather better in compression. Figure 21.7b shows the variation of σ_n with λ for the same elastomer as used in the tensile experiment, but in this case deformed into the compressive region ($\lambda < 1$). Good agreement between experiment and theory (Equation 21.39) is found between $\lambda = 0.4$ and 1.3. In this regime, the Gaussian function would be expected to be a good approximation of the behaviour of the chains and crystallization does not occur.

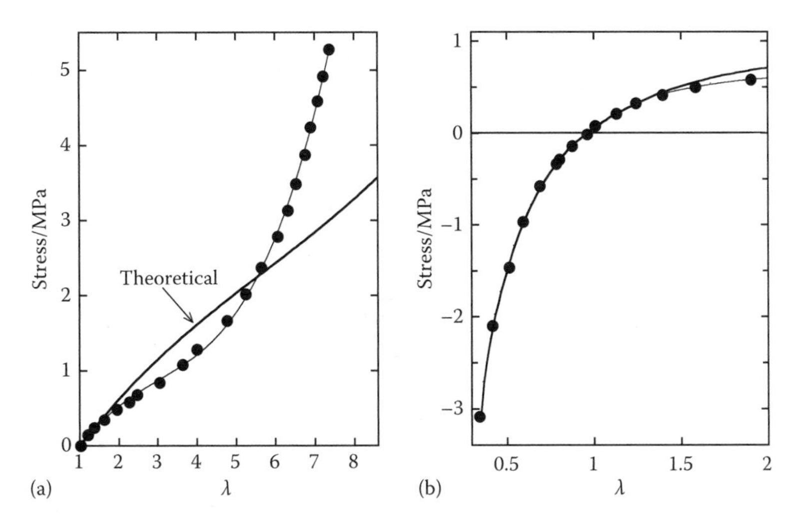

FIGURE 21.7 Relationship between nominal stress σ_n and extension ratio λ for vulcanized natural rubber. The theoretical curves for a value of $G = 0.4$ MPa are given as heavy lines: (a) extension and (b) compressive deformation. (Data taken from Treloar, L.R.G., *Physics of Rubber Elasticity*, 3rd edn., Clarendon Press, Oxford, 1975.)

21.5 FACTORS AFFECTING MECHANICAL BEHAVIOUR

Although it has been shown that the mechanical behaviour of elastomers is controlled principally by their network structure, there are number of other factors that affect their mechanical behaviour significantly. The first is swelling due to their interactions with solvents and the other is stress-induced crystallization.

21.5.1 SWELLING

It is well established that linear polymers are capable of being dissolved in suitable solvents to form polymer solutions. However, although crosslinked polymers are able to absorb large quantities of solvents up to an equilibrium concentration, they do not dissolve. Hence, in the presence of solvents, elastomers undergo a high degree of swelling that modifies their mechanical properties. Moreover, it is found rather surprisingly that they can undergo an increased amount of swelling upon the application of stress.

It can be assumed that, before swelling, the elastomer is in the form of a unit cube that contains N chains per unit volume. Upon swelling it is assumed that the increase in volume of the cube corresponds to the volume of liquid absorbed. The material therefore contains a volume fraction of liquid ϕ_1 and a volume fraction of polymer ϕ_2. The degree of swelling relative to the original unit cube can be defined in terms of the volume fraction of rubber and will be $1/\phi_2$. The corresponding change in the linear dimensions of the cube will be $\lambda_0 = 1/\phi_2^{1/3}$. This swelling involves isotropic expansion of the network that leads to a reduction in the network entropy. The deformation of the swollen elastomer will cause a further reduction in entropy. The mechanical properties of the swollen material are thus controlled by a combination of these two factors.

Equation 21.24 can be employed to evaluate the changes in entropy upon swelling and subsequent deformation. The change in entropy $\Delta S_0'$ upon deformation of the swollen elastomer is given by

$$\Delta S_0' = -\frac{1}{2} N\mathbf{k}\left(l_1^2 + l_2^2 + l_3^2 - 3\right) \tag{21.41}$$

where l_1, l_2 and l_3 are the lengths of the edges relative to the original unswollen unit cube. The entropy change involved with the initial swelling will be given by Equation 21.24 as

$$\Delta S_0 = -\frac{1}{2} N\mathbf{k}\left(\lambda_0^2 + \lambda_0^2 + \lambda_0^2 - 3\right) = -\frac{1}{2} N\mathbf{k}\left(3\phi_2^{-2/3} - 3\right) \tag{21.42}$$

The difference between these two quantities will be the entropy of deformation of the swollen network given by

$$\Delta S' = \Delta S_0' - \Delta S_0 = -\frac{1}{2} N\mathbf{k}\left(l_1^2 + l_2^2 + l_3^2 - 3\phi_2^{-2/3}\right) \tag{21.43}$$

At this stage, it is convenient to define the extension ratios, λ_1, λ_2 and λ_3 with respect to the unstrained swollen network. In this case, $l_1 = \lambda_1\lambda_0 = \lambda_1\phi_2^{-1/3}$ etc., and so

$$\Delta S' = -\frac{1}{2} N\mathbf{k}\phi_2^{-2/3}\left(\lambda_1^2 + \lambda_2^2 + \lambda_3^2 - 3\right) \tag{21.44}$$

which is the entropy per unit volume of the unswollen elastomer. The entropy per unit volume of the swollen elastomer is given by

$$\Delta S = \phi_2 \Delta S' = -\frac{1}{2} N\mathbf{k}\phi_2^{1/3}\left(\lambda_1^2 + \lambda_2^2 + \lambda_3^2 - 3\right) \tag{21.45}$$

Following on from Equations 21.25 and 21.26, the isothermal reversible work of deformation per unit volume given by

$$w = -T\Delta S = \frac{1}{2} N\mathbf{k}T\phi_2^{1/3}(\lambda_1^2 + \lambda_2^2 + \lambda_3^2 - 3) \tag{21.46}$$

Comparing the above analysis with that of dry elastomers in Equation 21.26 shows that the main difference is the factor $\phi_2^{1/3}$ and this leads to a modulus of the swollen rubber, G' being given by

$$G' = G\phi_2^{1/3} = \frac{\rho\mathbf{R}T}{M_c}\phi_2^{1/3} \tag{21.47}$$

Hence the effect of the swelling is to reduce the modulus of the elastomer in proportion to the cube root of the swelling ratio; the form of the stress–strain curve, however, is controlled by the deformation of the polymer network and so remains unchanged.

The above analysis is valid for any level of swelling of the elastomer although in practice there will be an *equilibrium level of swelling* for a particular elastomer–solvent–temperature combination. This can estimated using a modified form of the Flory–Huggins (Section 10.2.2) analysis taking into account the configurational entropy of the elastomer network. The total free energy of dilution of an elastomer by a solvent is given by

$$\Delta G_1 = \Delta G_{1m} + \Delta G_{1e} \tag{21.48}$$

where

ΔG_{1m} is the free energy of the mixing of the polymer in the uncrosslinked state
ΔG_{1e} is the free energy change upon expanding the elastomer network.

The second term can be obtained form the isothermal reversible work per unit volume w given by Equation 21.29. When the network is expanded isotropically then $\lambda_1 = \lambda_2 = \lambda_3 = \phi_2^{-1/3}$, hence

$$w = \frac{\rho\mathbf{R}T}{2M_c}(\lambda_1^2 + \lambda_2^2 + \lambda_3^2 - 1) = \frac{3\rho\mathbf{R}T}{2M_c}(\phi_2^{-2/3} - 1) \tag{21.49}$$

The free energy change upon expanding the network is given by $\partial w/\partial n_1$, where n_1 is the number of moles of liquid expanding the elastomer. The relationship between ϕ_2 and n_1 is given by

$$\phi_2 = \frac{n_2V_2}{n_1V_1 + n_2V_2} \quad \text{or} \quad \frac{1}{\phi_2} = 1 + n_1V_1 \tag{21.50}$$

where V_1 and V_2 are the molar volumes of the solvent and polymer, respectively. Substituting this relationship into Equation 21.49 and differentiating with respect to n_1 leads to

$$\Delta G_{1e} = \frac{\rho\mathbf{R}T}{M_c}V_1\phi_2^{1/3} \tag{21.51}$$

The first term in Equation 21.48 is given by the Flory–Huggins expression for the solvent (Equation 10.37) so that the total free energy of dilution becomes

$$\Delta G_1 = \mathbf{R}T\left[\ln(1-\phi_2) + \left(1 - \frac{1}{x}\right)\phi_1 + \chi\phi_2^2 + \frac{\rho V_1}{M_c}\phi_2^{1/3}\right] \tag{21.52}$$

since the condition for equilibrium swelling is $\Delta G_1 = 0$ and assuming that the degree of polymerization x is high, this then leads to the *Flory–Rehner* equation

$$\ln(1-\phi_2)+\phi_2+\chi\phi_2^2+\frac{\rho V_1}{M_c}\phi_2^{1/3}=0 \tag{21.53}$$

The value of the Flory–Huggins polymer–solvent parameters χ for a large variety of elastomer/solvent combinations are tabulated in the literature. It is also possible to estimate χ from solubility parameters using the principles of additivity (Section 10.2.5). The equilibrium degree of swelling is determined by the value of ϕ_2 that satisfies Equation 21.53. This is an important equation in that it shows that if the values of χ and V_1 are known for an elastomer–solvent system, then the number-average molar mass of the chain lengths between crosslinks M_c can be determined from the value of the equilibrium degree of swelling defined by ϕ_2. Moreover, it is possible to relate ϕ_2 to the mechanical properties of the elastomer since Equation 21.31 shows that the term $\rho V_1/M_c$ is proportional to the elastomer modulus.

Equation 21.53 can be modified by expanding the logarithmic term up to terms in ϕ_2^2 to give

$$\left(\chi-\frac{1}{2}\right)\phi_2^2+\frac{\rho V_1}{M_c}\phi_2^{1/3}\approx 0 \quad \text{or} \quad \frac{\rho V_1}{M_c}\approx\left(\frac{1}{2}-\chi\right)\phi_2^{5/3} \tag{21.54}$$

Figure 21.8 shows a log-log plot of the modulus against the reciprocal of the equilibrium degree of swelling $1/\phi_2$ for samples of crosslinked butyl rubber with different degrees of crosslinking and swollen by cyclohexane. It can be seen that the data fall on a straight line with a slope of –5/3 as predicted by Equation 21.54.

The final issue that needs to be explained is that the ability to swell an elastomer increases when the material is subjected to tension. Although this may seem surprising, it can be readily explained in terms of the concept of three-dimensional stresses introduced in Section 19.2.1. It can be assumed that the shear stress has no effect upon the swelling, whereas hydrostatic pressure will produce a

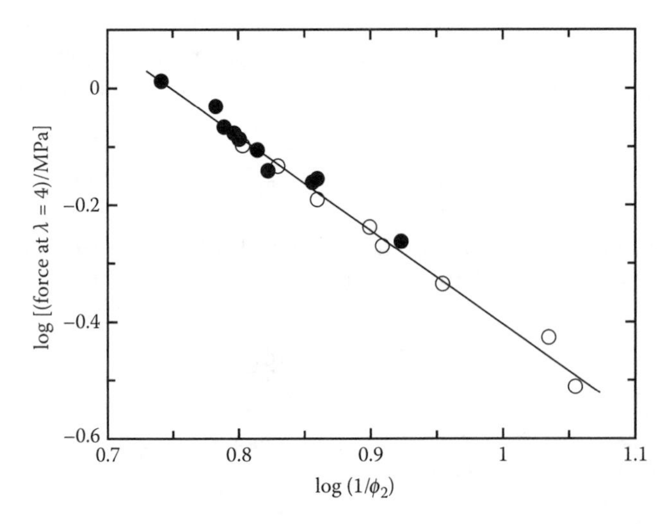

FIGURE 21.8 Log-log plot of the modulus (determined as the force at $\lambda = 4$) as a function of the reciprocal of the equilibrium degree of swelling $1/\phi_2$ for samples of crosslinked butyl rubber with different values of M_c and swollen with cyclohexane. (Data taken from Flory, P.J., *Ind. Engng. Chem.*, 38, 417, 1946.)

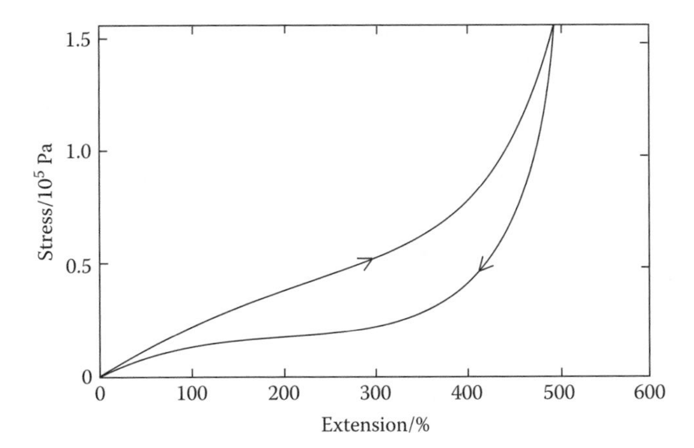

FIGURE 21.9 Illustration of mechanical hysteresis for a strain-crystallizing elastomer. (Data taken from Andrews, E.H., *Fracture in Polymers*, Oliver and Boyd Ltd, London, 1968.)

reduction in swelling. When the swollen elastomer is subjected to tension, there will be both a shear component of stress and an overall hydrostatic tension (the opposite of hydrostatic pressure), which will lead to an increase in ability to swell the material.

21.5.2 Strain-Induced Crystallization

It was pointed out in Section 21.1.1 that the crystallization of natural rubber takes place slowly under ambient conditions. When an elastomer is stretched, however, the polymer chains become aligned and these ordered chains may be capable of undergoing rapid crystallization to form crystallites in the amorphous elastomer. Figure 21.9 shows the stress–strain curve of a strain-crystallizing elastomer. It can be seen that on loading, the stress rises rapidly above about 400% strain ($\lambda = 3$) and is very different from the behaviour predicted by the theoretical curve in Figure 21.7a. This increase in stiffness of the material at high strain is due to two effects. First, the crystallites hinder the sliding of adjacent molecules and so act as physical crosslinks. Second, since crystalline polymers have a much higher modulus than amorphous polymers (Chapter 19) they can also act as reinforcing fillers.

Another aspect of the stress–strain behaviour found particularly in strain-crystallizing elastomers and not predicted by the statistical theory is the phenomenon of *mechanical hysteresis*, which is encountered when the deformed elastomer is allowed to relax. This type of behaviour is shown in Figure 21.9. In this case, although all the strain is recovered upon removing the load, the unloading curve does not follow the same path as the loading curve. The area between the two curves corresponds to the energy dissipated within the cycle. Hysteresis is found to be more prevalent in strain crystallizing and filled elastomers and in certain applications it can lead to undesirable consequences. For example, if an elastomer exhibiting a high hysteresis is cycled rapidly between low and high strains, the dissipation of energy can lead to a large heat buildup, which may cause a deterioration in the properties of the material.

PROBLEMS

21.1 Show that the integral in Equation 21.24 can be evaluated to give

$$\Delta S = -\frac{1}{2} N\mathbf{k}(\lambda_1^2 + \lambda_2^2 + \lambda_3^2 - 3)$$

21.2 The isothermal reversible work of deformation w per unit volume of an elastomer is given by the statistical theory of elastomer deformation as

$$w = \frac{1}{2}G(\lambda_1^2 + \lambda_2^2 + \lambda_3^2 - 3)$$

where λ_1, λ_2 and λ_3 are the principal extension ratios. Using this equation, deduce the stress–strain relations for the general two-dimensional strain of a sheet of material and hence show that the stress–strain relation for an equal two-dimensional extension of λ is

$$\sigma_t = G\left(\lambda^2 - \frac{1}{\lambda^4}\right)$$

where σ_t is the true stress acting on planes normal to the plane of stretch.

21.3 An experiment was undertaken to determine the effect of different liquids upon the swelling of a vulcanized sample of natural rubber at room temperature. It was found that the equilibrium swelling ratio of the natural rubber in carbon tetrachloride was 5.5 and in benzene it was 4.4. Given the information below determine the following parameters.

(a) The molar mass between crosslinks, M_c, for the natural rubber sample.

(b) The value of the Flory–Huggins interaction parameter, χ, for natural rubber in benzene.

(The density of natural rubber = 910 kg m^{-3}; Flory–Huggins interaction parameter χ for natural rubber in carbon tetrachloride = 0.290; molar volume of carbon tetrachloride = 97.1×10^{-6} m^3 mol^{-1}; and molar volume of benzene = 89.4×10^{-6} m^3 mol^{-1}.)

FURTHER READING

Andrews, E.H., *Fracture in Polymers*, Oliver and Boyd Ltd, London, U.K., 1968.

Bhowmick, A.K. and Stephens H.L. (eds), *Handbook of Elastomers*, Marcel Dekker, New York, 2001.

Flory, P.J., *Principles of Polymer Chemistry*, Cornell University Press, Ithaca, NY, 1953.

Mark, J.E., Erman B., and Eirich, F.R. (eds), *Science and Technology of Rubber*, 2nd edn., Academic Press, San Diego, CA, 1994.

Mark, J.E., Ngai, K., Graessley, W., Mandelkern, L., Samulski, E., Koenig, J., and Wignall, G., *Physical Properties of Polymers*, 3rd edn., Cambridge University Press, Cambridge, U.K., 2003.

Treloar, L.R.G., *The Physics of Rubber Elasticity*, 3rd edn., Clarendon Press, Oxford, U.K., 1975.

22 Yield and Crazing

22.1 INTRODUCTION

The determination of the stress–strain characteristics of a material up to failure is particularly useful as it gives information concerning important mechanical properties such as Young's modulus, yield strength and brittleness. These parameters are vital in design considerations when the material is used in a practical situation. Stress–strain curves can be readily obtained for polymers by subjecting a specimen to a tensile force applied at a constant rate of testing. The stress–strain behaviour of polymers is not fundamentally different from that of conventional materials, the main difference being that polymers show a marked time or rate dependence (i.e. viscoelasticity). The situation can be simplified greatly by making measurements at a fixed testing rate, but it is important to bear in mind the underlying time dependence that still exists.

22.2 PHENOMENOLOGY OF YIELD

When a material undergoes elastic deformation, as described in detail in Chapter 19, there is complete recovery of strain on the removal of stress. Yield, on the other hand, can be defined as the onset of permanent plastic deformation in a material which is not recovered when the stress is removed. It will be considered first of all from a phenomenological viewpoint.

22.2.1 DEFINITIONS

Figure 22.1 shows an idealized stress–strain curve for a ductile polymer sample. In this case, the nominal stress σ_n is plotted against strain e. The change in the cross-section of a parallel-sided specimen also is sketched schematically at different stages of the deformation. Initially the stress is proportional to the strain and Hooke's law is obeyed. The tensile modulus can be obtained from the initial slope. As the strain increases, the curve decreases in slope until it reaches a maximum. This is conventionally known as the *yield point* and the *yield stress* σ_y and *yield strain* e_y are indicated on the curve. The yield point for a polymer is rather difficult to define. It should correspond to the point at which permanent plastic deformation takes place, but for polymers a 'permanent set' can be found in specimens loaded to a stress, below the maximum, where the curve becomes non-linear. The situation is complicated further by the observation that even for specimens loaded well beyond the yield strain the plastic deformation can sometimes be recovered completely by annealing the specimen at elevated temperature. In practice, the exact position of the yield point is not of any great importance and the maximum point on the curve suffices as a definition of yield. The value of the yield strain for polymers is typically of the order of 5%–10%, which is very much higher than that of metals and ceramics; yield in metals normally occurs at strains below 0.1%.

During elastic deformation, the cross-sectional area of the specimen decreases uniformly as length increases, but an important change occurs at the yield point. The cross-sectional area starts to decrease more rapidly at one particular point along the gauge length as a 'neck' starts to form. The nominal stress falls after yield and settles at a constant value as the neck extends along the specimen. Eventually, when the whole specimen is necked, strain-hardening occurs and the stress rises until fracture eventually intervenes. The process whereby the neck extends is known as *cold-drawing*. It was originally thought that this was due to local heating of the specimen by the energy expended during deformation. Detailed measurements, however, have shown that a neck can form even at very low strain-rates when the heat is easily dissipated.

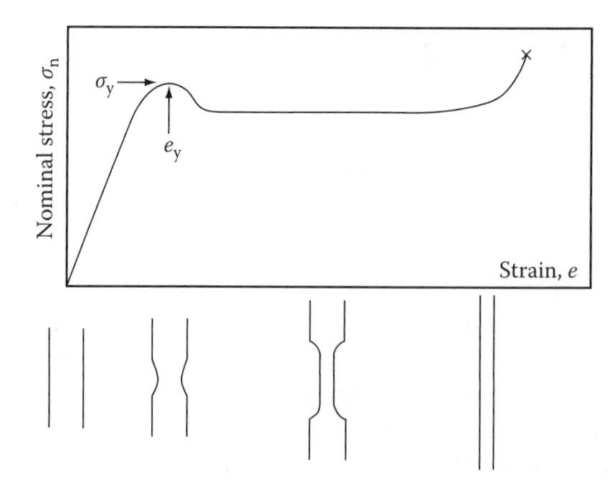

FIGURE 22.1 Schematic representation of the stress–strain behaviour of a ductile polymer and consequent change in specimen dimensions (× marks the point at which fracture occurs).

Polymers differ in their ability to form a stable neck and some do not neck at all. The form of the neck varies from polymer to polymer and with the testing conditions for a given polymer. The degree of necking is characterized by the 'draw ratio' which is defined as the length of a fully necked specimen divided by the original length. Since cold-drawing takes place at approximately constant volume it also is equal to the ratio of original specimen cross-sectional area to that of the drawn specimen. The exact form of the stress–strain curve for a particular polymer varies with both the temperature and rate of testing. A series of stress–strain curves for poly(methyl methacrylate) tested at different temperatures is presented in Figure 22.2. The modulus, as given by the initial slope of the curves, increases as the temperature is reduced. On the other hand, the ductility of the polymer decreases and below about 50°C the polymer does not draw and behaves in a brittle manner. The exact temperature at which this *ductile-to-brittle transition* takes place depends upon the rate of testing. The transition temperature decreases as the rate of testing increases. There is a similar interdependence of the properties upon temperature and rate as is found for viscoelasticity (Chapter 20). In general, it is found that the effect of increasing the rate of testing upon properties such as Young's modulus or yield stress is the same as reducing the testing temperature.

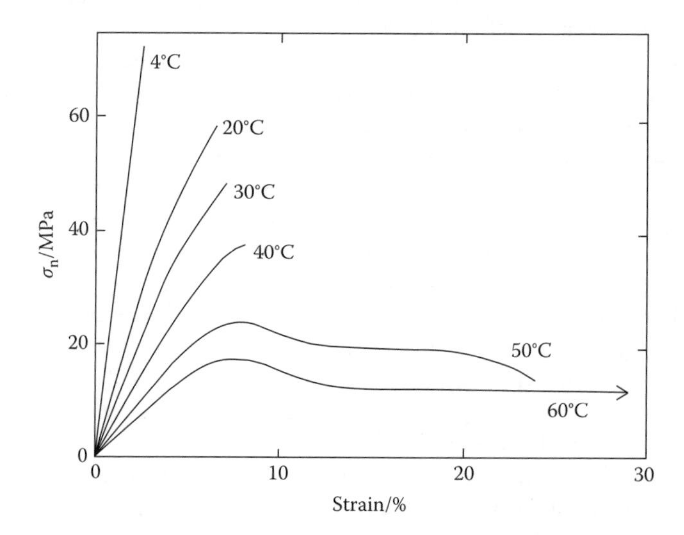

FIGURE 22.2 Variation of the stress–strain behaviour of poly(methyl methacrylate) with temperature. (Data taken from Andrews, E.H., *Fracture in Polymers*, Oliver and Boyd Ltd, London, U.K., 1968.)

22.2.2 Necking and the Considère Construction

It is clear that some polymers have the ability under certain conditions to form
undergo cold-drawing. This ability is shared by certain metals such as mild st
metals form an unstable neck, which continues to thin down until fracture ens
enon of necking and cold-drawing can best be considered using the *Considère construction*, which involves plotting the true stress σ against nominal strain e. This must be contrasted with the curves in Figure 22.1 and 22.2 where the nominal stress σ_n is plotted against the nominal strain e.

The *true stress* σ is defined as the load or force f divided by the instantaneous cross-sectional area A and so

$$\sigma = \frac{f}{A} \tag{22.1}$$

It may be assumed that deformation takes place at approximately constant volume and so A can be related to the original cross-sectional area A_0 through the equation

$$Al = A_0 l_0 \tag{22.2}$$

where l and l_0 are the instantaneous and original specimen lengths, respectively. The nominal strain is defined as

$$e = \frac{(l - l_0)}{l_0} \quad \text{so} \quad \frac{l}{l_0} = (1 + e) \tag{22.3}$$

and the nominal stress as

$$\sigma_n = \frac{f}{A_0} \tag{22.4}$$

Combining Equations 22.1 through 22.4 gives the relationship between the nominal and true stress as

$$\sigma_n = \frac{\sigma}{(1 + e)} \tag{22.5}$$

This means that for finite tensile strains σ will always be greater than σ_n.

When a neck starts to form, the load on the specimen ceases to rise as the strain is increased. This corresponds to the condition that $\mathrm{d}f/\mathrm{d}e = 0$ or $\mathrm{d}\sigma_n/\mathrm{d}e = 0$ (Figure 22.1). Applying this condition to Equation 22.5 gives

$$\frac{\mathrm{d}\sigma_n}{\mathrm{d}e} = \frac{1}{(1 + e)}\frac{\mathrm{d}\sigma}{\mathrm{d}e} - \frac{\sigma}{(1 + e)^2} = 0$$

or

$$\frac{\mathrm{d}\sigma}{\mathrm{d}e} = \frac{\sigma}{(1 + e)} \tag{22.6}$$

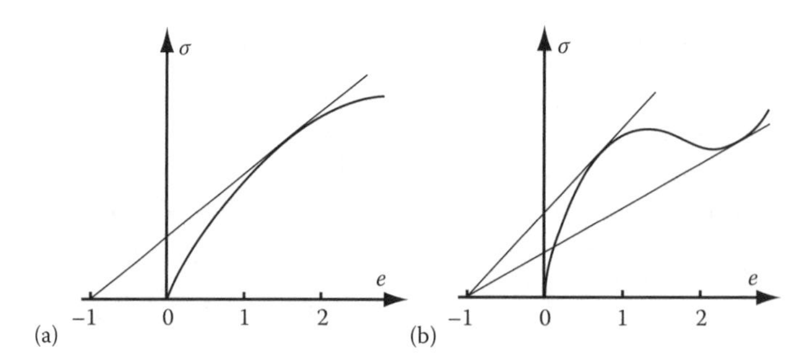

FIGURE 22.3 Schematic curves of true stress σ against nominal strain e for polymers, showing the Considère construction: (a) unstable neck and (b) stable neck.

as the condition for a neck to form. The use of this equation is illustrated in Figure 22.3. The point at which necking occurs can be determined by drawing a tangent to the curve of true stress against nominal strain from the point $e=-1$. Equation 22.6 will then be satisfied at the point at which the tangent meets the curve. This construction also can be used to determine whether or not the neck will be stable. In the curve drawn in Figure 22.3a, only one tangent can be drawn, and after the neck forms the sample continues to thin down in the necked region until fracture occurs. A stable neck will only form if a second tangent can be drawn to the curve, as shown in Figure 22.3b. The point where the second tangent meets the σ–e curve corresponds to a minimum in the nominal stress–nominal strain curve (Figure 22.1), which is necessary for the neck to be stable. The condition for both necking *and* cold-drawing to occur is therefore that $\mathrm{d}\sigma/\mathrm{d}e$ must be equal to $\sigma/(1+e)$ at *two* points on the σ–e curve.

So far the mechanisms whereby a stable neck forms in a polymer have not been considered. In metals, it is known to be due to the multiplication of dislocations that lead to a phenomenon known as strain-hardening. The metal becomes 'harder' as the strain is increased and, if the strain-hardening takes place to a sufficient extent, the material in the neck, which supports a smaller area, does not strain further. The material outside the neck is less deformed and hence 'softer' and continues to deform even though the cross-sectional area is larger and the true stress locally is lower. Strain-hardening in polymers is thought to be due principally to the effects of orientation. It was shown in Chapter 17 that X-ray diffraction reveals that the molecules are aligned parallel to the stretching direction in the cold-drawn regions of both amorphous and crystalline polymers. Since the anisotropic nature of the chemical bonding in the molecules causes an oriented polymer to be very much stronger and stiffer than an isotropic one, the material in the necked region is capable of supporting a much higher true stress than that outside the neck. Hence, polymers tend to form stable necks and undergo cold-drawing if they do not first undergo brittle fracture.

22.2.3 Rate and Temperature Dependence

Up to now, we have considered only deformation that takes place at constant rate and temperature, but plastic deformation, like other aspects of the mechanical behaviour of polymers, has a strong dependence upon the testing rate and temperature. Typical behaviour is illustrated in Figure 22.4 for a glassy thermoplastic deformed in tension. At a given strain-rate the yield stress drops as the temperature is increased and σ_y falls approximately linearly to zero at the glass transition temperature when the polymer glass becomes a rubber. If the strain-rate is increased and the temperature held constant, the yield stress increases (cf. time–temperature superposition, Section 20.7).

The behaviour of semi-crystalline polymers is rather similar with the main difference being that the yield stress drops to zero at the melting temperature of the crystals rather than at the T_g. Between T_g and T_m the non-crystalline regions are rubbery and the material gains its stiffness and strength from the crystalline regions, that reinforce the rubbery matrix. If the temperature of a

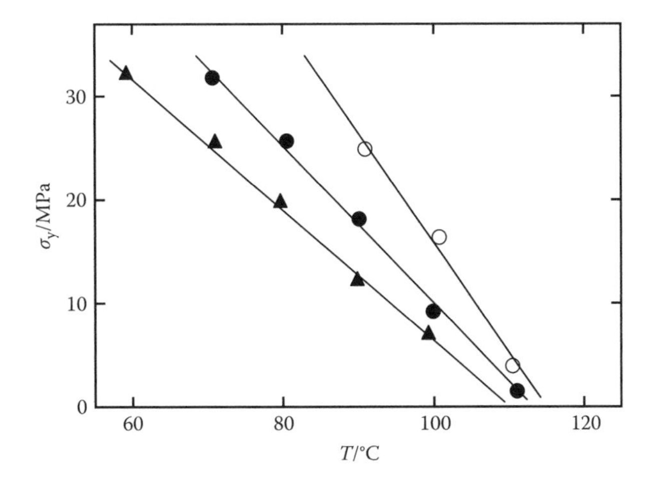

FIGURE 22.4 Variation of the yield stress of poly(methyl methacrylate) with temperature at strain-rates of: ▲, $0.002\,\text{min}^{-1}$; ●, $0.02\,\text{min}^{-1}$; and ○, $0.2\,\text{min}^{-1}$. (Data taken from Langford, G. et al., *Materials Research Laboratory Report No. R63-49*, Dept. Civil Eng., MIT, Cambridge, MA, 1963.)

semi-crystalline polymer is reduced below T_g it behaves more like a glassy polymer and the crystals do not have such a significant reinforcing effect.

22.3 YIELD CRITERIA

Only the yield behaviour of polymers in tensile deformation has been considered so far. In order to obtain a complete idea of the yield process, it is necessary to know under what conditions yield occurs for any general combination of stresses. For example, glassy polymers usually are brittle in tension when the temperature of testing is sufficiently below T_g whereas when they are deformed in compression at similar temperatures, they can undergo considerable plastic deformation. Also, knowledge of yield behaviour under general stress systems is important in engineering structures where components are subjected to a variety of tensile, compressive and shear stresses. A description of the conditions under which yield can occur under a general stress system is called a *yield criterion.* Several criteria have been suggested for materials in general, but before they can be described it is necessary to consider how the state of stress on a body can be represented. If the body is isotropic, then the stresses can in general be given in terms of a tensor (Section 19.2)

$$\sigma_{ij} = \begin{pmatrix} \sigma_{11}\,\sigma_{21}\,\sigma_{31} \\ \sigma_{21}\,\sigma_{22}\,\sigma_{23} \\ \sigma_{31}\,\sigma_{23}\,\sigma_{33} \end{pmatrix}$$

It is possible to choose a set of three mutually orthogonal axes such that the shear stress all are zero and the stress system can be described in terms of three normal stresses such that the tensor becomes

$$\begin{pmatrix} \sigma_1 & 0 & 0 \\ 0 & \sigma_2 & 0 \\ 0 & 0 & \sigma_3 \end{pmatrix}$$

These three *principal stresses* σ_1, σ_2 and σ_3 are used in the formulation of the different yield criteria outlined below.

22.3.1 Tresca Yield Criterion

This was one of the earliest criteria developed to describe the yield behaviour of metals and it is proposed that yield occurs when a critical value of the maximum shear stress σ_s is reached. If $\sigma_1 > \sigma_2 > \sigma_3$ then the criterion can be given as

$$\frac{1}{2}(\sigma_1 - \sigma_3) = \sigma_s \tag{22.7}$$

In a simple tensile test, σ_1 is equal to the applied stress and $\sigma_2 = \sigma_3 = 0$ and so at yield

$$\sigma_s = \frac{\sigma_1}{2} = \frac{\sigma_y}{2} \tag{22.8}$$

where σ_y is the yield stress in tension.

22.3.2 Von Mises Yield Criterion

Although the relative simplicity of the Tresca criterion is rather attractive, it is found that the criterion suggested by von Mises gives a somewhat better prediction of the yield behaviour of most materials. The criterion corresponds to the condition that yield occurs when the shear-strain energy in the material reaches a critical value and it can be expressed as a symmetrical relationship between the principal stresses of the form

$$(\sigma_1 - \sigma_2)^2 + (\sigma_2 - \sigma_3)^2 + (\sigma_3 - \sigma_1)^2 = \text{constant} \tag{22.9}$$

If the case of simple tension is considered, then $\sigma_2 = \sigma_3 = 0$ and if the tensile yield stress is σ_y, then the constant is equal to $2\sigma_y^2$. Equation 22.9 can, therefore, be expressed as

$$(\sigma_1 - \sigma_2)^2 + (\sigma_2 - \sigma_3)^2 + (\sigma_3 - \sigma_1)^2 = 2\sigma_y^2 \tag{22.10}$$

If the experiment is done in pure shear then $\sigma_1 = -\sigma_2$ and $\sigma_3 = 0$ and Equation 22.10 becomes

$$\sigma_1 = \frac{\sigma_y}{\sqrt{3}} \tag{22.11}$$

Hence the shear yield stress is predicted to be $1/\sqrt{3}$ times the tensile yield stress. This should be compared with the Tresca criterion, which predicts that the shear yield stress is $\sigma_y/2$. On the other hand, it can be readily seen that both criteria predict that the yield stresses measured in uniaxial tension and compression will be equal.

It is useful to represent the yield criteria graphically. This is done for the case of plane stress deformation ($\sigma_3 = 0$) in Figure 22.5 where the stress axes are normalized with respect to the tensile yield stress. In plane stress, the von Mises criterion defined by Equation 22.10 reduces to

$$\left(\frac{\sigma_1}{\sigma_y}\right)^2 + \left(\frac{\sigma_2}{\sigma_y}\right)^2 - \left(\frac{\sigma_1}{\sigma_y}\right)\left(\frac{\sigma_2}{\sigma_y}\right) = 1 \tag{22.12}$$

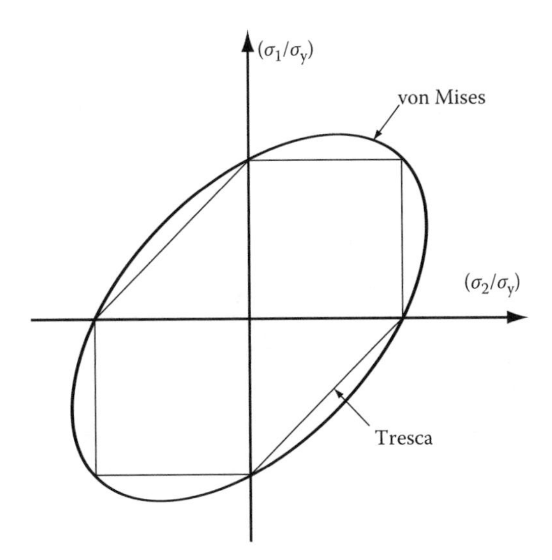

FIGURE 22.5 Schematic representation of the Tresca and von Mises yield criteria for plane stress deformation.

which describes the ellipse shown in Figure 22.5. It also shows that under certain conditions the applied tensile stress can be larger than σ_y. The Tresca criterion also can be represented on the same plot since when $\sigma_3 = 0$ Equations 22.7 and 22.8 can be combined to give

$$\frac{1}{2}(\sigma_1 - \sigma_2) = \sigma_s = \frac{1}{2}\sigma_y$$

or

$$\frac{\sigma_1}{\sigma_y} - \frac{\sigma_2}{\sigma_y} = 1 \quad (22.13)$$

This equation is applicable in the quadrants where σ_1 and σ_2 have different signs. Where they have the same signs the maximum shear stress criterion requires that

$$\frac{\sigma_1}{\sigma_y} = 1 \quad \text{and} \quad \frac{\sigma_2}{\sigma_y} = 1$$

It can be seen that the Tresca criterion gives a surface consisting of connected straight lines that inscribe the von Mises ellipse.

In practice, materials are subjected to triaxial stress systems and the form of the criterion in the three-dimensional principal stress space is of interest. They both plot out as infinite cylinders parallel to the [111] direction (cf. crystallography). The Tresca cylinder has a regular hexagonal cross-section and the von Mises one is circular in cross-section. It is found that metals tend to obey the von Mises criterion rather than Tresca, although the Tresca criterion often is assumed in calculations because of its simpler form.

22.3.3 Pressure-Dependent Yield Behaviour

In order to determine the most appropriate yield criterion for a particular material it is necessary to follow the yield behaviour by using a variety of different combinations of multiaxial stress. With

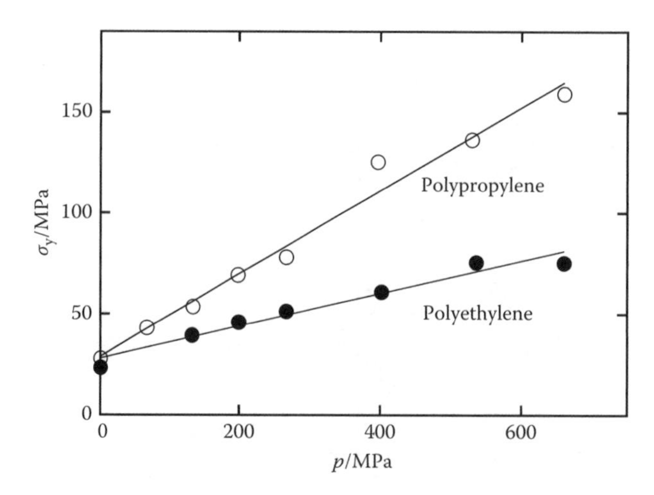

FIGURE 22.6 Variation of yield stress of polyethylene and polypropylene with hydrostatic pressure p. (Data taken from Mears, D.R. et al., *J. Appl. Phys.*, 40, 4229, 1969.)

polymers, however, rather unusual features are revealed when the yield stress of the same polymer is measured just in simple uniaxial tension and compression. Both the Tresca and von Mises criteria predict that the yield stress should be the same in both cases and this is what is found for metals. But for polymers, the compressive yield stress usually is higher than the tensile one. This difference between the compressive and tensile yield stress can be taken as an indication that the hydrostatic component of the applied stress is exerting an influence upon the yield process. This can be demonstrated more directly by measuring the tensile yield stress under the action of an overall hydrostatic pressure. The results of such measurements on different polymers are shown in Figure 22.6 and it can be seen that there is a clear increase in σ_y with hydrostatic pressure. In general, it is found that the yield stresses of amorphous polymers show larger pressure dependences than those of crystalline polymers. There is no significant change in the yield stresses of metals at similar pressures, but it is quite reasonable to expect that the physical properties of polymers will change at these relatively low pressures. They tend to have much lower bulk moduli than metals and so undergo significant volume changes on pressurization.

It is possible to modify the yield criteria described in the previous section to take into account the pressure dependence. The hydrostatic pressure p is given by

$$p = \frac{1}{3}(\sigma_1 + \sigma_2 + \sigma_3) \tag{19.2}$$

and the effect of hydrostatic stress can be incorporated into Equation 22.9 by writing it as

$$A[\sigma_1 + \sigma_2 + \sigma_3] + B[(\sigma_1 - \sigma_2)^2 + (\sigma_2 - \sigma_3)^2 + (\sigma_3 - \sigma_1)^2] = 1 \tag{22.14}$$

where A and B are constants. It is possible to define A and B in terms of the simple uniaxial tensile and compressive yield stresses, σ_{yt} and σ_{yc}, since when $\sigma_2 = \sigma_3 = 0$ then

$$A\sigma_{yt} + 2B\sigma_{yt}^2 = 1\,(\text{tension})$$

$$-A\sigma_{yc} + 2B\sigma_{yc}^2 = 1\,(\text{compression})$$

and so

$$A = \frac{(\sigma_{yc} - \sigma_{yt})}{\sigma_{yc}\sigma_{yt}} \quad \text{and} \quad B = \frac{1}{2\sigma_{yc}\sigma_{yt}} \tag{22.15}$$

Equation 22.14 can, therefore, be written as

$$2(\sigma_{yc} - \sigma_{yt})[\sigma_1 + \sigma_2 + \sigma_3] + [(\sigma_1 - \sigma_2)^2 + (\sigma_2 - \sigma_3)^2 + (\sigma_3 - \sigma_1)^2] = 2\sigma_{yc}\sigma_{yt} \tag{22.16}$$

If the magnitude of the tensile yield stress is the same as that for compression (i.e. $\sigma_{yt} = \sigma_{yc}$) the normal von Mises criterion is recovered.

It is possible to modify the Tresca maximum shear-stress criterion in several ways. The simplest way is to make the critical shear stress a function of the hydrostatic pressure p so that σ_s can be expressed by an equation of form

$$\sigma_s = \sigma_s^o - \mu p \tag{22.17}$$

where

σ_s^o is the shear yield stress in the absence of any overall hydrostatic pressure
μ is a material constant.

The hydrostatic pressure p is taken to be positive for uniaxial tensile loading and negative in compression. The yield stress in uniaxial tension is, therefore, given by combining Equations 19.2, 22.8 and 22.17 as

$$\sigma_{yt} = 2\sigma_s^o - \frac{2\mu\sigma_{yt}}{3}$$

or

$$\sigma_{yt} = \frac{2\sigma_s^o}{(1 + 2\mu/3)} \tag{22.18}$$

and in compression ($\sigma_1 = -\sigma_{yc}$) as

$$\sigma_{yc} = 2\sigma_s^o + \frac{2\mu\sigma_{yc}}{3}$$

or

$$\sigma_{yc} = \frac{2\sigma_s^o}{(1 - 2\mu/3)} \tag{22.19}$$

Combining these two equations leads to

$$\mu = \frac{3}{2}\left(\frac{\sigma_{yc} - \sigma_{yt}}{\sigma_{yc} + \sigma_{yt}}\right) \tag{22.20}$$

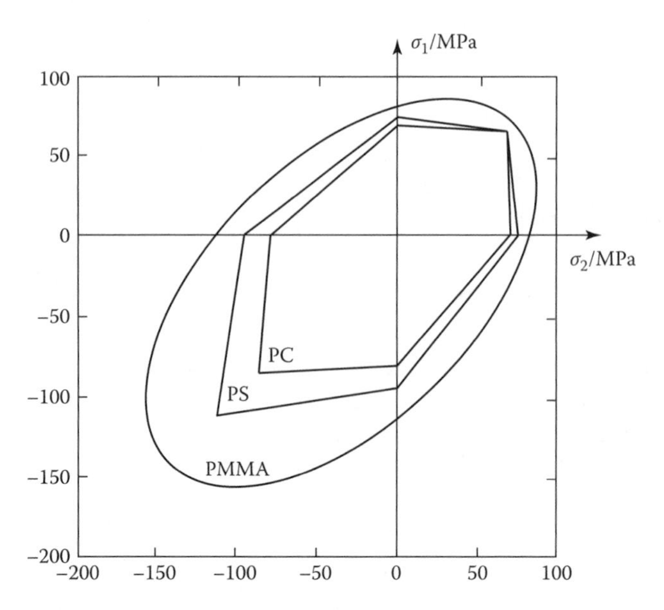

FIGURE 22.7 Modified von Mises and Tresca criteria fitted to experimental data upon the yield behaviour of polycarbonate (PC), polystyrene (PS) and poly(methyl methacrylate) (PMMA) at 20°C. (Data taken from Quinson, R. et al., *J. Mater. Sci.*, 32, 1371, 1997.)

which shows that $\mu > 0$ when the magnitude of the compressive yield stress is greater than that measured in tension.

In order to determine the most appropriate yield criterion for a particular polymer, it is necessary to follow the yield behaviour under a variety of states of stress. This is most conveniently done by working in plane stress ($\sigma_3 = 0$) and making measurements in pure shear ($\sigma_1 = -\sigma_2$) and biaxial tension (σ_1, $\sigma_2 > 0$) as well as in the simple uniaxial cases. The results of such experiments on different glassy polymers are shown in Figure 22.7. At room temperature, polycarbonate and polystyrene follow a modified Tresca criterion whereas poly(methyl methacrylate) obeys a modified von Mises one. It can be seen that the modified Tresca and von Mises criteria plot out as distorted hexagons and ellipses, respectively. The yield behaviour of the different glassy polymers can also change with temperature and it will be shown in the next section that the particular type of behaviour depends upon the deformation mechanisms involved.

In three-dimensional principal stress space, the modified Tresca cylinder becomes a hexagonal pyramid and the pressure-dependent von Mises cylinder becomes a cone. The significance of the apexes of the pyramid and cone is that they define the conditions for which there can be yielding under the influence of hydrostatic stress alone. This is something which cannot happen for materials that obey the unmodified criteria.

22.4 DEFORMATION MECHANISMS

So far the yield behaviour of polymers has been considered solely from a purely phenomenological viewpoint. It is important also to understand the effects of polymer structure and morphology, and the molecular processes that occur during yield in polymers.

22.4.1 Theoretical Shear Stress

In the consideration of the resistance of materials to plastic flow, it is useful to estimate theoretically the stress that may be required to shear the structure. It is necessary to do this before a sensible comparison can be made of the relative strengths of different types of materials such as metals and

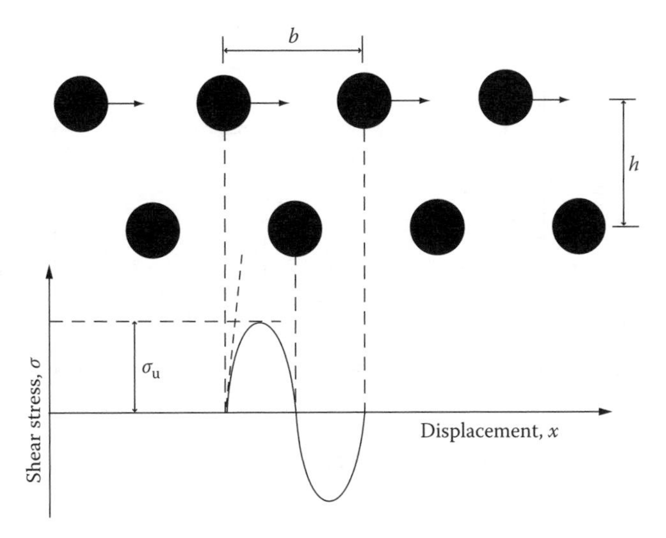

FIGURE 22.8 Schematic representation of the shear of layers of atoms in a crystal and the expected variation of shear stress with displacement. The parameters are defined in the text.

polymers. The general situation to be considered is shown in Figure 22.8 where one sheet of atoms or molecules is allowed to slip over another under the action of a shear stress σ. The separation of the sheets is h and the repeat distance in the direction of shear is b. It is envisaged that the shear stress will vary periodically with the shear displacement x in an approximately sinusoidal manner and that the maximum resistance to shear will be when $x=b/4$. The variation of σ with the displacement x can then be written as

$$\sigma = \sigma_u \sin\left(\frac{2\pi x}{b}\right) \tag{22.21}$$

The initial slope of the curve of σ as a function of x will be equal to the shear modulus G of the material and so at small strains

$$\sigma = \frac{\sigma_u 2\pi x}{b} = \frac{Gx}{h} \tag{22.22}$$

Thus the ideal shear strength is given by

$$\sigma_u = \frac{Gb}{2\pi h} \tag{22.23}$$

In a face-centred cubic metal $h = a/\sqrt{3}$ and $b = a/\sqrt{6}$ where a is the lattice parameter and so the theoretical shear strength is predicted to be $\sigma_u \approx G/9$. For chain direction slip on the (020) planes in a polyethylene crystal $b=0.254$ nm and h is equal to the separation of the (020) planes, which is 0.247 nm. Hence the theoretical shear stress for a polyethylene crystal would be expected to be of the order of $G/6$. More sophisticated calculations of Equation 22.23, however, lead to lower estimates of σ_u, which come out to be of the order of $G/30$ for most materials. Even so, this estimate of the stress required to shear the structure is very much higher than the values that normally are obtained experimentally and the discrepancy is due to the presence of defects, such as *dislocations*, within the crystals. The high values are realized only for certain crystal whiskers and other perfect crystals.

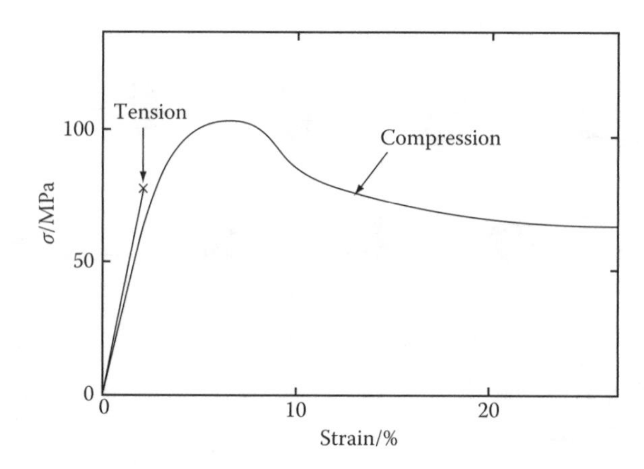

FIGURE 22.9 Stress–strain curves for an epoxy resin deformed in tension or compression at room temperature (× marks the point at which fracture occurs).

22.4.2 Shear Yielding in Glassy Polymers

Glassy polymers are generally thought of as being rather brittle materials, but they are capable of displaying a considerable amount of ductility below T_g especially when deformed under the influence of an overall hydrostatic compressive stress. This behaviour is illustrated in Figure 22.9 where true stress–strain curves are given for an epoxy resin tested in uniaxial tension and compression at room temperature. The T_g of the cured resin is 100°C and such crosslinked polymers are found to be brittle when tested in tension at room temperature. In contrast, they can show considerable ductility in compression and undergo shear yielding. Another important aspect of the deformation is that glassy polymers tend to show 'strain softening'. The true stress drops after yield, not because of necking, which cannot occur in compression, but because there is an inherent softening of the material.

The process of shear yielding in glassy polymers can either be homogeneous or inhomogeneous. Figure 22.10 shows an optical micrograph of a section of polystyrene deformed in plane-strain compression. Fine deformation bands are obtained in which the shear is highly localized. The formation of shear bands is associated with the strain softening in the material since once a small region starts to undergo shear yielding, it will continue to do so because it has a lower flow stress than the surrounding relatively undeformed regions. Other glassy polymers undergo uniform deformation with no evidence of any localization. It is found in general that polymers such as poly(methyl methacrylate) that undergo homogeneous deformation obey the modified von Mises

FIGURE 22.10 Polarized optical micrograph of shear bands in polystyrene deformed in plane-strain compression. The compression was applied in the vertical direction. (Courtesy of R.J. Oxborough, University of Cambridge, Cambridge, U.K.)

criterion whereas those that have localized shear bands such as polystyrene obey a modified Tresca (Figure 22.7).

There have been many attempts over the years to explain the process of shear yielding on a molecular level and some of the approaches are outlined below.

22.4.2.1 Stress-Induced Increase in Free Volume

It has been suggested that one effect of an applied strain upon a polymer is to increase the free volume (Section 16.2.2) and hence increase the mobility of the molecular segments. It follows that this would also have the effect of reducing T_g. The developments of this idea predict that during tensile deformation, the overall hydrostatic tension should cause an increase in free volume and so aid plastic deformation. However, there are problems in the interpretation of compression tests where plastic deformation can still take place even though the hydrostatic stress is compressive. Moreover, careful measurements have shown that there generally is a slight reduction in overall specimen volume during plastic deformation in both tension and compression. It would seem, therefore, that the concept of free volume is not very useful in the explanation of plastic deformation of glassy polymers.

22.4.2.2 Application of the Eyring Theory to Yield in Polymers

It is possible to think of yield and plastic deformation in polymers as a type of viscous flow, especially since glassy polymers are basically 'frozen' liquids that have failed to crystallize. Eyring developed a theory to describe viscous flow in liquids and it can be readily adapted to describe the behaviour of glassy polymers. The segments of the polymer chain can be thought of as being in a pseudo-lattice and for flow to occur a segment must move to an adjacent site. There will be a potential barrier to overcome, because of the presence of adjacent molecular segments, and the situation is shown schematically in Figure 22.11a. The height of the barrier is given by ΔG_0 and the frequency ν_0 at which the segments jump the barrier and move to the new position can be represented by an Arrhenius type equation

$$\nu_0 = B\exp\left(\frac{-\Delta G_0}{\mathbf{k}T}\right) \tag{22.24}$$

where

B is a temperature-independent constant
$\mathbf{k}$ is the Boltzmann constant.

The rate at which segments jump forward over the barrier will be increased by the application of a shear stress σ. This has the effect of reducing the energy barrier by an amount $(1/2)\sigma Ax$, that is

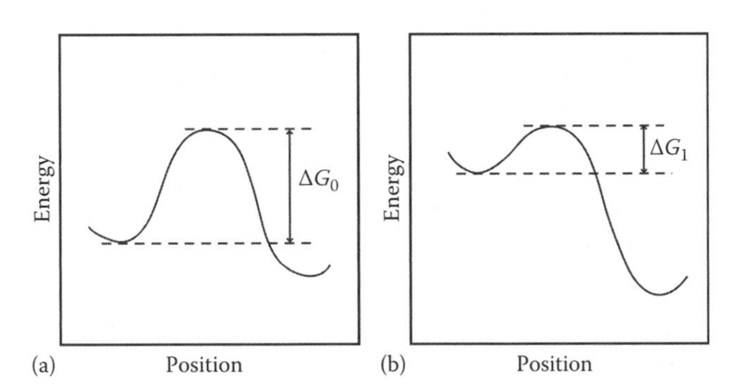

FIGURE 22.11 Schematic illustration of the potential barrier for flow in a glassy polymer. (a) Unstressed state with barrier of ΔG_0. (b) Stressed state with barrier reduced to $\Delta G_1 = \left(\Delta G_0 - \frac{1}{2}\sigma Ax\right)$.

the work done in moving the segment a distance x through the lattice of unit cross-sectional area A. The new situation is illustrated schematically in Figure 22.11b and the frequency of forward motion increases to become

$$\nu_f = B\exp\left[\frac{-(\Delta G_0 - (1/2)\sigma Ax)}{\mathbf{R}T}\right]$$

$$\text{or} \quad \nu_f = \nu_0 \exp\left(\frac{\sigma Ax}{2\mathbf{R}T}\right] \tag{22.25}$$

There is also the possibility of backward motion occurring. The frequency of backward motion ν_0 is given by

$$\nu_b = \nu_0 \exp\left(\frac{-\sigma Ax}{2\mathbf{k}T}\right) \tag{22.26}$$

but the application of stress makes backward motion more difficult. The resultant strain-rate $\dot{e}$ of the polymer will be proportional to the difference between the frequencies of forward and backward motion, i.e.

$$\dot{e} \propto (\nu_f - \nu_b) = \nu_0 \exp\left(\frac{\sigma Ax}{2\mathbf{k}T}\right) - \nu_0 \exp\left(\frac{-\sigma Ax}{2\mathbf{k}T}\right)$$

and this can be written as

$$\dot{e} = K\sinh\left(\frac{\sigma V}{2\mathbf{k}T}\right) \tag{22.27}$$

where K is a constant and V (=Ax) is called the 'activation volume' or 'Eyring volume' and is the volume of the polymer segment which must move to cause plastic deformation. If the stress is high and the frequency of backward motion is small then Equation 22.27 can be approximated to

$$\dot{e} = \left(\frac{K}{2}\right)\exp\left(\frac{\sigma V}{2\mathbf{k}T}\right) \tag{22.28}$$

This approach has met with considerable success in the prediction of the strain-rate dependence of the yield stress of polymers. It is tempting to relate V to structural elements in the polymer. For most glassy polymers, it comes out to be the volume of several polymer segments (typically between 2 and 10) although there is no clear correlation between the size of the activation volume and any particular aspect of the polymer microstructure.

22.4.2.3 Molecular Theories of Yield

Recent developments in the interpretation of yield behaviour of glassy polymers have involved attempts to relate plastic deformation to specific kinds of molecular motion. One particularly successful approach was suggested by Robertson for the case of a planar zig-zag polymer chain. He assumed that at any time, the molecular chain segments would be distributed between low-energy *trans* and high-energy *cis* conformations and that the effect of an applied stress would be to cause certain segments to change over from *trans* to *cis* conformations as illustrated in Figure 22.12. This

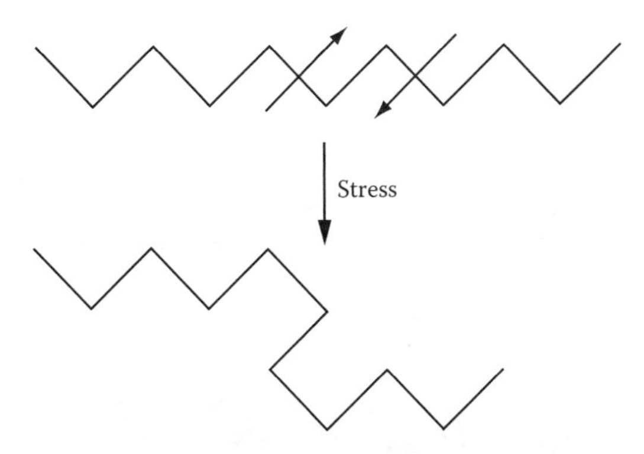

FIGURE 22.12 Schematic illustration of the conformational changes involved in the deformation of a planar zig-zag polymer molecule. (Adapted from Robertson, R.E., *J. Polym. Sci. Polym. Symp.*, 63, 173, 1978.)

and other similar molecular theories have led to accurate predictions of the variation of yield stress with strain-rate and temperature for glassy polymers. The reader is referred to more advanced texts for further details.

22.4.3 Plastic Deformation of Polymer Crystals

There is now clear evidence to suggest that polymer crystals are capable of undergoing plastic deformation in a similar way to other crystalline solids through processes such as slip, twinning and martensitic transformations. It is thought that similar deformation processes occur in both isolated single crystals and in the crystalline regions of semi-crystalline polymers. The main difference between the deformation of polymer crystals and crystals of atomic solids is that the deformation occurs in the polymer such that molecules are not broken and, as far as possible, remain relatively undistorted. This problem of molecular integrity does not arise in atomic solids. Deformation, therefore, generally involves only the sliding of molecules in directions parallel or perpendicular to the chain axis, which breaks only the secondary van der Waals bonds. The specific deformation mechanisms will be discussed separately.

22.4.3.1 Slip

The deformation of metal crystals takes place predominantly through slip processes that involve the sliding of one layer of atoms over the other through the motion of dislocations. In fact, the ductility of metals stems from there normally being a large number of slip systems (i.e. slip planes and directions) available in metal crystals. Slip also is important in the deformation of polymer crystals, but the need to avoid molecular fracture limits the number of slip systems available. The slip planes are limited to those that contain the chain direction (i.e. ($hk0$) planes when [001] is the chain direction). The slip directions can in general be any direction which lies in these planes, but normally only the slip that takes place in directions parallel or perpendicular to the chain direction is considered in detail. Slip in any general direction in a ($hk0$) plane will have components of these two specific types. *Chain direction slip* is illustrated schematically in Figure 22.13. The molecules slide over each other in the direction parallel to the molecular axis. It is known to occur during the deformation of both bulk semi-crystalline polymers and of solution-grown polymer single crystals. It can be readily detected from the change of orientation of the molecules within the crystals as shown in Figure 22.14 for polytetrafluoroethylene. Initially the molecules are approximately normal to the lamellar surface (Figure 17.20), but during deformation they slide past each other and become tilted within the crystal. Slip in directions perpendicular to the molecular axes is sometimes called *transverse slip*. It has been shown to occur during the deformation of both melt-crystallized polymers and solution-grown single crystals, but is rather more difficult to detect because it involves no change in molecular orientation within the crystals.

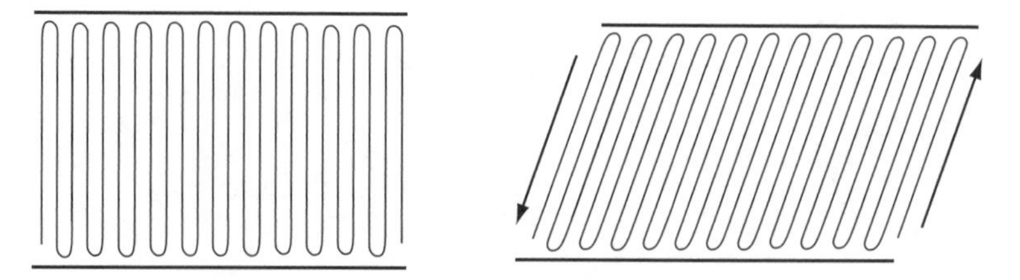

FIGURE 22.13 Schematic representation of chain direction slip in a chain-folded polymer crystal.

FIGURE 22.14 Replica of a fracture surface of polytetrafluoroethylene after compressive deformation in the vertical direction showing: (i) the tilt of the molecules in the crystals following chain direction slip and (ii) the breakup of some crystals. (cf. The undeformed polytetrafluoroethylene morphology in Figure 17.20.)

22.4.3.2 Dislocation Motion

It is well established that slip processes in atomic solids are associated with the motion of dislocations. The theoretical shear strength of a crystal lattice has been shown (Section 22.4.1) to be of the order of $G/10$ to $G/30$ where G is the shear modulus of the crystal. The measured shear yield strengths of many crystalline solids are typically between $10^{-4}\ G$ and $10^{-5}\ G$. The difference between the theoretical and experimentally determined values is due to deformation taking place by the glide of dislocations, which occurs at stresses well below the theoretical strength. The presence of dislocations in polymer crystals has been discussed previously (Section 17.6.2) and there is no fundamental reason why slip in polymer crystals cannot occur through dislocation motion.

The energy per unit length $U(l)$ of a dislocation in an isotropic medium can be shown to be given approximately by

$$U(l) = \frac{Gb^2}{4\pi}\ln\left(\frac{r}{r_0}\right) \tag{22.29}$$

where

b is the Burgers vector of the dislocation
r is the radius of the crystal
r_0 is the core radius of the dislocation (normally taken $\approx b$).

The shear modulus G in Equation 22.29 is that for an isotropic medium. Since the bonding in polymer crystals is highly anisotropic G must be replaced by an effective crystal modulus when the equation is applied to polymer crystals. The inspection of Equation 22.29 shows that for a crystal of a given size, dislocations will have low energy when the shear modulus and b are both small. The total energy of a dislocation also will be reduced if the length of the dislocation line is as short as possible. Since dislocations cannot terminate within a crystal, the dislocations with the shortest length can be obtained in lamellar polymer crystals by having the dislocation line perpendicular to lamellar surface and so parallel to the chain direction. The need to avoid excess deformation of the polymer chains means that b should also be parallel to the chain direction. The result of these considerations is that a particularly favourable type of dislocation has its Burgers vector parallel to the chain direction, as illustrated schematically in Figure 17.27a.

Such dislocations have been shown to be present in solution-grown polyethylene single crystals and their number is found to increase following deformation. The Burgers vector of such dislocations is the same as the chain direction repeat of 0.254 nm (Table 17.1) and the shear modulus for slip in the chain direction will be either $c_{44}=2.83$ GPa or $c_{55}=0.78$ GPa (Table 19.2) depending upon whether the slip plane is (010) or (100). It is found that at room temperature [001] chain direction slip in bulk polyethylene occurs at a stress of the order of 0.015 GPa (i.e. ~1/50 to 1/200 of the shear modulus) and calculations have shown that chain direction slip could occur through the thermal activation of the type of dislocations shown in Figure 17.27a. This mechanism is, however, only favourable when the crystals are relatively thin where the length of the dislocation line is small. It is expected that such a dislocation would be favourable in thin crystals of other polymers since they generally have low shear moduli. Most other polymer crystals have larger chain direction repeats than polyethylene (Table 17.1) especially when the molecule is in a helical conformation. This would give rise to larger values of $U(l)$ through higher Burgers vectors, but it would be expected that in this case, the full dislocations might split into partial dislocations in order to reduce their energy.

22.4.3.3 Twinning

Twinned crystals of many materials are produced during both growth and deformation. Many mineral crystals are found in a twinned form in nature and deformation twins often are obtained when metals are deformed, especially at low temperatures. The most commonly obtained types of twinned crystals are those in which one part of the crystal is a mirror image of another part. The boundary between the two regions is called the *twinning plane*. The particular types of twins that form in a crystal depend upon the structure of the crystal lattice and deformation twins can be explained in terms of a simple shear of the crystal lattice.

As in the case of slip, there is no fundamental reason why polymer crystals cannot undergo twinning and it is found that twinning frequently occurs when crystalline polymers are deformed. The types of twins that are obtained are formed in such a way that the lattice is sheared without either breaking or seriously distorting the polymer molecules. A good example of the formation of twins in polymer crystals is obtained by deforming solid-state polymerized polydiacetylene single crystals (Section 17.3.2). Figure 22.15a shows a micrograph of such a crystal viewed at 90° to the chain direction. The striations on the crystal define the molecular axis and it can be seen that the molecules kink over sharply at the twin boundary. Accurate measurements have shown that there is a 'mirror image' orientation of the crystal on either side of the boundary. This indicates clearly that the crystal contains a twin. A schematic diagram of the molecular arrangements on either side of the twin boundary is given in Figure 22.15b. The angle that the molecules can bend over and still retain the molecular structure can be calculated from knowledge of the crystal structure and exact agreement is found between the measured and calculated angles. The result of the twinning process is that the molecules on every successive plane in the twin are displaced by one unit of c parallel to the chain direction. This clearly is identical to the result of chain direction slip taking place and it now has become recognized that this type of twinning could be important in the deformation of many types of polymer crystals.

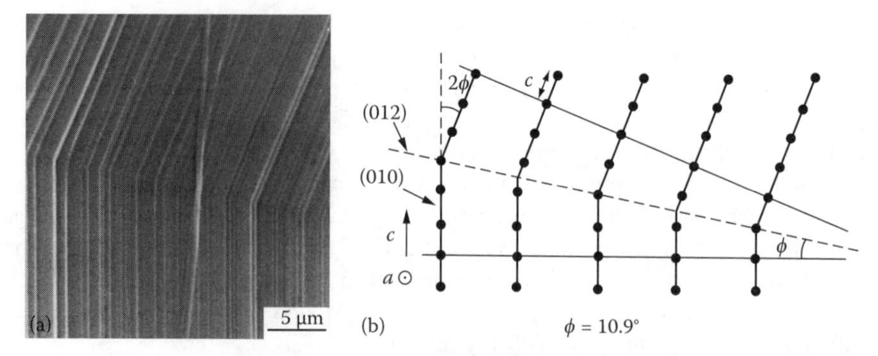

FIGURE 22.15 Large twin in a polydiacetylene single crystal: (a) scanning electron micrograph of twin and (b) schematic diagram of twin on a molecular level. The chain direction axis c is indicated, the a axis is perpendicular to the page and ϕ is the angle of rotation of the twin. (Adapted from Young, R.J. et al., *J. Mater. Sci.*, 13, 62, 1978.)

Another type of twinning that has been found to occur in polymer crystals involves a simple shear in directions perpendicular to the chain axis, which, in this case, does not bend the molecules. An example of this type of twinning is shown schematically in Figure 22.16 for the polyethylene unit cell projected parallel to the chain direction. Two particular twinning planes (110) and (310) have been found and the twinning has been reported after deformation of both solution-grown single crystals and melt-crystallized polymer. The occurrence of twinning can be detected from measurements of the rotation of the a and b crystal axes about the molecular c axis since the twinning causes a large rotation of the crystal lattice.

22.4.3.4 Martensitic Transformations

This is a general term which covers transformations which bear some similarity to twinning, but involve a change of crystal structure. The name comes from the well-known, important transformation which occurs when austenitic steel is rapidly cooled and forms the hard tetragonal metastable phase called 'martensite'. These transformations also can occur on the deformation or cooling of many crystalline materials and all of these martensitic transformations share the similar feature that they take place with no long-range diffusion.

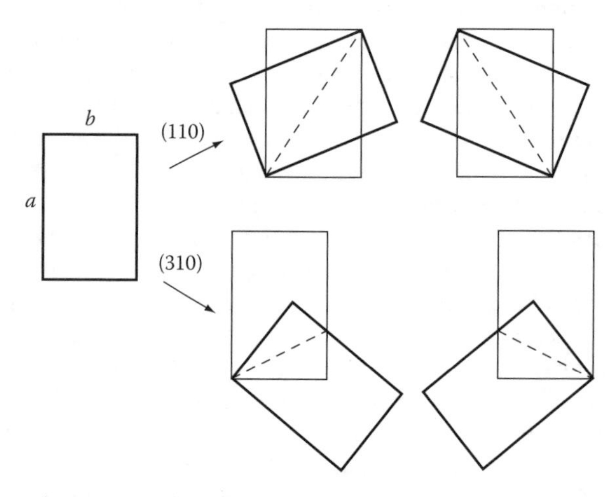

FIGURE 22.16 Schematic representation of (110) and (310) twinning in orthorhombic polyethylene. The unit cells are viewed projected parallel to the chain direction. (Adapted from Frank, F.C. et al., *Philos. Mag.*, 3, 64, 1958.)

Martensitic transformations have not been widely reported in polymers, but an important one occurs during the deformation of polyethylene. The normal crystal structure of both solution-grown single crystals and melt-crystallized polymer is orthorhombic but, after deformation, extra Bragg reflections are found in diffraction patterns and these have been explained in terms of the formation of a monoclinic form (Table 17.1). The transformation takes place by means of a simple two-dimensional shear of the orthorhombic crystal structure in a direction perpendicular to *c*. The chain axis length remains virtually unchanged by the transformation, at about 0.254 nm, but in the monoclinic cell (Table 17.1) it is conventionally indexed as *b*.

22.4.4 Plastic Deformation of Semi-Crystalline Polymers

Melt-crystallized polymers have a complex structure consisting of amorphous and crystalline regions in a spherulitic morphology (Figure 17.13). They display their most useful properties in the temperature range $T_g < T < T_m$ and they normally are designed to be used within this temperature range. Below the T_g they are brittle and above T_m they behave as viscous liquids or rubbers. The deformation of semi-crystalline polymers can be considered on several scales. Figure 22.17 shows how the spherulitic morphology changes with deformation. The spherulites become elongated parallel to the stretching direction as the strain is increased and at high strains the spherulitic morphology breaks up and a fibre-like morphology is obtained. Clearly the change in appearance of the spherulitic structure must be reflecting changes that are occurring on a smaller dimensional scale within the crystalline and amorphous areas. It is thought that when the spherulites are deforming homogeneously, during the early stages of deformation, the crystalline regions deform by a combination of slip, twinning, and martensitic transformations, as outlined in the last section. In addition, it is thought that because the amorphous phase is in a rubbery state, it allows deformation to take place by the shear of crystals relative to each other. During these early and intermediate stages of deformation, the crystals become distorted but remain essentially intact. Towards the later stages when the spherulitic morphology is lost, there appears to be a complete breakup of the original crystalline morphology and a new fibre-like structure is formed. The mechanisms whereby this happens are not understood fully but one possible mechanism by which this could take place is illustrated in Figure 22.18. It is assumed for simplicity that the undeformed polymer has crystals which are stacked with the molecules regularly folded. Deformation then takes place by slip and twinning etc., until eventually the crystals start to crack and chains are pulled out. At sufficiently high strains, the molecules and crystal blocks become aligned parallel to the stretching direction and a fibrillar structure is formed. Earlier theories of cold-drawing suggested that the structural

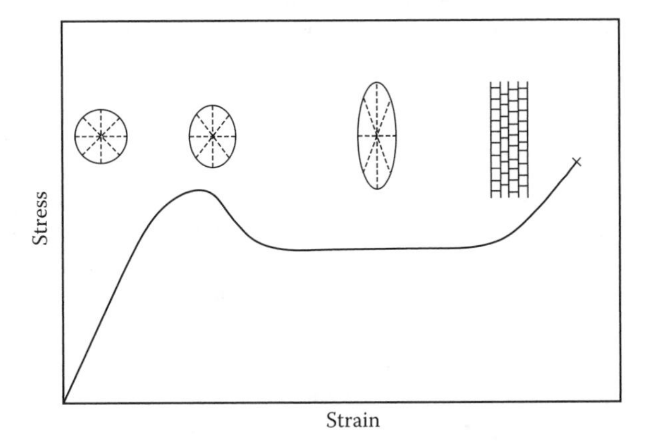

FIGURE 22.17 Schematic illustration of the change in spherulite morphology during the cold-drawing of a semi-crystalline polymer. (Adapted from Magill, J., in J.M. Schultz (ed.), *Treatise on Materials Science and Technology*, Vol. 10A, Academic Press, New York, 1977.)

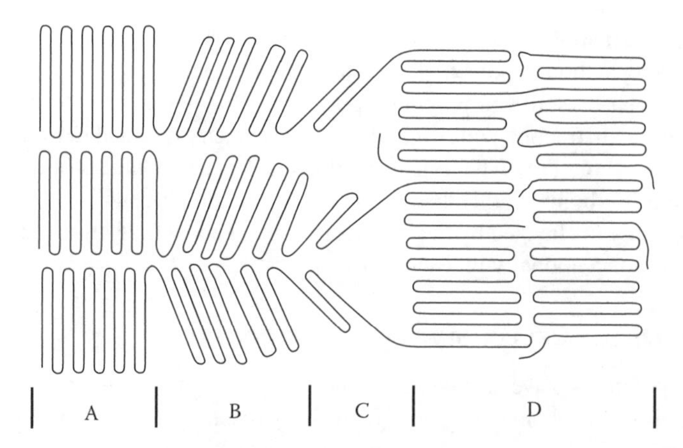

FIGURE 22.18 Schematic representation of the deformation which takes place on a molecular level during the formation of a neck in a semi-crystalline polymer: (A) idealized representation of undeformed structure; (B) crystals deforming by slip, twinning and martensitic transformations; (C) crystals break up and molecules are pulled out; and (D) fibrillar structure forms.

changes occurred because the heat generated during deformation caused melting to occur and the material subsequently re-crystallized on cooling. Although it is clear that there is a small rise in temperature during deformation, this is not sufficient to cause melting, and at normal strain-rates the heat is dissipated relatively quickly.

Polymers oriented by drawing are of considerable practical importance because they can display up to 50% of the theoretical chain direction modulus of the crystals. This short-fall in modulus is caused by imperfections within the crystals and the presence of folds and non-crystalline material. The value of modulus obtained depends principally upon the crystal structure of the polymer and the extent to which a sample is drawn (i.e. the draw ratio). The maximum draw ratio that can be obtained in turn depends upon the temperature of drawing and the average molar mass and molar mass distribution of the polymer. There is now considerable interest in optimizing these variables to obtain ultrahigh-modulus oriented polymers.

22.5 CRAZING

Consideration of the plastic deformation of polymers has, so far, been concerned only with 'shear yielding'. This occurs at constant volume and can take place uniformly throughout the sample. Certain polymers, particularly thermoplastics in the glassy state, are capable of undergoing a localized form of plastic deformation known as crazing.

22.5.1 Craze Yielding

Crazing is found to take place in a glassy polymer only when there is an overall hydrostatic tensile stress (i.e. $p > 0$) and the formation of crazes causes the material to undergo a significant increase in volume. The crazes appear as small crack-like entities that usually are initiated on the specimen surface and are oriented perpendicular to the tensile axis. Closer examination shows that they are regions of cavitated polymer and so are not true cracks, although the cracks that lead to eventual failure of the specimen usually nucleate within pre-existing crazes. Figure 22.19 shows a deformed specimen of polystyrene that has undergone crazing.

Crazing can be demonstrated quite simply by rubbing one's fingers over the surface of a sample of glassy polystyrene and then bending it to near its point of failure. The crazes appear as a fine surface mist on the tension surface. The crazes are very small, ~100 nm thick and several microns in lateral dimensions, but they can be seen by the naked eye because they are less dense than the

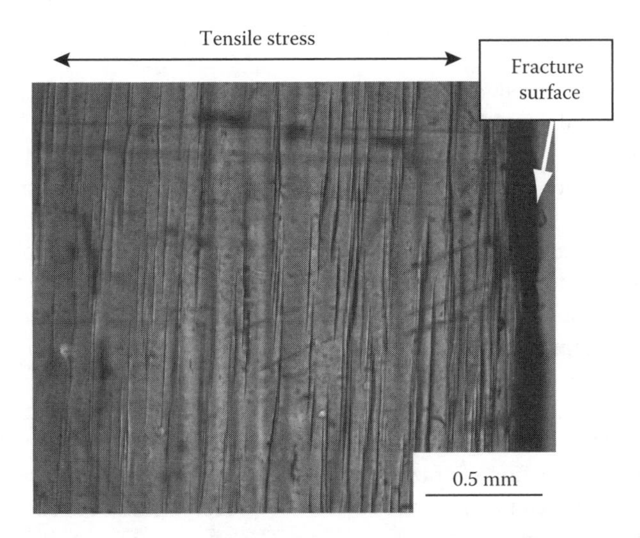

FIGURE 22.19 Optical micrograph of a sheet of polystyrene deformed to fracture showing crazes, aligned perpendicular to the axis of tensile stress, below the fracture surface. (Courtesy of S. Rannou, University of Manchester, Manchester, U.K.)

undeformed matrix and so reflect and scatter light. The demonstration also shows how crazing can be made easier by the presence of certain liquids ('crazing agents'), such as finger grease. In practice, the use of many glassy polymers often is limited by their tendency to undergo crazing at relatively low stresses in the presence of crazing agents. The crazes can mar the appearance of the specimen and lead to eventual catastrophic failure.

As with other mechanical properties, the formation of crazes in polymers depends upon testing temperature and the rate or time-period of loading. This is demonstrated clearly in Figure 22.20 where the stress required to cause crazing in simple uniaxial tensile specimens of polystyrene is plotted against testing temperature. The craze initiation stress drops as the testing temperature is increased and as the strain-rate is decreased. The data in Figure 22.20 are for dry crazing in air. The presence of crazing agents (e.g. finger grease or organic liquids with polystyrene) causes the craze initiation stress to drop further.

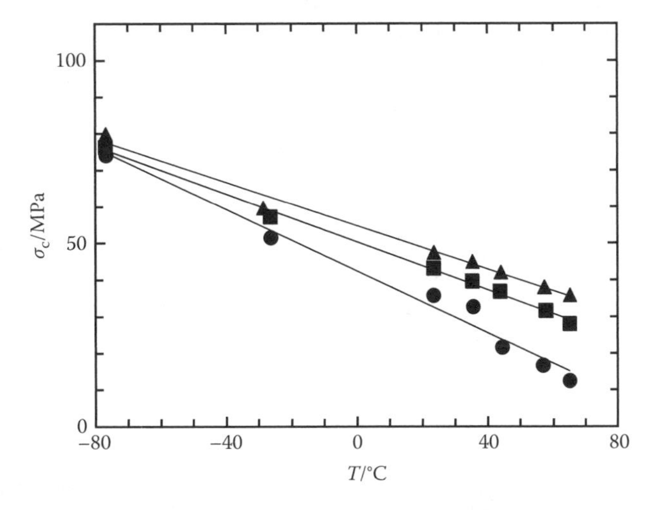

FIGURE 22.20 Dependence of crazing stress upon temperature for polystyrene deformed at different tensile strain-rates: ●, 0.00067 s^{-1}; ■, 0.0267 s^{-1}; and ▲, 0.267 s^{-1}.

22.5.2 Craze Criteria

As crazing is a yield phenomenon, there have been attempts to establish a *craze criterion* in the same way that Tresca and von Mises criteria have been used for shear yielding. Crazing occurs typically at about one-half of the yield stress of the polymer. This is in the 'elastic' region (Figure 22.3), but the occurrence of crazing does not usually cause any detectable change in slope of the stress–strain curve as the volume fraction of crazed material initially is very low. Early suggestions of craze criteria were that a critical uniaxial tensile stress or strain was required for craze formation, but these were inadequate to describe the behaviour in multiaxial stress systems. Observations upon the way in which polymers craze in uniaxial tension under the action of a superimposed hydrostatic pressure have been made in a similar way to the measurement of shear yielding under hydrostatic pressure described earlier (Figure 22.6). It is found that superimposed hydrostatic compression makes crazing more difficult and increases the tensile stress or strain needed to initiate crazes. Hydrostatic tension on the other hand produces a dilatation that helps to open up voids in the structure and hence aids the formation of crazes.

Since crazing involves the opening up of cavitated regions, it would be reasonable to expect that it might take place at a critical strain e_c for crazing. The effect of pressure, outlined above, could then be incorporated by writing, when there is an overall hydrostatic tension

$$e_c = C + \frac{D}{p} \tag{22.30}$$

where

p is the hydrostatic stress

C and D are time- and temperature-dependent parameters.

The maximum tensile strain in an isotropic body subjected to a general state of stress defined by the principal stresses σ_1, σ_2 and σ_3 (where $\sigma_1 > \sigma_2 > \sigma_3$) is given from Equation 19.26 as

$$e_1 = \frac{1}{E}(\sigma_1 - \nu\sigma_2 - \nu\sigma_3) \tag{22.31}$$

where ν is Poisson's ratio and this strain always acts in the direction of the maximum principal stress. The critical strain criterion requires that crazing occurs when e_1 reaches a critical value and so Equation 22.30 can be rewritten to define the criterion in terms of stresses only

$$\sigma_1 - \nu\sigma_2 - \nu\sigma_3 = X + \frac{Y}{(\sigma_1 + \sigma_2 + \sigma_3)} \tag{22.32}$$

where X and Y are new time- and temperature-dependent constants. This criterion is useful in that it gives both the conditions under which crazes form and the direction in which the crazes grow (i.e. perpendicular to σ_1). As with yield criteria, the easiest way to test craze criteria is by making measurements in plane stress ($\sigma_3 = 0$) where Equation 22.32 becomes

$$\sigma_1 - \nu\sigma_2 = X + \frac{Y}{(\sigma_1 + \sigma_2)} \tag{22.33}$$

This equation is plotted in Figure 22.21 and X and Y have been chosen by fitting the curve to experimental data for crazing in an amorphous polymer under biaxial stress. Crazing cannot take place

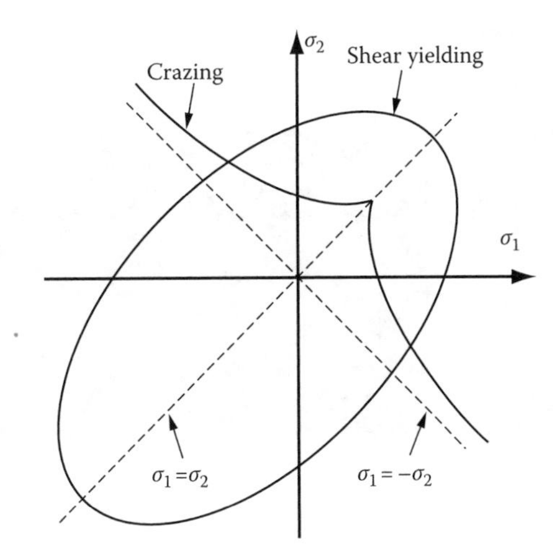

FIGURE 22.21 Envelopes defining crazing and yield for an amorphous polymer undergoing plane-stress deformation. (Adapted from Sternstein, S. and Ongchin, L., *ACS Polym. Prepr.*, 19, 1117, 1969.)

unless there is overall hydrostatic tension (i.e. $(\sigma_1 + \sigma_2 + \sigma_3) > 0$), which is necessary to allow cavitation to occur. The curves from Equation 22.33 will then be asymptotic to the line where $\sigma_1 = -\sigma_2$. It is instructive to consider the criteria for shear yielding at the same time as craze criteria are examined and so the pressure-dependent von Mises curve (Figure 22.7) for the amorphous polymer tested under similar conditions also is given in Figure 22.21. Having both the yield and craze envelopes on the same plot allows prediction of the type of yielding that may occur under any general state of stress.

22.5.3 Crazing in Glassy Polymers

The phenomenology of crazing was discussed earlier, but as yet the mechanisms by which crazing occurs have not been considered. Crazes normally nucleate on the surface of a specimen, probably at the site of flaws such as scratches or other imperfections. They tend to be lamellar in shape and grow into the bulk of the specimen from the surface, as shown in Figure 22.22. The flaws on the surface tend to raise the magnitude of the applied stress locally and allow craze initiation to occur.

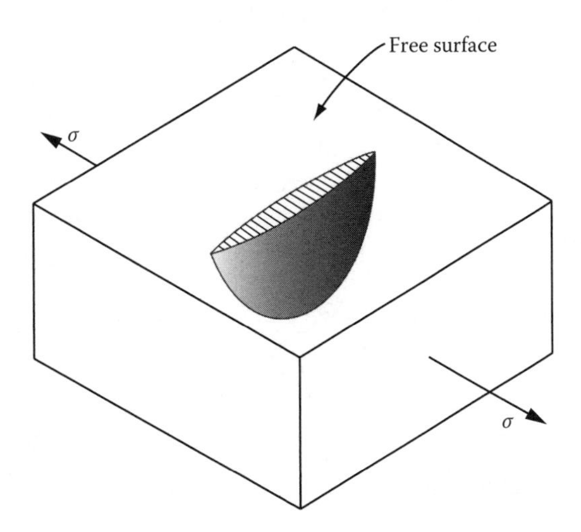

FIGURE 22.22 Schematic diagram of a craze nucleating at the surface of a polymer.

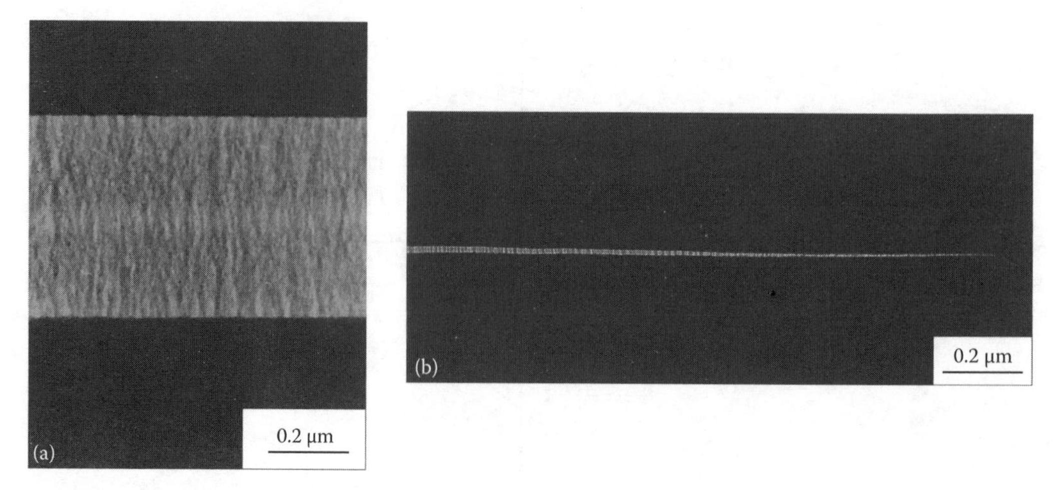

FIGURE 22.23 Transmission electron micrographs of a tapering craze in a thin film of polystyrene taken at the same magnification: (a) thick centre of the craze showing a mid-rib region and (b) thin craze tip region. (Courtesy of E.J. Kramer, Cornell University, Ithaca, NY.)

The basic difference between crazes and cracks is that crazes contain polymer, typically about 50% by volume, within their bulk whereas cracks do not. The presence of polymer within crazes was deduced originally from measurements of the refractive index of crazed material, but it can be demonstrated more directly by using electron microscopy. Figure 22.23 shows a section through a craze in polystyrene viewed at high magnification in a transmission electron microscope. It can be clearly seen that there is a sharp boundary between the craze and the polymer matrix and that the craze itself contains many cavities and is bridged by a dense network of fine fibrils of the order of 20 nm in diameter. If a craze is followed along its length to the tip it is found to taper gradually until it eventually consists of a row of voids and microfibrils on the 5 nm scale.

Microscopic examination has given the most important clues as to how crazes nucleate and grow in a glassy polymer. The basic mechanism is thought to be a local-yielding and cold-drawing process, which takes place in a constrained zone of material. It is known that the local stresses at the tip of a surface flaw or a growing craze are higher than the overall tensile stress applied to the specimen. It is envisaged that yielding takes place locally at these stress concentrations at a stress below that which is required to cause general yielding in the rest of the specimen. If the bulk tensile uniaxial stress is σ_0 then there will be an overall bulk tensile hydrostatic stress of $p=\sigma_0/3$ but this hydrostatic component will be very much higher at the tip of the flaw or growing craze. These conditions of stress then favour the initiation of a craze at a flaw or the propagation of a pre-existing craze. It is envisaged that the fibrils within the craze form by a local-yielding and cold-drawing process and the fibrils remain intact because they contain oriented molecules and consequently are stronger than the surrounding uncrazed polymer. The thickening of the craze is thought to take place by uncrazed material at the craze/matrix boundary being deformed into fibrils and so transformed into craze matter. The crazes form as very thin lamellae in a plane perpendicular to the tensile axis because the tensile stress is highest across this plane and the elastic constraint of the surrounding uncrazed material allows the craze to thicken and grow only laterally.

PROBLEMS

22.1 The tensile stress–strain curve of a polymer may be represented by $\sigma=ae^b$ where σ is the true stress and e is the nominal strain, and a and b are constants. Show that necking will occur when

$$e = \frac{b}{(1-b)}$$

Determine also the tensile yield stress of the material in terms of a and b.

22.2 A polymer obeys a pressure-dependent von Mises yield criterion of the form

$$A[\sigma_1 + \sigma_2 + \sigma_3] + B[(\sigma_1 - \sigma_2)^2 + (\sigma_2 - \sigma_3)^2 + (\sigma_3 - \sigma_1)^2] = 1$$

where

σ_1, σ_2 and σ_3 are the principal stresses
A and B are constants

Show that the material will yield under the action of hydrostatic stress alone and calculate the hydrostatic stress required to cause yield, in terms of the yield stresses of the material in tension and compression

22.3 It is thought that polyethylene is capable of undergoing twinning through a kinking of the molecules in a similar way to which twinning takes place in polydiacetylene single crystals. The most likely planes on which the kinking can occur are thought to be (200), (020) and (110). Draw molecular sketches of each of these twins viewed perpendicular to the plane of shear and hence or otherwise determine

(a) The twinning plane in each case
(b) The angle through which the molecules bend
(c) The direction of shear for each twin
(d) The twin involving the lowest shear

(Polyethylene is orthorhombic with a=0.740 nm, b=0.493 nm, c=2.537 nm).

22.4 Polystyrene is capable of undergoing either crazing or shear yielding during deformation depending upon the state of stress imposed. It is found that it undergoes shear yielding at a stress of 73 MPa in uniaxial tension and at 92 MPa in uniaxial compression at a particular strain-rate and temperature. Under the same conditions, it is found to undergo crazing at a stress of 47 MPa in uniaxial tension and under a biaxial tensile stress of 45 MPa.

Assuming that polystyrene obeys a pressure-dependent von Mises yield criterion and a critical strain crazing criterion, determine the conditions of stress at which crazing and shear yielding will take place simultaneously for plane stress deformation. (Poisson's ratio for polystyrene may be taken as 0.33.)

FURTHER READING

Andrews, E.H., *Fracture in Polymers*, Oliver and Boyd Ltd, London, U.K., 1968.

Haward, R.N. (ed.), *The Physics of Glassy Polymers*, Applied Science Publishers, London, 1973.

Haward, R.N. and Young, R.J. (eds.), *The Physics of Glassy Polymers*, 2nd edn., Chapman & Hall, London, 1997.

Kausch, H.H. and Michler, G.H., The effect of time on crazing and fracture, *Advances in Polymer Science*, **187**, 33 (2005).

Kinloch, A.J. and Young, R.J., *Fracture Behaviour of Polymers*, Applied Science, London, 1983.

Monnerie, L., Halary, J.L., and Kausch, H.H., Intrinsic molecular mobility and toughness of polymers 1, *Advances in Polymer Science*, **187**, 215 (2005).

Ward I.M., *Mechanical Properties of Solid Polymers*, 2nd edn., Wiley-Interscience, New York, 1983.

23 Fracture and Toughening

23.1 INTRODUCTION

The failure of engineering components through fracture can have catastrophic consequences and much effort has been put in by materials scientists and engineers in developing materials and designing structures that are resistant to premature failure through fracture. It is possible to classify materials as being either *brittle* like glass which shatters readily or *ductile* like pure metals such as copper or aluminium which can be deformed to high strains before they fail. Polymers are found to display both types of behaviour depending upon their structure and the conditions of testing. In fact, the same polymer can sometimes be made either brittle or ductile by changing the temperature or rate of testing. Over the years, some polymers have been advertised as being "unbreakable." Of course, this is very misleading as every material can only withstand limited stresses. On the other hand, domestic containers made from polymers, such as polypropylene or low-density polyethylene, are remarkably tough and virtually indestructible under normal conditions of use.

Polymers are capable of undergoing fracture in many different ways and the stress–strain curves of several different polymers tested to failure were shown schematically in Figure 19.4. Glassy polymers tend to be rather brittle, failing at relatively low strains with very little plastic deformation. On the other hand, semi-crystalline polymers normally are more ductile, especially between T_g and T_m and undergo cold-drawing before ultimate failure. Crosslinked rubbers (elastomers) are capable of being stretched elastically to high extensions, but a tear will propagate at lower strains if a pre-existing cut is present in the sample. Polymer single-crystal fibres break at very high stresses, but because their moduli are high, the stresses correspond to strains of only a few percent. Because of this wide variation in fracture behaviour, it is convenient to consider the different fracture modes separately.

23.2 FUNDAMENTALS OF FRACTURE

Fracture can be defined simply as the creation of new surfaces within a body through the application of external forces. In this section, the fundamental molecular aspects of polymer fracture will be considered before moving on to consider the mechanics of fracture in subsequent sections.

23.2.1 Theoretical Tensile Strength

A parameter of interest in consideration of the mechanical behaviour of materials is the stress that is expected to be required to cause cleavage through the breaking of bonds. The situation is visualized in Figure 23.1 in which a bar of unit cross-sectional area, made of a perfectly elastic material such as a polymer crystal, is pulled in tension. Work is done in stressing the material and since fracture involves the creation of two new surfaces, thermodynamic considerations demand that the work cannot be less than the surface energy of the new surfaces. The expected dependence of the stress σ upon the displacement x is shown in Figure 23.1 and it is envisaged that this can be approximated to a sine function of wavelength λ which takes the form

$$\sigma = \sigma_t \sin\left(\frac{2\pi x}{\lambda}\right) \tag{23.1}$$

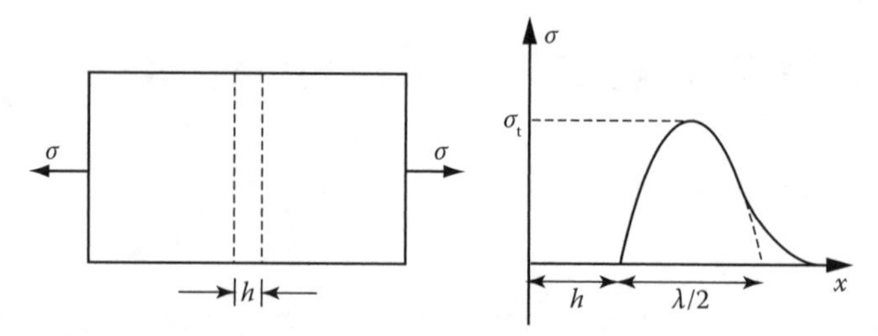

FIGURE 23.1 Model used to calculate the theoretical cleavage stress of a polymer crystal. The parameters are defined in the text.

where σ_t is the maximum tensile stress. This can be written for low strains as

$$\sigma = \sigma_t \frac{2\pi x}{\lambda} \tag{23.2}$$

but another equation can be derived relating σ to x since the material is assumed to be perfectly elastic. If the equilibrium separation of the atomic planes in the crystal perpendicular to the tensile axis is h then for small displacements Hooke's law gives

$$\sigma = \frac{Ex}{h} \tag{23.3}$$

and comparing Equations 23.2 and 23.3 leads to

$$\sigma_t = \frac{\lambda E}{2\pi h} \tag{23.4}$$

The work done when the bar is fractured is given approximately by the area under the stress–displacement curve up to the point of separation and equating this to the energy required to create the two new surfaces leads to

$$\int_0^{\lambda/2} \sigma_t \sin\left(\frac{2\pi x}{\lambda}\right) dx = 2\gamma$$

or at low strains

$$\gamma = \frac{\lambda \sigma_t}{2\pi}$$

Substituting for $\lambda/2\pi$ from Equation 23.4 then gives

$$\sigma_t = \sqrt{\frac{E\gamma}{h}} \tag{23.5}$$

Putting typical values of E, γ and h into this equation for a variety of materials shows that it is generally expected that

$$\sigma_t \approx \frac{E}{10} \tag{23.6}$$

This result is particularly interesting in the context of polymers because, although their moduli generally are low, there are certain circumstances in which E can be made large. Oriented amorphous and semi-crystalline polymers have relatively high moduli and the moduli of polymer crystals in the chain direction are comparable to those of metals. High tensile strengths might, therefore, be expected to be displayed when polymers are deformed in ways which exploit these high moduli.

In general, it is found that the tensile strengths of materials are very much lower than $E/10$ because brittle materials fracture prematurely due to the presence of flaws and ductile materials undergo plastic deformation through the motion of dislocations. However, fine whiskers of glass, silica and certain polymer crystals that do not contain any flaws and are not capable of plastic deformation, have values of fracture strength that are close to the theoretically-predicted values.

23.2.2 Molecular Failure Processes

In this section, the processes that take place on a molecular level when a polymer is fractured will be examined. It is clear that, in general, failure will occur through a combination of molecular rupture and the slippage of molecules past each other, but the extent to which these two processes occur is not easy to determine except for specific polymer structures and morphologies.

23.2.2.1 Bond Rupture

The rupture of primary bonds is likely to be the principal mechanism of failure during the fracture of polymer single crystals deformed in tension parallel to the chain direction. Figure 23.2 shows a micrograph of a polydiacetylene single-crystal fibre (see Figure 17.11a) that has been fractured in tension. The stress–strain curve of a similar crystal was given in Figure 19.6. The examination of the fractured end of the crystal shows clearly that molecular fracture has taken place. Since the covalent bonds along the polymer chain are very strong, the material has a very high fracture strength. Figure 23.3 shows a plot of the fracture strength of the polymer as a function of fibre diameter. The fracture strength increases as the diameter is reduced. This is because the finer fibres contain fewer strength-reducing flaws. The finest fibres tested had a strength of the order of 1.5 GPa. The modulus of the polymer is about 60 GPa and so the maximum fracture strength is of the order of $E/40$. Theoretical considerations (Section 23.2.1) show that the highest fracture strength might be expected to be about $E/10$ and most materials are, because of flaws, well below this value. The values of fracture strengths of the polydiacetylene single crystals are, therefore, reasonably close to the theoretical estimates and compare favourably with those of other high-strength materials.

Another group of materials in which molecular fracture must occur during crack growth are crosslinked polymers such as elastomers. The network junction points do not allow molecular pull-out to occur and bond breakage must then take place. It can be demonstrated that this is indeed

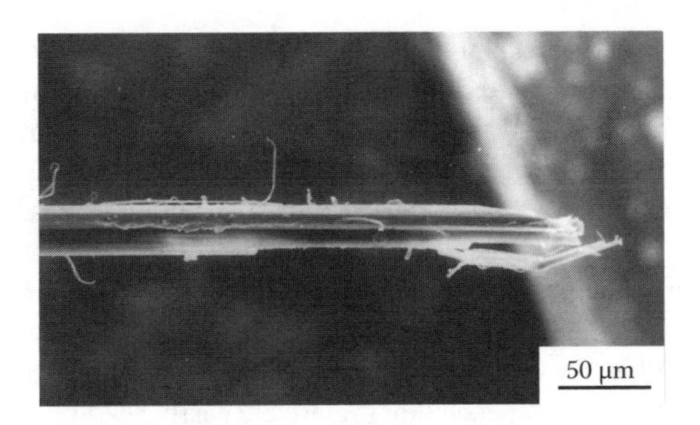

FIGURE 23.2 Scanning electron micrograph of a fractured polydiacetylene single-crystal fibre.

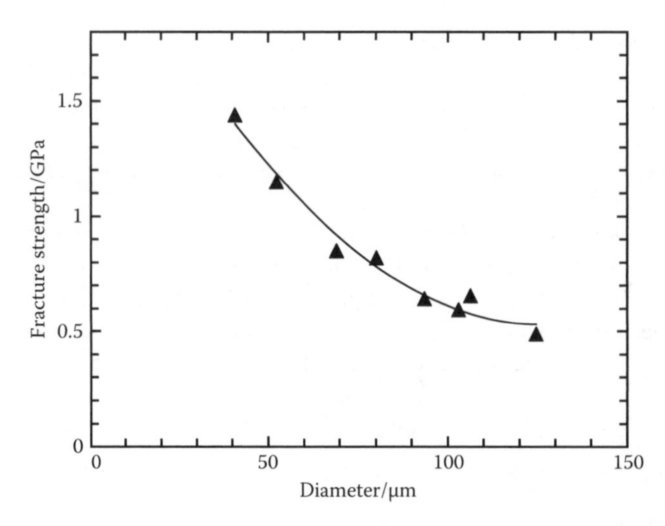

FIGURE 23.3 Dependence of fracture strength of polydiacetylene single crystal fibres upon fibre diameter.

the failure mechanism by determining the tearing energy of an elastomer in the absence of any viscoelastic energy losses. This is found to be close to that expected if failure takes place through bond breakage.

The technique of electron spin resonance spectroscopy (ESR) (Section 15.2.1) can be used to detect the presence of radicals generated by the fracture of covalent bonds. Specimens can be fractured within the cavity of the spectrometer and molecular fracture is manifested by the generation of an ESR signal. The technique suffers from certain limitations such as difficulty in being able to get a sufficiently large volume of material into the cavity to produce a strong enough signal and the drop in signal strength due to the decay of radicals through combination and other reactions. The radicals are found to be more stable at low temperature where the mobility of the chains is reduced. For many polymers, only weak signals are obtained because very little molecular fracture takes place and failure occurs primarily through molecular slippage. The strongest signals are obtained from crosslinked polymers and crystalline polymer fibres where the crosslinks and crystals act as anchorage points and reduce the amount of slippage that can occur, thereby increasing failure by the fracture of covalent bonds.

Figure 23.4 shows how the radical concentration increases with strain as nylon 6 fibres are deformed in tension. The molecules in the fibre were aligned and strained by drawing to a certain extent before deformation. An appreciable signal starts to build up above a strain of about 10%, well before the specimen fractures. The build up in ESR signal is thought to be due to failure of the most highly stressed molecules within the fibres before overall fracture of the specimen takes place.

23.2.2.2 Effect of Molar Mass

Since the failure of a polymer depends upon the balance between the fracture of primary covalent bonds and slippage of molecules past each other, it might be expected that there will be a strong dependence of the fracture behaviour upon molar mass. In general, it is found that polymers of low molar mass have very low tensile strengths and that their strength increases significantly as molar mass increases. It is found that eventually the strength of the polymer reaches a constant level at a critical molar mass above which any further increase in molar mass has little effect.

This type of behaviour is demonstrated in Figure 23.5 for brittle polymers, such as poly(methyl methacrylate) and polystyrene, and it can be rationalised in terms of molecular entanglements (see Sections 16.4 and 20.8). When the chains have low molar mass and are relatively short, they are not entangled and so are able to slide past each other relatively easily, leading to low strength. When the molar mass is increased and the chains are sufficiently long to form entanglements (i.e. above the

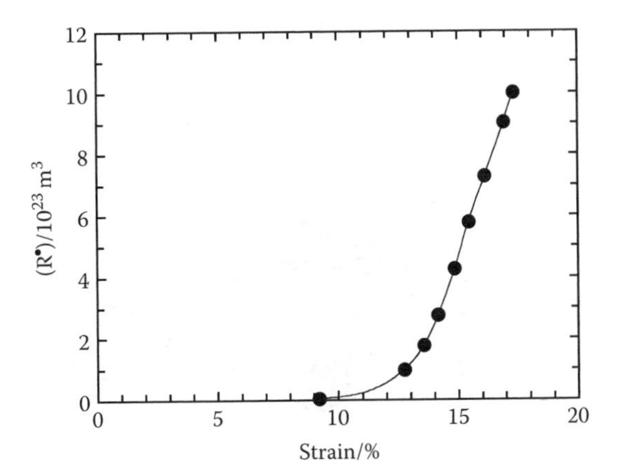

FIGURE 23.4 Dependence of radical concentration, [R•], upon strain during the tensile deformation of nylon 6 fibres at room temperature. (Data taken from Becht, J. and Fischer, H., *Kolloid-Zeitschrift*, 229, 167, 1969.)

effective molar mass between entanglements M_e, see Section 16.4.1) chain sliding is more difficult and the strength increases. The examples shown in Figure 23.6 are for glassy polymers in which stable crazing (Section 22.5.3) will also occur at sufficiently high molar mass when chain sliding is suppressed. These findings highlight the need to ensure that a polymer for use in structural applications has a molar mass well above M_e.

23.3 MECHANICS OF FRACTURE

Having considered the fundamental aspects of fracture on the molecular level, it is now necessary to look at fracture as a macroscopic phenomenon and the mechanics of the process.

23.3.1 Brittle Fracture and Flaws

One of the most useful approaches that can be used to explain the fracture of brittle polymers is the theory of brittle fracture developed by Griffith to account for the fracture behaviour of glass.

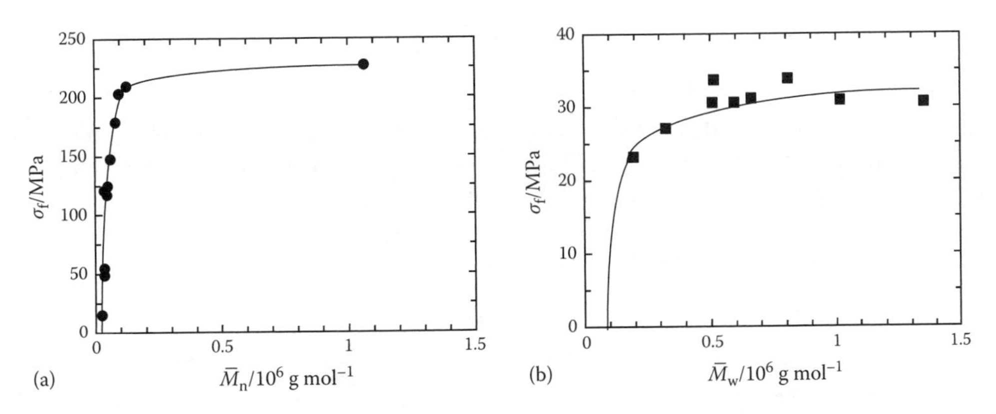

FIGURE 23.5 Dependence of the fracture stress σ_f upon molar mass for different glassy polymers. (a) Variation of σ_f, with number average molar mass $\bar{M}_n$ for poly(methyl methacrylate) at −196°C. (Data taken from Vincent, P.I., *Polymer*, 1, 425, 1960.) (b) Variation of σ_f with weight average molar mass $\bar{M}_w$ for polystyrene at room temperature. (Data taken from Martin, J.R. et al., *J. Macromol. Sci. Rev. Macromol. Chem.*, C8, 57, 1972.)

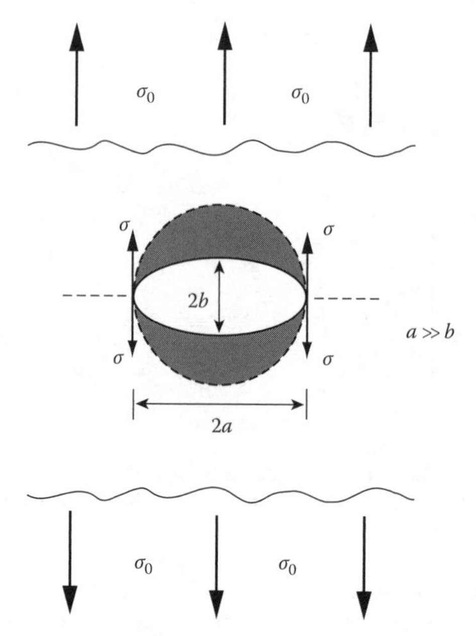

FIGURE 23.6 Model of an elliptical crack of length $2a$ in a uniformly loaded infinite plate. The parameters are defined in the text. The stress-free zone above and below the crack is shaded.

The theory is concerned with the thermodynamics of initiation of cracks from pre-existing flaws in a perfectly-elastic brittle solid. These flaws could be either scratches or cracks, both of which have the effect of causing a *stress concentration*. This means that the local stress in the vicinity of the crack tip is higher than that applied to the body as a whole. The theory envisages that fracture starts at this point and the crack then continues to propagate catastrophically.

The effect of the presence of a crack in a body can be determined by considering an elliptical crack of major and minor axes a and b in a uniformly loaded infinite plate as illustrated in Figure 23.6. If the stress at a great distance from the crack is σ_0 then the stress σ at the tip of the crack is given by

$$\sigma = \sigma_0\left(1+\frac{2a}{b}\right) \tag{23.7}$$

The ratio σ/σ_0 defines the stress-concentrating effect of the crack and is three for a circular hole ($a=b$). It will be even larger when $a>b$. The equation can be put in a more convenient form using the radius of curvature ρ of the sharp end of the ellipse, which is given by

$$\rho = \frac{b^2}{a}$$

Equation 23.7 then becomes

$$\sigma = \sigma_0\left[1+2\left(\frac{a}{\rho}\right)^{1/2}\right] \tag{23.8}$$

If the crack is long and sharp ($a \gg \rho$), σ will be given approximately by

$$\sigma = \sigma_0 2\left(\frac{a}{\rho}\right)^{1/2} \tag{23.9}$$

Inspection of these equations shows that the effect of a sharp crack will be to cause a large concentration of stress that is a maximum at the tip of the crack, which is, therefore, the point from which the crack propagates.

Griffith's contribution to the argument concerned a calculation of the energy released by putting the sharp crack into the plate and relaxing the surrounding material, and relating this to the energy required to create new surface. The full calculation requires an integration over the stress field around the crack, but an estimate of the energy released in the creation of new surface can be made by considering that the crack completely relieves the stresses in the shaded circular zone of material of radius a around the crack, as shown in Figure 23.6. If the plate has unit thickness then the volume of this region is $\sim\pi a^2$. Since the material is assumed to be linearly elastic then the strain energy per unit volume in an area deformed to a stress of σ_0 is $(1/2)\sigma_0^2/E$, where E is the modulus. If it is assumed that the circular area is completely relieved of stress when the crack is present then the energy released is $(1/2)\pi a^2\sigma_0^2/E$ per unit thickness. The surface energy of the crack is $4a\gamma$ per unit thickness, where γ is the surface energy of the material. The extra factor of 2 appears because the crack of length $2a$ has two surfaces. It is postulated that the crack will propagate if the energy gained by relieving stress is greater than that needed to create new surface and this can be expressed mathematically as

$$\frac{\partial}{\partial a}\left(-\frac{\pi\sigma_0^2 a^2}{2E} + 4a\gamma\right) > 0 \tag{23.10}$$

or at the point of fracture

$$\sigma_0 = \sigma_f = \left(\frac{4E\gamma}{\pi a}\right)^{1/2} \tag{23.11}$$

where σ_f is the fracture stress. This equation is similar to that obtained by Griffith who used a more rigorous calculation of the stresses around the crack to show that

$$\sigma_f = \left(\frac{2E\gamma}{\pi a}\right)^{1/2} \tag{23.12}$$

This equation is the usual form in which the Griffith criterion is expressed, but it is strictly only applicable when plane stress conditions prevail. In plane strain it becomes

$$\sigma_f = \left[\frac{2E\gamma}{\pi(1-v^2)a}\right]^{1/2} \tag{23.13}$$

where v is Poisson's ratio.

The Griffith theory was tested initially for glass and it has since been applied to the fracture of other brittle solids. The main predictions are borne out in practice. The theory predicts that $\sigma_f \propto a^{-1/2}$

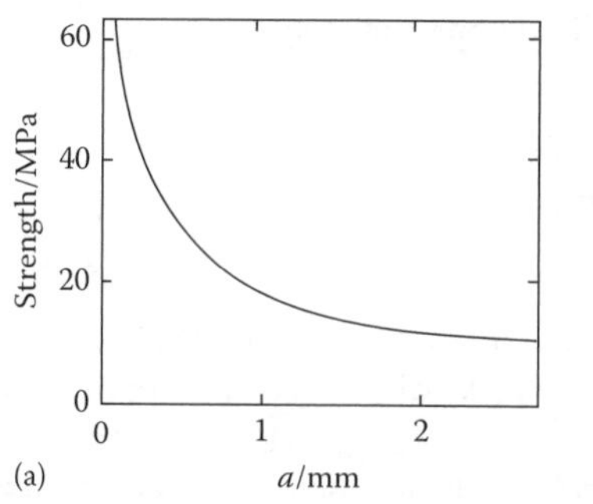

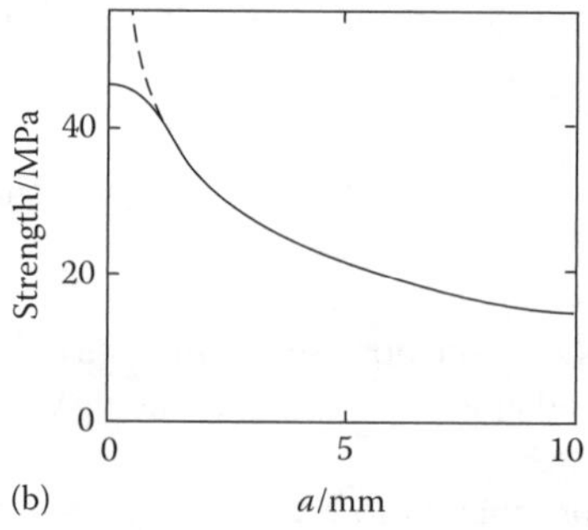

FIGURE 23.7 Dependence of the uniaxial tensile fracture strength of (a) poly(methyl methacrylate) and (b) polystyrene upon the crack length, a. (Data taken from Berry, J.P., in *Fracture VII*, H. Liebowitz (ed.), Academic Press, New York, 1972.)

and so once a crack starts to propagate in a uniform stress field it should continue to do so because the stress required to drive it drops as the crack becomes longer.

The theory provides a good qualitative prediction of the fracture behaviour of brittle polymers such as polystyrene and poly(methyl methacrylate). Figure 23.7 shows how the strength of specimens of these two polymers deformed in tension varies with the length of artificially induced cracks. There is a clear dependence of σ_f upon $a^{-1/2}$ as predicted by the Griffith theory, but detailed investigations have shown that there are significant discrepancies. For example, since Young's moduli of the polymers are known, the values of γ for the two polymers can be calculated from the curves in Figure 23.7 using Equation 23.12. The value of γ for poly(methyl methacrylate) is found to be ~210 J m^{-2} and that for polystyrene ~1700 J m^{-2}. These values are very much larger than the surface energies of these polymers which are thought to be of the order of 1 J m^{-2}. The values of γ determined for other materials also are larger than the theoretical surface energies (see, e.g. Table 23.1). The discrepancy arises because the Griffith approach assumes that the material behaves elastically and does not undergo plastic deformation. It is known that even if a material appears to behave in a brittle manner and does not undergo general yielding, there is invariably a small amount of plastic deformation, perhaps only on a very local level, at the tip of the crack. The energy absorbed during such plastic deformation is very much higher than the surface energy and this is reflected in the measured values of γ for most materials being much larger than the theoretical predictions.

The Griffith approach assumes that for any particular material, the fracture stress is controlled by the size of the flaws present in the structure. The strength of any particular sample can be increased by reducing the size of these flaws. This can easily be demonstrated for glass in which

TABLE 23.1
Approximate Values of Fracture Surface Energy γ Determined Using the Griffith Theory

Material	γ/J m^{-2}
Inorganic glass	7–100
MgO	8–10
High-strength aluminium	~8000
High-strength steel	~25000
Poly(methyl methacrylate)	200–400
Polystyrene	1000–2000
Epoxy resins	50–200

flaws are introduced during normal handling and contributes to its low fracture stress. The strength can be increased greatly by etching the glass in hydrofluoric acid, which removes most of the surface flaws. The reduction of the size of artificially-induced flaws (*notches*) clearly causes an increase in the strength of brittle polymers (Figure 23.7). This rise in strength, however, does not go on indefinitely and when the flaw size is reduced below a critical level, about 1 mm for polystyrene and 0.07 mm for poly(methyl methacrylate) at room temperature, σ_f becomes independent of flaw size. The materials, therefore, behave as if they contain natural flaws with these critical sizes and so the introduction of smaller artificial flaws does not affect the strength. The exact size of the natural or inherent flaws can be determined from the fracture strength of an unnotched specimen, the modulus and the value of γ measured from curves such as those shown in Figure 23.7. The relatively low inherent flaw size of poly(methyl methacrylate) accounts for the observation that, in the unnotched state, it has a higher strength than polystyrene, even though polystyrene has a higher value of γ (Table 23.1).

Since polystyrene behaves as if it contains natural flaws of the order of 1 mm, it might be expected that they could be detected in undeformed material but this is not the case. The inherent flaws are in fact crazes that form during deformation (Section 22.5). When an unnotched specimen of polystyrene is deformed in tension, large crazes of the order of 1 mm in length are seen to nucleate (Figure 22.19) whereas in the case of poly(methyl methacrylate), the crazes are much smaller. Fracture then takes place through the breakdown of one of these crazes that becomes transformed into a crack. On the other hand, when specimens containing cracks of greater than the inherent flaw size are deformed, general crazing does not occur and failure takes place at the artificial flaws.

23.3.2 Linear Elastic Fracture Mechanics

23.3.2.1 Definitions

The derived value of γ is usually greater than the true surface energy because other energy-absorbing processes occur and so it is normal to replace 2γ in Equation 23.12 with G_c which then represents the total work of fracture, i.e. the *fracture energy* which is also known as the *critical strain-energy release-rate*. The equation then becomes

$$\sigma_f = \left(\frac{EG_c}{\pi a}\right)^{1/2} \tag{23.14}$$

and this equation can be thought of as a prediction of the critical stress and crack length at which a crack will begin to propagate in an unstable manner. Failure would be expected to occur when the term $\sigma\sqrt{\pi a}$ reaches a value of $\sqrt{EG_c}$. The term $\sigma\sqrt{\pi a}$ can be considered as the driving force for crack propagation and this leads to the stress intensity factor K being defined as

$$K = \sigma\sqrt{\pi a} \tag{23.15}$$

The condition for crack propagation to occur is that K reaches a critical value K_c given by

$$K_c = \sqrt{EG_c} \tag{23.16}$$

The *critical stress intensity factor* K_c is sometimes called the *fracture toughness* since it characterizes the resistance of a material to crack propagation.

K_c and G_c are the two main parameters used in *linear elastic fracture mechanics*. Equations 23.15 and 23.16 strictly can be applied only if the material is linearly elastic and any yielding is confined to a small region in the vicinity of the crack tip. These restrictions generally are appropriate for glassy polymers undergoing brittle fracture and the application of linear elastic fracture mechanics to the failure of brittle polymers has allowed significant progress to be made in the understanding of polymer fracture processes. In the case of more ductile polymers and elastomers, generalized versions of the Griffith equation must be employed.

23.3.2.2 Fracture Mechanics Testing

The understanding of brittle fracture also is improved greatly by using more sophisticated test pieces than simple uniaxial tensile specimens. It is possible to design test pieces such that cracks can be grown in a stable manner at controlled velocities. In the case of the uniaxial tensile specimen, once a crack starts to grow it propagates very rapidly. This can be overcome by having a specimen in which the effective stress driving the crack drops as the crack grows longer. Figure 23.8 shows typical fracture mechanics test pieces. In the double-torsion specimen, the crack is driven by twisting the two halves of one end of the specimen in opposite directions and the stress distribution is such that, for a given load applied at the end of the specimen, the stress-intensity factor K is independent of the crack length. This must be contrasted with the unaxial tensile specimen where, for a given applied stress, K increases rapidly with increasing crack length. It is possible to have specimens where K actually drops as the crack length increases, such as in the case of the double cantilever beam which is also shown in Figure 23.8. The effect of this variation of K with crack length upon load–displacement curves when the specimen ends are displaced at a constant rate are shown schematically in Figure 23.8. In all three cases, the load is initially approximately proportional to the displacement. For the simple tensile specimen (SEN), the load drops immediately, the crack starts

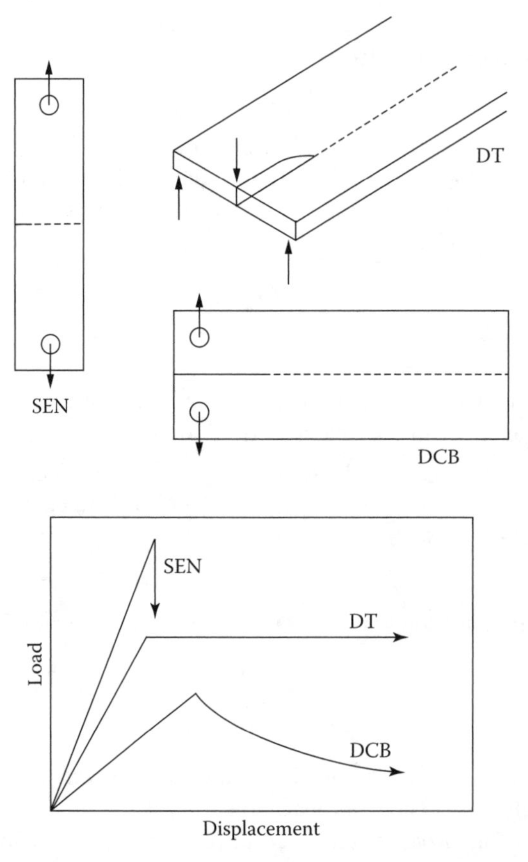

FIGURE 23.8 Schematic representation of different fracture mechanics specimens used with brittle polymers and load–displacement curves for each specimen for a constant displacement rate: SEN, single-edge notched; DT, double torsion; and DCB, double-cantilever beam.

to grow and the crack propagates rapidly in an uncontrolled manner to the other side of the specimen. When the crack starts to grow in the double-torsion specimen, it propagates at a constant load (and also constant velocity) until it reaches the end of the specimen. Finally, in the double cantilever beam, the crack propagates at a decreasing load and velocity when the specimen ends are displaced at a constant rate.

The value of K_c at which crack propagation occurs in a specimen of finite size depends upon the applied load, crack length and specimen dimensions. Equation 23.16 gives K_c only for a crack of length $2a$ in an infinite plate. For other specimen geometries, it must be determined from an elastic analysis of the particular specimen used and this so-called K-calibration has been done only for certain specific specimens. The formulae for K which have been determined for the three specimen types shown in Figure 23.8 are given in Table 23.2. It should be noted that for the double-torsion specimen the crack length a does not appear in the formula.

It sometimes is more convenient to use G_c rather than K_c to characterize the fracture behaviour of a material, especially since G_c is an energy parameter and so can be related more easily to particular energy-absorbing fracture processes. If the value of K_c is known, then G_c can be obtained through Equation 23.16 but care is required with polymers since E depends upon the rate of testing and the appropriate strain-rate for the fracture experiment may not be obvious. It is possible to calculate G_c without prior knowledge of K_c or E for any general type of fracture specimen using a compliance calibration technique. This involves measurement of $(\partial C/\partial a)$, the variation of specimen compliance C with crack length, the compliance in this case being defined as the ratio of the displacement of the ends of the specimens to the applied load or force f (see, e.g. Figure 23.20c). When crack propagation occurs, G_c is given by

$$G_c = \frac{f^2}{2t}\left(\frac{\partial C}{\partial a}\right) \tag{23.17}$$

where t is the specimen thickness.

TABLE 23.2
Expression for the Fracture Toughness K for Different Fracture Mechanics Specimens Used in Studying the Fracture of Brittle Polymers

Geometry of Specimen	$K =$
Single-edge notched (SEN)	$\frac{fa^{1/2}}{bt}[1.99 - 0.41(a/b) + 18.7(a/b)^2 - 38.48(a/b)^3 + 53.85(a/b)^4]$
Double torsion (DT)	$fW_m\left[\frac{3(1+v)}{Wt^3t_n}\right]^{1/2}$ $(W/2 \gg t)$
Double cantilever beam (DCB)	$\frac{2f}{t}\left[\frac{3a^2}{h^3}+\frac{1}{h}\right]^{1/2}$

Source: Adapted from Young, R.J., in *Developments in Polymer Fracture*, E.H. Andrews (ed.), Applied Science, London, 1979.

f, applied load; b, specimen breadth; t, sheet thickness; a, crack length; h, distance of specimen edge from fracture plane; t_n, thickness of sheet in plane of crack; W, specimen width; W_m, length of moment arm; v, Poisson's ratio.

23.3.2.3 Crack Propagation in Poly(Methyl Methacrylate)

The slow growth of cracks in poly(methyl methacrylate) is an ideal application of linear elastic fracture mechanics to the failure of brittle polymers. Cracks grow in a very well-controlled manner when stable test pieces such as the double-torsion specimen are used. In this case, the crack will grow steadily at a constant speed if the ends of the specimen are displaced at a constant rate. The values of K_c and G_c at which a crack propagates depends upon both the crack velocity and the temperature of testing, another result of the rate dependence and temperature dependence of the mechanical properties of polymers. This behaviour is demonstrated clearly in Figure 23.9. The value of K_c is a unique function of crack velocity at a given temperature with K_c increasing as the crack velocity increases and the temperature of testing is reduced. The data in Figure 23.9 are presented in the form of a log-log plot and can be fitted to an empirical equation of the form

$$K_c = A\left(\frac{da}{dt}\right)^n \tag{23.18}$$

where

A is a temperature-dependent constant
n is an exponent equal to the slope of the lines (~0.07).

At a given temperature, the value of K_c does not continue to rise indefinitely but reaches a maximum value, K_c^*. This corresponds to a point of instability above which the crack grows rapidly at a lower value of K_c. This instability has been explained in terms of an adiabatic-isothermal transition. It is thought that at velocities below the instability, cracks grow isothermally because the heat can be dissipated from the tip of the growing crack. At velocities above the transition, the crack grows too rapidly for heat to be dissipated completely and partially adiabatic conditions prevail at the crack tip. The consequent local rise in temperature softens the polymer in the vicinity of the crack tip and so reduces the value of K_c for crack propagation.

A consequence of the dependence of K_c upon crack velocity for poly(methyl methacrylate) is that the polymer is prone to time-dependent failure during periods of prolonged loading. Since polymers often are used under stress for long periods of time this can be a serious problem. A typical use might be as windows in the pressurized cabins of aircraft. The time-to-failure t_f for a specimen

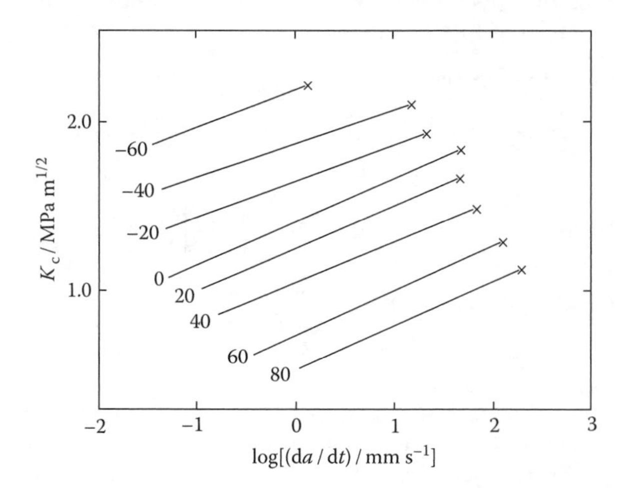

FIGURE 23.9 Dependence of K_c upon crack velocity for crack growth in poly(methyl methacrylate) at different temperatures (in °C). The values of K_c^* in each case are indicated by a cross (×). (Data taken from Marshall, G.P. et al., *J. Mater. Sci.*, 9, 1409, 1974.)

under stress can be calculated by integrating Equation 23.18. If the specimen contains flaws of size a_0 and is subjected to a constant stress σ_0 then a crack will start to grow slowly at a velocity governed by Equation 23.18. If it is postulated that failure occurs when the crack reaches a critical length a^* then the equation can be rearranged to give

$$\int_0^{t_f} dt = \int_{a_0}^{a^*} \frac{A^{1/n}}{K_c^{1/n}} da \tag{23.19}$$

If the flaw is small and can be considered to be in an infinite plate, then since K is given by $\sigma\sqrt{\pi a}$, Equation 23.19 can be written as

$$t_f = \int_{a_0}^{a^*} A^{1/n} \sigma_0^{-1/n} \pi^{-1/2n} a^{-1/2n} da \tag{23.20}$$

which upon integration gives

$$t_f = \frac{A^{1/n}(\sigma_0^2 \pi)^{-1/2n}}{(1-1/2n)}\left[a^{*(1-1/2n)} - a_0^{(1-1/2n)}\right]$$

Since $a^* \gg a_0$ and $n \ll 1$ (Figure 23.9), then the term in a^* can be neglected and so the time-to-failure will be given approximately by

$$t_f = \frac{A^{1/n} a_0^{(1-1/2n)}}{(\sigma_0^2 \pi)^{1/2n}(1/2n - 1)} \tag{23.21}$$

The time-to-failure can, therefore, be determined from knowledge of σ_0, a_0 and the variation of K_c with crack velocity. This particular equation is for a small crack in an infinite plate. Similar equations can be derived for other specimen geometries.

So far, the mechanisms of crack propagation have not been considered but they are important since the observed fracture behaviour is a reflection of deformation processes taking place at the crack tip. There is accumulated evidence to show that in poly(methyl methacrylate), polystyrene and probably many other glassy polymers, crack propagation takes place through the breakdown of a craze at the tip of the growing crack. One way that this can be demonstrated indirectly is from the presence of craze debris on the polymer fracture surfaces that gives rise to interference colours when the surfaces are viewed in reflected light. Figure 23.10 shows the craze debris on the fracture surface of a poly(methyl methacrylate) specimen and the geometry of the crack/craze entity. It is envisaged that propagation occurs through the growth of a crack with a single craze at its tip and that a steady-state situation is reached whereby the rate of craze growth equals that of craze breakdown with the crack/craze entity propagating gradually through the specimen. The breakdown can take place either down the centre of the craze or at the craze-polymer interface. In the example shown in Figure 23.10, failure has occurred at the polymer-craze interface. The formation of a craze and its subsequent breakdown absorbs a considerable amount of energy and it is these processes that account for the fracture energies being much higher than the energy required simply to create new surface (Table 23.1).

The mechanisms of crack propagation in poly(methyl methacrylate) are particularly amenable to analysis. The situation is not so simple for other polymers. For example, in polystyrene there is usually multiple crazing in the vicinity of the crack tip. In tougher polymers, such as polycarbonate,

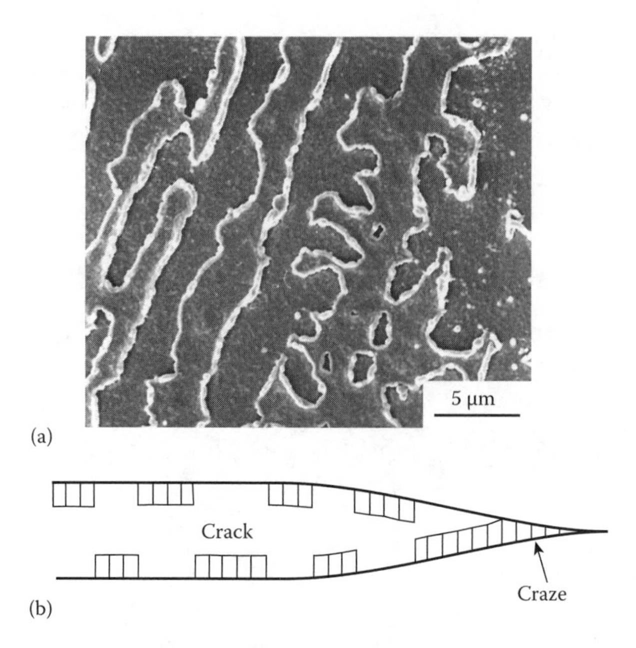

FIGURE 23.10 (a) Scanning electron micrograph of the fracture surface of a sample of poly(methyl methacrylate) showing craze debris on the surface. (Courtesy of M.N. Sherratt, University of Manchester, Manchester, U.K.) (b) A schematic representation of a crack growing through a craze with craze breakdown at the craze-polymer interface.

shear yielding as well as crazing often takes place at the crack tip. In these cases, crack propagation does not occur in such a well-controlled manner as in poly(methyl methacrylate) and it is more difficult to analyse.

23.3.3 Tearing of Elastomers

It is impossible to apply linear elastic fracture mechanics to polymers that fail at high strains and when the deformation is not linear elastic. There are, however, certain cases where the Griffith approach can be generalized and failure criteria established. One such case is in the tearing of elastomers for which the crack growth can be considered as being a balance between strain energy in the material and the tearing energy of the material. This case is particularly easy to analyse because the material behaves in a reasonably (non-linear) elastic manner and the energy losses are confined to regions in the vicinity of the crack tip. The situation can best be analysed by taking a specific example such as the pure shear test piece of Rivlin and Thomas which is shown in Figure 23.11. It is possible to calculate for this type of specimen the "energy release rate" $-\mathrm{d}U/\mathrm{d}A$ which is the amount of energy released per unit increase in crack area A.

When the specimen is deformed, the strain energy will not be the same in the different zones of the specimen and so they must be considered separately. Region A is unstressed and so the strain energy U_{A} will be zero. In region B around the crack tip, the stress field is complex and the strain energy is difficult to determine. Region C will be in pure shear and so the energy in this region will be given by $U_{\mathrm{C}}=wV_{\mathrm{C}}$ where w is the energy density (i.e. strain energy per unit volume) of the material in pure shear at the relevant strain and V_C is the volume of region C. Region D will not be in pure shear because of its proximity to the ends of the specimen and so U_{D} will be difficult to calculate. The strain energy in the different regions can be summarized as follows

$$U_{\mathrm{A}}=0, \quad U_{\mathrm{B}}=?, \quad U_{\mathrm{C}}=wV_{\mathrm{C}}, \quad U_{\mathrm{D}}=?$$

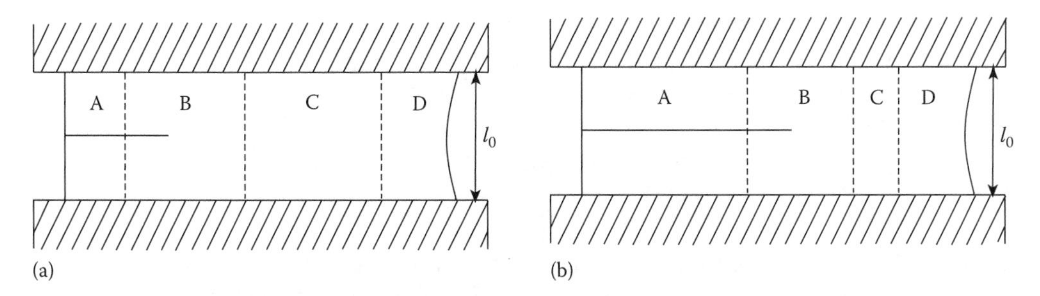

FIGURE 23.11 Pure shear test piece used in the study of the tearing of elastomers: (a) specimen before crack propagation and (b) after crack propagation. The parameters and labels are defined in the text. (Adapted from Andrews, E.H., in *Polymer Science*, Jenkins (ed.), North-Holland, Amsterdam, the Netherlands, 1972.)

The lack of knowledge of U_B and U_D can be overcome by calculating only the *change* in energy when crack propagation occurs. If the specimen grips are held at constant displacement and the crack propagates a distance Δa then the net result is to increase region A at the expense of region C and so, since region A is unstrained, the change in specimen strain energy ΔU is

$$\Delta U = -w\Delta V_C \tag{23.22}$$

where ΔV_C is the reduction in the volume of C due to crack growth. The size of regions B and D are unchanged and so the fact that their strain energies are not known does not matter. Holding the grips at constant displacement ensures that no work is done by or on the specimen and so the overall change in stored energy is given by Equation 23.22. Since elastomers shear at approximately constant volume, ΔV_C can be determined from either the unstrained or strained specimen dimensions and is

$$\Delta V_C = l_0 t_0 \Delta a \tag{23.23}$$

where l_0 and t_0 are the initial specimen length and thickness. The energy change then is given by

$$\Delta U = -w l_0 t_0 \Delta a \tag{23.24}$$

and so

$$\frac{-\partial U}{\partial a} = w l_0 t_0 \tag{23.25}$$

Since the crack has two surfaces then the change in crack area ΔA is given by

$$\Delta A = 2t_0 \Delta a$$

and so

$$\frac{-\partial U}{\partial A} = \frac{1}{2} w l_0 \tag{23.26}$$

The Griffith criterion for brittle solids assumes that crack propagation occurs when the release of elastic strain energy during an increment of crack growth is greater than the corresponding increase in surface energy due to the creation of new surface. This can be expressed in a similar way for the case of an elastomer

$$\frac{-\partial U}{\partial A} \geq \mathfrak{I} \tag{23.27}$$

where $\mathfrak{I}$ is the *tearing energy* or the characteristic energy for unit area of crack propagation in the material (analogous to γ in the Griffith theory or $G_c/2$ in linear elastic fracture mechanics). In terms of a critical stored energy density w_c the criterion for crack growth in the pure shear test piece is

$$w_c = \frac{2\mathfrak{I}}{l_0} \tag{23.28}$$

It is possible to determine w_c from the state of strain in the sample and the stress/strain characteristics of the material.

The quantity $-\partial U/\partial A$ can be determined for testing geometries other than the pure shear specimen. The values of $-\partial U/\partial A$ for a variety of other specimens commonly used to study the tearing of elastomers are given with schematic diagrams of the specimens in Figure 23.12. It is found that cracks will propagate at a given crack velocity and temperature at the same value of $-\partial U/\partial A$ in all of these types of specimen. This is strong evidence that $\mathfrak{I}$ can be thought of as a true material property. It also is found that $\mathfrak{I}$ varies with crack velocity and temperature for elastomers in a similar way that K_c depends upon these variables for glassy polymers such as poly(methyl methacrylate) (Figure 23.9). This is summarized conveniently in the three-dimensional plot in Figure 23.13 for the tearing of a vulcanized styrene–butadiene rubber (SBR). The value of $\mathfrak{I}$ increases with increasing crack velocity and decreasing temperature. This variation with rate and temperature can be represented in the form of a single WLF plot (Section 20.7) that enables the value of $\mathfrak{I}$ to be determined for any velocity or temperature. At any given temperature, the variation of $\mathfrak{I}$ with crack velocity can be represented by an equation of the form

$$\mathfrak{I} = B\left(\frac{\mathrm{d}a}{\mathrm{d}t}\right)^n \tag{23.29}$$

where

the exponent n in this case has a value of about 0.25 for the SBR used
B is a temperature-dependent constant.

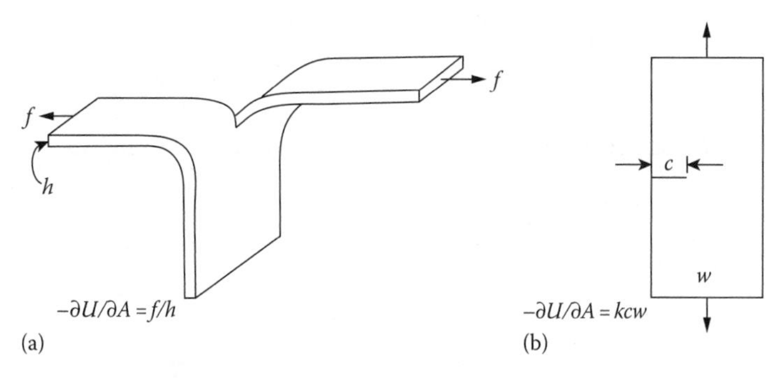

FIGURE 23.12 (a) Trouser and (b) edge-crack test pieces used to study the tearing of elastomers. Expressions for $-\partial U/\partial A$ are given in each case. The parameters are defined in the text. (Adapted from Andrews, E.H., in *Polymer Science*, Jenkins (ed.), North-Holland, Amsterdam, the Netherlands, 1972.)

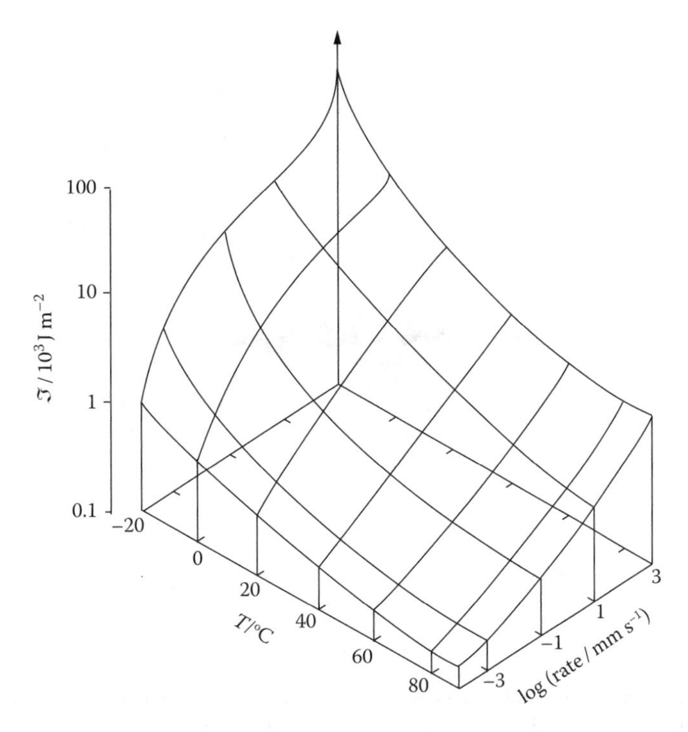

FIGURE 23.13 Dependence of the tearing energy $\mathfrak{I}$ upon rate and temperature for vulcanized styrene–butadiene rubber. (Data taken from Greensmith, H.W. and Thomas, A.G., *J. Polym. Sci.*, 18, 189, 1955.)

This equation can be used to predict the time-dependent failure of the SBR in the same way that Equation 23.18 is used with poly(methyl methacrylate).

The values of $\mathfrak{I}$ for elastomers are typically between 10^3 and 10^5 J m^{-2} which is very much higher than the energies required to break bonds. The extra energy is not used up in crazing or plastic deformation but is dissipated when material in the vicinity of the crack tip is deformed and relaxed viscoelastically as the crack propagates. If the cracks are propagated very slowly at high temperatures, these losses can be reduced and the values of $\mathfrak{I}$ become closer to those expected for bond breaking. The behaviour shown in Figure 23.13 is encountered only if the elastomer behaves as an entangled network is unfilled and does not crystallize. In such more complex materials crack propagation tends to be unstable with the crack moving in a series of jumps and the plots of $\mathfrak{I}$ against crack velocity and temperature tend to be more complicated than that in Figure 23.13. This difference in behaviour is due to the presence of filler particles and crystallization modifying the viscoelastic properties of the elastomer.

23.3.4 Ductile Fracture

It has been demonstrated that it is possible to apply extensions and developments of the Griffith fracture approach to brittle polymers when any yielding is localised to the vicinity of the crack tip, and to elastomers, where large elastic deformations are encountered without yielding. One of the main challenges in fracture mechanics has been the ability to model and analyse the behaviour of ductile polymers where there is a large plastic zone in the vicinity of the crack tip, as shown schematically in Figure 23.14. This problem has recently been solved using the concept of the *essential work of fracture*. In this approach, it is proposed that when a ductile material is deformed, the fracture process and plastic deformation take place in two different regions of the specimen, namely, the inner process zone and the outer plastic zone. As a crack propagates through a ductile material, much of the fracture work dissipated in growing the plastic zone is not associated with the fracture process.

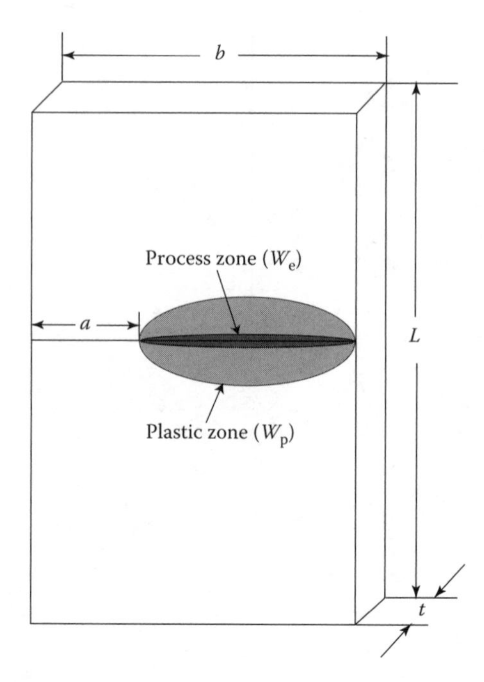

FIGURE 23.14 Schematic illustration of a ductile fracture specimen. The parameters are defined in the text. (Adapted from Wu, J. and Mai, Y.W., *Polym. Sci. Eng.*, 36, 2275, 1996.)

This work involved in plastic deformation will be a function of the specimen geometry (e.g. specimen width and thickness) and not a material constant. It is only the work that goes into the fracture process zone that is a material constant and a parameter of interest.

The total work of fracture W_f can be partitioned into two separate components, the essential work of fracture W_e and the non-essential work of fracture W_p. Hence it can be written that

$$W_f = W_e + W_p \tag{23.30}$$

The parameter W_e can be thought of as a fracture energy term and will be proportional to the area of material being fractured, i.e. the ligament length l ($=b-a$, Figure 23.14) multiplied by the specimen thickness t. The W_p term, on the other hand, is a volume energy and so will be proportional to l^2 and t. Hence Equation 23.30 can be rewritten as

$$W_f = w_e tl + \beta w_p tl^2 \tag{23.31}$$

and so the specific (i.e. per unit specimen area) work of fracture w_f is given by

$$w_f = \left(\frac{W_f}{tl}\right) = w_e + \beta w_p l \tag{23.32}$$

where

w_e is the specific essential work of fracture
w_p is the specific non-essential work of fracture (or the specific plastic work).

The parameter β is a geometrical factor related to the shape of the plastic zone.

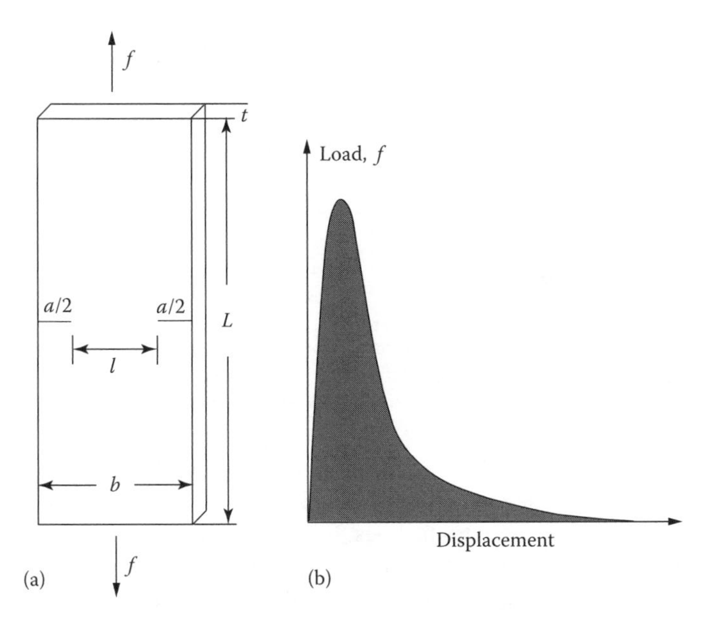

FIGURE 23.15 (a) Double-edge-notched specimen geometry. (b) Schematic load–displacement curve showing how the total fracture energy is determined. The parameters are defined in the text. (Adapted from Wu, J. and Mai, Y.W., *Polym. Sci. Eng.*, 36, 2275, 1996.)

Examination of Equation 23.32 shows that if w_e is a material constant, then measuring w_f as a function of l should yield a straight line plot, with w_e as the intercept as $l \rightarrow 0$. Figure 23.15 shows how this can be done for a ductile polymer. A typical specimen that can be employed is the double-edge-notched tensile specimen (Figure 23.15a) where two sharp notches are cut in the edges of the specimen to give a ligament of length l. The specimen then is loaded to failure in a tensile testing machine and the load plotted as a function of cross-head displacement. The total work of fracture W_f is given by the area under the load-displacement curve as shown schematically in Figure 23.15b. This measurement is repeated for a series of similar specimens with different ligament lengths, and W_f determined in each case. An example of this analysis is shown in Figure 23.16 for a tough

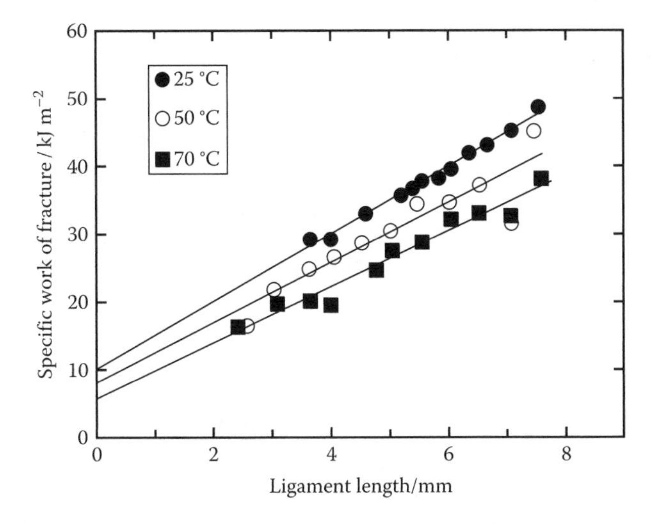

FIGURE 23.16 Plot of specific work of fracture against ligament length determined at the different indicated temperatures for a toughened polymer blend. (Data taken from Wu, J. and Mai, Y.W., *Polym. Sci. Eng.*, 36, 2275, 1996.)

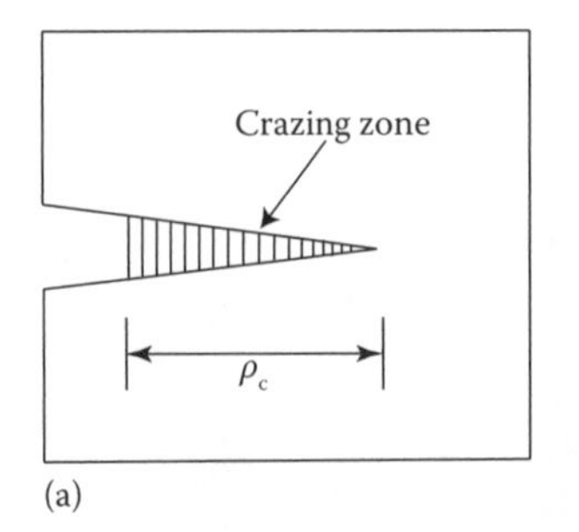

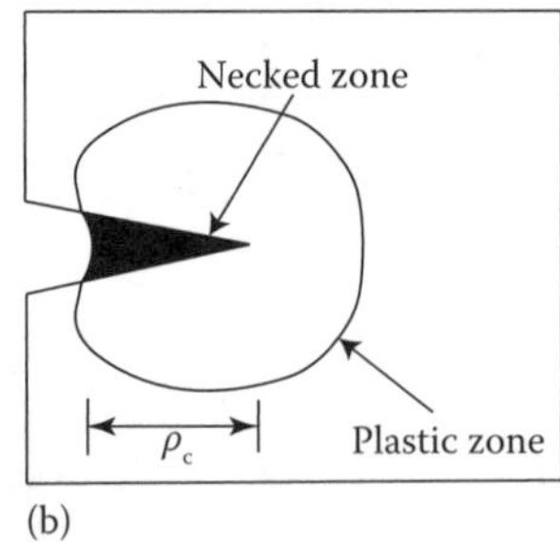

FIGURE 23.17 Schematic diagrams of the fracture process zones in (a) a glassy polymer and (b) a ductile polymer. The parameter ρ_c represents the lengths of (a) the crazing zone in a glassy polymer and (b) the necked zone in a ductile polymer. (Adapted from Wu, J. and Mai, Y.W., *Polym. Sci. Eng.*, 36, 2275, 1996.)

poly(butylene terephthalate)/polycarbonate impact-modified blend fractured at different temperatures. The specific work of fracture w_f ($=W_f/tl$) is plotted against ligament length l from which it can be seen that the data points fall close to the lines for each temperature. The intercept, which is the specific work of fracture ($=w_e$), decreases with increasing temperature.

Finally, it is worth considering the significance of the essential work of fracture of a ductile polymer in relation to the brittle fracture process in a glassy polymer. The process zones in the vicinity of the crack tip in each case are shown schematically in Figure 23.17. In the case of the glassy polymer, there is a localised crazing zone in the vicinity of the crack tip but no overall crazing or yielding away from the crack tip. In contrast, there is a local necked fracture zone for the ductile polymer and a large plastic zone around the crack. The essential work of fracture approach allows the fracture energy of the local necked zone to be determined and the energy expended in growing the plastic zone to be eliminated. Hence w_e for a ductile polymer can be considered to be the counterpart of the fracture energy G_c of a brittle polymer.

23.4 FRACTURE PHENOMENA

The phenomenon of fracture manifests itself in a number of different forms depending upon the material and the conditions of testing, such as rate and temperature. These aspects will be considered now.

23.4.1 Ductile–Brittle Transitions

Many materials are ductile under certain testing conditions, but when these conditions are changed such as, for example, by reducing the temperature they become brittle. This type of behaviour also is encountered in steels and is very common with polymers. Figure 23.18a shows how the strength of poly(methyl methacrylate) varies with testing temperature. At low temperatures, the material fails in a brittle manner, but when the testing temperature is raised to just above room temperature, the polymer undergoes general yielding and brittle fracture is suppressed. This type of ductile–brittle transition has been explained in terms of brittle fracture and plastic deformation being independent processes which have a different dependence on temperature, as shown schematically in Figure 23.18b. Since the yield stress σ_y increases more rapidly as the temperature is reduced, at a critical temperature it becomes higher than the stress σ_b required to cause brittle fracture. The temperature at which this occurs corresponds to that of the *ductile–brittle transition* T_b, and it is envisaged that the process which occurs during deformation is that which can take place at the lowest stress, i.e. brittle fracture at $T<T_b$ and general yielding at $T>T_b$.

In comparison with metals, polymers have the added complication of there being a strong dependence of the mechanical properties upon strain rate as well as upon temperature. For example,

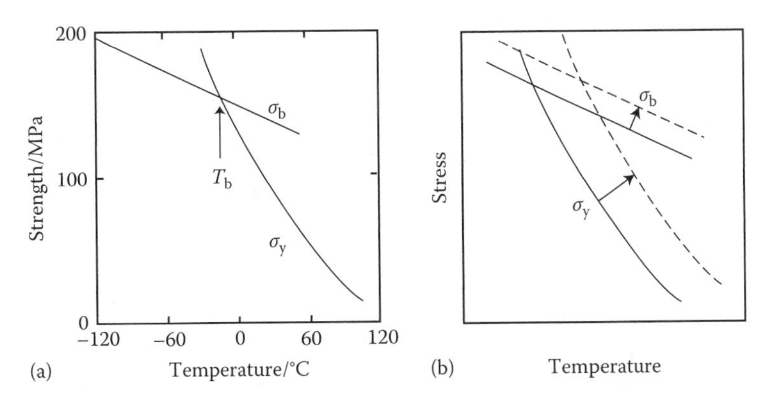

FIGURE 23.18 Ductile-to-brittle transitions. (a) Variation with temperature of brittle strength σ_b measured in flexure and yield strength σ_y measured in tension for poly(methyl methacrylate). (b) Effect of strain-rate upon the transition, where curves for a higher strain-rate are drawn as dashed lines. (Adapted from Ward, I.M., *Mechanical Properties of Solid Polymers*, 2nd edn., Wiley-Interscience, New York, 1983.)

nylons can be cold-drawn at room temperature when relatively low strain-rates are used, but become brittle as the strain-rate is increased. It is thought that this is because the curves of yield stress and brittle stress against temperature are moved to the right by increasing the strain-rate and so the temperature of the ductile–brittle transition is increased (Figure 23.18b).

It is known that the temperature at which the ductile–brittle transition occurs for a particular polymer is sensitive to both the structure of the polymer and the presence of surface flaws and notches. The temperature of the transition generally is raised by crosslinking and increasing the crystallinity. Both of these factors tend to increase the yield stress without affecting the brittle stress significantly. The addition of plasticizers to a polymer has the effect of reducing yield stress and so lowers T_b. Plasticization is used widely in order to toughen polymers such as poly(vinyl chloride), which would otherwise have a tendency to be brittle at room temperature.

The presence of surface scratches, cracks or notches can have the effect of both reducing the tensile strength of the material and also increasing the temperature of the ductile–brittle transition. In this way, the presence of a notch can cause a material to fail in a brittle manner at a temperature at which it would otherwise be ductile, a phenomenon known as *notch brittleness*. As one might expect, this phenomenon can play havoc with any design calculations which assume the material to be ductile. This type of *notch sensitivity* normally is explained by examining the state of stress that exists at the root of the notch. When an overall uniaxial tensile stress is applied to a body there will be a complex state of triaxial stress at the tip of a crack or notch. This can cause a large increase (up to three times) in the yield stress of the material at the crack tip, as shown in Figure 23.19. The brittle fracture stress will be unaffected and so the overall effect is to raise T_b. If the material is used between T_b for the bulk and T_b for the notched specimen it will be prone to suffer from notch brittleness.

It is tempting to relate the ductile–brittle transition to either the glass transition or secondary transitions (Section 20.6) occurring within the polymer. Indeed for some polymers, such as natural rubber or polystyrene, T_b and T_g occur at approximately the same temperature, but many other polymers are ductile below the glass transition temperature. When $T_b < T_g$ it sometimes is possible to relate T_b to the occurrence of secondary lower-temperature relaxations. However, more extensive investigations have shown that there is no general correlation between the brittle–ductile transition and molecular relaxations. This is not entirely unexpected since these relaxations are detected at low strains, whereas T_b is measured at high strains and depends upon factors such as the presence of notches which do not affect molecular relaxations.

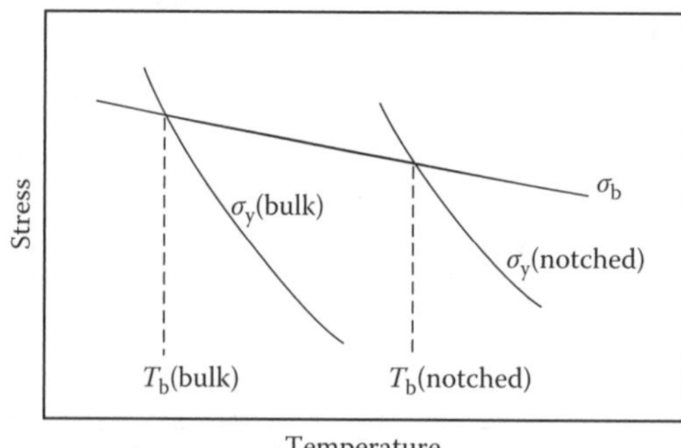

FIGURE 23.19 Schematic representation of the effect of notching upon the ductile-to-brittle transition. (Adapted from Andrews, E.H., *Fracture in Polymers*, Oliver and Boyd, London, U.K., 1968.)

23.4.2 Impact

It was explained in the previous section that brittle behaviour could be induced in a polymer through the presence of a notch or by testing at high rates. These effects can be evaluated though the use of impact testing. The exact behaviour of a polymer subjected to impact loading depends upon the geometry of the test and so it is essential to be aware of the different impact tests that have been developed. Two of the most widely used and easily analysed test methods are the *Charpy* and *Izod* tests shown schematically in Figures 23.20a and b respectively. In both cases, the specimen usually is notched and held against rigid supports as shown in the diagrams. The specimen is then struck by a pendulum hammer at the bottom of its arc of swing, the pendulum then continuing to a measured maximum height after fracturing the specimen. Since the pendulum is released from a fixed height, the energy absorbed in breaking the specimen can be deduced from the difference in release height and maximum height at the end of its first oscillation. Other methods of impact testing may be employed for specimens that are too thin or flexible. In such cases, a *falling weight* method often is employed on a circular polymer disc clamped around its periphery. A weight, often in the form of a dart, is then dropped on the specimen that may or may not break. The behaviour is analysed in terms of the weight or height of the dart needed to break the polymer specimen, i.e. the conversion of the potential energy of the dart into kinetic energy on impact.

It is possible to employ linear elastic fracture mechanics (Section 23.3.2) to analyse the Charpy and Izod impact tests if the specimens have a sharp notch and if they exhibit bulk linear elastic

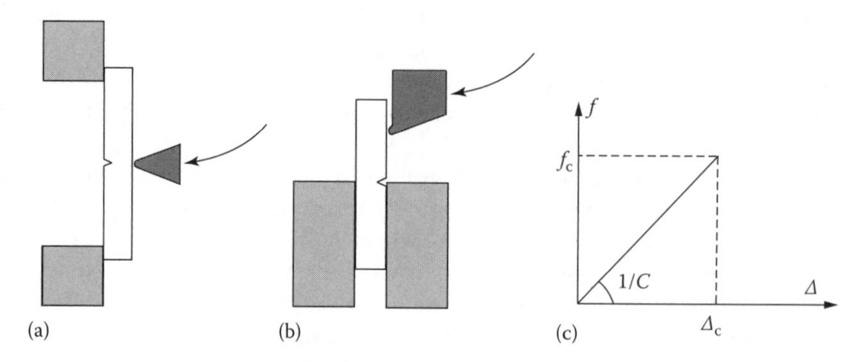

FIGURE 23.20 Geometry of testing and support arrangements for (a) the Charpy impact test and (b) the Izod impact test. It is normal to use single-edge-notched specimens, as shown in the diagrams. (c) The variation of load f with displacement Δ for linear elastic fracture behaviour, showing the critical values (subscript c) at failure. The parameter C is the specimen compliance.

behaviour. It can be assumed that the elastic energy absorbed as strain energy U_S by the specimen is given by

$$U_S = U_L - U_T \tag{23.33}$$

where

U_L is the energy lost by the pendulum

U_T is the kinetic (or tossing) energy lost when the specimen breaks.

The value of U_T normally is determined by carefully rejoining the broken pieces of the specimen in the impact tester and repeating the test without breaking them. The value of U_L can be determined from the force–displacement curve of the test shown schematically in Figure 23.20c. The elastic strain energy absorbed by the specimen in this case is simply the area under the triangle given by

$$U_S = \frac{1}{2} f_c \Delta_c \tag{23.34}$$

which can be expressed in terms of the compliance C of the specimen (Figure 23.20c) as

$$U_S = \frac{1}{2} f_c^2 C \tag{23.35}$$

It was shown in Section 23.3.2.1 that the fracture energy G_c can be expressed in terms of the maximum load and the change of specimen compliance with crack length ($\partial C/\partial a$) through Equation 23.17 and so substituting this equation into Equation 23.35 gives

$$U_S = G_c t \frac{C}{\partial C/\partial a} \tag{23.36}$$

If the specimen width is defined as b then

$$U_S = G_c tbZ \tag{23.37}$$

where Z is a dimensionless geometry factor given by

$$Z = \frac{C}{\partial C/\partial (a/b)} \tag{23.38}$$

In this way, the parameter Z can be considered to be a calibration factor for the particular impact test geometry used and its value has been tabulated for various Charpy and Izod specimen sizes. Figure 23.21 shows a typical plot of U_S versus tbZ for Charpy and Izod tests conducted upon a series of specimens of medium density polyethylene containing sharp notches. A linear relationship is found that gives a value of G_c from the slope of the line of around 8.1 kJ m^{-2}, independent of specimen geometry and test method.

23.4.3 Fatigue

Many materials suffer failure under cyclic loading at stresses well below those they can sustain during static loading. This phenomenon is known as *fatigue* and the development of fatigue cracks

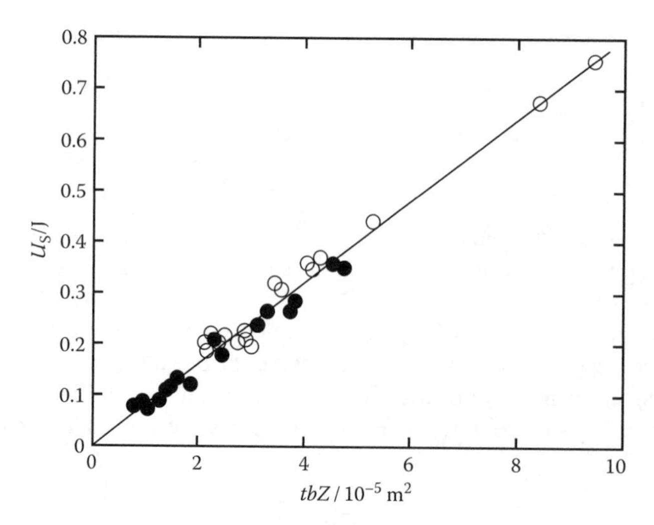

FIGURE 23.21 Impact data for medium density polyethylene (• = Charpy test, ○ = Izod test). The elastic strain energy U_S absorbed by the specimen is plotted against the parameter tbZ according to Equation 23.37. (Data taken from Plati, E. and Williams, J.G., *Polym. Eng. Sci.*, 15, 470, 1975.)

in the metallic components of aircraft and other structures has led to disastrous consequences. Polymers also can undergo fatigue fracture. Figure 23.22 gives the dependence of the failure stress of a variety of polymers upon the number N of cycles during fatigue loading. The polymers had been cycled at a frequency of 0.5 Hz between zero stress and the stress indicated. The curves have a characteristic sigmoidal shape and tend to flatten out when N becomes sufficiently large, suggesting a fatigue limit. This is stress at which the material can be cycled without causing failure. Similar so-called S–N curves have been found for metals, but one important difference with polymers is the strong dependence of the fatigue behaviour upon the testing frequency. There often is a large rise in specimen temperature, especially at high stresses and high testing frequencies, that causes failure to take place through thermal softening. The temperature rise is due to the energy dissipated by the viscoelastic energy loss processes not being conducted away sufficiently rapidly because of the low thermal conductivity of the polymer.

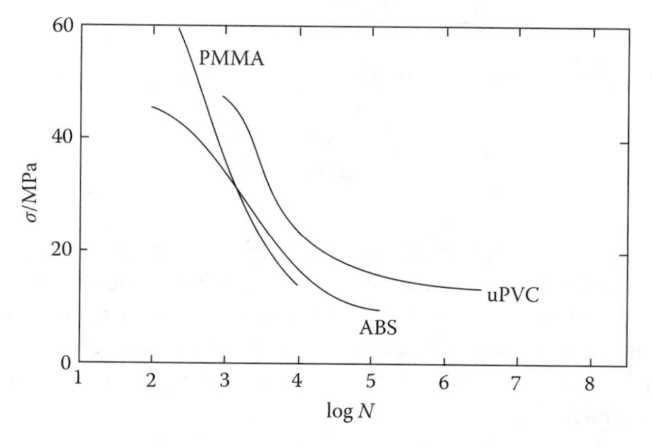

FIGURE 23.22 Dependence of failure stress of a variety of engineering polymers upon the number of cycles accumulated during a dynamic fatigue test at 20°C. The polymers used were poly(methyl methacrylate) (PMMA), unplasticized poly(vinyl chloride) (uPVC) and an acrylonitrile–butadiene–styrene copolymer (ABS). (Data taken from Bucknall, C.B. et al., in *Polymer Science*, Jenkins (ed.), North-Holland, Amsterdam, the Netherlands, 1972.)

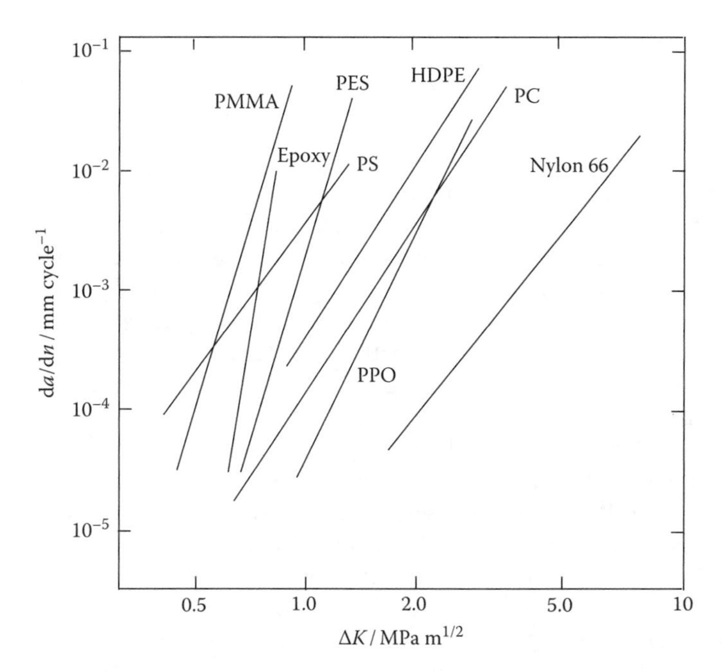

FIGURE 23.23 Fatigue crack propagation behaviour of different amorphous and semi-crystalline polymers (PMMA = poly(methyl methacrylate); PES = polyethersulphone; PS = polystyrene; HDPE = high-density polyethylene; PC = polycarbonate; PPO = poly(2,6-dimethyl-1,4-phenylene oxide)). The fatigue crack growth rate da/dn is plotted against ΔK for each material. (Data taken from Hertzberg, R.W. and Manson, J.A., *Fatigue of Engineering Plastics*, Academic Press, New York, 1980; Hertzberg, R.W. et al., *Polym. Eng. Sci.*, 15, 25, 1975.)

Fatigue failure occurs through a process of initiation and propagation of cracks. In the propagation phase the fatigue crack grows by a small amount during each cycle. This behaviour can be monitored by measuring the amount of crack growth per cycle da/dN as a function of the applied stress. It is found that for many brittle polymers the propagation rate is related to the range of stress-intensity factor ΔK by an equation of the form

$$\frac{da}{dN} = C(\Delta K)^m \tag{23.39}$$

where the exponent m and C is a temperature-dependent constants for the particular polymer. This type of behaviour is shown in Figure 23.23 and similar relationships also are found for many metals. The exact form of the fatigue crack propagation behaviour for a particular polymer is found to depend upon the testing conditions, the mean value of K used and material variables. In general, tougher polymers are more fatigue resistant and have low values of m of the order of 4, whereas more brittle ones have higher values of m and show steeper slopes in the plot in Figure 23.23.

Analogous behaviour also is found for non-crystallizing elastomers where the crack growth rate can be related to $\mathfrak{I}$ through the equation

$$\frac{da}{dN} = D\mathfrak{I}^n \tag{23.40}$$

where D and n are constants which depend on the type of material and testing conditions used.

23.4.4 Environmental Fracture

The Achilles heel of many polymers is their tendency to fail at relatively low stress levels through the action of certain hostile environments. This type of failure falls into several different categories.

1. One of the most widely studied forms of environmental failure is that of vulcanized natural rubber in the presence of *ozone*. The ozone reacts at the surface of unsaturated hydrocarbon elastomers to form ozonides that lead to chain scission and even when the materials are subjected to relatively low stresses, cracks can nucleate and grow catastrophically through the material. It is found that ozone cracking takes place above a critical value of tearing energy $\mathfrak{J}_0$ which is of the order of 0.1 J m^{-2}. This is close to the true surface energy of the polymer and well below the tearing energies of elastomers in the absence of a hostile environment.
2. Another type of environmental failure that occurs through surface interactions is the failure of polyolefins such as high density polyethylene, in the presence of *surfactants* such as certain detergents. It is found that when this material is held under constant stress in these environments its behaviour changes from ductile failure at high stresses over short time periods to brittle fracture at low stresses after longer periods of time.
3. Glassy polymers, in particular, are prone to failure through the action of certain organic liquids which act as *crazing* and *cracking* agents. Failure can occur at stresses and strains well below those required in the absence of the agents. The detailed behaviour is rather complex and it is difficult to generalize from one system to another. It is thought that in most cases the failure occurs through the initial formation of crazes, although the crazes may not be particularly stable and may transform directly into cracks. Two particular mechanisms have been proposed to account for the action of crazing and cracking agents in glassy polymers and in many cases it is thought that both mechanisms may apply. One suggestion is that the presence of the liquid lowers the surface energy of the polymer and makes the formation of new surface during crazing easier. Another is that the organic liquid swells the polymer and lowers its T_g allowing deformation and crazing to take place at lower stresses and strains.

The effect of certain liquids with different solubility parameters δ upon the critical strain for crazing and cracking poly(2,6-dimethyl-1,4-phenylene oxide) is shown in Figure 23.24. In certain cases, the crazes are not stable and they transform directly into cracks. The minimum in the curve corresponds to liquids which have the highest miscibilities in the polymer, when the values of δ are matched for the polymer and solvent (Section 10.2.5) and the polymer is essentially swollen by the solvent. It is clear, therefore, that the crazing and cracking in this particular system is due principally to a reduction in the T_g by solvent plasticization. This is not, however, the whole story since in the liquids with high or low values of δ, that produce no swelling at all, the polymer fails by crazing, whereas when tested at constant strain-rate in air it undergoes cold-drawing with very little crazing. It is thought that these liquids tend to reduce the surface energy and make the formation of voids during crazing easier.

23.5 TOUGHENED POLYMERS

Glassy polymers have properties that are adequate for many applications but they are prone to brittle fracture particularly when notched or subjected to high strain-rate impact loading (Section 23.4.2). Because of such problems, toughened polymers have been developed in which a glassy polymer matrix is toughened through the incorporation of a second phase consisting of particles which usually are spherical and of a rubbery polymer above its glass transition temperature, T_g. This can lead to significant improvements in the mechanical behaviour of the matrix polymer and

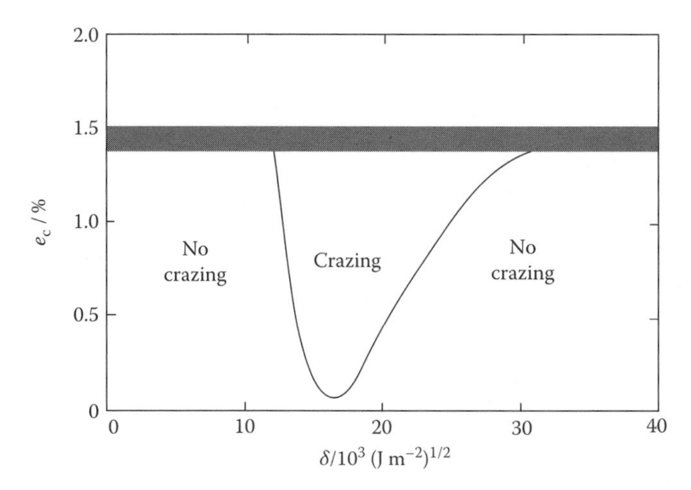

FIGURE 23.24 Dependence of critical strain e_c for crazing upon solubility parameter δ for poly(2,6-dimethyl-1,4-phenylene oxide) (PPO) in various liquids. The shaded area gives e_c for the polymer in air. (Data taken from Kambour, R.P. and Robertson, R.E., in *Polymer Science*, Jenkins (ed.), North-Holland, Amsterdam, the Netherlands, 1972.)

is often referred to as *rubber toughening*. There are two general ways of adding the second phase to form multi-component systems. The first is through blending an incompatible rubbery polymer with the matrix and the second is through copolymerization (Sections 18.2 and 18.3). Block copolymers sometimes are employed in which one part of the molecule provides the glassy matrix and the other part is an incompatible rubber which produces the dispersed phase. The covalent bonding between the blocks ensures a good interface between the two phases.

Some of the different geometries of rubber particles that have been employed are shown in Figure 23.25. As well as using simple solid rubber particles, nowadays particles with more complex morphologies often are employed. The presence of a low-modulus rubber in a polymer also has the effect of reducing the overall Young's modulus of the material which sometimes is a problem. The advantage of having a glassy core or glassy inclusions within the rubber particle is that good toughening can be obtained without reducing the modulus of the polymer significantly. A strong interface between the rubbery and glassy phases usually is required for the material to have good mechanical properties and this can be facilitated through copolymerization between the different components in the particles, and sometimes also with the matrix.

23.5.1 Mechanical Behaviour of Rubber-Toughened Polymers

The incorporation of rubber particles into a brittle polymer has a profound effect upon the mechanical properties as shown in the stress–strain curves in Figure 23.26. This can be seen in

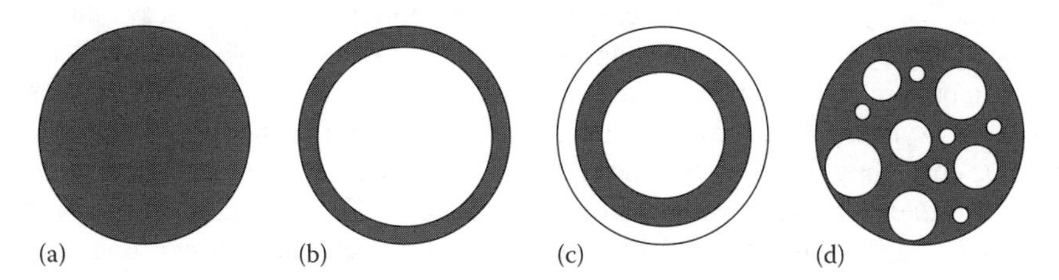

FIGURE 23.25 Schematic diagram of the different forms of rubber particles employed to toughen brittle polymers: (a) solid rubber particle; (b) two-layer core-shell particle with a glassy core; (c) three-layer core-shell particle with a glassy outer layer, rubber interlayer and glassy core; and (d) 'salami-type' rubber particle with spherical glassy inclusions. The rubber phase is shaded dark in each case.

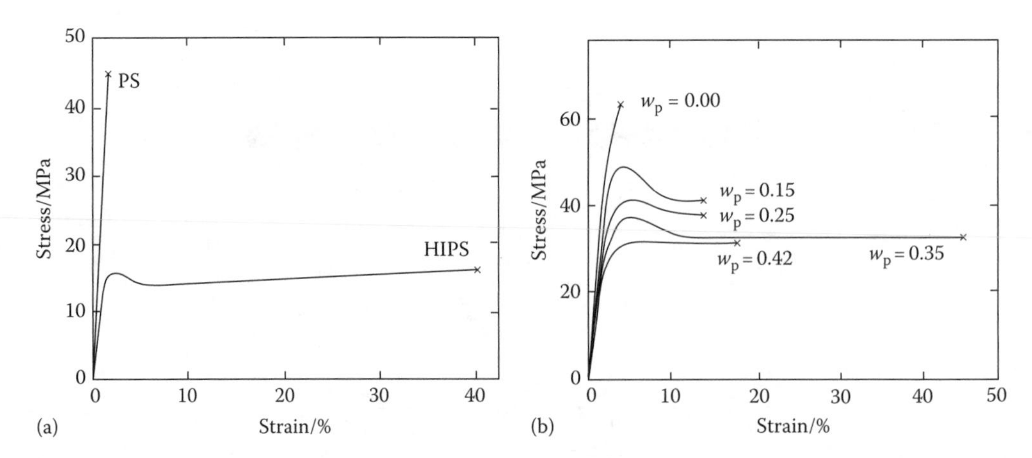

FIGURE 23.26 Stress–strain curves for rubber-toughened polymers. (a) Polystyrene (PS) and high-impact polystyrene (HIPS). (Data taken from Bucknall, C.B., *Toughened Plastics*, Applied Science, London, U.K., 1977.) (b) Rubber-toughened poly(methyl methacrylate) showing the effect of the weight fraction w_p of rubber particles. (Data of D.E.J. Saunders, P.A. Lovell, and R.J. Young, University of Manchester, Manchester, U.K.)

Figure 23.26a for high-impact polystyrene (HIPS) which usually is a blend of polystyrene and polybutadiene, the interfaces in which are stabilized by graft copolymer. The stress–strain curve for polystyrene shows brittle behaviour, whereas the inclusion of the rubbery phase causes the material to undergo yield and the sample to deform plastically to about 40% strain before eventually fracturing. The plastic deformation is accompanied by *stress-whitening*, whereby the necked region becomes white in appearance during deformation. As will be explained later, this is due to the formation of a large number of crazes around the rubber particles in the material.

Figure 23.26b shows a series of stress–strain curves for samples of rubber-toughened poly(methyl methacrylate) (RTPMMA) containing different weight fractions w_p of toughening particles. It can be seen that PMMA is relatively brittle but that following the addition of rubbery particles the material is able to undergo yield and deform plastically. A maximum elongation of over 45% is obtained for $w_p = 0.35$. It is found that, as with HIPS, stress-whitening is obtained following yield, although in the case of RTPMMA this is thought to be due to the particles undergoing voiding rather than due to crazing taking place. The reduction in elongation for $w_p > 0.35$ is because at high values of w_p the particles become so close that they touch and interact with each other and thereby reduce the efficiency of toughening.

As well as toughening thermoplastics such as polystyrene and PMMA, it is also possible to toughen brittle network polymers such as cured epoxy resins (Section 3.3.1.2) by the addition of a rubbery second phase. Figure 23.27 shows the effect of the volume fraction V_p of rubber particles upon G_c, the fracture energy for the rubber-toughened epoxy resin. The samples were prepared by the addition of 8.7% by weight of a carboxyl-terminated poly(butadiene-*co*-acrylonitrile) (CTBN) rubber to the epoxy prepolymer and curing agent. The rubber is soluble in and reactive towards the epoxy resin but precipitates in the form of particles as the resin increases in molar mass during curing. The carboxylic acid end-groups of the rubber react with the epoxy resin and so the rubber chains become covalently linked to the epoxy matrix, ensuring a strong interface between the rubber particles and the matrix. The variation of V_p in Figure 23.27 was obtained by varying the cure conditions employed. Although the weight of rubber used in each case was constant, it can be seen from Figure 23.27 that it is the volume fraction of rubber precipitated in the form of particles that controls the toughness of the material. Table 23.3 gives typical values of K_c and G_c for three different rubber-toughened polymer systems.

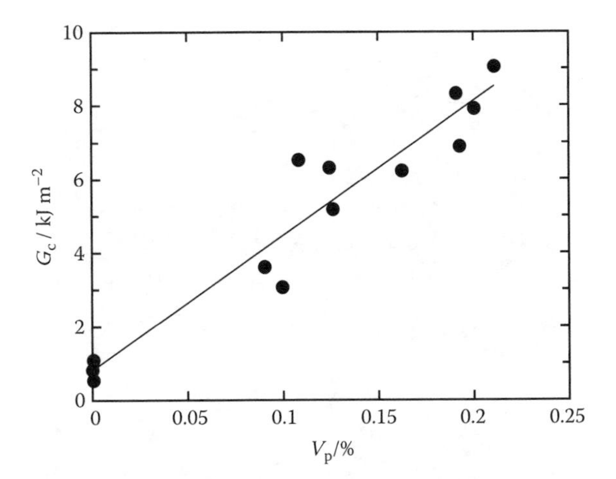

FIGURE 23.27 Dependence of G_c upon rubber phase volume fraction for a rubber-toughened epoxy resin. (Data taken from Bucknall, C.B. and Yoshi, T., *Brit. Polym. J.*, 10, 53, 1978.)

TABLE 23.3
Typical Values of Young's Modulus *E*, Fracture Toughness K_c and Fracture Energy G_c for Different Polymers

Polymer	*E*/GPa	K_c/MPa m$^{1/2}$	G_c/kJ m^{-2}
Polystyrene	3.0	1.1	0.4
HIPS	2.1	5.8	16
Poly(methyl methacrylate)	3.0	1.2	0.5
RTPMMA	2.1	2.4	2.6
Epoxy resin	2.8	0.5	0.1
Rubber-toughened epoxy resin	2.4	2.2	2.0
Silica-filled epoxy resin	7.5	1.4	0.3

23.5.2 Mechanisms of Rubber-Toughening

Both crazing and shear yielding involve the adsorption of energy and most methods of toughening brittle polymers involve modifying the polymer such that controlled high levels of crazing or shear yielding are able to take place. The second-phase spherical particles of the rubber have a Young's modulus about 3 orders of magnitude lower than that of the glassy matrix (Table 19.1). This leads to a stress concentration at the equators of the particles during mechanical deformation which is similar to the stress concentration found around holes and notches in plates (Section 23.3.1). The presence of the stress concentration can lead to shear yielding or crazing around every particle and hence throughout a large volume of material rather than just at the crack tip. Hence, the polymer adsorbs a large amount of energy during deformation and is toughened.

23.5.2.1 Transmission Electron Microscopy

The exact mechanisms of deformation around the rubber particles depend upon the type of polymer and the conditions of testing (rate and temperature). The morphology of rubber-toughened polymers generally is visualized by transmission electron microscopic (TEM) examination of thin

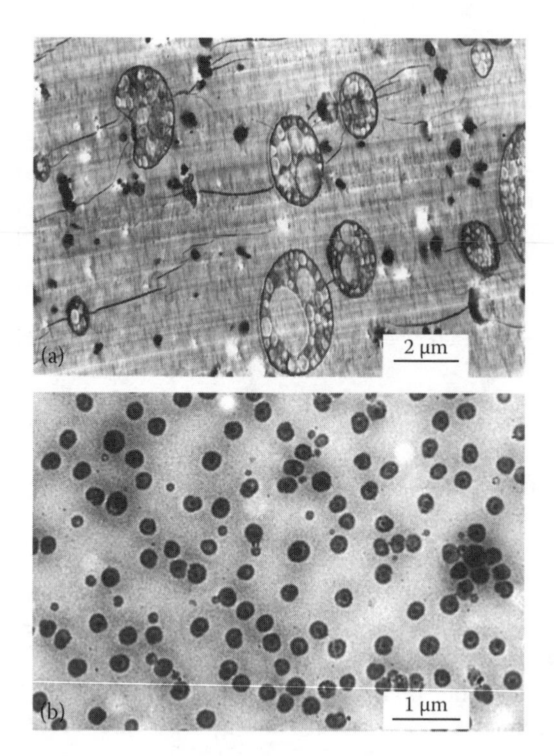

FIGURE 23.28 Transmission electron micrographs of OsO_4-stained sections of rubber-toughened polymers. (a) A sample of commercial HIPS deformed in tension (approximately vertically). The crazes around the particles can be seen clearly and the small black particles are pigment. (Courtesy of X. Hu, University of Manchester, Manchester, U.K.) (b) Section of RTPMMA containing four-layer core-shell rubber particles. (Courtesy of M.N. Sherratt, University of Manchester, Manchester, U.K.)

microtomed sections, such as those shown in Figure 23.28. It is sometimes possible to follow the deformation processes by the examination of such sections. Figure 23.28a shows an OsO_4-stained section of HIPS after tensile deformation. The OsO_4 stains the rubbery domains dark by depositing osmium in areas containing the unsaturated molecules (i.e. the polybutadiene). It can be seen that relatively large crazes emanate from the equators of the rubber particles. It should be noted that the rubber particles are not simple rubber spheres but have inclusions that have not been stained and so must be polystyrene. It has been found that this complex salami-type rubber particle morphology (Figure 23.25d) can lead to a given volume fraction of rubber V_r giving rise to a larger number of effective toughening particles than when there are no inclusions. Hence it makes the toughening process significantly more efficient.

The behaviour of HIPS can be contrasted with that of rubber-toughened poly(methyl methacrylate) (RTPMMA) shown in Figure 23.28b. In this case, a large number of smaller particles generally are employed. The particles are typically of the order of 0.2–0.3 μm in diameter and again have a complex morphology. They are made by emulsion polymerization (Section 4.4.4) and have a "core-shell" structure with up to four alternating layers of rubbery and glassy polymers. The glassy layers are essentially identical to the matrix polymer whereas the rubber is poly[(*n*-butyl acrylate)-*co*-styrene] with a composition chosen to give a refractive index identical to that of the PMMA matrix at room temperature. Since the particles also are smaller than the wavelength of light, the toughened polymer remains transparent and the excellent optical properties of the matrix PMMA are virtually unaltered. The presence of high modulus glassy material within the particles again ensures more efficient toughening than for pure rubber spheres. The micrograph of the RTPMMA

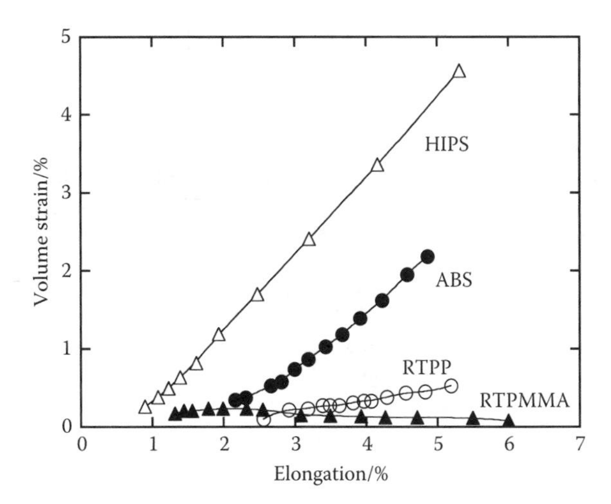

FIGURE 23.29 Volume-strain measurements as a function of axial strain for a series of different rubber-toughened polymers, HIPS, ABS (acrylonitrile–butadiene–styrene copolymer), toughened PP (polypropylene) and RTPMMA. (Data taken from Bucknall, C.B. et al., *J. Mater. Sci.*, 19, 2064, 1984.)

in Figure 23.28b shows the core-shell structure of the four-layer particles employed in this material. In this case the rubbery phase was stained with OsO_4 to give it a higher electron density than the glassy phase and hence it appears dark. It should be noted that only three layers can be visualized since the outer layer of the particles is PMMA and blends with the matrix. There is chemical graftlinking between the different layers in the particles and hence strong interfaces are obtained both between the particles and matrix and within the particles. It is thought that the particles in the RTPMMA toughen the material mainly by inducing the formation of shear bands around the particles that cannot be resolved easily by TEM, although crazing is sometimes seen in the high hydrostatic stress fields around crack tips.

23.5.2.2 Volume Change Measurements

A further insight into the deformation mechanisms that take place in rubber-toughened polymers can be obtained by measuring the change in specimen volume ΔV during deformation, as shown in Figure 23.29, for a variety of materials. It can be seen that in the case of HIPS, there is a significant increase in specimen volume following yield. Plastic deformation normally occurs at constant volume and so this large increase in volume shows that crazing takes place on a large scale during deformation of the material. The behaviour of HIPS in Figure 23.29 can be contrasted with that of RTPMMA for which there is little change in specimen volume during deformation. Since shear yielding takes place at approximately constant volume, this is further confirmation that the toughening mechanism in the tensile deformation of RTPMMA is mainly shear yielding. The volume-strain behaviour of other polymers in Figure 23.29 is between that of HIPS and RTPMMA showing that for these materials both crazing and shear yielding are probably taking place. Hence it can be seen that the deformation mechanisms in rubber-toughened polymers can be readily monitored using volume-strain measurements.

PROBLEMS

23.1 The fracture strength of an un-notched parallel-sided sheet of a brittle polymer of thickness 5 mm and breadth 25 mm is found to be 85 MPa. If the critical stress intensity factor K_c for the polymer is determined from a separate experiment on a notched specimen to be 1.25 MPa $m^{1/2}$, calculate the inherent flaw size for the polymer. (The value of K_c for a single-edge notched sheet specimen may be determined from the appropriate expression in Table 23.2.)

23.2 The data tabulated below were obtained from an experiment upon polystyrene at room temperature where the fracture stress σ_f was measured as a function of crack length, a. Determine for polystyrene: (a) G_c, (b) K_c and (c) the inherent flaw size. (Assume the Young's modulus of polystyrene is 3 GPa.)

σ_f/MPa	a/mm
17.2	10
17.5	9
17.9	8
18.4	7
19.0	6
20.0	5
22.0	4
25.0	3
31.0	2
40.0	1
41.0	0.5
42.0	0.1

23.3 It is found that the rate of growth of a crack da/dt in a non-crystallizing elastomer can be given by an equation of the form

$$\frac{da}{dt} = q\mathfrak{I}^n$$

where

$\mathfrak{I}$ is the tearing energy
q and n are constants for a given set of testing conditions

Derive an equation for the time-to-failure t_f for a sheet of elastomer held at constant extension and containing a small edge-crack of initial length a_0, given that for the sheet

$$\mathfrak{I} = Kaw$$

where

K is a constant
w is the stored energy density in the material at the strain imposed

FURTHER READING

Andrews, E.H., *Fracture in Polymers*, Oliver and Boyd, London, U.K., 1968.
Bucknall, C.B., *Toughened Plastics*, Applied Science, London, 1977.
Haward, R.N. and Young R.J. (eds.), *The Physics of Glassy Polymers*, 2nd edn., Chapman & Hall, London, U.K., 1997.
Hertzberg, R.W. and Manson, J.A., *Fatigue of Engineering Plastics*, Academic Press, New York, 1980.
Kausch, H.H., *Polymer Fracture*, Springer-Verlag, Berlin, 1988.

Kausch, H.H. and Michler, G.H., The effect of time on crazing and fracture, *Advances in Polymer Science*, **187**, 33 (2005).
Kinloch, A.J. and Young, R.J., *Fracture Behaviour of Polymers*, Applied Science, London, 1983.
Ward I.M., *Mechanical Properties of Solid Polymers*, 2nd edn., Wiley-Interscience, New York, 1983.
Ward, I.M. and Sweeney, J., *An Introduction to the Mechanical Properties of Solid Polymers*, 2nd edn., Wiley, New York, 2004.

24 Polymer Composites

24.1 INTRODUCTION TO COMPOSITE MATERIALS

The development of composites has been driven by the need for materials with a specific combination of properties beyond those obtainable from a single material. These aims can be achieved in polymers through the use of copolymers or blends, but the specific aspect that characterizes composite materials is that they are made up of distinct phases with very different physical properties. They are often, but not exclusively, found to consist of a relatively soft flexible matrix reinforced by a stiffer, often fibrous component. Sometimes, however, a softer phase is used to improve properties such as when rubber particles are added to a rigid polymer as in the case of a rubber-toughened polymer (Section 23.5). In practice, a large number of materials can be classified as being composites and nature, in particular, provides a number of celebrated examples. Wood is made up of polymeric cellulose fibres reinforcing a continuous organic lignin matrix and bone consists of collagen, a natural polymer, reinforced with a mineral phase of hydroxyapatite. The most widely studied synthetic composites are based upon polymers reinforced with stiff fibres and their use currently is increasing at the rate of 5%–10% per year across a range of different applications.

Although polymers are being employed more and more as structural materials, their use often is limited by their relatively low levels of stiffness and strength compared, for example, with metals (Table 24.1). Because of the need for light, stiff and strong materials for applications as diverse as aerospace structures and sporting goods, polymer-based high-performance composites have been developed over recent years with extremely impressive mechanical properties.

In some cases, a second phase is added to a polymer to improve properties other than just mechanical behaviour. In the case of a rubber tyre, the elastomer is reinforced with mineral and carbon-black particles with the aim of improving wear resistance, as well as increasing stiffness and strength. The tyre also is reinforced on a larger scale with cords made from polymer fibres or steel wires to enable it to withstand the high forces due to inflation and cornering. Carbon black also can be added to polymers to modify their electrical properties and, in particular to render them conductive (Chapter 25).

The need for property improvement is not the only reason for the development of composite materials. For instance, polymers often are employed in low-cost high-volume applications where the addition of a cheap inert mineral filler may reduce the quantity of relatively expensive polymer used with no sacrifice in mechanical properties (and in some cases an improvement in properties, such as increased stiffness). Mineral fillers also are employed for functional reasons, such as pigments to colour the polymer or render it opaque, or additives to improve the fire resistance or reduce the thermal expansion coefficient of the polymer.

There are a number of broad classifications of polymer composites based upon their geometries, as illustrated in Figure 24.1. Within the three main classifications there are a number of different sub-categories. In *particulate composites*, the particles may be spherical but often have non-uniform shapes and are sometimes elongated or even plate like. In *discontinuous fibre composites*, the fibres may be aligned or randomly placed and in practice their alignment can vary with position in the component. In *continuous fibre composites*, the fibres can be aligned uniaxially (i.e. along only one axis) as shown in the diagram. However, the different layers of aligned fibres can be stacked in different directions (as in a cross-ply composite) or groups of aligned fibres may even be woven or knitted into complex structures.

A characteristic feature of fibre-reinforced composites is that, because of the complex arrangement of fibres, their mechanical properties are anisotropic (i.e. they vary significantly when

TABLE 24.1
Stiffness and Strength of a Number of Different Engineering Materials

Material	Density, ρ/Mg m^{-3}	Young's Modulus, E/GPa	Tensile Strength, σ^*/MPa
Rubber (polyurethane)	1.2	0.01	20
Engineering thermoplastic (nylon)	1.1	2.5	80
Thermosetting resin (epoxy)	1.25	3.5	50
Metal (mild steel)	7.8	208	400
Engineering ceramic (alumina)	3.9	380	500
Wood (spruce)			
(load // grain)	0.6	16	80
(load ⊥ grain)	0.6	1	2
Advanced polymer composite (APC-2)			
(load // fibres)	1.6	200	1500
(load ⊥ fibres)	1.6	3	50

Source: Data taken from Hull, D. and Clyne, T.W., *An Introduction to Composite Materials*, 2nd edn., Cambridge University Press, Cambridge, U.K., 1996.

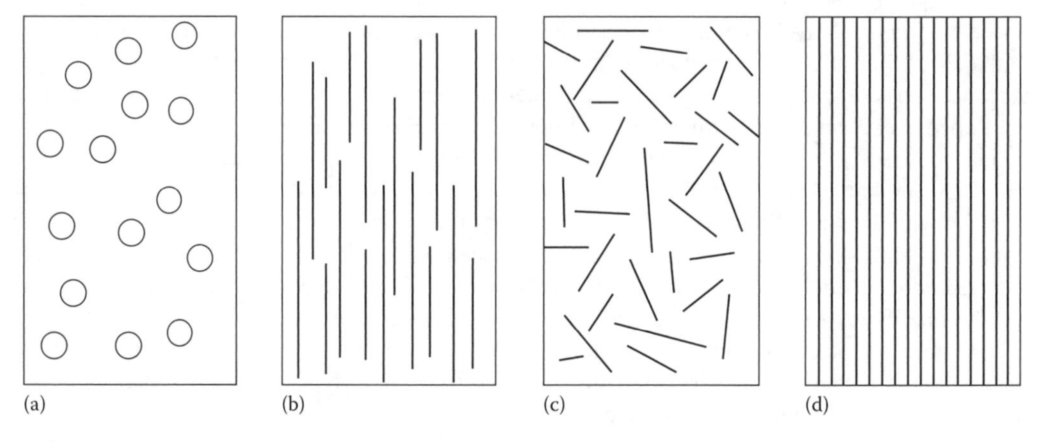

FIGURE 24.1 Schematic diagrams of the geometries of different types of composites: (a) particulate, (b) discontinuous aligned, (c) discontinuous random and (d) continuous fibre reinforced.

measured in different directions). This is because the fibres aligned in particular directions invariably have higher stiffness and strength than the matrix. For a similar reason, anisotropy also can be found for particulate-reinforced composites if the elongated particles or platelets are preferentially aligned in particular directions. This anisotropy makes the analysis of the mechanical properties of composites potentially rather complex but it offers significant scope for the ability to tailor the mechanical behaviour of composite components. For example, large differences in stiffness can be obtained in different directions and this can be incorporated into the design of a component by ensuring that the reinforcing fibres are aligned with the direction of maximum stress.

24.2 MATRIX MATERIALS

Historically, the first types of synthetic polymer composites developed were formaldehyde-based resins filled with mineral particles or sawdust. The resins now most commonly used as a matrix

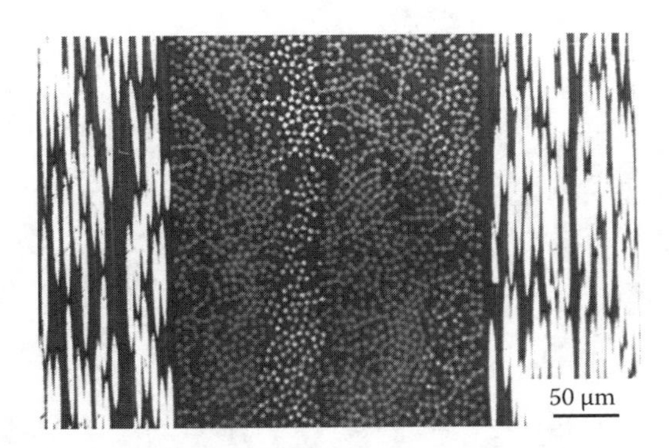

FIGURE 24.2 Optical micrograph of a polished section of an APC-2 carbon-fibre/PEEK cross-ply laminate. (Courtesy of A.J. Ball, University of Manchester, Manchester, U.K.)

materials in fibre-reinforced composites are thermosetting network polymers, such as polyester (Section 4.6.2.1) and epoxy resins (Section 3.3.1.2). Polyester resins tend to be used in glass-reinforced plastics (GRPs) for large structures such as boats, tanks, pipes etc., and epoxy resins are employed where their higher cost can be justified in terms of their superior mechanical properties. Carbon-fibre-reinforced epoxy resin composites are employed extensively for the bodies of racing cars and military aircraft, and increasingly in civilian aircraft.

In more recent years, there have been developments in the field of thermoplastic matrix composites using different types of nylon or polypropylene as matrix materials. When short fibres are employed, such as chopped glass fibres, the material can be processed using conventional techniques such as injection moulding. Care must be taken, however, that the stress exerted on the fibres during processing does not break them and reduce their length making them less effective for reinforcement. High-performance thermoplastic matrix composites reinforced with continuous carbon fibres have been developed using polyetheretherketone (PEEK, Table 3.2) as the matrix as shown in Figure 24.2. These types of materials have been targeted at aerospace applications with the aim of replacing conventional aluminium alloys.

24.3 TYPES OF REINFORCEMENT

A wide variety of different types of materials for reinforcement have been employed in polymer composites. Here, the materials used for both particulate and fibre reinforcement are discussed.

24.3.1 PARTICLES

Mineral particles have been used as cheap fillers and additives since synthetic polymers were first developed. Silicate-based minerals such as talc and mica have layer-type structures and the particles are used in the form of thin platelets. They produce composites that are relatively easy to process and the shape of the particles gives rise to platelet reinforcement, with significant increases in stiffness and strength over those of the pure polymer.

A whole variety of equiaxed particles are added to polymers for a variety of reasons. Carbon black is added to polymers for a number of different reasons; as a black pigment, to enhance UV stability, to modify electrical properties and to give reinforcement as in elastomers. Solid spherical glass particles also can be used to reinforce glassy polymers (Figure 24.3a) and hollow glass spheres are sometimes employed if a lower density is required. Figure 24.3b shows the microstructure of an epoxy resin with silica particles added to improve the dielectric properties of the resin. Sub-micron-sized particles of rutile TiO_2 often are added to polymers as a white pigment or as an opacifier. In many cases, however, the properties of the particles and the effect they have upon the matrix polymer are not important, especially in the case of cost reduction.

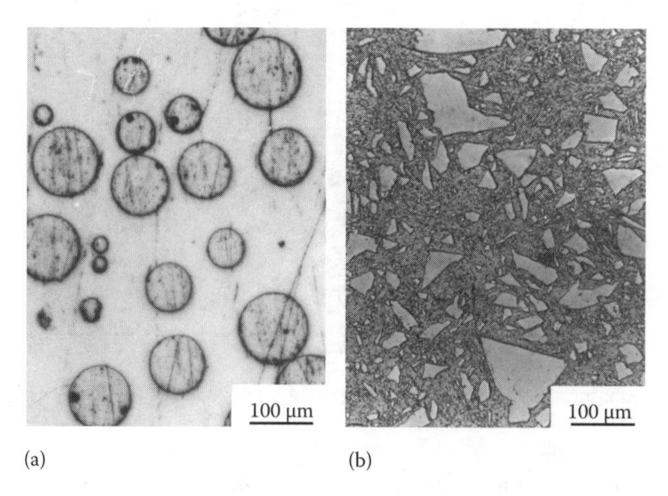

FIGURE 24.3 Optical micrographs of polished sections of particulate reinforced composites: (a) glass spheres in an epoxy resin and (b) silica particles in an epoxy resin.

24.3.2 Fibres

24.3.2.1 Glass Fibres

Glass fibres are the most commonly used type of fibre employed for reinforcing polymers, due to their low cost and good mechanical properties (Table 24.2). These are inorganic fibres based upon silica mixed with other oxides and a number of different composites are used. The most common type of fibre is E-glass that has a composition of SiO_2 (52%), CaO (17%), Al_2O_3 and Fe_2O_3 (14.0%), B_2O_3 (11%) and MgO (5%). Other types of silica-based fibres with somewhat different compositions are employed; the more expensive S-glass for higher modulus and strength, and C-glass for better corrosion resistance in acids and water.

The glass fibres are manufactured by extruding molten glass under gravity at high temperature through a series of holes in a platinum plate or "bushing." The fine filaments are drawn mechanically downwards and wound at high speed on to a drum. By controlling the processing conditions the fibres produced have a diameter typically between 8 and 15 μm. The structure of the glass fibres consists of a non-crystalline network of silica-based tetrahedra. Rapid cooling through their T_g during processing ensures that they do not crystallize since any crystals in the structure would cause a significant reduction in fibre strength. The fibres invariably are coated during manufacture with a "size" which is an emulsified polymer that is applied to the fibres for a number of reasons. Glass fibres are very susceptible to surface damage that causes a significant reduction in their strength.

TABLE 24.2
Mechanical Properties of Reinforcing Fibres Used in High-Performance Polymer Matrix Composites

Fibre	Density ρ_f/Mg m^{-3}	Tensile Modulus E_f/GPa	Tensile Strength σ_f^*/GPa
E-glass	2.55	76	1.5
Aramid (Kevlar 49)	1.45	125	3.0
PBO (Zylon HM)	1.56	270	5.8
Carbon (high strength)	1.77	230	3.3
Carbon (high modulus)	1.90	360	2.5

The size both protects the fibres from damage and provides a linkage between the fibres and matrix to improve the bond strength in composites.

24.3.2.2 Carbon Fibres

Carbon fibres are being used increasingly as reinforcement for polymer composites. They were developed in the United Kingdom in the 1960s, can be produced in a number of ways but the main method now employed in the manufacture of carbon fibres is through the pyrolysis of the polymer polyacrylonitrile (PAN). The PAN is drawn into fibres in which the molecules are aligned along the fibre axis. The fibres are then heated and the PAN converts into a ladder polymer as shown in Figure 24.4.

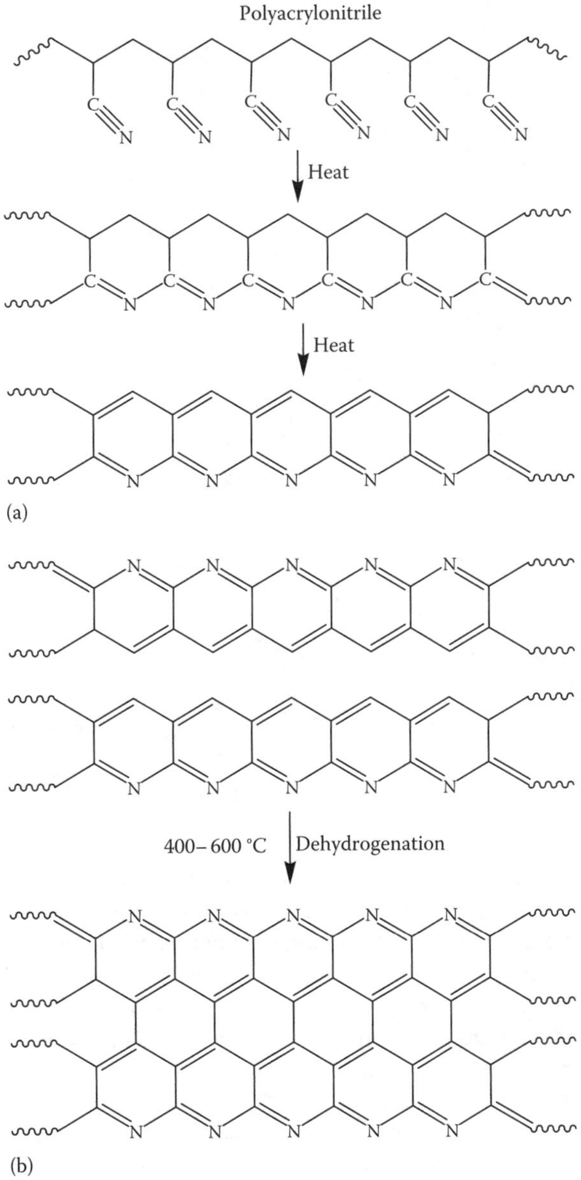

FIGURE 24.4 Schematic illustration of the formation of a graphitic ladder polymer by the pyrolysis of PAN: (a) initial heating, (b) dehydrogenation.

(continued)

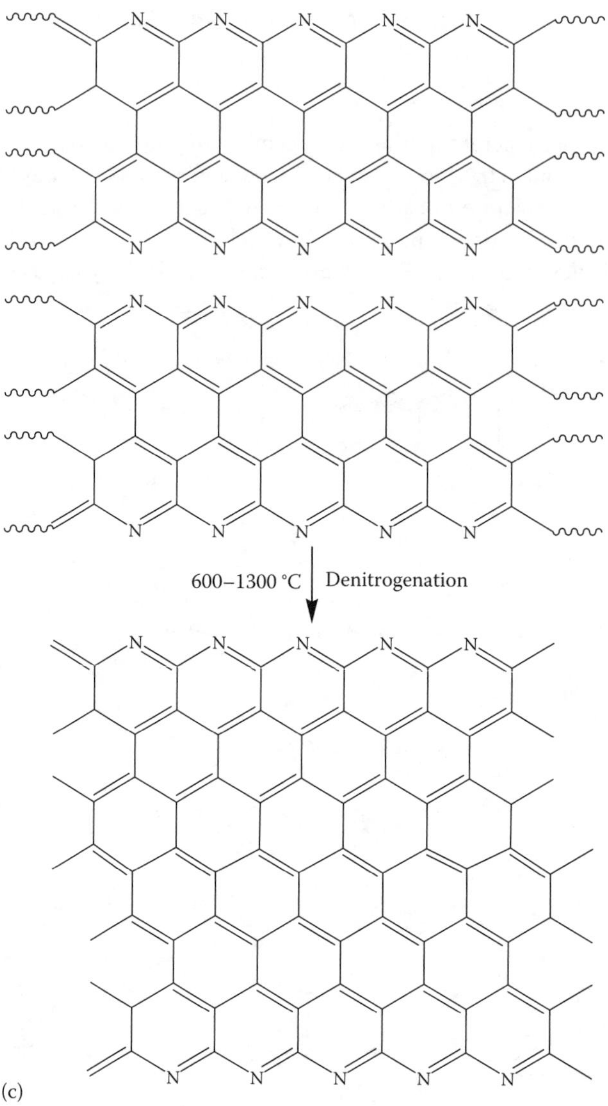

600–1300 °C Denitrogenation

(c)

FIGURE 24.4 (continued) (c) denitrogenation and aromatization.

Increasing the temperature and heating the fibres under tension in an oxygen-containing atmosphere causes further reaction in which the ladder molecules form crosslinks with each other.

The material is then reduced and heated at high temperatures to give a "turbostratic" graphite structure, an allotropic form of carbon, in which the graphene basal planes are not packed regularly as in a graphite single crystal. They are, however, aligned approximately parallel to each other along the fibre axis, leading to fibres with high levels of axial stiffness and strength. The mechanical properties of the carbon fibres are, therefore, dependent upon the processing conditions that control the microstructure of the fibres. Table 24.2 lists the mechanical properties of different types of carbon fibres (high strength fibres and high modulus fibres). Carbon fibres are inert and have high electrical and thermal conductivities. It should be noted, however, that the turbostratic structure leads to the fibres having anisotropic properties with a much lower modulus transverse to the fibre axis.

24.3.2.3 High-Modulus Polymer Fibres

The development of high-modulus polymer fibres (Section 19.3) has opened up the opportunity to produce all-polymer high-performance composites. Fibres such as PPTA aramids and PBO (Tables 3.2 and 24.2) are being used increasingly, in competition with carbon fibres. Their relatively low densities mean that when specific properties are considered (e.g. fibre modulus divided by fibre density, E_f/ρ_f) they are very competitive with carbon fibres, but it should be noted that both carbon and high-modulus polymer fibres are more expensive than glass fibres.

24.4 COMPOSITE COMPOSITION

The properties of composites depend upon their composition, which often is discussed in terms of the volume fraction of the reinforcement ϕ_r. If the total mass of the composite is W_c and the mass of the reinforcement and matrix are W_r and W_m, respectively, then

$$W_c = W_r + W_m \tag{24.1}$$

Similarly if V_c is the composite volume and V_r and V_m are the volumes of the reinforcement and the matrix, then assuming that there are no voids in the material

$$V_c = V_r + V_m \tag{24.2}$$

The volume fraction of reinforcement is given by

$$\phi_r = \frac{V_r}{V_c} \tag{24.3}$$

and that of the matrix by

$$\phi_m = \frac{V_m}{V_c} \tag{24.4}$$

It also follows from Equation 24.2 that $\phi_r + \phi_m = 1$. The density ρ_c ($=W_c/V_c$) of the composite can be obtained by dividing Equation 24.1 through by V_c and using Equations 24.3 and 24.4 to give

$$\rho_c = \phi_r\rho_r + \phi_m\rho_m \tag{24.5}$$

where ρ_r and ρ_m are the densities of the reinforcement and matrix, respectively. Table 24.1 shows that polymers have lower densities than metals or ceramics. In addition, Table 24.2 shows that the reinforcements employed to reinforce polymer matrices also have low densities such that the composite density given by Equation 24.5 remains relatively low for polymer-matrix composites.

24.5 PARTICULATE REINFORCEMENT

The level of reinforcement offered by particles depends upon both the geometry of packing and the mechanical properties of the particles and polymer. These factors will now be considered in detail.

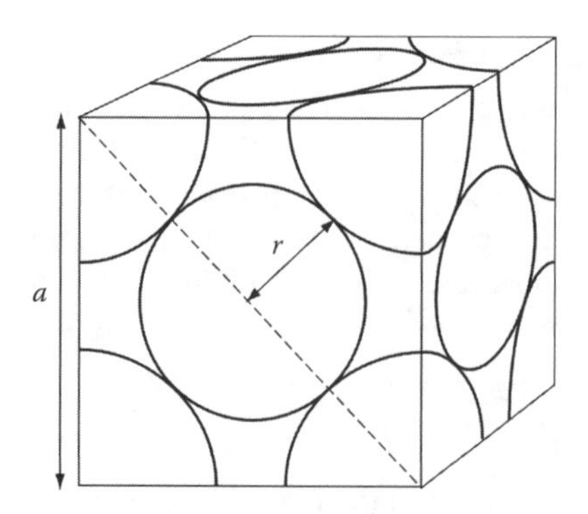

FIGURE 24.5 Schematic illustration of the unit cell for face-centred cubic (FCC) close packing of spherical particles in a composite.

24.5.1 Packing Geometries

As with fibre-reinforced composites, the properties of particulate-reinforced composites depend upon the volume fraction ϕ_p of particles within the composite. It is possible to calculate the maximum volume fraction of particles of the same size that can be packed into a composite using the calculations developed for the close-packing of spheres that have been employed in the analysis of atomic solids. The most efficient form of packing is the close-packed structure of spheres of radius r in a face-centred cubic (FCC) unit cell, as shown in Figure 24.5. If the cell edge dimension is a then the volume of the cell is a^3. The geometry of the right-angle triangle in the face gives

$$a^2 + a^2 = (4r)^2 \tag{24.6}$$

and so solving for a gives

$$a = 2r\sqrt{2} \tag{24.7}$$

The FCC unit cell volume V_c is given by

$$V_c = a^3 = (2r\sqrt{2})^3 = 16r^3\sqrt{2} \tag{24.8}$$

The volume fraction of particles ϕ_p is given by the total volume of particles in the cell divided by the volume of the cell. Inspection of Figure 24.5 shows that there are a total of 4 particles in an FCC cell ($8\times\frac{1}{4}$ particles and $4\times\frac{1}{2}$ particles). Since each particle has a volume of $4\pi r^3/3$ then the volume of particles in the cell V_p will be $16\pi r^3/3$. Hence the maximum volume fraction of particles is given by

$$\phi_p = \frac{V_p}{V_c} = \frac{(16/3)\pi r^3}{16r^3\sqrt{2}} = 0.74 \tag{24.9}$$

An important assumption in the above derivation is that the particles are all the same size. If there is a range of particle sizes present then the maximum volume fraction could be higher. In reality, however, most practical particulate composites have particle volume fractions that are lower than that given by Equation 24.9 and in which the particles are not touching.

24.5.2 Elastic Deformation

It is relatively easy to predict the Young's modulus of particulate-reinforced polymers, although in practice only upper and lower bounds can be determined rather than exact values. It is necessary to determine the Young's modulus for the situations where the particles and matrix (and hence the composite) are subjected to either *uniform strain* or *uniform stress*.

In the case of uniform strain

$$e_p = e_m = e_c \tag{24.10}$$

where e_p, e_m and e_c are the strain on the particle, matrix and composite, respectively. Assuming that the deformation of both the particles and matrix is elastic then it follows that

$$\sigma_p = E_p e_p \quad \text{and} \quad \sigma_m = E_m e_m \tag{24.11}$$

where

E_p is the Young's modulus of the particles
E_m is the Young's modulus of the matrix.

Assuming that, under the condition of uniform strain, the overall stress on the composite is the weighted sum of the stresses σ on the two phases leads to

$$\sigma_c = \phi_p \sigma_p + \phi_m \sigma_m = \phi_p E_p e_p + \phi_m E_m e_m \tag{24.12}$$

where

ϕ_p is the volume fraction of particles
ϕ_m is the volume fraction of the matrix ($\phi_p + \phi_m = 1$).

Dividing through by e_c and using Equation 24.10 gives

$$\frac{\sigma_c}{e_c} = E_c = \phi_p E_p + \phi_m E_m \tag{24.13}$$

and so the Young's modulus of the particulate composite is predicted to be

$$E_c = \phi_p E_p + (1 - \phi_p) E_m \tag{24.14}$$

which represents an upper bound of the predicted particulate-composite modulus in the case of *uniform strain*.

A similar calculation can be undertaken to predict the composite modulus for the situation of uniform stress. In this case

$$\sigma_p = \sigma_m = \sigma_c \tag{24.15}$$

where σ_p, σ_m and σ_c are the stresses on the particle, matrix and composite, respectively. Since the particles and matrix are again assumed to deform in an elastic manner it follows that

$$e_p = \frac{\sigma_p}{E_p} \quad \text{and} \quad e_m = \frac{\sigma_m}{E_m} \tag{24.16}$$

Assuming that under the condition of uniform stress, the overall strain on the composite is the weighted sum of the strains in the individual phases leads to

$$e_c = \phi_p e_p + \phi_m e_m \tag{24.17}$$

Applying Equation 24.16, this equation then becomes in terms of stresses

$$\frac{\sigma_c}{E_c} = \frac{\phi_p \sigma_p}{E_p} + \frac{\phi_m \sigma_m}{E_m} \tag{24.18}$$

Cancelling through the stresses using Equation 24.15 leads to

$$\frac{1}{E_c} = \frac{\phi_p}{E_p} + \frac{\phi_m}{E_m} \tag{24.19}$$

which can be rearranged to give

$$E_c = \frac{E_p E_m}{\phi_m E_p + \phi_p E_m} \tag{24.20}$$

This equation represents the lower bound of the Young's modulus of particulate composites in the case of *uniform stress.*

Some experimental data upon the determination of the Young's modulus of an epoxy resin containing different volume fractions of silica particles are shown in Figure 24.6. The predictions of Equations 24.14 and 24.20 are shown and it can be seen that the data fall closer to the (lower) uniform stress bound indicating that the high modulus particles tend to carry a similar level of stress as the low modulus matrix. The behaviour shown in Figure 24.6 is typical of particulate composites and it demonstrates that high levels of Young's modulus are difficult to attain with particulate reinforcement unless high filler loadings are employed. As will be seen later, better levels of reinforcement are achieved with fibre-reinforced composites where the fibres are subjected to much higher levels of stress than the matrix.

24.5.3 Fracture

It is interesting to note that if brittle silica particles are added to a brittle polymer such as an epoxy resin then, as well as increasing the stiffness of the polymer, the resulting composite can

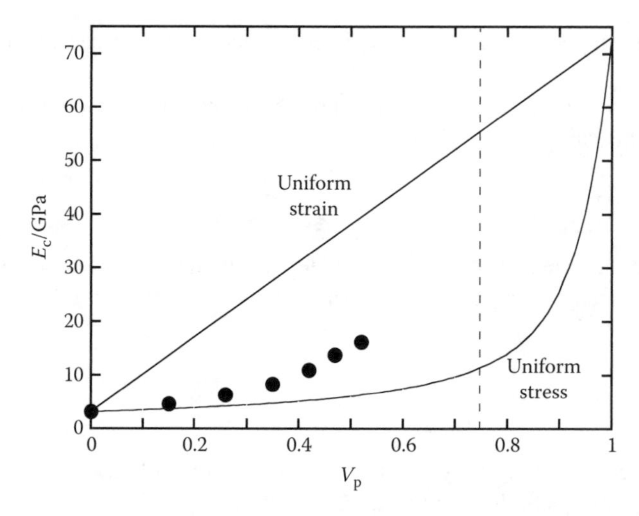

FIGURE 24.6 Young's modulus of a silica-particle-filled epoxy resin as a function of particle volume fraction. The two bounds for the cases of uniform strain (Equation 24.14) and uniform stress (Equation 24.20) are indicated. The vertical dashed line shows the maximum volume fraction of particles given by Equation 24.9.

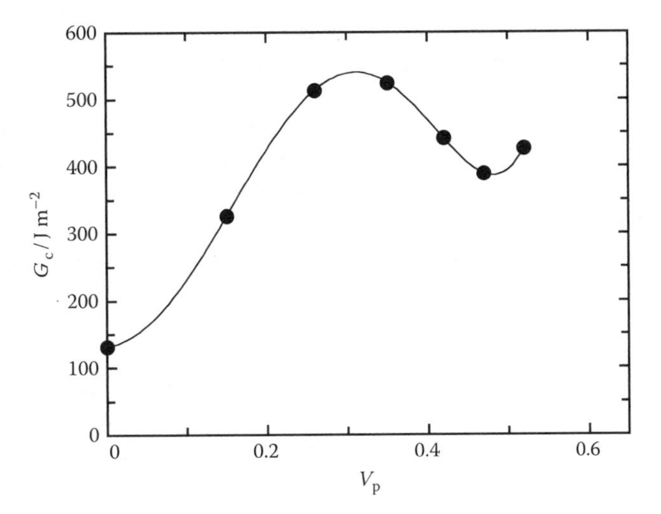

FIGURE 24.7 Dependence of the fracture energy G_c of a silica-particle-filled epoxy resin upon the volume fraction of particles V_p.

have improved toughness, as shown in Figure 24.7. It can be seen that the fracture energy G_c first increases with the addition of silica particles and then goes through a maximum value. The addition of further particles causes a reduction in toughness.

This toughening behaviour has been explained in terms of *crack pinning*, which is due to an interaction between the moving crack front and the second-phase dispersion, as shown in Figure 24.8. During crack propagation both a new fracture surface is formed and the length of the crack front is increased due to its change in shape between the pinning positions. As the crack moves through the composite, it is first pinned by the array of particles and bows around the particles until it breaks away. Strong evidence for the process is obtained from the micrograph in Figure 24.8a where it can be seen that as the crack goes around each side of the particles it joins up on different levels leading to tails in the wake of the particles on the fracture surface.

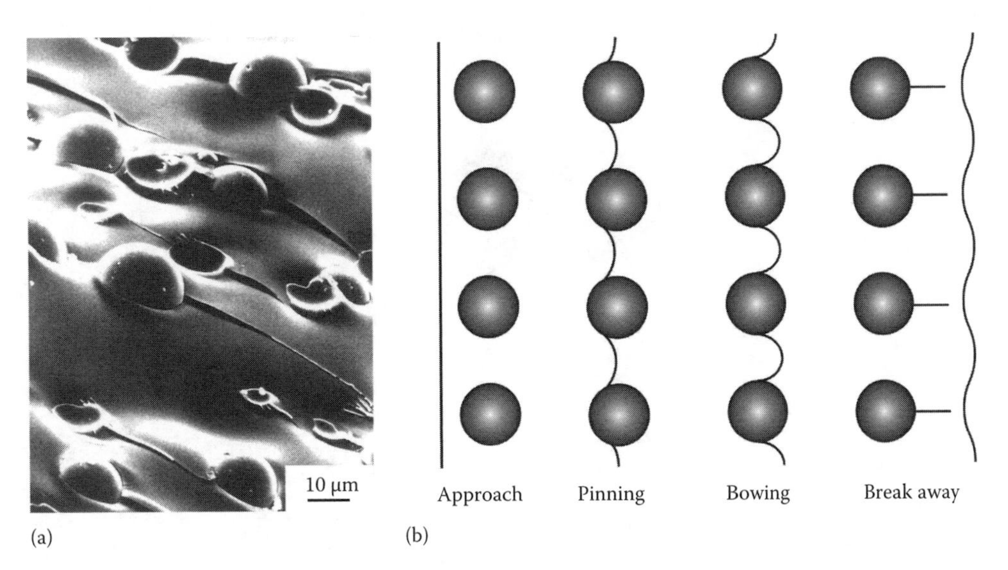

FIGURE 24.8 (a) Scanning electron micrograph of a fracture surface of a glass-particle-filled epoxy resin. (b) Schematic diagram of the crack pinning process for a crack propagating through a row of glass particles embedded in a polymer matrix.

The fractional increase in crack-front length per unit crack extension will depend upon the inter-particle spacing (controlled by the volume fraction of particles and their size). It is assumed that energy not only is required to create new fracture surface but, by analogy with dislocation theory, energy must be supplied to the newly-formed length of crack front which is assumed to possess line energy. The length of the crack front will increase as the inter-particle separation decreases and so it is predicted that the fracture toughness G_c increases with increasing particle volume fraction V_p. The peak in the fracture energy at high volume fraction is thought to be due to the filler particles becoming too close at high loadings for effective crack-front bowing to occur.

24.6 FIBRE REINFORCEMENT

The most impressive levels of reinforcement in polymer composites are obtained through the use of high-modulus fibres. The reinforcement is controlled by both the geometry of fibre packing and the mechanical properties of the fibres and matrix polymer. These aspects are now considered in detail.

24.6.1 Composite Geometry

24.6.1.1 Fibre Packing

The properties of fibre-reinforced composites depend crucially upon the geometry of fibre packing. The situation is shown schematically in Figure 24.9 for the relatively simple case of a parallel array of fibres. Two idealised examples are given of the packing of fibres of the same diameter $2r$ with a circular cross-section: a hexagonal array of fibres and a square array. If the closest separation of the fibre centres in the two cases is $2R$, then the volume fraction for the hexagonal array of fibres can be easily calculated as

$$\phi_f = \frac{\pi}{2\sqrt{3}}\left(\frac{r}{R}\right)^2 \tag{24.21}$$

and for the square array as

$$\phi_f = \frac{\pi}{4}\left(\frac{r}{R}\right)^2 \tag{24.22}$$

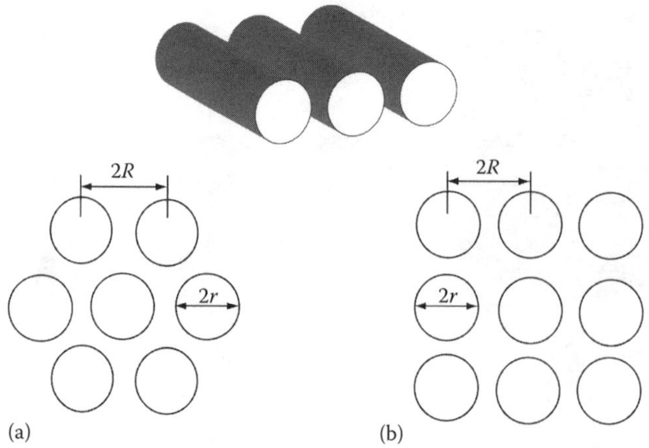

2R
2r
(a)
2R
2r
(b)

FIGURE 24.9 Schematic diagrams of the packing of reinforcing fibres in a resin matrix: (a) hexagonal array and (b) square array.

The maximum volume fraction of fibres in these packing arrangements can be determined as the cases when the fibres are touching, i.e. $r=R$. This gives maximum values of $\phi_f=0.907$ for the hexagonal array and $\phi_f=0.785$ for the square array.

The situation shown in Figure 24.9 clearly shows an idealised case. In practice, the fibres may not all be of the same diameter and they may not have circular cross-sections. Their spacing may not be uniform and clustering can often occur as can be seen in Figure 24.2. Further analysis of the model shows that the inter-fibre separation is relatively small even at low volume fractions of fibres. For example, it is easy to show for both models that the separation is less than one fibre diameter at a volume fraction of only about 0.3. This means that the polymer matrix can be constrained by the presence of fibres and that, at high ϕ_f, its morphology and mechanical behaviour may be different from that of the bulk matrix material.

24.6.1.2 Fibre Arrangements

One of the advantages of using fibre-reinforced composites is that it is possible to tailor different fibre arrangements to take advantage of the anisotropic properties of the fibre reinforcement. With layers of continuous fibres (Figure 24.1d) it is possible to stack them in unidirectional (stacked parallel), angle-ply (stacked at angles) and cross-ply (stacked at 90°) arrangements. The microstructure of a cross-ply laminate was shown in Figure 24.2. The highest axial stiffness and strength are obtained with the unidirectional fibres whereas crossing fibres give the composite more uniform properties in different directions in the angle-ply and cross-ply configurations.

Other arrangements of continuous fibres include knitted or woven structures, such as that shown in Figure 24.10, that can be draped into complex shapes, such as a helmet, and then impregnated with a thermosetting resin. The off-axis fibres tend to reduce the reinforcement efficiency and lower the composite modulus compared with composites containing straight fibres. The composites, however, tend to have more uniform properties and the interlacing of the fibres leads to composites with better impact and penetration resistance.

Different fibre arrangements can be obtained with short discontinuous fibres, as shown in Figure 24.1b and c. The morphology generally is controlled by the processing conditions, but the presence of short fibres in a liquid resin or molten polymer greatly increases the viscosity and makes

FIGURE 24.10 Optical micrograph of a woven aramid PPTA fabric that can be impregnated with an epoxy resin to make a composite.

processing more difficult. The two general random arrangements of reinforcing fibres are random in-plane and random three-dimensional, although in practice both completely random arrangements are difficult to achieve. The complex polymer flow patterns encountered with processing techniques such as injection moulding lead to the local alignment of fibres in different positions in the moulded object. It is possible, however, in certain cases to take advantage of this and design the material flow in the mould such that fibres are aligned in specific directions in the most highly stressed positions that the moulded object will be subjected to in service, to give reinforcement where it is most required.

24.6.2 Continuous Fibres

24.6.2.1 Axial Stiffness

It is possible to predict the stiffness of an aligned composite along its axis using the relatively simple model shown in Figure 24.11. If the fibres are considered to be infinitely long, then it can be assumed that the aligned fibres and matrix can be approximated as the equivalent to two slabs of material bonded together at an interface. When deformed in the axial direction, x_1, the two slabs of material are constrained to be under the same strain

$$e_{1f} = e_{1m} = e_1 \tag{24.23}$$

If the deformation of both the fibres and the matrix is elastic, then the axial stresses in the fibres and matrix will be given by

$$\sigma_{1f} = E_f e_1 \quad \text{and} \quad \sigma_{1m} = E_m e_1 \tag{24.24}$$

where

E_f is the fibre Young's modulus

E_m is that for the matrix.

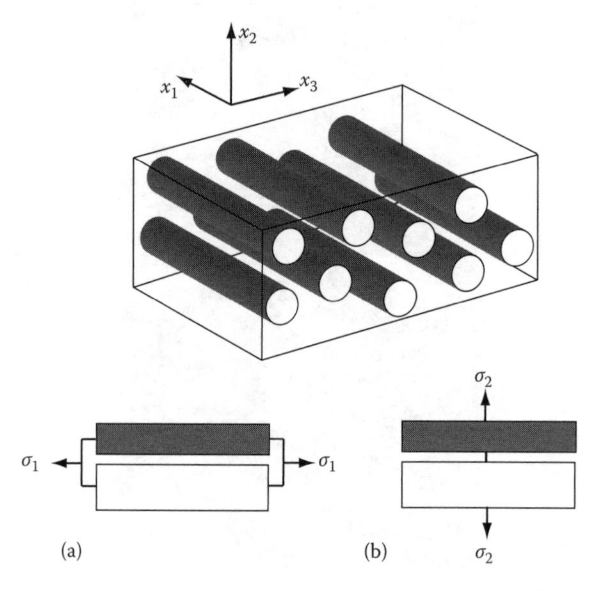

FIGURE 24.11 Unidirectionally aligned fibre-reinforced composite: (a) slab model for axial deformation and (b) slab model for transverse deformation.

Since normally $E_f \gg E_m$ then Equation 24.24 shows that $\sigma_{1f} \gg \sigma_{1m}$. This is the basis of fibre reinforcement where the fibres are subjected to the same strain as the matrix but, since they have a much higher modulus, they take a very much higher stress than the matrix.

It also is relatively easy to determine the axial Young's modulus of the composites E_1 using this model. The overall stress on the composite is given by the weighted sum of the stresses on the individual phases

$$\sigma_1 = \sigma_{1f}\phi_f + \sigma_{1m}\phi_m \tag{24.25}$$

where ϕ_f and ϕ_m are the volume fractions of the fibre and matrix, respectively. Dividing through by e_1 and using Equations 24.23 and 24.24 gives

$$E_1 = E_f\phi_f + E_m\phi_m = E_f\phi_f + E_m(1-\phi_f) \tag{24.26}$$

This equation, commonly known as the *rule of mixtures*, gives an accurate prediction of the axial Young's modulus of a unidirectional composite providing the fibres are sufficiently long to ensure that there are no end effects. Also if the volume fraction of high modulus fibres is sufficiently high, then Equation 24.26 can be written as

$$E_1 \cong E_f\phi_f \tag{24.27}$$

24.6.2.2 Transverse Stiffness

In many cases, the response of a composite to deformation in a direction transverse to the fibres axis can be critical in determining its suitability in load-bearing applications. Because of this, it is essential to know the transverse properties, in particular, the value of the transverse Young's modulus E_2. It is again possible to use a version of the slab model illustrated in Figure 24.11, where the slab is subjected to a stress σ_2 in the x_2 direction. In this case, the two slabs are subjected to the same stress

$$\sigma_{2f} = \sigma_{2m} = \sigma_2 \tag{24.28}$$

The transverse strains in the fibres and matrix are given by

$$e_{2f} = \frac{\sigma_2}{E_f} \quad \text{and} \quad e_{2m} = \frac{\sigma_2}{E_m} \tag{24.29}$$

The strains in the fibre and matrix will be additive and so assuming that the overall transverse strain is the weighted sum of the strains in the two slabs then

$$e_2 = e_{2f}\phi_f + e_{2m}\phi_m \tag{24.30}$$

Dividing through by σ_2 and using Equations 24.28 and 24.29 gives

$$\frac{1}{E_2} = \frac{\phi_f}{E_f} + \frac{\phi_m}{E_m} = \frac{\phi_f}{E_f} + \frac{(1-\phi_f)}{E_m} \tag{24.31}$$

This equation gives a prediction of the transverse mechanical properties of a uniaxially-aligned composite and if $E_f \gg E_m$ then it can be approximated to

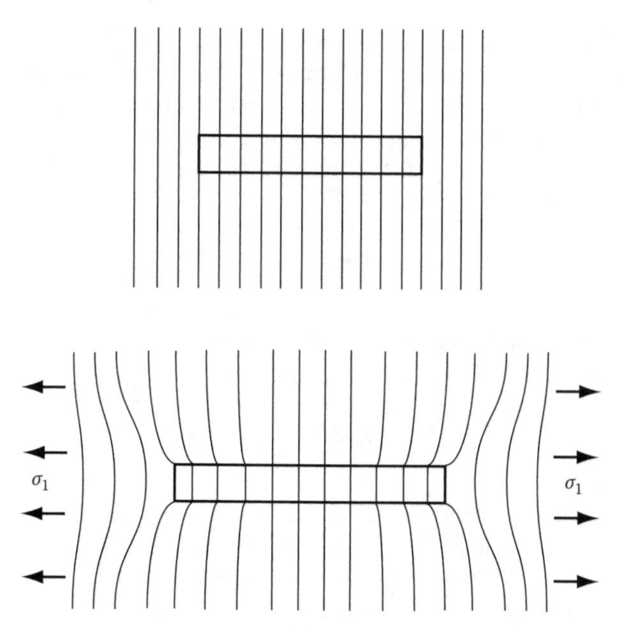

FIGURE 24.12 Deformation patterns for a discontinuous fibre in a polymer matrix.

$$E_2 \cong \frac{E_m}{(1-\phi_f)} \tag{24.32}$$

If the composite has an identical structure in the x_3 direction then $E_2 = E_3$.

In practice, this analysis of the transverse mechanical properties does not give as accurate a representation of the Young's modulus as in the case of the axial properties. This is because the strains in the matrix are not uniform and the matrix in the vicinity of the fibres can be constrained from deforming by the presence of the high-modulus fibres, a situation not found for axial deformation.

24.6.3 Discontinuous Fibres

In the case of discontinuous fibres reinforcing a composite matrix, stress transfer from the matrix to the fibre takes place through a shear stress at the fibre–matrix interface as shown in Figure 24.12. Before deformation, parallel lines perpendicular to the fibre can be drawn from the matrix through the fibre. When the system is subjected to an axial stress σ_1 parallel to the fibre axis, the lines become distorted since the Young's modulus of the matrix is much smaller than that of the fibre. This induces a shear stress at the fibre–matrix interface. The axial stress in the fibre builds up from zero at the fibre ends to a maximum value in the middle of the fibre. The uniform strain assumption means that, if the fibre is long enough, in the middle of the fibre Equation 24.23 is obeyed and the strain in the fibre equals that in the matrix. Since the fibres generally have a much higher Young's modulus, the fibres then carry most of the stress in the composite.

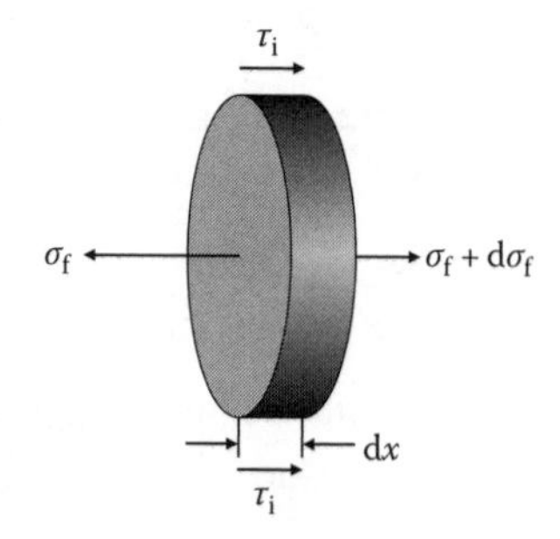

FIGURE 24.13 Balance of stresses acting on an element of the fibre of thickness dx in the composite.

The relationship between the interfacial shear stress τ_i near the fibre ends and the fibre stress σ_f can be determined by using a balance of the shear forces at the interface and the tensile forces in a fibre element, as shown in Figure 24.13. The main assumption is that the force due to the shear stress τ_i at the interface is balanced by the force due to the variation of axial stress $d\sigma_f$ in the fibre such that

$$2\pi r\tau_i dx = -\pi r^2 d\sigma_f \tag{24.33}$$

and so

$$\frac{d\sigma_f}{dx} = -\frac{2\tau_i}{r} \tag{24.34}$$

24.6.3.1 Elastic Stress Transfer

The behaviour of a discontinuous fibre in a matrix can be modelled using shear lag theory, developed initially by Cox to model the mechanical properties of paper. In the theory it is assumed that the fibre is surrounded by a cylinder of resin extending to a radius ρ from the fibre centre, as show in Figure 24.14. It is assumed that both the fibre and matrix deform elastically and that the fibre–matrix interface remains intact. If u is the displacement of the matrix in the fibre axial direction at a radius ρ then the shear strain γ at that position is given by

$$\gamma = \frac{du}{d\rho} \tag{24.35}$$

The shear modulus of the matrix is defined as $G_m = \tau/\gamma$, hence

$$\frac{du}{d\rho} = \frac{\tau}{G_m} \tag{24.36}$$

The shear force per unit length carried by the matrix cylinder surface is $2\pi\rho\tau$. It is transmitted to the fibre surface though the layers of resin and so the shear stress at radius ρ is given by

$$2\pi\rho\tau = 2\pi r\tau_i \tag{24.37}$$

and so

$$\tau = \left(\frac{r}{\rho}\right)\tau_i \tag{24.38}$$

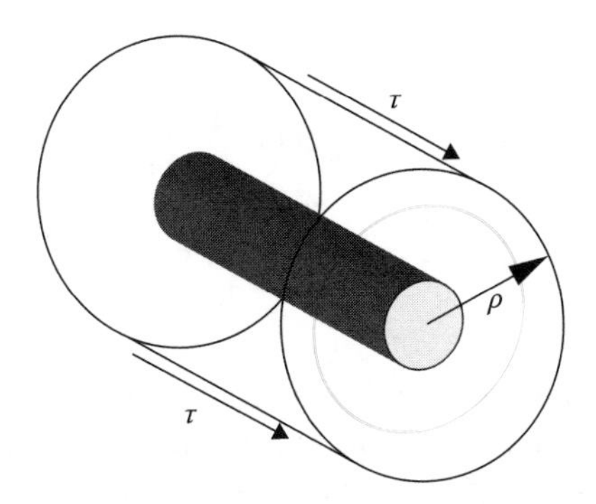

FIGURE 24.14 Model of a fibre within a resin used in shear-lag theory. The shear stress τ acts at a radius ρ from the fibre centre.

Using Equation 24.36, it follows that

$$\frac{du}{d\rho} = \left(\frac{r}{\rho}\right)\frac{\tau_i}{G_m} \tag{24.39}$$

This equation can be integrated using the limits of the displacement at the fibre surface ($\rho = r$) of $u = u_f$ and the displacement at $\rho = R$ of $u = u_R$

$$\int_{u_f}^{u_R} du = \left(\frac{r\tau_i}{G_m}\right)\int_r^R \frac{d\rho}{\rho} \tag{24.40}$$

hence

$$u_R - u_f = \left(\frac{r\tau_i}{G_m}\right)\ln\left(\frac{R}{r}\right) \tag{24.41}$$

It is possible to convert these displacements into strain since the fibre strain e_f and matrix strain e_m can be approximated as $e_f \approx du_f/dx$ and $e_m \approx du_R/dx$. It should be noted that this shear-lag analysis is not rigorous but it serves as a simple illustration of the process of stress transfer from the matrix to a fibre in a short-fibre composite. In addition, τ_i is given by Equation 24.34 and so differentiating Equation 24.41 with respect to x leads to

$$e_f - e_m = -\frac{r^2}{2G_m}\left(\frac{d^2\sigma_f}{dx^2}\right)\ln\left(\frac{R}{r}\right) \tag{24.42}$$

Multiplying through by E_f gives

$$\frac{d^2\sigma_f}{dx^2} = -\frac{n^2}{r^2}(\sigma_f - e_m E_f) \tag{24.43}$$

where

$$n = \sqrt{\frac{2G_m}{E_f \ln(R/r)}}$$

This differential equation has the general solution

$$\sigma_f = E_f e_m + C\sinh\left(\frac{nx}{r}\right) + D\cosh\left(\frac{nx}{r}\right)$$

where C and D are constants of integration. Equation 24.43 can be simplified and solved by double differentiation of the general solution, if it is assumed that the boundary conditions are that there is no stress transmitted across the fibre ends, i.e. if $x = 0$ in the middle of the fibre where $\sigma_f = E_f e_m$, then $\sigma_f = 0$ at $x = \pm l/2$ where l is the length of the fibre. This leads to $C = 0$ and comparing terms gives

$$D = -\frac{E_f e_m}{\cosh\left(nl/2r\right)}$$

The final equation for the distribution of fibre stress as a function of distance, x along the fibre is then

$$\sigma_f = E_f e_m \left[1 - \frac{\cosh(nx/r)}{\cosh(nl/2r)}\right]$$

Finally it is possible to determine the distribution of interfacial shear stress along the fibre using Equation 24.34 which leads to

$$\tau_i = \frac{n}{2} E_f e_m \frac{\sinh(nx/r)}{\cosh(nl/2r)}$$

It is convenient at this stage to introduce the concept of fibre aspect ratio s $(=l/2r)$ so that the two equations above can be rewritten as

$$\sigma_f = E_f e_m \left[1 - \frac{\cosh\left(ns\,2x/l\right)}{\cosh(ns)}\right] \tag{24.44}$$

for the axial fibre stress and as

$$\tau_i = \frac{n}{2} E_f e_m \frac{\sinh\left(ns\,2x/l\right)}{\cosh(ns)} \tag{24.45}$$

for the interfacial shear stress. The effect of the different parameters upon the buildup of stress in a fibre is demonstrated in Figure 24.15 for different values of the product ns. It can be seen that the fibre is most highly stressed, i.e. the most efficient fibre reinforcement is obtained, when the product ns is high. This implies that a high aspect ratio s is desirable along with a high value of n.

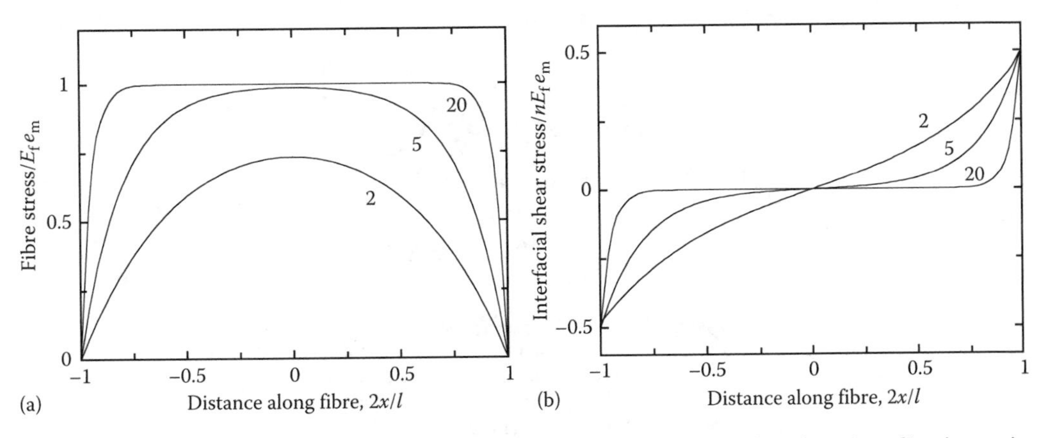

FIGURE 24.15 (a) Predicted variation of fibre stress with distance along the fibre for a short fibre in matrix. (b) Predicted variation of interfacial shear stress with distance along the fibre for a short fibre. (The values of the product ns are indicated in each case.)

If there are a number of aligned discontinuous fibres reinforcing a composite then it is a simple task to determine the mean fibre stress from the integral

$$\bar{\sigma}_{\mathrm{f}} = \frac{1}{l}\int_{-l/2}^{+l/2} \sigma_{\mathrm{f}}\mathrm{d}x \tag{24.46}$$

Substitution of Equation 24.44 into this integral gives

$$\bar{\sigma}_{\mathrm{f}} = E_{\mathrm{f}}e_{\mathrm{m}}\left[1-\left(\frac{\tanh(ns)}{ns}\right)\right] \tag{24.47}$$

This equation again shows that the best reinforcement is obtained when the product *ns* is very large.

24.6.3.2 Experimental Determination of Fibre Stress Distributions

Experimental measurements of the fibre stress in discontinuous fibres in a composite under stress have recently become available through the use of Raman spectroscopy (Section 15.5). The fibre stress or strain can be determined from the stress-induced shift of the Raman bands obtained from the fibre using a laser beam focused onto an individual fibre inside the matrix resin. Figure 24.16 shows the variation of fibre strain e_{f} ($=\sigma_{\mathrm{f}}/E_{\mathrm{f}}$) along a PBO fibre in an epoxy resin subjected to different levels of matrix strain, e_{m}. The data have been fitted to Equation 24.44 by choosing appropriate values of n and it can be seen that there is a close correlation between the theoretical curves and experimental data points. It should be noted, however, that Equation 24.43 cannot be used to determine n since the value of $\ln(R/r)$ is essentially indeterminate. It is more appropriate to think of n as a fitting parameter that characterises the efficiency of stress transfer between the matrix and fibre (Figure 24.15).

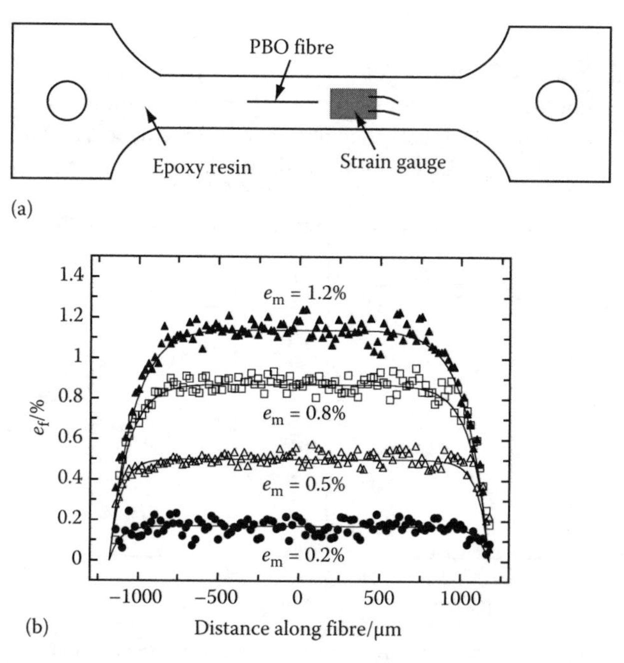

FIGURE 24.16 (a) Single fibre composite specimen. (b) Variation of fibre strain measured e_{f} along a short PBO fibre at different levels of matrix strain e_{m} in an epoxy resin matrix.

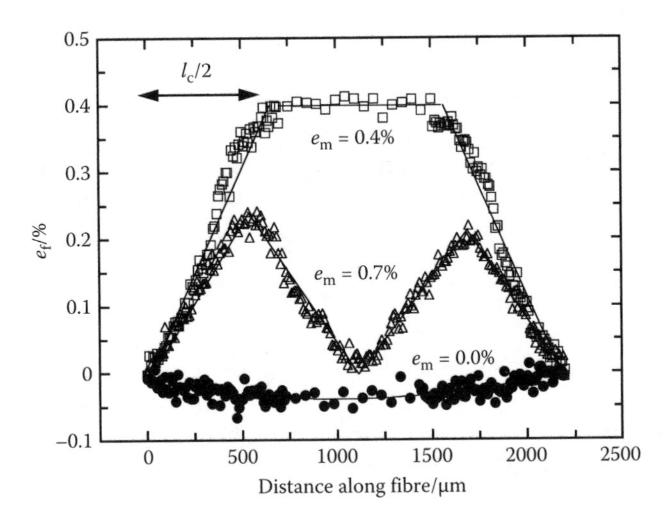

FIGURE 24.17 Variation of fibre strain measured along a high-modulus carbon fibre at different levels of epoxy matrix strain e_m.

The analysis of fibre reinforcement has so far only considered the situation where there is good adhesion across the fibre–matrix interface. Figure 24.17 shows the distribution of fibre strain, again determined by Raman spectroscopy, for a high-modulus carbon fibre embedded in an epoxy resin dumbbell specimen subjected to different levels of deformation. There is a slight compression in the fibre in the undeformed state ($e_m = 0\%$) as a result of shrinkage of the resin during the curing process. When the specimen is subjected to a matrix strain of 0.4%, the strain in the fibre builds up approximately linearly over a distance $l_c/2$ from the ends and then becomes equal to the matrix strain in a plateau region along the middle of the fibre. When the matrix strain is increased to 0.7%, the fibre breaks into two fragments with triangular strain distributions and in which the maximum strain is much less than the matrix strain. The interfacial shear stress τ_i can be determined from the slope of the lines at the fibre ends in Figure 24.17 by the application of Equation 24.34. If there is a weak interface between the fibre and matrix, it is controlled by the strength of the bonding between the fibre and matrix. For a strong interface τ_i will be equal to the shear yield strength of the matrix σ_y.

The parameter l_c (twice the distance over which the stress builds up to the plateau value from the fibre ends $= 2 \times l_c/2$) is known as the *critical length* since for fibres below this length, such as the fragments at $e_m = 0.7\%$ in Figure 24.17, the fibre strain always is less than the matrix strain and so the fibres will not offer much reinforcement to the matrix. It also follows that for discontinuous fibres, good levels of reinforcement are obtained only for fibres of length $l > 10l_c$ for which the ends are less important. The value of l_c can be determined by rearranging and integrating Equation 24.34 with the limits determined by the maximum value of $l_c/2$ occurring when the fibre stress σ_f is equal to the fibre fracture strength σ_f^*, such that

$$\int_0^{\sigma_f^*} d\sigma_f = \frac{2\tau_i}{r} \int_0^{l_c/2} dx \tag{24.48}$$

and so

$$l_c = \frac{\sigma_f^* r}{\tau_i} \tag{24.49}$$

Hence it follows that l_c is controlled by a balance between the strength of the fibre σ_f^*, the strength of the interface τ_i and the fibre radius r.

Another factor that affects the efficiency of reinforcement in discontinuous fibre-reinforced composites is the orientation of the fibres, as shown in Figure 24.1. In the case of long aligned fibres, the Young's modulus of the composite subjected to axial deformation is given by Equation 24.26. In short-fibre composites in which there is misalignment of the fibres, the modulus E_c is reduced and can be approximated to

$$E_c = K_e E_f \phi_f + E_m(1-\phi_f) \tag{24.50}$$

where K_e can be thought of as a fibre efficiency factor, which takes into account the effects of both fibre misorientation and finite length. Clearly, K_e will be less than unity and is typically between 0.6 and 0.1. The lower limit for E_c in a random discontinuous-fibre composite will be the same as that of a particulate composite and so be given by Equation 24.20.

24.6.4 Fracture

24.6.4.1 Continuous and Aligned Fibres

The first situation to be considered will be the fracture of continuous and aligned fibres deformed parallel to the fibre axis. The mechanical responses of the fibres and matrix are shown schematically in Figure 24.18a for the typical case of high-modulus high-strength brittle fibres in a ductile matrix. Both the failure stresses σ_f^* and σ_m^* and failure strains e_f^* and e_m^* of the fibre and matrix are indicated. The associated mechanical response of a composite is indicated in Figure 24.18b. The initial slope is between the value for the fibre and matrix and the slope of the stress–strain curve changes at the yield strain of the matrix above which the fibres take a higher proportion of the load. The onset of composite fracture occurs at the failure strain of the fibres e_f^* ($< e_m^*$), although this is generally not catastrophic as the fibres do not all break at the same strain and the matrix remains intact. It is possible to estimate the fracture stress of the composite, σ_c^*, in this case by adapting Equation 24.25 to give

$$\sigma_c^* = \sigma_m'(1-\phi_f) + \sigma_f^* \phi_f \tag{24.51}$$

where σ_m' is the stress in the matrix at the point of fibre failure.

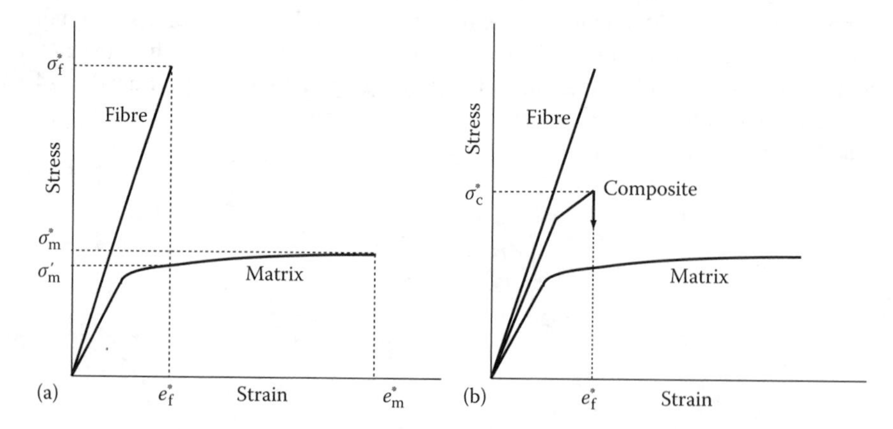

FIGURE 24.18 Schematic stress–strain curves (a) for a typical brittle fibre and ductile matrix and (b) additionally showing the behaviour of a unidirectional composite deformed parallel to the fibre axis.

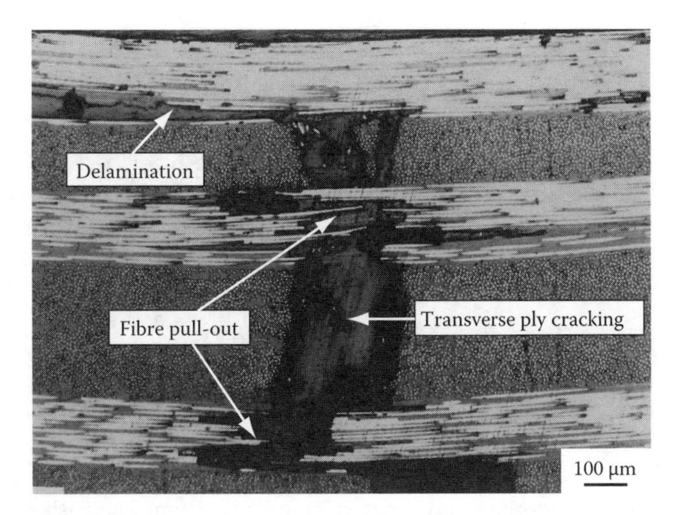

FIGURE 24.19 Optical micrograph of an impacted cross-ply APC-2 carbon-fibre in PEEK matrix laminate showing the different mechanisms of failure. (Courtesy of A.J. Ball, University of Manchester, Manchester, U.K.)

Composites with aligned fibres are designed to be subjected to axial stresses and can be very weak when subjected to stress transverse to the axis of fibre alignment. Their strength can even be lower than that of the matrix such that the reinforcement actually reduces the strength of the material. In contrast to their behaviour under axial deformation, where their strength is controlled by the strength of the fibres (Equation 24.51), the transverse strength depends upon factors such as the strength of the matrix material and/or fibre–matrix interface.

The resistance of composites to fracture can be improved by arranging layers of continuous aligned fibres into cross-ply laminates (Figure 24.2). In this case, a number of different failure mechanisms can be induced that add to the fracture resistance of the material. An example of a cross-ply laminate of carbon-fibres in a PEEK matrix which has been subject to impact from above in the vertical direction is shown in Figure 24.19. It can be seen that a number of different failure mechanisms have been induced: delamination, fibre pull-out and transverse ply cracking. Each of these processes involves the creation of new surfaces that leads to the absorption of energy, imparting high levels of toughness on the material. This means that fibre-reinforced composites are some of the toughest materials available and they often are employed in critical applications, such as being used as protective armour.

24.6.4.2 Discontinuous Fibres

Even though composites based upon discontinuous fibres have properties that are inferior to those of composites with continuous fibres they are finding increasing use due to their relative ease of processing and improvement of properties over those of the base polymer. Composites based upon chopped glass fibres are in widespread use and short carbon fibre and short aramid fibre composites are being developed.

If the discontinuous fibres in the composite are aligned (Figure 24.1b) and the fibre length l is greater than the critical length l_c then the strength of the composite, when deformed in the axial direction, can be determined by modifying Equation 24.51 to account for the less-stressed fibre ends to give

$$\sigma_c^* = \sigma_m'(1-\phi_f) + \sigma_f^*\phi_f\left(1-\frac{l_c}{2l}\right) \tag{24.52}$$

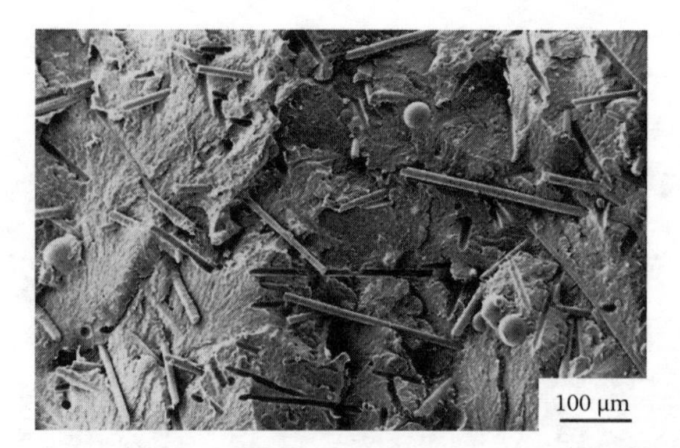

FIGURE 24.20 Scanning electron micrograph of the fracture surface of a polymer reinforced with both glass fibres and glass spheres showing fibre pull-out.

The fracture surface of a short-glass-fibre-reinforced polymer composite is shown in Figure 24.20 where fibres can be seen sticking out of the surface. Holes are also present and are characteristic of fibre pull-out having taken place. During the fracture of a composite, there will be a balance between fibre fracture and pull-out depending upon the relative strength of the fibres and interface; pull-out will be favoured for strong fibres and when there is a weak fibre–matrix interface.

24.7 NANOCOMPOSITES

The particles and fibres described so far for the reinforcement of polymers have had dimensions on the scale of microns (10^{-6} m) or larger. Over recent years there has been an increasing interest in the reinforcement of polymers by nanoparticles which have dimensions on the nanometre scale (10^{-9} m) and are capable of forming nanocomposites. There are a number of reasons why they are of interest. Firstly, as the dimensions of materials decreases the size of flaws diminishes and they become stronger (Section 23.3.1) offering better prospects for reinforcement. Secondly as the reinforcement becomes smaller in size, the surface-to-volume ratio increases and so the area of the interface between the reinforcement and the matrix is much larger in a nanocomposite, leading to potentially better stress transfer. It is possible to break down the types of reinforcement of that can be used in nanocomposites into three main classes, *nanoparticles*, *nanoplatelets* and *nanotubes* which will be described separately in the following sections.

24.7.1 Nanoparticles

24.7.1.1 Carbon Black

Carbon black is a material that has been known since ancient times and it has been employed for over 100 years as an additive to enhance the mechanical performance of elastomers. It can increase the stiffness, strength and wear resistance of elastomers significantly in practical applications, such as in vehicle tyres. Carbon black is a form of elemental carbon that is produced as black soot from a smoky flame during the combustion of hydrocarbons. The structure of carbon black is shown schematically in Figure 24.21. It consists of primary particles of the order of 20 nm in diameter (depending upon the grade) that invariably group together to form larger aggregates a factor of 10 times larger. The primary particles of carbon have a turbostratic graphite structure, made up of small imperfect crystallites, similar to that found in carbon fibres (Section 24.3.2.2).

Figure 24.22 shows the effect of the addition of carbon black upon the Young's modulus of natural rubber where it can be seen that the modulus is doubled upon the addition of 20% by volume of the carbon black. It was shown in Section 24.5.2 that particulate reinforcement usually takes place under the condition approximating to uniform stress. In this case the modulus of the composite is

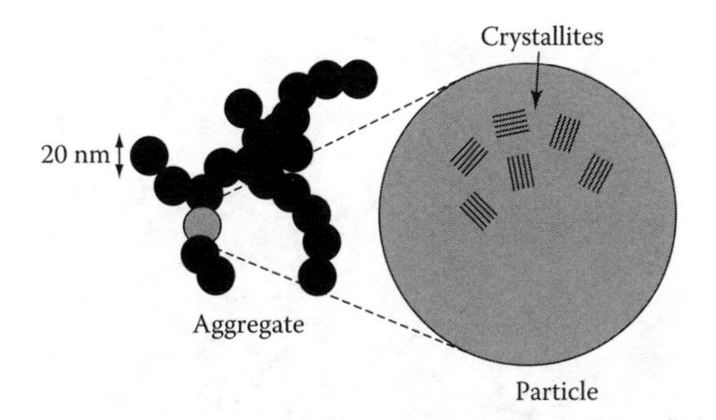

FIGURE 24.21 Schematic diagram of the structure of carbon black showing the aggregates of primary particles and the random arrangement of small imperfect crystallites in the particles.

given by Equation 24.20 and this is represented by the dashed curve in Figure 24.22 obtained using a Young's modulus of 10 GPa (i.e. similar to that of polycrystalline graphite) for the carbon black. It can be seen that the experimental data fall well above this curve showing that the carbon black gives better-than-expected reinforcement and it is found that the behaviour can be fitted by an equation of the form

$$\frac{E_r}{E_0} = 1 + 2.5V_p + 14.1V_p^2 \tag{21.55}$$

where

E_r is the Young's modulus of the reinforced elastomer
E_0 is the Young's modulus of the unreinforced material
V_p is the volume fraction of the reinforcement.

This equation was derived originally to describe the viscosity of liquids containing solid particles and is now used widely for hard fillers in soft elastomers. It is plotted in Figure 24.22 from which it can be seen that there is a good fit of the equation to the experimental data.

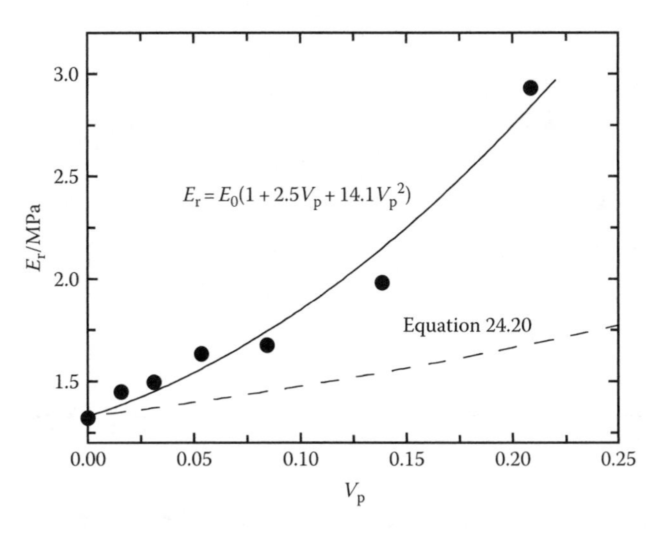

FIGURE 24.22 Effect of the volume fraction V_p of carbon black filler particles upon the Young's modulus of vulcanized natural rubber E_r. The dashed line represents the lower bound for the composite modulus predicted using Equation 24.20 for uniform stress. The value of E_0 was taken to be 1.33 MPa. (Data taken from Mullins, L. and Tobin, N.R., *J. Appl. Polym. Sci.*, 9, 2993, 1965.)

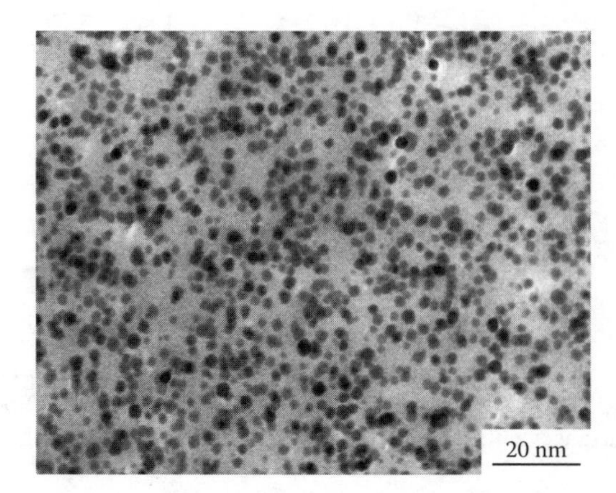

FIGURE 24.23 Transmission electron micrograph of a section through an epoxy resin containing nanosilica particles. (Courtesy of A.J. Kinloch and A.C. Taylor, Imperial College, London, U.K.; R. Ruggerone, EPFL, Lausanne, Switzerland.)

24.7.1.2 Nanosilica

Good dispersions of isolated 20 nm diameter silica (SiO_2) particles with a relatively uniform diameter as shown in Figure 24.23 can be produced in epoxy resins using a sol-gel technique. This can be contrasted with the dispersions of the much larger glass and silica particles shown in Figure 24.3. Upon curing, for a given volume fraction of silica, the epoxy resins containing the nanoparticles are found to have significantly better mechanical properties than those containing the larger particles.

Figure 24.24 shows the dependence of the Young's modulus and fracture energy of the epoxy-based nanocomposite upon the volume fraction of the nanosilica particles. It can be seen in Figure 24.24a that the modulus data lie well above the curve given by Equation 24.20 assuming uniform

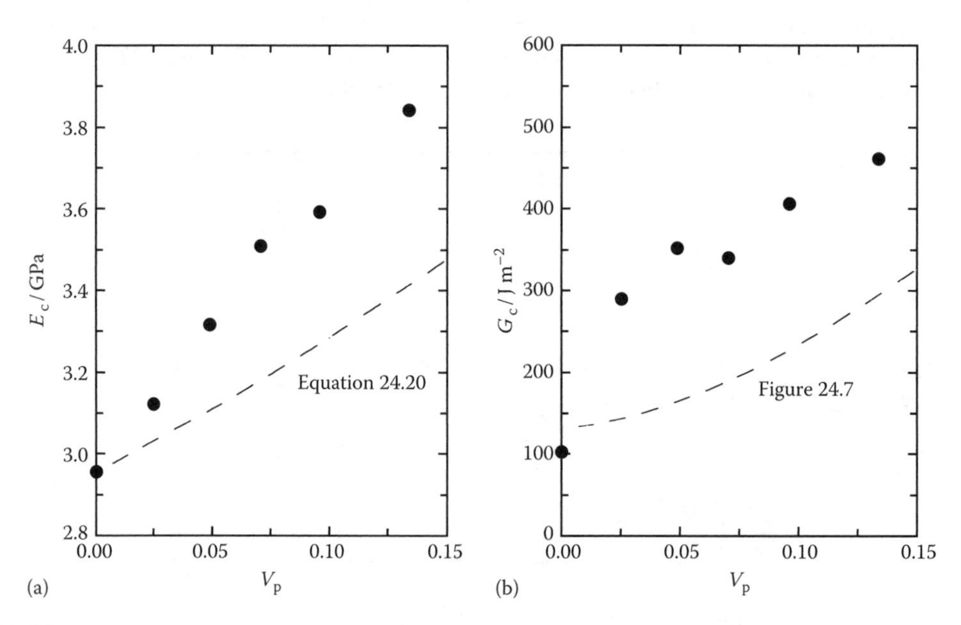

FIGURE 24.24 Mechanical properties of an epoxy resin reinforced with different volume fractions V_p of nanosilica particles. (a) Young's modulus—the dashed line represents Equation 24.20 using a value of 70 GPa for the modulus of silica. (b) Fracture energy—the dashed line is the curve from Figure 24.7 for the epoxy resin reinforced with macroscopic silica particles. (Data taken from Johnsen, B.B et al., *Polymer*, 48, 530, 2007.)

stress, in contrast to the behaviour of macroscopic silica particles shown in Figure 24.6, where the modulus data fall close to the uniform stress lower bound, especially at low volume fractions. Hence it appears that the nanosilica particles give rise to much better reinforcement. The improvement is even more apparent in the case of fracture energy as shown in Figure 24.24b. This increased by a factor of 4 for only 10% ($V_p=0.1$) of nanoparticles, whereas a volume fraction of nearly 30% of macroscopic silica is need for a similar level of improvement in toughness (Figure 24.7).

24.7.2 Nanoplatelets

24.7.2.1 Clays

Clay minerals have structures consisting of layers of alumino-silicate platelets of the order of 1 nm thick held together by electrostatic forces. One type of clay that has been investigated extensively as an additive for polymers is montmorillonite which has sodium ions in the gaps between the silicate layers. In its natural state monmorillonite is not miscible with organic monomers or polymers and does not disperse well, but it can be modified by exchanging the sodium cations with organic cations to give a modified organoclay suitable for dispersion in a polymer matrix.

Dispersion of the organoclay in a polymer matrix involves two distinct steps, as shown schematically in Figure 24.25. The first step is known as *intercalation* in which organic species, such as monomer or polymer molecules, diffuse into the gaps between the silicate layers. At this stage the layers remain parallel to each other while their separation increases. The interaction between the polymer and the organoclay can be usefully monitored using wide-angle X-ray diffraction (Section 17.2.1) as shown in Figure 24.25. The maximum in the scattering angle 2θ moves to a lower value as the d-spacing between the layers increases through intercalation of the polymer molecules. The

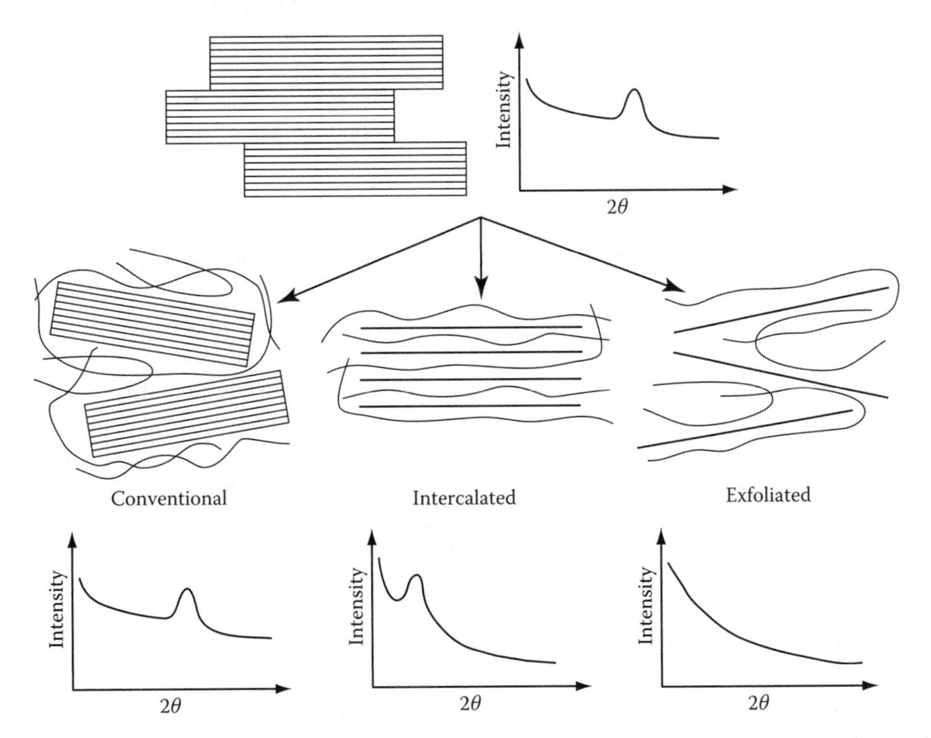

FIGURE 24.25 Schematic diagrams of the structures of conventional, intercalated and exfoliated clay nanocomposites showing schematic wide-angle X-ray scattering patterns from the materials. The straight lines are the edge-on views of the clay platelets. (Adapted from Sperling, L.H., *Introduction to Physical Polymer Science*, 4th edn., Wiley-Interscience, New York, 2006; Paul, D.R. and Robeson, L.M., *Polymer*, 49, 3187, 2008.)

TABLE 24.3
Mechanical and Thermal Properties of Clay–Nylon-6 Composites

Composite of Nylon 6 with	Weight % Clay	Young's Modulus/MPa	Tensile Strength/MPa	Heat Distortion Temperature/°C
No clay	0	1.1	69	65
Unmodified clay	5.0	1.0	61	85
Exfoliated clay	4.2	2.1	107	145

Source: Data taken from LeBaron, P.C. et al., *Appl. Clay Sci.*, 15, 11, 1999.

second step is known as *exfoliation* or delamination where the clay layers become completely separated from each other. In this case the peak in the X-ray diffraction pattern corresponding to the *d*-spacing disappears as long-range order in the structure no longer exists. It is generally only when exfoliation has occurred that significant improvements in the properties of the nanocomposites over those of the matrix are found. This is demonstrated in Table 24.3 for the clay-nylon-6 system. Addition of unmodified clay particles leads to a degradation of the Young's modulus and tensile strength where a similar loading of a fully-exfoliated clay causes significant increases in the modulus and strength of the nylon 6 (Table 24.3).

It is not only the mechanical properties that are improved by the formation of polymer nanocomposites with exfoliated clay platelets. Table 24.3 shows also that there is a significant increase in the heat distortion temperature of the nylon 6 enabling the nanocomposites to be used at higher temperatures than the original polymer, such as, for example, in under-bonnet applications in the automobile industry. Other properties that may be improved are flame retardance and barrier properties—the dispersion of nano-sized exfoliated platelets leads to a tortuous path for gas molecules to diffuse through the polymer structure, hence decreasing gas diffusion coefficients. This improvement in barriers properties is being exploited with clay-modified polymers now being used in increasingly food and drink packaging applications.

24.7.2.2 Graphene

Graphite is a naturally-occurring form of carbon that has been used since the Middle Ages. It has a layered structure consisting of stacks of covalently-bonded, sp^2-hybridised 0.34 nm thick graphene sheets held together by van der Waals forces. Interest in this material has grown rapidly since the discovery by Novoselov and Geim in Manchester in 2004 that individual one-atom thick graphene layers, with exceptional mechanical and extraordinary electronic transport properties, could be prepared by mechanical cleavage of graphite using adhesive tape. This discovery led to them being awarded the 2010 Nobel Prize for physics. Figure 24.26a shows an optical micrograph of a cleaved monolayer of graphene on a polymer substrate and the structure of the single graphene layer is shown in Figure 24.26b. It is possible to see the graphene since sp^2-hybridised carbon is a strong absorber of light and the monolayer absorbs 3% of the visible light which gives it sufficient contrast for a single layer of carbon atoms to be observed on the substrate in an optical microscope.

Graphene can be prepared by a number of other routes, such as chemical vapour deposition or the thermal reduction of graphene derivatives such as graphene oxide which can be produced by the exfoliation of graphite using strong acids. The cleavage route, however, currently produces the most structurally-perfect material. Graphene is of interest for nanocomposites since it is chemically and electronically similar to carbon nanotubes and structurally analogous to the clay minerals. Measurements of the mechanical properties of individual graphene sheets have shown the material to have an in-plane Young's modulus of the order of 1000 GPa and a strength in excess of 100 GPa, hence approaching the theoretical value for a material (Section 23.2.1). Such impressive mechanical properties have led to high levels of reinforcement being reported even using relatively low loadings

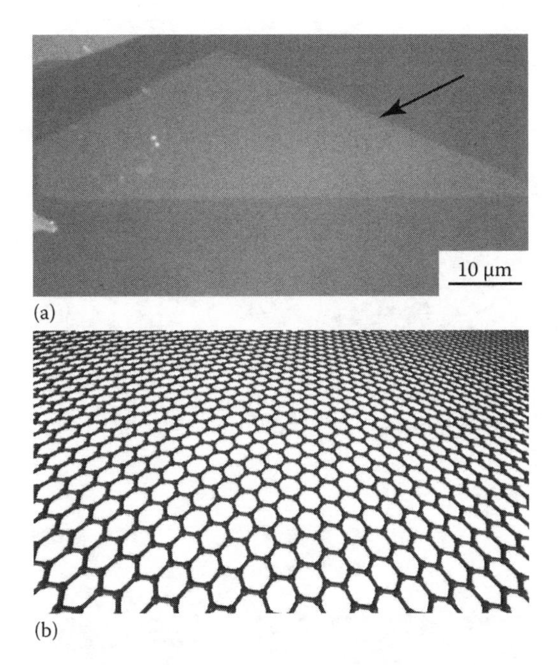

FIGURE 24.26 (a) Optical micrograph of a graphene monolayer (indicated by an arrow) prepared by mechanical cleavage and deposited on a polymer substrate. (Courtesy of K.S. Novoselov, University of Manchester, Manchester, U.K.) (b) Schematic representation of a graphene monolayer on the atomic level. (Courtesy of F. Ding, Hong Kong Polytechnic University, Hong Kong.)

of graphene, as shown in Table 24.4. Such improvements in a wide range of mechanical properties are found to be significantly better than those obtained with similar loadings of either carbon nanotubes or exfoliated clay.

24.7.3 Carbon Nanotubes

In the mid-1980s interest in different forms of carbon was opened up by the discovery of fullerene molecules such as C_{60} which is a cage-like structure made up of hexagonal and pentagonal rings of carbon atoms. Within 10 years, new tubular forms of carbon had been synthesised and isolated, firstly multi-walled carbon nanotubes (MWNTs) and then single-walled carbon nanotubes (SWNTs), the structures of which are shown in Figure 24.27. SWNTs can be considered to be single sheets of graphene rolled up seamlessly into a cylinder. They are essentially ladder polymers made up only of carbon atoms with no side groups and so are found to have excellent

TABLE 24.4
Mechanical Properties of Epoxy Resin Matrix Nanocomposites Reinforced with 0.1% by Weight of Graphene Platelets Produced by the Reduction of Graphene Oxide

Cured Epoxy Resin Containing	Young's Modulus/GPa	Tensile Strength/MPa	Fracture Toughness K_c/MPa m$^{1/2}$	Fracture Energy G_c/J m^{-2}
No graphene	2.85	55	1.0	260
Graphene platelets	3.74	78	1.5	580

Source: Data taken from Rafiee, M.A. et al., *ACS Nano*, 3, 3884, 2009.

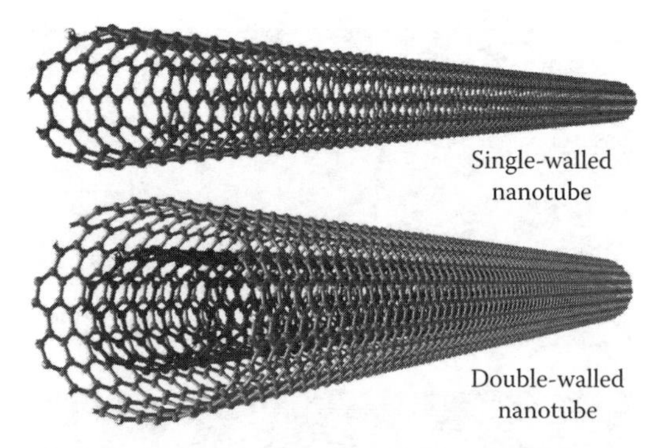

FIGURE 24.27 Schematic representations of single- and double-walled carbon nanotubes. Multiwalled nanotubes can have up to 20 layers of nested tubes. (Courtesy of F. Ding, Hong Kong Polytechnic University, Hong Kong.)

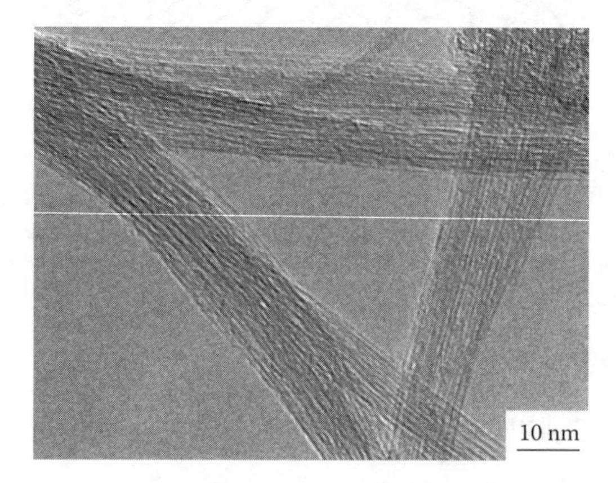

FIGURE 24.28 Transmission electron micrograph of bundles (or ropes) of single-walled carbon nanotubes. (Courtesy of U. Bangert, University of Manchester, Manchester, U.K.)

mechanical properties; values of Young's modulus of up to 1 TPa have been reported along with strengths in excess of 100 GPa. SWNTs may be either semiconducting or metallic depending upon their exact molecular structure and so can be employed as an additive to render polymers conductive (Section 25.3.2). MWNTs consist of two or more nested SWNT cylinders held together by van der Waals forces and are found also to have good mechanical properties. Their levels of strength and stiffness are found to be somewhat inferior to those of SWNTs, however, since during deformation there is the possibility of slippage between the weakly-bonded layers in the structure.

From experience with carbon fibres, carbon nanotubes are obvious candidates as additives for reinforcing polymers in the form of nanocomposites. One issue that arises with both MWNTs and particularly SWNTs is that the nanotubes tend to be entangled, as shown in Figure 24.28, and this can make their separation and dispersion in polymer matrices difficult. Nevertheless there have been a number of reports of improvements in the mechanical properties of polymers such as the data for polystyrene shown in Table 24.5. In this case nanocomposites were prepared using 1% of two different types of MWNTs which had a similar diameter but different aspect ratios s. It can be seen that the addition of the nanotubes increases both the strength and stiffness of the polystyrene and that the highest level of Young's modulus is found for the MWNTs with the higher aspect ratio. This would be expected from considerations of composite micromechanics (Section 24.6.3) and it is illuminating that the micromechanics approach can still be applied at the nano level. The data

TABLE 24.5
Mechanical Properties of Polystyrene Reinforced with 1% by Weight of Two Types of MWNTs with Different Aspect Ratios, s ($=l/2r$)

Polystyrene Containing	Nanotube Diameter $2r$/μm	Nanotube Aspect Ratio s	Tensile Modulus/GPa	Tensile Strength/MPa
No MWNTs	—	—	1.19	12.8
MWNT-1	33.6	446	1.62	16
MWNT-2	30	1167	1.69	16

Source: Data taken from Qian, D. et al., *Appl. Phys. Lett.*, 76, 2868, 2000.

shown in Table 24.5 are for a random dispersion of MWNTs and alignment of the nanotubes has the potential to produce even more impressive levels of reinforcement.

Carbon nanotubes are smooth-sided molecules that do not bond well with polymer matrices and tend to stick together in bundles or ropes. Hence they do not disperse well in polymer matrices and the interface between the nanotube and polymer tends to be weak. These issues have led to the drive to functionalize nanotubes with covalently bonded groups anchored to the surface of the nanotubes. This can be done using highly reactive intermediates that first of all attack the nanotubes and then introduce the functional groups on the surface. The levels of functionalization can be as high as 1 group for every 10 carbon atoms. The functionalized nanotubes are found to disperse much better in solvents or polymers and have a stronger interface with the matrix in nanocomposites. Figure 24.29 shows the effect of adding different loadings of functionalized SWNTs to a poly(dimethyl siloxane) (PDMS) elastomer. The SWNTs had short –OH-terminated polymer chains attached to their surface and the PDMS molecules were amine terminated. It can be seen that upon the addition of the SWNTs, there is nearly a 10-fold increase in the tensile modulus of the PDMS whereas the elongation-to-break remains relatively unchanged. Such levels of reinforcement are very impressive and much higher than those found in the case of the addition of carbon black to natural rubber shown in Figure 24.22.

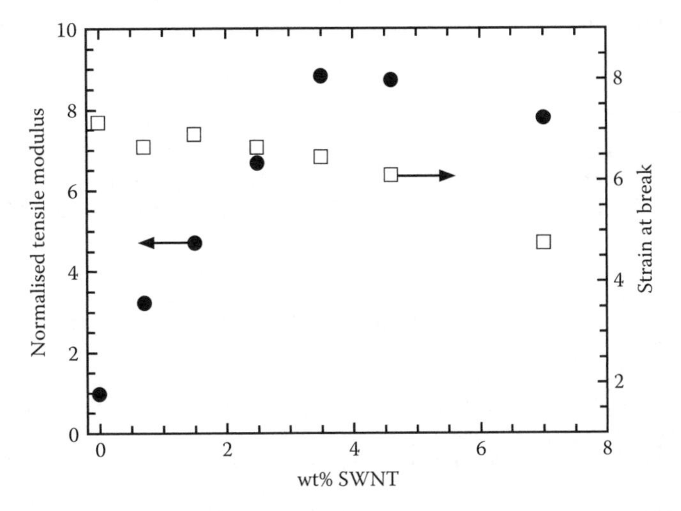

FIGURE 24.29 Dependence of the normalized tensile modulus E_c/E_m and strain at break of PDMS upon the weight percentage addition of functionalized single-walled carbon nanotubes. (Data taken from Dyke, C.A. and Tour, J.M., *J. Phys. Chem.*, 51, 11151, 2004.)

PROBLEMS

24.1 Calculate the mass of each of the constituents required per unit mass to prepare fibre-reinforced composites with the following compositions:

(a) Polypropylene reinforced by 30% glass fibres (by volume); and

(b) Epoxy resin reinforced by a combination of 20% carbon fibres and 30% aramid fibres (by volume).

(You can assume that the densities (in Mg/m^3) of the materials are as follows: polypropylene 0.90; glass fibres is 2.54; epoxy resin 1.30; carbon fibres 1.79; and aramid fibres 1.45).

24.2 Show that volume fraction of fibres, ϕ_f, in a fibre-reinforced composite is given by the following equations for the different geometrical arrangements:

(a) Hexagonal array $\phi_f = \frac{\pi}{2\sqrt{3}}\left(\frac{r}{R}\right)^2$

(b) Square array $\phi_f = \frac{\pi}{4}\left(\frac{r}{R}\right)^2$

In each case, determine also the maximum volume fraction of fibres in the composite. (Assume that the diameter of the fibres is $2r$ and the separation of the fibre centres is $2R$).

24.3 The Young's moduli in the longitudinal and transverse directions of a continuous and oriented fibre-reinforced composite are 20.0 and 3.7 GPa, respectively. Determine the Young's moduli of the fibre and matrix polymer given that the volume fraction of the fibres is 0.26.

24.4 A fibre-reinforced polymer composite consists of 50% by volume of continuous and aligned glass fibres with a Young's modulus of 70 GPa and 50% by volume of a polymer matrix resin with a Young's modulus of 3.3 GPa.

(a) What is the Young's modulus of the composite in the longitudinal direction parallel to the fibre axis?

(b) Determine the strain in each phase when the stress of 60 MPa is applied. Comment on the values you determine.

(c) Calculate the load carried by each of the fibre and matrix phases if the cross-sectional area of the specimen is 300 mm^2 and a stress of 60 MPa is applied in the longitudinal direction. Comment again on the difference between the loads in the fibres and matrix.

FURTHER READING

Chawla, K.K., *Composite Materials: Science and Engineering*, Springer, New York, 1998a.

Chawla, K.K., *Fibrous Materials*, Cambridge University Press, New York, 1998b.

Donnet, J.B., Wang, T.K., Peng, J.C.M. and Rebouillat, S. (eds), *Carbon Fibers*, 3rd edn., Marcel Dekker, New York, 1998.

Friedrich, K., Fakirov, S., and Zhang, Z., *Polymer Composites: From Nano-to-Macro-Scale*, Springer, New York, 2005.

Hull, D. and Clyne, T.W., *An Introduction to Composite Materials*, 2nd edn., Cambridge University Press, Cambridge, U.K., 1996.

Mai, Y.W. and Yu, Z.Z., *Polymer Nanocomposites*, CRC Press, Boca Raton, FL, 2006.

McCrum, N.G., Buckley C.P., and Bucknall C.B., *Principles of Polymer Engineering*, 2nd edn., Oxford Science Publications, Oxford, U.K., 1997.

Nielsen, L.E. and Landel R.F., *Mechanical Properties of Polymers and Composites*, Marcel Dekker, New York, 1994.

Sperling, L.H., *Introduction to Physical Polymer Science*, 4th edn., Wiley-Interscience, New York, 2006.

Usuki, A., Hasegawa, N., and Kato, M., Polymer-clay nanocomposites, *Advances in Polymer Science*, **179**, 135 (2005).

25 Electrical Properties

25.1 INTRODUCTION TO ELECTRICAL PROPERTIES

Polymers have been employed historically as electrical components because of their excellent behaviour as insulators. Indeed, the development of the electrical and electronic industries has been facilitated by the availability of synthetic polymers with high-quality insulation, good chemical and physical properties combined with their ease of fabrication. The characteristic of electrical insulation is an inherent property of saturated polymer molecules allowing the material to stop the flow of current and sustain high levels of electrical field without breakdown. High levels of electrical insulation can, however, give rise to problems with the polymers in practical applications. There can be a buildup of static electricity on a component that can give rise to problems such as attracting dust and giving electric discharge shocks. Such problems can be overcome by modifying the polymer to have intermediate levels of conductivity by the addition of a conductive second phase to make a conductive composite or the addition of antistatic agents (often just in the surface layer) that rely upon ionic conduction. This then allows the leakage of the charge to earth. Such property improvements, however, often are accompanied by a degradation of the mechanical properties of the polymer.

The virtues of polymers, such as toughness and flexibility, along with good processibility led to the long-held desire to obtain conducting polymers that could be fabricated easily. The discovery of metallic conduction in polyacetylene in the 1970s represented a major breakthrough in the preparation of inherently conducting polymers and led to the award of the 2000 Nobel Prize for Chemistry (Section 6.4.7). The initial work upon polyacetylene has been followed by the discovery of a whole family of related conducting polymers that are finding use in a number of applications such as electrodes in batteries. Even though they can have conductivities per unit mass better than copper, it has been found that they tend to lack the long-term stability needed for signal and power transmission. A further breakthrough, however, occurred in the 1990s with the discovery that these conducting polymers can undergo electroluminescent light emission, opening up the possibility of using them in a new generation of displays and other electronic devices.

25.2 DIELECTRIC PROPERTIES

The dielectric properties of polymers control their performance as insulators. Additionally, the analysis of the dielectric behaviour can yield valuable information upon polymer structure. In this section, dielectric properties will be considered firstly from a molecular viewpoint, and it then will be demonstrated how the analysis of the dielectric behaviour of polymers can be related to structural relaxations in the material.

25.2.1 Molecular Polarizability

When a molecule is subjected to an electric field, a dipole moment is induced, the magnitude of which depends upon the strength of the electric field and the polymer structure. This phenomenon is known as *polarization*. The polarizability of the molecule α is defined as the magnitude of the dipole moment induced divided by the strength of the electric field and is related to the relative permittivity ε (or dielectric constant) of the material. The polarizability depends both upon the chemical structure of the molecules and their physical environment. The analysis of the response of

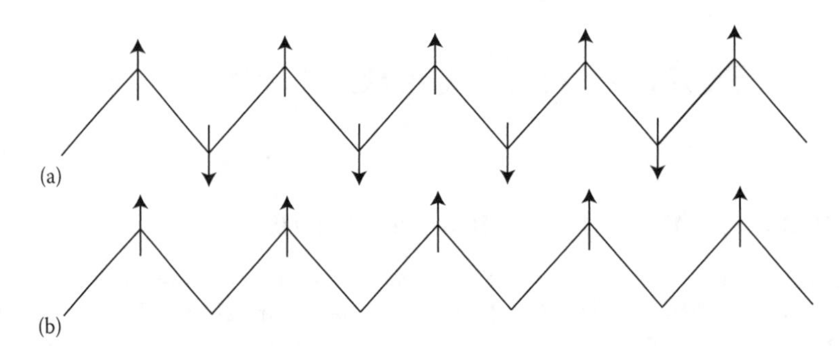

FIGURE 25.1 Schematic diagram of the arrangement of polar groups on a planar zig-zag polymer chain. (Adapted from Blythe, A.R. and Bloor, D., *Electrical Properties of Polymers*, 2nd edn., Cambridge University Press, Cambridge, U.K., 2005.)

a polymer molecule to an applied electric field (invariably an oscillating one) can, therefore, be used to characterize the material. A number of aspects of the polarization of molecules by an electric field have been identified and are described below.

Electronic polarization—This is where an electric field causes a small displacement of the electrons in an atom relative to the nucleus. The effect is relatively small as the size of the electric field usually is small relative to the size of the electric fields within the atom. It can, however, be induced at very high frequencies and is responsible for a number of optical properties of materials, such as the refraction of light.

Atomic polarization—In this case, the electric field distorts the arrangement of the atoms in the molecule. It is activated at lower frequencies than electronic polarization, typically at infrared frequencies. It is generally quite a weak effect but can be significant in ionic compounds.

Orientational polarization—This phenomenon is important in the analysis of the dielectric properties of polymers. Molecules that possess a permanent dipole moment will tend to become aligned in the applied electric field. The rate at which they move will depend upon the structure of the material and the molecular environment of the dipole. The process is relatively slow and is observed at frequencies of the order of 10^6 Hz or less.

A convenient way of analysing the polarizability of polymer molecules is to consider the polarizability of the individual repeat units. Some polymer molecules have permanent dipole moments, but the resulting dipole moment of the whole molecule will depend upon whether or not the individual dipole moments cancel each other out. This behaviour is shown schematically for a planar zig-zag polymer chain in Figure 25.1. In the case of a polymer such as polytetrafluoroethylene (PTFE), the situation is as shown in Figure 25.1a and the dipoles will cancel each other out in the normal helical conformation of the PTFE molecules (Table 17.1). Any small dipole orientation effects will be found only from defects in the structure, which invariably exist. In contrast, the situation shown in Figure 25.1b is typical of a vinyl polymer, such as poly(vinyl chloride), which adopts a planar zig-zag conformation (Table 17.1), where the dipoles from the C–Cl groups will be additive and the material has a high dielectric constant as a result of the high level of polarizability.

25.2.2 Dielectric Measurements

Viscoelastic phenomena in polymers often are probed using mechanical stressing, as was demonstrated in Chapter 20, but in some cases it is possible also to investigate perturbations in the material using an oscillating electric field that addresses the polarizability of the molecules. Although this technique is most applicable to polar polymers, it has a number of advantages over mechanical methods such as enabling polymers to be tested over a very wide range of frequencies, typically from 10^{-4} to 10^{10} Hz. Moreover, cells can be employed in which temperature or pressure is varied

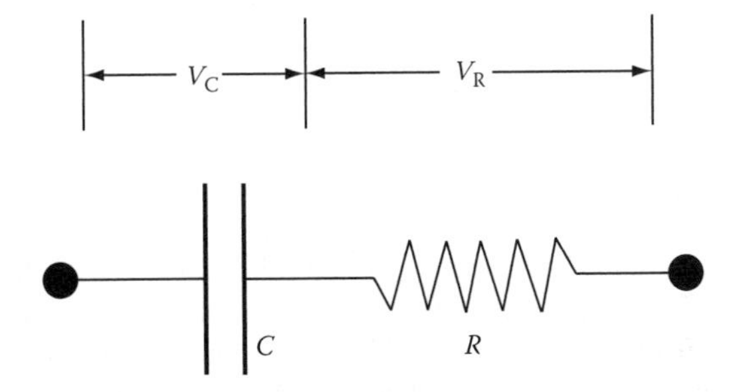

FIGURE 25.2 Schematic representation of a simple RC circuit.

and changes that occur during chemical reactions, such as polymerization, crosslinking and degradation, can be followed.

The phenomenon can be best considered initially in terms of an electric circuit consisting only of a capacitor of capacitance C and a resistor of resistance R as shown in Figure 25.2. The resistance is defined by Ohm's law as

$$R = \frac{V}{I} \tag{25.1}$$

where
V is the voltage
I is the current.

The capacitor of capacitance C_0, which consists ideally of two parallel electrodes separated by a vacuum, is capable of holding a charge Q when a voltage V is applied across the electrodes such that

$$Q = C_0 V \tag{23.2}$$

When an insulating material such as a polymer, which is capable of holding more charge due to polarization in the dielectric, is used instead of a vacuum, the capacitance becomes

$$C = \varepsilon C_0 \tag{23.3}$$

where ε is the dielectric constant. The correct term for this parameter is the relative permittivity, the ratio of the permittivity of the material to that of free space ε_0. Its value often is quoted as being a constant for different materials (e.g. $\varepsilon = 1.00054$ for air, $= 3.5$ for paper and $= 3.7$ for silica), but it should be remembered that ε is a function of frequency and temperature. This is the basis for undertaking measurements of the dielectric behaviour of polymers, as the frequency or temperature dependence of the dielectric constant carries information about molecular motions in the polymer.

The behaviour of a capacitor and resistor in series in the electrical circuit shown in Figure 25.2 will now be considered. In this case, the voltage across each component in the circuit will be additive such that

$$V = V_C + V_R \tag{25.4}$$

and the current, which is the same in each component, is defined by

$$I = \frac{dQ}{dt} \tag{25.5}$$

Combining Equations 25.1 through 25.4 gives the following general relationship:

$$\frac{dQ}{dt}+\frac{Q}{RC}=\frac{V}{R} \tag{25.6}$$

The form of this equation is remarkably similar to that of Equation 20.16 for the Voigt mechanical model for viscoelasticity (Section 20.2.2). In this case, however, the charge is the electrical analogue of strain and the voltage the analogue of stress. The capacitance can be considered to be equivalent to the compliance (reciprocal of the modulus, $1/E$) and R the viscosity. Equation 25.6 is for the case of a resistor and capacitance in parallel, whereas the Voigt model is for a spring and dashpot in series. The analogy works because the stresses are additive for the Voigt model, whereas the voltages are additive in the electrical circuit.

As in the case of the mechanical model, Equation 25.6 can be integrated using a constant voltage $V=V_0$ and this leads to

$$Q=CV_0\left[1-\exp\left(\frac{-t}{RC}\right)\right] \tag{25.7}$$

Or, since the product RC has dimensions of time, we can set this as equal to τ_0 and the equation becomes

$$Q=CV_0\left[1-\exp\left(\frac{-t}{\tau_0}\right)\right] \tag{25.8}$$

This again is of a similar form to Equation 20.18 and is the electrical analogue of mechanical creep when stress (rather than voltage) is held constant.

The above analysis has assumed that the capacitance is independent of time, which will only be true for a vacuum. Since the capacitance of a material is related to the dielectric constant through Equation 25.3, the capacitance of all materials will show some time dependence through the time dependence of the dielectric constant ε, which carries information about molecular motions in the material. The equations derived above can be used to describe an equivalent circuit in analysis of the behaviour of polymers, since a capacitor containing a dielectric polymer will behave like the RC circuit and the same set of equations can be employed to analyse it.

The analysis of the dielectric properties of polymers invariably is undertaken using alternating current (AC) circuits and the behaviour of the material can be analysed using the concept of equivalent circuits, two examples of which are shown in Figure 25.3. Because of the phase difference between the voltage and current, it is conventional to use a complex number notation in the analysis of AC circuits and thus a sinusoidal voltage is give by

$$V=V_0e^{i\omega t} \tag{25.9}$$

where ω is the angular frequency and $i=(-1)^{1/2}$. From Equations 25.2 and 25.3, when the material is subjected to an oscillating voltage

$$I=\frac{dQ}{dt}=\varepsilon^{*}C_0\frac{dV}{dt} \tag{25.10}$$

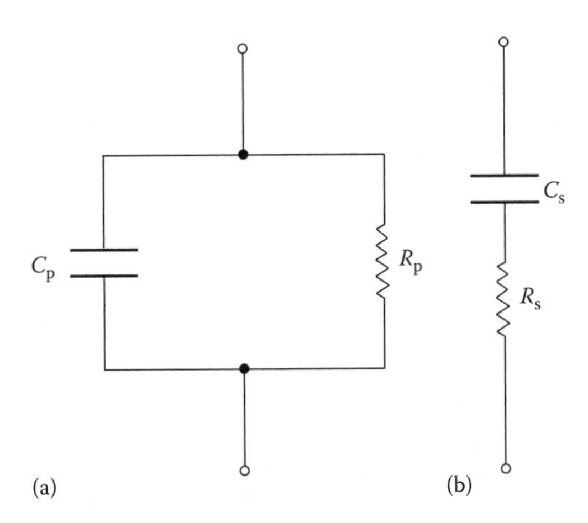

FIGURE 25.3 Equivalent electrical circuits of dielectric specimens: (a) parallel and (b) series. (Adapted from Blythe, A.R. and Bloor, D., *Electrical Properties of Polymers*, 2nd edn., Cambridge University Press, Cambridge, U.K., 2005.)

In AC circuits, both the voltage and current are complex quantities and so the dielectric constant will also be a complex quantity ε^*, given conveniently by

$$\varepsilon^* = \varepsilon' - \mathrm{i}\varepsilon'' \tag{25.11}$$

Substituting for dV/dt in Equation 25.10 and differentiating Equation 25.9 gives

$$I = \mathrm{i}\omega\varepsilon^* C_0 V = \omega C_0(\mathrm{i}e' + e'')V \tag{25.12}$$

which implies that there are two components of the current, a capacitative component

$$I_C = \mathrm{i}\omega C_0 \varepsilon' V \tag{25.13}$$

out of phase by 90° with the voltage and a resistive component

$$I_R = \omega C_0 \varepsilon'' V \tag{25.14}$$

in phase with the voltage.

The two quantities in the complex permittivity are linked by the relation

$$\frac{\varepsilon''}{\varepsilon'} = \tan\delta \propto \frac{\text{energy dissipated per cycle}}{\text{energy stored per cycle}} \tag{25.15}$$

where

$\tan\delta$ is, in this case, called the *dielectric loss tangent*
ε'' is known as the *dielectric loss factor.*

All of these parameters, ε', ε'' and $\tan\delta$, are readily observed experimentally over a range of frequencies and the variation in their behaviour can be related to the structure and properties of the material.

Since both the current and voltage are complex quantities, the resistance also is complex for an AC circuit. The concept of impedance Z is introduced to account for this; for a single resistance, $Z=R$ and for a single capacitance $Z=1/\mathrm{i}\omega C$. In the case of a resistance and capacitance in parallel (Figure 25.3a), the impedances are additive and total impedance is given by

$$\frac{1}{Z}=\frac{1}{R_{\mathrm{P}}}+\mathrm{i}\omega C_{\mathrm{P}} \tag{25.16}$$

Experimentally, a sinusoidal voltage is often applied to the material and this will give rise to an out-of-phase (imaginary) current

$$I_C=\text{Imaginary part of}\left[\frac{V}{Z}\right]=\mathrm{i}\omega C_{\mathrm{P}}V \tag{25.17}$$

and an in-phase (real) current

$$I_R=\text{Real part of}\left[\frac{V}{Z}\right]=\frac{V}{R_{\mathrm{P}}} \tag{25.18}$$

Comparison of Equation 25.17 with 25.13 gives

$$\varepsilon'=\frac{C_{\mathrm{P}}}{C_0} \tag{25.19}$$

and comparing Equation 25.18 with 25.14 gives

$$\varepsilon''=\frac{1}{R_{\mathrm{P}}C_0\omega} \tag{25.20}$$

Equation 25.15 then leads to

$$\tan\delta=\frac{\varepsilon''}{\varepsilon'}=\frac{1}{R_{\mathrm{P}}C_{\mathrm{P}}\omega} \tag{25.21}$$

If the specimen is analysed as a series circuit, as shown in Figure 25.3b, then the impedances are additive such that

$$Z=R_{\mathrm{S}}+\frac{1}{\mathrm{i}\omega C_{\mathrm{S}}} \tag{25.22}$$

An equivalent set of equations for ε', ε'' and $\tan\delta$ can be determined. Simple parallel or series *RC* circuits are often adequate to model the behaviour of polymers, although in practice extra components often are introduced to account for the actual dielectric response of the material.

25.2.3 Dielectric Relaxations

Dielectric measurements often are undertaken upon polymers over a wide range of frequency. The simplest method for use up to about 10^5 Hz is to employ a cell based upon a parallel plate capacitor with a thin disc of solid or liquid polymer between the electrodes. Certain corrections have to be made to account for experimental problems such as stray electric fields at the edge of the capacitor; otherwise, the behaviour can be analysed using the appropriate equations for an equivalent circuit. At frequencies higher than 10^5 Hz, the wavelength of electromagnetic radiation becomes similar to the dimensions of the polymer sample and a parallel plate capacitor can no longer be used. In this case, the dielectric properties have to be followed using alternative techniques that depend upon the attenuation of electromagnetic radiation by the polymer.

Measurement of the dielectric behaviour of polymers over a wide range of frequency and temperature is often referred to as *dielectric spectroscopy*, and this can yield a wealth of information about molecular motions and relaxation processes in the material. Non-polar polymers, such as polyethylene, polypropylene and polytetrafluoroethylene, have relatively low values of ε' (in the range of 2–3) because of the absence of dipoles and ions (assuming they are relatively pure). They also show very low values of the loss factor ε'' (typically $< 10^{-3}$) over a very wide range of frequency. Hence, these materials are employed widely as dielectric insulators in a range of applications, such as electric cables and capacitors in electrical circuits. Polar polymers, such as poly(vinyl chloride), poly(methyl methacrylate), or aromatic polyesters, tend to have low values of ε' and ε'' when they are in the glassy or semi-crystalline state at low temperatures where the dipoles are immobilized. For polar polymers in the liquid or elastomeric state above the T_g, the dipoles are much more mobile and larger values of ε' and ε'' are found.

It is of interest to compare the results of experiments upon dielectric relaxations in polymers with mechanical relaxation data (Section 20.6) obtained on the same material. Figure 25.4 shows a plot of the mechanical loss tangent and dielectric loss tangent as a function of temperature for poly(methyl methacrylate). It can be seen that there are two peaks in both sets of data, but that the β-relaxation is much more prominent than the α-relaxation in the case of the dielectric relaxation analysis. It is known that the α-relaxation corresponds to the glass transition corresponding to the onset of general molecular motion, giving rise to a high value of $\tan\delta$ for the mechanical relaxation. In contrast, the β-relaxation is associated with motion of the relatively polar $-COOCH_3$ ester side group on the non-polar polymer backbone and so is more prominent in the dielectric relaxation analysis.

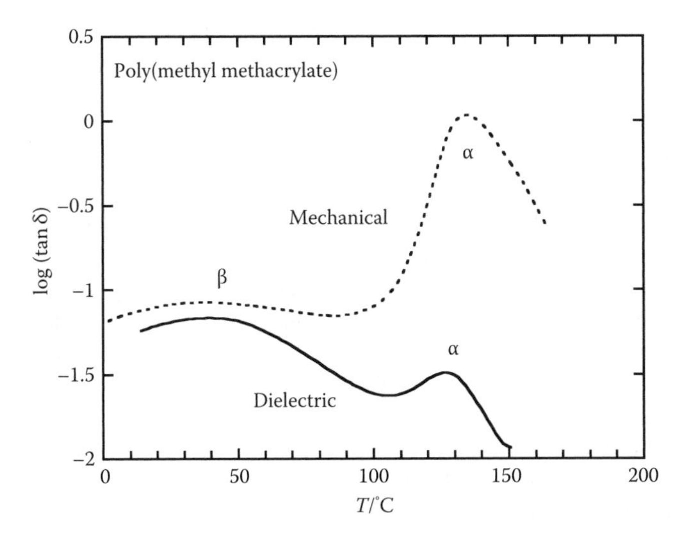

FIGURE 25.4 Comparison of measured values of the dependence of $\tan\delta$ upon temperature for PMMA measured by mechanical and dielectric analysis. (Data taken from Shaw, M.T. and MacKnight, W.J., *Introduction to Viscoelasticity*, 3rd edn., John Wiley & Sons, New York, 2005.)

25.2.4 Dielectric Breakdown

If the strength of the electric field across a polymeric insulator is increased sufficiently, there will come a point at which the insulating properties of the material are lost and dielectric breakdown occurs. This failure can be sudden and catastrophic and can sometimes be accompanied by burning of the material. Attempts to evaluate the dielectric strength of materials are fraught with difficulty because of the random nature of the breakdown process. It often is caused by voids and impurities in the material and attempts to determine it, even using apparently identical samples, have encountered considerable irreproducibility in the data.

A number of possible processes have been identified as leading to dielectric breakdown in polymers. Every material will have an intrinsic electrical breakdown strength at which electrons will be forced to move in the structure at a sufficiently high voltage, although it is unlikely that this has ever been reached for a polymer before other processes have occurred. Thermal breakdown can take place if there is sufficient conductivity in the material to allow Joule heating to occur. It may also be exacerbated in an AC field through heat dissipated in the dielectric relaxation processes described in the previous section. Since the dielectric strength of gases is much less than that of solid insulators, gas-filled voids are potential sources of dielectric breakdown in a polymer. Such voids may be present in the polymer as a result of processing or may develop through degradation of the material as a result of gas discharges in the high electric field. Because polymers are relatively soft materials, electromechanical degradation, due to a combination of mechanical and electrical stress (i.e. high voltages), has been suggested as another source of breakdown, particularly in the case of elastomers.

The random nature of the electrical breakdown of polymers is a challenge for design engineers and statistical methods often are employed to predict the behaviour of the materials over the long term. It has to be borne in mind that general ageing or deterioration of polymers due to degradation or cracking under mechanical stress will also reduce the resistance of a material to electrical failure. Moreover, the sustained application of high electrical fields over long periods of time can lead to progressive and cumulative ageing in the material, thereby inducing further electrical weakness.

25.3 CONDUCTION IN POLYMERS

The conductivity of polymers can vary over many orders of magnitude which makes them extremely versatile materials for a variety of electrical and electronic applications. This section covers the measurement of conductivity, conducting polymer composites, ionic conduction and, finally, the recently-identified inherently-conducting polymers.

25.3.1 Measurement of Conductivity

In the study of electrical conduction in polymers, it is essential to have an accurate and reliable method to determine the *conductivity* or its reciprocal, known as *resistivity*. Although instruments will allow the resistance of a sample to be determined experimentally, the material property of interest is the volume resistivity ρ_V, which is the resistance between the opposite faces of a unit cube of the material and has units of Ω m. The volume conductivity σ_V is defined similarly and has units of Ω^{-1} m^{-1} or S m^{-1}. The conductivity of materials varies over a wide range covering many orders of magnitude; Figure 25.5 shows typical values of the conductivity of a number of different materials, including polymers. Insulators have conductivities typically less than 10^{-12} S m^{-1}, semi-conductors are in the range 10^{-12} S m^{-1} to around 10 S m^{-1} and the conductivity of metals is greater than about 10 S m^{-1}. It can be seen that polymers are now available that span the full range of conductivities found in other materials.

The simplest way to determine the conductivity or resistivity of a polymer is to measure the flow of current between two electrodes at the ends of a strip of material, as shown in Figure 25.6a. In

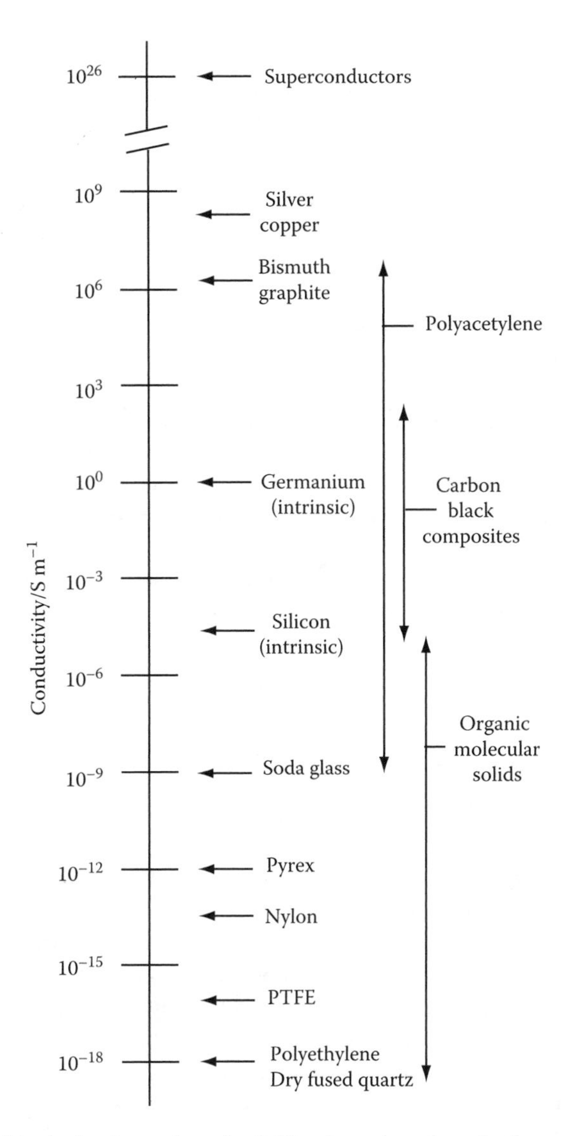

FIGURE 25.5 Chart of typical values of conductivities for polymers and other materials. (Adapted from Blythe, A.R. and Bloor, D., *Electrical Properties of Polymers*, 2nd edn., Cambridge University Press, Cambridge, U.K., 2005.)

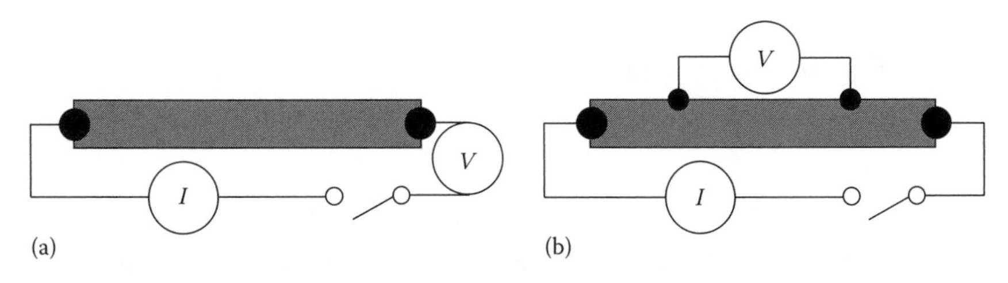

FIGURE 25.6 Electrode configurations for resistivity measurements: (a) two-terminal and (b) four-terminal in line. (Adapted from Blythe, A.R. and Bloor, D., *Electrical Properties of Polymers*, 2nd edn., Cambridge University Press, Cambridge, U.K., 2005.)

the case of a polymer with high resistivity, it may be more appropriate to use a thin disc of material between electrodes and measure the complex impedance. The resistance R of the sample is related to the current I and voltage V through Equation 25.1 and the volume resistivity ρ_V is given by

$$\rho_V = \frac{RA}{l} \tag{25.23}$$

where
A is the cross-sectional area of the specimen
l is the thickness if it is a disc or the distance between the electrodes for a strip of material.

One problem that arises in determination of the conductivity or resistivity of a polymer is contact resistance between the electrodes and the polymer for the two-terminal configuration shown in Figure 25.6a. This can be reduced by using conductive paints at the electrodes rather than metal pressure contacts. A better way, however, is to use a four-point contact method as shown in Figure 25.6b. In this case, there is a current I running through the sample and the potential difference ΔV is determined between two point electrodes separated by a distance Δl along the surface. In this way, contact resistance at the two surface electrodes is not an issue and the resistivity of the polymer is given by

$$\rho_V = \frac{\Delta V/\Delta l}{I/A} \tag{25.24}$$

25.3.2 Conducting Composites

The simplest way to render a polymer conductive is to add a conductive filler, such as a metal powder, to make a conductive composite. The issue that arises, however, is to attain sufficient contact between the metal particles to allow a conductive pathway to be established in the material. There clearly will be a dependence of the conductivity upon the volume fraction ϕ_p of the filler particles. In practice it is found that there is little increase in conductivity with the addition of small quantities of filler, but then a sudden rise of conductivity occurs at a critical level of filler loading. Unfortunately, the addition of the filler usually is accompanied by deterioration in the mechanical properties of the polymer, such as becoming stiff and brittle, and so a careful balance has to be struck between the conductivity and mechanical properties. Nevertheless, conductive metal-containing polymers, such as silver-loaded epoxy resins, are used widely in the electronics industry in a variety of applications, such as cold-solder materials.

In practice it is found that as well as being controlled by the volume fraction of the filler, the conductivity of particulate composites depends upon a range of other factors including particle size and shape and filler distribution. The sudden onset of a rapid rise in conductivity at a critical loading can be analysed using percolation theory, developed initially to model the permeation of fluids through porous solids. This approach predicts a relationship between the conductivity of the material and the volume fraction of the filler of the form

$$\sigma_V \propto (\phi_p - \phi_p^*)^t \tag{25.25}$$

where ϕ_p^* is the critical volume fraction of filler at which the conductivity rises rapidly, the percolation threshold, corresponding to the formation of a three-dimensional conductive network of filler particles. The exponent t generally reflects the dimensionality of the system (e.g. the particle shape) and has values typically between 1.3 and 2.0 for two- and three-dimensional systems, respectively.

Computer simulations of the percolation process have predicted a value of $\phi_p^* \sim 0.16$ for systems containing spherical particles.

In practice, systems demonstrating percolation can be obtained for lower loading levels, significantly less than the 0.16 volume fraction threshold. Carbon black, used widely to reinforce elastomers, can be added to many polymers at low levels of loading to give composites with significant levels of conductivity. An example of this is shown in Figure 25.7a, where the conductivity of a carbon black filled high-density polyethylene is plotted on a log–log scale as a function of the particle volume fraction. It can be seen that there is a rapid rise in conductivity of more than 10 orders of magnitude above the percolation threshold ϕ_p^* corresponding to a carbon-black loading of about 0.10 by volume. The maximum conductivity for the highest loading of carbon black is around 100 S m^{-1}. Carbon black is a useful additive to polymers in that it disperses well and has a low density, typically of the order of 1.8 Mg m^{-3}. The individual particles are approximately spherical and very small with dimensions in the range 20–300 nm, but the highest levels of conductivity are found for the so-called furnace blacks that have *structure*. This terminology refers to the aggregation of individual particles into chains so that they are more like fibrous fillers rather than particulate ones (see Figure 24.21). Analysis of the morphology of the composites shows that this structure persists even when the material is compounded into a polymer. There are a number of techniques that can be used to lower the percolation threshold further. One method involves coating small polymer particles with carbon black and then processing the material by avoiding significant levels of shear so that the particles remain essentially intact and a conductive network is established in the morphology of the moulded object.

Even better levels of conductivity at low loadings can be found using carbon nanotubes, which are fibrous fillers of nano-dimensions (Figure 24.27). They typically are of the order of 1–10 nm in diameter and up to microns in length with a density of the order of 1.4 Mg m^{-3}. Figure 25.7a also shows the dependence of the conductivity upon volume fraction for well-dispersed and aligned carbon nanotubes in an epoxy resin matrix. It can be seen that for this system, the percolation threshold ϕ_p^* is of the order of 0.0025%, several orders of magnitude lower than that obtained with

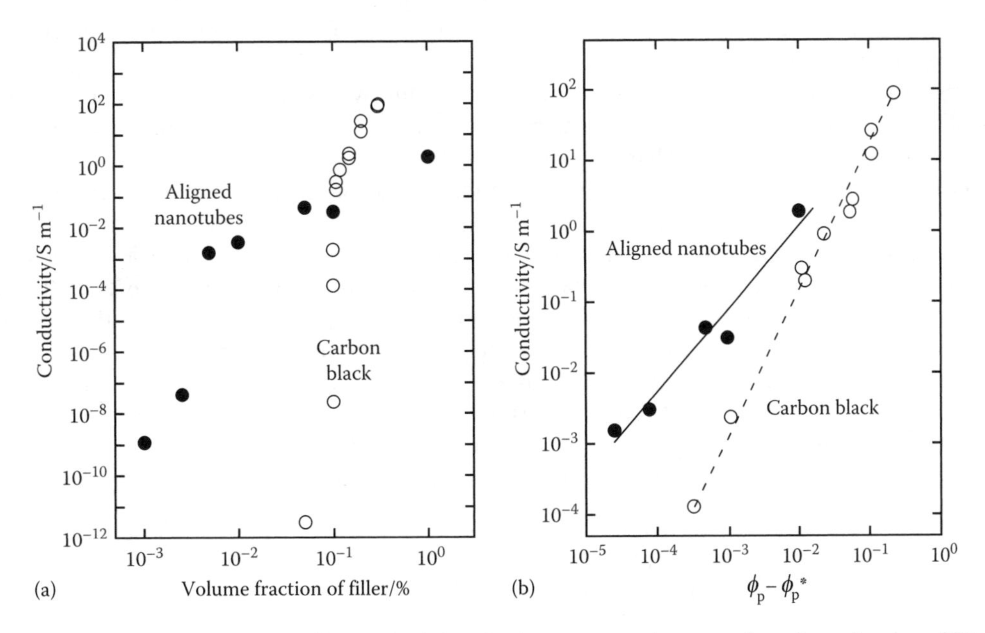

FIGURE 25.7 (a) Dependence of the conductivity of polymer composites upon the volume fraction of filler: aligned carbon nanotubes in an epoxy resin (Data taken from Sandler, J.K.W. et al., *Polymer*, 44, 5893, 2003.) and carbon black in polyethylene (Data taken from Hindermann-Bischkoff, M. and Ehrburger-Dolle, F., *Carbon*, 39, 375, 2001.). (b) Data in Figure 25.7a replotted as a function of $\phi_p - \phi_p^*$.

carbon-black particles. This system offers the possibility of obtaining a polymer with a reasonable level of conductivity at a low level of loading of nanotubes (~0.01%) that does not affect the bulk mechanical properties. The fibrous nature of this nano-filler allows a percolation network to be established at a significantly lower volume fraction than for the particulate carbon-black systems, whilst still remaining transparent.

The behaviour of the carbon-black and nanotube systems is compared in Figure 25.7b where the conductivity in the region of the percolation threshold is plotted against $(\phi_p - \phi_p^*)$ in a log–log plot. It can be seen that the data points for the nanotube composite lie above those for the carbon-black composite, showing the better performance with nanotubes. The exponent t for the two systems can be determined from the slope by applying Equation 25.25. A value of t of about 2 is found for the carbon back and a value of 1.2 for the nanotube composite, reflecting the difference in dimensionality between the two systems.

There are number of practical applications of conducting composites that exploit their response to external variables. The conductivity of conducting composites is sensitive to temperature but can either increase or decrease with increasing temperature. It also is dependent upon mechanical deformation and swelling due to the uptake of solvents. It would be expected that conduction in these systems might be thermally activated and so increase with temperature, but thermal expansion of the matrix polymer can break up a percolating conductive network leading to a reduction in conductivity with increasing temperature. The latter phenomenon, which is found in many carbon-black-based systems, may be used in polymer-based heating elements in which their resistance increases with increasing temperature. This facilitates good temperature control and prevents thermal runaway.

25.3.3 Ionic Conduction

In polymers that exhibit very low levels of conductivity, such as polyethylene, any conductivity measured is usually due to the presence of ionic impurities. In fact, such polymers can have problems with static charges being unable to leak away, giving rise to the attraction of dust and the possible generation of electric discharge shocks. Antistatic agents are employed widely as additives for polymers to avoid such problems and they invariably are based upon ionic conduction allowing the leakage of charge way to earth.

The family of polymers known as *ionomers* have chains with some repeat units that have ionizable groups. When these groups are ionized, one group of ions (which may either be anions or cations) remains fixed to the polymer chain, whilst the counter-ions are able to move under the influence of an electric field. For example, polyamides will show appreciable levels of conductivity (~10^{-8} S m^{-1}) at elevated temperatures thought to be due to the production of protons from the dissociation of the amide groups. A characteristic feature of ionomers is that their behaviour depends strongly upon the water content of the material; they often show no ionic conductivity in the dry state. There has been an increase in demand for polymer-based ionic conductors of this type for use in applications such as hydrogen-based fuel cells in which a membrane will both separate the electrodes and at the same time allow the transfer of protons. One of the most successful ranges of materials for this application is based upon perfluorinated copolymers with side groups terminated with sulphonate groups, such as Nafion®. The chemical structure of this material is shown below and values of the number of repeat units of groups in the structure are as follows: $x=5$–15; $y=1$; $m \geq 1$; and $n=2$.

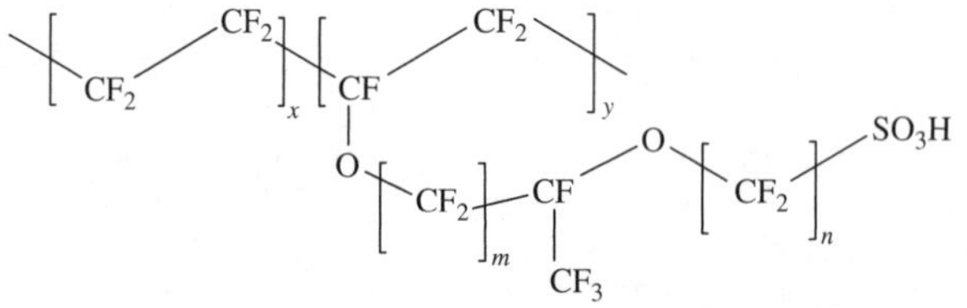

The polymer is a processible thermoplastic that has good mechanical properties combined with excellent thermal and chemical stability. It becomes conductive on the uptake of water and can achieve a conductivity of the order of 1 S m^{-1}.

There is now an increasing interest in using solid polymer electrolytes for applications, such as separators in lithium ion batteries. These materials are based upon the discovery that alkali metal salts, such as $LiClO_4$, could be dissolved in polymers such as poly(ethylene oxide), $(–CH_2–CH_2–O–)_n$, with a maximum loading of 1 mole of the metal salt to 4 moles of the polymer repeat units. This is found to yield a material with a reasonable level of conductivity, up to 10^{-2} S m^{-1}, with the conductivity being due to the rapid movement of the metal ions through the structure.

25.3.4 Inherently Conducting Polymers

Polymers with saturated chemical structures were developed originally as insulating materials, but it became increasingly apparent that there could be considerable potential for polymer-based conductors with high levels of conductivity, combined with the other advantages of polymers such as ease of processibility and light weight. There was, therefore, a rapid growth of interest in inherently conducting polymers following the recognition in 1974 that a conductive material could be made from a conjugated polymer such as polyacetylene (PA) (Section 6.4.7). Since that time, there has been a worldwide interest in developing a very wide range of conducting polymers and the chemical structures of a number of the more important ones are listed in Figure 25.8.

The development of inherently conducting polymers has involved a balance between synthesizing polymers with electronic structures that are capable of undergoing electrical conduction and making materials that are chemically stable and can be processed. Polyphenylenes are clear candidates as conducting polymers and exemplify the difficulties involved. Their properties depend upon their linkage, i.e. *meta-* or *para-*, and although they can be synthesized by a number of techniques and can be around 50% crystalline, they generally are intractable and insoluble coloured powders. Difficulties in processing are highlighted in Figure 25.9, which shows the dependence of the melting point T_m of both poly(*p*-phenylene) and poly(*m*-phenylene) as a function of the number n of chain repeats. Benzene ($n = 1$) has a melting point of 6 °C and the melting point of the oligomers increases as n increases, much more rapidly for the *para-* than the *meta*-derivative. Poly(*p*-phenylene) is soluble in toluene only up to $n = 6$ and poly(*m*-phenylene) up to $n = 16$. The polymer powders can be processed by sintering under pressure at around 400 °C, but poly(*p*-phenylene) degrades at temperatures in excess of 500 °C in air; poly(*m*-phenylene) tends to degrade at an even lower temperature (~300 °C). Although poly(*p*-phenylene) can be doped with either electron donors or electron acceptors and conductivities of up to 10^4 S m^{-1} can be achieved, processing difficulties have limited its applications.

25.3.5 Polyacetylene

25.3.5.1 Structure of Polyacetylene

Polyacetylene is the simplest example of a polymer in which metallic levels of conductivity are found and so it is worth investigating its electronic structure to see how this conductivity occurs. It might be expected that the structure would be similar to that of benzene with the electrons delocalized along the polymer chain as shown in Figure 25.10a. Each carbon atom would contribute one electron to the continuous band of π-electrons in this case, and the structure would be expected to be a one-dimensional metal. This issue was, in fact, considered a number of years earlier than the discovery of how to prepare conductive PA in a useful form. It was shown then that this structure would be unstable to a Peierls distortion whereby it would need to have alternating single and double bonds of different length, as shown in Figure 25.10b rather than delocalized electrons. This is because the energy gained in introducing an energy gap into the electronic structure is less than that needed

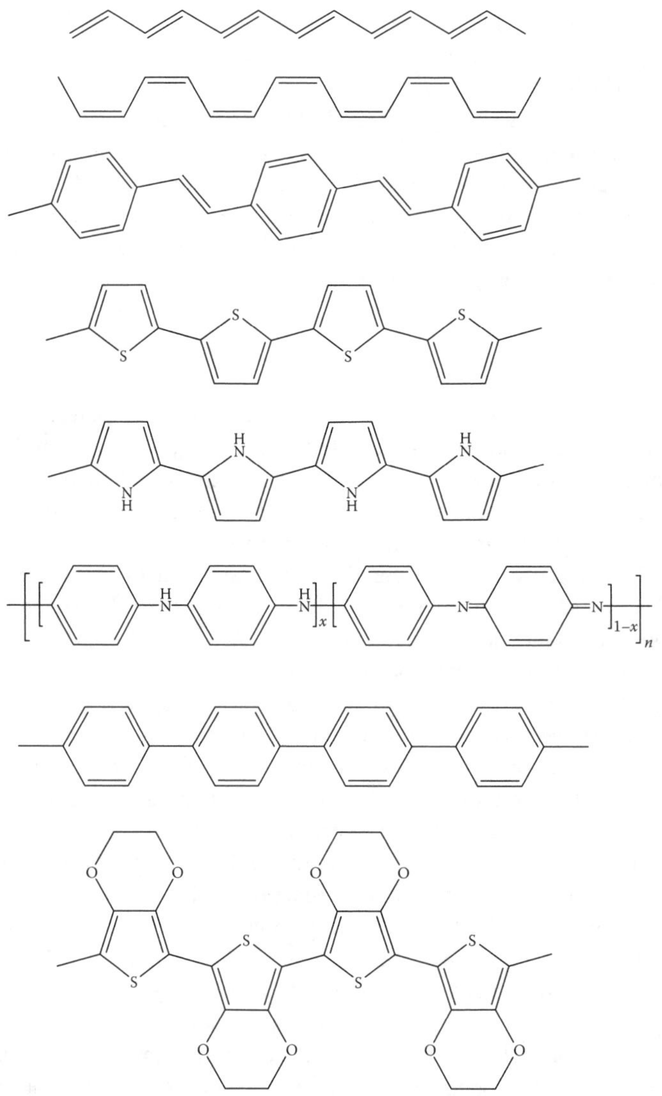

trans-polyacetylene (*trans*-PA)

cis-polyacetylene (*cis*-PA)

poly(*p*-phenylene vinylene) (PPV)

polythiophene (PT)

polypyrrole (PPY)

polyaniline (PANI)

poly(*p*-phenylene) (PPP)

poly(ethylene dioxythiophene) (PEDOT)

FIGURE 25.8 Chemical structures of some important conducting polymers. (Adapted from Fischou, D. and Horowitz, G., *Encyclopaedia of Materials: Science and Technology*, Elsevier, Amsterdam, the Netherlands, 2001.)

to distort the carbon–carbon bonds to equal length. This means that the length of the chain repeat unit in the structure is increased from that in $(–CH–)_n$ by a factor of two in $(–CH{=}CH–)_n$ and the polymer has two degenerate ground states, as shown in Figure 25.10c. Experimental measurements upon pristine PA have shown it to be a semi-conductor with a band gap of around 1.5 eV.

25.3.5.2 Preparation of Conducting Polyacetylene

Polyacetylene had been synthesized as early as 1958 by Natta and co-workers using a conventional type of Ziegler–Natta catalyst (see Section 6.4) in an inert solvent, but was obtained as an intractable powder. A breakthrough came in 1974 when Shirakawa and co-workers synthesized *cis*-PA (Figure 25.8) directly in the form of flexible, fibrillar films (Section 6.4.7), which typically have a conductivity of 1.7×10^{-7} S m^{-1}. The *cis*-PA can be transformed by heating at 150–200 °C to material with the *trans*-PA conformation (Figure 25.8), which is silver in colour and has a conductivity

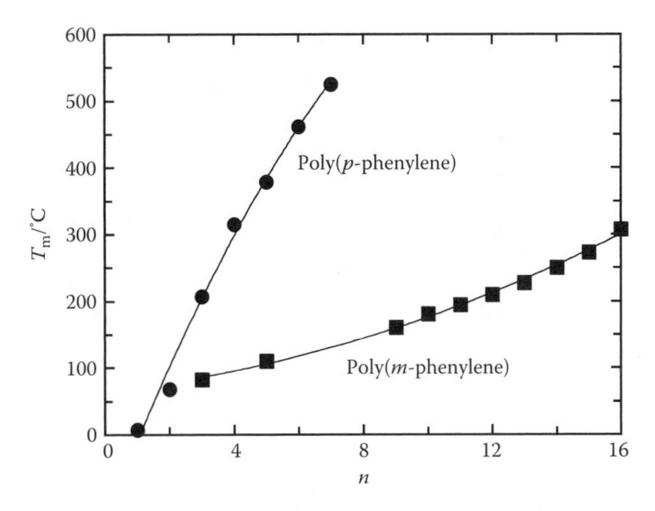

FIGURE 25.9 Dependence of the melting temperature of different isomers of polyphenylene upon the number n of repeat units. (Data taken from Naarmann, H., in *Ullmann's Encyclopedia of Industrial Chemistry*, Wiley-VCH, New York, 2005.)

of about 4.4×10^{-3} S m^{-1}. This material can, therefore, be classified as a semi-conductor, but it can be transformed into a metallic-level conductor by oxidation on exposing the *trans*-PA to iodine or arsenic pentafluoride in a process known as "doping." This process of doping is not the same as the conventional doping of semi-conductors. It is essentially a redox reaction in which a salt is formed between the polymer and doping ion. An example of this is shown in Figure 25.11, where it can be seen that the conductivity increases by 9 orders of magnitude on exposure to AsF_5. The doped film has a conductivity of the order of 10^5 S m^{-1}, which is approaching the value of 10^8 S m^{-1} for metals such as copper (Figure 25.5). The oxidation reaction with iodine, otherwise known as p-doping, is shown in the following text:

$$(CH)_n + \frac{3x}{2} I_2 \rightarrow (CH)_n^{x+} + xI_3^-$$

Metallic levels of conduction also can be obtained by undertaking n-doping (reductive doping) of PA with an alkali metal such as sodium

$$(CH)_n + xNa \rightarrow (CH)_n^{x-} + xNa^+$$

The doping ions have a very low mobility and so are virtually stationary, whilst charges on the polymer can move more easily along the chain backbone due to the presence of delocalized π-electrons.

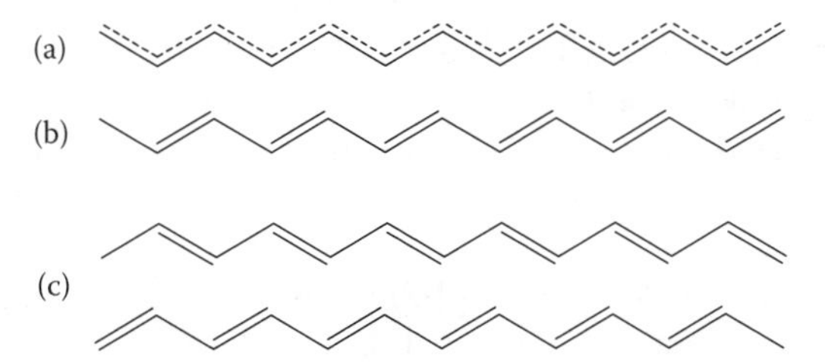

FIGURE 25.10 Polyacetylene: (a) structure with delocalized electrons; (b) structure with alternating single and double bonds due to the Peierls instability; and (c) degenerate A and B phases in *trans*-polyacetylene. (Adapted from Heeger, A.J., *Synthetic Metals*, 125, 23, 2002.)

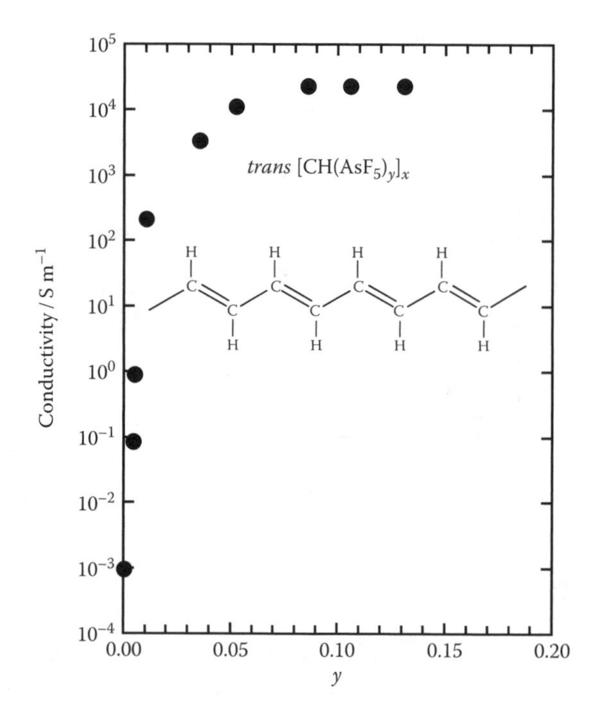

FIGURE 25.11 Increase in room temperature conductivity of *trans*-PA after exposure to arsenic pentafluoride. (Data taken from Heeger, A.J., *Synthetic Metals*, 125, 23, 2002.)

Although it is possible to prepare materials with high levels of conductivity by chemical doping, the process can be somewhat difficult to control and inhomogeneous material often results. Electrochemical doping has now been developed to overcome some of these problems and has proven to be easier to control. As with many other physical properties of polymers, the conductivity of PA is affected by molecular orientation. Stretching the film produces a material in which the conductivity is further increased along the orientation direction to over 2×10^5 S m^{-1} and considerably less in the transverse direction. Well-defined X-ray diffraction patterns can be obtained from such films and the crystal structure of the Shirakawa *trans*-PA has been determined to be in the form of a orthorhombic unit cell (Table 17.1) with $a = 0.733$ nm, $b = 0.409$ nm and $c = 0.246$ nm.

Although PA was the first example of a conducting polymer, it was found that it is inherently unstable under ambient atmospheric conditions and so there has been an upsurge of interest over the past 30 years to find other conjugated polymers that can exhibit high levels of conductivity with better stability and which are more easily processed. Examples of some of these materials are shown in Figure 25.8 and their synthesis is described in Sections 3.2.1.3, 8.3.2 and 8.4.

25.3.5.3 Electronic Structure of Doped Polyacetylene

The electronic structure of PA subjected to oxidative doping with iodine is shown in Figure 25.12a. The vacancy (hole) created does not delocalize completely but a radical-cation is produced. There is a localized distortion of the polymer chain and the radical-cation associated with this local deformation is called a *polaron*. A polaron can carry either positive (cation) or negative (anion) charges.

The formation of a positive polaron (spin = 1/2) by oxidation is shown schematically in Figure 25.12b. The positive charge normally is immobilized in the vicinity of the counter-ion (Figure 25.12a), whereas the unpaired electron (radical) is mobile and capable of moving freely along the chain. The effect of further oxidation also is shown in Figure 25.12b. This yields a positive *bipolaron* (spin = 0) which consists of two separate charges coupled through lattice distortion.

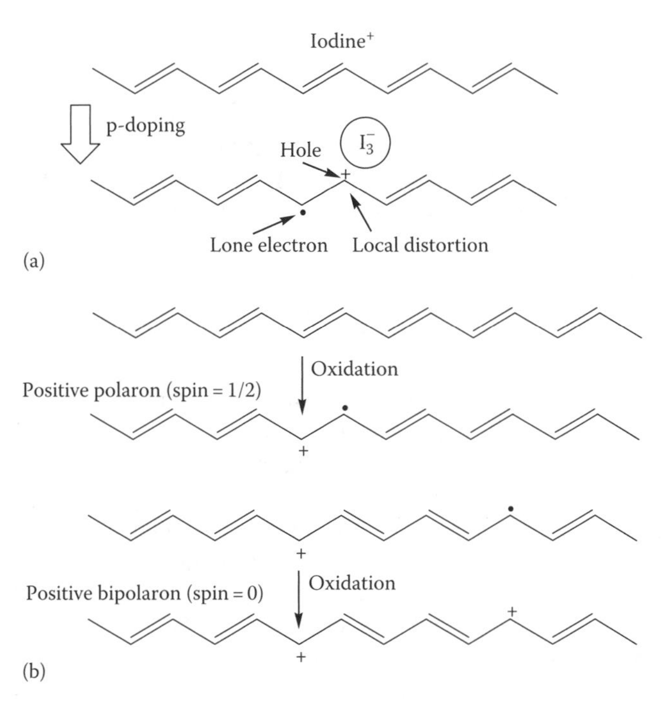

FIGURE 25.12 (a) Oxidation of PA by iodine showing the formation of a radical-cation and a counter anion. (b) Formation of a polaron and bipolaron by the oxidation of PA.

The presence of bond alternation in PA (Figure 25.10a) naturally leads to the possibility of the presence of bond alternation defects capable of moving freely along the chain, as shown schematically in Figure 25.13. This defect is in the form of a single unpaired electron that marks the transition between two regions of mirror-image bond alternation. The defect is termed a *soliton* and they can be neutral, as shown in Figure 25.13, or may carry charge. Neutral solitons in PA are highly mobile as the polymer chains on either side have the same energy. Charged solitons are subject to electrostatic attraction so that their mobility depends upon the number of counter-ions in the vicinity of the polymer chain.

Theoretical investigations of such defects in PA have shown that the sudden change in bond length required for the localized soliton represented in Figure 25.13 would require too much energy and that the distortion of the structure would spread out over a number of carbon atoms, as depicted at the bottom of Figure 25.13. Experimental evidence points to such defects extending over approximately seven carbon atoms. Doping will clearly affect the electronic structure of PA; if strong electron-accepting or electron-donating species are employed, charged solitons may be produced that are also capable of moving charges along the chain.

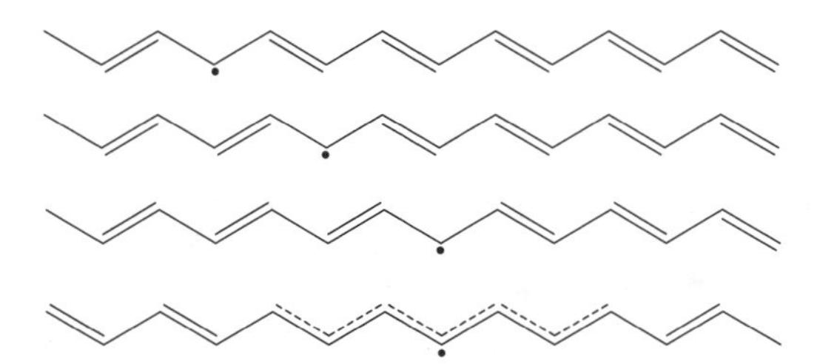

FIGURE 25.13 Motion of solitons in PA also showing delocalization of the bonding (bottom structure).

25.4 POLYMER ELECTRONICS

There has been a rapid upsurge of activity in recent years in using organic materials in electronic applications. Electroluminescence in organic materials was discovered in the 1960s and organic light-emitting diode (OLED) prototype devices were developed within 20 years. The discovery of conjugated conducting polymers has led to further interest in such devices, although the polymers tend to be utilized in their semi-conducting, rather than conducting, states.

25.4.1 POLYMER-BASED LIGHT-EMITTING DIODES

Figure 25.14a shows schematically the structure of a polymer light-emitting diode (LED). The semiconducting polymer is sandwiched between an electron-injecting cathode and a hole-injecting anode. The cathode has to be made of a low work-function conductor such as calcium, aluminium or alloys such as MgAg to provide a low potential barrier to allow electrons to be injected in the conduction band of the polymer. The anode is a high work-function conductor that also has to be transparent to allow the light to come out of the device. Indium tin oxide (ITO) is used widely as the anode material since it is transparent to visible light and has a high work function, which enables holes to be injected into the polymer. The electrons and holes recombine in the device, but this does not lead directly to the emission of light. Initially, an *exciton* (an excited electron-hole state) is formed and then this decays to its ground state and emits light in the visible region.

The first polymer that was used in this application was poly(*p*-phenylene vinylene) PPV (Figure 25.8), the synthesis of which was described in Sections 3.2.1.3 and 8.4. A schematic energy level diagram for an ITO/PPV/Al LED device is shown in Figure 25.14b. A small barrier ΔE_h exists for the injection of holes from the ITO electrode into the valence band states (the highest occupied molecular orbital, HOMO). When Al is used as the cathode, there is a larger barrier ΔE_e for electron injection into the conduction band of the PPV (or the lowest unoccupied molecular orbital, LOMO).

PPV produces a yellow-green luminescence as shown in Figure 25.15. The original devices produced were relatively inefficient in producing light, but devices able to produce light of different

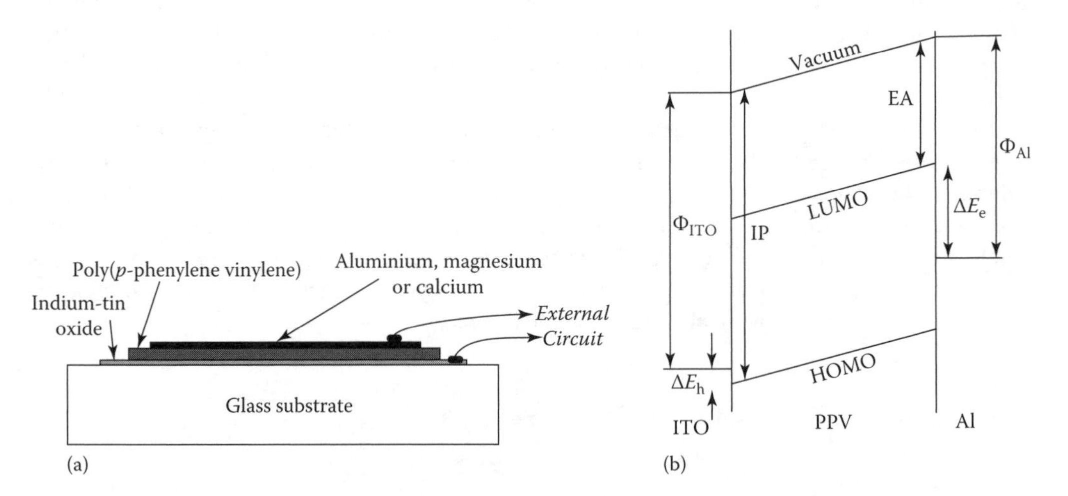

FIGURE 25.14 Single-layer poly(*p*-phenylene vinylene) (PPV) polymer electroluminescent diode: (a) schematic structure of the device and (b) schematic energy level diagram for the device: IP, ionization potential of the PPV; EA, electron affinity of the PPV; Φ_{ITO}, work function of ITO; Φ_{Al}, work function of Al; ΔE_e, barrier to injection of electrons; and ΔE_h, barrier to injection of holes. (Adapted from Friend, R.H. et al., *Nature*, 397, 121, 1999.)

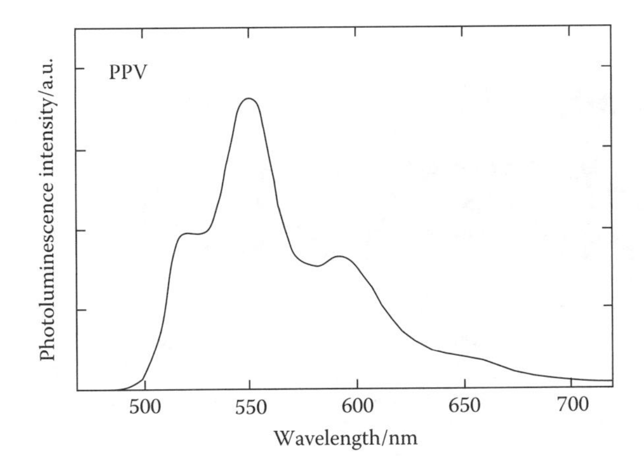

FIGURE 25.15 Photoluminescence spectrum of a PPV-based device of the type indicated in Figure 25.14a, showing the optically-excited luminescence for a thin film on a glass substrate. (Data taken from Friend, R.H. et al., *Nature*, 397, 121, 1999.)

wavelengths with much higher levels of efficiency (up to 25%) have now been developed. Polymers are attractive for this application because of their good mechanical properties and the possibility of undertaking solution processing and spin-coating or ink-jet printing of thin polymer films at relatively low cost. Polymer LEDs are particularly useful for electronic displays because, unlike conventional liquid crystal displays, they do not have to be back lit and allow a wide range of viewing angles. The display can even be flexible if a polymer substrate is used.

25.4.2 Polymer-Based Solar Cells

The polymer LED relies upon the recombination of holes and electrons to produce light, but the process can be inverted such that electrons and holes are produced when light falls on the device. In this case, it acts as a photovoltaic cell, most commonly called a solar cell. The efficiency in such an application, however, is not yet as good as silicon-based solar cells. The impinging photon is absorbed by the molecule which generates an exciton, but this exciton does not decay efficiently into a hole and electron, unless the electric field is very high. This cannot be achieved in the device and therefore the application of conjugated polymers in solar cells has not received as much attention as polymer LEDs.

25.4.3 Polymer-Based Transistors

Field-effect transistors (FET) based upon conjugated polymers have been investigated along with polymer LEDs, but their development lags behind that of polymer LEDs because they do not yet compete successfully with silicon-based electronics. A schematic diagram of a polymer-based FET is shown in Figure 23.16a. It is essentially a capacitor in which one of the plates has been replaced by a semi-conducting layer, with the other plate acting as the gate electrode. The conducting channel of the device is through the polymer semi-conductor film between two electrodes, the source and the drain. If a bias is applied between the source and gate, an equal amount of charge with an opposite sign is induced at each side of the insulating layer. The charge induced at the interface between the insulator and polymer semi-conductor can be driven by a second bias that is applied across the source and drain. The device performs essentially as a variable resistance, with the magnitude of its resistance being controlled though the gate voltage.

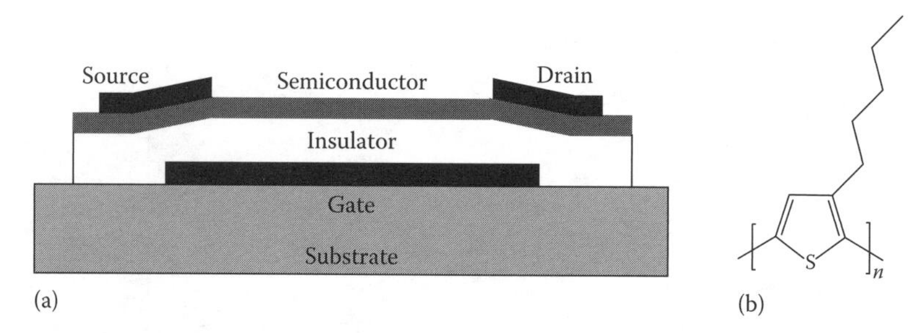

FIGURE 25.16 (a) Schematic diagram of a thin-film polymer transistor (Adapted from Fischou, D. and Horowitz, G., *Encyclopaedia of Materials: Science and Technology*, Elsevier, Amsterdam, the Netherlands, 2001). (b) The structure of semi-conducting polymer poly(3-pentylthiophene) that can be used in polymer-based transistors.

The performance of the FET is judged by the lowest "off" current and the highest "on" current that can be achieved in the device, known as the ON/OFF ratio. This behaviour is controlled strongly by the mobility of the carriers in the semi-conductor, and the organic semi-conductors used in the original devices had mobilities many orders of magnitude less than that of silicon. A great deal of effort has since gone into improving their mobilities, and factors such as purity and structural order are found to be crucial in controlling mobility. It is found that improvements can be obtained by using semi-conducting polymers with pendant side groups. One class of polymers that has proved to be useful are poly(3-alkylthiophene)s (P3HT), an example of which is shown in Figure 25.16b, that has enabled devices with ON/OFF ratios of better than 10^6 to be produced, which starts to rival silicon-based FETs. The pendant alkyl group not only improves solubility, allowing solution processing of the material, but also helps to induce better structural order in films of the material.

Even though polymer-based FETs have their limitations and may not yet be able to rival silicon-based devices, silicon is a brittle material and its technology is somewhat expensive. Polymer-based FETs could be used in all-polymer multilayer integrated optoelectronic circuits. Such devices could be inexpensive, easily processible and used in applications such as flexible polymer-based displays.

25.4.4 Polymer-Based Sensors

It is possible to envisage a number of different applications of conjugated polymer in sensors. For example, it was shown earlier that the conductivity of polyacetylene changes on exposure to different chemical species through doping. Hence, an electrochemical cell could be constructed in which the variation in resistance of a conjugated polymer is determined on exposure to an active chemical. A more sophisticated approach is to use a chemically-sensitive FET in which the polymer gate electrode is exposed to the species under analysis. This can cause a change in the gate electrode potential, which in turn affects the drain current. Some of the earliest devices of this type were base upon electrolytically-deposited films of polypyrrole (Figure 25.8), but the advent of soluble conjugated polymers has enabled solution processing to be used to fabricate sensors.

Doping of the conjugated polymers with specific species can be used to make them responsive to particular volatile organic compounds enabling them to function as an *electronic nose*, although difficulties have been encountered in making them sensitive enough to specific analytes. Developments also have taken place in the field of biosensors by, for example, using conjugated polymers as a host matrix for enzymes, antibodies and cells. Applications have been investigated such as developing devices for the detection of molecules such as glucose and cholesterol, in environmental monitoring and in the food industry.

PROBLEMS

25.1 The following data* were obtained upon the conductivities and densities of a range of metals and two different types of polyacetylene (PA). Determine the specific conductivity (i.e. conductivity per unit mass) for each material and comment upon the values you obtain.

Metal or PA	Conductivity/S m^{-1}	Density/Mg m^{-3}
Mercury	1,040,000	13.54
Platinum	10,150,000	21.45
Chromium	7,750,000	7.19
Iron	10,300,000	7.87
Gold	47,000,000	19.3
Silver	67,100,000	10.49
Copper	64,500,000	8.94
PA—doped with $FeCl_3$	90,000	1.10
PA—oriented and doped	12,000,000	1.15

25.2 The following data were measured for the conductivity σ_V of polymer nanocomposites with different volume fractions of nanoparticles ϕ_p. For each composite determine the percolation threshold ϕ_p^* and the exponent t in the equation relating the conductivity of the material to the volume fraction of filler

$$\sigma_V \propto (\phi_p - \phi_p^*)^t$$

Comment upon the results you obtain.

ϕ_p/%	Conductivity/S m^{-2}
Carbon black in polyethylene	
0.049830	3.1623e–12
0.099308	2.4342e–08
0.099769	0.00013689
0.099769	0.0019745
0.10894	0.16876
0.10844	0.30010
0.11951	0.73053
0.14923	1.7783
0.14786	2.4342
0.19883	12.991
0.19883	28.480
0.29742	85.471
0.30298	94.901
Carbon nanotubes in epoxy resin	
0.0010000	1.1788e–09
0.0025161	4.2170e–08
0.0049333	0.0016379
0.0100000	0.0035774
0.050160	0.045784
0.10083	0.032950
1.0168	1.9307

* Data of Naarmann, in *Ullmann's Encyclopedia of Industrial Chemistry*, Wiley-VCH, New York, 2005.

25.3 *Trans*- and *cis*-polyacetylene (PA) have orthorhombic unit cells with packing of the polymer chains similar to that in polyethylene unit cells.

(a) Given the following information calculate the densities of *trans*- and *cis*-PA crystals, where c is the polymer chain direction.
Trans-PA: $a=0.733$ nm, $b=0.409$ nm and $c=0.246$ nm
Cis-PA: $a=0.762$ nm, $b=0.444$ nm and $c=0.438$ nm

(b) Explain why the values of c differ by a factor of almost 2 in the two crystal structures.

(The Avogadro constant is 6.022×10^{23} mol^{-1} and the molar mass of C and H atoms may be taken as 12 g mol^{-1} and 1 g mol^{-1}, respectively.)

FURTHER READING

Blythe, A.R. and Bloor, D., *Electrical Properties of Polymers*, 2nd edn., Cambridge University Press, Cambridge, U.K., 2005.

Fischou, D. and Horowitz, G., *Encyclopaedia of Materials: Science and Technology*, Elsevier, Amsterdam, the Netherlands, 5748, 2001.

Friend, R.H., Gymer, R.W., Holmes, A.B., Burroughes, J. H., Marks, R.N., Taliani, C., Bradley, D.D.C., Santos, D.A. Dos, Brédas, J.L., Lögdlund, M., and Salaneck, W.R., Electroluminescence in conjugated polymers, *Nature*, **397**, 121 (1999).

Heeger, A.J., Nobel Prize Lecture, *Synthetic Metals*, **125**, 23 (2002).

Jang, J., Conducting polymer nanomaterials and their applications, *Advances in Polymer Science*, **199**, 189 (2005).

Pron, A. and Rannou, P., Processible conjugated polymers, *Progress in Polymer Science*, **27**, 135 (2002).

Shaw, M.T. and MacKnight, W.J., *Introduction to Viscoelasticity*, 3rd edn., John Wiley & Sons, New York, 2005.

Answers to Problems

Note that answers are provided for all numerical problems and for non-numerical problems where simple indications of answers can be given.

PART I Concepts, Nomenclature and Synthesis of Polymers

Chapter 1: Concepts and Nomenclature

1.1 $\bar{x}_n = 860$ and $\bar{M}_w = 134{,}160\ \text{g mol}^{-1}$

1.2 $M_0^{\text{cop}} = 30.67\ \text{g mol}^{-1}$ and $\bar{x}_n = 1300$

1.3 (a) $\bar{M}_n = 46{,}667\ \text{g mol}^{-1}$, $\bar{M}_w = 78{,}571\ \text{g mol}^{-1}$ and $\bar{M}_w/\bar{M}_n = 1.68$
(b) $\bar{M}_n = 20{,}930\ \text{g mol}^{-1}$, $\bar{M}_w = 46{,}667\ \text{g mol}^{-1}$ and $\bar{M}_w/\bar{M}_n = 2.23$
(c) $\bar{M}_n = 43{,}384\ \text{g mol}^{-1}$, $\bar{M}_w = 86{,}950\ \text{g mol}^{-1}$ and $\bar{M}_w/\bar{M}_n = 2.00$

Chapter 2: Principles of Polymerization

2.1 Synthesis of poly(vinyl chloride) from vinyl chloride proceeds by chain polymerization, whereas synthesis of poly(ethylene terephthalate) from 1:1 stoichiometric reaction of ethylene glycol with terephthalic acid proceeds via step polymerization.

2.2 The data show that increasing the length of the methylene chain in the acid has no effect on the reactivity of the $-CO_2H$ group except at very short lengths ($x = 1$–3) and provide evidence to justify the principle of equal reactivity of functional groups, that is, that the inherent reactivity of a reactive species is independent of its degree of polymerization.

2.3 This would lead to the formation of branches and, ultimately, network polymer.

Chapter 3: Step Polymerization

3.1 (a) 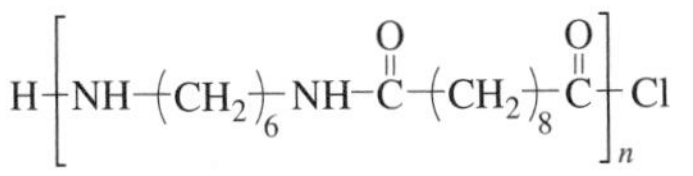 $H\!\left[NH\!-\!(CH_2)_6\!-\!NH\!-\!\overset{O}{\overset{\|}{C}}\!-\!(CH_2)_8\!-\!\overset{O}{\overset{\|}{C}}\right]_n\!Cl$

(b) $O{=}C{=}N\!-\!(CH_2)_6\!-\!NH\!-\!\overset{O}{\overset{\|}{C}}\!\left[O\!-\!(CH_2)_6\!-\!O\!-\!\overset{O}{\overset{\|}{C}}\!-\!NH\!-\!(CH_2)_6\!-\!NH\!-\!\overset{O}{\overset{\|}{C}}\right]_{n-1}\!O\!-\!(CH_2)_6\!-\!OH$

(c) HO–C(=O)–C6H4–[benzobisthiazole (S, N)–C6H4]$_{n-1}$–benzothiazole (S, N) bearing SH and NH_2

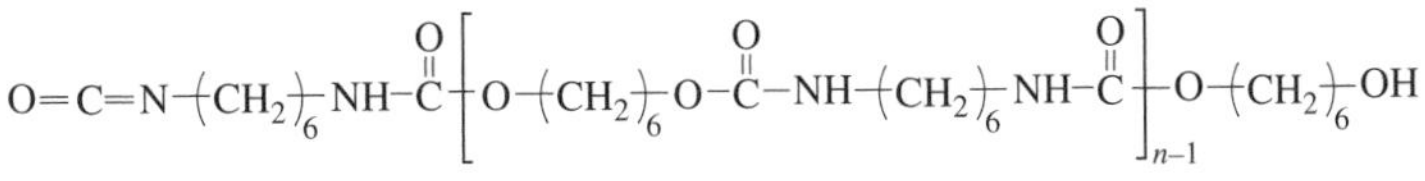

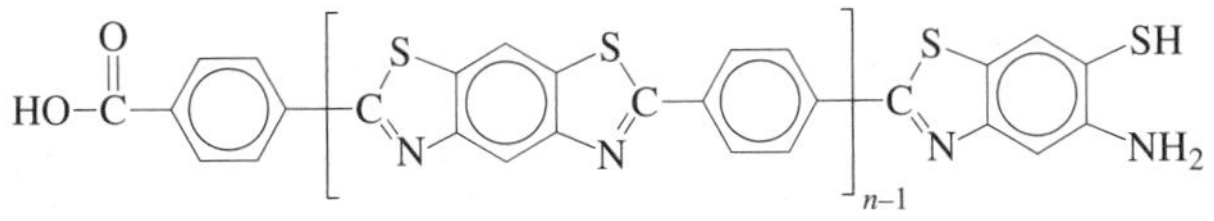

3.2 (I) $n\,O{=}C{=}N{-}C_6H_4{-}CH_2{-}C_6H_4{-}N{=}C{=}O + n\,HO{-}\!\!\left(CH_2\right)_4\!\!{-}OH \longrightarrow$

(II) $n\,HO{-}\overset{O}{\overset{\|}{C}}{-}C_6H_4{-}\overset{O}{\overset{\|}{C}}{-}OH + n\,HO{-}\!\!\left(CH_2\right)_6\!\!{-}OH \longrightarrow$

(III) $n\,H_2N{-}\!\!\left(CH_2\right)_7\!\!{-}NH_2 + n\,ClOC{-}\!\!\left(CH_2\right)_5\!\!{-}COCl \longrightarrow$

3.3 Plots of second- and third-order kinetics equations show that the reaction is second order, indicating that a catalyst was used, in which case the rate coefficient $k = 2.42 \times 10^{-4}$ $dm^3\ mol^{-1}\ s^{-1}$. After 1 h $p = 0.731$ and after 5 h $p = 0.931$.

3.4 Percentage conversion = 99.0% (i.e., $p = 0.990$).

3.5 $\bar{x}_n = 75.9$, $\bar{M}_n = 8.58$ kg mol^{-1}, and $\bar{M}_w = 17.07$ kg mol^{-1}.

3.6 Mixture before reaction: $\bar{M}_n = 15{,}000$ g mol^{-1}, $\bar{M}_w = 40{,}000$ g mol^{-1}. The product from the ester interchange reaction has: $\bar{M}_n = 15{,}000$ g mol^{-1}, $\bar{M}_w = 30{,}000$ g mol^{-1}.

3.7 $n\,H_2N{-}\!\!\left(CH_2\right)_4\!\!{-}NH_2 + n\,ClOC{-}\!\!\left(CH_2\right)_8\!\!{-}COCl \longrightarrow H{-}\!\!\left[NH{-}\left(CH_2\right)_4{-}NH{-}\overset{O}{\overset{\|}{C}}{-}\left(CH_2\right)_8{-}\overset{O}{\overset{\|}{C}}\right]_n\!\!{-}Cl$

(b) Nylon 4.10 or poly(tetramethylene terephthalamide) or poly(butylene terephthalamide)

(c) $p = 0.995$

(d) The monomer mixture has 1:1 NH_2:COCl stoichiometry and so inclusion of the tetrafunctional amine will lead to the formation of a network polyamide.

3.8 Flory statistical theory gives $p_c = 0.816$ whereas Carothers theory gives $p_c = 0.917$.

3.9 No short indicative answer.

3.10 Branching factors are: (a) 0.71 and (b) 0.86.

Chapter 4: Radical Polymerization

4.1 Half-life = 24.07 h. The initiator concentration will have decreased by a factor of 0.972 (i.e., to $0.972[I]_0$).

4.2 (i) (a) No effect; (b) increases by a factor of 2.
(ii) (a) Increases by a factor of 4; (b) increases by a factor of 2.
(iii) (a) Increases by a factor of 4; (b) decreases by a factor of 2.

4.3 Combination.

4.4 (a) $[\text{benzoyl peroxide}]_0 = 3.336 \times 10^{-3}$ mol dm^{-3}.
(b) (i) $[M^\bullet] = 1.0 \times 10^{-8}$ mol dm^{-3} and (ii) $R_p = 2.887 \times 10^{-5}$ mol dm^{-3} s^{-1}.

4.5 $R_p = 1.33 \times 10^{-5}$ mol dm^{-3} s^{-1} and $\bar{M}_n = 796.2$ kg mol^{-1}.

4.6 R_p is increased by a factor of 2.64, and $\bar{x}_n$ is reduced by a factor of 0.699.

4.7 $R_p = 3.43 \times 10^{-5}$ mol dm^{-3} s^{-1}.

4.8 (a) [styryl-TEMPO] $= 4.846 \times 10^{-2}$ mol dm^{-3}; and (b) time to 60% conversion = 25 h 56 min.

4.9 (a) Free-radical emulsion polymerization; (b) reversible-addition-fragmentation chain-transfer solution polymerization; (c) free-radical solution polymerization; (d) free-radical bulk polymerization (for sheet materials) or suspension polymerization (for moulding materials); (e) free-radical catalytic chain transfer bulk or solution polymerization; (f) free-radical solution polymerization; and (g) free-radical suspension polymerization.

Chapter 5: Ionic Polymerization

5.1 $\bar{M}_n = 5470$ g mol^{-1}.

5.2 $R_p = 0.165$ mol dm^{-3} s^{-1} and $\bar{M}_n = 1560$ kg mol^{-1}.

5.3 No short indicative answer.

Chapter 6: Stereochemistry and Coordination Polymerization

6.1 (a) Tactic polymer with two stereochemical forms of the repeat unit.
(b) The repeat unit has one form only (because there is no asymmetric carbon atom).
(c) The repeat units from 1,2- and 3,4-addition are tactic and have two stereochemical forms each. The repeat units from 1,4-addition have *cis*- and *trans*-forms.

6.2 $\theta_M = 0.95$ and $C_p^* = 3.6 \times 10^{-6}$ mol dm^{-3}.

6.3 (a) Ziegler–Natta or metallocene polymerization of but-1-ene.
(b) Anionic polymerization of methyl methacrylate at low temperature (e.g., −78 °C) using a bulky initiator that provides a small counter-ion (e.g., fluorenyl lithium) in a hydrocarbon (e.g., benzene) that is a solvent for both the monomer and polymer.
(c) Metallocene polymerization of ethylene using a diimine Pd complex as catalyst.

Chapter 7: Ring-Opening Polymerization

7.1 (a) cyclo-$(Si(CH_3)(Ph)—O)_3$ anionic

(b) $CF_2—CF(CF_3)$, O (epoxide ring) anionic

(c) cyclopentene metathesis

(d) O=C, O (five-membered lactone ring) anionic

(e) $CH_2{=}CH—CH—CCl_2$, CH_2 (cyclopropane ring) free-radical

7.2 No short indicative answer.

7.3 (i) O=C, NH (five-membered lactam ring) anionic

(ii) $CH_2{=}C$, O—CH_2—CH_2—NH (ring) free radical

Chapter 8: Specialized Methods of Polymer Synthesis

8.1 No short indicative answer.

8.2 A copolymer with a blocky structure would form first, but with time it will become more and more random because the mechanism involves attack at any point along an existing chain.

8.3 No short indicative answer.

Chapter 9: Copolymerization

9.1 (i) The mol% styrene repeat units in the copolymers formed are: (a) 42.7, (b) 51.0, (c) 59.7, (d) 57.4, (e) 50.5, (f) 94.6, (g) 98.2; (ii) the mol% styrene required in comonomer mixtures are: (a) 57.2, (b) 48.5, (c) 32.9, (d) 24.0, (e) any value in the range 0.001–99.999, (f) 3.3, (g) 1.3. The mol% azeotropic compositions with respect to styrene are: (b) 52.9, (c) 76.6, (d) 61.5, (e) 50.5 (the other copolymerizations are not azeotropic).

9.2 No short indicative answer.

9.3 $r_S = 0.40$ and $r_{AN} = 0.04$.

9.4 The question describes synthesis of a polystyrene-*block*-polyisoprene-*block*-polystyrene tri-block copolymer which would be suitable for use as thermoplastic elastomer. The polystyrene end-blocks have $\bar{x}_n = 144$ and $\bar{M}_n = 15.0$ kg mol^{-1} and the central polyisoprene block has $\bar{x}_n = 735$ and $\bar{M}_n = 100.0$ kg mol^{-1} (neglecting the methylene linking group).

PART II CHARACTERIZATION OF POLYMERS

Chapter 10: Theoretical Description of Polymers in Solution

10.1 $\chi = 0.426$.

10.2 No short indicative answer.

10.3 (a) 1320 nm; (b) 1078 nm; (c) 20.17 nm.

Chapter 11: Number-Average Molar Mass

11.1 $\bar{M}_n = 1.5$ kg mol^{-1}.

11.2 $\bar{M}_n = 143.6$ kg mol^{-1} and $\chi = 0.463$.

11.3 $\bar{M}_n = 2.0$ kg mol^{-1}.

11.4 (a) Of the methods available for the evaluation of $\bar{M}_n$ in this range, end-group analysis would be best and could be carried out via acetylation of the hydroxyl groups or by NMR spectroscopy using as solvents pyridine and deuterium oxide, respectively. (b) Membrane osmometry with a highly polar solvent, such as dimethylformamide, is required.

Chapter 12: Scattering Methods

12.1 (a) $\bar{M}_w = 735.8$ kg mol^{-1}; (b) $\left\langle \overline{s^2} \right\rangle_z^{1/2} = 48.4\,\text{nm}$; (c) $A_2 = 4.19 \times 10^{-4}$ m^3 mol kg^{-2}.

12.2 $\left\langle \overline{R_h} \right\rangle_{hz} = 24.1\,\text{nm}$.

12.3 No short indicative answer.

Chapter 13: Frictional Properties of Polymers in Solution

13.1 Sample B: $[\eta]=26.16$ cm^3 g^{-1}; sample E: $[\eta]=127.28$ cm^3 g^{-1}; $K=8.15\times10^{-2}$ cm^3 g^{-1} (g mol^{-1})$^{-a}$ and $a=0.50$.
13.2 $\bar{M}_v=65.6$ kg mol^{-1}.
13.3 $K_\theta=5.74\times10^{-2}$ cm^3 g^{-1} (g mol^{-1})$^{1/2}$. α_η has the following values in order of increasing molar mass: 1.15; 1.23; 1.26; 1.28; and 1.31 (i.e., α_η increases with molar mass due to the increased volume exclusion as chain length increases).

Chapter 14: Molar Mass Distribution

14.1 $\chi=1.66$.
14.2 $M=119.15$ kg mol^{-1}.
14.3 $\lambda=8.75\times10^{-3}$ and $D=8.93\times10^{-12}$ m^2 s^{-1}.
14.4 $M=21{,}185$ g mol^{-1}.

Chapter 15: Chemical Composition and Molecular Microstructure

15.1 $\varepsilon=258$ dm^3 mol^{-1} cm^{-1}.
15.2 **I**: C≡N stretch; **II**: N–H stretch, but no C=O stretch; **III**: C=O stretch and O–H stretch; **IV**: C=O stretch and 1,4-disubstituted benzene C–H bend; **V**: 1,4-disubstituted benzene C–H bend, but no C=O stretch.
15.3 No short indicative answer.
15.4 Sample A approximates to Bernoullian statistics and is predominantly syndiotactic, indicating preparation by radical polymerization. Sample B is predominantly isotactic and not Bernoullian, so has most likely been prepared by living anionic polymerization at low temperature using a non-polar solvent with Li$^+$ as the counterion.

Sample	Mole Fractions of Triads				Bernoullian Test
	x_{mm}	x_{mr}	x_{rr}	P_m	$\frac{4x_{mm}x_{rr}}{x_{mr}^2}$
A	0.04	0.33	0.63	0.205	0.93
B	0.95	0.04	0.01	0.970	23.8

15.5 Since MALDI is to be used, it is reasonable to assume that each *m/z* species detected will have only a single charge. (a) The series of *m/z* peaks will have values corresponding to either $18+44p$ Da **e**$^{-1}$ (two OH end-groups) or $32+44p$ Da **e**$^{-1}$ (one methoxy and one hydroxyl end-group. (b) The separation of the *m/z* peaks will be 128 Da **e**$^{-1}$ for poly(*t*-butyl acrylate) and 72 Da **e**$^{-1}$ for poly(acrylic acid). (c) The *m/z* peaks should have values corresponding to $114(1+x)$ Da **e**$^{-1}$.

PART III Phase Structure and Morphology of Bulk Polymers

Chapter 16: The Amorphous State

16.1 The glass transition temperature $T_g=89$ °C.
16.2 Assuming the rule of mixtures, the copolymer composition should be 82% (by weight) *n*-butyl acrylate and 18% (by weight) styrene. The material would only be transparent at room temperature since $\bar{n}$ will vary differently with temperature for each component.
16.3 (a) Polyethylene (flexible molecule) lower than polypropylene (side group hinders rotation). (b) Poly(methyl acrylate) lower than poly(vinyl acetate) (bulky side group further from main chain).

(c) Poly(but-1-ene) (long bulky side group) higher than poly(but-2-ene) (shorter side groups).
(d) Poly(ethylene oxide) (flexible molecule) lower than poly(vinyl alcohol) (–OH side groups).
(e) Poly(ethyl acrylate) (long flexible side group) lower than poly(methyl methacrylate) (shorter bulky side groups).
NB All pairs are isomers.

16.4 The copolymer has $T_g = 246$ K.

16.5 The average number of branches per molecule = 3.15

Chapter 17: The Crystalline State

17.1 The planes giving rise to the reflections are (110), (200) and (020).

17.2 (a) Chain repeat, $c = 0.254$ nm
(b) Higher order lines cannot be obtained since $\sin\phi \leq 1$ (Figure 17.3). Second order lines could be obtained if X-rays with wavelength $\lambda < 0.127$ nm were used.
(c) The specimen would have to be tilted by 17.65°.

17.3 (a) Density of the polypropylene crystal = 945.04 kg m^{-3}
(b) Volume fraction of crystals = 27.92%
(c) Polymer single crystals—fold regions reduce crystallinity by 20% as the folds are unable to crystallize.

17.4 Volume fraction and weight fraction of crystals $\phi_c = 43.75\%$ and $x_c = 46.70\%$

17.5 (a) $K_{DSC} = 1.02$ and (b) $x_c = 68.9\%$

17.6 Equilibrium melting temperature, $T_m^o = 349\,\text{K}$. The fold surface energy, $\gamma_e = 0.048$ J m^{-2}

17.7 The equilibrium melting temperature, $T_m^o = 79.5$ °C.

17.8 (a) The behaviour is due to the "odd-even" effect. Polyamide crystals with even values of x can undergo H-bonding more easily than with odd x.
(b) The presence of the p-phenylene groups will make the polymer chains much stiffer and so increase T_m greatly.

Chapter 18: Multicomponent Polymer Systems

18.1 (a) and (b) No short indicative answers.
(c) The value of ϕ_{1c} at the critical point when the two polymers have the same number of segments is 0.5. The corresponding critical value of the Flory–Huggins interaction parameter χ_c is $2/x$

18.2 The T_g of the blend with $\phi_{PVDF} = 0.8$ is −18 °C. Since the T_g varies linearly with ϕ_{PVDF}, the blend appears to be compatible and so the microstructure will be single phase.

18.3 NR/NBR 50/50: Incompatible blend.
NR/CR 50/50: Incompatible blend.
NR/BR 50/50: Compatible blend.

18.4 The thickness of the polyisoprene layer is approximately 8.30 nm.

PART IV Properties of Bulk Polymers

Chapter 19: Elastic Deformation

19.1 (a) and (b) No short indicative answers
(c) There will be no volume change when $\nu = 1/2$, which implies that $K = \infty$. (In practice $\nu < 1/2$ and $K \gg E$)
(d) $$\Delta V = (\sigma_1 + \sigma_2 + \sigma_3)\left[\frac{(1-2\nu)}{E}\right]$$

(e) This combination of stresses gives no volume change, equivalent to a uniaxial stress of $\sigma_1 = 0$ MPa

19.2 (a) $E = 175.2$ GPa

(b) The analysis underestimates the modulus of the crystal as it does not consider interchain forces

19.3 The chain direction Young's modulus is approximately 300 GPa

Chapter 20: Viscoelasticity

20.1 (a) Maxwell model: $\frac{de}{dt} = \frac{R}{E} + \frac{Rt}{\eta}$

Voigt Model: $e = \frac{R}{E}\left\{t - \tau_0\left[1 - \exp\left[-\frac{t}{\tau_0}\right]\right]\right\}$

where constant stressing rate, $R = d\sigma/dt$

(b) Maxwell model: $\sigma = S\eta\left[1 - \exp\left[-\frac{t}{\tau_0}\right]\right]$

Voigt model: $\sigma = ESt + S\eta$

where constant straining rate, $S = \frac{de}{dt}$

20.2 Time before seal leaks = 208 days.

20.3 The total strain at $t = 200$ s is 0.006

20.4 (a) The strain after 100 s (before the stress is reversed) = 0.02361

(b) The residual strain when the stress falls to zero = 0.02515

20.5 $\Delta U = \sigma_0 e_0 \pi \sin\delta$

Chapter 21: Elastomers

21.1 $\Delta S = -\frac{1}{2}N\mathbf{k}(\lambda_1^2 + \lambda_2^2 + \lambda_3^2 - 3)$

21.2 $\sigma_t = G(\lambda^2 - 1/\lambda^4)$

21.3 (a) $M_c = 5308$ g mol^{-1}

(b) $\chi = 0.41$ for natural rubber in benzene

Chapter 22: Yield and Crazing

22.1 Necking occurs when $e = b/(1-b)$

Tensile stress at yield, $\sigma_t = a\left[\frac{b}{(1-b)}\right]^b = ab^b(1-b)^{-b}$

22.2 The hydrostatic pressure to cause yielding, $p_c = \frac{\sigma_{yc}\sigma_{yt}}{3(\sigma_{yc} - \sigma_{yt})}$

22.3 See Bevis, M., *Colloid and Polymer Science*, **256**, 234 (1978).

Molecular Plane	(a) Twinning Plane	(b) Angle of Bend	(c) Direction of Shear
(200)	(202)	37.8°	$[20\bar{2}]$
(020)	(022)	54.5°	$[02\bar{2}]$
(110)	(112)	34.4°	~$[11\bar{2}]$*

* Irrational.

22.4 $\sigma_1 = 59$ MPa and $\sigma_2 = -30$ MPa
or $\sigma_1 = -30$ MPa and $\sigma_2 = 59$ MPa

Chapter 23: Fracture and Toughening

23.1 Inherent flaw size, $a_0 = 55$ μm
23.2 (a) Fracture energy, $G_c = 1.3$ kJ m^{-2}
(b) Fracture toughness, $K_c = 2.0$ MPa $m^{-1/2}$
(c) Inherent flaw size, $a_0 = 1$ mm
23.3 The time to failure is given by

$$t_f = \frac{1}{qK^n w^n (n-1) a_0^{n-1}}$$

Chapter 24: Composites

24.1 (a) $W_m = 0.4526$ and $W_f = 0.5474$
(b) $W_m = 0.4505$, $W_c = 0.2481$ and $W_a = 0.3015$
24.2 (a) Hexagonal array $\phi_f = \frac{\pi}{2\sqrt{3}}\left(\frac{r}{R}\right)^2$; Maximum 0.9069
(b) Square array $\phi_f = \frac{\pi}{4}\left(\frac{r}{R}\right)^2$; Maximum 0.7854
24.3 $E_m = 2.775$ GPa and $E_f = 69.025$ GPa
24.4 (a) $E_1 = 36.65$ GPa
(b) $e_1 = e_f = e_m = 0.1637\%$ (uniform strain situation)
(c) The load on the fibres is 17,190 N; the load on the matrix is 810.3 N

(Hence the fibres are carrying more than 20× the load on the matrix)

Chapter 25: Electrical Properties

25.1

Metal or PA	Specific Conductivity / S m^2 Mg^{-1}
Mercury	76,809
Platinum	473,193
Chromium	1,077,885
Iron	1,308,767
Gold	2,435,233
Silver	6,396,568
Copper	7,214,765
PA—doped with $FeCl_3$	81,818
PA—oriented and doped	10,434,782

Oriented PA has a much better conductivity than non-oriented material.
The specific conductivity of the PA is as good as the metals due to its low density.
25.2 The percolation thresholds: carbon black, $\phi_p^* \sim 10^{-1}\%$, nanotubes, $\phi_p^* \sim 3\times10^{-3}\%$.
(Nanotubes act as fibres and give a better percolation network at lower volume fraction)
Exponents: nanotubes $t \sim 1.2$, carbon black $t \sim 2$ (fibres versus particles).
25.3 Density of the *trans*-polyacetylene crystal = 1170.8 kg m^{-3}
Density of the *cis*-polyacetylene crystal = 1165.4 kg m^{-3}
The value of c differs by nearly a factor of 2 due to the different conformations of the repeat units.

Index

C

D

E

G

H

I

K

L

M

N

O

P

T